Investigating Astronomy
A Conceptual View of the Universe

SECOND EDITION

Investigating Astronomy

A Conceptual View of the Universe

SECOND EDITION

TIMOTHY F. SLATER

University of Wyoming

ROGER A. FREEDMAN

University of California, Santa Barbara

W.H. Freeman and Company
A Macmillan Higher Education Company

Publisher: Kate Parker
Acquisitions Editor: Alicia Brady
Development Editors: Blythe Robbins; Kharissia Pettus
Senior Media and Supplements Editor: Amy Thorne
Assistant Editor: Tue Tran
Marketing Manager: Taryn Burns
Marketing Assistant: Samantha Zimbler
Project Editor: Kerry O'Shaughnessy
Production Manager: Julia DeRosa
Cover and Text Designer: Blake Logan
Illustration Coordinator: Janice Donnola
Illustrations: Imagineering
Photo Editors: Sheena Goldstein, Nick Ciani
Photo Researcher: Feldman & Associates
Composition: Northeastern Graphic, Inc.
Printing and Binding: RR Donnelley

Credit is given to the following sources for photos in the **Cosmic Connections** and **Visual Literacy Tasks** and on the cover: NASA/SDO/Goddard Space Flight Center; **Cosmic Connections** and **Visual Literacy Tasks:** p. 6: Earth: JPL/NASA; Sun: NOAA National Climatic Data Center; starry sky: Space Telescope Science Institute (hubblesite.org); nebula: NASA,ESA, M. Robberto (STScI/ESA) and the Hubble Space Telescope Orion Treasury Project Team; supernova: NASA, ESA, J. Hester and A. Loll (Arizona State University); Milky Way: NASA/JPL-Caltech/R. Hurt (SSC); galaxy clusters: National Optical Astronomy Observatory; observable universe: 2MASS, IPAC/Caltech, and University of Massachusetts; pp. 98–99: NASA/JPL/Space Science Institute; p. 110: NASA/JPL/Space Science Institute; p.154: NASA Jet Propulsion Laboratory; p. 158: NASA Jet Propulsion Laboratory; p. 164: NASA/JPL; p. 177: Earth: NASA; all others: NASA/JPL; p. 208: Sun: SOHO (ESA & NASA); stars: NASA and H. Richer (University of British Columbia); p. 291: Sun: SOHO/MDI (ESA & NASA); p. 319: spiral galaxy: NASA/ESA; globular cluster: S. Fakfa and K. Honeycutt, Indiana University/WIYN/NOAO/NSF; p. 343: colliding galaxies #1: Bob and Bill Twardy/Adam Block/NOAO/AURA/NSF; colliding galaxies #2: NASA, ESA, and the Hubble Heritage Team (STScI/AURA)-ESA/Hubble Collaboration; colliding galaxies #3: NASA/JPL-Caltech/Z. Wang (Harvard-Smithsonian CfA); colliding galaxies #4: John Dubinski, University of Toronto.

Library of Congress Control Number: 2013956580

ISBN-13: 978-1-4641-4085-3
ISBN-10: 1-4641-4085-5

Printed in the United States of America

Second printing

W. H. Freeman and Company
41 Madison Avenue, New York, NY 10010
Houndmills, Basingstoke RG21 6XS, England
www.whfreeman.com

TIM: To Stephanie Jean Slater, my wife and partner in all things; and to my children, who I hope never stop wondering about what floats beyond Earth

ROGER: To my wife Caroline Robillard, for expanding my universe

ABOUT THE AUTHORS

Timothy F. Slater is the University of Wyoming Excellence in Higher Education Endowed Chair of Science Education and holds faculty appointments in the College of Education and the College of Science. Internationally known for his seminal work in astronomy education research, he is a Senior Fellow at the CAPER Center for Astronomy & Physics Education Research, where his research focuses on uncovering learners' conceptual models when engaging in science. Before becoming a chaired professor at the University of Wyoming, Dr. Slater was a tenured professor in the Astronomy Department at the University of Arizona, where he constructed the first Ph.D. program focusing on astronomy education research. Winner of numerous teaching awards, Dr. Slater has been elected to the Council and Boards of Directors for the American Astronomical Society, the Astronomical Society of the Pacific, the Society of College Science Teachers, and the National Science Teachers Association. He has served on, and chaired, education committees for the American Association of Physics Teachers, the American Physical Society, and the American Geophysical Union. Dr. Slater and his wife, Stephanie, also a noted discipline-based education researcher, spend their evenings in search of the perfect location to watch the Sun set and count the stars as they appear. *(Photo courtesy of UW Photo Service)*

Roger A. Freedman is on the faculty of the Department of Physics at the University of California, Santa Barbara. He grew up in San Diego, California, and was an undergraduate at the University of California campuses in San Diego and Los Angeles. He did his doctoral research in nuclear theory and its astrophysical applications at Stanford University under the direction of Professor J. Dirk Walecka. Dr. Freedman came to UCSB in 1981 after three years of teaching and doing research at the University of Washington. Dr. Freedman holds a commercial pilot's license, and when not teaching or writing he can frequently be found flying with his wife, Caroline. He has piloted aircraft across the United States and Canada. *(Photo courtesy of Caroline J. Robillard)*

BRIEF CONTENTS

CONTENTS

A LETTER FROM THE AUTHORS

Astronomy is first and foremost about standing outside and gazing up at the night sky with wonder. Although it is certainly true that modern astronomers use giant telescopes, high-tech space probes, and supercomputers, astronomy is at its core a quest for knowledge to help better understand our place in this vast universe. This new edition of *Investigating Astronomy* has been carefully designed to highlight astronomy not as a long list of disconnected and therefore meaningless facts, but instead as a deeply human enterprise.

As one of humanity's first sciences, astronomy might initially seem to be stagnant "old news." Perhaps surprisingly, because of rapid advances in technology and computers, astronomers are taking enormous strides and increasing our knowledge of the universe. In other words, what we understand about the universe expands each and every day.

In much the same way, what we know about how students learn develops and grows each day. This new edition uniquely brings the latest astronomical discoveries together with cutting-edge teaching and learning strategies in a partnership designed specifically to improve the depth of students' learning. The scientific concepts selected for this edition are presented with a focus on making tight connections to the experiences of today's students and helping them develop flexible understandings of scientific processes. Overall, the goal of this edition is to leverage students' innate enthusiasm for learning about outer space to help them more deeply understand and value the scientific and technical enterprises that shape twenty-first–century society.

Timothy F. Slater

Roger A. Freedman

PREFACE

TEXT FEATURES

For the second edition of *Investigating Astronomy*, we continue to provide instructors and students with the tools and framework they need to successfully tackle the one-semester astronomy course. Throughout the text, we've updated the science as needed as well as changed the illustrations and photos to include the most recent images.

Careful Selection of Topics, Concise Coverage

This book contains 15 chapters to match more closely the typical one-semester astronomy course. While an instructor comparing the concepts in Slater/Freedman with any other text will find the same core concepts that are covered in most introductory astronomy courses, we have limited the coverage of higher-level topics. Further, this textbook has far fewer bold-faced words than found in most textbooks; we purposely strive to describe and explain the universe in the most common terms possible. As such, our strategy is not simply to tell students long lists of names and facts about the universe, but to engage students in a conversation about how astronomers study and decipher the universe's underlying fundamental processes. As such, *Investigating Astronomy* helps students to build their own understanding of astronomy.

A Conceptual Approach to Comparative Planetology

Emphasizing how the formation of the solar system relates to the composition of the planets, we introduce the solar system as a whole in Chapter 4, including planetary orbits and physical composition. The fundamental concepts of Earth science are presented in Chapter 5, laying the groundwork for studying the entire solar system. Chapter 6 covers the terrestrial planets, not planet by planet, but by focusing on the processes that shaped them and how astronomers have discerned these processes. Chapter 7 applies the same approach to the specifics of the giant planets of our solar system. Students are asked to understand the logic behind what we know scientifically, and not just memorize facts. The result is a cumulative and contextual understanding.

Building Mental Models

We have constructed this textbook using mental models. For example, when we introduce the nature of stars and galaxies to students, we emphasize observational data and inference before listing facts and formulae. In the same way, when we talk about evolution of stars and galaxies, we focus on what we see rather than on what things are called.

Conceptual Approach to Quantitative Topics

In favor of a more effective use of mathematical reasoning, we have largely removed arithmetic calculations. Instead of a stream of formulas and equations, numbers are used to show relationships and comparisons as concepts are explained and evaluated.

Scientific Process as an Obvious Theme

The text focuses on the continuing process of scientific discovery, on how astronomers know what they know and how they embark on future discoveries.

LEARNING TOOLS

In constructing *Investigating Astronomy* we've taken advantage of the long-standing strengths of the more comprehensive *Universe* book, now in its 10th edition. We've enhanced the text's writing and pedagogical approach by building on the newest results of modern cognitive science on how students learn best.

Immediate Assessment and Application of Concepts Allow Students to Go Beyond Passive Reading

• **ConceptChecks:** Our experience teaching thousands of students has shown us that continuous feedback and conceptual reinforcement are necessary to create comprehensive and lasting understanding. Included at the end of each section, these thought-provoking questions go beyond reading comprehension, often asking students to draw conclusions informed by, but not explicitly mentioned, in the text, calling for applied thinking and synthesis of concepts. Answers are provided at the end of each chapter.

ConceptCheck 4-5: If planets reflect the Sun's light rather than emitting light of their own, how can spectroscopy reveal information about a planet's atmosphere?

ConceptCheck 5-5: What would happen to the temperature in the stratosphere if there was an absence of ozone?

• **CalculationChecks:** Similar to ConceptChecks and focusing on mathematics, these give students the opportunity to test themselves by solving different mathematical problems associated with the chapter concepts. They appear only within sections where relevant mathematical reasoning is presented. Answers are provided at the end of each chapter.

CalculationCheck 3-3: How fast is the space shuttle traveling 3 s after launch if it is accelerating at a rate of about 20 m/s^2?

VISUAL LITERACY TASK

Kepler's Laws

B
A
C
Sun
Perihelion
Aphelion
D

PROMPT: What would you tell a fellow student who said, **"Planets move fastest when farthest from the Sun because there is more gravity at great distances"?**

ENTER RESPONSE:

Guiding Questions:

1. Planets move about our Sun
   a. fastest when farthest from our Sun.
   b. fastest when closest to our Sun.
   c. at the same speed no matter their distance.
   d. slower for relatively smaller orbits.
2. The gravitational attraction between an orbiting planet and our Sun is greatest for
   a. longer distances.
   b. shorter distances.
   c. the more rapidly spinning planets.
   d. planets with the thickest atmospheres.
3. For a given amount of time, a line drawn between a planet and the Sun will sweep out
   a. the same size area regardless of distance.
   b. a greater area when close to the Sun.
   c. a smaller area when close to the Sun.
   d. an area that always covers the ellipse's second focus point.
4. With an average distance of about 150 million km (93 million mi), the difference between Earth's closest approach and most distant positions in its elliptical orbit about the Sun is roughly
   a. 5 km (4 mi).
   b. 5 million km (4 million mi).
   c. 50 million km (40 million mi).
   d. 500 million km (400 million mi).

• **Visual Literacy Tasks:** At the end of each chapter, students are presented with a figure and a question that they must defend or refute. Writing space is provided to underscore the value of writing to learning. Prompt questions are supplied to help students determine their answers. Specific answers are not provided in the text.

Text Boxes Key into Student Misconceptions and Link the Material to Everyday Life

• **Cosmic Connections:** Full-page illustrations summarize key concepts.

• **Cautions—Confronting Misconceptions:** Special "caution" paragraphs alert students to conceptual pitfalls.

• **Analogies—Bringing Astronomy Down to Earth:** Analogy paragraphs relate new ideas to more familiar experiences on Earth.

• **Margin Notes:** Text in the margin of the book highlights key details in the chapter for students.

• **Tools of the Astronomer's Trade:** Worked examples follow a logical sequence of steps called S.T.A.R.: assess the Situation, select the Tools, find the Answer, and Review the answer and explore its significance.

• **The Heavens on Earth:** Astronomical concepts are brought down to Earth.

Chapter Opening and Closing Features Are Study Aids

• **Chapter Learning Objectives:** Found at the start of each chapter, these objectives provide the most benefit when students use them in conjunction with the notes they take while reading.

• **Key Ideas and Terms:** Each chapter closes with a bulleted outline of topics, including embedded key terms to reinforce their context.

• **Chapter Review and Web Chat Questions, and Collaborative Exercises:** These exercises offer additional opportunities for students to apply the material, both on their own and collaboratively.

• **Observing Questions:** For use with the *Starry Night*™ planetarium software, written by Marcel Bergman, T. Alan Clark, and William J. F. Wilson, University of Calgary.

NEW TO THE SECOND EDITION

Further Emphasis on Scientific Method

Recognizing how important the scientific method is to the study of any science, we have placed further emphasis throughout the chapters on the scientific method as an important process in the study of astronomy. We intend to emphasize a contemporary view of the nature of science, moving away from its more traditional description as a linear approach and more toward science as a human enterprise that is tied to social and technological issues that students can directly relate to.

Improved Design and Illustrations

To ensure greater readability, the second edition includes an improved design that allows for easier navigation through text, features, and figures. Illustrations have also been improved and refined to highlight the key information in the figure. Figures with complex information have been broken down or amended to present the content in easier-to-handle chunks.

QR Codes

Responding to rapid advances in technology and computers, we have placed QR codes throughout the chapters to lend a "just-in-time" teaching component to the concepts covered in the text. Every day, astronomers are increasing our knowledge of the universe, and now with just a click of a button, students can quickly access more information on various topics from the speed of astronomical objects to tectonics on Mars and estimating the age of the universe.

Chapter Learning Objectives

Found at the start of each chapter, these objectives provide students with a clear list of learning outcomes that they can focus on as they read through the chapter. This list also provides students with a means of self-assessing their understanding of the material presented in the chapters.

CHAPTER LEARNING OBJECTIVES By reading the sections of this chapter, you will learn

2-1 Light travels through empty space at a speed of nearly 300,000 km/s

2-2 Glowing objects, like stars, emit an entire spectrum of light

2-3 An object's temperature is revealed by the most intense wavelength of its spectrum of light

2-4 An object's chemical composition is revealed by the unique pattern of its spectrum of light

2-5 An object's motion through space is revealed by the precise wavelength positions of its spectrum of light

2-6 Telescopes use lenses, mirrors, and electronics to concentrate and capture incoming light for study

Chapter Changes

- **Chapter 1:** Modernized treatment of the scientific method emphasizing that there are multiple approaches to doing science instead of a single, step-by-step scientific method of testing hypotheses.
- **Chapter 2:** Increased emphasis on a conceptual understanding of how light is produced, balanced by a moderate reduction in historical accounts.
- **Chapter 3:** Changed emphasis from a historical "How *do* we know?" to a more active "How *can* we know?" view of the history of astronomy.
- **Chapter 4:** Modernized explanations of our search for extrasolar planets to emphasize the transit method over the radial velocity method.
- **Chapter 5:** Rearranged Earth science chapter to emphasize the role of water throughout the Earth system.
- **Chapter 6:** Updated discussion of water in the solar system to include the newest observations of Mercury and the Moon.
- **Chapter 7:** Updated art and illustrations to emphasize the most modern images from the world's most high-tech telescopes and advanced space probes.
- **Chapter 8:** Provided a more contemporary and scientifically accurate discussion of the possibility of life on Mars and other solar system bodies.
- **Chapter 9:** Updated traditional images with new, higher-resolution, space-based solar telescope observations of the Sun.

- **Chapter 10:** Purposefully adjusted the reading comprehension level of the text to make it more accessible for the growing diversity of students.
- **Chapter 11:** Included new telescope images that more clearly illustrate concepts described in the text.
- **Chapter 12:** Expanded the presentation of recent Chandra X-Ray Observatory discoveries.
- **Chapter 13:** Included data and images from Spitzer that better illustrate the nature of our Galaxy.
- **Chapter 14:** Emphasized a conceptual view of the structure and evolution of galaxies over a historical accounting of thought.
- **Chapter 15:** Reduced the reliance on mathematical explanations to focus instead on spatial representations.

MEDIA AND SUPPLEMENTS

Investigating Astronomy was designed to support a wide variety of teaching styles and course environments. A variety of multimedia and supplemental materials provides an array of choices for students and instructors in their use of these materials, which were created based on input and contributions from a large number of faculty.

Electronic Versions

Investigating Astronomy is offered in two electronic versions. One is an interactive e-Book as part of the LaunchPad, and the other is a PDF-based e-Book from CourseSmart. These options are provided to offer students and instructors flexibility in their use of course materials.

CourseSmart e-Book

The *Investigating Astronomy* CourseSmart e-Book offers the complete text in an easy-to-use, flexible format. Students can choose either to view the CourseSmart e-Book online or download it to their computer or to a portable media player, such as an iPhone. To help students study and to mirror the experience of a printed textbook, CourseSmart e-Books feature note-taking, highlighting, and bookmarking features.

ONLINE LEARNING OPTIONS

Investigating Astronomy supports instructors with a variety of online learning preferences. Its rich array of resources and platforms provides solutions according to each instructor's teaching method. Students can also access the resources through the Student Companion Web Site.

LaunchPad: Because Technology Should Never Get in the Way

W. H. Freeman and Company is committed to providing online instructional materials that meet the needs of instructors and students in powerful yet simple ways—powerful enough to dramatically enhance teaching and learning, yet simple enough to use right away.

We've taken what we've learned from thousands of instructors and hundreds of thousands of students and created a new generation of technology—featuring LaunchPad. LaunchPad offers our acclaimed content curated and organized for easy assignability in a breakthrough user interface in which power and simplicity go hand in hand.

- **Curated LaunchPad Units Make Class Prep a Whole Lot Easier:** Combining a curated collection of videos, tutorials, animations, projects, multimedia activities and exercises, and e-Book content, LaunchPad's interactive units give you a building block to use as is or as a starting point for your own learning units. An entire unit's worth of work can be assigned in seconds, drastically saving the amount of time it takes for you to have your course up and running.
- **Interactive e-Book:** The Interactive e-Book is a complete online version of the textbook with easy access to rich multimedia resources. All text, graphics, tables, boxes, and end-of-chapter resources are included in the e-Book. Features include:
    - Quick, intuitive navigation to any section or subsection
    - Full text search, including the glossary and index

- Sticky notes, which allow users to place notes anywhere on the screen and to choose the note color for easy categorization.
- "Top notes," which allow users to place a prominent note at the top of the page to provide a more significant alert or reminder.
- Text highlighting, down to the level of individual phrases, in a variety of colors

- **LearningCurve:** Powerful adaptive quizzing, a gamelike format, direct links to the e-Book, instant feedback, and the promise of better grades make LearningCurve an ideal student resource. Customized quizzing tailored to the text adapts to students' responses and provides material at different difficulty levels and different topics based on student performance. Students love the simple yet powerful system and instructors can access class reports to help refine lecture content.

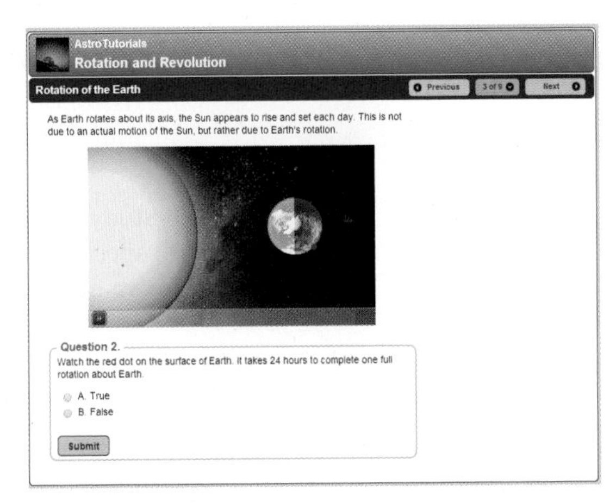

- **Interactive Tutorials:** Developed by prominent astronomy education researchers, the Interactive Tutorials present astronomy topics in a flexible multimedia environment. They take advantage of the best means to illustrate each topic, using a blend of text, review questions, animations, videos, and quizzes to produce a thorough understanding that students can carry with them. The tutorial topics were chosen after careful analysis of the most commonly taught subjects, with particular emphasis on which topics were most often misunderstood by students. The tutorials have been shown to increase student understanding and produce a meaningful, memorable learning experience.

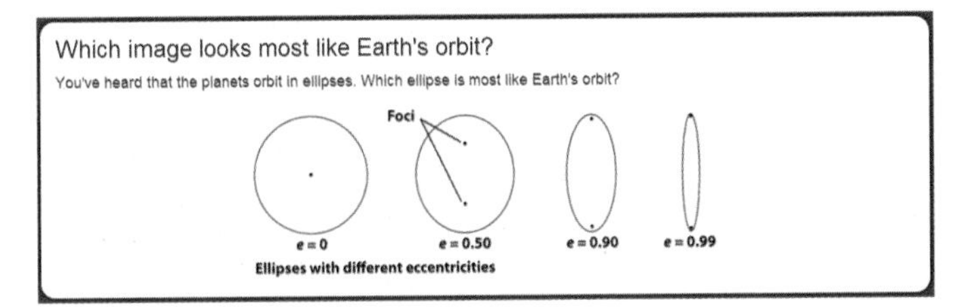

- **Image Map Activities:** These activities use figures and photographs from the text to assess key ideas, helping students to develop their visual literacy skills. Students must click the appropriate section(s) of the image and answer corresponding questions.
- **Other Resources:** Animations, videos, interactive exercises, flashcards, and other resources highlight key concepts in introductory astronomy for students to explore at their own pace.
- **Assignments for Online Quizzing, Homework, and Self-Study:** Instructors can create and assign automatically graded homework and quizzes from the complete test bank, which is preloaded in LaunchPad. All quiz results feed directly into the instructor's gradebook.

  The Gradebook quickly and easily allows you to look up performance metrics for your whole class, for individual students, or for individual assignments. Having ready access to this information can help with lecture prep and in making office hours more productive and efficient for both professors and students.

- ***Scientific American* Newsfeed:** To demonstrate the continued process of science and the exciting new developments in the field, the *Scientific American* Newsfeed delivers regularly updated material from the well-known magazine. Articles, podcasts, news briefs, and videos on subjects related to astronomy are selected for inclusion by *Scientific American's* editors. The newsfeed provides several updates per week, and instructors can archive or assign the content they find most valuable.

Sapling Learning

www.saplinglearning.com

Developed by educators with both online expertise and extensive classroom experience, Sapling Learning provides highly effective interactive homework and instruction that improve student learning outcomes for the problem-solving disciplines. Sapling Learning offers an enjoyable teaching and effective learning experience that is distinctive in three important ways:

- **Ease of Use:** Sapling Learning's easy-to-use interface keeps students engaged in problem solving, not struggling with software.
- **Targeted Instructional Content:** Sapling Learning increases student engagement and comprehension by delivering immediate feedback and targeted instructional content.

• **Unsurpassed Service and Support:** Sapling Learning makes teaching more enjoyable by providing a dedicated Masters- and Ph.D.-level colleague to serve instructors' unique needs throughout the course, including help with content customization.

We offer bundled packages that include Sapling Learning Online Homework with all versions of our texts.

STUDENT COMPANION WEB SITE

The *Investigating Astronomy* Student Companion Web Site, accessed through www.whfreeman.com/slater2e, provides a range of tools for student self-study and review. They include:

- **Online Self-Study Quizzes** offering randomized questions and answers with instant feedback referring to specific sections in the text, to help students study, review, and prepare for exams. Instructors can access results through an online database or they can have them e-mailed directly to their accounts.
- **Animations** of key concepts
- **NASA Videos** highlight important processes and phenomena
- **Vocabulary** and concept-review flashcards
- **Interactive Exercises** based on text illustrations

Starry Night™

Starry Night™ is a brilliantly realistic planetarium software package produced by Simulation Curriculum Corp. It is designed for easy use by anyone with an interest in the night sky, and particularly college students. See the sky from anywhere on Earth or lift off and visit any solar system body or any location up to 20,000 light years away. View 2,500,000 stars along with more than 170 deep-space objects like galaxies, star clusters, and nebulae. You can travel 15,000 years in time, check out the view from the International Space Station, and see planets up close from any one of their moons. Included are stunning OpenGL graphics. You can also print handy star charts to explore outside. This version of *Starry Night™* contains student exercises specific to the Freeman version for use with *Investigating Astronomy*. *Starry Night™* is available via online download using an access code packaged with the text at no extra charge upon instructor request.

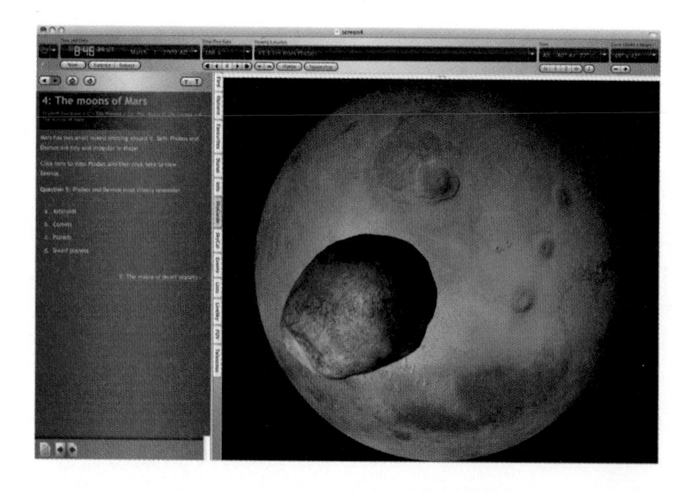

Observing Projects Using Starry Night™

ISBN: 1-4641-2502-3

by Marcel Bergman, T. Alan Clark, and William J. F. Wilson, University of Calgary

Available for packaging with the text, and compatible with both PC and Mac, this book contains a variety of comprehensive lab activities for *Starry Night™*.

Test Bank CD-ROM

Windows and Mac versions on one disc, ISBN 1-4641-6889-X

1500 multiple-choice questions are referenced by section. The easy-to-use CD-ROM version includes Windows and Mac versions on a single disc, in a format that lets you add, edit, re-sequence, and print questions to suit your needs.

Online Course Materials (Blackboard, Desire2Learn, Moodle, Canvas)

As a service for adopters, we will provide content files in the appropriate online course format, including the instructor and student resources for this text. The files can be used as is or can be customized to fit specific needs. Course outlines, prebuilt quizzes, links, activities, and a whole array of materials are included.

PowerPoint Lecture Presentations

A set of online lecture presentations created in PowerPoint allows instructors to tailor their lectures to suit their own needs, using images and notes from the textbook. These presentations are available on the instructor portion of the companion Web site, and within LaunchPad.

ACKNOWLEDGMENTS

This textbook would not have been possible were it not for the exceptional dedication and pool of talent at W. H. Freeman and Company. Our deepest gratitude goes to all those involved in the development of the first edition of this book: Valerie Raymond, who shared our vision of this text when first approached and brought considerable skill to the developmental process of the first edition in order to achieve the clarity and readability that are so essential; Anthony Palmiotto, acquisitions editor, for guiding the overall publishing effort; and Amy Thorne for adeptly managing the extensive array of multimedia and supplemental resources that support students and instructors.

The second edition of this text also included a dedicated team that we extend our thanks to: in particular, our publisher, Jessica Fiorillo, and acquisitions editor, Alicia Brady, as well as development editors Blythe Robbins and Kharissia Pettus. We extend our appreciation to Kerry O'Shaughnessy, our project editor, who carefully shepherded the book through many stages of development and proofs, as well as Julia DeRosa for coordinating the vast array of resources necessary to produce such a book. Designer Blake Logan brought extensive creativity and enormous talent to the task of establishing and implementing the look and feel of the text. Photos and illustrations are extremely important to the study of astronomy, and we were very fortunate to work with Jennifer Macmillan, Sheena Goldstein, Nick Ciani, and Feldman & Associates, for photo permissions. In addition, Janice Donnola expertly managed the art rendering process. Imagineering Art is to thank for producing the excellent artwork. We commend copyeditor Louise Ketz for her diligent work and precise feedback as we strove to make this text as accurate as possible.

Kerri Russini, our market development manager for the first edition, was extremely impressive in her long-term dedication and attention to every detail as she ensured that astronomy instructors nationwide learned about the book throughout its development. She managed extensive reviews, surveys, focus groups, events, class tests, and the development of preview materials and multimedia to help us share our vision with our colleagues. For the second edition, marketing manager Debbie Clare brought outstanding innovation and vision to the promotion effort, and we thank her for developing and implementing a fantastic marketing strategy.

We thank Marcel Bergman, T. Alan Clark, and William J. F. Wilson for developing exceptional activities and questions using *Starry Night*™, and we value the very fruitful partnership between W. H. Freeman and the team at Simulation Curriculum Corp., the producers of *Starry Night*™.

On a personal note, Tim Slater notes that his contribution to this book would not have happened had it not been for the unwavering support and love of his dear wife, Stephanie. When he would feel down, she helped him gaze upward to the stars. When he was tired, Stephanie reminded him that tomorrow the Sun would rise, bringing a new day. And when the stars were twinkling most brightly, she was always by his side to share with him the beauty of the universe.

Roger Freedman would like to thank his late father, Richard Freedman, for first cultivating an interest in space many years ago, and for his father's personal contributions to the exploration of the universe as an engineer for the Atlas and Centaur launch vehicle programs. Most of all, Roger thanks his charming wife, Caroline, for putting up with the many hours he devotes to writing textbooks!

Although we have made a concerted effort to make this textbook error free, some mistakes may have crept in. We would appreciate hearing from anyone who finds an error or wishes to comment on the text.

We are deeply grateful to the astronomers and teachers who reviewed the manuscript. This is a stronger and better book because of their conscientious efforts.

Reviewers

Kurt S. J. Anderson (emeritus), *New Mexico State University*
Stuart Anderson, *Lake-Sumter State College*
Dixie L. Androes, *Northwest Arkansas Community College*
Douglas Arion, *Carthage College*
Pauline Barmby, *University of Western Ontario*
S. Leslie Blatt, *Clark University*
Daniel C. Boice, *San Antonio College*
Lowell M. Boone, *Evansville University*

Sukanta Bose, *Washington State University*
Robert Braunstein, *Northern Virginia Community College, Loudoun*
Bill Briscoe, *The George Washington University*
Allison I. Bruce, *El Paso Community College*
Gary Burk, *Otterbein College*
Juan Cabanela, *Minnesota State University, Moorhead*
Joseph Caprioglio, *Hofstra University*
Kwang-Ping Cheng, *California State University, Fullerton*
Michol Christopher, *Mt. San Antonio College*
Josh Colwell, *University of Central Florida*
James Cooney, *University of Central Florida*
John Cowan, *University of Oklahoma*
Matthew Craig, *Minnesota State University, Moorhead*
Peter Detterline, *Kutztown University*
Bryan Dunne, *University of Illinois, Urbana-Champaign*
Joanna Eisberg, *Chaffey College*
Rebecca Ericson, *George Mason University*
Jason Ferguson, *Wichita State University*
Yan Fernandez, *University of Florida*
Efrain J. Ferrer, *State University of New York, Fredonia*
Kent D. Fisher, *Columbus State Community College*
Terrence Flower, *St. Catherine University*
Anthony J. George, Jr., *Columbia Basin College*
Satyajit P. Ghosh, *University of Scranton*
Elaine Gibb, *Middle High School*
Henry S. Greenside, *Duke University*
Erika Grundstrom, *Vanderbilt University*
Jim Hamm, *Big Bend University*
Andy Hollerman, *University of Louisiana, Lafayette*
Richard Ignace, *East Tennessee State University*
Bruce E. Ivey, *Pacific Union College*
Francine Jackson, *Framingham State University*
Fred Jaquin, *Onondaga Community College*
John Michael Kalko, *Fullerton College*
Kishor Kapale, *Western Illinois University*
Arthur Ketterer, *Raritan Valley Community College*
Patrick Koehn, *Eastern Michigan University*
Lucy Kulbago, *John Carroll University*
Andrew Layden, *Bowling Green State University*
Denis Leahy, *University of Calgary*
Kevin Lee, *University of Nebraska, Lincoln*
Douglas Leonard, *San Diego State University*
Karina Leppik, *Oberlin College*
Ran Li, *Kent State University, Stark Campus*
Bernhard Lee Lindner, *College of Charleston*
Arthur H. Litka, *Seminole State College of Florida*
Ian Littlewood, *California State University, Stanislaus*
Michael LoPresto, *Henry Ford Community College*
Vera E. Margoniner, *California State University, Sacramento*
Christopher Martin, *California Institute of Technology*
Eduardo Martin, *University of Central Florida*
Danielle Lynn Martino, *Santiago Canyon College*
Janet McLarty-Schroeder, *Cerritos College*
Benjamin Mendelsohn, *West Valley College*
Armando Miccoli, *Adairsville High School*
Milan Mijic, *California State University, Los Angeles*
Scott T. Miller, *Sam Houston State University*
Jeffrey R. Miller, *St. Lawrence University*
Anatoly Miroshnichenko, *University of North Carolina, Greensboro*
Terry R. Mitchell, *Lake Sumter Community College*
Michele Montgomery, *University of Central Florida*
Windsor Morgan, *Dickinson College*
Peter Newbury, *University of British Columbia*
Kris Ochwat, *Wilbur Wright University*
Douglas O'Handley, *Santa Clara University*
Richard Olenick, *University of Dallas*
John P. Oliver, *University of Florida*
Michelle Ouellette, *California Polytechnic State University*
Christopher Palma, *Penn State University*
Bruce Palmquist, *Central Washington University*
Nicolas Pereryra, *University of Texas–Pan American*
Delphine Perrodin, *Franklin and Marshall College*
Dale Pleticha, *Gordon College*
Barton Pritzl, *University of Wisconsin, Oshkosh*
Jeff W. Robertson, *Arkansas Tech University*
Carl Rosenzweig, *Syracuse University*
Louis Joseph Rubbo, *Coastal Carolina University*
Jeffrey Sabby, *Southern Illinois University, Edwardsville*
Victoria Alten Sahami, *Metropolitan State College of Denver*
Ronald Samec, *Bob Jones University*
Gregory R. Schultz, *University of California, Berkeley*
Teresa Schulz, *Lansing Community College*
Kendra Sibbernsen, *Metropolitan Community College*
John Sievers, *Mesa College*
Murray Silverstone, *University of Alabama*
Allyn Smith, *Austin Peay State University*
Daniel Snowman, *Rhode Island College*
James R. Sowell, *Georgia Institute of Technology*
Don Sparks, *Pierce College*
James Stickler, *Allegany College of Maryland*
Jeff J. Sudol, *West Chester University of Pennsylvania*
Ben E. K. Sugerman, *Goucher College*
Michael E. Summers, *George Mason University*
Jonathan Tan, *University of Florida*
Christopher Taylor, *California State University, Sacramento*
Christian Thomas, *Belmont University*
Jamey Thompson, *Hudson Valley Community College*
Toshiya Ueta, *University of Denver*
Colin Wallace, *Colorado College*
Edward White, *St. Charles Community College*
Rob Wilson, *Muskingum University*
John Wayne Wooten, *Pensacola State College*
Todd Young, *Wayne State College*
Nicolle Zellner, *Albion College*
Yan Zeng, *Savannah State*

Special thanks to Inge Heyer, Joint Astronomy Centre, who volunteered to carefully read and comment on every chapter. Her eagle eye and attention to detail have made the book much better.

R I V U X G The Milky Way Galaxy. *(A. Fujii/ESA/NASA)*

Predicting the Motions of the Stars, Sun, and Moon

CHAPTER LEARNING OBJECTIVES **By reading the sections of this chapter, you will learn:**

1-1 Astronomy is both an ancient cultural practice and a cutting-edge science

1-2 The stars are grouped by constellations

1-3 All of the observed celestial motions can be described if our planet Earth spins once each day while it orbits around our Sun each year

1-4 The Sun appears to change position over the day and throughout the year, and these changes result in Earth's seasons

1-5 The Moon appears to change its position in the sky every hour and its phase throughout each month

1-6 Eclipses occur only when the Sun, Moon, and Earth are perfectly aligned

Imagine yourself in the desert on a clear, dark, moonless night, far from the glare of city lights. As you gaze upward, you see a panorama that no poet's words can truly describe and that no artist's brush could truly capture. Literally thousands of stars are scattered from horizon to horizon, many of them grouped into a luminous band called the Milky Way (which extends across the middle of the photograph on the previous page). As you watch, the entire spectacle swings slowly overhead from east to west as the night progresses.

For thousands of years people have looked up at the heavens and contemplated the universe. Like our ancestors, we find our thoughts turning to profound questions as we gaze at the stars. How was the universe created? Where did Earth, the Moon, and the Sun come from? What are the planets and stars made of? And how do we fit in? What is our place in the cosmic scope of space and time?

Wondering about the universe is a key part of what makes us human. Our curiosity, our desire to explore and discover, and, most important, our ability to reason about what we have discovered are qualities that distinguish us from other animals. The study of the stars transcends all boundaries of culture, geography, and politics. In a literal sense, astronomy is a universal subject—its subject is the entire universe.

In this chapter, we describe how astronomers predict the motions of the stars, Sun, and Moon as they move across the sky. What we will discover is that these apparent motions are due to a spinning Earth against the backdrop of our Sun, Moon, and the distant stars. Understanding why we see what we see is the first step in exploring our dynamic universe. ■

1-1 Astronomy is both an ancient cultural practice and a cutting-edge science

Wondering about the night sky has a rich heritage that dates back to the myths and legends of antiquity. Centuries ago, the heavens were thought to be populated with demons and heroes, gods and goddesses. Astronomical phenomena were often explained as the result of supernatural forces and divine intervention and were used to remind Earth-bound inhabitants that their lives are intimately connected with the goings-on in the sky.

The course of civilization has been greatly affected by a profound realization: *The universe is comprehensible.* This awareness is one of the great gifts to come to us from the great thinkers of ancient Greece. Greek astronomers discovered that by observing the heavens and carefully reasoning about what they saw, they could uncover something about how the universe operates. For example, as we will see, they measured the size of Earth and were able to understand and predict eclipses without imagining supernatural forces. Modern science is a direct descendant of the work of these ancient Greek philosophers.

There is no single sequence of steps followed in all scientific investigations. All scientific methods use *multifaceted observations* and *logical inference* to *pursue questions about the natural world.*

Variety of Scientific Methods

Like musicians, philosophers, or advertisers, scientists are people who make use of creativity, intuition, and experience. Although many people learn how to scientifically study the universe in college, there is no such thing as an official scientist's license to be purchased or a formal scientist certificate to be granted before someone can call him/herself a scientist. Instead, individuals call themselves scientists when they follow a *scientific method* and agree to its code of ethics.

What is a scientific method? Throughout much of our schooling, we have often learned that the scientific method is a boring, nonimaginative, step-by-step way of studying the world by first making predictive *hypotheses* and then designing *experiments* to prove that a hypothesis is correct. We often read that this scientific method is supposedly free from any subjective human emotions and is based entirely on observed facts. While this sounds appealing, science does not actually work this way.

You might be surprised to learn that a seemingly simple story of scientists following the specific step-by-step approach of a scientific method is not how the science of astronomy is actually done. Instead, the truth of the scientific enterprise is actually far more interesting. The real process of humans conducting scientific exploration is filled with twists and wrong turns, epic failures and glorious triumphs, surprising insights, and sometimes catastrophic accidents. Let's consider some of characteristics of how science is done.

There is no single sequence of steps followed in all scientific investigations. In other words, there is no single scientific method with specific steps that everyone follows identically. There are, however, several attributes that are common across scientific studies: All **scientific methods** use multifaceted observations to pursue consistent evidence when trying to answer questions about the natural world. All scientific investigations are centered on pursuing a question or an unexpected observation, but they do not necessarily test a hypothesis.

A **hypothesis** is traditionally defined as a collection of thoughtfully proposed mechanisms for how the world operates or a comprehensive and complete explanation for why particular observations are seen. Hypotheses are weighed by the degree to which they can accurately make testable predictions. However, astronomers rarely devise and test hypotheses through experimentation; instead, astronomy is generally an observational science.

Let's start with an example. Imagine you are growing fruit and you want to know how much fertilizer to put on apple trees in order to make them grow as much fruit as possible. One scientific approach to studying fruit trees and fertilizer is to make a hypothesis that apples trees grow better with fertilizer than without fertilizer. The next step would be to plant 50 trees without any fertilizer and 50 trees with fertilizer as an experiment. Finally, after waiting for a few months, you could count which of the two groups of trees produced the most apples and determine if your hypothesis about apple trees with fertilizer producing more apples was correct. This is an example of a traditional, two-group comparison approach to the scientific method.

There are, however, other approaches to pursuing the question of how to grow more apples. A different scientific method of studying the relationship between apple tree production and the use of fertilizer could be to plant 100 apple trees, put slightly increasing amounts of fertilizer on each tree, and then count the number of apples produced in terms of how much fertilizer was applied. Because some trees would have too little fertilizer and other trees would have too much fertilizer, this scientific study result would be a graph revealing the optimal amount of fertilizer for growing the most apples. Like the initially described study, this second scientific study described the responses of apple trees to fertilizer. Although the first study tested a hypothesis and the second study did not, both are completely valid scientific methods.

Scientists Develop and Test Theories

Rather than hypotheses, astronomers deal largely with developing and refining *theories*. A body of related observations can be pieced together into a comprehensive, self-consistent explanatory description of nature called a **theory.** In fact, many astronomers go their entire career and never create and test a single hypothesis. From the above example about how to best produce lots of apples, the theory being studied was that there exists an optimum amount of fertilizer to maximize an apple tree's production.

An astronomy example from later in this book is the theory that the planets are held in their orbits around the Sun by a gravitational force between the Sun and planets (**Figure 1-1**). Without theories that make broad statements about the natural world, there is no understanding and no science, only long collections of disconnected facts. In fact, it is when widely encompassing ideas come together to form a theory that one can predict the outcome of experiments and observations that the scientific method is at its peak performance.

CAUTION In everyday language the word "theory" is often used to mean an idea that looks good on paper, but has little to do with reality. In science, however, a good theory is one that explains reality and that can be applied to explain new observations. An excellent example is the theory of gravitation, which was devised by the English scientist Isaac Newton in the late 1600s to explain the orbits of the six planets known at that time. When astronomers of later centuries discovered the planets Uranus and Neptune and objects like Pluto, they found that these objects also moved in accordance with Newton's theory. The same theory describes the motions of satellites around Earth as well as the orbits of planets around other stars.

FIGURE 1-1 Planets Orbiting the Sun An example of a scientific theory is the idea that Earth and the other planets orbit the Sun because of the Sun and planets' gravitational attraction. This theory is universally accepted because it makes predictions that have been tested and confirmed by observation. *(Calvin J. Hamilton and NASA/JPL)*

The most widely accepted scientific theories are ones that make accurate predictions that can be independently tested by other scientists. If the predictions are verified by observation, this lends wider support to a proposed theory and suggests that it might be a reasonably accurate description of how nature operates. Alternatively, if a theory's predictions are not verified, the theory needs to be modified or completely replaced.

For example, an old theory held that the Sun and planets orbit around a stationary Earth. This theory led to certain predictions that could be checked by observation, as we will see later in this book. In the early 1600s the Italian scientist Galileo Galilei used one of the first telescopes to show that these predictions were incorrect. As a result, the theory of a stationary Earth was rejected, eventually to be replaced by the modern picture shown in Figure 1-1 in which Earth and other planets orbit the Sun.

CAUTION Sometimes, scientists refer to universal laws. Some people mistakenly think that once a scientific theory is proven, it is elevated to the status of a law, such as the *law of gravity.* In science today, some things are referred to as "laws" for historical reasons because people used to believe the laws were completely true and would stand forever and we have used that name for a long time. However, one of the important characteristics of science is that scientific knowledge and theories are always open to being revised with new observations—and the term *law* is rarely used any more. A comprehensive theory, not a law, is what many scientists aspire to create.

In astronomy, a theory that cannot be tested by observation or experiment does not qualify as a scientific theory. An example is the idea that there is a little man living in your refrigerator who turns the inside light on or off when you open and close the door. The little man is invisible, weightless, and makes no sound, so you cannot detect his presence. While this is an amusing idea, it cannot be tested and so cannot be considered science.

Scientific Theories Can Change

Like observation, informed skepticism is an essential part of all scientific methods. New theories must be able to withstand the close scrutiny of other scientists. The more radical the theory, the more skepticism and critical evaluation it will receive from the scientific community, because the general rule in science is that extraordinary claims require extraordinary evidence. That is why scientists as a rule do not accept claims that people have been abducted by aliens and taken aboard UFOs. The evidence presented for these claims is unverifiable.

At the same time, scientists must be open-minded. They must be willing to discard long-held ideas if these ideas fail to agree with new observations and experiments, provided the new data have survived evaluation. That is why scientific knowledge is always subject to change. If, for example, an alien spacecraft really did land on Earth, scientists would be the first to accept that aliens existed—provided they could take a careful look at the spacecraft and its occupants.

The most productive scientists are diligently open to considering alternative ideas, particularly when presented with new evidence. This is true because one of the characteristics of science is that the specific way a particular observation is made can dramatically influence the results. Consider, for example, that a large telescope on Earth's surface obtains very different views of our Sun than much smaller space telescopes. Orbiting our planet above Earth's obscuring atmosphere, space telescopes have special instruments that provide very different views of the same object, as can be seen in different ground-based and space-based telescope images of our Sun. These views are valuable, but show different aspects of the Sun because the different methods obtain different results. Moreover, even two astronomers looking at the exact same set of images may decide that the data reveal very different things. This is not about making an error; but it is about people arriving at different conclusions based on the same data.

Ethics in Science

Earlier, we hinted at the idea that science is fundamentally a human endeavor and, as such, is subject to much of the same celebration, conflict, and missteps as any other human activity. For example, scientists often compete to be the first person to make a new discovery. Just as many Olympic athletes try to be the fastest record holder in a racing event, scientists who are first to make an important discovery are widely celebrated. For example, many people have tried to discover new comets, but one comet hunter is probably more famous than all others, Edmund Halley of the famed Comet Halley.

Celebrated scientists are often rewarded with high-paying jobs and are invited to travel to distant locations to present the results of their research. At the same time, sometimes the desire to be the first scientist to announce a new discovery or to be proven correct can result in unethical behavior—just as it occasionally happens in sports, where the drive to win at all costs overcomes a moral judgment.

People who do science informally agree to an unwritten code of ethics. This code generally includes working to benefit the world without causing harm. It also means that scientists should be honest and transparent about their observational and experimental strategies and publicize their results for public review. Scientists also allocate time to reviewing the scientific work of other scientists. Perhaps most importantly, scientists agree to publicly acknowledge the intellectual and financial contributions of all others from whose work they have drawn while not falsifying scientific data or being otherwise deceitful in their work. Many scientists also accept responsibility for training the next generation of upcoming scientists and future teachers. As stated by Bruce Alberts, former president of the National Academy of Sciences, "honesty, generosity, a respect for evidence, and openness to all ideas and opinions" can be considered a summary statement about the scientific code of ethics.

While we recognize that there is not a single scientific method, we also must recognize that not all questions are appropriate for scientific investigations. Important questions about what it means to be human—such as what is love, how valuable is truth, and what is the purpose of life—are tremendously important intellectual pursuits but are not necessarily the types of questions science can answer. Rather, the questions that science can best address are those about physical processes governing the natural world—questions for which numerous observations and carefully planned predictive tests can be made and evaluated using logical inference, and a bit of informed skepticism.

Technology in Science

Go to Video 1-1

As we will see in Chapter 2, in recent years astronomers have constructed telescopes that can detect such nonvisible forms of light (**Figure 1-2**). These instruments give us views of the universe vastly different from anything our eyes can see. These new views have allowed us to see through the atmospheres of distant planets, study the thin but incredibly violent gas that surrounds our Sun, and even observe new solar systems being formed around distant stars. Aided by high-technology telescopes, today's astronomers carry on the program of careful observation and logical analysis begun thousands of years ago by their ancient Greek predecessors.

Scientific advancements often depend on the development of new technology. For example, it was not until the use of telescopes in the seventeenth century that astronomers came to widely agree that the planets orbit our Sun. The new technology of the telescope provided observations about the

FIGURE 1-2 R I V U X G **A Telescope in Space** Because it orbits above most of Earth's obscuring atmosphere, the Hubble Space Telescope (HST) can better detect nonvisible forms of light absorbed by our atmosphere that are difficult or impossible to detect with a telescope on Earth's surface. *(NASA)*

changing appearances of planets and motivated a new scientific theory to replace one that was deeply entrenched in society.

As our technology increases, with computers squeezing vast amounts of digital memory in smaller and smaller places and cell phones having longer and longer battery lives, new tools for research and new techniques of observation dramatically impact astronomy. For example, tiny digital cameras are commonly integrated into cell phones, and this same technology can be used to make ultrasensitive cameras that work with telescopes. As another example, until fairly recently everything we knew about the distant universe was based on a narrow band of visible light. By the end of the nineteenth century, however, scientists had begun discovering forms of light invisible to the human eye: X-rays, gamma rays, radio waves, microwaves, and ultraviolet and infrared radiation.

A Quick Guide to Objects in the Sky

The science of astronomy allows our intellects to voyage across the cosmos. We can think of three stages in this voyage: from Earth through the solar system, from the solar system to the stars, and from the stars to galaxies and the grand scheme of the universe (see **Cosmic Connections: Size and Structure of the Universe** on page 6).

The star we call the Sun and all the celestial bodies that orbit our particular star—including Earth, the other seven planets and all their various moons, and smaller bodies such as dwarf planets, asteroids, and comets—make up the solar system (Chapter 4).

The nearest star to Earth is the Sun. All stars emit energy in the form of light, so all stars shine. Stars change over time and have life stages that we will discuss later (Chapters 10, 11, and 12). A few key points about stars help set the stage for us here:

- Huge clouds of interstellar gas, called *nebulae,* are scattered across the sky. Stars are born from the material of the nebula itself.
- Some stars that are far more massive than the Sun end their lives with a spectacular detonation called a *supernova* that blows the star apart, forming new nebulae.
- The most massive stars end their lives as almost inconceivably dense objects called *black holes,* whose gravity is so powerful that nothing—not even light—can escape.

Stars are not spread uniformly across the universe but are grouped together in huge assemblages called *galaxies* (Chapters 13 and 14). A typical galaxy, like the Milky Way, of which our Sun is part, contains several hundred billion stars.

We will begin our voyage from Earth to the distant reaches of the universe just as our ancestors did, by looking up from Earth into the sky and carefully studying the objects we can see with our eyes.

> **ConceptCheck 1-1: Which is held in higher regard by professional astronomers, a hypothesis or a theory? Explain your answer.**
>
> ***Answer appears at the end of the chapter.***

1-2 The stars are grouped by constellations

Go to Video 1-2

Looking at the sky on a clear, dark night, you might think that you can see millions of stars. Actually, the unaided human eye can detect only about 6000 stars. Because half of the sky is below the horizon at any one time, you can see at most about half of these stars. But how does one make sense of so many stars scattered about?

Stars Are Organized into Patterns Called *Asterisms* and the Sky Is Divided into Sections Called *Constellations*

When ancient people looked at these thousands of stars, they imagined that groupings of stars traced out pictures in the sky, often by creating figures by connecting stars from dot to dot. These figures, which make somewhat recognizable shapes, are called **asterisms.** You may already be familiar with some of these pictures or patterns in the sky, such as the asterism of the Big Dipper. To make it easier to describe particular parts of the sky, today's astronomers divide up the

COSMIC CONNECTIONS Size and Structure of the Universe

The Sun, Earth, and other planets are members of our solar system.

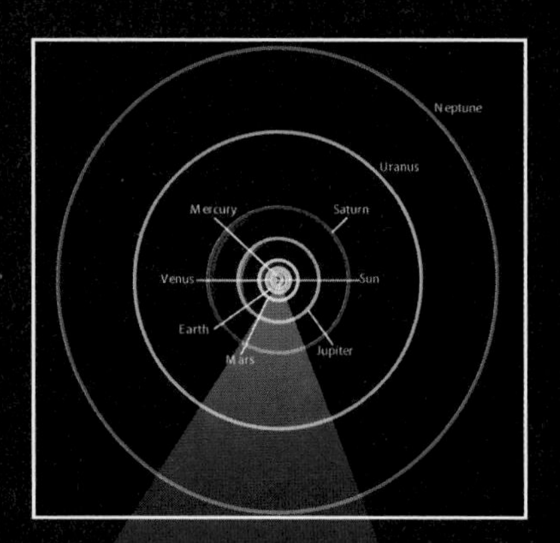

All the planets except Mercury and Venus have **satellites.** Ours is the Moon.

Smaller objects also orbit the Sun:

Asteroids: rocky, planetlike objects found in the inner solar system

Trans-Neptunian objects: bodies of rock and ice that orbit beyond Neptune

Comets: small objects of rock and ice from the outer solar system that can venture close to the Sun

Our Sun is a typical star.

Stars are so numerous they can look like glowing grains of sand scattered across the sky.

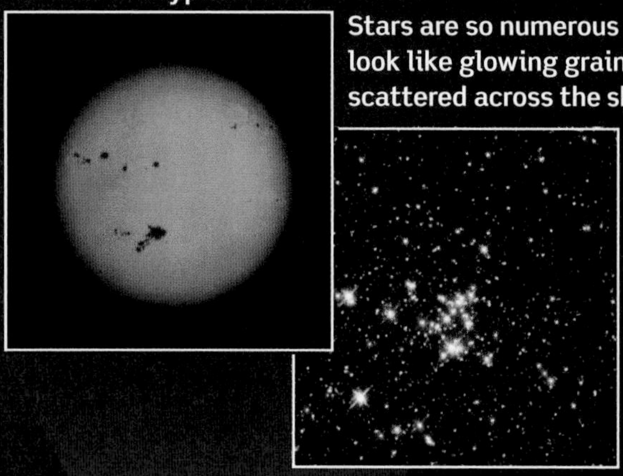

Stars are born in huge clouds of interstellar dust and gas called **nebulae.**

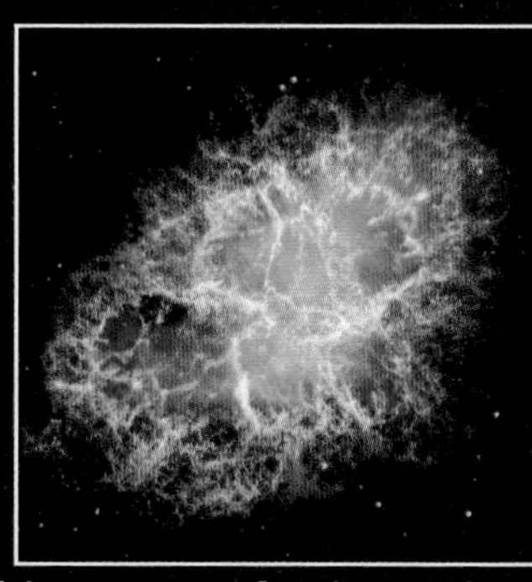

Some stars end their lives with a spectacular detonation, called a **supernova,** that blows the star apart.

Some dead stars become dizzily spinning pulsars; others become inconceivably dense objects called **black holes,** from which not even light can escape.

Stars are grouped together in huge assemblages called *galaxies.* Our Sun is only one of the stars in the Milky Way Galaxy.

Galaxies are grouped into clusters.

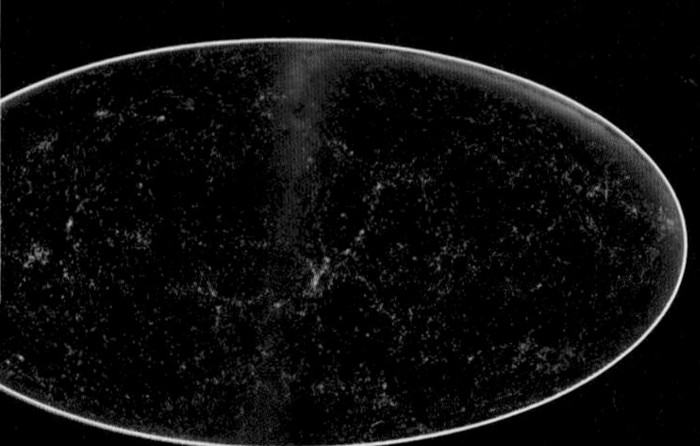

Each of the dots in this map of the entire sky represents a relatively nearby galaxy. This is a tiny fraction of the number of galaxies in the observable universe.

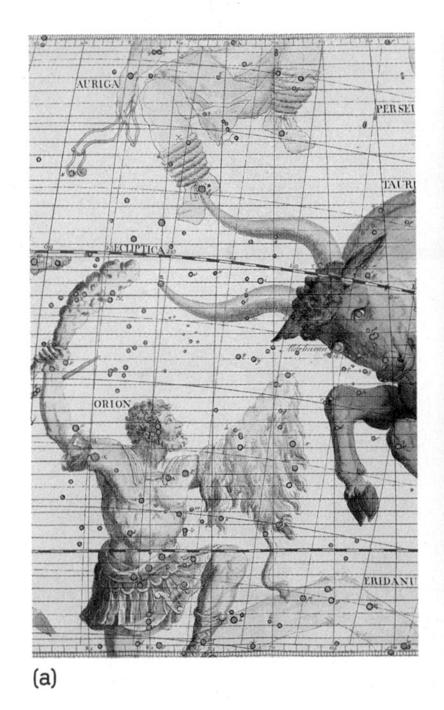

(a)

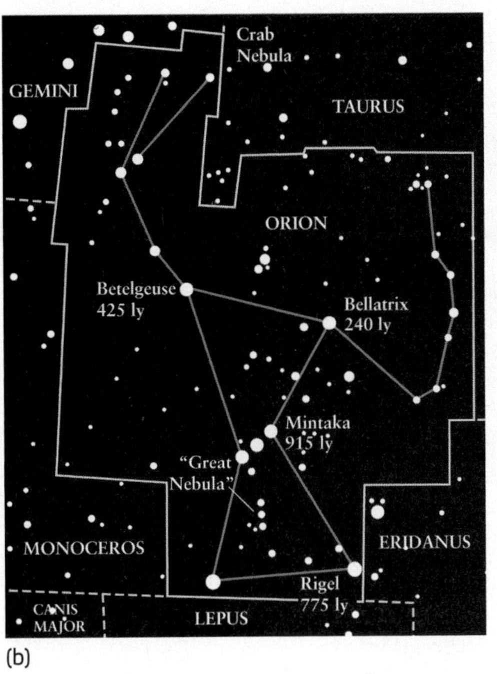

(b)

(c)

FIGURE 1-3
R I V U X G
Three Views of Orion The constellation Orion is easily seen on nights from December through March. (a) This fanciful drawing from a star atlas published in 1835 shows Orion the Hunter as well as other celestial creatures. (b) A portion of a modern star atlas shows some of the stars in Orion. The yellow lines show the constellation borders between Orion and its neighboring constellations (labeled in capitals), and the blue lines outline a more easily recognizable asterism. (c) This photograph of Orion shows many more stars than can be seen with the naked eye. *(a: The Stapleton Collection/The Bridgeman Art Library; c: Luke Dodd/Science Photo Library/Science Source)*

sky into semirectangular regions called **constellations** (from the Latin for "group of stars"). Many constellations, such as Orion in **Figure 1-3,** have names derived from the myths and legends of antiquity. Although some star groupings vaguely resemble the figures for which they are named (see Figure 1-3*a*), most do not. Constellations rarely, if ever, actually look like pictures of any kind.

The term "constellation" has a broader definition in present-day astronomy. On modern star charts, the entire sky is divided into 88 regions, each of which is called a constellation. For example, the constellation Orion is now defined to be an irregular patch of sky whose borders are shown in Figure 1-3*b*. When astronomers refer to the "Great Nebula" in Orion (see Figure 1-3*c*), they mean that as seen from Earth this nebula appears to be within Orion's patch of sky. Some constellations cover large areas of the sky (Ursa Major being one of the biggest) and others very small areas (Crux, the Southern Cross, being the smallest). But because the modern constellations cover the entire sky, every star lies in one constellation or another.

When you look at a constellation's star pattern, it is tempting to conclude that you are seeing a group of stars that are all relatively close together. In fact, most of these stars are nowhere near one another. As an example, in Figure 1-3*b,* although Bellatrix (Latin for "female warrior") and Mintaka (Arabic for "the belt") appear to be close to each other, Mintaka is actually farther away from us. The two stars only *appear* to be close because they are in nearly the same direction as seen from Earth. The same illusion often appears when you see an airliner's lights at night. It is very difficult to tell how far away a single bright light is, which is why you can mistake an airliner a few kilometers away for a star trillions of times more distant.

Many of the star names shown in Figure 1-3*b* are drawn from the Arabic language. For example, Betelgeuse is sometimes translated as "armpit," which makes sense when you look at the star atlas drawing in Figure 1-3*a*. Other types of names are also used for stars. For example, the brightest star in the night sky, Sirius, is also known as α Canis Major because it is also the brightest star in Canis Major (α, or alpha, is the first letter in the Greek alphabet).

CAUTION A number of commercial firms offer to name a star for you for a fee. The money that they charge you for this "service" is real, but the star names are not; none of these names is recognized by professional astronomers. It can be a fun gift, but if you want to use astronomy to formally commemorate your name or the name of a friend or relative, consider making a donation to your local planetarium or science museum. The money will be put to much better use!

Constellations Are a Map in the Night Sky

Easily recognizable groups of stars can help you find your way around the sky. For example, if you live in the northern hemisphere, you can use the asterism of the Big Dipper in the constellation of Ursa Major to find the north direction by drawing a straight line through the two stars at the front of the Big Dipper's bowl (**Figure 1-4**). The first moderately bright star you come to is Polaris, also called the North Star because it is located almost directly over Earth's north pole. If you draw a line from Polaris straight down to the horizon, you will find the north direction.

Astronomers group the thousands of stars visible in the night sky into semirectangular regions called constellations and recognizable connect-the-dot shapes called asterisms.

As Figure 1-4 shows, by following the handle of the Big Dipper you can locate the bright reddish star Arcturus in Boötes (the Shepherd) and the prominent bluish star Spica in Virgo (the Virgin). The saying "Follow the arc to Arcturus

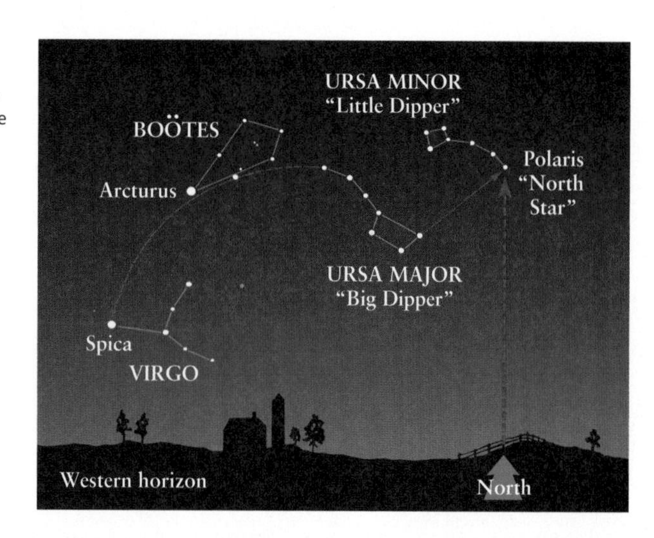

FIGURE 1-4 The Big Dipper as a Guide Polaris can be seen from anywhere in the northern hemisphere on any night of the year. This star chart shows how the Big Dipper (Ursa Major) can be used to point out Polaris as well as the brightest stars in two other constellations. The chart shows the sky at around 11 P.M. (daylight saving time) on August 1.

and speed on to Spica" may help you remember these stars, which are conspicuous in the evening sky during the spring and summer.

During winter in the northern hemisphere, you can see some of the brightest stars in the sky. Many of them are in the vicinity of the "winter triangle" (**Figure 1-5**), which connects bright stars in the constellations of Orion (the Hunter), Canis Major (the Large Dog), and Canis Minor (the Small Dog).

A similar feature, the "summer triangle," graces the summer sky in the northern hemisphere. This triangle connects the brightest stars in Lyra (the Harp), Cygnus (the Swan), and Aquila (the Eagle) (**Figure 1-6**). A conspicuous portion of the Milky Way forms a beautiful background for these constellations, which are nearly overhead during the middle of summer at midnight.

Note that all the star charts in this section and at the end of this book are drawn for an observer in the northern hemisphere. If you live in the southern hemisphere, you can see constellations that are not visible from the northern hemisphere, and vice versa. In the next section we will see why this is so.

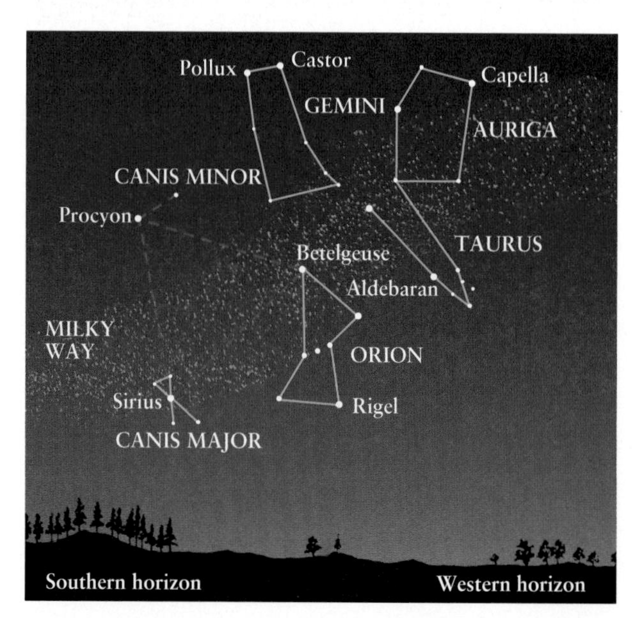

FIGURE 1-5 The Winter Triangle This star chart shows the view toward the southwest on a winter evening in the northern hemisphere (around midnight on January 1, 10 P.M. on February 1, or 8 P.M. on March 1). Three of the brightest stars in the sky make up the winter triangle. In addition to the constellations involved in the triangle, the chart shows the prominent constellations Gemini (the Twins), Auriga (the Charioteer), and Taurus (the Bull).

FIGURE 1-6 The Summer Triangle This star chart shows the eastern sky as it appears in the evening during spring and summer in the northern hemisphere (around 1 A.M. daylight savings time on June 1, around 11 P.M. on July 1, and around 9 P.M. on August 1). The constellations Sagitta (the Arrow) and Delphinus (the Dolphin) are much fainter than the three constellations that make up the triangle.

ConceptCheck 1-2: If Jupiter is reported to be in the constellation of Taurus the Bull, does Jupiter need to be within the outline of the bull's body? Why or why not?

Answer appears at the end of the chapter.

1-3 All of the observed celestial motions can be described if our planet Earth spins once each day while it orbits around our Sun each year

Go outdoors soon after dark, find a spot away from bright lights, and note the positions of stars in the sky. Do the same a few hours later. You will find that all of stars—as well as the Moon, if it is visible—have shifted position. New constellations will have risen above the eastern horizon, and others will have disappeared below the western horizon. If you look again before dawn, you will see that the stars that were just rising in the east when the night began are now low in the western sky. This daily motion of the stars can also be seen when looking toward the northern horizon where stars near a stationary North Star do not appear to rise and set, but move clockwise. These apparent motions are visible in time-exposure photographs like the one shown in **Figure 1-7.**

If you repeat your observations on the following night, you will find that the positions of stars are almost but not quite the same. The same constellations rise in the east and set

FIGURE 1-7 R I V U X G **Motions of the Stars** Earth's rotation makes stars appear to trace out circles in the sky. *(Peter Michaud/Gemini Observatory/AURA, NSF)*

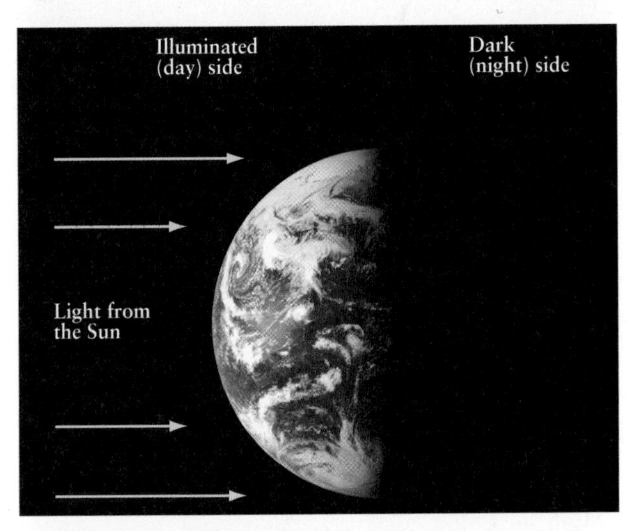

FIGURE 1-8 R I V U X G **Day and Night on Earth** At any moment, half of Earth is illuminated by the Sun. As Earth rotates from west to east, your location moves from the dark (night) hemisphere into the illuminated (day) hemisphere and back again. This image was recorded in 1992 by the *Galileo* spacecraft as it was en route to Jupiter. *(JPL/NASA)*

in the west, but they do so a few minutes earlier than on the previous night. If you look again after a month, the constellations visible at a given time of night (say, midnight) will be noticeably different, and after six months you will see an almost totally different set of constellations rising and setting. Only after a year has passed will the night sky have the same appearance as when you began your observations a year earlier.

Why does the sky go through this daily motion? Why do the constellations slowly shift from one night to the next? As we will see, the answer to the first question is that Earth spins once a day around an axis from the north pole to the south pole, while the answer to the second question is that Earth also orbits once a year around the Sun.

Daily Motion and Earth's Rotation

To understand daily motion, note that at any given moment it is daytime on the entire half of Earth illuminated by the Sun and nighttime on the other half (**Figure 1-8**). Earth rotates from west to east, making one complete rotation about every 24 hours, which is why there is a daily cycle of day and night. Because of this rotation, most of the stars appear to us to rise in the east and set in the west, as do the Sun and Moon.

Situated directly above Earth's north pole, Polaris (the North Star) is the only star that does not appear to move. The stars near Polaris appear to move because Earth spins beneath it, but they do not rise and set. These stars near Polaris appear to move from east to west for part of the night, and then, instead of setting, appear to move from west to east as they return to their original positions. In other words, these stars move clockwise around Polaris, never reaching the horizon.

Figure 1-9 helps to further explain daily motion. It shows two views of Earth as seen from a point above Earth's north pole. At the instant shown in Figure 1-9*a*, it

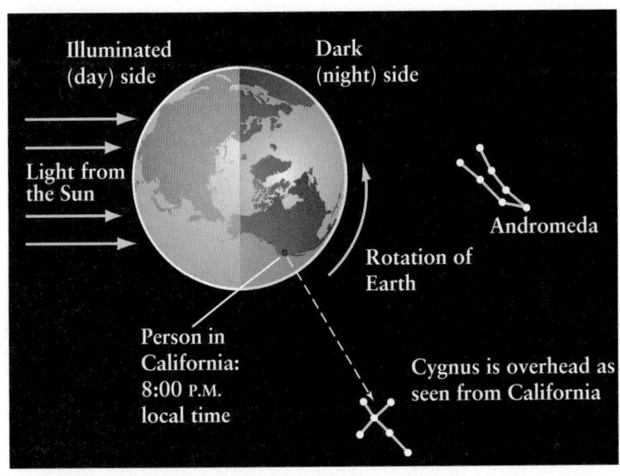

(a) Earth as seen from above the north pole

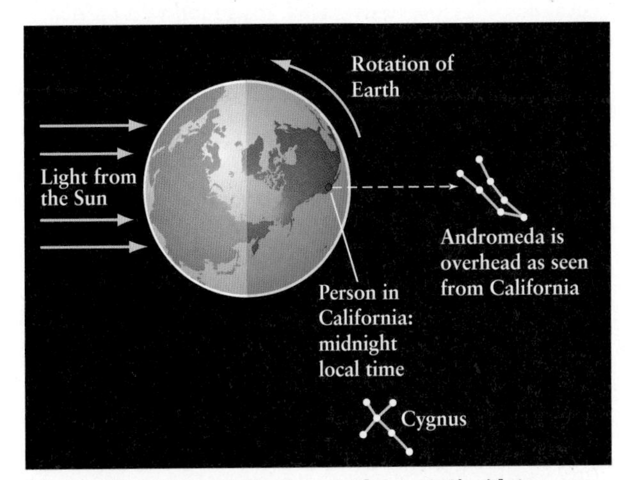

(b) Four hours (one-sixth of a complete rotation) later

FIGURE 1-9 Why Daily Motion Happens The apparent daily motion of the stars, Sun, and Moon from east to west across the sky is a consequence of Earth's rotation. (a) This drawing shows Earth from a vantage point above the north pole, with very distant stars in the background. In this drawing, imagine the red dot is a person in California, the local time is 8:00 P.M., and the constellation Cygnus is visible high overhead. (b) Four hours later, Earth has made one-sixth of a complete rotation to the east. As seen from Earth, the entire sky appears to have rotated to the west by one-sixth of a complete rotation. It is now midnight in California, and the constellation high over the observer in California is now Andromeda.

is daytime in Asia but nighttime in most of North America and Europe. Figure 1-9*b* shows Earth four hours later. Four hours is one-sixth of a complete 24-hour day, so Earth has made one-sixth of a rotation between Figures 1-9*a* and 1-9*b*. Europe is now in the illuminated half of Earth (the Sun has risen in Europe), while Alaska has moved from the illuminated half to the dark half of Earth (the Sun has set in Alaska). For a person in California, in Figure 1-9*a* the time is 8:00 P.M. and the constellation Cygnus (the Swan) is directly overhead. Four hours later, the constellation over California is Andromeda (named for a mythological princess). Because Earth rotates from west to east, it appears to us on Earth that the entire sky rotates around us in the opposite direction, from east to west.

ConceptCheck 1-3: **People in which one of the following cities in North America experience sunrise first: New York, San Francisco, Chicago, or Denver?**

CalculationCheck 1-1: **If the constellation Cygnus rises along the eastern horizon at sunset, at what time will it be highest above the southern horizon?**

Yearly Motion and Earth's Orbit

In addition to the daily motion of the stars in the sky, the specific constellations visible in the night sky also change slowly over the course of a year. This happens because Earth orbits the Sun (**Figure 1-10**). Over the course of a year, Earth makes one complete orbit, and the darkened, nighttime side of Earth gradually turns toward different parts of the heavens. For example, as seen from the northern hemisphere, at midnight in late July the constellation Cygnus is close to overhead; at midnight in late September the constellation Andromeda is close to overhead; and at midnight in late November the constellation Perseus (commemorating a mythological hero) is close to overhead. If you follow a particular star on successive evenings, you will find that it rises approximately four minutes earlier each night, or two hours earlier each month.

ConceptCheck 1-4: **If Earth suddenly rotated on its axis three times faster than it does now, then how many times would the Sun appear to rise and set each year?**

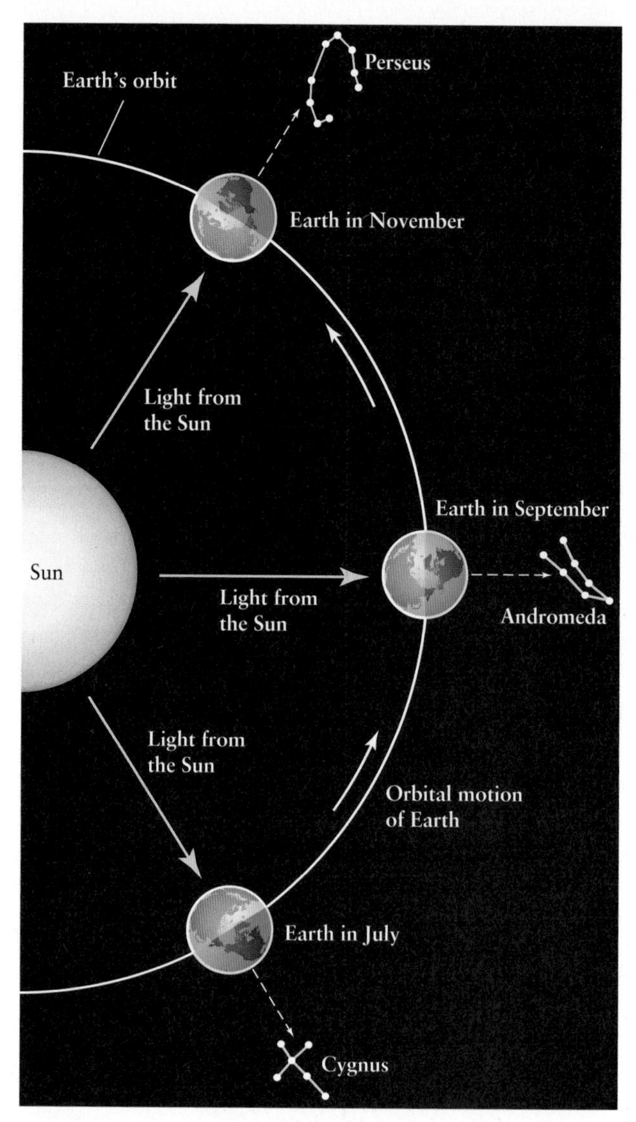

FIGURE 1-10 Why the Night Sky Changes During the Year As Earth orbits around the Sun, the nighttime side of Earth gradually turns toward different parts of the sky. Hence, the particular stars that you see in the night sky are different at different times of the year. This figure shows which constellation is overhead at midnight local time—when the Sun is on the opposite side of Earth from your location—during different months for observers at midnorthern latitudes (including the United States). If you want to view the constellation Andromeda, the best time of the year to do it is in late September, when Andromeda is nearly overhead at midnight.

Skywatchers and Astronomers Often Imagined the Sky to Be a Giant Celestial Sphere Surrounding Earth

As you look up into the night sky, one can imagine that all of the stars might be attached to a giant dome that stretches across the sky. Ancient skywatchers often imagined the stars to be bits of fire imbedded into the inner surface of an immense hollow sphere, called the **celestial sphere,** with Earth at its center, and all the stars at the same distance from Earth. In this picture of the universe, Earth was fixed and could not rotate. Instead, the entire celestial sphere rotated once a day around Earth from east to west, thereby causing the daily motion of the sky. This imaginary picture of a rotating celestial sphere fits well with how the sky appears to move and was a useful model of how the universe works.

Today's astronomers know that this simple model of the universe is not correct. The apparent changing sky over the course of a night is due to Earth's spin, not the rest of the universe moving around a stationary Earth. Furthermore, the stars are not all at the same distance from Earth. Indeed, the nearby stars you can see without a telescope can

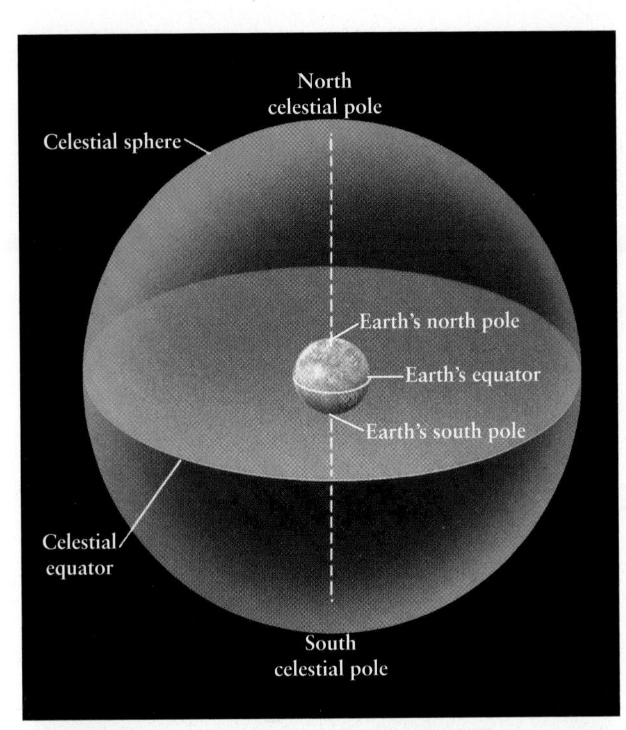

FIGURE 1-11 The Celestial Sphere The celestial sphere is the apparent sphere of the sky. The view in this figure is from the outside of this (wholly imaginary) sphere. Earth is at the center of the celestial sphere, so our view is always of the *inside* of the sphere. The celestial equator and poles are the projections of Earth's equator and axis of rotation out into space. The celestial poles are therefore located directly over Earth's poles.

be more than 1000 times closer than the most distant ones. And telescopes allow us to see objects even billions of times farther away.

Thus, the celestial sphere is an *imaginary* object that has no basis in physical reality. Nonetheless, the celestial sphere model is a useful framework for the study and tracking of objects in the sky. If we imagine, as did the ancients, that Earth is stationary and that the celestial sphere rotates around us, it is relatively easy to specify the directions to different objects in the sky and to visualize the motions of these objects.

Figure 1-11 depicts the celestial sphere, with Earth at its center. (A truly proportional drawing would show the celestial sphere as being millions of times larger than Earth.) We picture stars as tiny points of bright light that are fixed on the inner surface of a giant celestial sphere. If we project Earth's equator out into space, we obtain the **celestial equator** splitting the sky into two halves. The celestial equator divides the sky into northern and southern hemispheres, just as Earth's equator divides Earth into two hemispheres.

At the same time, if we project Earth's north and south poles into space, we obtain the points in the sky called the **north celestial pole** and the **south celestial pole.** The two celestial poles are where Earth's axis of rotation (extended out into space) intersects the celestial sphere (see Figure 1-11). Polaris is also known as the North Star because it is almost precisely above Earth's north pole.

In addition to the celestial pole positions, there is another important reference point in the sky. The point in the sky directly overhead an observer anywhere on Earth is called that observer's **zenith.** The zenith and celestial sphere are shown in **Figure 1-12** for an observer located at 35° north latitude—that is, at a location on Earth's surface 35° north of the equator. The zenith is shown at the top of Figure 1-12, so Earth and the celestial sphere appear "tipped" compared to Figure 1-11. At any time, an observer can see only half of the celestial sphere; the other half is below the horizon, hidden by the body of Earth. The hidden half of the celestial sphere is darkly shaded in Figure 1-12.

If you are standing at Earth's north pole, your zenith point is the same point as the north celestial pole on the celestial sphere.

Motions of the Celestial Sphere

For an observer anywhere in the northern hemisphere, including the observer in Figure 1-12, the north celestial pole is always above the northern horizon. As Earth turns from west to east, the celestial sphere seems to turn from east to west. Stars sufficiently near the north celestial pole revolve around the pole, never rising or setting. Such stars are called **circumpolar.** For example, as seen from North America or Europe, Polaris is a circumpolar star and can be seen at any time of night on any night of the year. Figure 1-7 shows the circular trails of stars around the north celestial pole. Stars near the south celestial pole revolve around that pole but always remain below the horizon of an observer in the northern hemisphere because Earth's body is in the way. Hence, these stars can never be seen by the observer in Figure 1-12. Stars between those two limits rise in the east and set in the west.

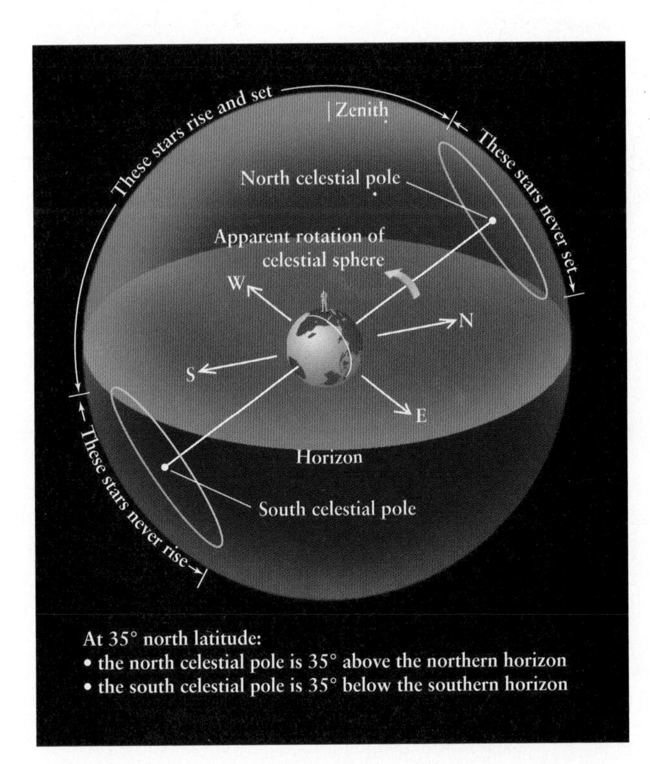

FIGURE 1-12 The View from 35° North Latitude To an observer at 35° north, the north celestial pole is always 35° above the horizon. Stars within 35° of the *north* celestial pole are circumpolar; they trace out circles around the north celestial pole during the course of the night and are always above the horizon on any night of the year. Stars within 35° of the *south* celestial pole are always below the horizon and can never be seen from this latitude. Stars that lie between these two extremes rise in the east and set in the west.

CAUTION Which stars are never-rising and never-setting circumpolar stars and which stars rise and set depends on the latitude from which you view the heavens. As an example, for an observer at 35° south latitude (roughly the latitude of Sydney, Cape Town, and Buenos Aires), the roles of the north and south celestial poles are the opposite of those shown in Figure 1-12. Objects close to the *south* celestial pole are circumpolar, that is, they revolve around that pole and never rise or set. For an observer in the southern hemisphere, stars close to the *north* celestial pole are always below the horizon and can never be seen. Hence, astronomers in Australia, South Africa, and Argentina never see Polaris but are able to see other stars that are forever hidden from North American and European observers.

BOX 1-1 TOOLS OF THE ASTRONOMER'S TRADE

Measuring Positions in the Sky

If you were standing outside and wanted to tell a friend about a particular star you are interested in, how would you do this? Simply saying, "The dim one over there sort of next to the bright one" probably isn't very helpful. Another strategy is to describe how far apart two objects seem to be using angles. An **angle** is the opening between two lines that meet at a point. The basic unit for angles is the **degree**, designated by the symbol ° (**Figure B1-1.1**).

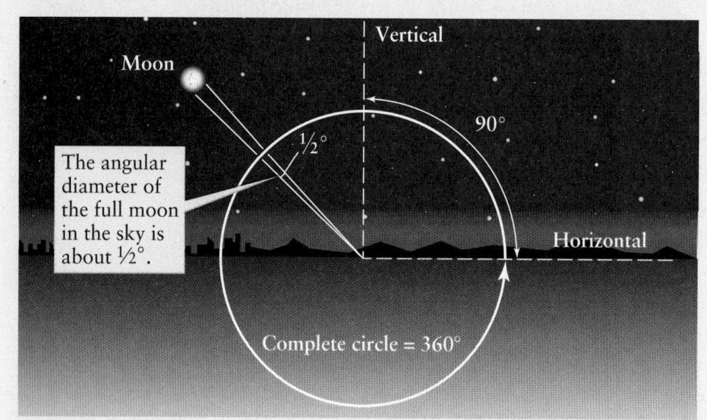

(a) Measuring angles in the sky

(b) Angular distances in the northern hemisphere

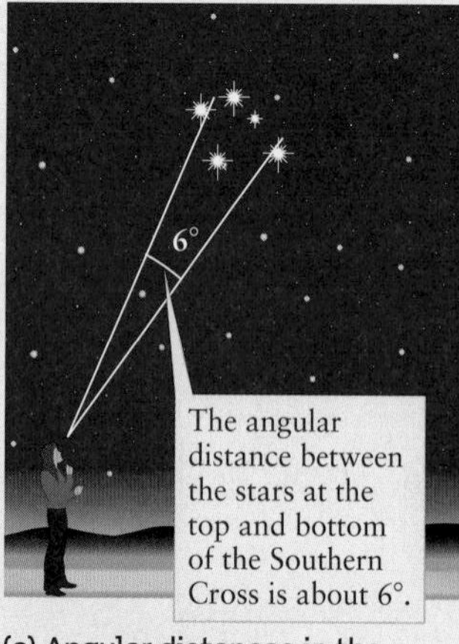

(c) Angular distances in the southern hemisphere

FIGURE B1-1.1 Angles and Angular Measure (a) Angles are measured in degrees (°). There are 360° in a complete circle and 90° in a right angle. For example, the angle between the vertical direction (directly above you) and the horizontal direction (toward the horizon) is 90°. The angular diameter of the full moon in the sky is about ½°. (b) The seven bright stars that make up the Big Dipper can be seen from anywhere in the northern hemisphere. The angular distance between the two "pointer stars" at the front of the Big Dipper is about 5°. (c) The four bright stars that make up the Southern Cross can be seen from anywhere in the southern hemisphere. The angular distance between the stars at the top and bottom of the cross is about 6°.

FIGURE B1-1.2 Estimating Angles with Your Hand The adult human hand extended to arm's length can be used to estimate angular distances and angular sizes in the sky.

Astronomers also use angles to describe the apparent size of a celestial object—that is, what fraction of the sky that object seems to cover. For example, the angle covered by the diameter of the full moon is about ½° (Figure B1-1.1*a*). We therefore say that the **angular diameter** of the Moon is ½° across or is ½° in size.

The adult human hand held at arm's length provides a means of estimating angles, as **Figure B1-1.2** shows. For example, your fist covers an angle of 10°, whereas the tip of your finger is about 1° wide. You can use various segments of your index finger, when your hand is extended to arm's length, to estimate angles a few degrees across.

To talk about smaller angles, we subdivide the degree into 60 **arcminutes** (also called minutes of arc), which is commonly abbreviated as 60 arcmin or 60′. An arcminute is further subdivided into 60 **arcseconds** (or seconds of arc), usually written as 60 arcsec or 60″. Thus,

$$1° = 60 \text{ arcmin} = 60'$$

$$1' = 60 \text{ arcsec} = 60''$$

(a) At middle northern latitudes

(b) At the north pole

(c) At the equator

FIGURE 1-13 R I V U X G **The Apparent Motion of Stars at Different Latitudes** As Earth rotates, stars appear to rotate around us along paths that are parallel to the celestial equator. (a) As shown in this long time exposure, at most locations on Earth the rising and setting motions are at an angle to the horizon that depends on the latitude. (b) At the north pole (latitude 90° north) the stars appear to move parallel to the horizon. (c) At the equator (latitude 0°) the stars rise and set along vertical paths. *(a: © David Miller/DMI; b: NPS Photo/Jacob W. Frank; c: O Chul Kwon)*

For observers at most locations on Earth, stars rise in the east and set in the west at an angle to the horizon (**Figure 1-13*a***). To see why this is so, notice that the rotation of the celestial sphere carries stars across the sky in paths that are parallel to the celestial equator. If you stand at the north pole, the north celestial pole is directly above you at the zenith (see Figure 1-11) and the celestial equator lies all around you at the horizon. Hence, as the celestial sphere rotates, the stars appear to move parallel to the horizon (Figure 1-13*b*). If instead you stand on the equator, the celestial equator passes from the eastern horizon through the zenith to the western horizon. The north and south celestial poles are 90° away from the celestial equator, so they lie, respectively, on the northern and southern horizons. As the celestial sphere rotates around an axis from pole to pole, the stars rise and set straight up and down—that is, in a direction perpendicular to the horizon (Figure 1-13*c*). At any location on Earth between the equator and either pole, the rising and setting motions of the stars are at an angle intermediate between Figures 1-13*b* and 1-13*c*. The particular angle depends on the latitude.

ConceptCheck 1-5: Where would you need to be standing on Earth for the celestial equator to pass through your zenith?

1-4 The Sun appears to change position over the day and throughout the year, and these changes result in Earth's seasons

As we travel with Earth around the Sun, we experience an annual cycle of seasons. But why are there seasons? Furthermore, the seasons are opposite in the northern and southern hemispheres. For example, February is midwinter in North America but midsummer in Australia. Why is this?

The Origin of the Seasons

The reason why we have seasons, and why they are different in different hemispheres, is that Earth's rotation axis is not perpendicular to the plane of Earth's orbit—in other words, Earth's axis does not appear to be straight up and down but tipped to one side. As **Figure 1-14** shows, Earth's rotational axis is tilted about 23½° away to the side. Earth maintains this tilt toward the north celestial pole as it orbits the Sun.

During part of the year, when Earth is in the part of its orbit shown on the left side of Figure 1-14, the northern hemisphere is tilted toward the Sun. As Earth spins on its axis, a point in the northern hemisphere spends more than

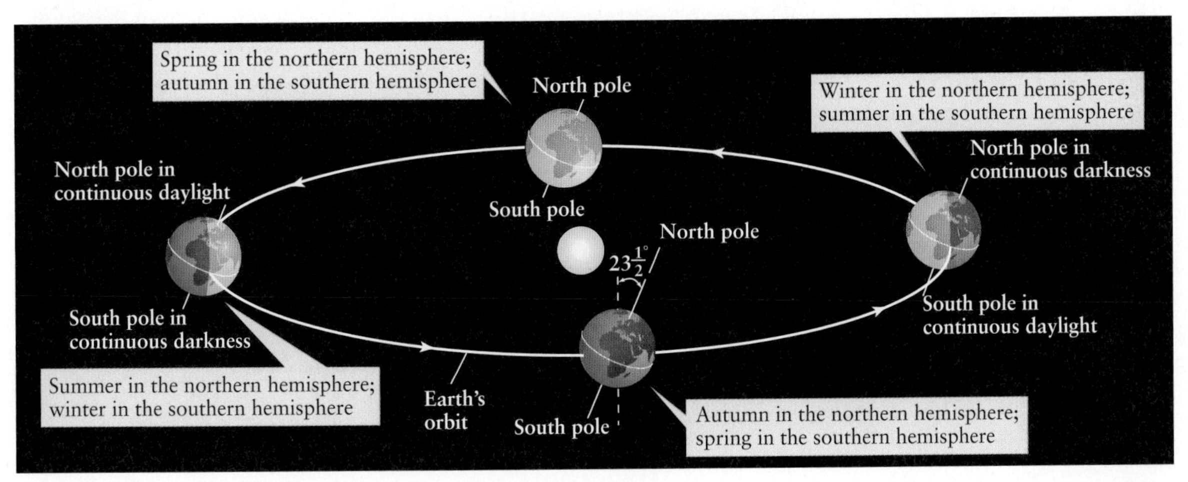

FIGURE 1-14
The Seasons Earth's axis of rotation is inclined 23½° away from the perpendicular to the plane of Earth's orbit. The north pole is aimed at the north celestial pole, near the star Polaris. Earth maintains this orientation as it orbits the Sun. Consequently, the amount of solar illumination and the number of daylight hours at any location on Earth vary in a regular pattern throughout the year. This is the origin of the seasons.

12 hours in the sunlight. Thus, the days there are long and the nights are short, and it is summer in the northern hemisphere. The summer is hot not only because of the extended daylight hours but also because the Sun is high in the northern hemisphere's sky. As a result, sunlight strikes the ground from a point near overhead that heats the ground efficiently (**Figure 1-15*a***). During this same time of year in the southern hemisphere, the days are short and the nights are long, because a point in this hemisphere spends fewer than 12 hours a day in the sunlight. The Sun is low in the sky, so sunlight strikes the surface at a grazing angle that causes little heating (Figure 1-15*c*), and it is winter in the southern hemisphere.

It is hotter in the summertime because the Sun is in the sky longer and reaches a higher altitude, not because Earth is any closer to the Sun than at other times of the year.

Half a year later, Earth is in the part of its orbit shown on the right side of Figure 1-14. Now the situation is reversed, with winter in the northern hemisphere and summer in the southern hemisphere. It is important to notice that Earth's axis is still pointed in the same direction. But which of Earth's hemispheres is receiving the most sunlight changes? During spring and autumn, the two hemispheres receive roughly equal amounts of illumination from the Sun, and daytime and nighttime are of equal length everywhere on Earth.

CAUTION A common misconception is that the seasons are caused by variations in the distance from Earth to the Sun. According to this idea, Earth is closer to the Sun in summer and farther away in winter. But in fact, Earth's orbit around the Sun is very nearly circular, and the Earth-Sun distance varies only about 3% over the course of a year. (Earth's orbit only *looks* elongated in Figure 1-14 because this illustration shows an oblique side view.) Earth is slightly closer to the Sun in January than in July, but this small variation has little influence on the cycle of the seasons. Also, if the seasons were really caused by variations in Earth-Sun distance, the seasons would be the same in both hemispheres!

ConceptCheck 1-6: If Earth's axis were not tilted, but rather was straight up and down compared to the path of Earth's orbit, would observers at Earth's north pole still observe periods where the Sun never rises and the Sun never sets?

How the Sun Moves on the Celestial Sphere

Each morning, the Sun appears above the eastern horizon and slowly and continuously moves up into the southern sky. After about half of the day, the Sun, having reached its highest point in the sky at about noon, slowly makes its way down to the western horizon and sets. This rising and setting pattern occurs day after day with seemingly little change.

However, upon closer inspection, one can observe that our Sun does not follow the same path every day nor does it always cover the same stars with its brilliant light. Rather, the Sun slowly appears to cover first one constellation, then another, then another over the course of a year, sometimes high in the sky at noon and sometimes low in the sky. This circular path that the Sun appears to trace out against the background of stars over the course of a year is called the **ecliptic** (**Figure 1-16*a***). The plane of this path is the same as the ecliptic plane (Figure 1-16*b*). (The name *ecliptic* suggests that the path traced out by the Sun has something to do with eclipses. We will discuss the connection later in this chapter.) Because there are 365¼ days in a year and 360° in a circle, the Sun appears to move along the ecliptic at a rate of about 1° per day. This motion is from west to east, that is, in the direction opposite to the apparent motion of the celestial sphere.

ANALOGY Note that at the same time that the Sun is making its yearlong trip around the ecliptic, the entire celestial sphere is rotating around us once per day. You can envision the celestial sphere as a merry-go-round rotating clockwise, and the Sun as a restless child who is walking slowly around the merry-go-round's rim in the counterclockwise direction. During the time it takes the child to make a round trip, the merry-go-round rotates 365¼ times.

The ecliptic plane is *not* the same as the plane of Earth's equator, thanks to the 23½° tilt of Earth's rotation axis, shown in Figure 1-14. As a result, the ecliptic and the celestial equator are inclined to each other by that same 23½° angle (**Figure 1-17**).

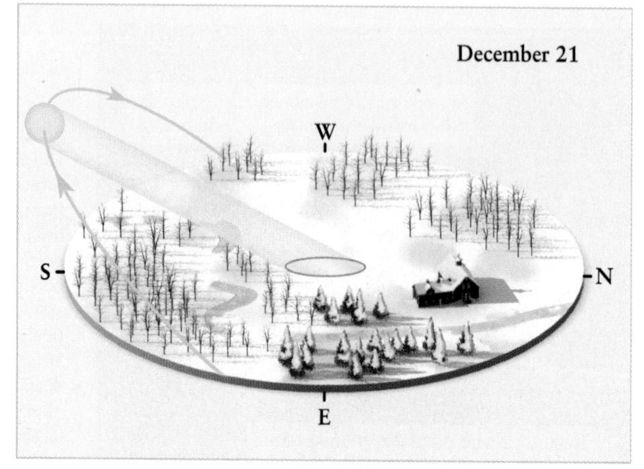

(a)

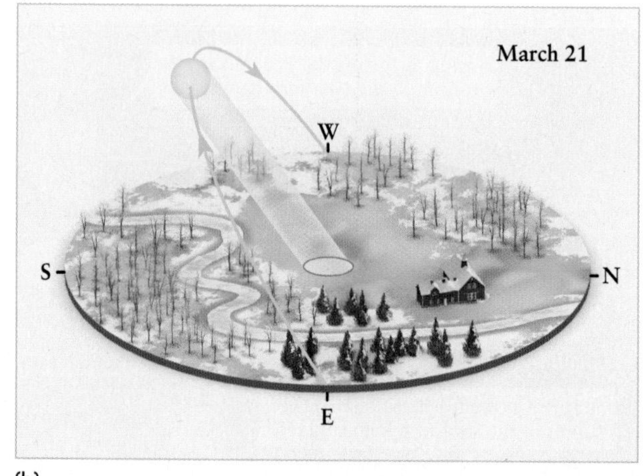

(b)

FIGURE 1-15 Solar Energy in Summer and Winter At different times of the year, sunlight strikes the ground at different angles. As illustrated for northern mid-latitudes, we see that: (a) and (b) In winter, the sunlight is less concentrated, the days are short, and little heating of the ground takes place. This accounts for the low temperatures in winter. (c) and (d) In summer, sunlight is concentrated and the days are also longer, which further increases the heating.

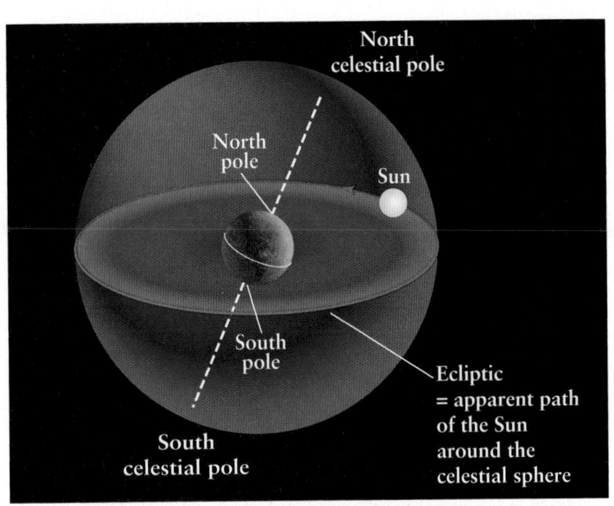

(a) It appears from Earth that the Sun travels around the celestial sphere once a year

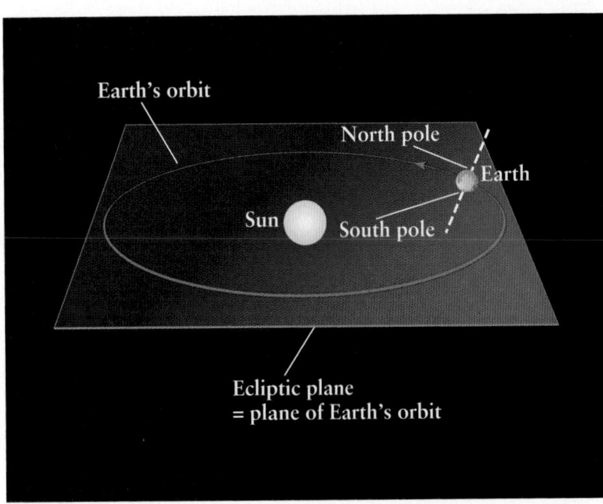

(b) In reality Earth orbits the Sun once a year

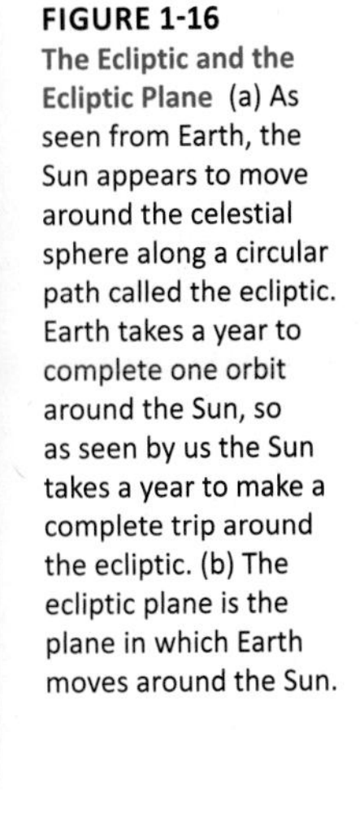

FIGURE 1-16 The Ecliptic and the Ecliptic Plane (a) As seen from Earth, the Sun appears to move around the celestial sphere along a circular path called the ecliptic. Earth takes a year to complete one orbit around the Sun, so as seen by us the Sun takes a year to make a complete trip around the ecliptic. (b) The ecliptic plane is the plane in which Earth moves around the Sun.

ConceptCheck 1-7: How long does the Sun take to move from being next to a bright star, all the way around the celestial sphere, and back to that same bright star?

Equinoxes and Solstices

The ecliptic and the celestial equator intersect at only two points, which are exactly opposite each other on the celestial sphere (see Figure 1-17). Each point is called an **equinox** (Latin for "equal night"), because when the Sun appears to cover either of these points, day and night are each about 12 hours long at all locations on Earth. The term "equinox" is also used to refer to the date on which the Sun passes in front of one of these special points on the ecliptic.

On about March 21 of each year, the Sun crosses northward across the celestial equator at a point on the sky known as the **March equinox.** This marks the beginning of spring in the northern hemisphere (it was once known as the "vernal equinox," from the Latin for "spring," but we more often use the term March equinox because it is less Eurocentric). On about September 22 the Sun moves southward across

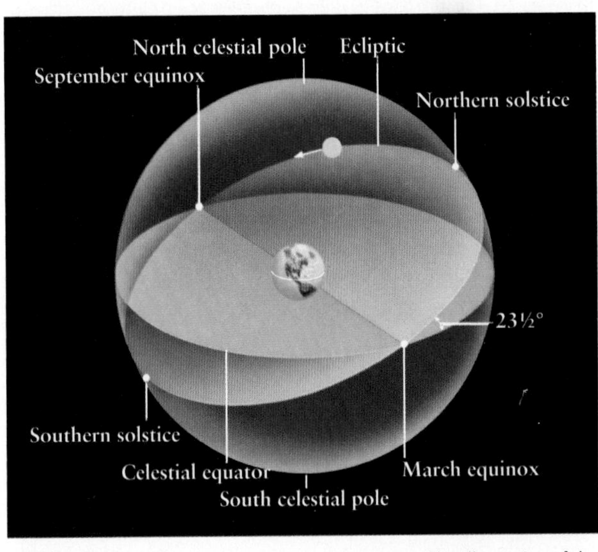

FIGURE 1-17 The Ecliptic, Equinoxes, and Solstices This illustration of the celestial sphere is similar to Figure 1-16*b*, but it is drawn with the north celestial pole at the top and the celestial equator running through the middle. The ecliptic is inclined to the celestial equator by 23½° because of the tilt of Earth's axis of rotation. It intersects the celestial equator at two points, called equinoxes. The northernmost point on the ecliptic is the northern solstice, and the southernmost point is the southern solstice. The Sun is shown in its approximate position for August 1.

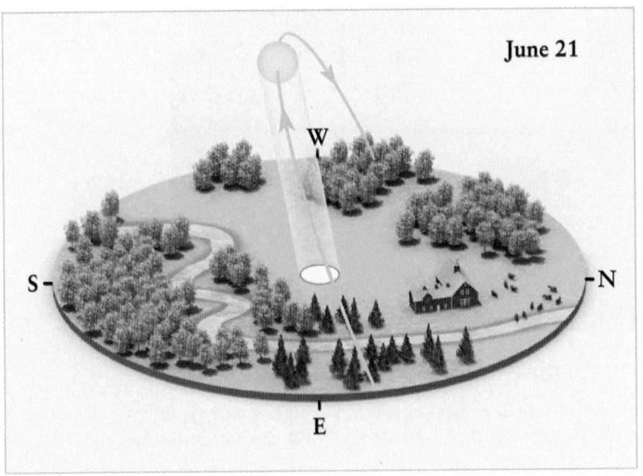

(c)

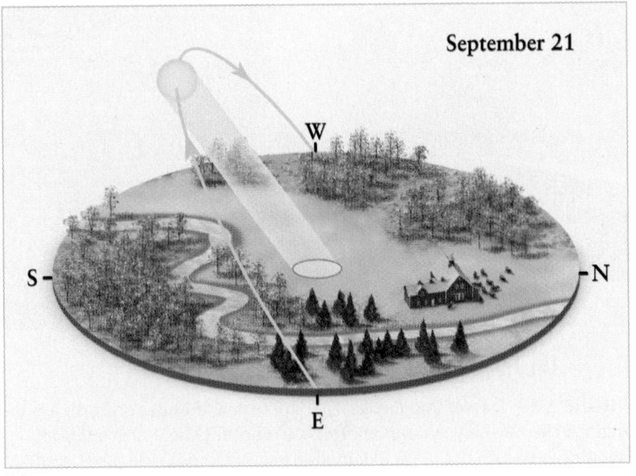

(d)

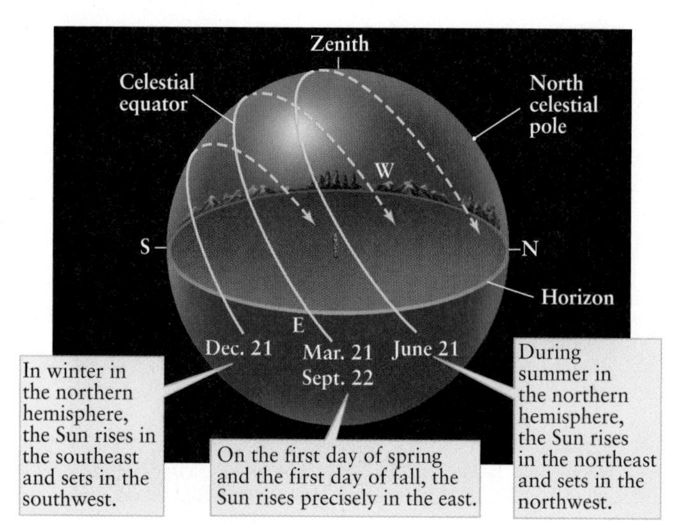

FIGURE 1-18 The Sun's Daily Path Across the Sky This drawing shows the apparent path of the Sun during the course of a day on four different dates. As in Figure 1-16*a*, this drawing is for an observer at 35° north latitude.

the celestial equator at the **September equinox,** marking the moment when autumn (or fall) begins in the northern hemisphere. Since the seasons are opposite in the northern and southern hemispheres, for Australians and South Africans the March equinox actually marks the beginning of autumn.

Between the March and September equinoxes, two other significant locations lie along the ecliptic. The point on the ecliptic farthest north of the celestial equator is called the **northern solstice.** The word "solstice" is from the Latin for "solar standstill," and it is at the northern solstice that the Sun stops moving northward on the celestial sphere. At this point, the Sun is as far north of the celestial equator as it can get. It marks the location of the Sun at the moment summer begins in the northern hemisphere—about June 21—and has traditionally been called the summer solstice, even though it marks the start of winter in the southern hemisphere. At the beginning of the northern hemisphere's winter—about December 21—the Sun is farthest south of the celestial equator at a point called the **southern solstice,** which, again, has traditionally been known as the winter solstice, even though it marks the beginning of summer in the southern hemisphere.

Because the Sun's position on the celestial sphere varies slowly over the course of a year, its daily path across the sky (due to Earth's rotation) also varies with the seasons (**Figure 1-18**). On the first day of spring or the first day of autumn, when the Sun is at one of the equinoxes, the Sun rises directly in the east and sets directly in the west.

When the northern hemisphere is tilted away from the Sun and it is winter in the northern hemisphere, the Sun rises in the southeast. Daylight lasts for fewer than 12 hours as the Sun skims low over the southern horizon and sets in the southwest. Northern hemisphere nights are longest when the Sun is at the southern solstice.

If you were to travel from Earth's equator toward the north pole, you would find that as you get farther north, the winter days get progressively shorter and the winter nights get progressively longer. In fact, anywhere within 23½° of Earth's north pole (that is, north of latitude 90° − 23½° = 66½° north latitude), the Sun is below the horizon for 24 continuous hours at least one day of the year. The circle around Earth at 66½° north latitude is called the **Arctic Circle** (**Figure 1-19**). The corresponding region around the south pole is bounded by the **Antarctic Circle** at 66½° south latitude. At the time of the winter solstice, explorers south of the Antarctic Circle enjoy the "midnight sun," or 24 hours of continuous daylight.

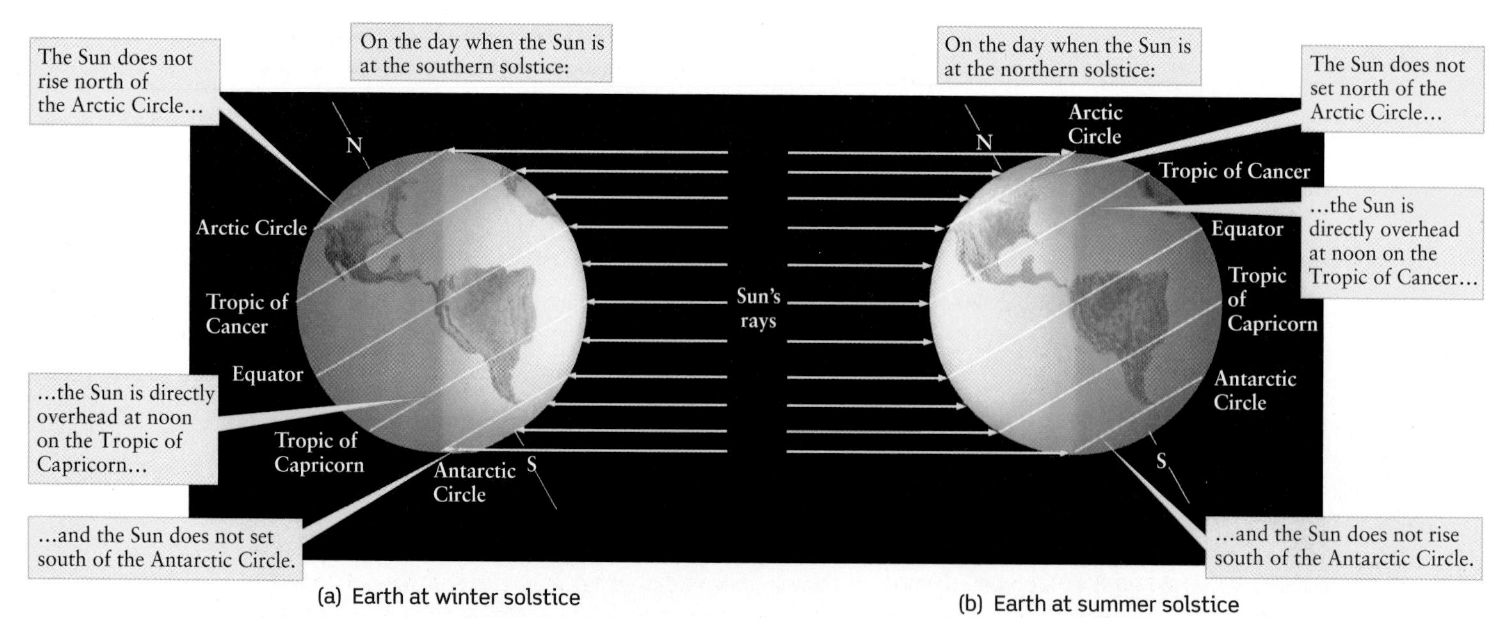

FIGURE 1-19 Tropics and Circles Four important latitudes on Earth are the Arctic Circle (66½° north latitude), Tropic of Cancer (23½° north latitude), Tropic of Capricorn (23½° south latitude), and Antarctic Circle (66½° south latitude). These drawings show the significance of these latitudes when the Sun is (a) at the southern solstice and (b) at the northern solstice.

Go to Video 1-3

The variations of the seasons are much less pronounced close to the equator. Between the **Tropic of Capricorn** at 23½° south latitude and the **Tropic of Cancer** at 23½° north latitude, the Sun is directly overhead—that is, at the zenith—at high noon at least one day a year. Outside of the tropics, the Sun is never directly overhead, but it is always either south of the zenith (as seen from locations north of the Tropic of Cancer) or north of the zenith (as seen from south of the Tropic of Capricorn).

ConceptCheck 1-8: **How often each year does an observer standing on Earth's equator experience no shadow during the noontime Sun?**

CalculationCheck 1-2: **Approximately how many days are there between the northern solstice and the March equinox?**

1-5 The Moon appears to change its position in the sky hourly and its phase throughout each month

Go to Video 1-4

As seen from Earth, both the Sun and the Moon appear to move from west to east on the celestial sphere—that is, relative to the background of stars. They move at very different rates, however. The Sun takes one year to make a complete trip around the imaginary celestial sphere along the path we call the *ecliptic*. By comparison, the Moon takes only about four weeks. In the past, these similar motions led people to believe that both the Sun and the Moon orbit around Earth. We now know that only the Moon orbits Earth, while the Earth-Moon system as a whole (**Figure 1-20**) orbits the Sun. (In Chapter 4 we will learn how this was discovered.)

One key difference between the Sun and the Moon is the nature of the light that we receive from them. The Sun emits its own light. So do the stars, which are objects like the Sun but much farther away, and so does an ordinary lightbulb. By contrast, the light that we see from the Moon is reflected light. This is sunlight that has struck the Moon's surface, bounced off, and ended up in our eyes here on Earth.

CAUTION You probably associate *reflection* with shiny objects like a mirror or the surface of a still lake. In science, however, the term refers to light bouncing off any object. You see most objects around you by reflected light. When you look at your hand, for example, you are seeing light from the Sun (or from a light fixture) that has been reflected from the skin of your hand and into your eyes. In the same way, moonlight is really sunlight that has been reflected by the Moon's surface.

FIGURE 1-20 RIVUXG
Earth and the Moon This picture of Earth and the Moon was taken by the *Galileo* spacecraft on its way toward Jupiter. The Sun, which provides the illumination for both Earth and the Moon, was far to the right and out of the camera's field of view when this photograph was taken. *(JPL/NASA)*

Understanding the Moon's Phases

Figure 1-20 shows both Earth and the Moon as seen from a spacecraft. When this image was recorded, the Sun was far off to the right. Hence, only the right-hand hemispheres of both worlds were illuminated by the Sun; the left-hand hemispheres were in darkness and are not visible in the picture. In the same way, when we view the Moon from Earth, we see only the half of the Moon that faces the Sun and is illuminated. However, not all of the illuminated half of the Moon is necessarily facing us. As the Moon moves around Earth, from one night to the next we see different amounts of the illuminated half of the Moon. These different appearances of the Moon are called **lunar phases.**

Figure 1-21 shows the relationship between the lunar phase visible from Earth and the position of the Moon in its orbit. For example, when the Moon is at position A, we see it in roughly the same direction in the sky as the Sun. Hence, the dark hemisphere of the Moon faces Earth. This phase, in which the Moon is barely visible, is called **new moon.** Since a new moon is near the Sun in the sky, it rises around sunrise and sets around sunset.

As the Moon continues around its orbit from position A in Figure 1-21, more of its illuminated half becomes exposed to our view. The result, shown at position B, is a phase called **waxing crescent moon** ("waxing" is a synonym for "increasing"). About a week after new moon, the Moon is at position C; we then see half of the Moon's illuminated hemisphere and half of the dark hemisphere. This phase is called **first quarter moon.**

As seen from Earth, a first quarter moon is one-quarter of the way around the celestial sphere from the Sun. It rises and sets about one-quarter of an Earth rotation, or six hours, after the Sun does: Moonrise occurs around noon, and moonset occurs around midnight.

One half of our Moon is always lit by the Sun. The Moon's phases are a result of our seeing more or less of the illuminated side.

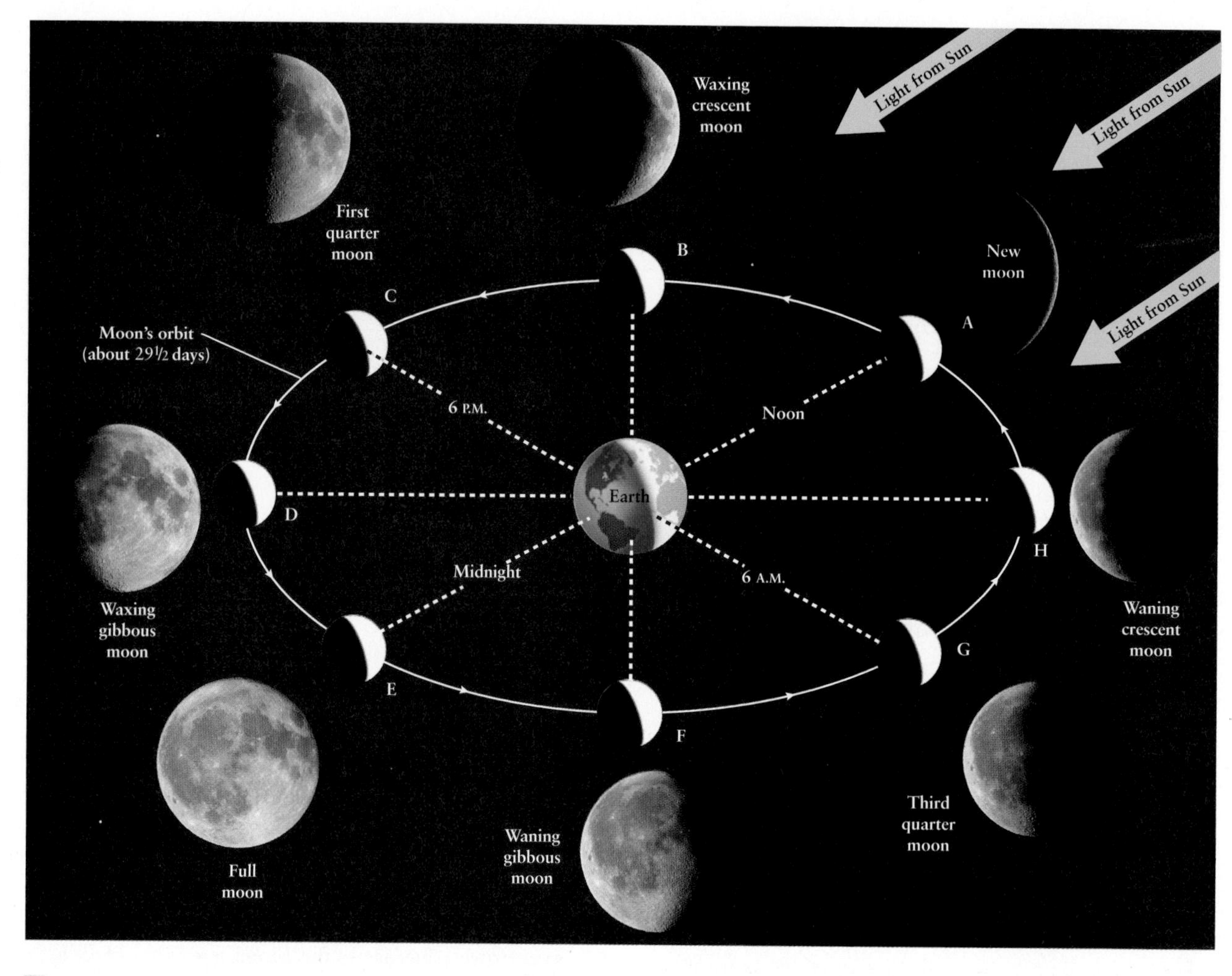

FIGURE 1-21
Why the Moon Goes Through Phases This figure shows the Moon at eight positions on its orbit, along with photographs of what the Moon looks like at each position as seen from Earth. The changes in phase occur because light from the Sun illuminates one half of the Moon, and as the Moon orbits Earth, we see varying amounts of the Moon's illuminated half. It takes about 29½ days for the Moon to go through a complete cycle of phases. *(Photos courtesy of Larry Landolfi/ Science Source)*

CAUTION Despite the name, a first quarter moon appears to be *half* illuminated, not one-quarter illuminated! The name means that this phase is one-quarter of the way through the complete cycle of lunar phases.

About four days later, the Moon reaches position D in Figure 1-21. Still more of the illuminated hemisphere can now be seen from Earth, giving us the phase called **waxing gibbous moon** ("gibbous" is another word for "swollen"). When you look at the Moon in this phase, as in the waxing crescent and first quarter phases, the illuminated part of the Moon is toward the west. Two weeks after new moon, when the Moon stands opposite the Sun in the sky (position E), we see the fully illuminated hemisphere. This phase is called **full moon.** Because a full moon is opposite the Sun on the celestial sphere, it rises at sunset and sets at sunrise.

FIGURE 1-22
R I V U X G
The Moon During the Day The Moon can be seen during the daytime as well as at night. The time of day or night when it is visible depends on its phase. *(Katrina Brown/ Shutterstock)*

Over the following two weeks, we see less and less of the Moon's illuminated hemisphere as it continues along its orbit, and the Moon is said to be *waning* ("decreasing"). While the Moon is waning, its illuminated side is toward the east. The phases are called **waning gibbous moon** (position F), **third quarter moon** (position G, also called *last quarter moon),* and **waning crescent moon** (position H). A third quarter moon appears one-quarter of the way around the celestial sphere from the Sun, but on the opposite side of the celestial sphere from a first quarter moon. Hence, a third quarter moon rises and sets about one-quarter of an Earth rotation, or six hours, *before* the Sun: Moonrise is around midnight and moonset is around noon.

The Moon takes about four weeks to complete one orbit around Earth, so it likewise takes about four weeks for a complete cycle of phases from new moon to full moon and back to new moon. Since the Moon's position relative to the Sun on the celestial sphere is constantly changing, the times of moonrise and moonset are different on different nights. On average, the Moon rises and sets about an hour later each night.

Figure 1-21 also explains why the Moon is often visible in the daytime, as shown in **Figure 1-22.** From any location on Earth, about half of the Moon's orbit is visible at any time.

For example, if it is midnight at your location, you are in the middle of the dark side of Earth that faces away from the Sun. At that time you can easily see the Moon at positions C, D, E, F, or G. If it is midday at your location, you are in the middle of Earth's illuminated side, and the Moon will be easily visible if it is at positions A, B, C, G, or H. (The Moon is so bright that it can be seen even against the bright blue sky.) You can see that the Moon is prominent in the midnight sky for about half of its orbit and prominent in the midday sky for the other half.

CAUTION A very common misconception about lunar phases is that they are caused by the shadow of *Earth* falling on the Moon. As Figure 1-21 shows, this is not the case at all. Instead, phases are simply the result of our seeing the illuminated half of the Moon at different angles as the Moon moves around its orbit. To help you better visualize how this works, Box 1-2: Phases and Shadows describes how you can simulate the cycle shown in Figure 1-21 using ordinary objects on Earth. (As we will learn in Section 1-6, Earth's shadow does indeed fall on the Moon on rare occasions. When this happens, we see a lunar eclipse.)

Although the phase of the Moon is constantly changing, one aspect of its appearance remains the same: It always keeps essentially the same hemisphere, or face, toward Earth. Thus, you will always see the same craters and mountains on the Moon, no matter when you look at it; the only difference will be the angle at which these surface features are illumi-

BOX 1-2 THE HEAVENS ON EARTH

Phases and Shadows

Figure 1-21 shows how the relative positions of Earth, the Moon, and the Sun explain the phases of the Moon. You can visualize lunar phases more clearly by doing a simple experiment here on Earth. All you need are a small round object, such as an orange or a baseball, and a bright source of light, such as a street lamp or the Sun.

In this experiment, you play the role of an observer on Earth looking at the Moon, and the round object plays the role of the Moon. The light source plays the role of the Sun. Hold the object in your right hand with your right arm stretched straight out in front of you, with the object directly between you and the light source (position A in **Figure B1-2.1**). In this orientation the illuminated half of the object faces away from you, like the Moon when it is in its new phase (position A in Figure 1-21).

Now, slowly turn your body to the left so that the object in your hand "orbits" around you (toward positions C, E, and G in the illustration). As you turn, more and more of the illuminated side of the "moon" in your hand becomes visible, and it goes through the same cycle of phases—waxing crescent, first quarter, and waxing gibbous—as does the real Moon. When you have rotated through half a turn so that the light source is directly behind you, you will be looking face on at the illuminated side of the object in your hand. This corresponds to a full moon (position E in Figure 1-21). Make sure your body does not cast a shadow on the "moon" in your hand—that would correspond to a lunar eclipse!

As you continue turning to the left, more of the unilluminated half of the object becomes visible as it moves through the phases of waning gibbous, third quarter, and waning crescent. When your body has rotated back to the same orientation that you were in originally, the unilluminated half of your handheld "moon" is again facing toward you, and its phase is again new. If you continue to rotate, the object in your hand repeats the cycle of "phases," just as the Moon does as it orbits around Earth.

The experiment works best when there is just one light source around. If there are several light sources, such as in a room with several lamps turned on, the different sources will create multiple shadows and it will be difficult to see the phases of your handheld "moon." If you do the experiment outdoors using sunlight, you may find that it is best to perform it in the early morning or late afternoon, when shadows are most pronounced and the Sun's rays are nearly horizontal.

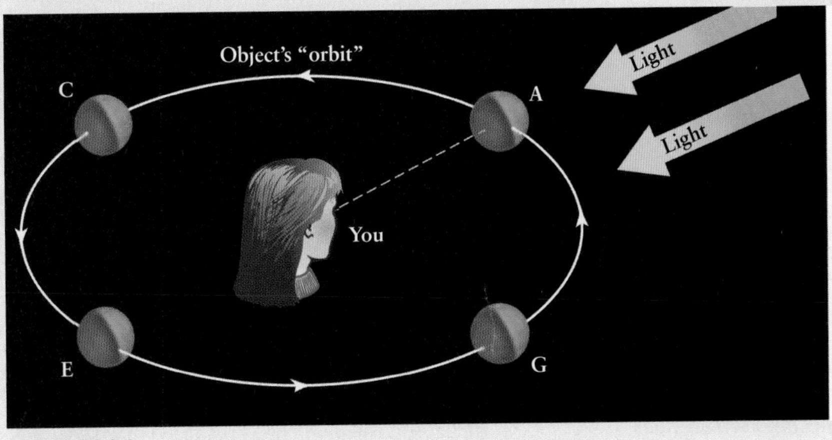

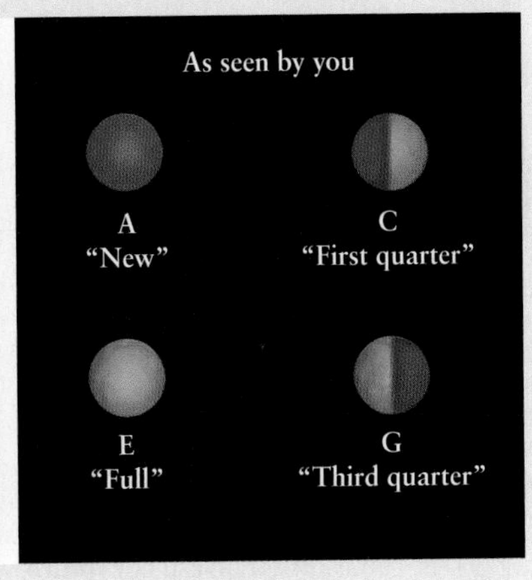

FIGURE B1-2.1

nated by the Sun. (You can verify this by carefully examining the photographs of the Moon in Figure 1-21.)

ConceptCheck 1-9: **If an observer on Earth sees just a tiny sliver of the crescent moon, how much of the Moon's total surface is being illuminated by the Sun?**

ConceptCheck 1-10: **If the Moon appears in its waxing crescent phase, how will it appear in two weeks?**

The Moon's Synchronous Rotation

Why is it that we only ever see one face of the Moon? You might think that it is because the Moon does not rotate (unlike Earth, which rotates around an axis that passes from its north pole to its south pole). To see that this *cannot* be the case, consider **Figure 1-23.** This figure shows Earth and the orbiting Moon from a vantage point far above Earth's north pole. In this figure two craters on the lunar surface have been colored, one in red and one in blue. If the Moon did not rotate on its axis, as in Figure 1-23*a,* at some times the red crater would be visible from Earth, while at other times the blue crater would be visible. Thus, we would see different parts of the lunar surface over time, which does not happen in reality.

In fact, the Moon always keeps the same face toward us because it *is* rotating, but in a very special way: It takes exactly as long to rotate on its axis as it does to make one orbit around Earth. This situation is called **synchronous rotation.** As Figure 1-23*b* shows, this keeps the crater shown in red always facing Earth, so that we always see the same face of the Moon.

An astronaut standing at the spot shown in red in Figure 1-23*b* would spend two weeks (half of a lunar orbit) in darkness, or lunar nighttime, and the next two weeks in sunlight, or lunar daytime. Thus, as seen from the Moon, the Sun rises and sets, and no part of the Moon is perpetually in darkness. This means that there really is no "dark side of the Moon." The side of the Moon that constantly faces away from Earth is properly called the *far side.* The Sun rises and sets on the far side just as on the side toward Earth. Hence, the blue crater on the far side of the Moon in Figure 1-23*b* is in sunlight for half of each lunar orbit.

ConceptCheck 1-11: **If astronauts landed on the Moon near the center of the visible surface at full moon, how many Earth days would pass before the astronauts experienced darkness on the Moon?**

Sidereal and Synodic Months

Go to Video 1-5

The time for a complete lunar "day"—the same as the time that it takes the Moon to rotate once on its axis—is about four weeks. (Because the Moon's rotation is synchronous, it takes the same time for one complete lunar orbit.) It also takes about four weeks for the Moon to complete one cycle of its phases as seen from Earth. This regular cycle of phases inspired our ancestors to invent the concept of a month. For historical reasons, none of which has much to do with the heavens, the calendar we use today has months of differing lengths. Astronomers find it useful to define two other types of months, depending on whether the Moon's motion is measured relative to the stars or to the Sun. Neither corresponds exactly to the familiar months of the calendar.

The **sidereal month** is the time it takes the Moon to complete one full orbit of Earth, as measured with respect to the stars. This true orbital period is equal to about 27.32 days. The **synodic month,** or *lunar month,* is the time it takes the Moon to complete one cycle of phases (that is, from new moon to new moon or from full moon to full moon) and thus is measured with respect to the Sun rather than the stars. The length of the "day" on the Moon is a synodic month, not a sidereal month.

The synodic month is longer than the sidereal month because Earth is orbiting the Sun while the Moon goes through

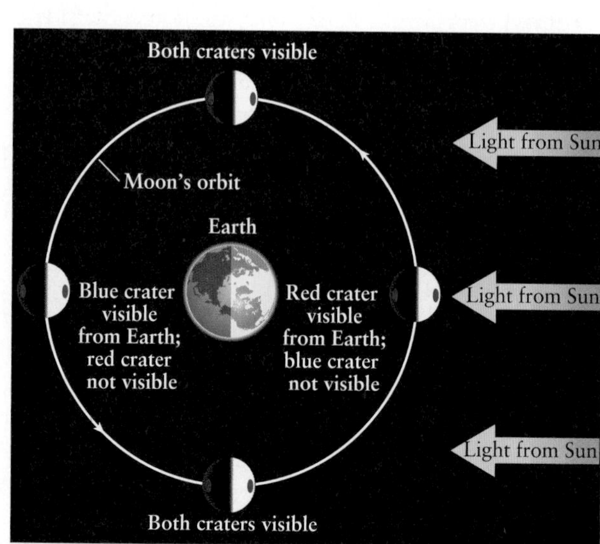

(a) If the Moon did not rotate, we could see all sides of the Moon

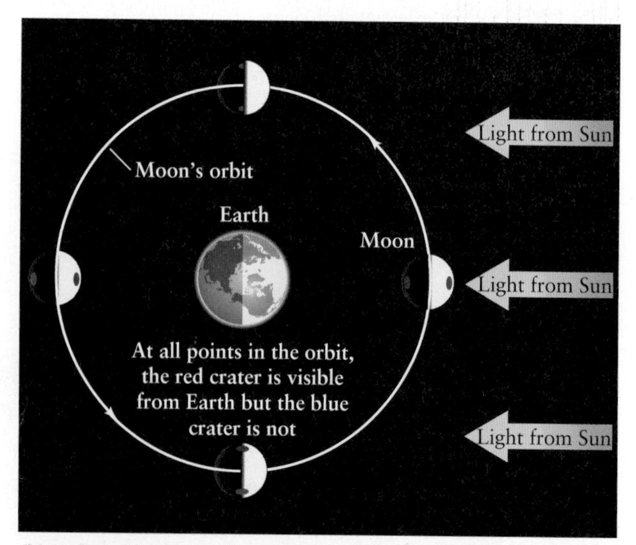

(b) In fact, the Moon does rotate, and we see only one face of the Moon

FIGURE 1-23
The Moon's Rotation These diagrams show the Moon at four points in its orbit as viewed from high above Earth's north pole. (a) If the Moon did not rotate, then at various times the red crater would be visible from Earth while at other times the blue crater would be visible. Over a complete orbit, the entire surface of the Moon would be visible. (b) In reality, the Moon rotates on its north-south axis. Because the Moon makes one rotation in exactly the same time that it makes one orbit around Earth, we only see one face of the Moon.

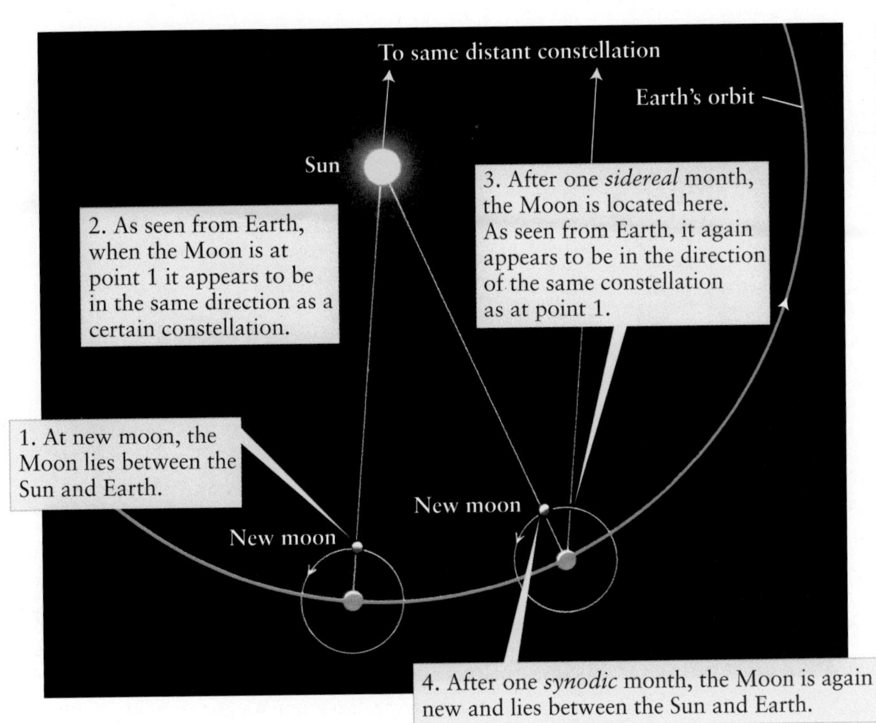

FIGURE 1-24 The Sidereal and Synodic Months The sidereal month is the time the Moon takes to complete one full revolution around Earth with respect to the background stars. However, because Earth is constantly moving along its orbit about the Sun, the Moon must travel through slightly more than 360° of its orbit to get from one new moon to the next. Thus, the synodic month—the time from one new moon to the next—is longer than the sidereal month.

its phases. As **Figure 1-24** shows, the Moon must travel *more* than 360° along its orbit to complete a cycle of phases (for example, from one new moon to the next). Because of this extra distance, the synodic month is equal to about 29.53 days, about two days longer than the sidereal month.

Both the sidereal month and synodic month vary somewhat from one orbit to another, the latter by as much as half a day. The reason is that the Sun's gravity sometimes causes the Moon to speed up or slow down slightly in its orbit, depending on the relative positions of the Sun, Moon, and Earth. Furthermore, the Moon's orbit changes slightly from one month to the next.

ConceptCheck 1-12: **If Earth was orbiting the Sun much faster than it is now, would the length of time between full moons increase, decrease, or stay the same?**

1-6 Eclipses occur only during rarely observed events when our Sun, Moon, and Earth are perfectly aligned

From time to time the Sun, Earth, and Moon all happen to lie along a straight line. When this occurs, Earth's shadow can fall on the Moon or the Moon's shadow can fall on Earth. Such phenomena are called **eclipses.** They are perhaps the most dramatic astronomical events that can be seen with the naked eye.

A **lunar eclipse** occurs when the Moon passes through Earth's shadow. This occurs when the Sun, Earth, and Moon are in a straight line, with Earth directly between the Sun and Moon so that the Moon is at full phase (position E in Figure 1-21). At this point in the Moon's orbit, the face of the Moon seen from Earth would normally be fully illuminated by the Sun. Instead, it appears quite dim because Earth casts a shadow on the Moon.

A **solar eclipse** occurs when Earth passes through the Moon's shadow. As seen from Earth, the Moon moves in front of the Sun. Once again, this can happen only when the Sun, Moon, and Earth are in a straight line. However, for a solar eclipse to occur, the Moon must be between Earth and the Sun. Therefore, a solar eclipse can occur only at new moon (position A in Figure 1-21).

CAUTION Both new moon and full moon occur at intervals of 29½ days. Hence, you might expect that there would be a solar eclipse every 29½ days, followed by a lunar eclipse about two weeks (half a lunar orbit) later. But in fact, there are only a few solar eclipses and lunar eclipses per year. Solar and lunar eclipses are so infrequent because the plane of the Moon's orbit and the plane of Earth's orbit are not exactly aligned, as **Figure 1-25** shows. The angle between the plane of Earth's orbit and the plane of the Moon's orbit is about

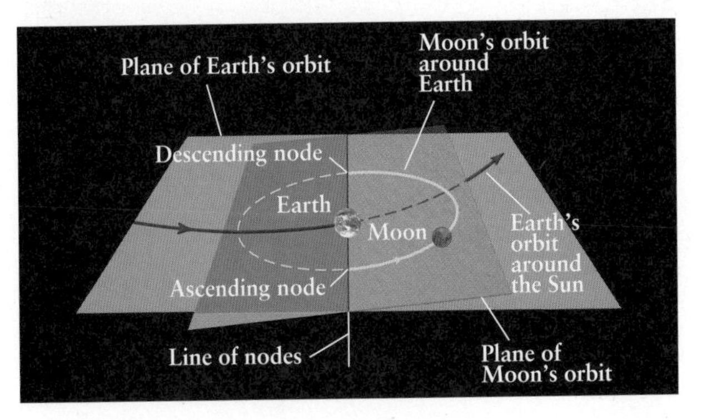

FIGURE 1-25 The Inclination of the Moon's Orbit This drawing shows the Moon's orbit around Earth (in yellow) and part of Earth's orbit around the Sun (in red). The plane of the Moon's orbit (shown in brown) is tilted by about 5° with respect to the plane of Earth's orbit, also called the plane of the ecliptic (shown in blue). These two planes intersect along a line called the line of nodes.

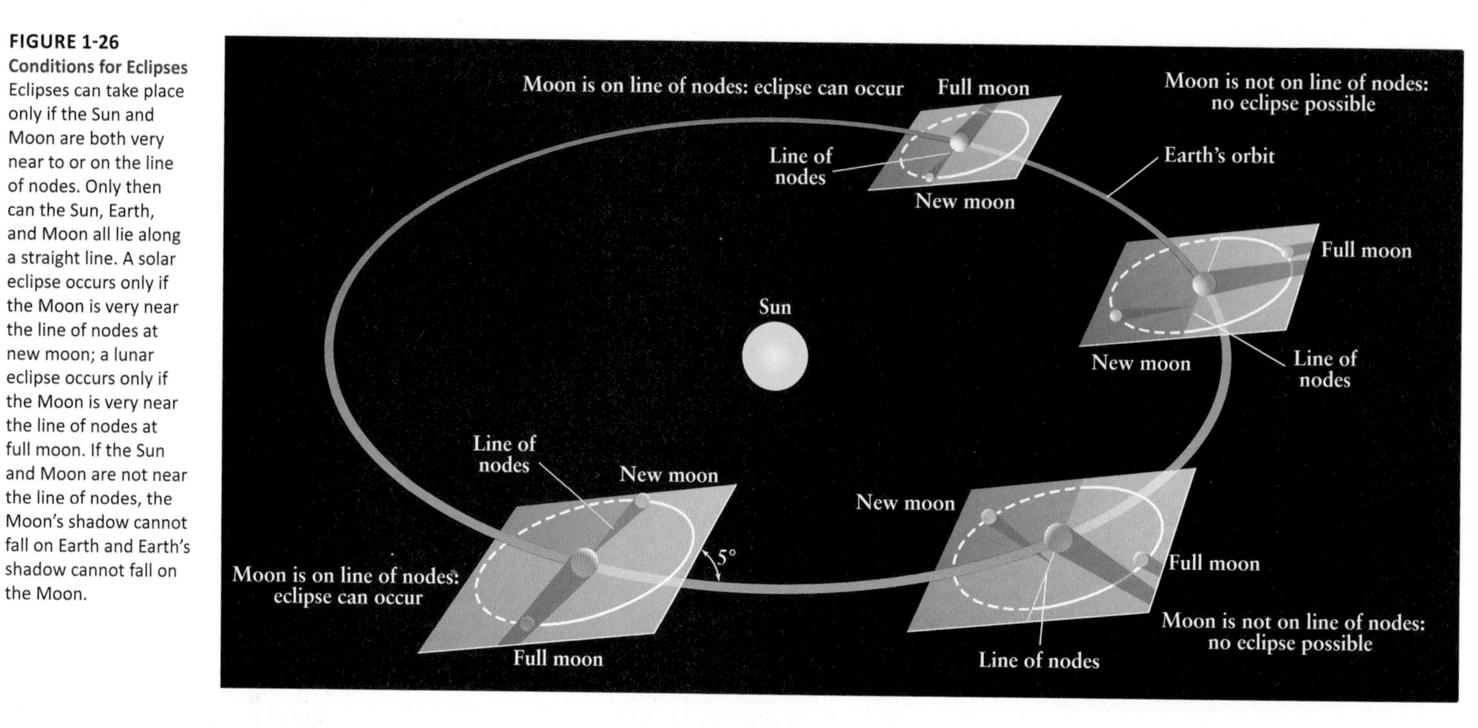

FIGURE 1-26
Conditions for Eclipses Eclipses can take place only if the Sun and Moon are both very near to or on the line of nodes. Only then can the Sun, Earth, and Moon all lie along a straight line. A solar eclipse occurs only if the Moon is very near the line of nodes at new moon; a lunar eclipse occurs only if the Moon is very near the line of nodes at full moon. If the Sun and Moon are not near the line of nodes, the Moon's shadow cannot fall on Earth and Earth's shadow cannot fall on the Moon.

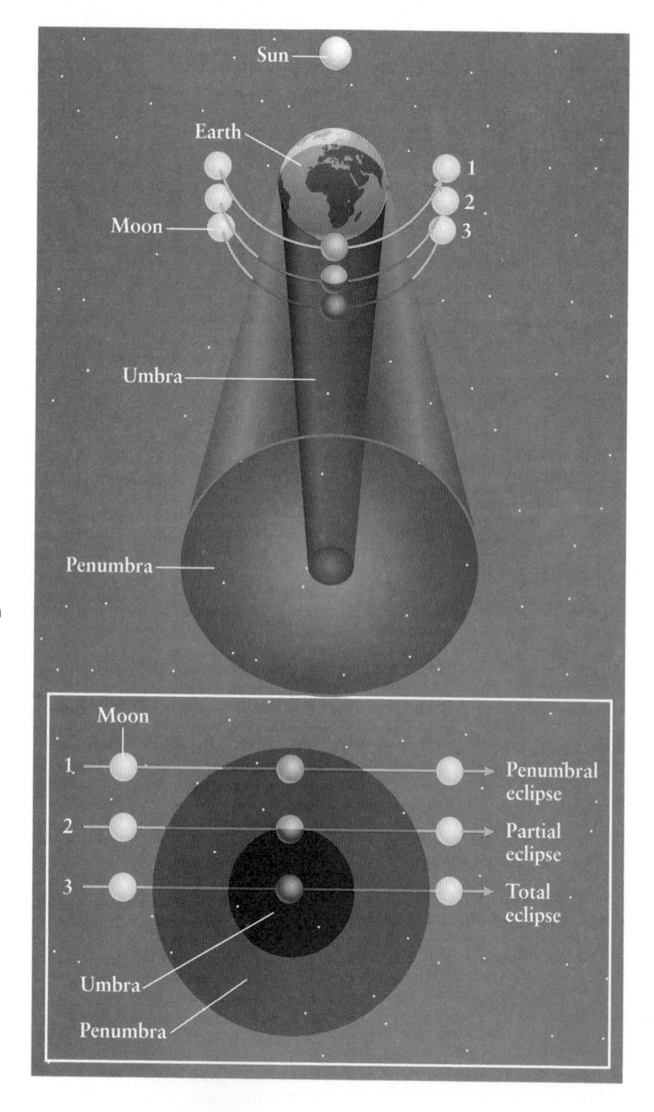

FIGURE 1-27
Three Types of Lunar Eclipses People on the nighttime side of Earth see a lunar eclipse when the Moon moves through Earth's shadow. In the umbra, the darkest part of the shadow, the Sun is completely covered by Earth. The penumbra is less dark because only part of the Sun is covered by Earth. The three paths show the motion of the Moon if the lunar eclipse is penumbral (Path 1), partial (Path 2), or total (Path 3). The inset shows these same paths, along with the umbra and penumbra, as viewed from Earth.

5°. Because of this tilt, new moon and full moon usually occur when the Moon is either above or below the plane of Earth's orbit. When the Moon is not in the plane of Earth's orbit, the Sun, Moon, and Earth cannot align perfectly, and an eclipse cannot occur.

In order for the Sun, Earth, and Moon to be lined up for an eclipse, the Moon must lie precisely in the same plane as Earth's orbit around the Sun. As we saw in Section 1-4, this plane is called the *ecliptic plane* because it is the same as the plane of the Sun's apparent path around the sky, or ecliptic (see Figure 1-17). Thus, when an eclipse occurs, the Moon appears from Earth to be on the ecliptic—which is how the ecliptic gets its name.

The planes of Earth's orbit and the Moon's orbit intersect along a line called the **line of nodes,** shown in **Figure 1-26.** The line of nodes passes through Earth and is pointed in a particular direction in space. Eclipses can occur only if the line of nodes is pointed toward the Sun—that is, if the Sun lies on or near the line of nodes—and if, at the same time, the Moon lies on or very near the line of nodes. Only then do the Sun, Earth, and Moon lie in a line straight enough for an eclipse to occur.

Anyone who wants to predict eclipses must know the orientation of the line of nodes. But the line of nodes is gradually shifting because of the gravitational pull of the Sun on the Moon. As a result, the line of nodes rotates slowly westward. Astronomers calculate such details to fix the dates and times of upcoming eclipses.

There are at least two—but never more than five—solar eclipses each year. The last year in which five solar eclipses occurred was 1935. The least number of eclipses possible (two solar, zero lunar) happened in 1969. Lunar eclipses occur just about as frequently as solar eclipses, but

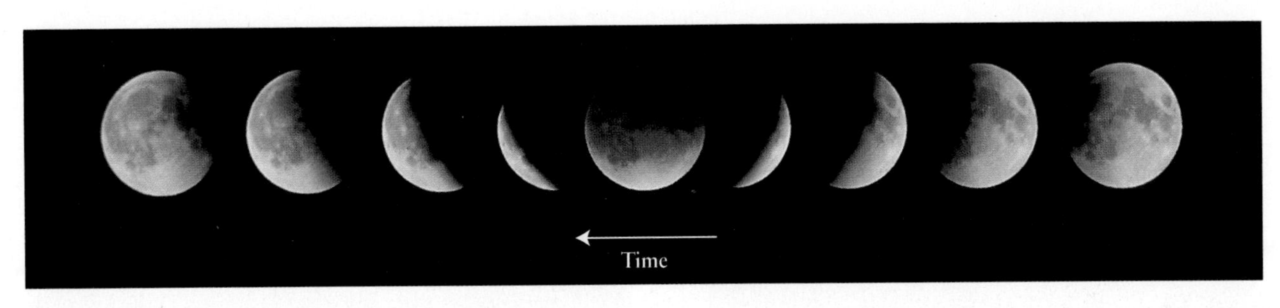

FIGURE 1-28 RIVUXG
A Total Lunar Eclipse This sequence of nine photographs was taken over a three-hour period during the lunar eclipse of January 20, 2000. The sequence, which runs from right to left, shows the Moon moving through Earth's umbra. During the total phase of the eclipse (shown in the center), the Moon has a distinct reddish color. *(© 2000 Fred Espenak, www.MrEclipse.com)*

the maximum possible number of eclipses (lunar and solar combined) in a single year is seven.

ConceptCheck 1-13: Why don't lunar eclipses occur each time the Moon reaches full moon phase?

Lunar Eclipses

The character of a lunar eclipse depends on exactly how the Moon travels through Earth's shadow. As **Figure 1-27** shows, the shadow of Earth has two distinct parts. In the **umbra,** the darkest part of the shadow, no portion of the Moon's surface can be seen. A portion of the Moon's surface is visible in the **penumbra,** which therefore is not quite as dark. Most people notice a lunar eclipse only if the Moon passes into Earth's umbra. As this umbral phase of the eclipse begins, a bite seems to be taken out of the Moon.

The inset in Figure 1-27 shows the different ways in which the Moon can pass into Earth's shadow. When the Moon passes through only Earth's penumbra (Path 1), we see a **penumbral eclipse.** During a penumbral eclipse, Earth blocks only part of the Sun's light and so none of the lunar surface is completely shaded. Because the Moon still looks full but only a little dimmer than usual, penumbral eclipses are easy to miss. If the Moon travels completely into the umbra (Path 2), a **total lunar eclipse** occurs. If only part of the Moon passes through the umbra (Path 3), we see a **partial lunar eclipse.**

If you were on the Moon during a total lunar eclipse, the Sun would be hidden behind Earth. But some sunlight would be visible through the thin ring of atmosphere around Earth, just as you would see sunlight through a person's hair whose head was between your eyes and the Sun. As a result, a small amount of light reaches the Moon during a total lunar eclipse, and so the Moon does not completely disappear from the sky as seen from Earth. Most of the sunlight that passes through Earth's atmosphere is red, and thus the eclipsed Moon glows faintly in reddish hues, as **Figure 1-28** shows.

Lunar eclipses occur at full moon, when the Moon is directly opposite the Sun in the sky. Hence, a lunar eclipse can be seen at any place on Earth where the Sun is below the horizon (that is, where it is nighttime). A lunar eclipse has the maximum possible duration if the Moon travels directly through the center of the umbra. The Moon's speed through Earth's shadow is roughly 1 kilometer per second (3600 kilometers per hour, or 2280 miles per hour), which means that **totality**—the period when the Moon is completely within Earth's umbra—can last for as long as 1 hour and 42 minutes.

Eclipses can only occur at new moon or full moon *AND* only if the Sun and Earth are in perfect alignment with the Moon's position.

On average, two or three lunar eclipses occur in a year. **Table 1-1** lists all 10 lunar eclipses from 2014 to 2018. Of all lunar eclipses, roughly one-third are total, one-third are partial, and one-third are penumbral.

ConceptCheck 1-14: Why does the eclipsing Moon spend more time in the penumbral shadow than the umbral shadow?

Table 1-1 Lunar Eclipses, 2014–2018

Date	Type	Where visible	Duration of totality (h = hours, m = minutes)
2014 April 15	Total	Australia, Pacific, Americas	1h 18m
2014 October 8	Total	Asia, Australia, Pacific, Americas	59m
2015 April 4	Total	Asia, Australia, Pacific, Americas	5m
2015 September 28	Total	Americas, Europe, Africa, western Asia	1h 12m
2016 March 23	Penumbral	Asia, Australia, Pacific, western Americas	—
2016 September 16	Penumbral	Europe, Africa, Asia, Australia, western Pacific	—
2017 February 11	Penumbral	Americas, Europe, Africa, Asia	—
2017 August 7	Partial	Europe, Africa, Asia, Australia	—
2018 January 31	Total	Asia, Australia, Pacific, western North America	1h 16m
2018 July 27	Total	South America, Europe, Africa, Asia, Australia	1h 43m

(a)

(b)

FIGURE 1-29 R I V U X G **A Total Solar Eclipse** (a) This photograph shows the total solar eclipse of August 11, 1999, as seen from Elâziğ, Turkey. The sky is so dark that the planet Venus can be seen to the left of the eclipsed Sun. (b) When the Moon completely covers the Sun's disk during a total eclipse, the solar corona is revealed. *(a: © 1999 Fred Espenak, www.MrEclipse.com; b: Timothy F. Slater)*

Solar Eclipses

As seen from Earth, the angular diameter of the Moon is almost exactly the same as the angular diameter of the far larger but more distant Sun—about 0.5°. Thanks to this coincidence of nature, the Moon just "fits" over the Sun during a **total solar eclipse.**

A total solar eclipse is a dramatic event. The sky begins to darken, the air temperature falls, and winds increase as the Moon gradually covers more and more of the Sun's disk. All nature responds: Birds go to roost, flowers close their petals, and crickets begin to chirp as if evening had arrived. As the last few rays of sunlight peek out from behind the edge of the Moon and the eclipse becomes total, the landscape around you is bathed in an eerie gray or, less frequently, in shimmering bands of light and dark. Finally, for a few minutes the Moon completely blocks out the dazzling solar disk and not much else (**Figure 1-29*a***). The Sun's thin, hot outer atmosphere, which is normally too dim to be seen—blazes forth in the darkened daytime sky (Figure 1-29*b*). It is an awe-inspiring sight.

CAUTION If you are fortunate enough to see a solar eclipse, keep in mind that the only time when it is safe to look at the Sun is during **totality**, when the solar disk is blocked by the Moon and only the Sun's outermost atmosphere is visible. Viewing this magnificent spectacle cannot harm you in any way. But you must *never* look directly at the Sun when even a portion of its intensely brilliant disk is exposed. *If you look directly at the Sun at any time without a special filter approved for solar viewing, you will suffer permanent eye damage or blindness.*

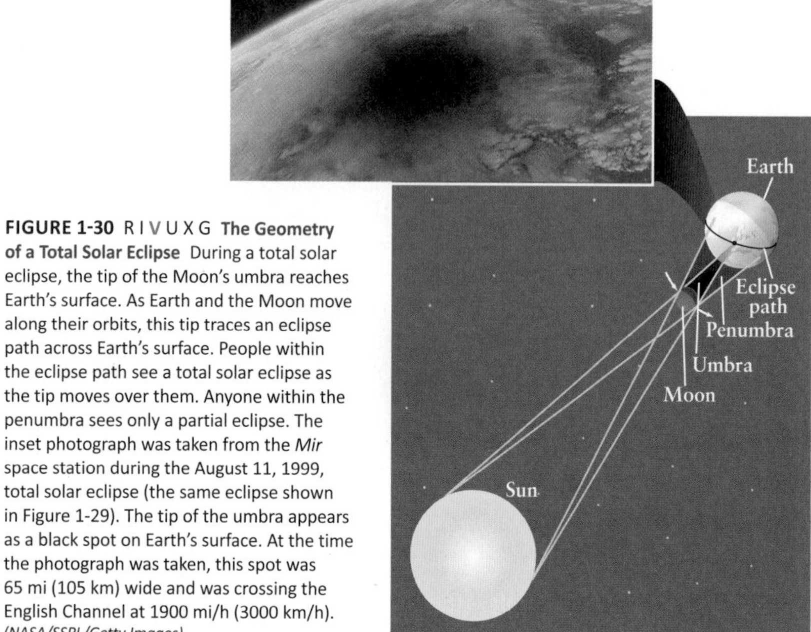

FIGURE 1-30 R I V U X G **The Geometry of a Total Solar Eclipse** During a total solar eclipse, the tip of the Moon's umbra reaches Earth's surface. As Earth and the Moon move along their orbits, this tip traces an eclipse path across Earth's surface. People within the eclipse path see a total solar eclipse as the tip moves over them. Anyone within the penumbra sees only a partial eclipse. The inset photograph was taken from the *Mir* space station during the August 11, 1999, total solar eclipse (the same eclipse shown in Figure 1-29). The tip of the umbra appears as a black spot on Earth's surface. At the time the photograph was taken, this spot was 65 mi (105 km) wide and was crossing the English Channel at 1900 mi/h (3000 km/h). *(NASA/SSPL/Getty Images)*

To see the remarkable spectacle of a total solar eclipse, you must be inside the darkest part of the Moon's shadow, also called the umbra, where the Moon completely blocks the Sun. Because the Sun and the Moon have nearly the same angular diameter as seen from Earth, only the tip of the Moon's umbra reaches Earth's surface (**Figure 1-30**). As Earth rotates, the tip of the umbra traces an **eclipse path** across Earth's surface. Only those locations within the eclipse path are treated to the spectacle of a total solar eclipse. The inset in Figure 1-30 shows the dark spot on Earth's surface produced by the Moon's umbra.

Immediately surrounding the Moon's umbra is the region of partial shadow called the penumbra. As seen from this area, the Sun's surface appears only partially covered by the Moon. During a solar eclipse, the Moon's penumbra covers a large portion of Earth's surface, and anyone standing inside the penumbra sees a **partial solar eclipse.** Such eclipses are much less interesting events than total solar eclipses, which is why astronomy enthusiasts strive to be inside the eclipse path. If you are within the eclipse path, you will see a partial eclipse before and after the brief period of totality.

The width of the eclipse path depends primarily on the Earth-Moon distance during totality. The eclipse path is widest if the Moon happens to be at **perigee,** the point in its orbit nearest Earth. In this case the width of the eclipse path can be as great as 270 kilometers (170 miles). In most eclipses, however, the path is much narrower.

> **ConceptCheck 1-15: Why can a total lunar eclipse be seen by people all over the world but total solar eclipses can only been seen from a very limited geographic location?**

Annular Solar Eclipses

In some eclipses the Moon's umbra does not reach all the way to Earth's surface. This can happen if the Moon is at or near **apogee,** its farthest position from Earth. In this case, the Moon appears too small to cover the Sun completely. The result is a third type of solar eclipse, called an **annular eclipse.** During an annular eclipse, a thin ring of the Sun is seen around the edge of the Moon (**Figure 1-31**). The length of the Moon's umbra is nearly 5000 kilometers (3100 miles) less than the average distance between the Moon and Earth's surface. Thus, the Moon's shadow often fails to reach Earth even when the Sun, Moon, and Earth are properly aligned for an eclipse. Hence, annular eclipses are slightly more common—as well as far less dramatic—than total eclipses.

Even during a total eclipse, most people along the eclipse path observe totality for only a few moments. Earth's rotation, coupled with the orbital motion of the Moon, causes the umbra to race eastward along the eclipse path at speeds in excess of 1700 kilometers per hour (1060 miles per hour). Because of the umbra's high speed, totality never lasts for more than 7½ minutes. In a typical total solar eclipse, the Sun-Moon-Earth alignment and the Earth-Moon distance are such that totality lasts much less than this maximum.

The details of solar eclipses are calculated well in advance. They are published in such reference books as the *Astronomical Almanac* and are available on the World Wide Web. **Figure 1-32** shows the eclipse paths for all

FIGURE 1-31 R I V U X G **An Annular Solar Eclipse** This composite of nine photographs was taken from the small Spanish town of Carrascosa del Campo on October 3, 2005. At maximum coverage, the Sun is still too bright to observe without special eye protection, even though 90% of the disk is covered. This type of eclipse occurs when the Moon happens to be too far from Earth to completely cover the Sun's disk. *(© 2005 by Fred Espenak, www.MrEclipse.com)*

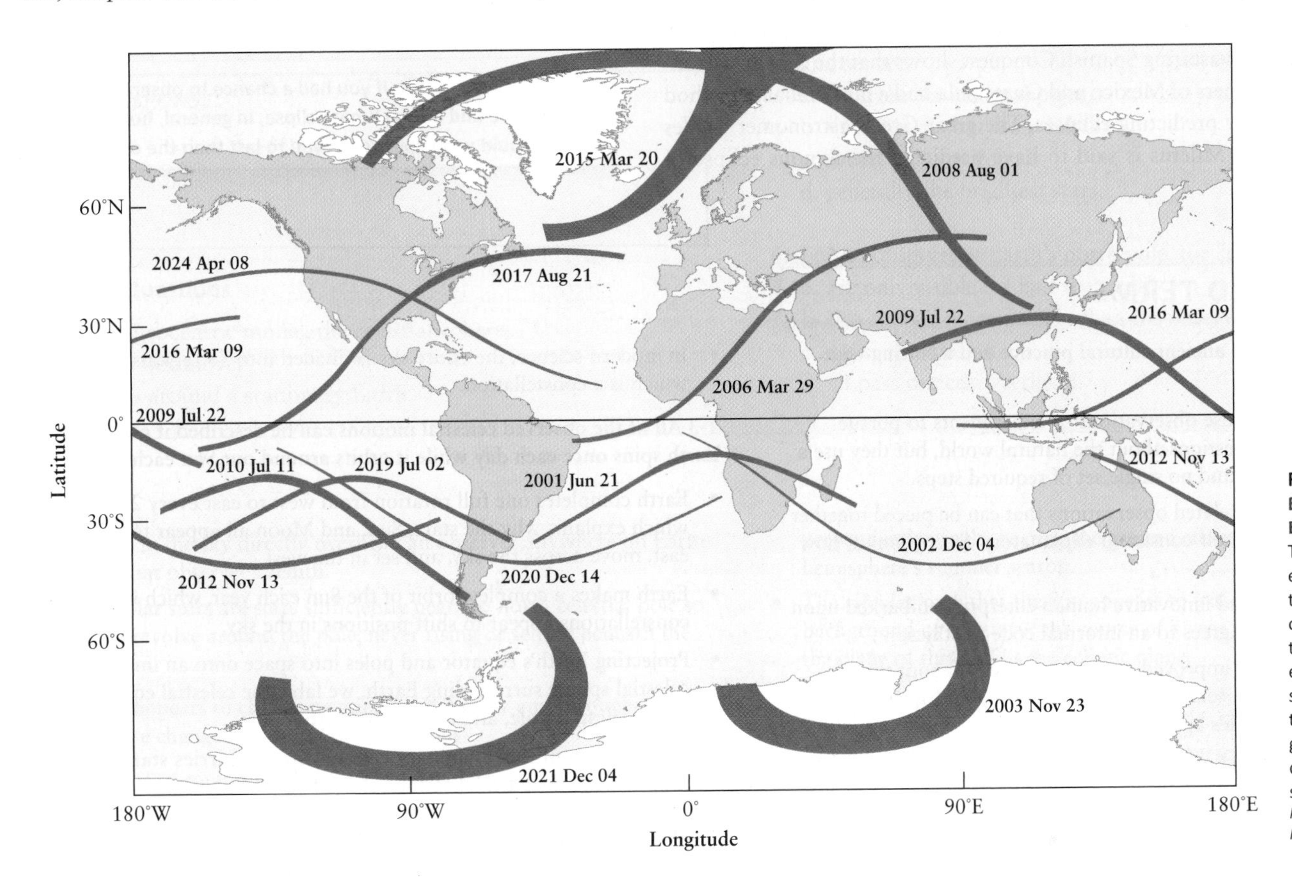

FIGURE 1-32 Eclipse Paths for Total Eclipses, 2001–2025 This map shows the eclipse paths for all total solar eclipses occurring from 2001 through 2025. In each eclipse, the Moon's shadow travels along the eclipse path in a generally eastward direction across Earth's surface. *(© Fred Espenak, NASA/Goddard Space Flight Center)*

- On the ecliptic between the March and September equinoxes lie the **northern solstice,** the point farthest north that the Sun reaches, and the **southern solstice,** the point farthest south.
- On the southern solstice at the **Arctic Circle,** the Sun is below the horizon for 24 continuous hours, and at the **Antarctic Circle,** there is 24 hours of continuous daylight, while on the northern solstice this is reversed.
- Between the **Tropic of Capricorn** at 23½° south latitude and the **Tropic of Cancer** at 23½° north latitude, the Sun is directly overhead—that is, at the zenith—at high noon at least one day a year.

1-5 The Moon appears to change its position in the sky hourly and its phase throughout each month

- Our Moon orbits Earth approximately every four weeks, while the whole Earth-Moon system orbits the Sun.
- Our Moon shines not from light of its own, but rather by reflecting sunlight. We only see the illuminated portion.
- The different amounts of the illuminated half of the Moon that we see over the course of its orbit are called **lunar phases.**
- The phases of the Moon are **new moon, waxing crescent moon, first quarter moon, waxing gibbous moon, full moon, waning gibbous moon, third quarter moon,** and **waning crescent moon.**
- The Moon takes exactly as long to rotate on its axis as it does to make one orbit around Earth, which is called **synchronous rotation** and is the reason why we always see the same face of the Moon.
- A **sidereal month** is the time it takes the Moon to complete one full orbit of Earth as measured with respect to the stars, while the **synodic month** is the time it takes the Moon to complete one cycle as measured with respect to the Sun.

1-6 Eclipses occur only during rarely observed events when our Sun, Moon, and Earth are perfectly aligned

- An **eclipse** happens when the shadow of Earth falls on the Moon—a **lunar eclipse**—or when the Moon moves in front of the Sun and the shadow of the Moon falls on Earth—a **solar eclipse.**
- Eclipses can only happen when the Sun, Earth, and Moon align on the same ecliptic plane along a line called the **line of nodes.**
- In a lunar eclipse, the shadow of Earth has two distinct parts: the **umbra,** the darkest part of the shadow where no portion of the Moon's surface can be seen, and the **penumbra,** which is not quite as dark and where a portion of the Moon's surface is visible.
- If the Moon travels completely into the umbra, we have a **total lunar eclipse**; if only *part* of the Moon travels through the umbra, we see a **partial lunar eclipse**; if the Moon travels only through the penumbra, we see a **penumbral eclipse.**
- Lunar eclipses occur at full moon, and **totality,** the period when the Moon is completely within Earth's umbra, can last for more than an hour.
- Because the angular diameter of the Moon is about the same as the larger but more distant Sun, the Moon just "fits" over the Sun during a **total solar eclipse.**
- During an eclipse, the Moon's penumbra covers a large portion of Earth's surface, and anyone inside the penumbra sees a **partial solar eclipse.**
- The **eclipse path** is widest if the Moon happens to be at **perigee,** the point in its orbit nearest Earth.
- If the Moon is at its farthest position from Earth, its **apogee,** the umbra may not reach Earth, resulting in an **annular eclipse.**

QUESTIONS

Review Questions

1. What is the difference between a hypothesis and a theory?
2. How are scientific theories tested?
3. How are constellations useful to astronomers? How many stars are not part of any constellation?
4. A fellow student tells you that only those stars in Figure 1-3*b* that are connected by blue lines are part of the constellation Orion. How would you respond?
5. Why are different stars overhead at 10:00 P.M. on a given night than two hours later at midnight? Why are different stars overhead at midnight on June 1 than at midnight on December 1?
6. What is the celestial equator? How is it related to Earth's equator? How are the north and south celestial poles related to Earth's axis of rotation?
7. Where would you have to look to see your zenith?
8. How do the stars appear to move over the course of the night as seen from the north pole? As seen from the equator? Why are these two motions different?
9. Using a diagram, explain why the tilt of Earth's axis relative to Earth's orbit causes the seasons as we orbit the Sun.
10. Give two reasons why it is warmer in summer than in winter.
11. What are the March and September equinoxes? What are the northern and southern solstices? How are these four points related to the ecliptic and the celestial equator?
12. How does the daily path of the Sun across the sky change with the seasons? Why does it change?
13. Describe how the seasons would be different if Earth's axis of rotation, rather than having its present 23½° tilt, were tilted (a) by 0° or (b) by 90°.
14. Explain the difference between sunlight and moonlight.
15. Explain why the Moon exhibits phases.
16. At approximately what time does the Moon rise when it is (a) a new moon; (b) a first quarter moon; (c) a full moon; and (d) a third quarter moon?
17. If you lived on the Moon, would you see Earth go through phases? If so, would the sequence of phases be the same as those of the Moon as seen from Earth, or would the sequence be reversed? Explain using Figure 1-21.
18. What is the difference between a sidereal month and a synodic month? Which is longer? Why?
19. What is the difference between the umbra and the penumbra of a shadow?
20. Why doesn't a lunar eclipse occur at every full moon and a solar eclipse at every new moon?
21. Which type of eclipse—lunar or solar—do you think more people on Earth have seen? Why?
22. How is an annular eclipse of the Sun different from a total eclipse of the Sun? What causes this difference?

Web Chat Questions

1. Scientists assume that "reality is rational." Discuss what this means and the thinking behind it.

2. All scientific knowledge is inherently provisional. Discuss whether this is a weakness or a strength of the scientific method.
3. Examine a list of the 88 constellations. Are there any constellations whose names obviously date from modern times? Where are these constellations located? Why do you suppose they do not have archaic names?
4. In William Shakespeare's *Julius Caesar* (act 3, scene 1), Caesar says:

   But I am constant as the northern star,
   Of whose true-fix'd and resting quality
   There is no fellow in the firmament.

   Translate Caesar's statement about the "northern star" into modern astronomical language. Is the northern star truly "constant"? Was the northern star the same in Shakespeare's time (1564–1616) as it is today?

Collaborative Exercises

1. A scientific theory is fundamentally different from the everyday use of the word "theory." List and describe any three scientific theories of your choice and creatively imagine an additional three hypothetical theories that are not scientific. Briefly describe what is scientific and what is nonscientific about each of these theories.
2. Using a bright light source at the center of a darkened room or a flashlight, use your fist held at arm's length to demonstrate the difference between a full moon and a lunar eclipse. (Use yourself or a classmate as Earth.) How must your fist "orbit" Earth so that lunar eclipses do not happen at every full moon? Create a simple sketch to illustrate your answers.
3. Imagine you are planning a trip to see a solar eclipse in the future. Using Table 1-2 showing when and where solar eclipses are visible, which solar eclipse would you most want to go see and why?

Observing Questions

1. If you have access to the *Starry Night*™ planetarium software, install it on your computer. There are several guides to the use of this software. As an initial introduction, you can run through the step-by-step basics of the program by clicking the **Sky Guide** tab to the left of the main screen and then clicking the **Starry Night basics** hyperlink at the bottom of the Sky Guide pane. A more comprehensive guide is available by choosing the **Student Exercises** hyperlink and then the **Tutorial** hyperlink. A User's Guide to this software is available under the **Help** menu. As a start, you can use this program to determine when the **Moon** is visible today from your location. If the viewing location in the Starry Night control panel is not set to your location, select **Set Home Location** in the **File** menu (on a Mac, this command is found under the **Starry Night** menu). Click the **List** tab in the Home Location dialog box; then select the name of your city or town and click the **Save As Home Location** button. Next, use the hand tool to explore the sky and search for the Moon by moving your viewpoint around the sky. (Click and drag the mouse to achieve this motion.) If the Moon is not easily seen in your sky at this time, click the **Find** tab at the top left of the main view. The **Find** pane that opens should contain a list of solar system objects. Ensure that there is no text in the edit box at the top of the Find pane. If the message "Search all Databases" is not displayed below this edit box, then click the magnifying glass icon in the edit box and select **Search All** from the dropdown menu that appears. Click the **+** symbol to the left of the listing for **Earth** to display **The Moon** and double-click on this entry in the list in order to center the view upon the Moon. (If a message is displayed indicating that "the Moon is not currently visible from your location," click on the **Best Time** button to advance to a more suitable time.) You will see that the Moon can be seen in the daytime as well as at night. Note that the **Time Flow Rate** is set to **1x**, indicating that time is running forward at the normal rate. Note also the phase of the Moon.
   a) Estimate how long it will take before the Moon reaches its full phase. Set the **Time Flow Rate** to **1 minute.**
   b) Find the time of moonset at your location.
   c) Determine which, if any, of the following planets are visible tonight: Mercury, Venus, Mars, Jupiter, and Saturn. (*Hint:* Use the **Find** pane and click on each planet in turn to explore the positions of these objects.) Feel free to experiment with the many features of *Starry Night*™.
2. Use *Starry Night*™ to observe the diurnal motion of the sky. First, set *Starry Night*™ to display the sky as seen from where you live, if you have not already done so. To do this, select **File > Set Home Location...** (on a Mac, **Starry Night > Set Home Location**) and click on the **List** tab to find the name of your city or town. Highlight the name and note the latitude of your location as given in the list and click the **Save As Home Location** button. Select **Options > Other Options > Local Horizon...** from the menu. In the Local Horizon Options dialog box, click on the radio button labeled **Flat** in the Horizon style section and click **OK.** For viewers in the northern hemisphere, press the "N" key (or click the **N** button in the **Gaze** section of the toolbar) to set the gaze direction to the northern sky. If your location is in the southern hemisphere, press the "S" key (or click the **S** button in the Gaze section of the toolbar) to set the gaze direction to the south. Select **Hide Daylight** under the **View** menu to view the present sky without daylight. Select **View > Constellations > Astronomical** and **View > Constellations > Labels** to display the constellation patterns on the sky. In the toolbar, click on the **Time Flow Rate** control and set the time step to **1 minute.** Then click the **Play** button to run time forward. (The rapid motions of artificial Earth-orbiting satellites can prove irritating in this view. You can remove these satellites by clicking on **View > Solar System** and turning off **Satellites**).
   a) Do the stars appear to rotate clockwise or counterclockwise? Explain this observation in terms of Earth's rotation.
   b) Are any of the stars circumpolar, that is, do they stay above your horizon for the full 24 hours of a day? If some stars at your location are circumpolar, adjust time and locate a star that moves very close to the horizon during its diurnal motion. Click the Stop button and right-click (Ctrl-click on a Mac) on the star and then select Show Info from the contextual menu to open the Info pane. Expand the Position in Sky layer and note the star's declination (its N-S position on the sky with reference to a coordinate system whose zero value is the projection of Earth's equator).
   c) How is this limiting declination, above which stars are circumpolar, related to your latitude, noted above?
      i) The limiting declination is equal to the latitude of the observer's location.
      ii) The limiting declination is equal to (90 − latitude).
      iii) There is no relationship between this limiting declination and the observer's latitude.
   d) Now center your field of view on the southern horizon (if you live in the northern hemisphere) or the northern horizon (if you live in the southern hemisphere) and click **Play** to resume time flow. Describe what you see. Are any of these stars circumpolar?
3. Use *Starry Night*™ to explore the concept of the celestial sphere. Open **Favourites > Explorations > Celestial Sphere.** The view shows Earth as it might look from space with time flowing at 300x. In the background are the stars of the Milky Way. Superimposed on Earth is

a grid of latitude and longitude with the equator shown in red. Earth's poles are also shown, the aqua line indicating the north pole and yellow line indicating the south pole. Superimposed on the background sky is a spherical grid pattern in red, which represents the celestial sphere. **Zoom** out as far as possible and use the location scroller tool to examine the celestial sphere. Now open **Favourites > Explorations > Surface View of CS** to examine the celestial sphere from the north pole of Earth. Notice that unlike the red gridlines on the celestial sphere, the gray meridian is fixed to the observing location on Earth since it does not move with respect to the horizon. Use the hand tool and **Zoom** controls to examine the celestial sphere from this perspective. Switch between these two perspectives as necessary to answer the questions that follow.

a) What is the relationship between the poles of Earth and the celestial poles?

b) What is the relationship between Earth's equator and the celestial equator?

4. Use the *Starry Night*™ program to measure angular spacing between stars in the sky. Open **Favourites > Explorations > N Pole** to display the northern sky from Calgary, Canada, at a latitude of 51°. This view shows several asterisms, or groups of stars, outlined and labeled with their common names. The stars in the Big Dipper asterism outline the shape of a "dipper," used for scooping water from a barrel. The two stars in the Big Dipper on the opposite side of the scoop from the handle, **Merak** and **Dubhe,** can be seen to point to the brightest star in the Little Dipper, the Pole Star. The Pole Star is close to the north celestial pole, the point in the sky directly above the north pole of Earth. It is thus a handy aid to navigation for northern hemisphere observers because it indicates the approximate direction of true north. Measure the spacing between the two "pointer stars" in the Big Dipper and then the spacing between the Pole Star and the closest of the pointer stars, Dubhe. (*Hint:* These measurements are best made by activating the angular separation tool from the cursor selection tool on the left side of the toolbar.)

a) What is the angular distance between the pointer stars Merak and Dubhe?

b) What is the angular spacing, or separation, between Dubhe (the pointer star at the end of the Big Dipper) and the Pole Star?

c) Approximately how many pointer-star spacings are there between Dubhe and the Pole Star?

Click the **Play** button in the toolbar. Notice that the Pole Star will appear to remain fixed in the sky as time progresses because it lies very close to the north celestial pole. Select **Edit > Undo Time Flow** or **File > Revert** from the menu to return to the initial view. Select **View > Celestial Guides > Celestial Poles** from the menu to indicate the position of the north celestial pole on the screen. Right-click on the Pole Star (Ctrl-click on a Mac) and select **Centre** from the dropdown menu to center the view on the Pole Star. **Zoom** in and use the angular separation tool to measure the angular spacing between the Pole Star and the north celestial pole.

d) What is the angular separation between the Pole Star and the north celestial pole?

Open **Favourites > Explorations > S Pole** to view the southern sky from Brisbane, Australia, at a southern latitude of 27½°. For observers in the southern hemisphere, there is no prominent star at or near to the position of the south celestial pole so its position is displayed on the view. The key asterism for estimating the position of the south celestial pole is the Southern Cross.

e) Use the angular separation tool to measure the angle between the fainter of the two stars making up the major axis of the cross, Gacrux, and the brighter star, Acrux. Extend this line toward the pole to demonstrate that this alignment is not exact.

f) Measure the angle between Acrux and the south celestial pole.

g) How many "pointer-star" spacings does this angle represent?

h) Measure the angle between the pole and the horizon at Brisbane. What is the relationship between this angle and the latitude of the observer? Click the **Play** button to demonstrate that the sky appears to rotate about this south celestial pole because of the rotation of Earth.

5. Use *Starry Night*™ to demonstrate the reason for seasonal variations on Earth at midlatitudes. A common misconception is that summertime is warmer because Earth is closer to the Sun in the summer. The real reason is that the tilt of the spin axis of Earth to its orbital plane places the Sun at a higher angle in the sky in the summer than in the winter. Thus, sunlight hits Earth's surface at a less oblique angle in summertime than in winter, thereby depositing greater heat. Open **Favourites > Explorations > Seasonal Variations** to view the southern sky in daylight from Calgary, Canada, at a latitude of 51°N, on December 21, 2013, at 12:38 P.M., local standard time, when the Sun is at its highest angle on that day. Form a table of values of Sun altitude and Sun-Earth distance as a function of date in the year. To find these values, move the cursor over the Sun to reveal the Info panel. (Ensure that the relevant information is displayed by opening **File > Preferences > Cursor Tracking (HUD)** and clicking on **Altitude** and **Distance from Observer.**) Note the Date, Sun Altitude, and Distance in your table. Advance the **Date** to March 21, 2014, and then to June 21, 2014, noting the values of these parameters.

a) On which of the three dates, in winter, spring and summer, respectively, is the Sun at the highest altitude in the Calgary sky?

b) From the values in your table, what is the Sun altitude at midday in December in Calgary?

c) How does the Sun's altitude at midday on March 21 relate to the latitude of Calgary?

i) The Sun's altitude is equal to the latitude of Calgary.

ii) The Sun's altitude is 90° minus the latitude of Calgary.

iii) The Sun's altitude is not related to the latitude of the location of the observer.

d) On which of these three observing dates is Earth closest to the Sun?

e) On which of these three dates is Earth farthest from the Sun?

6. Use the *Starry Night*™ program to observe the Sun's motion on the celestial sphere. Select **Favourites > Explorations > Sun** from the menu. The view shows the entire celestial sphere as if you were at the center of a transparent Earth on January 1, 2010. The view is centered upon the Sun and shows the ecliptic, the celestial equator, and the boundary and name of the constellation in which the Sun is located. With the **Time Flow Rate** set to **8 hours,** click the **Play** button. Observe the Sun for a full year of simulated time. The motion of Earth in its orbit causes this apparent motion.

a) How does the Sun appear to move against the background stars?

b) What path does the Sun follow and does it ever change direction?

c) Through which constellations does the Sun appear to move over the course of a full year? In the toolbar, click the **Now** button to return to the current date and time.

d) In which constellation is the Sun located today? The Sun (and therefore this constellation) is high in the sky at midday.

e) Approximately how long do you think it will take for this constellation to be high in the sky at midnight?

7. Use *Starry Night*™ to demonstrate the phases of the Moon. From the menu, select **Favourites > Explorations > Moon Phases.** In this view, you are looking down upon the southern hemisphere of Earth from a

location 69,300 kilometers above Earth's surface, with Earth centered in the view. The size of the Moon is greatly exaggerated in this view. The face of the Moon in this diagrammatic representation is configured to show its phase as seen from Earth. The inner green circle represents the Moon's orbit around Earth. The position of the Sun with respect to Earth is shown on the outer green circle, the ecliptic, near to the top of the view. In reality, the Sun would be in this direction from Earth but at a far larger distance. Click the **Play** button and observe the relative positions of the Sun, Earth, and Moon and the associated phases of the Moon.

a) Describe the geometry of the Sun, Moon, and Earth when the Moon is new.
b) Describe the geometry of the Sun, Moon, and Earth when the phase of the Moon is full.
c) When the phase of the Moon is at first or third quarter, what is the approximate angle at the Moon between the Moon-Sun and Moon-Earth lines?
d) What general term describes the phase of the Moon when the angle created by the Moon-Sun and Moon-Earth lines at the Moon is less than 90°?
e) What general term describes the phase of the Moon when the angle created by the Moon-Sun and Moon-Earth lines at the Moon is greater than 90°?

8. Use *Starry Night*™ to observe a lunar eclipse from a position in space near the Sun. From the main menu select **Favourites > Explorations > Lunar Eclipse.** This view from a position near to the Sun shows Earth and two tan-colored circles. The inner circle represents the umbra of Earth's shadow and the outer circle represents the boundary of the penumbra of Earth's shadow. (If these circles do not appear, click on **Options > Solar System > Planets-Moons...** and click on **Shadow Colour** to adjust the brightness of the umbral and penumbral outlines.)
   a) What is the Moon's phase? Click the **Play** button and watch as time flows forward and the Moon passes through Earth's shadow. Select **File > Revert** from the menu, move the cursor over the Moon, right-click the mouse, and click on **Centre** to center the view on the Moon. Zoom in to almost fill the field of view with the Moon and click the **Play** button again to watch the lunar eclipse in detail. Use the time controls to determine the duration of totality to the nearest minute.
   b) What is the duration of totality for this eclipse?

ANSWERS

ConceptChecks

ConceptCheck 1-1: While a hypothesis is a testable idea that seems to explain an observation about nature, a scientific theory represents a set of well-tested and internally consistent hypotheses that are able to successfully and repeatedly predict the outcomes of experiments and observations.

ConceptCheck 1-2: No, Jupiter does not need to be within the dot-to-dot asterism of the bull's body; rather, Jupiter would only need to be within the semirectangular boundaries of the Taurus constellation.

ConceptCheck 1-3: As locations on Earth rotate from the dark, nighttime side of the planet into the bright, daytime side of the planet, the easternmost cities experience sunrise first. New York is the farthest east of the cities listed, so the Sun rises there first.

ConceptCheck 1-4: The Sun would still only rise and set once each day, but each year would have 3 times more days ($365 \times 3 = 1095$ days each year).

ConceptCheck 1-5: The celestial equator is an extension of Earth's equator up into the sky. In order for this imaginary line to pass directly overhead, one would need to be standing somewhere on Earth's equator.

ConceptCheck 1-6: No, they would not; in this imaginary scenario, the Sun would rise and set every 12 hours all year long. Earth's tilted axis means that observers at Earth's north pole will experience six months where the Sun never rises (when it is tilted away from the Sun) and then six months when the Sun never sets (when it is tilted toward the Sun).

ConceptCheck 1-7: The Sun slowly moves toward the eastern part of the sky a little each day, taking 365¼ days to return to the same place it was one year earlier.

ConceptCheck 1-8: As the Sun's position on the celestial sphere slowly moves back and forth between the northern and southern solstice points over the course of a year, the noontime Sun will be directly overhead and cast no shadow for an observer at Earth's equator only twice each year, on the March and September equinoxes.

ConceptCheck 1-9: Because the Moon is always illuminated by the Sun, it is always half illuminated and half dark. Except in the rare events of eclipses, the Moon's apparent changing phases are due to observers on Earth seeing differing amounts of the Moon's half-illuminated surface.

ConceptCheck 1-10: The Moon goes through an entire cycle of phases in about four weeks, so if it is currently in the waxing crescent phase, in one week it will be in the waxing gibbous phase, and after two weeks it will be in the waning gibbous phase.

ConceptCheck 1-11: If the moon is full, then after one week, it would reach the third quarter phase, and the point that used to be in the center of the Moon's visible surface at full moon would now fall into darkness that would last for two weeks.

ConceptCheck 1-12: Increase. If Earth was moving around the Sun faster than it is now, Earth would move farther around the Sun during the Moon's orbit, and it would take longer for the Moon to reach the position where it was in line with the Sun and Earth, increasing its synodic period.

ConceptCheck 1-13: Lunar eclipses only occur when the Sun, Earth, and Moon are all exactly in a line, the line of nodes. Usually, the full moon is above or below the ecliptic plane of Earth's orbit around the Sun and misses being covered by Earth's shadow.

ConceptCheck 1-14: The eclipsing Moon spends more time in the penumbral shadow because Earth's penumbra is much larger than the umbra.

ConceptCheck 1-15: Total solar eclipses are only observed in the small region where the Moon's tiny umbral shadow lands on Earth, whereas when the Moon enters Earth's much larger umbral shadow, anyone on Earth who can see the Moon can observe it in Earth's shadow.

ConceptCheck 1-16: A total lunar eclipse can last for more than an hour, but a total solar eclipse lasts only a few minutes.

CalculationChecks

CalculationCheck 1-1: Earth rotates once in about 24 hours, so when the stars of Cygnus rise in the east at sunset, they take about 12 hours to go from one side of the sky to the other. As a result, it takes about half that time, or six hours, for stars of Cygnus to move halfway across the sky.

CalculationCheck 1-2: The northern solstice occurs on about June 21 and the March equinox occurs about March 21, so there are about nine months or 270 days between these two events.

RIVUXG Different telescopes capture different wavelengths of light from M81, a galaxy of several hundred billion stars. *(Hubble data: NASA, ESA, and A. Zezas [Harvard-Smithsonian Center for Astrophysics]; GALEX data: NASA, JPL-Caltech, GALEX Team, J. Huchra et al. [Harvard-Smithsonian Center for Astrophysics]; Spitzer data: NASA/JPL/Caltech/S. Willner [Harvard-Smithsonian Center for Astrophysics])*

Decoding the Hidden Messages in Starlight

CHAPTER LEARNING OBJECTIVES **By reading the sections of this chapter, you will learn:**

2-1 Light travels through empty space at a speed of nearly 300,000 km/s

2-2 Glowing objects, like stars, emit an entire spectrum of light

2-3 An object's temperature is revealed by the most intense wavelength of its spectrum of light

2-4 An object's chemical composition is revealed by the unique pattern of its spectrum of light

2-5 An object's motion through space is revealed by the precise wavelength positions of its spectrum of light

2-6 Telescopes use lenses, mirrors, and electronics to concentrate and capture incoming light for study

The sky overhead encourages us to wonder about the nature of the universe—places so distant that we will likely never visit them easily. Sunlight illuminates our day and twinkling stars decorate our nighttime sky. All the glowing objects in the sky reveal themselves to us, even though they are very far away, by sending their light our way. Because these objects are so distant, the only information we have about them comes from the light we can capture with our telescopes. Fortunately, the physical properties of light are predictable and well understood. In this chapter we will first consider the basic properties of light; then we will discover how astronomers use light to reveal temperature, chemical composition, and motion through space of all the glowing objects in the sky. We will conclude with a discussion of how astronomers capture and enhance starlight in modern telescopes, in order to study it. The technology developed for the most modern of telescopes influences other technologies, including the high-quality cameras in cell phones and the lightweight materials they are made of. ■

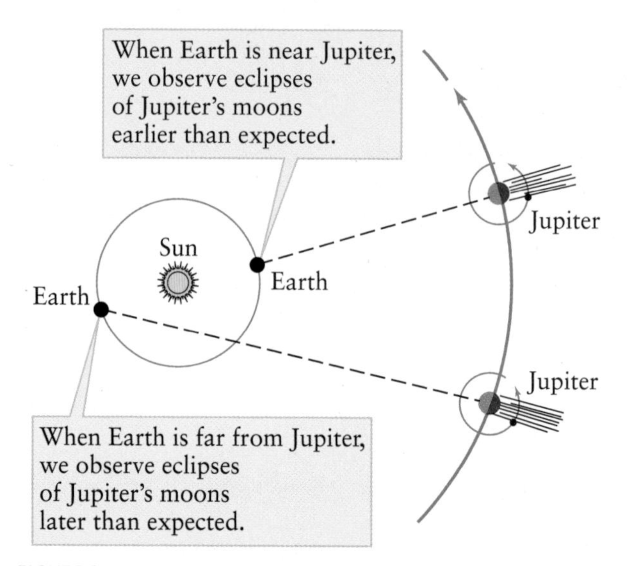

FIGURE 2-1 Rømer's Proof That Light Does Not Travel Instantaneously The timing of eclipses of Jupiter's moons as seen from Earth depends on the Earth-Jupiter distance. Rømer correctly attributed this effect to variations in the time required for light to travel from Jupiter to Earth.

2-1 Light travels through empty space at a speed of nearly 300,000 km/s

When you turn on a flashlight or a laser pointer, how long does it take for the beam to move across the room? Most people would agree that light moves quite quickly, but how would you actually measure this? Consider the following experiment: You and a partner stand on opposite sides of a room and point flashlights at one another. The agreement is that you are going to turn your flashlight on first, and when your partner receives your flashlight beam, she will turn on her flashlight aimed back at you. Your task is to measure how much time elapses between when you first send a beam of light across the room and when you see the other flashlight beam travel back to you. Does this seem possible?

In the early 1600s, Galileo tried to measure the speed of light using a similar approach. He and an assistant stood at night on two hilltops a known distance apart, each holding a shuttered lantern. First, Galileo opened the shutter of his lantern; as soon as his assistant saw the flash of light, he opened his own. Using his pulse as a timer, Galileo found that the measured time failed to increase noticeably, no matter how distant the assistant was stationed. Galileo therefore concluded—as you most likely did—that the speed of light is too high to be measured by slow human reactions.

Nearly 70 years after Galileo, the Danish astronomer Olaus Rømer was studying the orbits of the moons of Jupiter by carefully timing the moments when they passed into or out of Jupiter's shadow. To Rømer's surprise, the timing of these eclipses of Jupiter's moons seemed to depend on the relative positions of Jupiter and Earth. When Earth was far from Jupiter (**Figure 2-1**), the eclipses occurred several minutes later than when Earth was close to Jupiter.

Rømer realized that when Earth is closest to Jupiter, the image of a moon disappearing behind Jupiter arrives at our telescopes a little sooner than it does when Jupiter and Earth are farther apart. Rømer was not able to actually calculate the speed of light because he did not know the exact distances Earth was moving closer and farther from Jupiter, but using our modern understanding of these distances, Rømer's method yields an impressively correct value for the speed of light.

The speed of light in a vacuum is usually designated by the letter c (from the Latin *celeritas,* meaning "speed"). The current accepted value of the speed of light is c = 299,792.458 km/s (186,282.397 mi/s) but is slightly slower when traveling through a transparent substance such as air, water, or glass. The value in kilometers per second (km/s) is often most useful when comparing c to the speeds of objects in space, though the most convenient set of units to use for c is different in different situations. The speed of light in a vacuum is a universal constant: It has the same value everywhere in the cosmos. In most calculations, you can use:

$$c = 300{,}000 \text{ km/s}$$

Indeed, light moves almost unimaginably fast, fast enough to circle Earth almost 7 times in a single second. Light takes only a little more than a second to travel all the way to the Moon, 240,000 mi (384,000 km). Today, the speed of light in empty space is well known, and one of the most important numbers in modern physical science. This value appears in many equations that describe atoms, gravity, electricity, and magnetism. It is also an upper limit on speed: According to Einstein's special theory of relativity, nothing can travel faster than the speed of light.

ConceptCheck 2-1: Why has the speed of light been historically so difficult to measure?

Answer appears at the end of the chapter.

2-2 Glowing objects, like stars, emit an entire spectrum of light

Many people who have felt the warmth of the sunshine on a clear summer's day have wondered, "What is light?" We know that it has energy because light "feels" warm when it comes from the Sun. What else can we learn about light by studying it? As early as the 1600s, scientists started to reveal the important characteristics of light: that it is a mixture of all the colors of the rainbow; that it travels in waves and has electric and magnetic properties; and that light visible to the human eye is only one small part of the entire spectrum of light.

Sunlight Is a Mixture of All Colors

What color is sunlight? Observing the dusk at sunset might persuade you to believe that sunlight is reddish-pink. A quick (but potentially very harmful!) glance at the Sun through thin clouds might lead you to change your mind to believe that sunlight is actually white. If you hold a piece of white paper in a beam of sunlight, the paper appears to be bright white. Alternatively, if you hold a piece of blue paper in that same beam of sunlight, the paper appears bright blue. The same observation holds true with red paper, green paper, and yellow paper. In the end, capturing a ray of sunlight to determine its color is tricky indeed.

Another experimental approach to studying the nature of light is to observe what happens when light is separated into a rainbow of colors, when it reflects off an evenly scratched surface, such as a DVD. The same thing can happen when light passes through a raindrop, causing a rainbow of colors in the sky. However, we can study light more precisely when it passes through carefully designed materials, such as a glass prism (**Figure 2-2**). The first recorded experiment on passing light through a glass prism was performed by Isaac Newton around 1670. Newton was familiar with what he called the "celebrated Phenomenon of Colours," in which a beam of sunlight passing through a glass prism spreads out into the colors of the rainbow. This rainbow of all colors is called a **spectrum** (plural **spectra**).

Until Newton's time, it was thought that a prism somehow added colors to white light. To test this idea, Newton placed a second prism so that just one color of the spectrum passed through it (**Figure 2-3**). According to the old thinking, this should have caused a further change in the color of

FIGURE 2-2 A Prism and a Spectrum When a beam of sunlight passes through a glass prism, the light is broken into a rainbow-colored band called a spectrum.

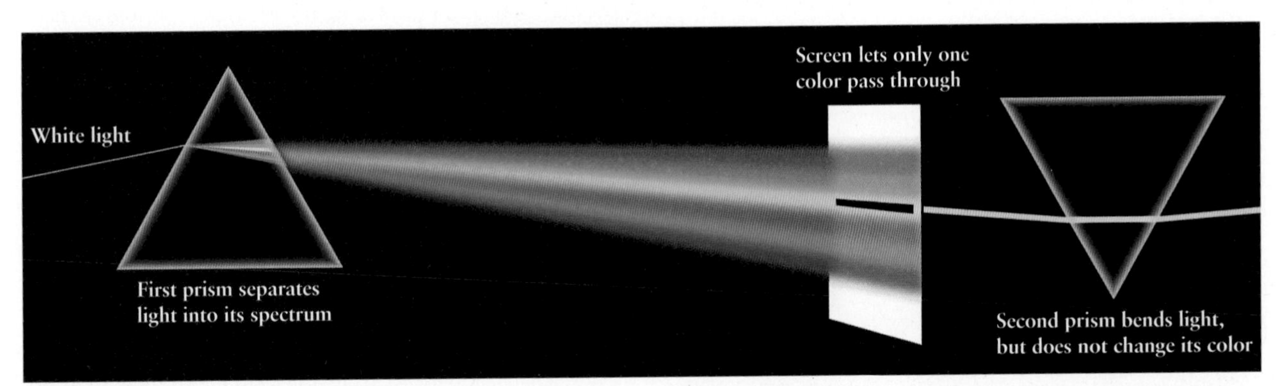

FIGURE 2-3 Newton's Experiment on the Nature of Light In a crucial experiment, Newton took sunlight that had passed through a prism and sent it through a second prism. Between the two prisms was a screen with a hole in it that allowed only one color of the spectrum to pass through. This same color emerged from the second prism. Newton's experiment proved that prisms do not add color to light but merely bend different colors through different angles. It also proved that white light, such as sunlight, is actually a combination of all the colors that appear in its spectrum.

the light. But Newton found that each color of the spectrum was unchanged by the second prism; red remained red, blue remained blue, and so on. These observations led him to conclude that a prism merely separates colors and does not add color. Hence, the spectrum produced by the first prism demonstrates that sunlight is a mixture of all the colors of the rainbow.

ConceptCheck 2-2: If you cover a white light with a specially designed green plastic gel so that only the green light passes through, which color plastic cover gel do you need to add to the pure green light to make it change to red?

Light Travels in Waves and Is Also Called Electromagnetic Radiation

Newton suggested that light is composed of particles too small to detect individually. In 1678, the Dutch physicist and astronomer Christiaan Huygens proposed a rival explanation. He suggested that light travels in the form of waves rather than as tiny particles. These two ideas are very different notions about the nature of light. So which is it?

Around 1801, Thomas Young in England carried out an experiment that convincingly demonstrated the wavelike aspect of light. He passed a beam of light through two thin, parallel slits in an opaque screen, as shown in **Figure 2-4.** On a white surface some distance beyond the slits, the light formed a pattern of alternating bright and dark bands. Young reasoned that if a beam of light was a stream of particles (as Newton had suggested), the two beams of light from the slits should simply form bright images of the slits on the white surface. The pattern of bright and dark bands he observed is just what would be expected, however, if light had wavelike properties. An analogy with water waves demonstrates why.

ANALOGY Imagine ocean waves pounding against a reef or breakwater that has two openings (**Figure 2-5**). A pattern of ripples is formed on the other side of the barrier as the waves come through the two openings and interfere with each other.

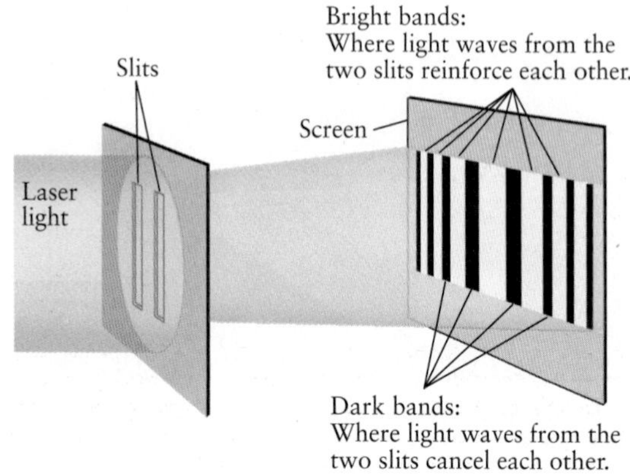

FIGURE 2-4 Young's Double-Slit Experiment Thomas Young's classic double-slit experiment can easily be repeated in the modern laboratory by shining light from a laser onto two closely spaced parallel slits. Alternating dark and bright bands appear on a screen beyond the slits.

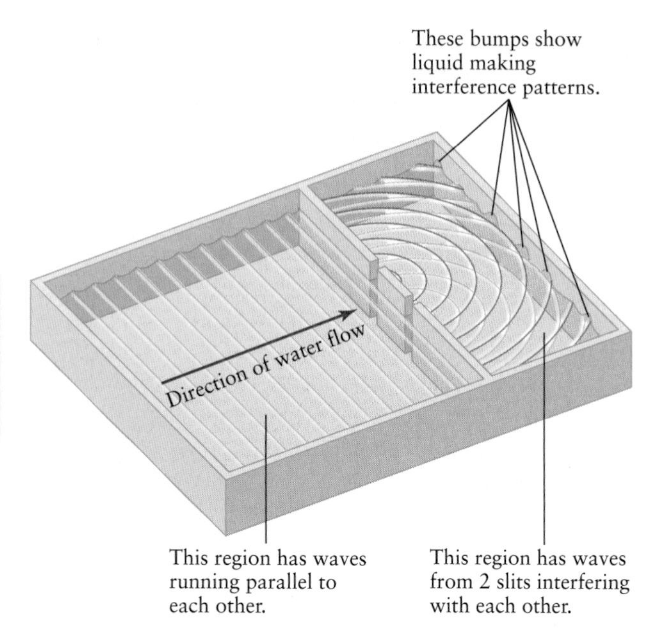

FIGURE 2-5 An Experiment with Water Waves The intensity of light on the screen shown in Figure 2-4 is analogous to the height of water waves that pass through a barrier with two openings. (The illustration shows this experiment with water waves in a small tank.) In certain locations, wave crests from both openings reinforce each other to produce extra high waves. At other locations a crest from one opening meets a trough from the other. The crest and trough cancel each other, producing still water. *(University of Colorado, Center for Integrated Plasma Studies, Boulder, CO)*

At certain points, wave crests arrive simultaneously from the two openings. These reinforce each other and produce high waves. At other points, a crest from one opening meets a trough from the other opening. These cancel each other out, leaving areas of still water. This process of combining two waves also takes place in Young's double-slit experiment: The bright bands are regions where waves from the two slits reinforce each other, while the dark bands appear where waves from the two slits cancel each other.

This demonstration of the wave nature of light poses some obvious questions. What exactly is "waving" in light? That is, what is it about light that goes up and down like water waves on the ocean? Because we can see light from the Sun, planets, and stars, light waves must be able to travel across empty space. Hence, whatever is "waving" cannot be any material substance. What, then, is it?

The answer is that light is energy. It is a particular kind of energy called *electromagnetic energy* that can radiate throughout the empty vacuum of outer space. It is a unique kind of energy because some kinds of energy require a material substance to travel through—water waves, for example, can only travel through water. Because of its ability to move by the process of radiation, light is sometimes given the long and somewhat clumsy name of electromagnetic radiation.

CAUTION You may associate the term *radiation* with radioactive materials like uranium, but this term refers to anything that radiates, or spreads away, from its source. For example, scientists sometimes refer to sound waves as "acoustic radiation." Radiation does not have to be related to radioactivity! Although some scientists make a distinction between light and electromagnetic radiation, for our purposes in this book, we use the two interchangeably.

Thinking of light as a wave of energy traveling through outer space allows us to describe light in of the same terms as we describe waves, like those shown in Figure 2-5. In addition to energy, we can also describe the distance between two successive wave crests as the **wavelength** of the light, usually designated by the Greek letter λ (lambda). No matter what the size of the wavelength or the amount of energy in a light wave, light always travels at the same speed $c = 300{,}000$ km/s through empty space.

> ConceptCheck 2-3: **If bigger and more water waves are generated more quickly by tapping the pan in Figure 2-5 more intensely and faster, what happens to the speed that the waves travel through the water?**

Go to Video 2-1

Our Eyes See Only Some of the Entire Spectrum of Light

The human eye is sensitive to only a very narrow range of wavelengths of light. Once scientists determined that there are no size restrictions on the possible wavelengths of light, they realized that light could and should exist with wavelengths both longer and shorter than the range of visible light seen across the entire spectrum of colors in a rainbow. Consequently, they began to look for *invisible* forms of light—forms of light to which the cells of the retina in a human eye do not respond. As it turns out, when one uses carefully designed sensors, light appears throughout our universe in a much wider variety of wavelengths than were ever imagined to exist.

One of the first kinds of invisible light was discovered around 1800 by the German-born British astronomer and acclaimed musician, Sir William Herschel. Trying to better understand the constituents of sunlight, he passed a beam of sunlight through a prism and held a thermometer just beyond the red end of the visible spectrum. The thermometer registered a temperature increase, indicating that it was being exposed to an invisible form of energy. It was only later that this invisible energy, now called **infrared light** (IR), was realized to be light with wavelengths somewhat longer than those of visible light.

In 1888, the German physicist Heinrich Hertz succeeded in identifying light with even longer wavelengths than infrared, having wavelengths of several centimeters, now known as **radio waves,** in experiments with electric sparks. At about the same time and studying light with much shorter wavelengths than visible light, another German physicist, Wilhelm Röntgen, invented a machine that created invisible light with incredibly small wavelengths, now known as **X-rays.** The X-ray machines in modern medical and dental offices are direct descendants of Röntgen's invention. Over the years, an entire spectrum of different wavelengths of light has been discovered.

An entire spectrum of light is emitted by glowing objects across the universe, much of which is invisible to human eyes and requires special telescopes to observe.

The surprisingly narrow range of wavelengths of light our human eyes can see, called **visible light,** occupies only a tiny fraction of the full range of possible wavelengths, collectively called the **electromagnetic spectrum** (Figure 2-6). To express such tiny distances conveniently, scientists use a

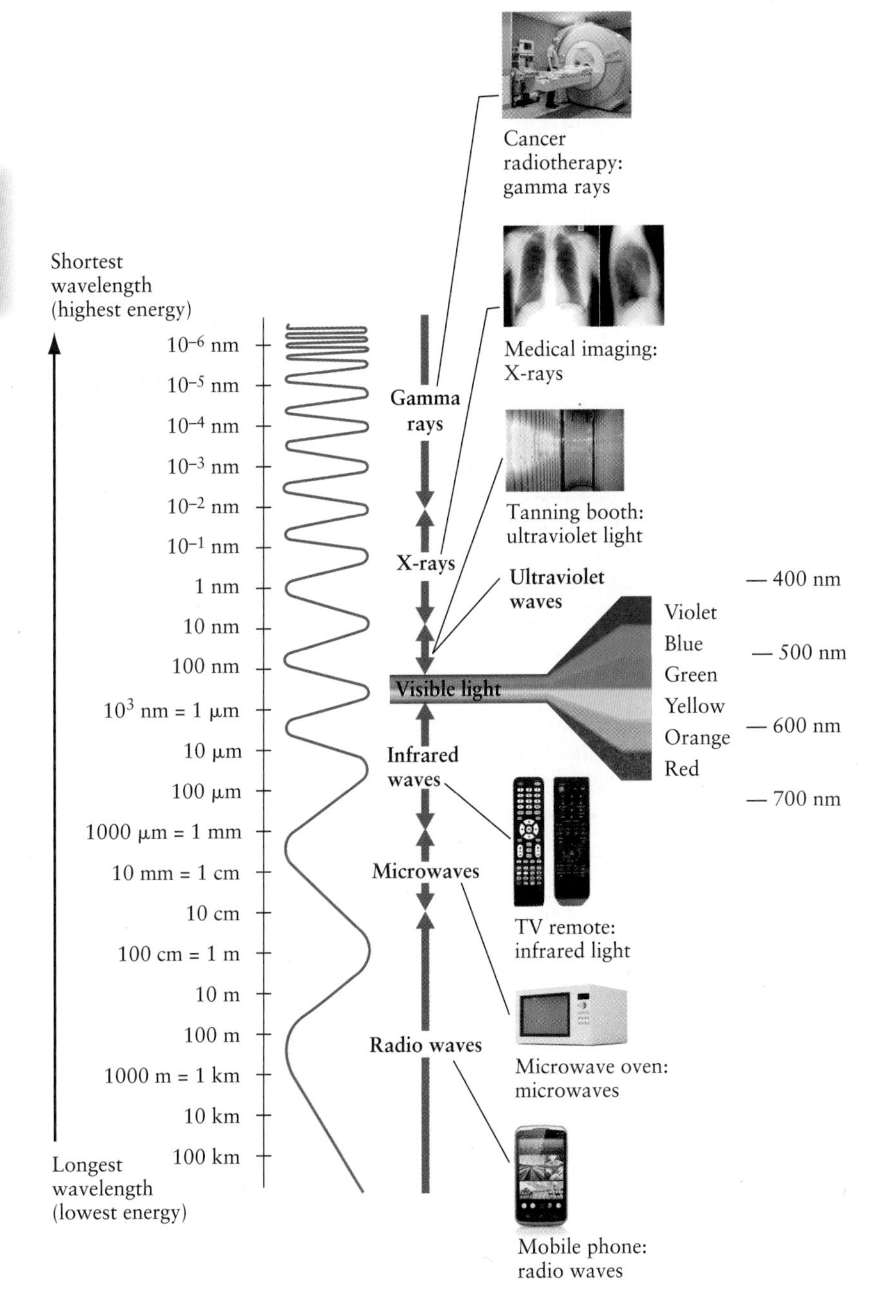

FIGURE 2-6 The Electromagnetic Spectrum of Light The full array of all types of light is called the electromagnetic spectrum. It extends from the longest-wavelength radio waves to the shortest-wavelength gamma rays. Visible light occupies only a tiny portion of the entire electromagnetic spectrum. *(from top: epstock/Shutterstock; Flik47/Shutterstock; Pavel L. Photo and Video/Shutterstock; Goran Bogicevic/Shutterstock; Kostsov/Shutterstock; Oleksiy Mark/Shutterstock)*

unit of length called the **nanometer** (abbreviated nm), where 1 nm is written as 10^{-9} m, which means there are nine digits to the right of the decimal place—a very small number indeed.[1] Some astronomers prefer to measure wavelengths in *angstroms*. One angstrom, abbreviated Å, is one-tenth of a nanometer: 1 Å = 0.1 nm or 10^{-10} m. Visible light has wavelengths covering the range from about 400 nm for violet light to about 700 nm for red light. Intermediate colors of the rainbow like yellow (550 nm) have intermediate wavelengths. In these units, the wavelengths of visible light shown in Figure 2-6 would extend from about 4,000 Å to about 7,000 Å, instead of 400 nm to 700 nm.

Gamma rays are the light waves that have the shortest wavelength. Only a little longer are X-rays, which have wavelengths between about 10 nm and 0.01 nm. At wavelengths just short of visible light, **ultraviolet light** (UV) extends from about 100 nm down to 10 nm.

For wavelengths of light longer than visible light, infrared light covers the range from about 700 nm to 1 mm. Astronomers interested in infrared light often express wavelength in *micrometers* or *microns,* abbreviated μm, where 1 μm = 10^{-3} mm = 10^{-6} m. **Microwaves** have wavelengths from roughly 1 mm to 10 cm, while radio waves have the longest wavelengths.

ConceptCheck 2-4: Which form of light has a wavelength similar to the diameter of your finger?

Frequency Relates to the Number of Wave Crests Passing by Each Second

Although wavelength is a common way to distinguish among the entire spectrum of light emitted by objects, light can also be described in terms of *frequency.* This is commonly done by astronomers working with radio telescopes who look at the largest wavelengths of light. The **frequency** (f) of a wave is the number of wave crests that pass a given point in 1 s. Equivalently, it is the number of complete *cycles* of the wave that pass per second (a complete cycle is from one crest to the next). The unit of frequency is the cycle per second, also called the *hertz* (abbreviated Hz) in honor of Heinrich Hertz, the physicist who first produced radio waves. For example, if 500 crests of a wave pass you in 1 second, the frequency of the wave is 500 cycles per second or 500 Hz.

In working with frequencies, it is often convenient to use the prefix *mega-* (meaning "million," or 10^6, and abbreviated M) or *kilo-* (meaning "thousand," or 10^3, and abbreviated k). For example, AM radio stations broadcast at frequencies between 535 kHz and 1605 kHz (kilohertz), while FM radio stations broadcast at frequencies in the range from 88 MHz to 108 MHz (megahertz).

[1] For a tutorial on scientific notation, go to Appendix 9.

The relationship between the frequency and wavelength of an electromagnetic wave is relatively straightforward. Because light in a vacuum moves at a constant speed c = 300,000 km/s, if the wavelength—distance from one crest to the next—is made shorter, then the frequency must increase because more of those closely spaced crests pass by you each second.

The frequency f of light is inversely related to its wavelength λ, such that when one value is large, the other value is small. This idea can also be expressed as a simple equation as

Frequency and wavelength of an electromagnetic wave

$$f = c/\lambda$$

f = frequency of an electromagnetic wave (in hertz)
c = speed of light (in meters/second) = 300,000,000 m/s
λ = wavelength of the wave (in meters)

That is, the frequency of a wave equals the wave speed divided by the wavelength. What this relationship means is that the longest wavelengths of light have the lowest frequencies.

For example, hydrogen atoms in space emit radio waves with a wavelength of 21.12 cm. To calculate the frequency of this electromagnetic wave using the equation $c = f \times \lambda$, we must first express the wavelength in meters rather than centimeters: λ = 0.2112 m. Then we can use the above formula to find the frequency f:

$$\begin{aligned} f = c/\lambda &= 3 \times 10^8 \text{ m/s} / 0.2112 \text{ m} \\ &= 1.42 \times 10^9 \text{ Hz} \\ &= 1420 \text{ MHz} \end{aligned}$$

Visible light has a much shorter wavelength and higher frequency than radio waves. You can use the above formula to show that for yellow-orange light of wavelength 600 nm, the frequency is 5×10^{14} Hz or, with this many zeros, 500 *million* MHz!

ConceptCheck 2-5: How do the frequencies of the longest wavelengths of light compare to the frequencies of the shortest wavelengths of light?

CalculationCheck 2-1: What is the wavelength of radio waves from a nearby FM radio station?

2-3 An object's temperature is revealed by the most intense wavelength of its spectrum of light

Understanding the properties of light is the key astronomers need to unlock the mysteries of the universe. In this section, we will see how astronomers use light to measure the temperature of distant objects.

Infrared Light Can Pass Through Interstellar Clouds that Visible Light Cannot

If light exists in a wide spectrum of forms, the question arises as to which objects actually emit light and which do not. Our eyes are sensitive to only a very small range of possible wavelengths of light, and we often do not realize that almost all objects in the universe emit light of one form or another (with the notable exception of dark matter, described elsewhere in this book), depending on how hot or cool they are.

Common objects, like your book or the shoes on your feet, are at room temperature, or about 20°C. (Ordinary room temperature is 293 K, 20°C, or 68°F; see **Box 2-1 Tools of the Astronomer's Trade: Temperatures and Temperature Scales**). They are emitting some light, but most likely infrared light, which our eyes do not detect. Infrared light is the most typical wavelength of light emitted by objects at

BOX 2-1 TOOLS OF THE ASTRONOMER'S TRADE

Temperatures and Temperature Scales

Three temperature scales are in common use. Throughout most of the world, temperatures are expressed in **degrees Celsius** (°C). The Celsius temperature scale is based on the behavior of water, which freezes at 0°C and boils at 100°C at sea level on Earth. This scale is named after the Swedish astronomer Anders Celsius, who proposed it in 1742.

Astronomers usually prefer the Kelvin temperature scale. This is named after the nineteenth-century British physicist Lord Kelvin, who made many important contributions to our understanding of heat and temperature. Absolute zero, the temperature at which atomic motion is at the absolute minimum, is 2273°C in the Celsius scale but 0 K in the Kelvin scale. Atomic motion cannot be any less than the minimum, so nothing can be colder than 0 K; hence, there are no negative temperatures on the Kelvin scale. Note that we do *not* use degree (°) with the Kelvin temperature scale.

A temperature expressed in kelvins is always equal to the temperature in degrees Celsius plus 273. On the Kelvin scale, water freezes at 273 K and boils at 373 K. Water must be heated through a change of 100 K or 100°C to go from its freezing point to its boiling point. Thus, the "size" of a kelvin is the same as the "size" of a Celsius degree. When considering temperature changes, measurements in kelvins and Celsius degrees are the same. For extremely high temperatures the Kelvin and Celsius scales are essentially the same: For example, the Sun's core temperature is either 1.55 × 107 K or 1.55 × 107 °C.

The now-archaic Fahrenheit scale, which expresses temperature in **degrees Fahrenheit** (°F), is used only in the United States. When the German physicist Gabriel Fahrenheit introduced this scale in the early 1700s, he intended 100°F to approximate the temperature of a healthy human body. On the Fahrenheit scale, water freezes at 32°F and boils at 212°F. There are 180 Fahrenheit degrees between the freezing and boiling points of water, so a degree Fahrenheit is only 100/180 = 5/9 as large as either a Celsius degree or a kelvin.

Two simple equations allow you to convert a temperature from the Celsius scale to the Fahrenheit scale and from Fahrenheit to Celsius:

$$T_F = \frac{9}{5}T_C + 32$$

$$T_C = \frac{5}{9}T_F - 32$$

T_F = temperature in degrees Fahrenheit
T_C = temperature in degrees Celsius

EXAMPLE: A typical room temperature is 68°F. We can convert this to the Celsius scale using the second equation:

$$T_C = \frac{5}{9}(68 - 32) = 20°\text{C}$$

To convert this to the Kelvin scale, we simply add 273 to the Celsius temperature. Thus,

68°F = 20°C = 293 K

Figure B2-1.1 below displays the relationships among these three temperature scales.

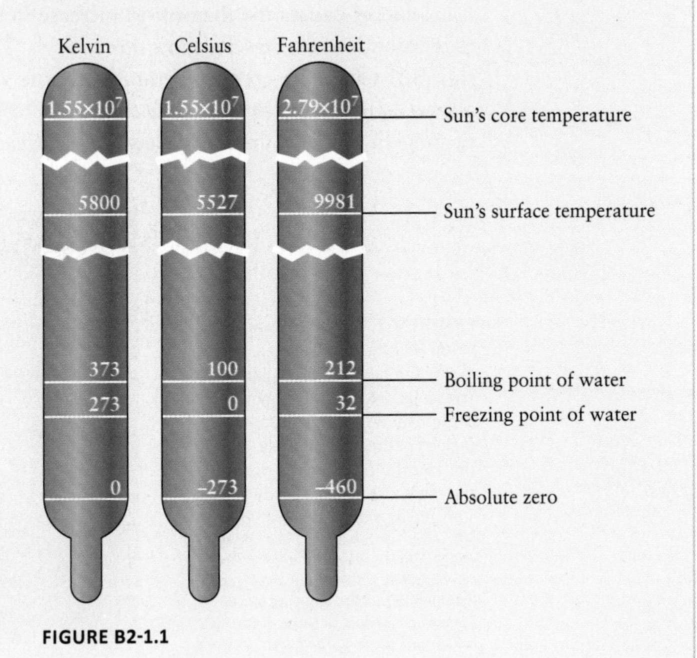

FIGURE B2-1.1

temperatures common on Earth's surface. The reason you can read your book and see your shoes is not that they are glowing with visible light; rather, they are reflecting visible light in the room into your eyes. In other words, your book and shoes are both emitting *and* reflecting, but different wavelengths of light.

Infrared light is an important part of astronomy because infrared light can successfully pass from distant stars through interstellar clouds, which visible light cannot penetrate. A visible light camera can detect only a very few stars emitting at visible wavelengths from the Orion Nebula, whereas the camera that is sensitive to infrared light can easily see the infrared light emitted by stars within the clouds (**Figure 2-7**).

CAUTION The right-hand image in Figure 2-7 is a **false-color** image. False-color images do not represent the true color of the stars shown. False color is often used when the image is made using wavelengths that the eye cannot detect, as with the infrared image in Figure 2-7. A different use of false color is to indicate the relative brightness of different parts of the image, as in the infrared image of a person in **Figure 2-8**.

ConceptCheck 2-6: Which form of light is being emitted most intensely by a frozen ice cube at 0°C?

Objects Emit Specific Amounts of Light, Revealing Their Temperatures

The simplest and most common way for any object to produce more light is to increase its temperature. The hot wire inside an ordinary filament lightbulb emits light because electrical energy causes the filament to increase in temperature to thousands of degrees and glow more visibly. This relationship between an object's temperature and the way it glows allows astronomers to determine the precise characteristics of some of the most distant objects in the universe.

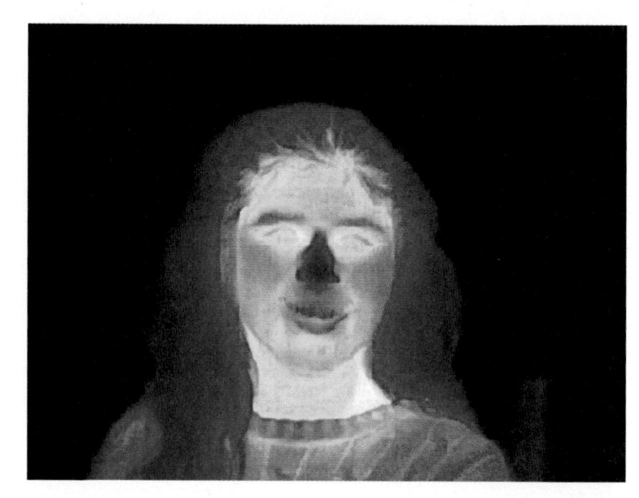

FIGURE 2-8 R I V U X G **An Infrared Portrait** In this image, made with a camera sensitive to infrared light, the different colors represent regions of different temperature. Red areas (like the woman's face) are the warmest and emit the most infrared light, while the bright blue areas (including the woman's nose and hair) are at the lowest temperatures and emit the least amount of infrared light. *(NASA/IPAC)*

Imagine a welder or blacksmith heating a bar of iron using a torch. The hottest flame glows blue, whereas the lower temperature flames glow more yellow and red-orange, as shown in **Figure 2-9*a–c*.** (You can see this same glow from the coils of a toaster, or from an electric range turned on "high.") As the temperature drops further, more of the light given off is from longer wavelengths and appears as reddish-orange light (Figure 2-9*b*). If the bar could be prevented from melting and vaporizing, at extremely high temperatures it would emit a dazzling blue-white light.

This experiment demonstrates that both the total amount of energy emitted by glowing objects and which wavelength of light is emitted most intensely depend directly on the temperature, where hotter objects emit shorter wavelength light that is more blue. The hotter the object, the brighter it is and the more energy it emits, and the shorter the

(a) R I V U X G

(b) R I V U X G

FIGURE 2-7 Newborn Stars in the Orion Nebula One light-year across in size and taken by the Hubble Space Telescope, (a) the visible-light view on the left shows the dusty Trapezium regions of the Orion Nebula, and (b) the infrared image on the right of the same area reveals newly formed stars hiding within the dust. *(a: NASA, C.R. O'Dell and S.K. Wong [Rice University /NASA Hubblesite.org]; b: NASA, K. L. Luhman [Harvard-Smithsonian Center for Astrophysics, Cambridge, MA], and G. Schneider, E. Young, G. Rieke, A. Cotera, H. Chen, M. Rieke, R. Thompson [Steward Observatory, University of Arizona, Tucson])*

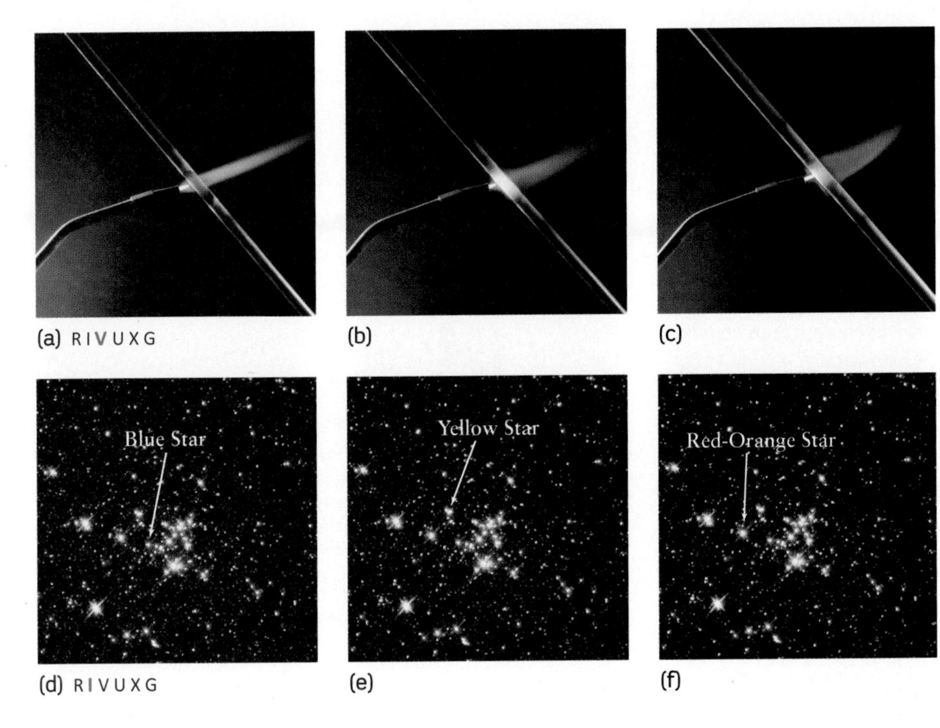

FIGURE 2-9 Heating a Bar of Iron (a–c) The sequence of photographs along the top row shows how the appearance of a flame changes with temperature. As the temperature of the torch flame is reduced, it radiates most of its energy at longer and longer wavelengths and becomes less blue and more red in color. (d–f) The stars shown have roughly the same surface temperatures as the flames above them. *(a–c: © Richard Megna Fundamental Photographs, NYC; d–f: Don Figer [Space Telescope Science Institute] and NASA)*

wavelength at which most of the energy is emitted. Colder objects emit relatively little energy, and the wavelength of light that has the maximum intensity emitted is primarily at the longer wavelengths. But what does this tell us about the stars? If you can measure the wavelength of light most intensely emitted from a star that is far away, you can determine its temperature.

However, if all objects, including stars, emit a wide range of wavelengths of light, then how can you determine the relative amounts of different forms of light that are being emitted? One strategy is to observe glowing objects with cameras designed to detect specific wavelengths of light and then graph the intensity of light emitted at various wavelengths for an object by plotting the wavelength emitted versus the brightness. The resulting graph of this spectrum is sometimes called a *blackbody curve* (**Figure 2-10**).

CAUTION A **blackbody** is an *idealized example of a dense object* that astronomers use as a model. A sample blackbody curve, showing brightness versus wavelength, is a graph of an object's spectrum that illustrates the intensity of light at every wavelength emitted at a particular temperature. How might this plot be different for different objects? In our experiment "heating the bar of iron," we noticed that as the bar increased in temperature, both its total amount of energy increased and the most intense wavelength emitted moved to shorter and bluer colors. As an object's temperature changes, the graph of the spectrum changes accordingly. The higher the temperature, the shorter the wavelength of maximum emission (at which the curve peaks) and the greater the amount of light emitted at every wavelength (see Figure 2-10).

This inversely proportional relationship between the temperature, measured in kelvins, of an emitting object and the maximum wavelength of light emitted, measured in meters, is known most commonly as **Wien's law,** after its originator, Wilhelm Wien.

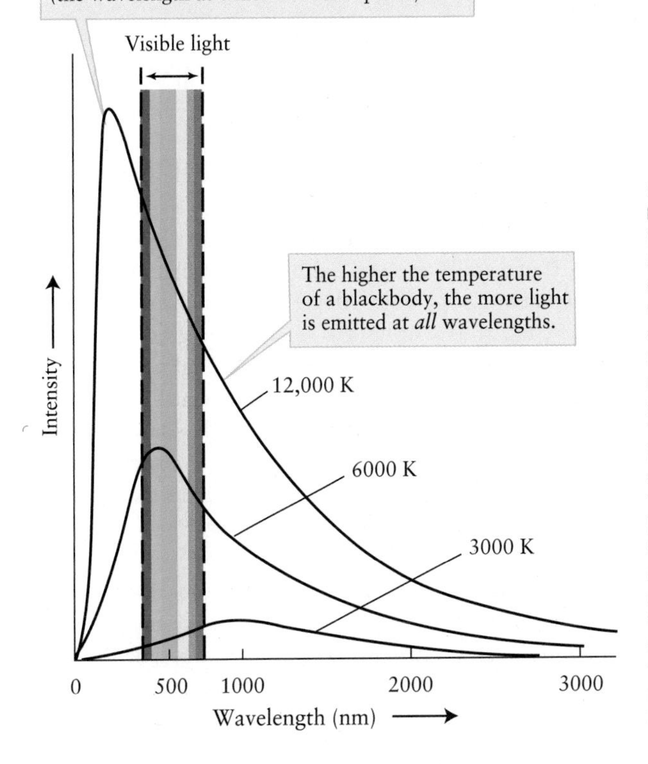

FIGURE 2-10 Blackbody Curves Each of these curves shows the intensity of light at every wavelength that is emitted by a blackbody (an idealized case of a dense object) at a particular temperature. The rainbow-colored band shows the range of visible wavelengths. The vertical scale has been compressed so that all three curves can be seen; the peak intensity for the 12,000-K curve is actually about 1000 times greater than the peak intensity for the 3000-K curve.

Our observations lead to a general rule that helps us measure the temperatures of celestial objects such as planets and stars:

The higher an object's temperature, the more intensely the object emits light and the shorter the wavelength at which it emits light most strongly.

We can express this general rule as a simple mathematical equation showing that the maximum wavelength is inversely related to the temperature.

Wien's law for a blackbody

$$\lambda = \frac{0.0029 \text{ K m}}{T}$$

λ_{max} = **wavelength of maximum emission of the object (in meters)**

T = **temperature of the object (in kelvins)**

In other words, the higher the temperature, the shorter the maximum wavelength emitted. Remember that Wien's law involves the wavelength of maximum emission in *meters*. If you want to convert the wavelength to nanometers, you must multiply the wavelength in meters by $(10^9 \text{ nm})/(1 \text{ m})$.

Figure 2-10 shows that for a dense object at a temperature of 3000 K, the wavelength of maximum emission is around 1000 nm (1 μm). Because this is an infrared wavelength well outside the visible range, you might think that you cannot see the light emitted from an object at this temperature. In fact, the glow from such an object *is* visible; the curve shows that this object emits plenty of light within the visible range, as well as at even shorter wavelengths. The 3000-K curve is quite a bit higher at the red end of the visible spectrum than at the violet end, so a dense object at this temperature will appear red in color. Similarly, the 12,000-K curve has its wavelength of maximum emission in the ultraviolet part of the spectrum, at a wavelength shorter than visible light. But such a hot, dense object also emits copious amounts of visible light (much more than at 6000 K or 3000 K, for which the curves are lower) and thus will have a very visible glow. The curve for this temperature is higher for blue light than for red light, and so the color of a dense object at 12,000 K is a brilliant blue or blue-white. These conclusions agree with the color changes of a heated rod shown in Figure 2-9. The same principles apply to stars: A star that looks blue has a high surface temperature, while a red star has a relatively cool surface.

The temperature of a star can be determined by measuring which color wavelength of light is being given off most intensely.

Figure 2-11 shows the spectral curve for a temperature of 5800 K. It also shows the intensity curve for light from the Sun, as measured from above Earth's atmosphere. (This is necessary because Earth's atmosphere absorbs certain wavelengths.) The peak of both curves is at a wavelength of about 500 nm, near the middle of the visible spectrum. This is a strong indication that the temperature of the Sun's glowing surface is about 5800 K—a temperature that we can measure across a distance of 150 million km! This is very valuable information for Earthbound astronomers trying to study the stars.

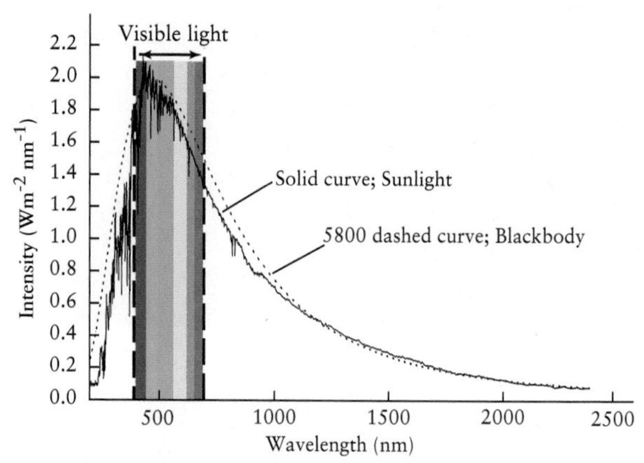

FIGURE 2-11 The Sun as a Blackbody This graph shows that the intensity of sunlight over a wide range of wavelengths (solid curve) is a remarkably close match to the intensity of light coming from a blackbody at a temperature of 5800 K (dashed curve). The measurements of the Sun's intensity were made above Earth's atmosphere (which absorbs and scatters certain wavelengths of sunlight). It is not surprising that the range of visible wavelengths includes the peak of the Sun's spectrum; the human eye evolved to take advantage of the most plentiful light available.

Wien's law is very useful for determining the surface temperature of stars. It is not necessary to know how far away the star is, how large it is, or how much energy it radiates into space. All we need to know is the dominant wavelength of the star's emitted light.

ConceptCheck 2-7: **What single piece of information do astronomers need to determine if a star is hotter than our Sun?**

CalculationCheck 2-2: **Which wavelength of light would our Sun emit most if its temperature were twice its current temperature of 5800 K?**

How Much Energy a Star Emits Is Determined by Both Temperature and Surface Area

Several variables in addition to temperature are involved in determining how much energy an object emits. They include the object's diameter, mass, and chemical composition. To understand how all these aspects work together, let us first consider how we measure energy. When we eat food to nourish our human bodies, we absorb energy. The amount of energy stored inside food is usually measured in calories in the United States. When we talk about stars, we more often use **joules** (J), named after the nineteenth-century English physicist James Joule. A joule is a standard unit for measuring energy and is equivalent to the kinetic energy of a 1-kg object moving at 1 m/s. The joule is a convenient unit of energy in astronomy because it is closely related to the more familiar **watt** (W): 1 W is 1 J/s[2]. For example, a 100-W lightbulb uses energy at a rate of 100 joules per second, or 100 J/s. As it turns out, the energy content of food can also often be measured in joules; in most of the world, diet soft drinks are labeled as "low joule" rather than "low calorie."

The amount of the star's total energy output, or its *luminosity,* depends both on its temperature and on its surface area. Just as a large burning log radiates much more heat than a burning match, even though the temperatures are the same, so do larger stars emit more energy than smaller stars, even at the same temperature. So, to understand how much energy a star emits, astronomers need to do more than measure temperature; they also need to consider the surface area of the star. This makes sense: To consider the effects of temperature alone, it is convenient to look at the amount of energy emitted from each square meter of an object's surface in a second. This quantity is called the **energy flux** (F). Flux means "rate of flow," and thus F is a measure of how rapidly energy is flowing out of the object.

Experimentally, the total amount of energy emitted from an object depends only on the temperature and its surface area. In fact, the energy flux from glowing objects is highly dependent on temperature—more so than anything else. Known as the **Stefan-Boltzmann law,** this idea can be represented using a simple mathematical relationship where the flux emitted by a star is proportional to the fourth power of the object's temperature (measured in kelvins).

Stefan-Boltzmann law for a blackbody

$$F = \sigma T$$

F = energy flux, in joules per square meter of surface per second[3]
σ = a constant = 5.67×10^{-8} W m^{-2} K^{-4}
T = object's temperature, in kelvins

The value of the constant σ (the Greek letter sigma) is known from laboratory experiments.

This relationship between temperature and flux emitted is incredibly powerful. If you double the temperature of an object (for example, from 300 K to 600 K), then the energy emitted from the object's surface each second increases by a factor of $2^4 = 16$. If you increase the temperature by a factor of 10 (for example, from 300 K to 3000 K), the rate of energy emission increases by a factor of $10^4 = 10{,}000$. Thus, a chunk of iron at room temperature (around 300 K) emits very little energy at visible wavelengths, but an iron bar heated to 3000 K glows quite intensely. When astronomers apply the Stefan-Boltzmann law to stars, we are able to infer a star's surface area, and thus its diameter. For example, there are some relatively cool stars that are intensely bright—the only way a cool star can be very bright is if these special stars themselves are gigantic, as deduced from these principles. As another example, we find some amazingly hot stars that are quite dim; again, the only way this can be is if the star itself is quite small. **Box 2-2 Tools of the Astronomer's Trade: Using the Laws of Blackbody Radiation** gives several examples of applying Wien's law and the Stefan-Boltzmann law to typical astronomical problems.

[2] 1 W = 1 J/s = 1 J s^{-1}. The superscript -1 means you are dividing by that quantity.
[3] It is measured in joules per square meter per second, usually written as J/m^2/s or J m^{-2} s^{-1}. Alternatively, because 1 W equals 1 J/s, we can express flux in watts per square meter (W/m^2, or W m^{-2}).

ConceptCheck 2-8: **If you observe a red star and a blue star in the sky, how do you distinguish which star is at a higher temperature?**

CalculationCheck 2-3: **How many times more energy flux comes from a star that is 3 times hotter than the Sun?**

Light Has Properties of Both Waves and Particles

To round out our understanding of why this works the way it does, we need to realize that the famous German-born physicist Albert Einstein had a great influence on our modern thinking about light. In 1905, Einstein realized that light beams are not continuous; rather, they come in discrete, particlelike packets known as light *quanta* (the plural of *quantum,* from a Latin word meaning "how much"). Further, Einstein proposed that the energy of each light quantum—today we usually call it a **photon**—is related to the wavelength of light: The greater the wavelength, the lower the energy of a photon associated with that wavelength. For example, a photon of red light (λ = 700 nm) has less energy than a photon of violet light (λ = 400 nm). In this picture, light has a dual personality; it behaves as a stream of particlelike photons, but each photon has wavelike properties. In this sense, the best answer to the question "Is light a wave or a stream of particles?" is "Yes!"

Imagining light photons as being both a particle and a wave explains a lot. For example, the photon view of light explains why only ultraviolet light causes suntans and sunburns. The reason is that tanning or burning involves a chemical reaction in the skin. High-energy, short-wavelength ultraviolet photons can trigger these reactions, but the lower-energy, longer-wavelength photons of visible light cannot. Similarly, normal photographic film is sensitive to visible light but not to infrared light; a long-wavelength infrared photon does not have enough energy to cause the chemical change that occurs when film is exposed to the higher-energy photons of visible light. To commemorate this great achievement, Einstein and other scientists working on

BOX 2-2 TOOLS OF THE ASTRONOMER'S TRADE

Using the Laws of Blackbody Radiation

The Sun and stars behave like nearly perfect blackbodies. Wien's law and the Stefan-Boltzmann law can therefore be used to relate the surface temperature of the Sun or a distant star to the energy flux and wavelength of maximum emission of its radiation. The following examples show how to do this.

EXAMPLE: The maximum intensity of sunlight is at a wavelength of roughly 500 nm = 5.0×10^{-7} m. Use this information to determine the surface temperature of the Sun.

Situation: We are given the Sun's wavelength of maximum emission λ_{max}, and our goal is to find the Sun's surface temperature, denoted by $T_{\odot}$. (The symbol $\odot$ is the standard astronomical symbol for the Sun.)

Tools: We use Wien's law to relate the values of λ_{max} and $T_{\odot}$.

Answer: As written, Wien's law tells how to find λ_{max} if we know the surface temperature. To find the surface temperature from λ_{max}, we first rearrange the formula, then substitute the value of λ_{max}:

$$T_{\odot} = \frac{0.0029 \text{ K m}}{\lambda} = \frac{0.0029 \text{ K m}}{5.0 \times 10^{-7} \text{m}} = 5800 \text{ K}$$

Review: This is a very high temperature by Earth standards, about the same as an iron welding arc.

EXAMPLE: Using detectors above Earth's atmosphere, astronomers have measured the average flux of solar energy arriving at Earth. This value, called the **solar constant,** is equal to 1370 W m^{-2}. Use this information to calculate the Sun's surface temperature. (This provides a check on our result from the preceding example.)

Situation: The solar constant is the flux of sunlight as measured at Earth. We want to use the value of the solar constant to calculate $T_{\odot}$.

Tools: It may seem that all we need is the Stefan-Boltzmann law, which relates flux to surface temperature. However, the quantity F in this law refers to the flux measured at the Sun's surface, *not* at Earth. Hence, we will first need to calculate F from the given information.

Answer: To determine the value of F, we first imagine a huge sphere of radius 1 AU with the Sun at its center, as shown in the figure. Each square meter of that sphere receives 1370 W of power from the Sun, so the total energy radiated by the Sun per second is equal to the solar constant multiplied by the sphere's surface area. The result, called the **luminosity** of the Sun and denoted by the symbol $L_{\odot}$, is 3.90×10^{26} W. That is, in 1 s the Sun radiates 3.90×10^{26} J of energy into space. Because we know the size of the Sun, we can compute the energy flux (energy emitted per square meter per second) at its surface. The radius of the Sun ($R_{\odot}$) is 6.96×10^{8} m, and the Sun's surface area is $4\pi R_{\odot}^{2}$. Therefore, its energy flux ($F_{\odot}$) is the Sun's luminosity (total energy emitted by the Sun per second) divided by the Sun's surface area (the number of square meters of surface):

$$F_{\odot} = \frac{L_{\odot}}{4\pi R_{\odot}^{2}} = \frac{3.90 \times 10^{26} \text{ W}}{4\pi(6.96 \times 10^{8} \text{ m})^{2}} = 6.41 \times 10^{7} \text{W m}^{-2}$$

Once we have the Sun's energy flux $F_{\odot}$, we can use the Stefan-Boltzmann law to find the Sun's surface temperature $T_{\odot}$:

$$T_{\odot}^{4} = \frac{F_{\odot}}{\sigma} = 1.13 \times 10^{15} \text{K}^{4}$$

Taking the fourth root (the square root of the square root) of this value, we find the surface temperature of the Sun to be 5800 K.

Review: Our result for $T_{\odot}$ agrees with the value we computed in the previous example using Wien's law. Notice that the solar constant of

the photon nature of light won Nobel Prizes for their contributions to understanding the nature of light.

ConceptCheck 2-9: **If a photon's wavelength is measured to be longer than the wavelength of a green photon, will it have a greater or lower energy than a green photon?**

2-4 An object's chemical composition is revealed by the unique pattern of its spectrum of light

The development of strategies to measure the temperature of a star so distant that we will probably never visit is exciting science in and of itself, but you might be wondering what these stars, galaxies, and giant interstellar clouds actually are. The astronomer and science popularizer Carl Sagan once quipped, "We are all star stuff," but what does that actually mean? How might we determine what star stuff really is?

One strategy would be to send a space probe to the stars and do a chemical analysis. Unfortunately, the next star to our solar system is trillions of miles (trillions of kilometers) away, too far to send a probe. And even if we could send a space probe to a star, it is impossible to get close to stars because, like our Sun, they are far too hot. Fortunately, just as light gives us the clues to understand temperature, it gives us clues to chemical composition and what sorts of atoms these distant objects are made of. However, to figure this out, we have to exploit what we now understand about the structure of atoms. Why do different atoms each emit their own distinctive wavelengths of light? Why is a particular atom only able to absorb specific wavelengths of light? Answers to these questions did not become clear until astronomers began to use their knowledge of the structure and properties of atoms.

1370 W m^{-2} is very much less than $F_{\odot}$, the flux at the Sun's surface. By the time the Sun's light reaches Earth, it is spread over a greatly increased area.

EXAMPLE: Sirius, the brightest star in the night sky, has a surface temperature of about 10,000 K. Find the wavelength at which Sirius emits most intensely.

Situation: Our goal is to calculate the wavelength of maximum emission of Sirius (λ_{max}) from its surface temperature *T*.

Tools: We use Wien's law to relate the values of λ_{max} and *T*.

Answer: Using Wien's law,

$$\lambda_{max} = \frac{0.0029 \text{ K m}}{T} = \frac{0.0029 \text{ K m}}{10{,}000 \text{ K}} = 2.9 \times 10^{7} \text{ m} = 290 \text{ nm}$$

Review: Our result shows that Sirius emits light most intensely in the ultraviolet. In the visible part of the spectrum, it emits more blue light than red light (like the curve for 12,000 K in Figure 2-10), so Sirius has a distinct blue color.

EXAMPLE: How does the energy flux from Sirius compare to the Sun's energy flux?

Situation: To compare the energy fluxes from the two stars, we want to find the *ratio* of the flux from Sirius to the flux from the Sun.

Tools: We use the Stefan-Boltzmann law to find the flux from Sirius and from the Sun, which from the preceding examples have surface temperatures 10,000 K and 5800 K, respectively.

Answer: For the Sun, the Stefan-Boltzmann law is $F_{\odot} = \sigma T_{\odot}^4$, and for Sirius we can likewise write $F_* = \sigma T_*^4$, where the subscripts ⊙ and * refer to the Sun and Sirius, respectively. If we divide one equation by the other to find the ratio of fluxes, the Stefan-Boltzmann constants cancel out and we get

$$\frac{F_*}{F_{\odot}} = \frac{T_*^4}{T_{\odot}^4} = \frac{(10{,}000 \text{ K})^4}{(5800 \text{ K})^4} = \left(\frac{10{,}000}{5800}\right)^4 = 8.8$$

Review: Because Sirius has such a high surface temperature, each square meter of its surface emits 8.8 times more energy per second than a square meter of the Sun's relatively cool surface. Sirius is actually a larger star than the Sun, so it has more square meters of surface area and hence its *total* energy output is *more* than 8.8 times that of the Sun.

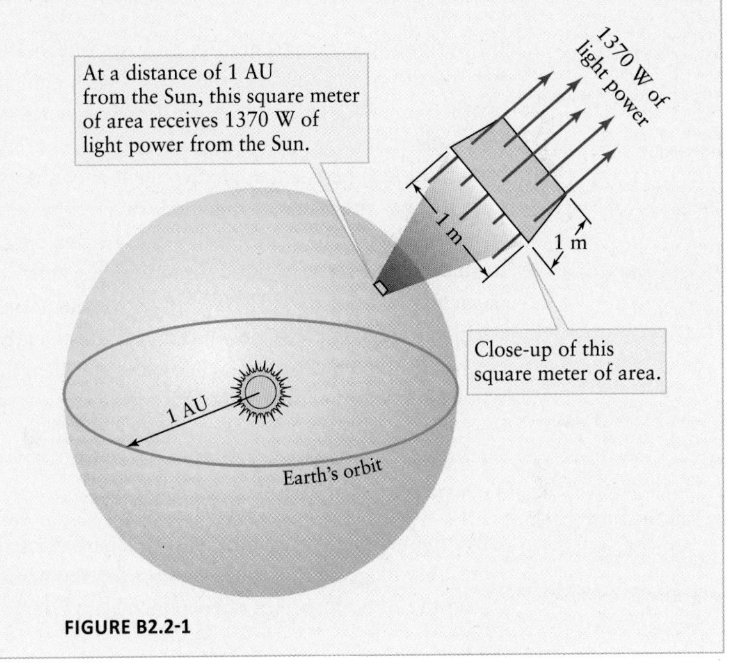

FIGURE B2.2-1

Electrons Can "Jump" Only to Specific Orbits Within Atoms, Creating the Particular Sets of Spectral Lines

You might be surprised to learn that only about 100 years ago, the inner workings of the atom were mostly unknown. Today, we understand that most of an atom's mass is located in its very center. Tiny electrons, nearly without mass, are found in the outer extremities, often called electron orbits, while the majority of the atom's volume is empty space. We know that the central mass of an atom contains two types of particles, protons and neutrons, but it is the behavior of the electrons in the outer extremities of the atom that generates light.

In the early 1900s, physicist Niels Bohr made the rather wild assumption that the electron in a hydrogen atom can orbit the nucleus only in certain specific orbits. (This was a significant break with the ideas of Newton, who predicted that any orbital path should be possible.) To be clear, the electrons in an atom do not actually move precisely in circular orbits, but thinking of it this way is an extremely handy conceptual tool.

Figure 2-12 shows a hydrogen atom as an example of this mental model of an atom. The four smallest orbital paths are labeled by the numbers $n = 1$, $n = 2$, $n = 3$, and so on. In this example, an electron can jump from one of these orbits to another. For an electron to do this, the hydrogen atom must gain or lose a specific amount of energy. **Figure 2-13** shows an electron jumping between the $n = 2$ and $n = 3$ orbits of the hydrogen atom as the atom absorbs or emits light. (In the case illustrated in Figure 2-13, the particular wavelength of light is 656.3 nm.) The atom must absorb energy for the electron to go from an inner to an outer orbit (Figure 2-13*a*); the atom must release energy for the electron to go from an outer to an inner orbit (Figure 2-13*b*). It is these transitions between electron orbit positions that help to explain

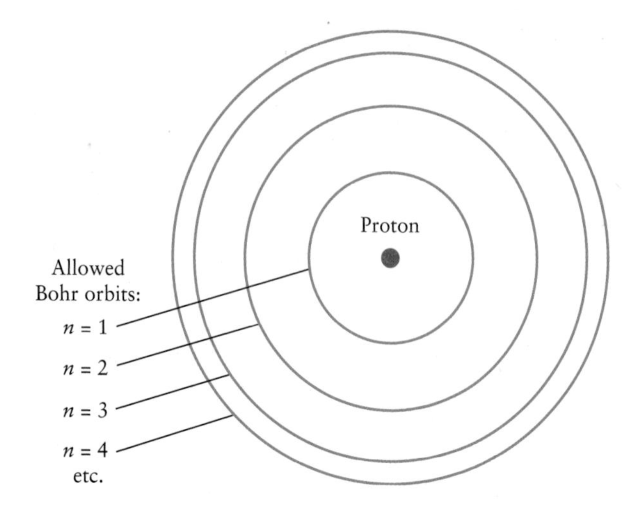

FIGURE 2-12 **The Bohr Model of the Hydrogen Atom** In this model, an electron circles the hydrogen nucleus (a proton) only in allowed orbits $n = 1, 2, 3$, and so forth. The first four Bohr orbits are shown here. This figure is not drawn to scale; in the Bohr model, the $n = 2, 3$, and 4 orbits are respectively 4, 9, and 16 times larger than the $n = 1$ orbit.

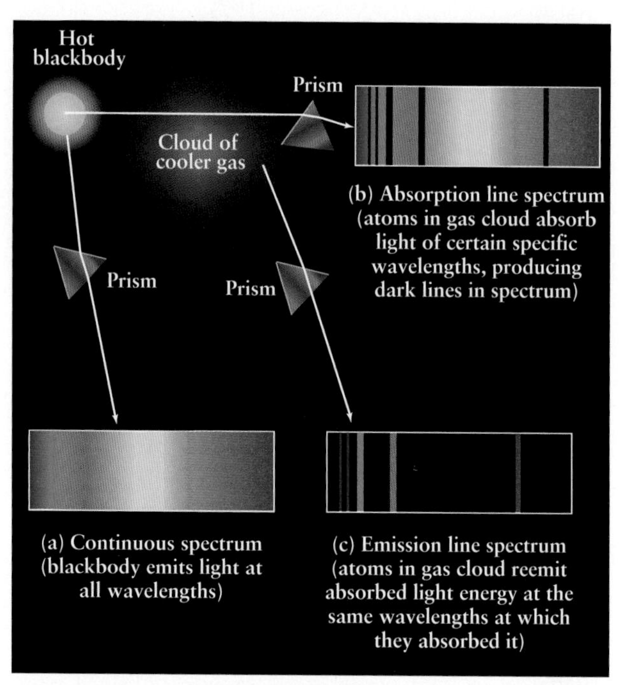

FIGURE 2-14 **Continuous, Absorption Line, and Emission Line Spectra** A hot, opaque body (like a blackbody) emits a continuous spectrum of light (spectrum **a**). If this light is passed through a cloud of a cooler gas, the cloud absorbs light of certain specific wavelengths, and the spectrum of light that passes directly through the cloud has dark absorption lines (spectrum **b**). The cloud does not retain all the light energy that it absorbs but radiates it outward in all directions. The spectrum of this reradiated light contains bright emission lines (spectrum **c**) with exactly the same wavelengths as the dark absorption lines in spectrum b. The specific wavelengths observed depend on the chemical composition of the cloud.

the various spectra astronomers observe and to answer our key questions: Why do different atoms each emit their own distinctive wavelengths of light? Why is a particular atom only able to absorb specific wavelengths of light? The most important idea here is that an atom will only absorb or emit light if it has the precisely matched wavelength necessary to move the electron to a new orbit. There are two consequences to this notion. First, in most circumstances, only light with particular wavelengths will be emitted from an atom. Second, when light hits an atom, only particular wavelengths of light can be absorbed by an atom, while other wavelengths of light will pass through without affecting the atom.

Kirchhoff's Laws Explain How Different Types of Spectra Are Produced

Once the nature of the atom was understood, astronomers were able to generalize three important statements about the nature of how spectra are produced, applying principles outlined years before by Gustav Kirchhoff, called **Kirchhoff's laws.** These laws, which are illustrated in **Figure 2-14,** are as follows:

Law 1 *A hot, opaque body or a hot, dense gas produces a* **continuous spectrum**—*a complete rainbow of colors without any spectral lines.*

Law 2 *A hot, transparent gas produces an* **emission line spectrum**—*a series of bright spectral lines against a dark background.*

Law 3 *A cool, transparent gas in front of a source of a continuous spectrum produces an* **absorption line spectrum**—*a series of dark spectral lines among the colors of the continuous spectrum. Furthermore, the dark lines in the absorption spectrum of a particular gas occur at exactly the same wavelengths as the bright lines in the emission spectrum of that same gas.*

Kirchhoff's laws imply that if a beam of white light is passed through a gas, the atoms of the gas somehow extract light of very specific wavelengths from the white light. This is a direct consequence of the behavior of electrons in an atom, as described in the previous section. Hence, an observer who

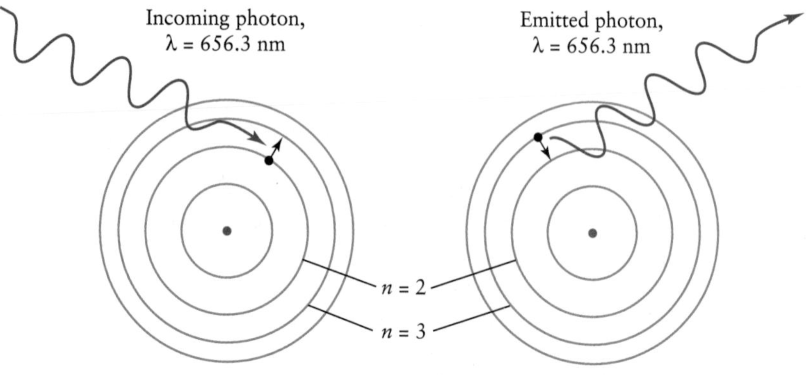

FIGURE 2-13 **The Absorption and Emission of an H_α Photon** This schematic diagram, drawn according to the Bohr model, shows what happens when a hydrogen atom (a) absorbs or (b) emits a photon whose wavelength is 656.3 nm.

looks straight through the gas at the white-light source will receive light whose spectrum has dark absorption lines superimposed on the continuous spectrum of the white light (Figure 2-14*b*). The gas atoms then radiate light of precisely these same wavelengths in all directions. An observer at an oblique angle (that is, one who is not sighting directly through the cloud toward the glowing object) will receive only this light radiated by the gas cloud; the spectrum of this light is bright emission lines on a dark background (Figure 2-14*c*).

CAUTION Figure 2-14 shows that light can either pass through a cloud of gas or be absorbed by the gas. But there is also a third possibility: The light can simply bounce off the atoms or molecules that make up the gas, a phenomenon called **light scattering**. In other words, photons passing through a gas cloud can miss the gas atoms altogether, be swallowed whole by the atoms (absorption), or bounce off the atoms like billiard balls colliding (scattering). **Box 2-3 The Heavens on Earth: Light Scattering** describes how light scattering explains the blue color of the sky and the red color of sunsets.

Whether an emission line spectrum or an absorption line spectrum is observed from a gas cloud depends on the relative temperatures of the gas cloud and its background. Absorption lines are seen if the background is hotter than the gas, and emission lines are seen if the background is cooler.

For example, if we drop sodium in the flame of a Bunsen burner in a darkened room, as illustrated in **Figure 2-15**, the flame will emit a characteristic orange-yellow glow. (This same glow is produced if we use ordinary table salt, which is a compound of sodium and chlorine.) If we look at the light from the flame through a prism, we observe an emission line spectrum with two closely spaced spectral lines at wavelengths of 588.99 nm and 589.59 nm in the orange-yellow part of the spectrum (see Figure 2-6).

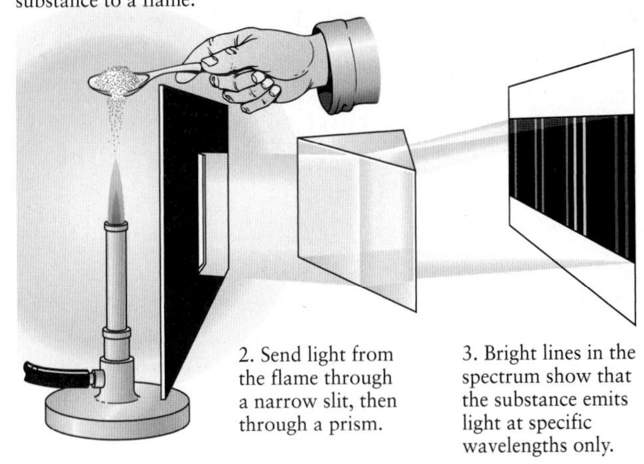

FIGURE 2-15 The Kirchhoff-Bunsen Experiment In the mid-1850s, Gustav Kirchhoff and Robert Bunsen discovered that when a chemical substance is heated and vaporized, the spectrum of the emitted light exhibits a series of bright spectral lines. They also found that each chemical element produces its own characteristic pattern of spectral lines. (In an actual laboratory experiment, lenses would be needed to focus the image of the slit onto the screen.)

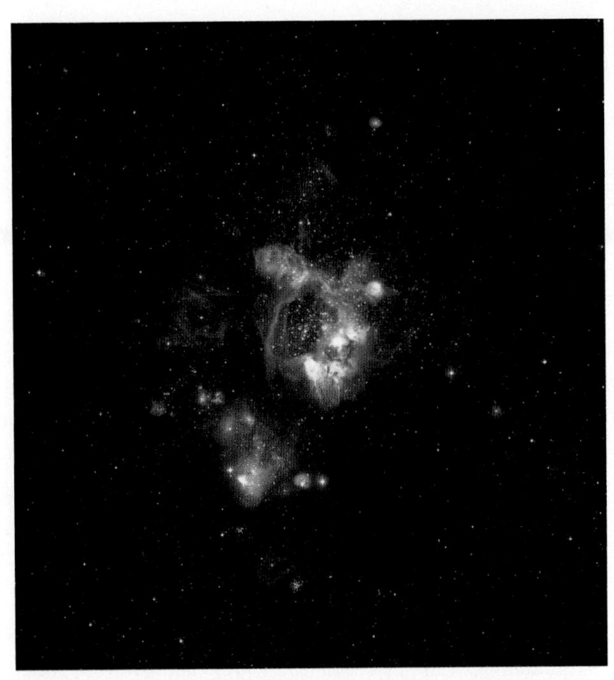

FIGURE 2-16 R I V U X G **The Nebula N44** Stretching about 1,000 light-years across, this glowing gas cloud shown here is not within our Milky Way Galaxy, but in the nearby Large Magellanic Cloud about 170,000 light-years away. Hot stars within the nebula emit high-energy, ultraviolet photons, which are absorbed by the surrounding gas and heat the gas to high temperature. This heated gas produces light with an emission line spectrum. The particular wavelength of red light emitted by the nebula is characteristic of hydrogen gas. *(WFI, MPG/ESO 2.2-m Telescope, La Silla, ESO)*

What is truly remarkable about spectroscopy is that it can determine chemical composition at any distance. The 656-nm red light produced by a sample of heated hydrogen gas on Earth is the same as that observed coming from the glowing emission nebula in **Figure 2-16,** located about 170,000 light-years away.

The unique features of a spectrum can be deciphered to determine precisely which atoms make up the glowing object, even when it is very far way.

Liquids, Solids, and Dense Gases All Can Produce Continuous Spectra

We have seen that changes to an electron's orbit explain the emission line spectra and absorption line spectra of gases. But how can that explain the continuous spectra produced by dense objects like the filament of a lightbulb or the heated coils of a toaster? These objects are also made of atoms, so why don't they emit light with an emission line spectrum characteristic of the particular atoms of which they are made?

The reason is directly related to the difference between a thin gas on the one hand and a dense gas on the other. In a thin gas, atoms are widely separated and can emit light without interference from other atoms. But in the case of a dense gas, atoms are so close that they almost touch, and thus these atoms interact strongly with each other. (In fact, a liquid and a solid can act like glowing, dense gas when heated to a high enough temperature.) These closely packed atom-to-atom

BOX 2-3 THE HEAVENS ON EARTH

Light Scattering

Light scattering is the process whereby photons bounce off particles in their path. These particles can be atoms, molecules, or clumps of molecules. You are reading these words using photons from the Sun or a lamp that bounced off the page—that is, were scattered by the particles that make up the page.

An important fact about light scattering is that very small particles—ones that are smaller than a wavelength of visible light—are quite effective at scattering short-wavelength photons of blue light, but less effective at scattering long-wavelength photons of red light. This fact explains a number of phenomena that you can see here on Earth.

The light that comes from the daytime sky is sunlight that has been scattered by the molecules that make up our atmosphere (see part (**a**) of **Figure B2-3.1**). Air molecules are less than 1 nm across, far smaller than the wavelength of visible light, so they scatter blue light more than red light—which is why the sky looks blue. Smoke particles are also quite small, which explains why the smoke from a cigarette or a fire has a bluish color.

Distant mountains often appear blue thanks to sunlight being scattered from the atmosphere between the mountains and your eyes. (The Blue Ridge Mountains, which extend from Pennsylvania to Georgia, and Australia's Blue Mountains derive their names from this effect.) Sunglasses often have a red or orange tint, which blocks out blue light. This cuts down on the amount of scattered light from the sky reaching your eyes and allows you to see distant objects more clearly.

Light scattering also explains why sunsets are red. The light from the Sun contains photons of all visible wavelengths, but as this light passes through our atmosphere the blue photons are scattered away from the straight-line path from the Sun to your eyes. Red photons undergo relatively little scattering, so the Sun always looks a bit redder than it really is. When you look toward the setting sun, the sunlight that reaches your eyes has had to pass through a relatively thick layer of atmosphere (part (**b**) of the figure). Hence, a large fraction of the blue light from the Sun has been scattered, and the Sun appears quite red.

The same effect also applies to sunrises, but sunrises seldom look as red as sunsets. The reason is that dust is lifted into the atmosphere during the day by the wind (which is typically stronger in the daytime than at night), and dust particles in the atmosphere help to scatter even more blue light.

If the small particles that scatter light are sufficiently concentrated, there will be almost as much scattering of red light as of blue light, and the scattered light will appear white. This explains the white color of clouds, fog, and haze, in which the scattering particles are ice crystals or water droplets. Whole milk looks white because of light scattering from tiny fat globules; nonfat milk has only a very few of these globules and so has a slight bluish cast.

Light scattering has many applications to astronomy. For example, it explains why very distant stars in our Galaxy appear surprisingly red. The reason is that there are tiny dust particles in the space between the stars, and this dust scatters blue photons. By studying how much scattering takes place, astronomers have learned about the tenuous material that fills interstellar space.

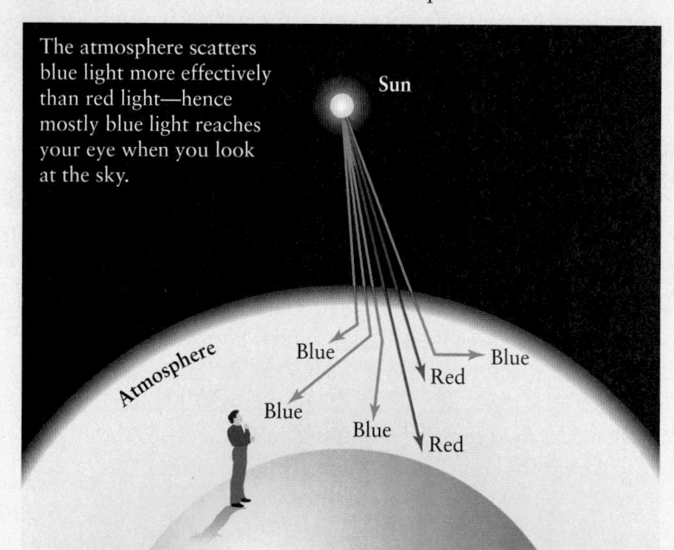

(a) Why the sky looks blue

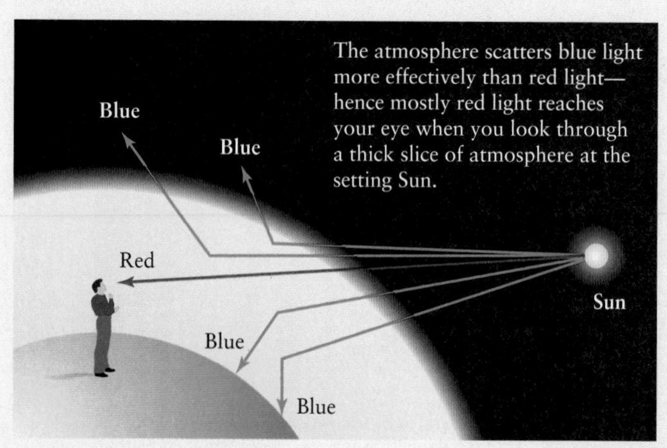

(b) Why the setting Sun looks red

FIGURE B2.3-1

interactions interfere with the process of emitting light. As a result, the pattern of distinctive bright spectral lines that the atoms would emit in isolation becomes "smeared out" into a continuous spectrum.

Go to Video 2-2

ConceptCheck 2-10: **What type of spectra would result from a glowing field of hot, dense lava as viewed by an orbiting satellite through Earth's atmosphere?**

Each Chemical Substance Has a Unique Pattern of Spectral Lines

Now we are in a position to talk specifically about how atoms can be detected within and among the distant stars. You might have noticed that fireworks displays often show highly varied, glowing colors. What would you see if you could pass these colored lights through a prism? Actually, scientists have been studying the light from burning chemicals for more than a century. In the mid-1850s, the German chemist Robert Bunsen and the Prussian-born physicist Gustav Kirchhoff discovered that the light emitted from burning chemicals has a specific appearance when passed through a prism. In fact, every chemical substance has its own specific color spectrum when burned and passed through a prism, showing *a pattern of thin, bright lines against a dark background* known as **spectral lines** (Figure 2-16). The identification of chemical substances by the unique patterns of lines in their spectrum is called **spectroscopy** and is a core aspect of contemporary astronomy. Spectral lines are tremendously important in astronomy, because they provide reliable evidence about the chemical composition of distant objects. We will use them again and again throughout this book.

2-5 An object's motion through space is revealed by the precise wavelength positions of its spectrum of light

Often, astronomical reports in the media describe the speed at which astronomical objects are moving. Or, Hollywood takes a turn at illustrating science, and impending doom is imminent because of the asteroid or meteor speeding our way. How might one go about determining the speed of a star moving through space? If the star is moving side to side, one could reasonably hope to simply measure how far it moves in a particular amount of time. But, what if the object is moving toward us or away from us? Fortunately, in addition to telling us about temperature and chemical composition, the spectrum of a planet, star, or galaxy can also reveal something about that object's motion through space.

Exploiting the Doppler Effect

Go to Video 2-3

In 1842, Christian Doppler, a professor of mathematics in Prague, pointed out that the observed wavelength of light must be affected by motion. **Figure 2-17** shows why. In this figure, a light source is moving from right to left, and the circles represent the crests of waves emitted from the moving source at various positions. Each successive wave crest is emitted from a position slightly closer to the observer on the left, so she sees a shorter wavelength—the distance from one crest to the next—than she would if the source were stationary. All the lines in the spectrum of an approaching source are shifted toward the short-wavelength (blue) end of the spectrum. This phenomenon is called a **blueshift.**

The source is receding from the observer on the right in Figure 2-17. The wave crests that reach him are stretched apart, so that he sees a longer wavelength than he would if the source were stationary. All the lines in the spectrum of a receding source are shifted toward the longer-wavelength (red) end of the spectrum, producing a **redshift.** In general, the effect of relative motion on wavelength is called the **Doppler effect.** Some radar guns use the Doppler effect to check for cars exceeding the speed limit: The radar gun sends a radio wave toward the car and measures the wavelength shift of the reflected wave—and thus the speed of the car.

ANALOGY You have probably noticed a similar Doppler effect for sound waves. When a police car is approaching, the sound waves from its siren have a shorter wavelength and

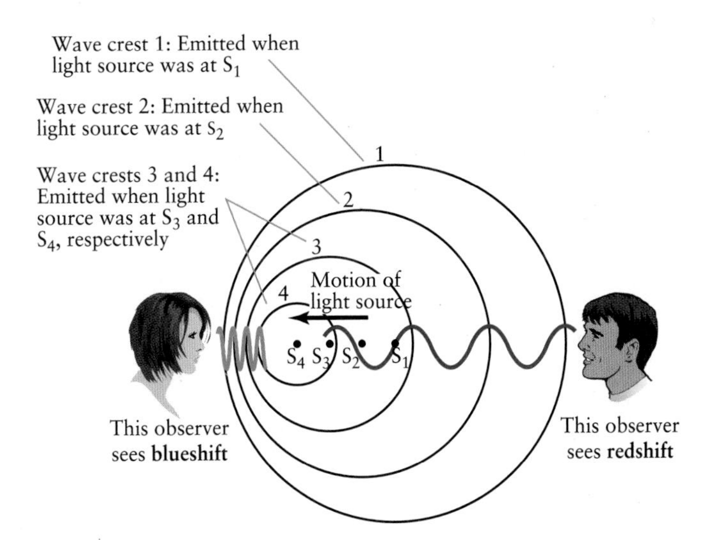

FIGURE 2-17
The Doppler Effect The wavelength of light is affected by motion between the light source and an observer. The light source shown here is moving, so wave crests 1, 2, etc., emitted when the source was at points S_1, S_2, etc., are crowded together in front of the source but are spread out behind it. Consequently, wavelengths are shortened (blueshifted) if the source is moving toward the observer and lengthened (redshifted) if the source is moving away from the observer. Motion perpendicular to an observer's line of sight does not affect wavelength.

higher frequency than if the moving siren were at rest, and hence you hear a higher pitch. After the police car passes you and is moving away, you hear a lower pitch from the siren because the sound waves have a longer wavelength and a lower frequency.

CAUTION The redshifts and blueshifts of stars visible to the naked eye, or even through a small telescope, are only a small fraction of a nanometer. These tiny wavelength changes are far too small to detect visually. (Astronomers were able to detect the tiny Doppler shifts of starlight only after they had developed highly sensitive equipment for measuring wavelengths. This was done around 1890, a half-century after Doppler's original proposal.) So, if you see a star with a red color, it means that the star really is red; it does *not* mean that it is moving rapidly away from us.

The Doppler effect is an important tool in astronomy because it uncovers basic information about the motions of planets, stars, and galaxies. For example, the rotation of the planet Venus was deduced from the Doppler shift of radar waves reflected from its cloud-shrouded surface. Small Doppler shifts in the spectrum of sunlight have shown that the entire Sun is vibrating like an immense gong. The back-and-forth Doppler shifting of the spectral lines of certain stars reveals that these stars are being orbited by unseen companions; from this astronomers have discovered planets around other stars and massive objects that may be black holes. This is only one of many examples of how Doppler's discovery has empowered astronomers in their quest to understand the universe.

Doppler proved that the wavelength shift ($\Delta\lambda$) is governed by the following simple equation for objects moving at speeds much less than the speed of light:

Doppler shift equation

$$\frac{\Delta\lambda}{\lambda_0} = \frac{v}{c}$$

$\Delta\lambda$ = wavelength shift
λ_0 = wavelength if source is not moving
v = velocity of the source measured along the line of sight
c = speed of light = 3.0×10^5 km/s

CAUTION The capital Greek letter Δ, or delta, is commonly used to denote a change in the value of a quantity. Thus, $\Delta\lambda$ is the change in the wavelength λ due to the Doppler effect. It is *not* equal to a quantity Δ multiplied by a second quantity λ.

So far in this chapter we have glimpsed how much can be learned by analyzing light from the heavens. To analyze this light, however, it is first necessary to collect as much of it as possible, because most light sources in space are very dim. Collecting the faint light from distant objects is a key purpose of telescopes.

ConceptCheck 2-11: **How is the spectrum changed when looking at an emission spectrum from an approaching cloud of interstellar gas as compared to a stationary cloud?**

CalculationCheck 2-4: **How fast and in what direction is a star moving if it has a line that shifts from 486.2 nm to 486.3 nm?**

2-6 Telescopes use lenses, mirrors, and electronics to concentrate and capture incoming light for study

Understanding what information is carried by light is only half the challenge of the enterprise of astronomy. The other half involves capturing this light so it can be decoded. In this section we will first cover the basic properties and functions of telescopes, as well as the new technologies that have moved us forward. Next, we will cover the telescopes in space. We will conclude with a discussion of the advances in our technological abilities to capture and record light.

Telescopes Gather Light

The most common belief about telescopes is that astronomers use telescopes to magnify stars so they can be studied. As it turns out, almost all telescopes show stars to be tiny pinpoints of light—regardless of the amount of magnification used. This is because stars are so very, very far away that magnification is not very helpful. On the other hand, telescopes are useful because they can capture light from very distant and dim objects and can collect and focus starlight from otherwise very dim stars into a central place to be studied. To aid in this effort of trying to make dim objects appear bright enough to be carefully studied, astronomers' telescopes are constructed to have as much light-gathering ability as possible.

The most important dimension of a telescope is its diameter. Compared with a small-diameter telescope, a large-diameter telescope captures more light, produces brighter images, and allows astronomers to detect fainter objects. It is for this same reason that the iris of your eye opens wider when you go into a darkened room, to allow you to better see dimly lit objects.

The **light-gathering power** of a telescope is directly related to the size of its objective lens, the gathering area. In other words, the wider across your telescope's objective lens is, the more light it can gather. Mathematically, the area of a telescope's objective lens is directly related to the square of the telescope's diameter (**Figure 2-18**). Thus, if you double the diameter of the telescope, the light-gathering power increases by a factor of $2^2 = 2 \times 2 = 4$. Because light-gathering power is so important for seeing faint objects, the telescope's diameter is almost always given in describing a telescope.

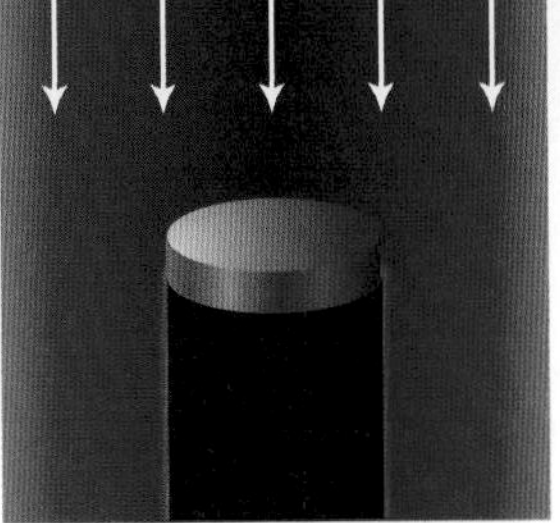

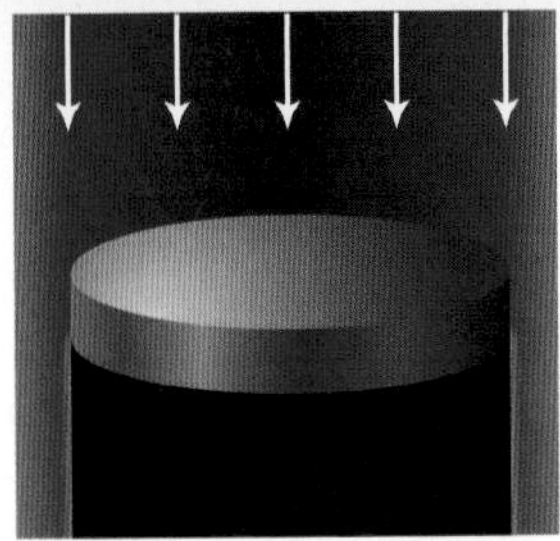

Small-diameter objective lens: dimmer image, less detail

Large-diameter objective lens: brighter image, more detail

FIGURE 2-18 RIVUXG **Light-Gathering Power** These two photographs of the galaxy M31 in Andromeda were taken using the same exposure time and at the same magnification, but with two different telescopes, which had objective lenses of different diameters. The right-hand photograph is brighter and shows more detail because it was made using the large-diameter lens, which intercepts more starlight than a small-diameter lens. This same principle applies to telescopes that use curved mirrors rather than lenses to collect light. *(Association of Universities for Research in Astronomy [AURA])*

For example, the Lick Observatory telescope on Mount Hamilton in California is 90 cm in diameter. By comparison, Galileo's telescope of 1610 was a much smaller 3 cm across. The Lick telescope is 30 times larger in diameter, and so has $30 \times 30 = 900$ times the light-gathering power of Galileo's instrument.

> **ConceptCheck 2-12: If someone says they are using an 8-inch telescope, which dimension of the telescope's size are they most likely referring to?**

Refracting Telescopes Use a Lens to Concentrate Incoming Light at a Focal Point

Which strategies can makers of telescopes use to gather and concentrate light to a central point? One is to use glass lenses, similar to those found in a common magnifying glass. In fact, the first telescopes recorded in history, those invented in the Netherlands in the early seventeenth century and the ones Galileo used for his groundbreaking astronomical observations, used *lenses* to make distant objects appear brighter and, in the case of the planets, somewhat larger.

The principle job of a telescope is to gather together light from a distant, dim object so it can be studied carefully, rather than just magnify it.

Lenses can be large or small, thick or thin, clear or coated, glass or plastic. Regardless of their exact specifications, lenses are characterized by how much they bend light. When a beam of light rays passes through a lens, all of the light converges to a single point. The distance from the lens to this special point is called the **focal length** of the lens. In order to view the image visually, a second lens to magnify the image is added. Such an arrangement of two lenses is called a **refracting telescope** (**Figure 2-19**). The large-diameter, long-focal-length lens at the front of the telescope, called the **objective lens,** forms the image; the smaller, shorter-focal-length lens at the rear of the telescope, called the **eyepiece lens,** magnifies the image for the observer.

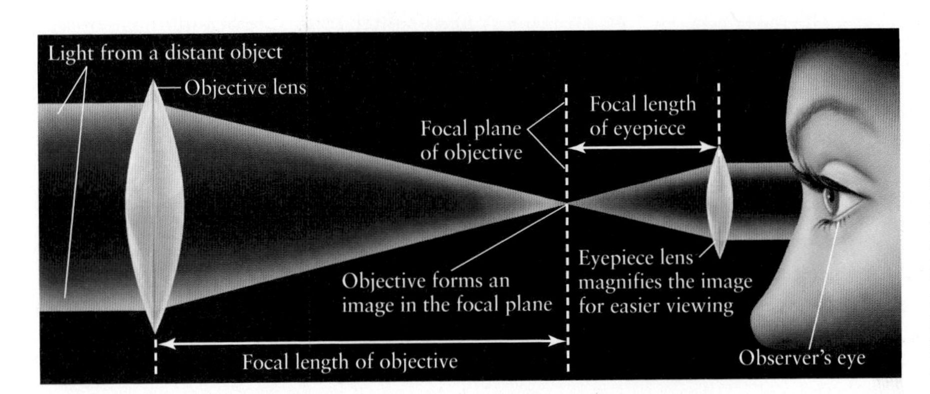

FIGURE 2-19 A Refracting Telescope A refracting telescope consists of a large-diameter objective lens with a long focal length and a small eyepiece lens of short focal length. The eyepiece lens magnifies the image formed by the objective lens in its focal plane (shown as a dashed line). To take a photograph, the eyepiece is removed and the film or electronic detector is placed in the focal plane.

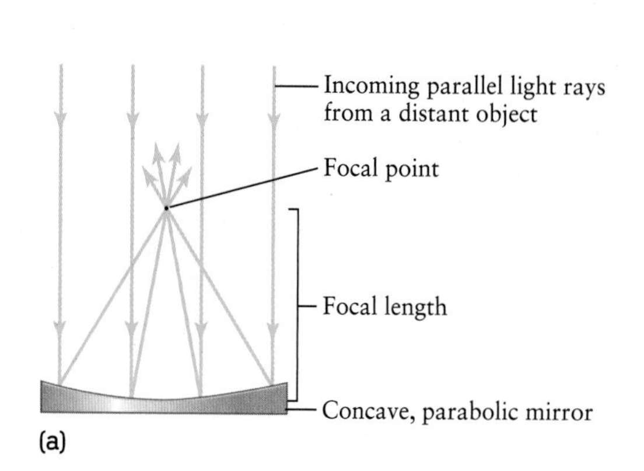

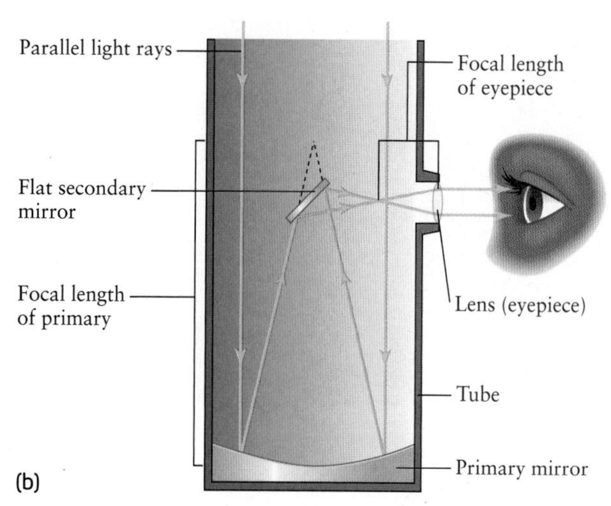

FIGURE 2-20 Reflecting Telescopes (a) A curved mirror causes parallel light rays from stars to converge and gather together at a focal point. The distance between the mirror and focal point is the focal length. (b) A Newtonian telescope uses a tiny flat mirror, called the secondary mirror, to send light outside the telescope. If small, this secondary mirror does not block too much of the incoming light. The light rays are made parallel again by passing through the magnifying eyepiece so the star can be observed. The dashed line shows where the focal point of this primary mirror would be if the secondary mirror were not in the way.

> ConceptCheck 2-13: **If a thick lens is able to bend light more than a thin lens, which lens has a greater focal length?**

Magnification The ability of the two lenses working in concert not only provides a brighter image, but also a slightly larger one. As an example, the Moon as viewed with the naked eye is much smaller across than when viewed with even a small telescope. When Galileo viewed the Moon through his telescope, the moon appeared large enough so that he could identify craters and mountain ranges. The magnifying power, or **magnification,** of a telescope is the ratio of how big an object is when seen through the telescope compared to its naked-eye size.

The magnification of a refracting telescope depends on the focal lengths of both of its lenses:

$$\text{Magnification} = \frac{\text{focal length of objective lens}}{\text{focal length of eyepiece lens}}$$

This formula shows that using a long-focal-length objective lens with a short-focal-length eyepiece gives a large magnification. **Box 2-4 Tools of the Astronomer's Trade: Magnification and Light-Gathering Power** illustrates how this formula is used.

> ConceptCheck 2-14: **How do the eyepieces with the largest focal length affect a telescope's overall magnification?**

Reflecting Telescopes Use a Mirror to Concentrate Incoming Light at a Focal Point

The second and more efficient approach of telescope makers to gather and focus light is to use curved mirrors rather than lenses. A telescope that uses a curved mirror to make an image of a distant object is called a **reflecting telescope.** Using terminology similar to that used for reflecting telescopes and illustrated in **Figure 2-20,** the mirror that forms the image is called the objective mirror, or **primary mirror.** In much the same way that a glass lens is able to concentrate light at a single point, a curved mirror also has a point at which light comes together at the focal length from the primary mirror. Telescope mirrors are commonly made from shaped glass with silver, aluminum, or a similar highly reflective substance applied to make them reflective.

Because light reflects off the surface of the glass rather than passing through it, defects within the glass—which would have very negative consequences for the objective lens of a refracting telescope—have no effect on the optical quality of a reflecting telescope. Telescopes can be made much larger when made from mirrors than when made from glass. More importantly, these giant mirrors can be made out of materials that are significantly lighter and more durable than heavy glass, making them easier to point and less delicate to clean.

Although a reflecting telescope has many advantages over a refracting telescope, the arrangement shown in Figure 2-20 is not ideal. One problem is that the focal point is in front of the objective mirror. If you try to view the image formed at the focal point, your head will block part or all of the light from reaching the mirror. To get around this, the light is usually directed out of the telescope after bouncing off the primary mirror, often by the use of additional, or secondary, mirrors.

For astronomers, the most important thing is that reflecting telescopes can be constructed to be extremely large and have tremendous power to gather light from the dimmest of objects. As of 2013, there were 15 optical reflectors in operation with primary mirrors between 8 m (26.2 ft) and 11 m (36.1 ft) in diameter. These are very large indeed! **Figure 2-21** shows the inner workings and seven enormous objective mirrors of the Giant Magellan Telescope, planned for completion in 2020 and being built in the Andes in Chile. These seven mirrors with diameters of 8.4 m (27.6 ft) add

BOX 2-4 TOOLS OF THE ASTRONOMER'S TRADE

Magnification and Light-Gathering Power

The magnification of a telescope is equal to the focal length of the objective lens divided by the focal length of the eyepiece lens. Telescopic eyepieces are usually interchangeable, so the magnification of a telescope can be changed by using eyepieces of different focal lengths.

EXAMPLE: A small refracting telescope has an objective lens of focal length 120 cm. If the eyepiece has a focal length of 4.0 cm, what is the magnification of the telescope?

Situation: We are given the focal lengths of the telescope's objective and eyepiece lenses. Our goal is to calculate the magnification provided by this combination of lenses.

Tools: We use the relationship that the magnification equals the focal length of the objective (120 cm) divided by the focal length of the eyepiece (4.0 cm).

Answer: Using this relationship,

$$\text{Magnification} = \frac{120\ \text{cm}}{4.0\ \text{cm}} = 30\ \text{(usually written as } 30\times)$$

Review: A magnification of 30× means that as viewed through this telescope, a large lunar crater that subtends an angle of 1 arcmin to the naked eye will appear to subtend an angle 30 times greater, or 30 arcmin (one-half of a degree). This makes the details of the crater much easier to see.

If a 2.0-cm-focal-length eyepiece is used instead, the magnification will be (120 cm)/(2.0 cm) = 60×. The shorter the focal length of the eyepiece, the greater the magnification.

The light-gathering power of a telescope depends on the diameter of the objective lens; it does not depend on the focal length. The light-gathering power is proportional to the square of the diameter. As an example, a fully dark adapted human eye has a pupil diameter of about 5 mm. By comparison, a small telescope whose objective lens is 5 cm in diameter has 10 times the diameter and $10^2 = 100$ times the light-gathering power of the eye. (Recall that there are 10 mm in 1 cm.) Hence, this telescope allows you to see objects 100 times fainter than you can see without a telescope.

EXAMPLE: The same relationships apply to reflecting telescopes. Each of the two Keck telescopes on Mauna Kea in Hawaii (see Figure 2-28) uses a concave mirror 10 m in diameter to bring starlight to a focus. How many times greater is the light-gathering power of either Keck telescope than that of the human eye?

Situation: We are given the diameters of the pupil of the human eye (5 mm) and of the mirror of either Keck telescope (10 m). Our goal is to compare the light-gathering powers of these two optical instruments.

Tools: We use the relationship that light-gathering power is proportional to the square of the diameter of the area that gathers light. Hence, the *ratio* of the light-gathering powers is equal to the square of the ratio of the diameters.

Answer: We first calculate the ratio of the diameter of the Keck mirror to the diameter of the pupil. To do this, we must first express both diameters in the same units. Because there are 1000 mm in 1 m, the diameter of the Keck mirror can be expressed as

$$10\ \text{m} \times \frac{1000\ \text{nm}}{1\ \text{m}} = 10{,}000\ \text{nm}$$

Thus, the light-gathering power of either of the Keck telescopes is greater than that of the human eye by a factor of

$$\frac{(10{,}000\ \text{mm})^2}{(5\ \text{mm})^2} = (2000)^2 = 4 \times 10^6 = 4{,}000{,}000$$

Review: Either Keck telescope can gather *4 million* times as much light as a human eye. When it comes to light-gathering power, the bigger the telescope, the better!

Drawbacks of Refracting Telescopes When building extremely large telescopes with enormous light-gathering power, there are several negative aspects of *reflecting telescopes* that even the finest optician cannot overcome. First, because faint light must readily pass through the objective lens first, the glass from which the lens is made must be totally free of defects, such as the bubbles that frequently form when molten glass is poured into a mold. Such defect-free glass is extremely expensive. Second, glass is opaque to certain kinds of light. Ultraviolet light is absorbed almost completely, and even visible light is dimmed substantially as it passes through the thick slab of glass that makes up the objective lens. Third, it is impossible to produce a large lens that bends light of all wavelengths in equal amounts. Fourth, because glass lenses are extremely heavy, yet can be supported only around their edges, they can sag and distort under their own weight. This has dramatically negative effects on the image clarity.

their light together and work as if they were one telescope with a total diameter of 22 m (72 ft) across. This telescope will have enough light-gathering power to capture light from some of the dimmest objects ever observed in the distant universe.

Several other reflecting telescopes around the world have objective mirrors between 3 m and 6 m in diameter—which is still quite large—and dozens of smaller but still powerful telescopes have mirrors in the range of 1 m to 3 m. There are thousands of professional astronomers, each of whom

FIGURE 2-21 R I V U X G **Giant Magellan Telescope** This artist's rendering shows the seven primary objective mirrors of the Giant Magellan Telescope, planned for completion in 2020. These seven mirrors with diameters of 8.4 m (27.6 ft) add their light together and work as if they were one telescope with an enormous diameter of 22 m (72 ft) across. The tiny secondary mirror is seen at the top and supported by thin bars above the primary mirrors. Notice that the center mirror has a tiny opening so that light reflected downward from the secondary mirror can reach the scientific instruments under the telescope. *(© Courtesy GMTO Corp.)*

has several ongoing research projects, and thus the demand for all of these telescopes is high. On any night of the year, nearly every research telescope in the world is being used to explore the universe.

> ConceptCheck 2-15: **In large sizes, which type of telescope can be made lightest and most inexpensively?**

Recent Advances in Technology Dramatically Improve Telescope Performance and Place Telescopes Above Earth's Atmosphere

If telescopes can make dim objects appear brighter so they can be studied more carefully, telescopes can also make objects clearer. Imagine one night that you are watching the headlights of very distant cars and motorcycles slowly moving toward you. For the most distant of cars, it is difficult to distinguish the twin headlights on a car from the single headlight of a motorcycle. In fact, it is only when the vehicles get relatively close to you that you can determine which are double headlighted cars and which are single headlighted motorcycles. Just how close do these cars need to be for you to resolve the headlights as single or double? Your ability to see two glowing objects as distinct and separate light sources is called **angular resolution,** and it gauges how well fine details can be seen. Because of the blurring effects of Earth's surrounding atmosphere, the resolution of a telescope can be limited when trying to observe two stars very close to one another or to see fine details on a planet's surface.

When you are asked to read the letters on an eye chart, essentially what is being measured is the angular resolution of your eye. If you have 20/20 vision, the angular resolution of your eye is about 1 arcmin, or 60 arcsec. (You may want to review Box 1-1 Tools of the Astronomer's Trade: Measuring Positions in the Sky.) Hence, with the naked eye it is impossible to distinguish two stars less than 1 arcmin apart or to see details on the Moon with an angular size smaller than this. All the planets have angular sizes (as seen from Earth) of less than 1 arcmin, which is why they appear as featureless points of light to the naked eye. One factor limiting angular resolution is the tendency of light waves to naturally spread out when they are confined to a small area like the lens or mirror of a telescope. This effect can cause star images to blur together. In astronomy, poor angular resolution causes star images to be fuzzy and blurred together.

Which characteristics are most important in determining a telescope's angular resolving power? Diameter is almost everything in terms of importance. The specific wavelength of light also plays a role because shorter wavelengths of light are easier to resolve than longer wavelengths.

In practice, however, ordinary optical telescopes cannot achieve the highest levels of resolution due to the blurring effects of Earth's atmosphere. In many cases the angular resolution of a telescope can be augmented by rapidly changing

FIGURE 2-22 R I V U X G **Creating an Artificial "Star"** A laser beam shines upward from Gemini North Observatory, near the summit of Mauna Kea on the island of Hawaii. The beam strikes sodium atoms that lie about 56 mi (90 km) above Earth's surface, causing them to glow and make an artificial "star." Tracking the twinkling of this "star" makes it possible to undo the effects of atmospheric turbulence on telescope images. *(© Gemini Observatory/Association of Universities for Research in Astronomy)*

FIGURE 2-23 R I V U X G **Using Adaptive Optics to "Unblur" Telescope Images** These two false-color images show the same view of our Milky Way Galaxy's core at infrared wavelengths as observed with the 10.0-m Keck II telescope near the summit of Mauna Kea, Hawaii (see Figure 2-28). Without adaptive optics, it is nearly impossible to distinguish individual stars in this region. With adaptive optics turned on, numerous stars can be distinguished and studied. *(Dr. Andrea Ghez)*

a mirror's shape using a technique called **adaptive optics.** The goal of adaptive optics is to compensate for blurring atmospheric turbulence, which causes the image of a star to "dance" around erratically. Optical sensors monitor this dancing motion 10 to 100 times per second, and a powerful computer rapidly calculates the mirror shape needed to compensate. Fast-acting mechanical devices called *actuators* then deform the mirror accordingly. In some adaptive optics systems, the actuators deform a small secondary mirror rather than the large objective mirror.

One difficulty with adaptive optics is that a fairly bright star must be in or near the field of the telescope's view to serve as a "target" for the sensors that track atmospheric turbulence. This is seldom the case, because the field of view of most telescopes is rather narrow. Astronomers get around this limitation by shining a laser beam toward a spot in the sky near the object to be observed (**Figure 2-22**). The laser beam causes atoms in the upper atmosphere to glow, making an artificial "star." The light that comes down to Earth from this "star" travels through the same part of our atmosphere as the light from the object being observed, so its image in the telescope will "dance" around in the same erratic way as the image of a real star.

Figure 2-23 shows the dramatic improvement in angular resolution possible with adaptive optics. Images made with adaptive optics are nearly as sharp as if the telescope were in the vacuum of space, where there is no atmospheric distortion whatsoever and the only limit on angular resolution is diffraction. A number of large telescopes are now being used with adaptive optics systems.

ConceptCheck 2-16: **If astronomers are using an adaptive optics system on a night where the atmosphere is unusually turbulent, will the adaptive optics actuators be deforming the telescope's mirror more rapidly or less rapidly than on a typical night?**

Telescopes in Orbit Around Earth Detect and Resolve Light that Does Not Penetrate the Atmosphere

In addition to the distortion caused by Earth's atmosphere, another challenge to astronomers is that Earth's atmosphere blocks much of the light emitted by stars in ways that even the better telescopes under Earth's atmosphere cannot fix. **Figure 2-24** shows the transparency of Earth's atmosphere to different wavelengths of light. The atmosphere is most transparent in two wavelength regions—visible light and radio waves. There are also several relatively transparent regions at infrared wavelengths between 1 μm and 40 μm. Infrared light within these wavelength intervals can penetrate Earth's atmosphere somewhat and can be detected with ground-based telescopes. This wavelength range is called the *near-infrared,* because it lies just beyond the red end of the visible light regime.

One factor over which astronomers have absolutely no control is the weather. Optical telescopes cannot see through clouds, so it is important to build observatories where the weather is usually good. Mountaintop observatories in particular benefit from most of the clouds forming at altitudes below the observatory, giving astronomers a better chance of having clear skies. Regardless of how high your mountaintop perch is, in many ways, the best location for a telescope is

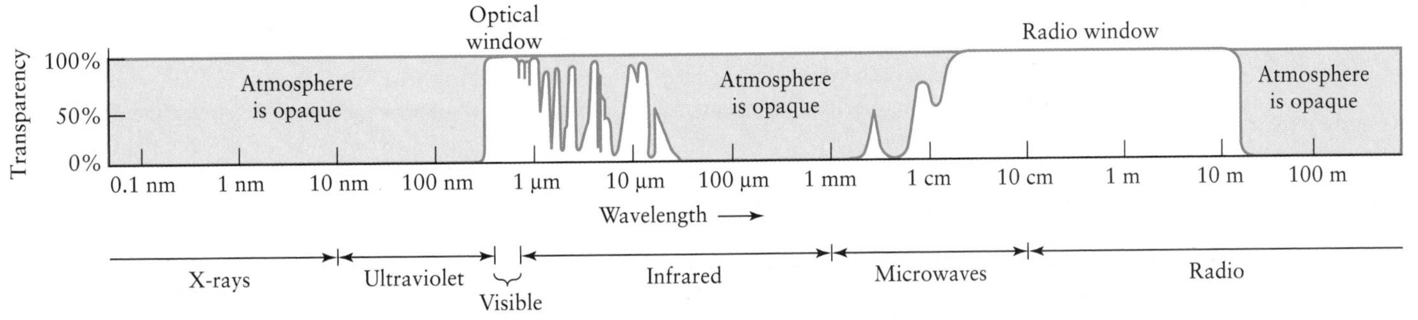

FIGURE 2-24 The Transparency of Earth's Atmosphere This graph shows the percentages of light that can penetrate Earth's atmosphere at different wavelengths. Regions in which the curve is high are called "windows," because the atmosphere is relatively transparent at those wavelengths. There are also three wavelength ranges in which the atmosphere is opaque and the curve is near zero: (1) at wavelengths less than about 290 nm, which are absorbed by atmospheric oxygen and nitrogen; (2) between the optical and radio window, due to absorption by water vapor and carbon dioxide; and (3) at wavelengths longer than about 20 m, which are reflected back into space by ionized gases in the upper atmosphere.

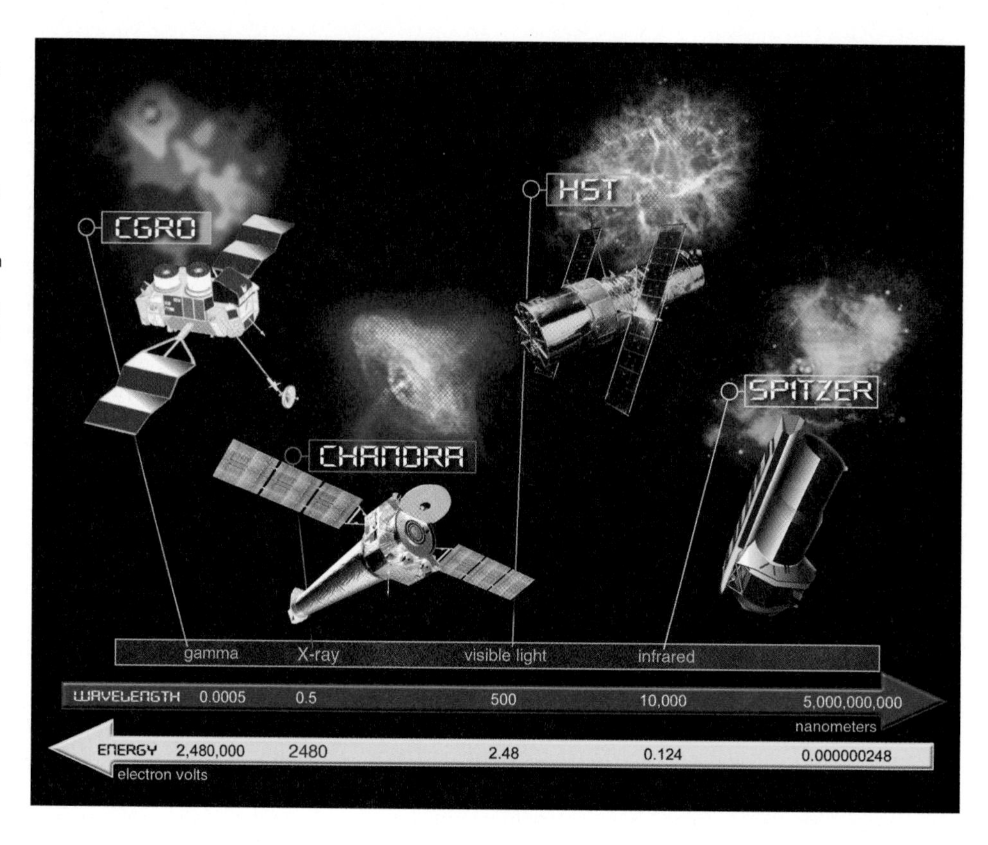

FIGURE 2-25
NASA's Great Observatories Space Telescopes Orbiting above most of Earth's obscuring atmosphere, space observatories can be configured to capture light across a wide range of wavelengths. From shortest wavelength light across to the longer wavelengths of light, this illustration shows the Compton Gamma Ray Observatory (CRGO), the Chandra X-ray Observatory, the Hubble Space Telescope (HST), and the Spitzer Space Telescope (SST). *(© NASA/CXC/M. Weiss)*

orbiting around Earth in outer space, where it is unaffected by weather, light pollution, or atmospheric turbulence.

Figure 2-25 illustrates four telescopes of NASA's Great Observatories program. Each mission is specially designed to look at a particular region of the electromagnetic spectrum. The Compton Gamma Ray Observatory (CGRO) was designed to capture the highly elusive, shortest wavelength gamma rays and discovered 3000 intense energy bursts from the formation of black holes and the merging of neutron stars (discussed in Chapter 12). The Chandra X-ray Observatory was launched because X-rays from outer space rarely penetrate Earth's protective atmosphere. Chandra was the first telescope to make false-color images of high temperature gas near exotic neutron stars and black holes.

CAUTION X-ray telescopes work on a very different principle from the X-ray devices used in medicine and dentistry. If you have your foot X-rayed to check for a broken bone, a piece of photographic film (or an electronic detector) sensitive to X-rays is placed under your foot and an X-ray beam is directed at your foot from above. The X-rays penetrate through soft tissue but not through bone, so the bones cast an "X-ray shadow" on the film. A fracture will show as a break in the shadow. X-ray telescopes, by contrast, do *not* send beams of X-rays toward astronomical objects in an attempt to see inside them. Rather, these telescopes detect X-rays that the objects emit on their own.

The Hubble Space Telescope (HST) made observations in the ultraviolet, visible, and infrared light regimes and was able to precisely measure the rate at which our universe is expanding (described in Chapter 15) and obtain images of the most distant galaxies in the universe as well as unprecedented views of our solar system's planets. At the longer wavelength end, the Space Infrared Telescope Facility launched the Spitzer Space Telescope, which is able to peer inside interstellar dust clouds to observe the youngest stars just as they are forming. There are other space telescopes, but these four Great Observatories illustrate some of the wide range of what astronomers have been able to accomplish by observing from space.

The success of the Hubble Space Telescope has inspired plans for its larger successor, the James Webb Space Telescope (JWST). Planned for a 2016 launch, JWST will observe at visible and infrared wavelengths from 600 nm to 28 μm. With its 6.5-m objective mirror—2.5 times the diameter of the HST objective mirror, with 6 times the light-gathering power—JWST will study faint objects such as planetary systems forming around other stars and galaxies near the limit of the observable universe. Unlike HST, which is in a relatively low-altitude orbit around Earth, JWST will orbit the Sun some 1.5 million km beyond Earth. In this orbit the telescope's view will not be blocked by Earth. Furthermore, by remaining far from the radiant heat of Earth, it will be easier to keep JWST at the very cold temperatures required by its infrared detectors.

The advantages and benefits of Earth-orbiting observatories cannot be overemphasized. We are no longer limited to the narrow ranges of whatever wavelengths manage to leak through our shimmering, hazy atmosphere (**Figure 2-26**). For the first time, we are really *seeing* the universe.

ConceptCheck 2-17: **Look at Figure 2-26, which shows the transparency of Earth's atmosphere. Would astronomers most prefer to have a new ground-based telescope constructed that is most sensitive in the X-ray region, the ultraviolet wavelength region, or the microwave region?**

ConceptCheck 2-18: **What is the primary advantage of an orbiting space telescope, compared to a ground-based telescope?**

Charge-Coupled Devices Record Very Fine Image Details

Until this point, we have been targeting our discussion on how telescopes capture light. The other half of the story is how the captured light is recorded for later study. For the past hundred years, astronomers primarily used photographic film or photographic plates. Unfortunately, photographic film is not a very efficient light detector. Only about 2% of the light striking photographic film is able to trigger the chemical reaction needed to produce an image. Thus, roughly 98% of the light falling onto photographic film is wasted.

The most sensitive light detector currently available to astronomers is the **charge-coupled device** (**CCD**). At the heart of a CCD is a semiconductor wafer divided into an array of small light-sensitive squares called picture elements or, more commonly, **pixels.** For example, each of the 40 CCDs shown in **Figure 2-27*a*** has more than 9.4 million pixels arranged in 2048 rows by 4608 columns. They have about 1000 times more pixels per square centimeter than a typical computer screen has, which means that a CCD of this type can record very fine image details. CCDs with smaller numbers of pixels are used in digital cameras, scanners, and fax machines.

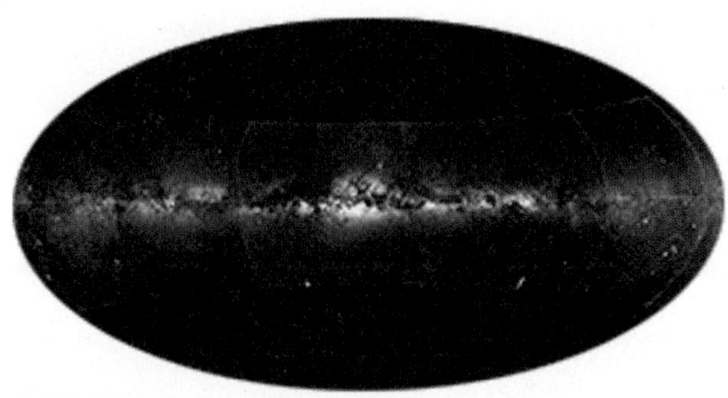

(a) RIVUXG

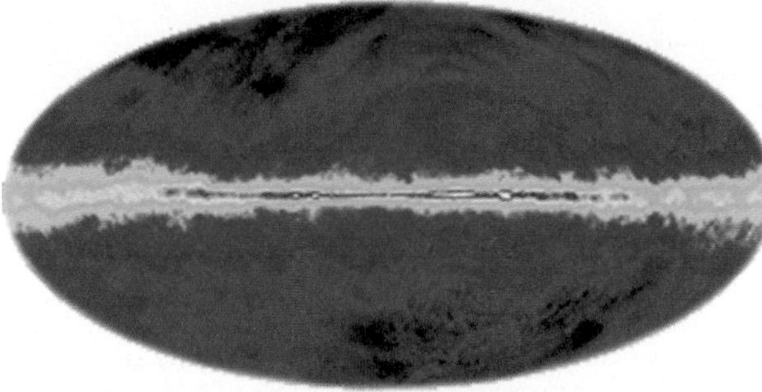

(b) RIVUXG

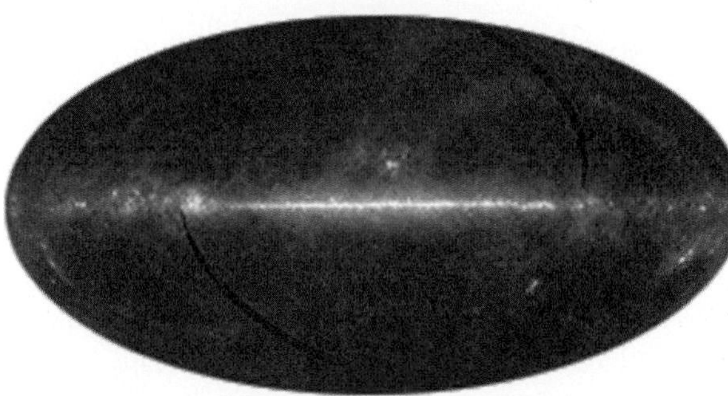

(c) RIVUXG

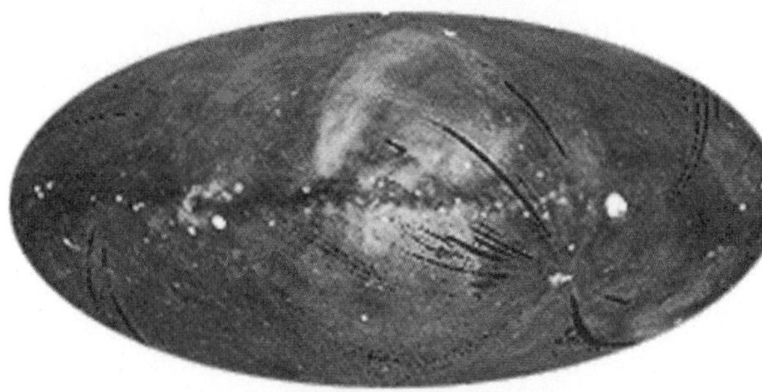

(d) RIVUXG

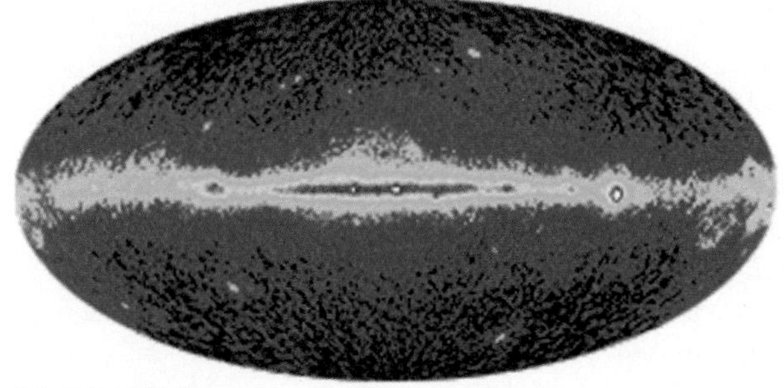

(e) RIVUXG

FIGURE 2-26 The Entire Sky at Five Wavelength Ranges These five views show the entire sky at visible, radio, infrared, X-ray, and gamma-ray wavelengths. The entire celestial sphere is mapped onto an oval, with the Milky Way stretching horizontally across the center. The black crescents in the infrared and X-ray images are where data are missing. (a) In the visible view the constellation Orion is at the right, Sagittarius in the middle, and Cygnus toward the left. Many of the dark areas along the Milky Way are locations where interstellar dust is sufficiently thick to block visible light. (b) The radio view shows the sky at a wavelength of 21 cm. This wavelength is emitted by hydrogen atoms in interstellar space. The brightest regions (shown in red) are in the plane of the Milky Way, where the hydrogen is most concentrated. (c) The infrared view from Infrared Astronomical Satellite (IRAS) shows emission at 100 μm, 60 μm, and 12 μm. Most of the emission is from dust particles in the plane of the Milky Way that have been warmed by starlight. (d) The X-ray view from the ROSAT observatory shows wavelengths of 0.8 nm (blue), 1.7 nm (green), and 5.0 nm (red), corresponding to photon energies of 1500 eV, 750 eV, and 250 eV. Extremely high temperature gas emits these X-rays. The white regions, which emit strongly at all X-ray wavelengths, are remnants of supernovae. (e) The gamma-ray view from the Compton Gamma Ray Observatory includes all wavelengths less than about 1.2×10^{-5} nm (photon energies greater than 10^8 eV). The diffuse light from the Milky Way is emitted when fast-moving subatomic particles collide with the nuclei of atoms in interstellar gas clouds. The bright spots above and below the Milky Way are distant, extremely energetic galaxies. *(NASA Goddard Space Flight Center)*

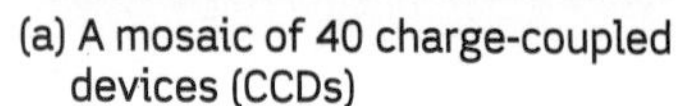

(a) A mosaic of 40 charge-coupled devices (CCDs)

(b) An image made with photographic film

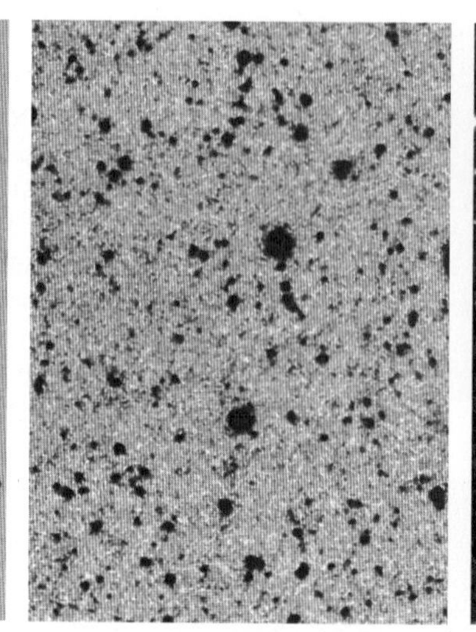

(c) An image of the same region of the sky made with a CCD

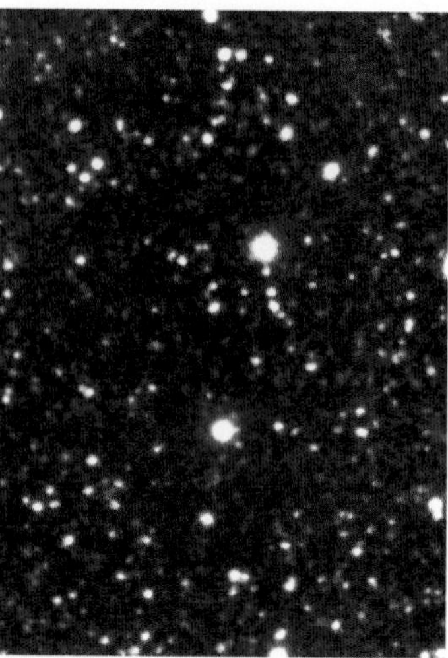

(d) Combining several CCD images made with different color filters

FIGURE 2-27 R I V U X G **Charge-Coupled Devices (CCDs) and Imaging** (a) The 40 CCDs in this mosaic are used to record the light gathered by the Canada-France-Hawaii Telescope (see Figure 2-28). After an exposure, data from each of the 377 million light-sensitive pixels in the mosaic are transferred to a waiting computer. (b) This negative print (black stars and white sky) shows a portion of the sky as imaged with a 4-m telescope and photographic film. (c) This negative image of the same region of the sky was made with the same telescope, but with the photographic film replaced by a CCD. Many more stars and galaxies are visible. (d) To produce this color positive view of the same region, sequences of CCD images were made using different color filters. These were then combined using a computer image-processing program. *(a: J. C. Cuillandre/ Canada-France-Hawaii Telescope; b, c, d: Patrick Seitzer, National Optical Astronomy Observatories)*

When an image from a telescope is focused on the CCD, an electric charge builds up in each pixel in proportion to the number of photons falling on that pixel. When the exposure is finished, the amount of charge on each pixel is read into a computer, where the resulting image can be stored in digital form and either viewed on a monitor or printed out. Compared with photographic film, CCDs are some 35 times more sensitive to light (they commonly respond to 70% of the light falling on them, versus 2% for film), can record much finer details, and respond more uniformly to light of different colors. Figures 2-27*b, c,* and *d* show the dramatic difference between photographic and CCD images. The great sensitivity of CCDs also makes them useful for measuring the brightnesses of stars and other astronomical objects.

In the modern world of CCD astronomy, astronomers no longer need to spend the night in the unheated dome of a telescope on a windy mountaintop far from civilizations. Instead, they operate the telescope electronically from a separate control room, where the electronic CCD images can be viewed on a computer monitor. The control room need not even be adjacent to the telescope. Although the Keck I and II telescopes (**Figure 2-28**) are at an altitude of 13,600 ft (4145 m), astronomers can now make observa-

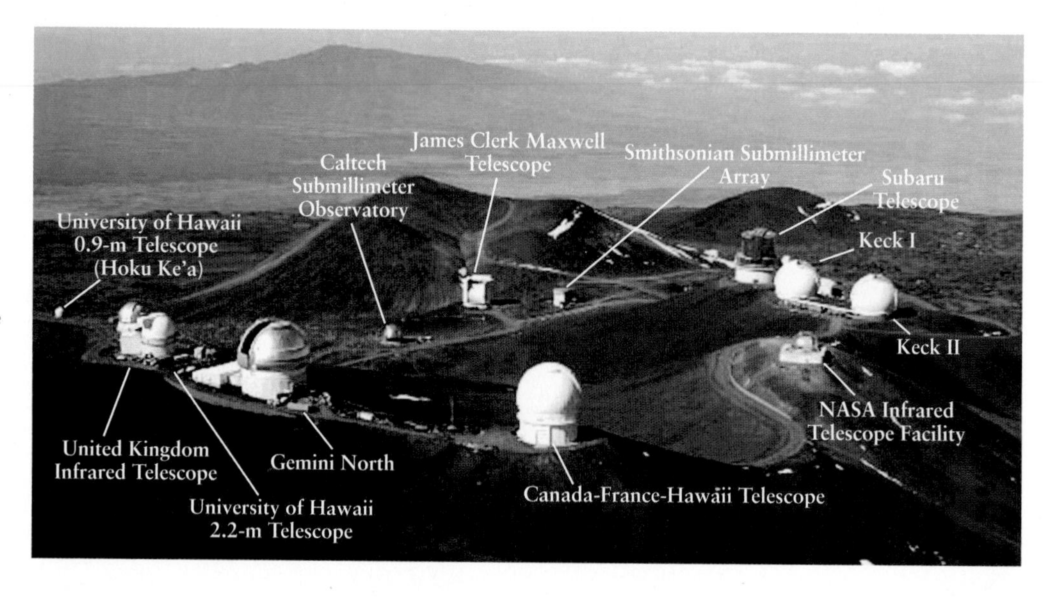

FIGURE 2-28 R I V U X G **The Telescopes of Mauna Kea** Mauna Kea—an extinct Hawaiian volcano that reaches more than 13,600 ft (4145 m) above the waters of the Pacific—has nighttime skies that are unusually clear, still, and dark. To take advantage of these superb viewing conditions, Mauna Kea has become the home to some of the world's most advanced telescopes. *(Richard Wainscoat/University of Hawaii Institute of Astronomy/AP Photo)*

tions from a facility elsewhere on the island of Hawaii that is much closer to sea level. This saves the laborious drive up to Mauna Kea and eliminates the need for astronomers to acclimate to the high altitude.

Most of the images that you will see in this book were made with CCDs. Because of their extraordinary sensitivity and their ability to be used in conjunction with computers, CCDs have attained a role of central importance in astronomy. For scientists and engineers on the cutting edge of technology, the next advancements are all focused on making CCDs faster and yet with more pixels, but on smaller platforms. In a very short time, this technology will filter down quickly to make better cell phone cameras, higher-definition video cameras, and high quality Web cams at unbelievably low prices.

ConceptCheck 2-19: Why can CCDs more efficiently observe faint stars than photographic film or photographic plates?

VISUAL LITERACY TASK

Blackbody Curves

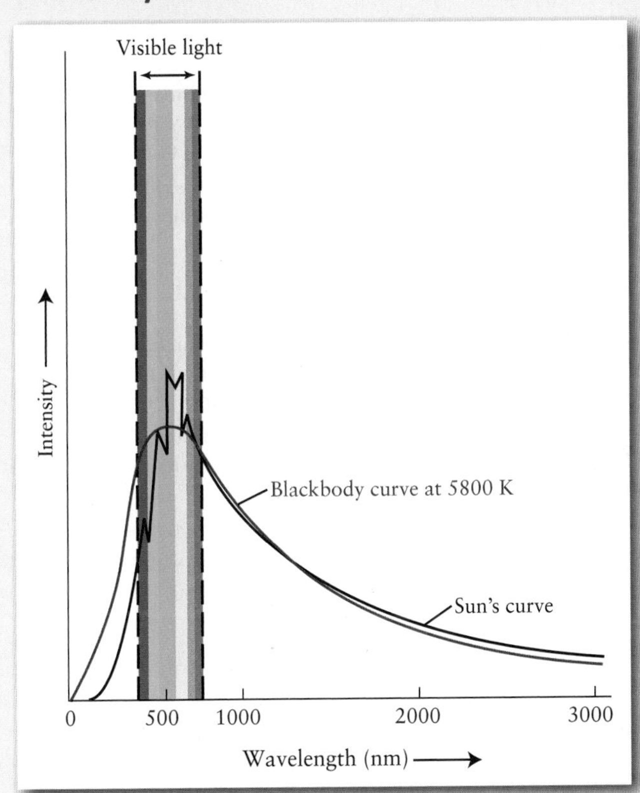

PROMPT: What would you tell a fellow student who said, **"All stars have the same intensity, but a star hotter than the Sun would have a peak in the red region and its graph would be spread out more equally on both sides of the visible region of the spectrum"?**

ENTER RESPONSE:

Guiding Questions:

1. On a graph like this, the specific wavelength being emitted with the most intensity is
   a. the highest point on the curve.
   b. where the line crosses the vertical axis.
   c. measured by the area under the curve.
   d. in the red region of the wavelength axis.

2. The stars with the greatest energy output
   a. have more intense visible colors.
   b. have the highest intensity.
   c. are closest to Earth.
   d. have the greatest densities.

3. The hottest stars have a blackbody curve with an intensity that
   a. peaks at longest wavelengths.
   b. peaks at shortest wavelengths.
   c. is nearly identical to cooler stars.
   d. peaks in the blue-green region.

4. Compared to much cooler stars, the shape of the curve for the hottest stars would be
   a. similar but much taller.
   b. similar but much shorter.
   c. more evenly distributed around the visible region.
   d. about the same shape and size for all stars.

KEY IDEAS AND TERMS

2-1 Light travels through empty space at a speed of nearly 300,000 km/s

- The speed of light, abbreviated as *c*, was historically measured over large, interplanetary distances.

2-2 Glowing objects, like stars, emit an entire spectrum of light

- Sunlight is a mixture of all colors that can be spread out into a **spectrum.**
- Light travels in waves and is characterized by **wavelength** and its **frequency.**
- The **electromagnetic spectrum** of light includes **radio waves, microwaves, infrared, visible light, ultraviolet, X-rays,** and **gamma rays.**
- Light has properties of both waves and particles where individual packets are called **photons.**

2-3 An object's temperature is revealed by the most intense wavelength of its spectrum of light

- **Wien's law** specifies the dominant wavelength of light emitted by objects and reveals their temperatures.
- Astronomers use the **Stefan-Boltzmann law** to determine the temperature of an object by carefully measuring the **energy flux** of light emitted.

2-4 An object's chemical composition is revealed by the unique pattern of its spectrum of light

- A hot, dense object emits a **continuous spectrum** covering all wavelengths.
- A hot, transparent gas produces an **emission line spectrum** containing bright lines.
- A cool, transparent gas in front of a light source that itself has a continuous spectrum produces an **absorption line spectrum.**
- **Spectroscopy** is the study of the unique pattern of **spectral lines** emitted by every chemical substance.
- Electrons can "jump" only to specific orbits within atoms, creating the particular sets of spectral lines.

2-5 An object's motion through space is revealed by the precise wavelength positions of its spectrum of light

- The **Doppler effect** describes how an observed wavelength can be shifted due to the relative motion of an energy-emitting object.
- An observed spectrum is **redshifted** when the distance between an object and an observer is increasing.
- An observed spectrum is **blueshifted** when the distance between an object and an observer is decreasing.

2-6 Telescopes use lenses, mirrors, and electronics to concentrate and capture incoming light for study

- The ability of a telescope to capture light is a telescope's **light-gathering power** and is closely related to the telescope's diameter.
- A **refracting telescope** uses an **objective lens** to concentrate incoming light to a point that is a **focal length** from the lens.
- A telescope using a **primary mirror** to concentrate incoming light is called a **reflecting telescope.**
- The **magnification** achieved by a telescope is accomplished by the **eyepiece lens.**
- **Angular resolution** is the ability of a telescope to see two glowing objects as distinct and separate light sources.
- The new telescope technology of **charged-coupled devices (CCDs), adaptive optics,** and **interferometry** dramatically improves modern telescope performance.
- Telescopes in orbit around Earth detect and resolve light that does not penetrate the atmosphere.

QUESTIONS

Review Questions

1. For each of the following wavelengths, state whether it is in the radio, microwave, infrared, visible, ultraviolet, X-ray, or gamma-ray portion of the electromagnetic spectrum and explain your reasoning:
   a) 2.6 μm
   b) 34 m
   c) 0.54 nm
   d) 0.0032 nm
   e) 0.620 μm
   f) 310 nm
   g) 0.012 m
2. A cellular phone is actually a radio transmitter and receiver. You receive an incoming call in the form of a wave of frequency 880.65 MHz. What is the wavelength (in meters) of this wave?
3. Using Wien's law and the Stefan-Boltzmann law, explain the color and intensity changes that are observed as the temperature of a hot, glowing object increases.
4. If you double the Kelvin temperature of a hot piece of steel, how much more energy will it radiate per second?
5. The bright star Bellatrix in the constellation Orion has a surface temperature of 21,500 K. What is its wavelength of maximum emission in nanometers? What color is this star?
6. The bright star Antares in the constellation Scorpius (the Scorpion) emits the greatest intensity of light at a wavelength of 853 nm. What is the surface temperature of Antares? What color is this star?
7. Explain how Bohr's model of the atom accounts for spectra.
8. Why do different elements display different patterns of lines in their spectra?
9. What is the Doppler effect? Why is it important to astronomers?
10. If you see a blue star, what does its color tell you about how the star is moving through space? Explain your answer.
11. With the aid of a diagram, describe a refracting telescope.
12. With the aid of a diagram, describe a reflecting telescope.
13. Which dimensions of the telescope determine its light-gathering power?
14. What is the purpose of a telescope eyepiece?
15. Quite often advertisements appear for telescopes that extol their magnifying power. Is this a good criterion for evaluating telescopes? Explain your answer.
16. Explain some of the disadvantages of refracting telescopes compared to reflecting telescopes.
17. What is the angular resolution of a telescope?
18. What is adaptive optics?
19. What is a charge-coupled device (CCD)? Why have CCDs replaced photographic film for recording astronomical images?

20. Why can radio astronomers make observations at any time during the day, whereas optical astronomers are mostly limited to observing at night? (*Hint:* Does your radio work any better or worse in the daytime than at night?)
21. Why must astronomers use satellites and Earth-orbiting observatories to study the heavens at X-ray and gamma-ray wavelengths?

Web Chat Questions

1. The human eye is most sensitive over the same wavelength range at which the Sun emits the greatest intensity of light. Suppose creatures were to evolve on a planet orbiting a star somewhat hotter than the Sun. To what wavelengths would their vision most likely be sensitive?
2. Why do you suppose that ultraviolet light can cause skin cancer but ordinary visible light does not?
3. If you were in charge of selecting a site for a new observatory, what factors would you consider important?
4. Discuss the advantages and disadvantages of using a small telescope in Earth's orbit versus a large telescope on a mountaintop.

Collaborative Exercises

1. The Doppler effect describes how relative motion impacts wavelength. With a classmate, stand up and demonstrate each of the following: (a) a blueshifted source for a stationary observer; (b) a stationary source and an observer detecting a redshift; and (c) a source and an observer both moving in the same direction, but the observer is detecting a redshift. Create simple sketches to illustrate what you and your classmate did.
2. Stand up and have everyone in your group join hands, making as large a circle as possible. If a telescope mirror were built as big as your circle, what would be its diameter?
3. Are there enough students in your class to stand and join hands and make two large circles that recreate the sizes of the two Keck telescopes? Explain how you determined your answer.

Observing Questions

1. You can use the *Starry Night*™ program to measure the speed of light by observing a particular event, in this case one of Jupiter's moons emerging from the planet's shadow, from two locations separated by a known distance. These two locations are at the north poles of Earth and the planet Mercury respectively. Open **Favourites > Explorations > Io from Earth.** The view shows Jupiter as seen from the north pole of Earth at 9:12:00 P.M. Standard Time on September 25, 2010. You will see the label for Io to the left of the planet. Keep the field of view about 11 arcmin wide. Click the **Play** button and observe Io suddenly brighten as it emerges from Jupiter's shadow. Depending upon your computer monitor, you may need to **Zoom** out slightly so that Io's transition from being invisible to visible occurs instantaneously (at high zoom levels, Io will brighten gradually). Next, use the time controls to **Step time backward** and **forward** in 1-s intervals to determine the time to the nearest second at which Io brightens. Open the **Status** pane and expand the **Time** layer. Record the **Universal Time** for this event. Then open the **Info** pane and be sure that **Io Info** appears at the top of the **Info** pane. Expand the **Position in Space** layer and record the value given for **Distance from Observer.** Now select **Favourites > Explorations > Io from Mercury.** This view once again shows Io labeled to the left of Jupiter but in this instance you are viewing the scene from the north pole of the planet Mercury. With the field of view set to 11 arcmin wide, click the **Play** button. **Stop time flow** as soon as you see Io suddenly brighten as it emerges from eclipse. Again, it may be necessary to adjust the **Zoom** level to make this event appear instantaneous rather than gradual. Use the time controls to find the time to the nearest second at which Io brightens. Open the **Status** pane and record the **Universal Time** for this event as seen from Mercury. Then open the contextual menu for Io and select **Show Info.** Record the **Distance from Observer** of Io from the **Position in Space** layer. Use your observations to calculate the speed of light. First, calculate the difference in the time between the two observations in seconds. Then calculate the difference in the **Distance from Observer** in AU for each of the locations. Divide the difference in the distance by the difference in time to calculate the speed of light in AU per second. Finally, convert this value to kilometers per second by multiplying the result by 1.496×10^8 (the number of kilometers in one AU). How does your result compare to the accepted value of the speed of light of 2.9979×10^5 km/s? Explain the difference between your calculated value from these observations and the accepted value for the speed of light.
2. Use *Starry Night*™ to examine the temperatures of several relatively nearby stars. Select **Favourites > Explorations > Atlas.** Use the **File** menu (**Starry Night** menu on a Mac) and select **Preferences . . .** to open the Preferences dialog window. Click the box in the top left of this dialog window and choose **Cursor Tracking (HUD).** Scroll through the **Show** list and click the checkbox next to **Temperature** to turn this option on. Then close the Preferences dialog window. Next, open the **Find** pane, click the magnifying glass icon in the edit box at the top of this pane and select **Star** from the dropdown menu. To locate each of the following stars—(i) Altair; (ii) Procyon; (iii) Epsilon Indi; (iv) Tau Ceti; (v) Epsilon Eridani; (vi) Lalande 21185—type the name of the star in the edit box and then press the **Enter** (**Return**) key. Use the **HUD** to find and record the star's temperature. Then answer the following questions:
   a) Which of the stars have a longer wavelength of maximum emission λ_{max} than the Sun?
   b) Which of the stars have a shorter λ_{max} than the Sun?
   c) Which of the stars will have a reddish color?
3. Use the *Starry Night*™ program to compare the brightness of two similarly sized stars in the constellation Auriga. Select **Favourites > Explorations > Auriga.** The two stars Capella and Delta Aurigae are labeled in this view. Select **Preferences** from the **File** menu (Windows) or **Starry Night** menu (Mac) and set **Cursor Tracking/HUD** options so that **Temperature** and **Radius** are shown in the HUD display. You will notice that these two stars have the same radius but differ in temperature. From these data, which of these stars is intrinsically brighter and by what proportion?
4. Use *Starry Night*™ to examine the celestial objects listed below. Select **Favourites > Explorations > Atlas** to show the whole sky as would be seen from the center of a transparent Earth. Ensure that deep space objects are displayed by selecting **View > Deep Space > Messier Objects** and **View > Deep Space > Bright NGC Objects** from the menu. Also, select **View > Deep Space > Hubble Images** and ensure that this option is turned **off.** To display each of the objects listed below, open the **Find** pane and then type the name of the selected object in the edit box followed by the **Enter** (**Return**) key. This object will be centered in the view. Use the **Zoom** controls to adjust your view until you can see the object in detail. For each object, decide whether you think it will have a continuous spectrum, an absorption line spectrum, or an emission line spectrum, and explain your reasoning. The objects to observe are:
   a) The Lagoon Nebula in Sagittarius. (With a field of view of about 6° × 4°, you can compare and contrast the appearance of the Lagoon Nebula with the Trifid Nebula just to the north of it.)

b) M31, the great galaxy in the constellation Andromeda. (*Hint:* The light coming from this galaxy is the combined light of hundreds of billions of individual stars.)

c) The Moon (*Hint:* Moonlight is simply reflected sunlight.)

ANSWERS

ConceptChecks

ConceptCheck 2-1: The speed of light is incredibly fast, which made it very difficult to measure its speed precisely before modern technology, except over enormous distances.

ConceptCheck 2-2: As Newton found when passing sunlight through a series of prisms, when one color is isolated from white light, there are no longer any other colors present in the remaining light. As a result, you cannot turn pure green into any other color by passing it through a plastic gel.

ConceptCheck 2-3: A more vigorous shaking of the pan would make more waves and make them crest higher, but the speed of waves across the water would stay the same, in much the same way that light can only travel so fast through outer space.

ConceptCheck 2-4: The width of your finger is about 1 cm, which falls in the range of the wavelength of microwaves (1 mm to 10 cm).

ConceptCheck 2-5: Because the relationship between wavelength and frequency is written as $c = \lambda \times f$, wavelength and frequency are inversely related; as one increases the other decreases. Thus, the longest wavelengths have the lowest frequencies and the shortest wavelengths have the highest frequencies.

ConceptCheck 2-6: Infrared light is the most typical wavelength of light emitted by objects at temperatures common on Earth's surface.

ConceptCheck 2-7: A star is hotter than our Sun if the dominant wavelength emitted is shorter than the light from our Sun.

ConceptCheck 2-8: The shorter the wavelength, the hotter the star, so the blue star will be at a higher temperature.

ConceptCheck 2-9: Photons with longer wavelengths will have lower energy than those with shorter wavelengths because the greater the wavelength, the lower the energy of a photon associated with that wavelength.

ConceptCheck 2-10: An absorption spectra results when the light from a hot, dense object passes through the cooler, transparent gas of our atmosphere.

ConceptCheck 2-11: When the distance between an object and a source is decreasing, the emissions lines will be shifted toward shorter wavelengths; alternatively, when the distance between an object and a source is decreasing, the emissions lines will be shifted toward longer wavelengths.

ConceptCheck 2-12: The diameter, because it is the width of the opening that determines how much light can be captured by the telescope.

ConceptCheck 2-13: The thin lens bends light less, so, according to Figure 2-21, it must focus light at a more distant location.

ConceptCheck 2-14: The larger the eyepiece focal length, the smaller the magnification.

ConceptCheck 2-15: Reflecting telescopes use mirrors that can be made to be lightweight rather than glass lenses, which are quite heavy and difficult to construct.

ConceptCheck 2-16: Adaptive optics actuators slightly deform the telescope's mirror to match the apparent movement of a star due to distortions in Earth's atmosphere. The actuators must deform the mirror more on nights when the stars appear more distorted due to a rapidly fluctuating atmosphere.

ConceptCheck 2-17: Given these three choices, astronomers would much prefer to have a new telescope in the microwave region because X-rays and ultraviolet wavelengths rarely pass through Earth's atmosphere to the ground.

ConceptCheck 2-18: Orbiting space telescopes are placed far above most of Earth's atmosphere, which blocks many wavelengths of light that astronomers want to see.

ConceptCheck 2-19: CCDs are able to detect 70% of the light that falls on them as compared to photographic film, which only captures 2% of the light, making CCDs far more efficient.

CalculationChecks

CalculationCheck 2-1: A popular radio station in Phoenix is 90.5 MHz. To calculate the wavelength of these radio waves, we rearrange the equation $c = f \div \lambda$ to get $\lambda = 3 \times 10^8$ m/s $\div$ 90.5×10^6 Hz $= 3.31$ m.

CalculationCheck 2-2: Wien's law can be rearranged to calculate the temperature of a star as $T = 0.0029$ K m $\div$ (5800 K $\times$ 2) $=$ 250 nm, which is ultraviolet.

CalculationCheck 2-3: $F = \sigma T^4$, so if temperature is 3 times greater, then the energy flux is 3^4, or 81 times greater than the flux from the Sun.

CalculationCheck 2-4: $v = c \times \Delta\lambda \div \lambda_0 = 3 \times 10^5$ km/s $\times$ (486.3 nm $-$ 486.2 nm) $\div$ 486.2 nm $=$ 61.7 km/s, and because it is moving toward longer wavelengths, the distance between the observer and the star must be increasing.

R I V U X G An astronaut orbits Earth attached to the International Space Station's manipulator arm. *(NASA)*

3 Analyzing Scales and Motions of the Universe

CHAPTER LEARNING OBJECTIVES **By reading the sections of this chapter, you will learn:**

3-1 Astronomers of antiquity used observation and reasoning to develop astonishing advances in the study of astronomy

3-2 Nicolaus Copernicus devised the first comprehensive Sun-centered model

3-3 Galileo's discoveries of moons orbiting Jupiter and the phases of Venus strongly supported a heliocentric model

3-4 Johannes Kepler proposed that planets orbit the Sun in elliptical paths, moving fastest when closest to the Sun, with the closest planets moving at the highest speeds

3-5 Isaac Newton formulated three laws relating force and motion to describe fundamental properties of physical reality

3-6 Newton's description of gravity accounts for Kepler's laws and explains the motions of the planets

Sixty years ago the idea of humans orbiting Earth or sending spacecraft to other worlds was regarded as science fiction. Today, science fiction has become commonplace reality. Literally thousands of artificial satellites orbit our planet to track weather, relay signals for communications and entertainment, and collect scientific data about Earth and our universe. Humans live and work in Earth orbit, have ventured as far as the Moon, and have sent dozens of robotic spacecraft to explore all the planets of the solar system. Given these amazing feats of science, you might ask, how did we get here?

We will start our discussion by asking a seemingly simple question that took our ancestors a thousand years to answer: How big is Earth? We will look at how astronomers learned to measure distances, starting with the early astronomers who made bold predictions about the distances to the Sun and Moon, moving on to later generations of astronomers who plotted the distances to stars and discovered the astonishing surprise that our Sun is not the center of our Galaxy.

Then we will learn how Nicolaus Copernicus, Johannes Kepler, and Galileo Galilei helped us understand that Earth is itself one of several planets orbiting the Sun. We will learn, too, about Isaac Newton's revolutionary discovery of why the planets move in the way that they do. This was just one aspect of Newton's immense body of work, which included formulating the fundamental laws of physics and developing a precise mathematical description of the force of gravitation—the force that holds the planets in their orbits and holds galaxies of stars together. Gravitation proves to be a truly universal force: It guides spacecraft as they journey across the solar system, keeps satellites and astronauts in their orbits, and moderates how the very galaxies of our universe move. ■

3-1 Astronomers of antiquity used observation and reasoning to develop astonishing advances in the study of astronomy

We sometimes perpetuate a myth that the Italian explorer Christopher Columbus proved that Earth is a sphere when he "sailed the seas of blue in the year 1492" in search of a western sailing route to the Orient for trading. Perhaps surprising, most formally educated people knew long before Columbus's voyages that Earth is round. How did they know so long ago, before satellites could photograph Earth, that Earth is round or how large the planet is that we live on? They had come to this conclusion using a combination of observation and logical deduction, much like modern scientists.

The achievements of our ancestors still stand as impressive applications of observation and reasoning and important steps toward the development of the scientific method, particularly the calculations used to determine the size of Earth, the distances to the Sun and Moon, and the reasons for the movement of objects across the sky.

Eratosthenes Measures Earth's Size

The earliest evidence of a round Earth is found in ancient Greek writings. No one knows specifically how Earth's roundness was first discovered, but many travelers would have noticed that the stars, Polaris (the North Star) in particular, appeared in different positions as they traveled north and south. This observation could only occur if Earth was spherical, rather than flat.

In much the same way, it was common knowledge among the Greek philosophers that during lunar eclipses, when the Moon passes through Earth's shadow (see Figure 1-28), the edge of the shadow is always circular. Because a sphere is the only shape that always casts a circular shadow from any angle, they correctly concluded that Earth itself is spherical.

Around 240 B.C., the Greek astronomer Eratosthenes devised a way to measure Earth's size. It was known that on the first day of summer (see Section 1.4), in the town of Syene in Egypt, near present-day Aswan, the Sun shone directly down the vertical shafts of water wells. Hence, at local noon on that day, the Sun was directly overhead as seen from Syene. Eratosthenes knew that the Sun never appeared directly overhead at his home in the Egyptian city of Alexandria, about 500 mi due north of Syene. Rather, on the summer solstice in Alexandria, the noontime Sun cast a short shadow, showing the Sun's position was about 7° south of the zenith (**Figure 3-1**). If we set up a simple ratio where the 7° portion of a 360° circle is the same as a 500-mi distance, and a portion of Earth's circumference, we get:

$$\frac{7^\circ}{360^\circ} = \frac{500\,\text{mi}}{\text{circumference}} \rightarrow \text{circumference} = \frac{500\,\text{mi} \times 360^\circ}{7^\circ} \rightarrow$$
$$\text{circumference} = 25{,}700\,\text{mi}$$

This calculation of 25,700 mi (41,300 km) is remarkably close to the actual value we know today of 24,901 mi (40,075 km).

Eratosthenes was only one of several brilliant astronomers to emerge from the Alexandrian school, which by his time had a distinguished tradition. Another of the Alexandrian astronomers was Aristarchus of Samos.

Aristarchus and Distances in the Solar System

Imagine how you might determine the distance to the Moon with none of our modern technologies—no satellite measurements, no laser radar ranging, not even a telescope. Aristarchus proposed a method of determining the relative distances to the Sun and Moon, perhaps as long ago as 280 B.C.

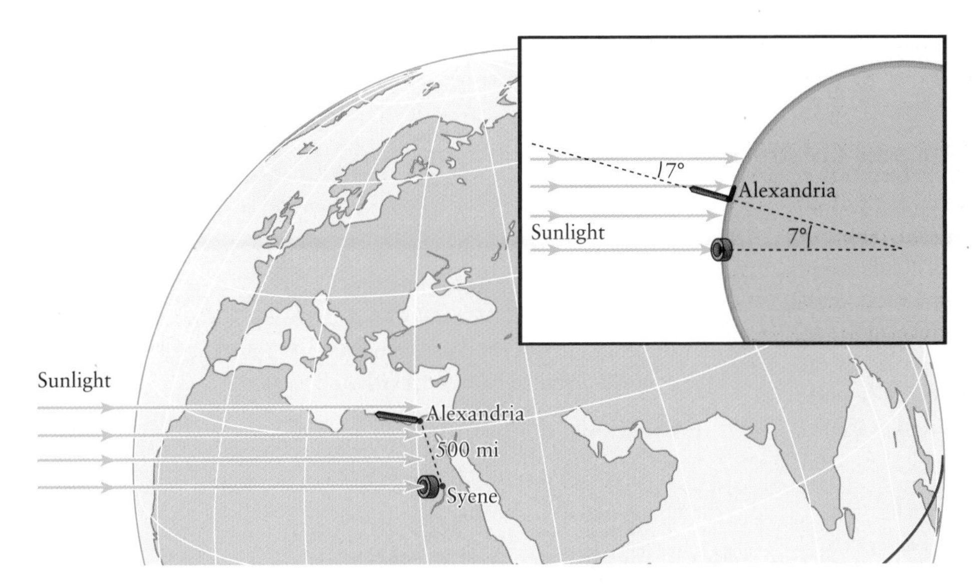

FIGURE 3-1 Eratosthenes's Method of Determining the Diameter of Earth Around 240 B.C., Eratosthenes used observations of the Sun's position at noon on the summer solstice to show that Alexandria and Syene were about 7° apart on the surface of Earth. The ratio of 7°/360° is the same as the distance between the cities divided by Earth's circumference.

Aristarchus knew that the Sun, Moon, and Earth form a right triangle at the moment of first or third quarter moon, with the right angle at the location of the Moon (**Figure 3-2**). He estimated that, as seen from Earth, the angle between the Moon and the Sun at first and third quarters is 87°, or 3° less than a 90° right angle. Using the rules of geometry, Aristarchus concluded that the Sun is about 20 times farther from us than is the Moon. In hindsight, he was not very close, but it is impressive that Aristarchus was able to determine a reasonable method for the measurement that demonstrated that the Sun is quite a bit more distant than the Moon. Today we know that he erred in measuring angles and that the average distance to the Sun is about 390 times greater than the average distance to the Moon. Fortunately, astronomers have more precise ways to measure large distances and use various units of length, including the **astronomical unit** (AU) and the **light-year** (ly). See **Box 3-1 Tools of the Astronomer's Trade: Modern Astronomical Distances Are Often Measured in Astronomical Units or Light-Years** for further explanation.

Aristarchus also made an equally bold attempt to determine the relative sizes of the Sun, Moon, and Earth. From his observations of how long the Moon takes to move through Earth's shadow during a lunar eclipse, Aristarchus estimated the diameter of Earth to be about 3 times larger than the diameter of the Moon. To determine the diameter of the Sun, Aristarchus simply pointed out that the Sun and the Moon have the same angular size in the sky. Therefore, their

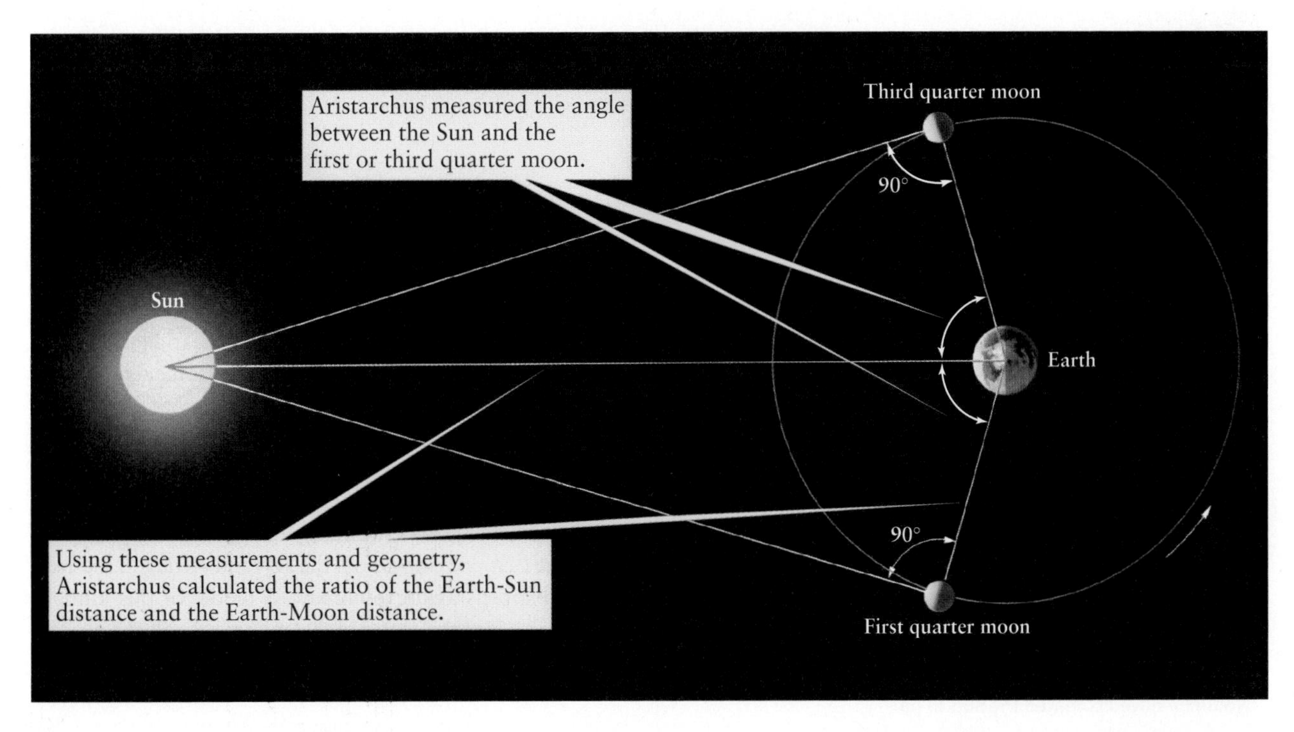

FIGURE 3-2 Aristarchus's Method of Determining Distances to the Sun and Moon Aristarchus knew that the Sun, Moon, and Earth form a right triangle at first and third quarter phases. Using geometrical arguments, he calculated the relative lengths of the sides of these triangles, thereby obtaining the distances to the Sun and Moon.

BOX 3-1 TOOLS OF THE ASTRONOMER'S TRADE

Modern Astronomical Distances Are Often Measured in Astronomical Units or Light-Years

Astronomers use many of the same units of measurement as do other scientists. They often measure lengths in meters (abbreviated m), masses in kilograms (kg), and time in seconds (s). Like other scientists, astronomers often find it useful to combine these units with **powers of ten** (see Appendix 9) and create new units using prefixes. As an example, the number 1000 (= 10^3) is represented by the prefix "kilo," and so a distance of 1000 m is the same as 1 kilometer (1 km). Here are some of the most common prefixes, with examples of how they are used:

one-billionth meter = 10^{-9} m = 1 nanometer

one-millionth second = 10^{-6} s = 1 microsecond

one-thousandth arcsecond = 10^{-3} arcsec = 1 milliarcsecond

one-hundredth meter = 10^{-2} m = 1 centimeter

one thousand meters = 10^3 m = 1 kilometer

one million tons = 10^6 tons = 1 megaton

In principle, we could express all sizes and distances in astronomy using units based on the meter. Indeed, we will use kilometers to give the diameters of Earth and the Moon, as well as the Earth-Moon distance. But, while a kilometer (roughly equal to three-fifths of a mile) is an easy distance for humans to visualize, a megameter (10^6 m) is not. Consider this: If the length of a shoe can be measured in centimeters, does it also make sense to describe the distance between two cities in centimeters? Similarly, if we measure the distance between two cities in kilometers, does it make sense to measure the great distances between stars in kilometers? Indeed, in describing the distances between objects in the cosmos, the particular units of length used depend on just how large the distances actually are. For this reason, astronomers have devised units of measure that are more appropriate for the tremendous distances between the planets and the far greater distances between the stars.

Measuring Inside our Solar System

When discussing distances across the solar system, astronomers use a unit of length called the **astronomical unit** (abbreviated AU). This is the average distance between Earth and the Sun:

$$1 \text{ AU} = 1.496 \times 10^8 \text{ km} = 92.96 \text{ million mi}$$

Thus, the average distance between the Sun and Jupiter can be conveniently stated as 5.2 AU.

Measuring Distances to the Stars

To talk about distances to the stars, astronomers use two different units of length. The light-year (abbreviated ly) is the distance that light travels in one year. This is a useful concept because the speed of light in empty space always has the same value, 3.00 × 105 km/s (kilometers per second) or 1.86×10^5 mi/s (miles per second). In terms of kilometers or astronomical units, 1 ly is given by

$$1 \text{ ly} = 9.46 \times 10^{12} \text{ km} = 63{,}240 \text{ AU}$$

This distance is roughly equal to 6 trillion mi.

CAUTION Keep in mind that despite its name, the light-year is a unit of distance and *not* a unit of time. As an example, Proxima Centauri, the nearest star other than the Sun, is a distance of 4.2 ly from Earth. This means that light takes 4.2 years to travel to us from Proxima Centauri.

Physicists often measure interstellar distances in light-years because the speed of light is one of nature's most important numbers.

diameters must be in the same proportion as their distances. In other words, because Aristarchus thought the Sun to be 20 times farther from Earth than the Moon, he concluded that the Sun must be 20 times larger than the Moon. Once Eratosthenes had measured Earth's circumference, astronomers of the Alexandrian school could use Aristarchus's method to estimate the diameters of the Sun and Moon as well as their distances from Earth.

ConceptCheck 3-1: If Eratosthenes had found that the Sun's noontime summer solstice altitude at Alexandria was much closer to directly overhead, would he then assume that Earth was larger, smaller, or about the same size?

CalculationCheck 3-1: If Aristarchus had estimated the Sun to be 100 times farther from Earth than the Moon, how large would he have estimated the Sun to be?

The Greek Geocentric Model

Go to Video 3-1

The easiest assumption to make about the cosmos is that we observers are actually stationary and everything else moves around us. This makes some sense in that you might think that if we were moving through space, we could feel it, much like we can sense the movement of a car we are riding in or a boat we are sailing in. Most Greek scholars also assumed that the Sun, the Moon, the stars, and the planets revolve about a stationary Earth, with everything rising in the east, moving across the sky, and setting in the west until it happens again tomorrow. A model of this kind, in which Earth is at the center of the universe, is called a **geocentric model.** Like ancient thinkers in many parts of the world, the ancient Greeks concluded that stars were brilliant points of light attached to an immense celestial sphere surrounding

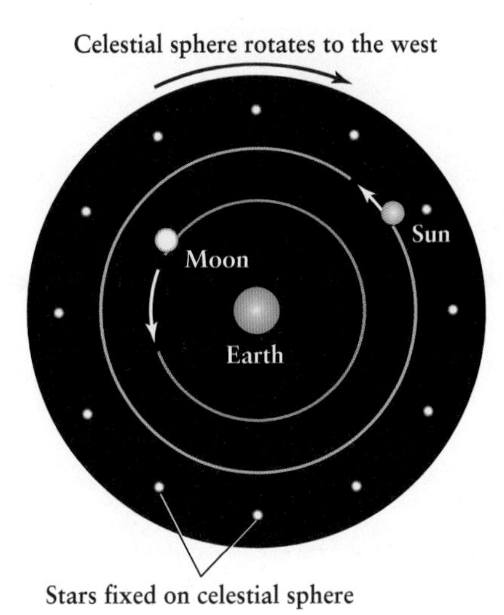

FIGURE 3-3 Earth-Centered (Geocentric) Model of the Universe The ancient Greeks imagined that the Sun and Moon move around the rotating celestial sphere with its fixed stars. Thus, the Sun and Moon would move from east to west across the sky every day and also move slowly eastward from one night to the next relative to the background of stars, matching what we see in the sky perfectly.

Earth. To explain the daily rising and setting of the stars, they assumed that the celestial sphere was *real,* and that it moved around the stationary Earth once a day.

In this Earth-centered model, the Sun and Moon both participated in this daily rotation of the sky, which explained their rising and setting motions. However, in order to be accurate, this model of how the universe works also needs to explain why the Sun and the Moon move at slightly different speeds than the stars. To explain why the Sun and Moon both slowly change their positions from night to night, compared to the fixed positions of stars, the ancient Greeks imagined that both of these objects orbit around Earth as shown in **Figure 3-3.**

The ancient Greeks and other cultures of that time knew of five other planets: Mercury, Venus, Mars, Jupiter, and Saturn, each of which is a bright object in the night sky. For example, when Venus is at its maximum brilliancy, it is 16 times brighter than the brightest star. (By contrast, Uranus and Neptune are quite dim and were not discovered until after the invention of the telescope.)

Like the Sun and Moon, all of the planets rise in the east and set in the west once a day. And like the Sun and Moon, from night to night the planets slowly move on the celestial sphere, that is, with respect to the background of stars. However, the character of this motion on the celestial sphere is quite different for the planets. Any model of how the universe works needs to account for all of these objects, which move at different rates.

Both the Sun and the Moon always move from west to east on the celestial sphere, that is, opposite the direction in which the celestial sphere appears to rotate. The Sun follows the path called the ecliptic (see Section 1-4), while the Moon follows a path that is slightly inclined to the ecliptic (see Section 1-5). Furthermore, the Sun and the Moon each move at relatively constant speeds around the celestial sphere. (The Moon's speed is faster than that of the Sun: It travels all the way around the celestial sphere in about a month while the Sun takes an entire year.) The planets, too, appear to move along paths that are close to the ecliptic. The difference is that each of the planets appears to wander back and forth on the celestial sphere with varying speed. As an example, **Figure 3-4** shows the wandering motion of Mars with respect to the background of stars during 2011 and 2012. (This figure shows that the name *planet* is well deserved; it comes from a Greek word meaning "wanderer.")

CAUTION On a map of Earth with north at the top, west is to the left and east is to the right. Why, then, is *east* on the left and *west* on the right in Figure 3-4? The answer is that a map of Earth is a view looking downward at the ground from above, while a star map like Figure 3-4 is a view looking upward at the sky. If the constellation Leo in Figure 3-4 were directly overhead, Virgo would be toward the eastern horizon and Cancer would be toward the western horizon.

Most of the time, planets drift slowly eastward relative to the stars, just as do the Sun and Moon. For example,

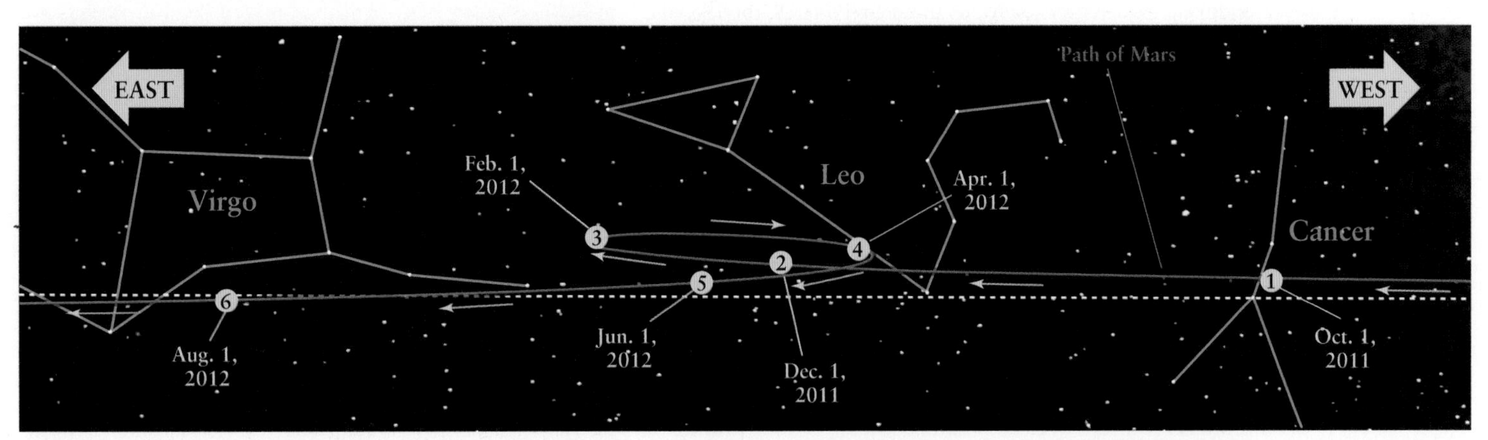

FIGURE 3-4 The Path of Mars in 2011–2012 From October 2011 through August 2012, Mars moved across the zodiac constellations Cancer, Leo, and Virgo. Mars's motion was direct (from west to east, or from right to left in this figure) most of the time but was retrograde (from east to west, or from left to right in this figure) during February and March 2011. Notice that the speed of Mars relative to the stars is not constant: The planet travels farther across the sky from October 1 to December 1 than it does from December 1 to February 1.

Figure 3-4 shows that Mars moved slowly eastward night after night from October 2011 through January 2012 and from April through August 2012. Occasionally, however, the planet will seem to stop and then back up for several weeks or months. This occasional westward movement is called **retrograde motion.** Mars underwent retrograde motion during February and March 2012 (see Figure 3-4), and will do so again about every 22½ months. All the planets go through retrograde motion, but at different intervals. This is quite a complication to the Earth-centered model shown in Figure 3-3, so the Greeks imagined the planets as going around, but once in a while having to stop and switch directions!

Whether a planet is in direct or retrograde motion, over the course of a single night you will see it rise in the east and set in the west. That is because both direct and retrograde motions are much slower than the apparent daily rotation of the sky. Hence, they are best detected by mapping the position of a planet against the background stars from night to night over a long period. Figure 3-4 is a map of just this sort.

ConceptCheck 3-2: If Mars is moving retrograde, will it rise above the eastern horizon or above the western horizon?

ConceptCheck 3-3: How fast is Earth spinning on its axis in the Greeks' geocentric model?

The Ptolemaic System

Although one could imagine that the stars, planets, Moon, and Sun generally moved in a simple motion around Earth, the alternating forward and retrograde motions of the planets were not so easily explained by imagining a celestial sphere simply moving around Earth. In fact, successfully proposing a mechanism for the nonuniform motions of the five planets was one of the main challenges facing the astronomers of antiquity.

The Greeks developed many different theories to account for retrograde motion and the loops that the planets trace out against the background stars. One of the most successful and enduring explanations was advanced by Ptolemy, the last of the great Greek astronomers, during the second century A.D. (**Figure 3-5*a***). In Ptolemy's model, each planet is assumed to move in a small circle, whose center in turn moves in a larger circle, which is centered approximately on Earth. Both the circles rotate in the same counterclockwise sense (looking down from the north), as shown in Figure 3-5*a*.

In this model, most of the time the planet will appear to drift slowly eastward as seen from Earth when compared to the background stars. However, when the planet is nearest Earth, the motion of the planet appears to slow down and halt its usual eastward movement among the constellations, and actually goes backward in retrograde (westward) motion for a few weeks or months (Figure 3-5*c*). Although complicated, this mechanism allowed the Greek astronomers to explain and accurately predict the retrograde loops of the planets. This incredibly complicated approach to predicting

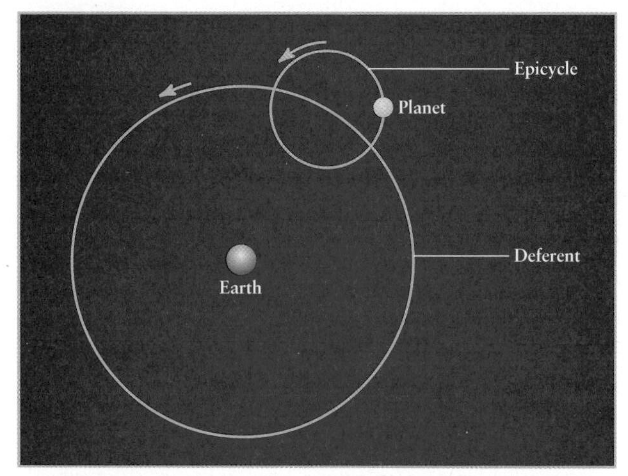

(a) Planetary motion modeled as a combination of circular motions

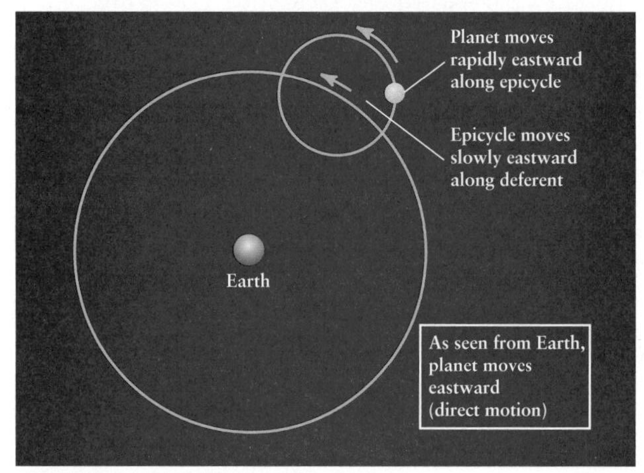

(b) Modeling direct motion

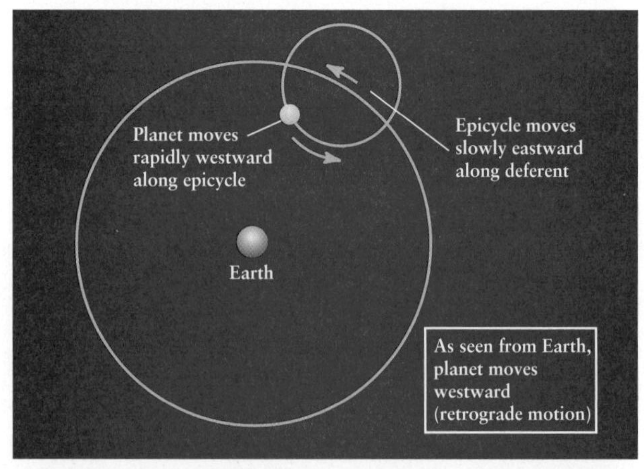

(c) Modeling retrograde motion

FIGURE 3-5 A Geocentric Explanation of Retrograde Motion (a) The ancient Greeks imagined that each planet moves along an epicycle, which in turn moves along a deferent centered approximately on Earth. The planet moves along the epicycle more rapidly than the epicycle moves along the deferent. (b) At most times the eastward motion of the planet on the epicycle adds to the eastward motion of the epicycle on the deferent. Then the planet moves eastward in direct motion as seen from Earth. (c) When the planet is on the inside of the deferent, its motion along the epicycle is westward. Because this motion is faster than the eastward motion of the epicycle on the deferent, the planet appears from Earth to be moving westward in retrograde motion.

the motions of the planets was so successful that it was used for more than a thousand years as the most useful description of the workings of the heavens.

ConceptCheck 3-4: What causes the planets to actually stop and change their direction of motion in the Ptolemaic model?

3-2 Nicolaus Copernicus devised the first comprehensive Sun-centered model

The idea that the most simple and most straightforward explanation of observations in nature is most likely to be correct is called **Occam's razor,** after William of Occam (or Ockham), the fourteenth-century English philosopher who first expressed it. (The "razor" refers to shaving extraneous details from an argument or explanation.) Although Occam's razor has no proof or verification, it appeals to a scientist's sense of beauty and elegance, and it has helped lead to the simple and powerful laws of nature that scientists use today. Clearly, Ptolemy's incredibly complicated mechanism for predicting the motions of planets is inconsistent with Occam's razor. The first astronomer to suggest a simpler and more straightforward model was the Greek astronomer Aristarchus, who in the third century B.C. suggested a Sun-centered, or heliocentric, model as a way to explain retrograde motion, but it was not widely accepted.

A Heliocentric Model Explains Retrograde Motion

Go to Video 3-2

Imagine riding on a fast racehorse. As you pass a slowly walking pedestrian, he appears to move backward, even though he is traveling in the same direction as you and your horse. This sort of simple observation inspired Aristarchus and others to formulate a **heliocentric (Sun-centered) model** in which all the planets, including Earth, revolve about the Sun. In this way of thinking, planets take different lengths of time to complete an orbit, so from time to time one planet will overtake another, just as a fast-moving horse overtakes a person on foot. When Earth overtakes Mars, for example, Mars appears to move backward in retrograde motion, as **Figure 3-6** shows. Thus, in the heliocentric picture, the occasional retrograde motion of a planet across the sky is an illusion due to Earth's motion.

Imagining a system where a spinning Earth orbits the Sun more easily describes observed retrograde motion.

In the last section, we described how Aristarchus demonstrated that the Sun is bigger than Earth. If the Sun is enormous in size compared to Earth, it becomes a bit more sensible to imagine a small Earth orbiting a larger Sun. He also imagined that Earth rotated on its axis once a day, which explained the daily rising and setting of the Sun, Moon, and planets and the diurnal motions of the stars. To explain why the apparent motions of the planets never take them far from the same path through the sky, Aristarchus proposed that the orbits of Earth and all the planets must lie in nearly the same plane.

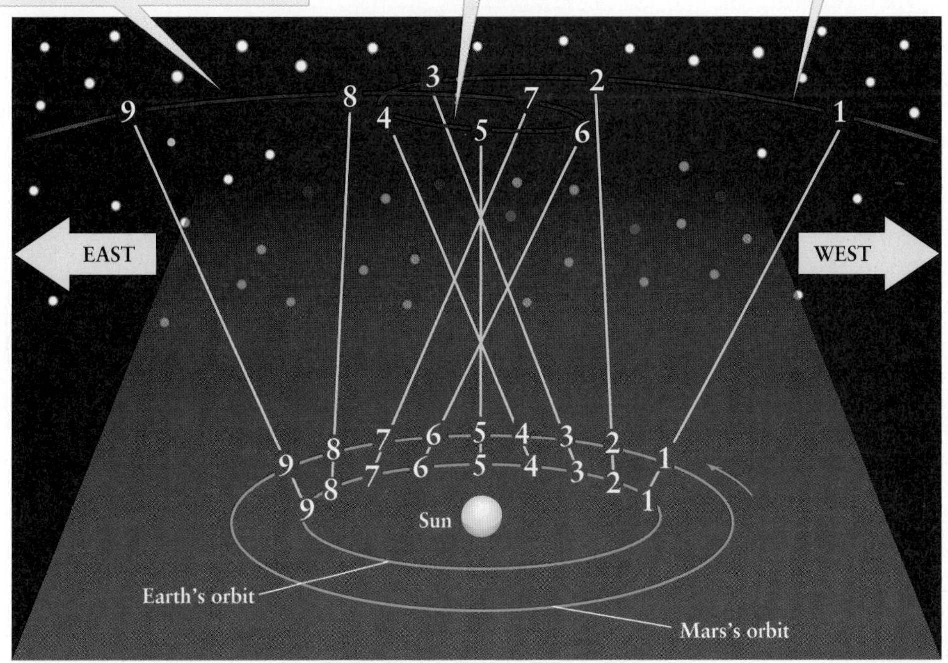

FIGURE 3-6 A Heliocentric Explanation of Retrograde Motion In the heliocentric model of Aristarchus, Earth and the other planets orbit the Sun. Earth travels around the Sun more rapidly than Mars. Consequently, as Earth overtakes and passes this slower-moving planet, Mars appears for a few months (from points 4 through 6) to fall behind and move backward with respect to the background of stars.

This Sun-centered model is conceptually much simpler than an Earth-centered system, such as that of Ptolemy, with all its "circles upon circles." In Aristarchus's day, however, the idea of an orbiting, rotating Earth was not widely accepted given Earth's apparent stillness and immobility.

ConceptCheck 3-5: In the heliocentric model, would an imaginary observer on the Sun look out and see planets moving in retrograde motion?

Copernicus and the Arrangement of the Planets

How can an observer determine which planets are closest to the Sun and which are farthest away? If you have a telescope, you can sometimes tell that some appear larger than others. But what if you do not have a telescope and planets appear to be shiny dots in the sky? For early astronomers, the order of the planets from closest to the Sun to most distant from the Sun was largely guesswork. During the first half of the sixteenth century, a Polish lawyer, physician, canon of the church, and gifted mathematician named Nicolaus Copernicus began to construct a new model of the universe. His model, which placed the Sun at the center, explained the motions of the planets in a more natural way than the Ptolemaic system, consistent with Occam's razor.

Copernicus came to realize that you could use the heliocentric model to determine the arrangement of the planets without ambiguity. Copernicus realized that because Mercury and Venus are always observed fairly near the Sun in the sky, their orbits must be smaller than Earth's. The other visible planets known at that time—Mars, Jupiter, and Saturn—are sometimes seen on the side of the celestial sphere opposite the Sun, so these planets appear high above the horizon at midnight (when the Sun is far below the horizon). When this happens, Earth must lie between the Sun and these planets. Copernicus therefore concluded that the orbits of Mars, Jupiter, and Saturn must be larger than Earth's orbit.

The heliocentric model also explains why planets appear in different parts of the sky on different dates. Both Mercury and Venus go through cycles as they orbit the Sun: They are seen in the west after sunset for several weeks or months, then for several weeks or months in the east before sunrise, and then in the west after sunset again.

Figure 3-7 shows the reason for this cycle. When Mercury or Venus is visible after sunset, it is near **greatest eastern elongation.** (The angle between the Sun and a planet as viewed from Earth is called the planet's **elongation.**) The planet's position in the sky is as far east of the Sun as possible, so it appears above the western horizon after sunset (that is, to the east of the Sun) and is often called an "evening star." At **greatest western elongation,** Mercury or Venus is as far west of the Sun as it can possibly be. It then rises before the Sun, gracing the predawn sky as a "morning star" in the east. When Mercury or Venus is at **inferior conjunction,** it is between us and the Sun, and it is moving from the evening sky into the morning sky. At **superior conjunction,** when the planet is on the opposite side of the Sun, it is moving back into the evening sky.

A planet farther from the Sun than Earth whose orbit is larger than Earth's, such as Mars, is best seen in the night sky when it is at **opposition.** At this point the planet is in the part of the sky opposite the Sun and is highest in the sky at midnight. This is also when the planet appears brightest, because it is closest to us. Alternatively, when a planet is located behind the Sun at superior conjunction, it is above the horizon during the daytime and thus is not well placed for nighttime viewing.

ConceptCheck 3-6: If Venus is visible high in the evening sky after sunset, is it leading Earth in its orbit or behind Earth?

ConceptCheck 3-7: How many times is Mars at inferior conjunction during one orbit around the Sun?

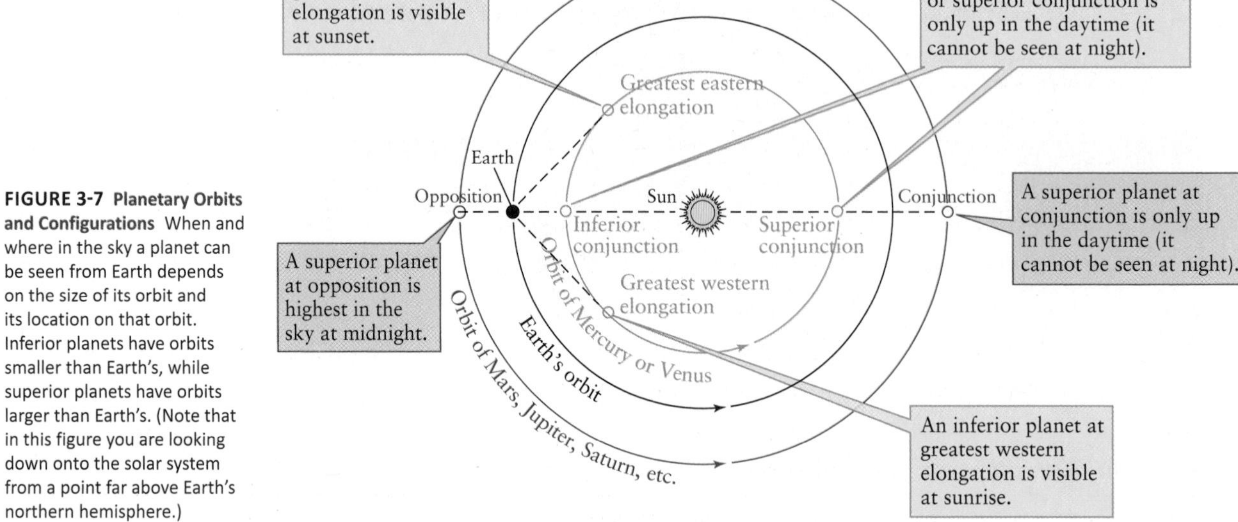

FIGURE 3-7 Planetary Orbits and Configurations When and where in the sky a planet can be seen from Earth depends on the size of its orbit and its location on that orbit. Inferior planets have orbits smaller than Earth's, while superior planets have orbits larger than Earth's. (Note that in this figure you are looking down onto the solar system from a point far above Earth's northern hemisphere.)

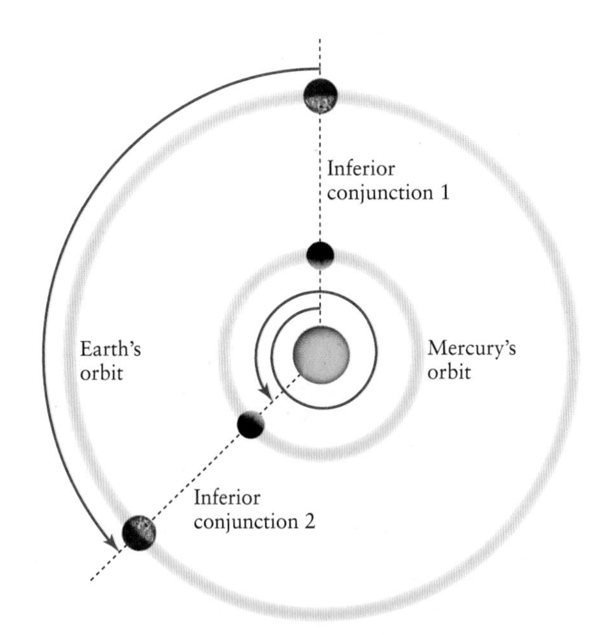

FIGURE 3-8 Synodic Period The time between consecutive conjunctions of Earth and Mercury is 116 days. As is typical of synodic periods for all planets, the location of Earth is different at the beginning and end of the period. You can visualize the synodic periods of the outer planets by putting Earth in Mercury's place in this figure and putting one of the outer planets in Earth's place.

Planetary Periods and Orbit Sizes

The Ptolemaic system has no simple rules relating the motion of one planet to another. But Copernicus showed that there *are* such rules in a heliocentric model. In particular, he found a correspondence between the time a planet takes to complete one orbit—that is, its **period**—and the size of the orbit.

Determining the period of a planet takes some care, because Earth, from which we must make the observations, is also moving. Realizing this, Copernicus was careful to distinguish between two different periods of each planet. The **synodic period** is the time that elapses between two successive identical configurations as seen from Earth—from one opposition to the next, for example, or from one conjunction to the next (**Figure 3-8**). The **sidereal period** is the true orbital period of a planet, the time it takes the planet to complete one full orbit of the Sun relative to the very distant background stars.

There is a definite relationship between the sidereal period of a planet and the size of its orbit, as listed in **Table 3-1.** One can see that the farther a planet is from the Sun, the longer it takes to travel around its orbit (that is, the longer its sidereal period). Today, we understand this is so for *two* reasons: (1) The larger the orbit, the farther a planet must travel to complete an orbit; and (2) the larger the orbit, the slower a planet moves. For example, Mercury, with its small orbit, moves at an average speed of 107,000 mi/h (47.9 km/s). Saturn travels around its large orbit much more slowly, at an average speed of 9.64 km/s (21,600 mi/h).

ConceptCheck 3-8: **What causes the planets to stop and change their direction of motion through the sky in the heliocentric model?**

ConceptCheck 3-9: **Why is Jupiter's sidereal period longer than its synodic period?**

The Shapes of Orbits in the Copernican Model

One might think that Copernicus's heliocentric model was more accurate than Ptolemy's geocentric model. Copernicus found that when he used perfectly circular orbits for a planet's path around the Sun, he could not accurately describe the paths of the other planets. As a result, he had to add extra tiny orbits to each planet's orbit, much like Ptolemy used. However, it needs to be emphasized that he added extra motions to improve the accuracy of his model, *not* to explain retrograde motion. Rather, these tiny orbits on orbits helped Copernicus account for slight variations in each planet's speed along its orbit.

Table 3-1 Relationship between Sidereal Period and a Planet's Orbit

	Synodic and Sidereal Periods of the Planets		Average Distances of the Planets from the Sun	
Planet	Synodic period	Sidereal period	Copernican value (AU*)	Modern value (AU)
Mercury	116 days	88 days	0.38	0.39
Venus	584 days	225 days	0.72	0.72
Earth		1.0 year	1	1
Mars	780 days	1.9 years	1.52	1.52
Jupiter	399 days	11.9 years	5.22	5.2
Saturn	378 days	29.5 years	9.07	9.55
Uranus	370 days	84.1 years		19.19
Neptune	368 days	164.9 years		30.07

*1 AU = 1 astronomical unit = average distance from Earth to the Sun

Copernicus compiled his ideas and calculations into a book entitled *De revolutionibus orbium coelestium* (*On the Revolutions of the Celestial Spheres*), which was published in 1543, the year of his death. For several decades after Copernicus, most astronomers saw little reason to change their allegiance from the older geocentric model of Ptolemy. The predictions that the Copernican model makes for the apparent positions of the planets are, on average, no better or worse than those of the Ptolemaic model. The test of Occam's razor does not really favor either model, because both use a combination of circles to describe each planet's motion.

More concrete evidence was needed to convince scholars to abandon the old, comfortable idea of a stationary Earth at the center of the universe.

ConceptCheck 3-10: Why was Copernicus's model more accurate than Ptolemy's model?

3-3 Galileo's discoveries of moons orbiting Jupiter and the phases of Venus strongly supported a heliocentric model

If you were a practicing scientist 400 years ago, what would you accept as "proof" that Earth orbited the Sun and not the other way around? When Dutch opticians invented the telescope during the first decade of the seventeenth century, astronomy was changed forever. The scholar who used this new tool to amass convincing evidence that the planets orbit the Sun, not Earth, was the Italian mathematician and physical scientist Galileo Galilei.

While Galileo did not invent the telescope, he was the first to point one of these new devices toward the sky and to publish his observations. Beginning in 1609, he saw sights of which no one had ever dreamed. He was the first to view mountains on the Moon, sunspots on the Sun, rings around Saturn, phases of Venus, moons orbiting Jupiter, and that the Milky Way is not a featureless band of light passing overhead but rather "a mass of innumerable stars."

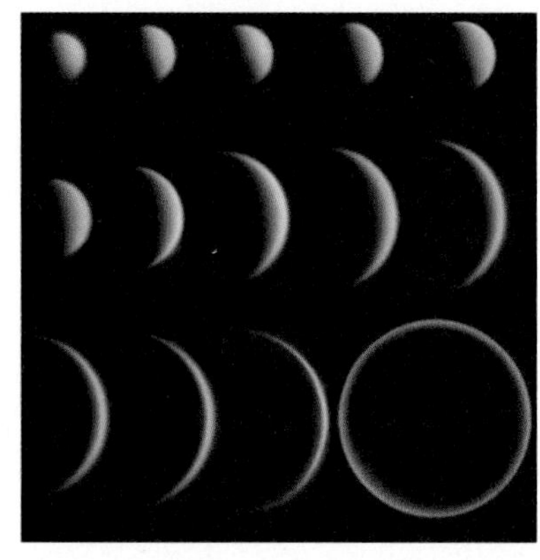

FIGURE 3-9 R I V U X G **The Phases of Venus** This series of photographs shows how Venus changes appearance and size as it orbits the Sun. When farthest from Earth, it appears quite small and nearly in a full phase as we see the entirety of its lit face. When closest to Earth, it appears much larger and we are able to see it in its new phase where the side opposite us is illuminated. Note on the largest image, that a thin ring can be seen around Venus as sunlight from the other side passes through Venus's atmosphere. *(Statis Kalyvas/European Southern Observatory)*

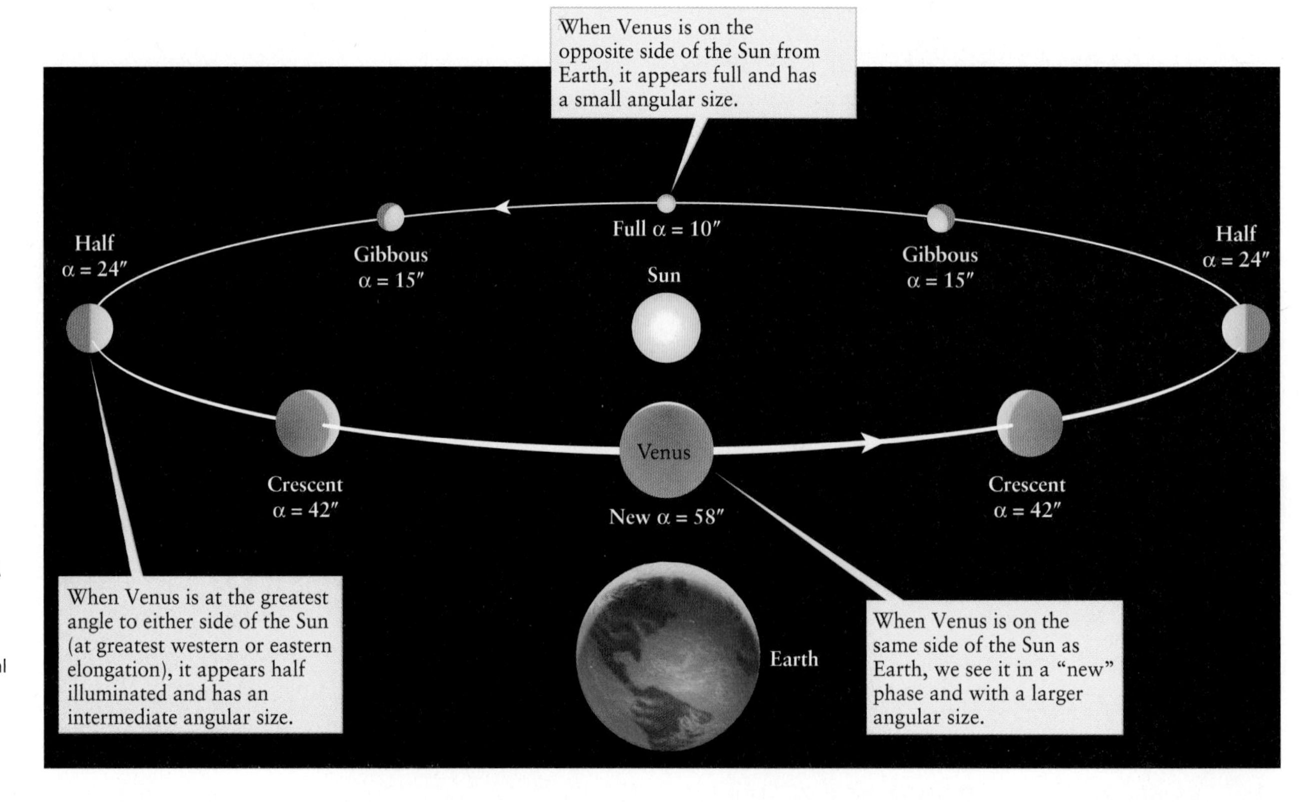

FIGURE 3-10 The Changing Appearance of Venus Explained in a Heliocentric Model A heliocentric model, in which Earth and Venus both orbit the Sun, provides a natural explanation for the changing appearance of Venus shown in this figure.

The Phases of Venus

One of Galileo's most important discoveries with the telescope was that Venus exhibits phases like those of the Moon (**Figure 3-9**). Galileo also noticed that the apparent size of Venus as seen through his telescope was related to the planet's phase. Venus appears small at gibbous phase and largest at crescent phase. **Figure 3-10** shows that these relationships are entirely compatible with a heliocentric model in which Earth and Venus both go around the Sun. They are also completely *incompatible* with the Ptolemaic system, in which the Sun and Venus both orbit Earth. To explain why Venus is never seen very far from the Sun, the Ptolemaic model had to assume that the orbits of Venus and of the Sun move together in lockstep, with Venus's path centered on a straight line between Earth and the Sun (**Figure 3-11**). In this model, Venus was never on the opposite side of the Sun from Earth, and so it could never have shown the gibbous phases that Galileo observed.

ConceptCheck 3-11: Which phase will Venus be in when it is at its maximum distance from Earth?

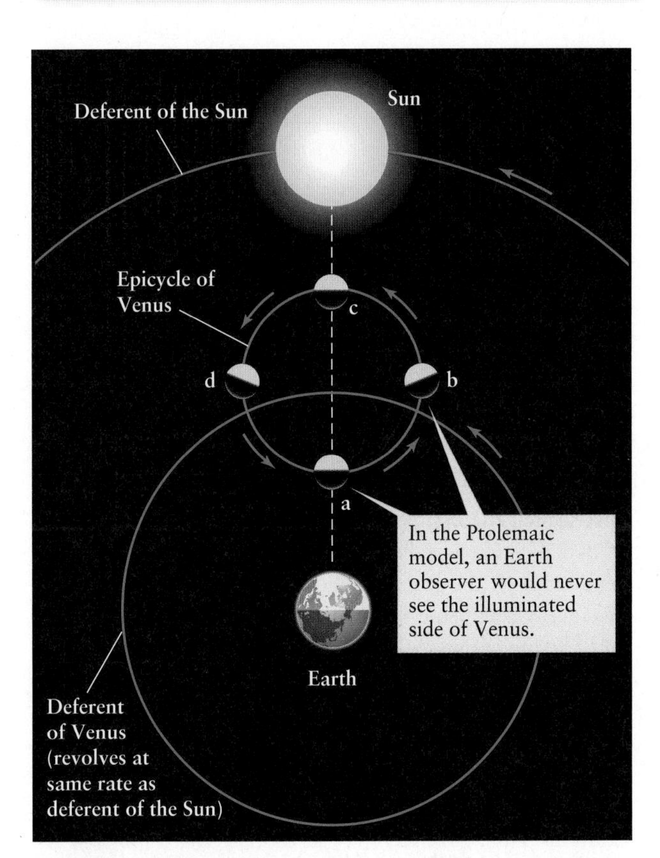

FIGURE 3-11 The Appearance of Venus in the Ptolemaic Model In the geocentric Ptolemaic model the deferents of Venus and the Sun rotate together, with the epicycle of Venus centered on a line (shown dashed) that connects the Sun and Earth. In this model an Earth observer would never see Venus as more than half illuminated. At positions a and c, Venus appears in a "new" phase; at positions b and d, it appears as a crescent. (Compare with Figure 3-2, which shows the phases of the Moon.) Because Galileo saw Venus in nearly fully illuminated phases, he concluded that the Ptolemaic model must be incorrect.

FIGURE 3-12 R I V U X G **Jupiter and Its Four Largest Moons** This photograph shows the four Galilean satellites alongside an overexposed image of Jupiter. Each satellite is bright enough to be seen with the unaided eye, were it not being overwhelmed by the glare of Jupiter. *(Rev. Ronald Royer/Science Source)*

The Moons of Jupiter

The guiding philosophy of most people in Galileo's day was that Earth was the center of all orbits. Imagine how surprised observers were in 1610 when he discovered four previously unknown moons orbiting Jupiter (**Figure 3-12**). He realized that they were orbiting Jupiter because they appeared to move back and forth from one side of the planet to the other. **Figure 3-13** shows confirming observations made by Jesuit observers in 1620. Astronomers soon realized that the larger the orbit of one of the moons around Jupiter, the slower that moon moves and the longer it takes

Jupiter's own orbiting moons and Venus's phases provide indirect evidence for a Sun-centered model.

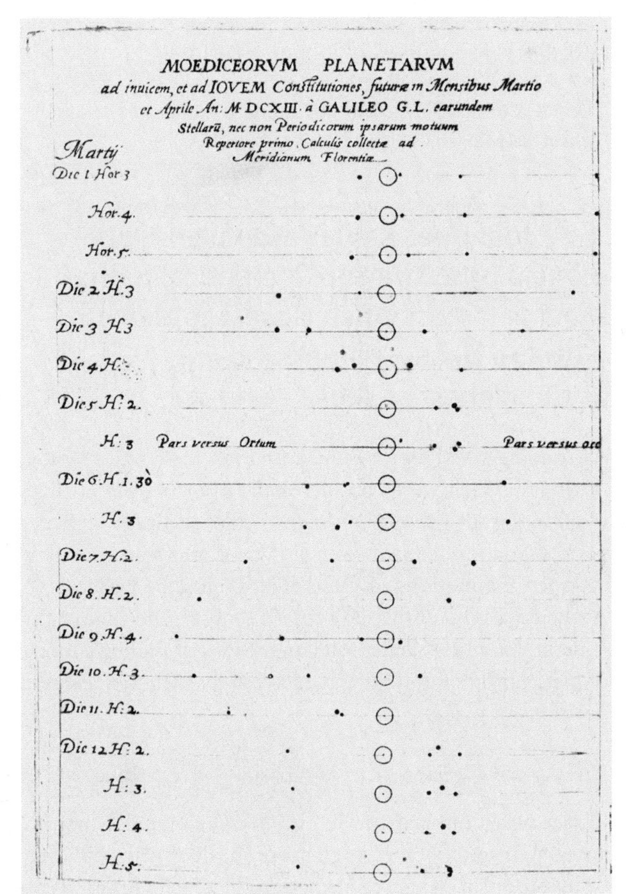

FIGURE 3-13 Early Observations of Jupiter's Moons In 1610 Galileo discovered four "stars" that move back and forth across Jupiter from one night to the next. He concluded that these four moons orbit Jupiter, much as our Moon orbits Earth. The circle represents Jupiter and the dots its moons. *(Royal Astronomical Society/Science Source)*

that moon to travel around its orbit. These are the same relationships that Copernicus deduced for the motions of the planets around the Sun. Thus, the moons of Jupiter behave like a Copernican system in miniature, demonstrating that Earth is not the center of all orbits.

Contradicting prevailing opinion and religious belief, Galileo's discoveries strongly suggested a heliocentric structure of the universe. The Roman Catholic Church of the day, which was a powerful political force in Italy and whose doctrine at the time placed Earth at the center of the universe, cautioned Galileo not to advocate a heliocentric model. He nonetheless persisted and was sentenced to spend the last years of his life under house arrest "for vehement suspicion of heresy." Nevertheless, there was no turning back. (The Roman Catholic Church lifted its ban against Galileo's heliocentric ideas in the 1700s and in 1992 apologized for its treatment of Galileo.)

While Galileo's observations showed convincingly that the Ptolemaic model was entirely wrong and that a heliocentric model is the more nearly correct one, he was unable to provide a complete explanation of why Earth should orbit the Sun and not vice versa. The first person who was able to provide such an explanation was the Englishman Isaac Newton, born on Christmas Day of 1642, the same year that Galileo died. While Galileo revolutionized our understanding of planetary motions, Newton's contribution was far greater: He deduced the basic laws that govern all motions on Earth as well as in the heavens. But before we can talk about Newton, we need to talk about a contemporary of Galileo, the mathematician Johannes Kepler.

ConceptCheck 3-12: Why had Jupiter's moons not been observed prior to Galileo's time?

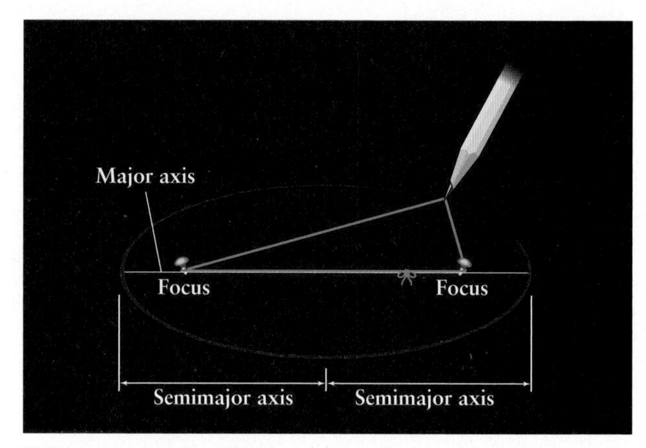

(a) The geometry of an ellipse

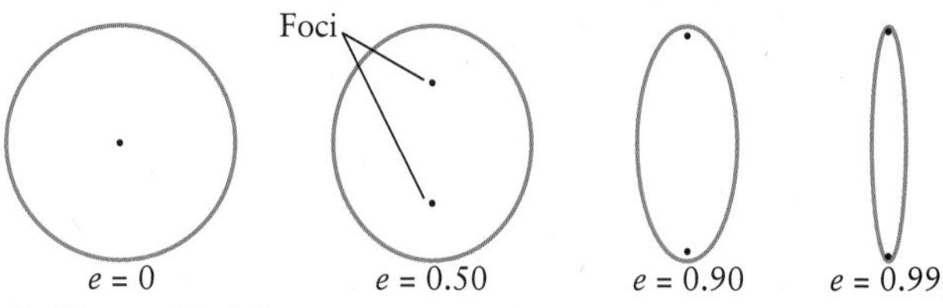

(b) Ellipses with different eccentricities

FIGURE 3-14 Ellipses (a) To draw an ellipse, use two thumbtacks to secure the ends of a piece of string, then use a pencil to pull the string taut. If you move the pencil while keeping the string taut, the pencil traces out an ellipse. The thumbtacks are located at the two foci of the ellipse. The major axis is the greatest distance across the ellipse; the semimajor axis is half of this distance. (b) This shows a series of ellipses with the same major axis but different eccentricities. An ellipse can have any eccentricity from $e = 0$ (a circle) to just under $e = 1$ (virtually a straight line).

3-4 Johannes Kepler proposed that planets orbit the Sun in elliptical paths, moving fastest when closest to the Sun, with the closest planets moving at the highest speeds

Astronomers had long assumed that heavenly objects move in circles, which were considered the most perfect and harmonious of all geometric shapes. Many believed that if a perfect God resided in heaven along with the stars and planets, then the motions of these objects must be perfect, too. At the beginning of the seventeenth century, and against this context, Johannes Kepler took on the task of finding a model of planetary motion that agreed precisely with the observed positions of planets.

Elliptical Orbits and Kepler's First Law

Kepler was a mathematician and had the unique benefit of access to incredibly precise planetary position data, collected by an eccentric Danish nobleman named Tycho Brahe. Using this data, Kepler dared to try to explain planetary motions with imperfect circles. In particular, he found that he had the best success with a particular shape called an **ellipse.**

You can draw an ellipse by using a loop of string, two thumbtacks, and a pencil, as shown in **Figure 3-14*a*.** Each thumbtack in the figure is at a **focus** (plural **foci**) of the ellipse; an ellipse has two foci. The longest diameter of an ellipse, called the *major axis,* passes through both foci. Half of that distance is called the **semimajor axis** and is usually designated by the letter a. Another way to think about the distance a is as being the average distance between the edge and a focus point. (A circle is a special case of an ellipse in which the two foci are at the same point. This corresponds to using only a single thumbtack in Figure 3-14*a*). The semimajor axis, a, of a circle is equal to its radius.

By assuming that planetary orbits are ellipses, Kepler found, to his delight, that he could make his theoretical calculations match precisely to observations. This important discovery, first published in 1609, is now called **Kepler's first law:**

The orbit of a planet about the Sun is an ellipse with the Sun at one focus.

The semimajor axis a of a planet's orbit is the average distance between the planet and the Sun.

CAUTION The Sun is at one focus of a planet's elliptical orbit, but there is *nothing* at the other focus. This "empty focus" has geometrical significance because it helps to define the shape of the ellipse, but it plays no other role.

Ellipses come in different shapes, depending on the elongation of the ellipse. The shape of an ellipse is described by its **eccentricity**, designated by the letter e. The value of e can range from 0 (a circle) to just under 1 (nearly a straight line). The greater the eccentricity, the more elongated the ellipse. Figure 3-14*b* shows a few examples of ellipses with different eccentricities. Because a circle is a special case of an ellipse, it is possible to have a perfectly circular orbit. But all of the objects that orbit the Sun have orbits that are at least slightly elliptical. The most circular of any planetary orbit is that of Venus, with an eccentricity of just 0.007; Mercury's orbit has an eccentricity of 0.206, and a number of small bodies called comets move in very elongated orbits with eccentricities just less than 1.

ConceptCheck 3-13: Which orbit is more circlelike—Venus's orbit, with $e = 0.007$, or Mars's orbit, with $e = 0.093$?

Orbital Speeds and Kepler's Second Law

Once Kepler knew the shape of a planet's orbit, he was ready to describe exactly *how* it moves on that orbit. As a planet travels in an elliptical orbit, its distance from the Sun varies. Kepler realized that the speed of a planet also varies along its orbit. A planet moves most rapidly when it is nearest the Sun, at a point on its orbit called **perihelion.** Conversely, a planet moves most slowly when it is farthest from the Sun, at a point called **aphelion** (**Figure 3-15**).

After much trial and error, Kepler found a way to describe just how a planet's speed varies as it moves along its orbit. Figure 3-15 illustrates this discovery, also published in 1609. Suppose that it takes 30 days for a planet to go from point A to point B. During that time, an imaginary line joining the Sun and the planet sweeps out a nearly triangular area. Kepler discovered that a line joining the Sun and the planet also sweeps out exactly the same area during any other 30-day interval. In other words, if the planet also takes 30 days to go from point C to point D, then the two shaded segments in Figure 3-15 are equal in area. **Kepler's second law** can be stated in this way:

A line joining a planet and the Sun sweeps out equal areas in equal intervals of time.

This relationship is also called the **law of equal areas.** In the idealized case of a circular orbit, a planet would have to move at a constant speed around the orbit in order to satisfy Kepler's second law.

ANALOGY An analogy for Kepler's second law is a twirling ice skater holding weights in each hand. If the skater moves the weights closer to her body by pulling her arms straight in, her rate of spin increases and the weights move faster; if she extends her arms so the weights move away from her body, her rate of spin decreases and the weights slow down. Just like the weights, a planet in an elliptical orbit travels at a higher speed when it moves closer to the Sun (toward perihelion) and travels at a lower speed when it moves away from the Sun (toward aphelion).

Kepler's laws explain equally well the orbits of planets around the Sun, moons around their own planets, and man-made satellites around Earth.

ConceptCheck 3-14: According to Kepler's second law, at what point in a communications satellite's orbit around Earth will it move the slowest?

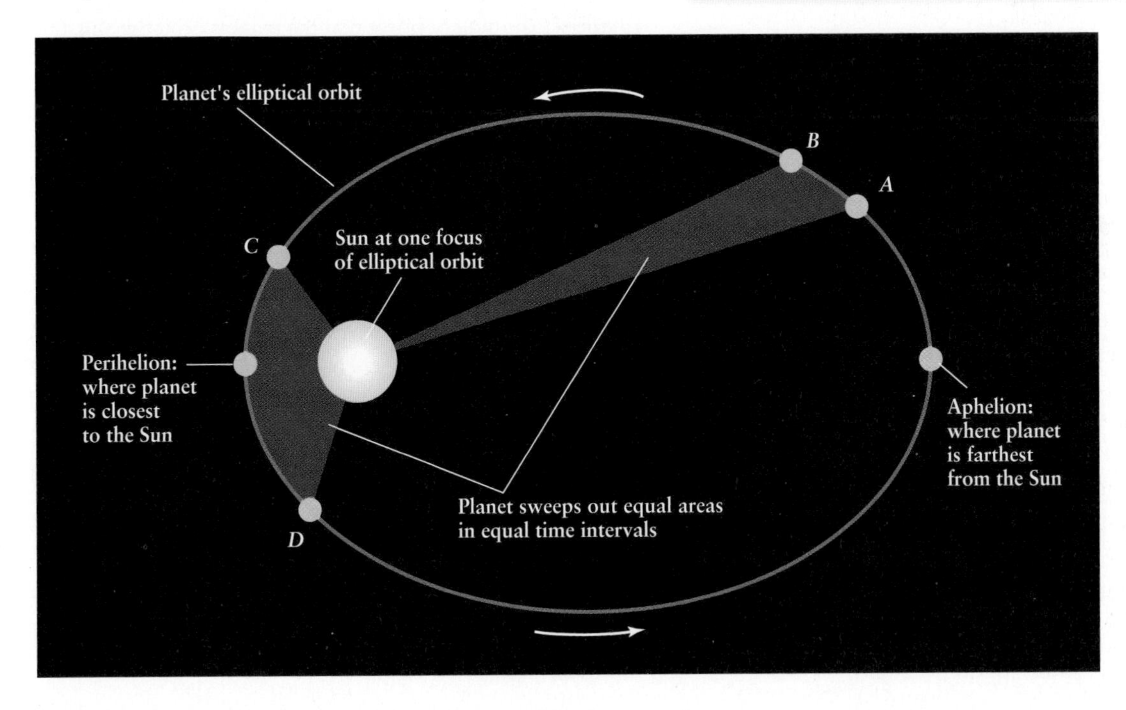

FIGURE 3-15 Kepler's First and Second Laws According to Kepler's first law, a planet travels around the Sun along an elliptical orbit with the Sun at one focus. According to his second law, a planet moves fastest when closest to the Sun (at perihelion) and slowest when farthest from the Sun (at aphelion). As the planet moves, an imaginary line joining the planet and the Sun sweeps out equal areas in equal intervals of time (from A to B or from C to D). By using these laws in his calculations, Kepler found a perfect fit to the apparent motions of the planets.

Orbital Periods and Kepler's Third Law

Kepler's second law describes how the speed of a given planet changes as it orbits the Sun. Kepler also deduced a relationship that can be used to compare the motions of *different* planets. Published later than his first two propositions and now called **Kepler's third law,** it states a relationship between the size of a planet's orbit and the time the planet takes to go once around the Sun:

The square of the sidereal period of a planet is directly proportional to the cube of the semimajor axis of the orbit.

Kepler's third law says that the larger a planet's orbit—that is, the larger the semimajor axis, or average distance from the planet to the Sun—the longer the sidereal period, which is the time it takes the planet to complete an orbit. From Kepler's third law one can show that the larger the semimajor axis, the slower the average speed at which the planet moves around its orbit. (By contrast, Kepler's *second* law describes how the speed of a given planet is sometimes faster and sometimes slower than its average speed.) This qualitative relationship between orbital size and orbital speed is just what Aristarchus and Copernicus used to explain retrograde motion. Kepler's great contribution was to make this relationship a quantitative one.

It is useful to restate Kepler's third law as an equation. If a planet's sidereal period P is measured in years and the length of its semimajor axis a is measured in astronomical units (AU), where 1 AU is the average distance from Earth to the Sun, then Kepler's third law is

Kepler's third law

$$P^2 = a^3$$

P = planet's sidereal period, in years
a = planet's semimajor axis, in AU

We can verify Kepler's third law for all of the planets, including those that were discovered after Kepler's death, using data from Table 3-1. If Kepler's third law is correct, for each planet the numerical values of P^2 and a^3 should be equal. This is indeed true to very high accuracy, as **Table 3-2** shows.

ConceptCheck 3-15: The space shuttle typically orbited Earth at an altitude of 300 km whereas the International Space Station orbits Earth at a higher altitude of 370 km. Although the space shuttle took less time to orbit Earth, which one actually moved at a faster rate?

CalculationCheck 3-2: If Pluto's orbit has a semimajor axis of 39.5 AU, how long does it take Pluto to orbit the Sun once?

The Significance of Kepler's Laws

Go to Video 3-3

Kepler's laws are a landmark in the history of astronomy. They made it possible to calculate the motions of the planets with better accuracy than any geocentric model ever had, and they helped to justify the idea of a Sun-centered model. Kepler's laws also pass the test of Occam's razor, for they are simpler in every way than the schemes of Ptolemy or Copernicus, both of which used a complicated combination of circles.

But the significance of Kepler's laws goes far beyond understanding planetary orbits. These same laws are also obeyed by spacecraft orbiting Earth, by two stars revolving about each other in a binary star system, and even by galaxies in their orbits about each other. Throughout this book, we will use Kepler's laws in a wide range of situations.

As impressive as Kepler's mathematical accomplishments were in describing when planets could be found where in the sky, he did not prove that the planets orbit the Sun, nor was he able to explain *why* planets move in accordance with his three laws. While we attribute the demonstration that planets orbit the Sun to Galileo, who worked independently

Table 3-2 A Demonstration of Kepler's Third Law ($P^2 = a^3$)

Planet	Sidereal Period P (years)	Semimajor axis a (AU)	P^2	a^3
Mercury	0.24	0.39	0.06	0.06
Venus	0.61	0.72	0.37	0.37
Earth	1.00	1.00	1.00	1.00
Mars	1.88	1.52	3.53	3.51
Jupiter	11.86	5.20	140.70	140.60
Saturn	29.46	9.55	867.90	871.00
Uranus	84.10	19.19	7,072.00	7,067.00
Neptune	164.86	30.07	27,180.00	27,190.00

Kepler's third law states that $P^2 = a^3$ for each of the planets. The last two columns of this table demonstrate that this relationship holds true to a very high level of accuracy.

from Kepler, Isaac Newton is credited as the luminary who figured out that gravity was responsible for why the planets move as they do.

ConceptCheck 3-16: **Do Kepler's laws of planetary motion apply only to the planets?**

3-5 Isaac Newton formulated three laws relating force and motion to describe fundamental properties of physical reality

Until the mid-seventeenth century, virtually all attempts to describe the motions of the heavens were empirical, or based directly on data and observations. From Ptolemy to Kepler, astronomers would adjust their ideas and calculations by trial and error until they ended up with answers that agreed with observation.

Isaac Newton introduced a new approach. He began with three quite general statements, now called **Newton's laws of motion.** These laws, deduced from experimental observation, apply to all forces and all objects. Newton then showed that Kepler's three laws follow logically from these laws of motion and from a formula for the force of gravity that he derived from observation.

In other words, Kepler's laws are not just an empirical description of the motions of the planets, but a direct consequence of the fundamental laws of physical matter. Using this deeper insight into the nature of motions in the heavens, Newton and his successors were able to accurately describe not just the orbits of the planets but also the orbits of the Moon and comets.

Newton's First Law of Motion

Newton's laws of motion describe objects on Earth as well as in the heavens. Thus, we can understand each of these laws by considering the motions of objects around us. We begin with **Newton's first law of motion:**

An object remains at rest, or moves in a straight line at a constant speed, unless acted upon by a net outside force.

By force we mean any push or pull that acts on the object. An *outside* force is one that is exerted on the object by something other than the object itself. The net, or total, outside force is the combined effect of all of the individual outside forces that act on the object.

Right now, you are demonstrating the first part of Newton's first law. As you sit in your chair reading this, there are two outside forces acting on you: The force of gravity between you and Earth pulls you downward and, at the same time, the chair pushes up on you. These two forces are of equal strength but of opposite direction, so their effects cancel—there is no *net* outside force. Hence, your body remains at rest as stated in Newton's first law. If you try to lift yourself out of your chair by grabbing your knees and pulling up, you will remain at rest because this force is not an outside force.

CAUTION The second part of Newton's first law, about objects in motion, may seem to go against common sense. If you want to make this book move across the floor in a straight line at a constant speed, you must continually push on it. You might therefore think that there *is* a net outside force, the force of your push. But another force also acts on the book—the force of friction as the book rubs across the floor. As you push the book across the floor, the force of your push exactly balances the force of friction, so again there is no net outside force. The effect is to make the book move in a straight line at constant speed, just as Newton's first law says. If you stop pushing, there will be nothing to balance the effects of friction. Then there will be a net outside force and the book will slow to a stop.

Newton's first law tells us that if no net outside force acts on a moving object, it can only move in a straight line and at a constant speed. This means that a net outside force must be acting on the planets since they do not move in straight lines, but instead move around elliptical paths. To see why, note that a planet would tend to fly off into space at a steady speed along a straight line if there were no other outside force acting on it to keep it moving around our Sun. Because this does not happen, Newton concluded that a force *must* act continuously on the planets to keep them in their elliptical orbits.

ConceptCheck 3-17: **If the 815 kg unmanned *Voyager 2* interplanetary space probe was traveling at 38,000 mph (62,000 kph) without any rocket engines firing in 2009, how fast will it be moving in 2015, still without engines?**

Newton's Second Law of Motion

Newton's second law describes how the motion of an object *changes* if there is a net outside force acting on it. To appreciate Newton's second law, we must first understand quantities that describe motion: speed, velocity, and acceleration.

Speed is a measure of how fast an object is moving. Speed and direction of motion together constitute an object's **velocity.** Compared with a car driving north at 100 km/h (62 mi/h), a car driving east at 100 km/h has the same speed but a different velocity. We can restate Newton's first law to say that an object has a constant velocity (its speed and direction of motion do not change) if no net outside force acts on the object.

Acceleration is the rate at which velocity changes. Because velocity involves both speed and direction, acceleration can result from changes in either. Contrary to popular use of the term, acceleration does not simply mean speeding up. A car is accelerating if it is speeding up, and it is also accelerating if it is slowing down or turning (that is, changing the direction in which it is moving).

You can verify these statements about acceleration if you think about the sensations of riding in a car. If the car is moving with a constant velocity (in a straight line at a constant speed), you feel the same as if the car were not moving at all. But you can feel it when the car accelerates in any way: You feel thrown back in your seat if the car speeds up, thrown forward if the car slows down, and thrown sideways if the car changes direction in a tight turn.

An apple falling from a tree is a good example of acceleration that involves only an increase in speed. Initially, at the moment the stem breaks, the apple's speed is zero. After 1 s, its downward speed is 9.8 meters per second, or 9.8 m/s (32 feet per second, or 32 ft/s). After 2 s, the apple's speed is twice this, or 19.6 m/s. After 3 s, the speed is 29.4 m/s. Because the apple's speed increases by 9.8 m/s for each second of free fall, the rate of acceleration is 9.8 meters per second per second, or 9.8 m/s^2 (32 ft/s^2). Thus, Earth's gravity gives the apple a constant acceleration of 9.8 m/s^2 downward, toward the center of Earth.

A planet revolving about the Sun along a perfectly circular orbit is an example of acceleration that involves change of direction only. As the planet moves along its orbit, its speed remains constant. Nevertheless, the planet is continuously being accelerated because its direction of motion is continuously changing.

Newton's second law of motion says that in order to give an object an acceleration (that is, to change its velocity), a net outside force *must* act on the object. To be specific, this law says that the acceleration of an object is proportional to the net outside force acting on the object. That is, the harder you push on an object, the greater the resulting acceleration. This law can be succinctly stated as an equation. If a net outside force F acts on an object of mass m, the object will experience an acceleration a described by the following mathematical equation:

Newton's laws provided a mechanism explaining why Kepler's much earlier laws worked so well.

Newton's second law

$$F = ma$$

F = net outside force on an object
m = mass of object
a = acceleration of object

The **mass** of an object is a measure of the total amount of material in the object. It is usually expressed in kilograms (kg). For example, the mass of the Sun is 2×10^{30} kg, the mass of a hydrogen atom is 1.7×10^{-27} kg, and the mass of an average adult is 75 kg. The Sun, a hydrogen atom, and a person have these masses regardless of where they happen to be in the universe.

CAUTION It is important not to confuse the concepts of mass and weight. Weight is the force of gravity that acts on an object and, like any force, is usually expressed in pounds or newtons (1 newton = 0.225 pound).

We can use Newton's second law to relate mass and weight. We have seen that the acceleration caused by Earth's gravity is 9.8 m/s^2. When a 50-kg swimmer falls from a diving board, the only outside force acting on her as she falls is her weight. Thus, from Newton's second law ($F = ma$), her weight is equal to her mass multiplied by the acceleration due to gravity:

$$50 \text{ kg} \times 9.8 \text{ m/s}^2 = 490 \text{ newtons} = 110 \text{ pounds}$$

Note that this answer is correct only when the swimmer is on Earth. She would weigh less on the Moon, where the gravitational attraction between the swimmer and the Moon is weaker because the Moon has less mass than Earth. Alternatively, she would weigh more on Jupiter because Jupiter has more mass and the gravitational attraction between the two is greater. Floating deep in space, she would have no weight at all; she would be "weightless." Nevertheless, in all these circumstances, she would always have exactly the same mass, because mass is an inherent property of matter regardless of where it is located. Whenever we describe the properties of planets, stars, or galaxies, we speak of their masses, never of their weights.

We have seen that a planet is continually accelerating as it orbits the Sun. From Newton's second law, this means that there must be a net outside force that acts continually on each of the planets. As we will see in the next section, these forces are the gravitational attractions between the planets and our Sun.

CalculationCheck 3-3: **How fast is the space shuttle traveling 3 s after launch if it is accelerating at a rate of about 20 m/s^2?**

Newton's Third Law of Motion

The last of Newton's general laws of motion, called **Newton's third law of motion**, is the famous statement about action and reaction:

Whenever one object exerts a force on a second object, the second object exerts an equal and opposite force on the first object.

For example, if you weigh 110 lb, when you are standing up you are pressing down on the floor with a force of 110 lb. Newton's third law tells us that the floor is also pushing up against your feet with an equal force of 110 lb. You can think of each of these forces as a reaction to the other force, which is the origin of the often used phrase "every action has an equal and opposite reaction."

Newton realized that because each planet is gravitationally attracted to the Sun, the Sun must also be attracted to the planets. However, the planets are much less massive than the Sun (for example, Earth has only 1/300,000 of the Sun's mass). Therefore, although the Sun's attraction to a planet is the same as the planet's attraction to the Sun, the planet's much smaller mass gives it a much larger acceleration and moves it significantly more than the Sun. This is why the planets circle the Sun instead of vice versa. Thus, Newton's laws reveal the reason for our heliocentric solar system.

ConceptCheck 3-18: **If a door on the International Space Station requires 100 newtons of force to be pushed open, and, according to Newton's third law, the door pushes back on an astronaut with an equal but opposite force of 100 newtons, why is it that an astronaut can successfully open the door?**

3-6 Newton's description of gravity accounts for Kepler's laws and explains the motions of the planets

Tie a ball to one end of a piece of string, hold the other end of the string in your hand, and whirl the ball around in a circle. As the ball "orbits" your hand, it is continuously accelerating because its velocity is changing. (Even if its speed is constant, its direction of motion is changing.) In accordance with Newton's second law, this can happen only if the ball is continuously acted on by an outside force—the pull of the string. The pull is directed along the string toward your hand. In the same way Newton saw the gravitational attractions between the planets and the Sun act as a pull toward the significantly more massive Sun. That pull is *gravity,* or *gravitational force.*

Newton's discovery about the forces that act on planets led him to suspect that the force of gravity pulling a falling apple straight down to the ground is fundamentally the same as the force on a planet that is always directed straight at the Sun. In other words, gravity is the force that shapes the orbits of the planets. What is more, he was able to determine how the gravitational attractions between the planets and the Sun depend on the distances between them. His result was a law of gravitation that could apply to the motion of distant planets as well as to the flight of a soccer ball on Earth. Using this law, Newton achieved the remarkable goal of deducing Kepler's laws from fundamental principles of nature.

To see how Newton reasoned, think again about a ball attached to a string. If you use a short string, so that the ball orbits in a small circle, and whirl the ball around your hand at a high speed, you will find that you have to pull fairly hard on the string (**Figure 3-16***a*). But if you use a longer string, so that the ball moves in a larger orbit, and if you make the ball orbit your hand at a slow speed, you only have to exert a light tug on the string (Figure 3-16*b*). The orbits of the planets behave in the same way; the larger the size of

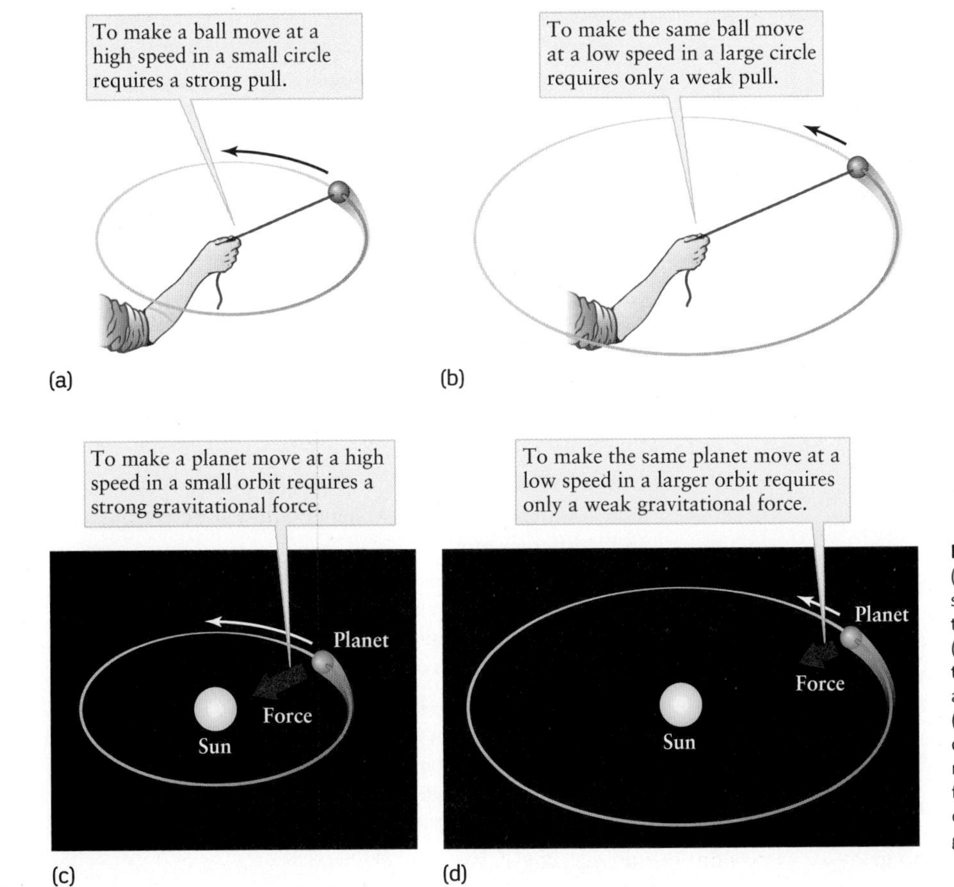

FIGURE 3-16 An Orbit Analogy
(a) To make a ball on a string move at high speed around a small circle, you have to exert a substantial pull on the string. (b) If you lengthen the string and make the same ball move at low speed around a large circle, much less pull is required. (c), (d) Similarly, a planet that orbits close to the Sun moves at high speed and requires a substantial gravitational force from the Sun, while a planet in a large orbit moves at low speed and requires less gravitational force to stay in orbit.

the orbit, the slower the planet's speed (Figure 3-16*c*, *d*). By analogy to the force of the string on the orbiting ball, Newton concluded that the force that attracts a planet toward the Sun must decrease with increasing distance from the Sun.

The Law of Universal Gravitation

Using his own three laws and Kepler's three laws, Newton succeeded in formulating a general statement that describes the nature of the gravitational force. Newton's **law of universal gravitation** is as follows:

Two objects attract each other with a force that is directly proportional to the mass of each object and inversely proportional to the square of the distance between them.

This law states that *any* two objects exert gravitational pulls on each other. Normally, you notice only the gravitational force that Earth exerts on you, otherwise known as your weight. In fact, you feel gravitational attractions to *all* the objects around you. For example, this book is exerting a gravitational force on you as you read it. But because the force exerted on you by this book is proportional to the book's mass, which is very small compared to Earth's mass, the force is too small to notice. (It can actually be measured with sensitive equipment.)

Consider two 1-kg objects separated by a distance of 1 m. Newton's law of universal gravitation says that the force is directly proportional to the mass, so if we double the mass of one object to 2 kg, the force between the objects will double. If we double both masses so that we have two 2-kg objects separated by 1 m, the force will be $2 \times 2 = 4$ times what it was originally (the force is directly proportional to the mass of *each* object). If we go back to two 1-kg masses, but double their separation to 2 m, the force will be only one-quarter its original value. This is because the force is inversely proportional to the square of the distance: If we double the distance, the force is multiplied by a factor of $\frac{1}{2^2} = \frac{1}{4}$.

Newton's law of universal gravitation can be stated more succinctly as an equation. If two objects have masses m_1 and m_2 and are separated by a distance r, then the gravitational force F between these two objects is given by the following equation:

Newton's law of universal gravitation

$$F = G\left(\frac{m_1 m_2}{r^2}\right)$$

F = gravitational force between two objects
m_1 = mass of first object
m_2 = mass of second object
r = distance between objects
G = universal constant of gravitation

If the masses are measured in kilograms and the distance between them in meters, then the force is measured in newtons. In this formula, G is a number called the universal constant of gravitation. Laboratory experiments have yielded a value for G of

$$G = 6.67 \times 10^{-11} \text{ newton} \bullet \text{m}^2/\text{kg}^2$$

We can use Newton's law of universal gravitation to calculate the force with which any two objects attract each other. For example, to compute the gravitational force that the Sun exerts on Earth, we substitute values for Earth's mass ($m_1 = 5.98 \times 10^{24}$ kg), the Sun's mass ($m_2 = 1.99 \times 10^{30}$ kg), the distance between them ($r = 1$ AU $= 1.5 \times 10^{11}$ m), and the value of G into Newton's equation. We get

$$F_{\text{Sun-Earth}} = 6.67 \times 10^{-11}\left[\frac{(5.98 \times 10^{24}) \times (1.99 \times 10^{30})}{(1.50 \times 10^{11})^2}\right]$$
$$= 3.53 \times 10^{22} \text{ newtons}$$

If we calculate the force that Earth exerts on the Sun, we get exactly the same result. (Mathematically, we just let m_1 be the Sun's mass and m_2 be Earth's mass instead of the other way around. The product of the two numbers is the same, so the force is the same.) This is in accordance with Newton's third law: Any two objects exert *equal* gravitational forces on each other.

Your weight is just the gravitational force that Earth exerts on you, so we can calculate it using Newton's law of universal gravitation. Earth's mass is $m_1 = 5.98 \times 10^{24}$ kg, and the distance r is the distance between the *centers* of Earth and you. This is just the radius of Earth, which is $r = 6378$ km $= 6.378 \times 10^6$ m. If your mass is $m_2 = 50$ kg, your weight is

$$F_{\text{Earth-you}} = 6.67 \times 10^{-11}\left[\frac{(5.98 \times 10^{24}) \times (50)}{(6.378 \times 10^6)^2}\right]$$
$$= 490 \text{ newtons}$$

This is the same as the weight of a 50-kg person that we calculated earlier. This example shows that your weight would have a different value on a planet with a different mass m_1 and a different radius r.

ConceptCheck 3-19: How much does the gravitational force of attraction change between two asteroids if the two asteroids drift 3 times closer together?

CalculationCheck 3-4: How much would a 75-kg astronaut, weighing about 165 pounds on Earth, weigh in newtons and in pounds if he was standing on Mars, which has a mass of 6.4×10^{23} kg and a radius of 3.4×10^6 m?

Gravitational Force and Orbits

Go to Video 3-4

Because there is a gravitational force between any two objects, Newton concluded that gravity is also the force that keeps the Moon in orbit around Earth. It is also the force that keeps artificial satellites in orbit. But if the force of gravity attracts two objects to each other, why don't satellites immediately fall to Earth? Why doesn't the Moon fall into Earth? And, for that matter, why don't the planets fall into the Sun?

Let's conduct an imaginary experiment. Suppose you were to imagine dropping a ball from a great height above Earth's surface, as in **Figure 3-17.** After you drop the ball, it, of course, falls straight down (path A in Figure 3-17). But if you *throw* the ball horizontally, it travels some distance across Earth's surface before hitting the ground (path B). If you throw the ball harder, it travels a greater distance (path C). If you could throw at just the right speed, the curvature of the ball's path will exactly match the curvature of Earth's surface (path E). Although Earth's gravity is making the ball fall, Earth's surface is moving away under the ball at the same rate. Hence, the ball does not get any closer to the surface, and we would say that "the ball is in circular orbit." So the ball in path E is in fact falling, but it is falling *around* Earth rather than *toward* Earth.

FIGURE 3-17 An Explanation of Orbits If a ball is dropped from a great height above Earth's surface, it falls straight down (A). If the ball is thrown with some horizontal speed, it follows a curved path before hitting the ground (B, C). If thrown with just the right speed (E), the ball goes into circular orbit; the ball's path curves but it never gets any closer to Earth's surface. If the ball is thrown with a speed that is slightly less (D) or slightly more (F) than the speed for a circular orbit, the ball's orbit is an ellipse.

A spacecraft is launched into orbit in just this way—by shooting it fast enough. The thrust of a rocket is used to give the spacecraft the necessary orbital speed. Once the spacecraft is in orbit, the rocket engines are turned off and the spacecraft falls continually around Earth.

CAUTION An astronaut on board an orbiting spacecraft feels "weightless." However, this is *not* because she is "beyond the pull of gravity." The astronaut is herself an independent satellite orbiting Earth, and Earth's gravitational pull is what holds her in orbit. She feels "weightless" because she and her spacecraft are falling *together* around Earth, so there is nothing pushing her against any of the spacecraft walls. You feel the same "weightless" sensation whenever you are falling, such as when you jump off a diving board into a swimming pool or ride one of the free-fall rides at an amusement park.

If the ball in Figure 3-17 is thrown with a slightly slower speed than that required for a circular orbit, its orbit will be an ellipse (path D). An elliptical orbit also results if instead the ball is thrown a bit too fast (path F). In this way, spacecraft can be placed into any desired orbit around Earth by adjusting the initial rocket thrust.

Just as the ball in Figure 3-17 will not fall to Earth if given enough speed, the Moon does not fall to Earth and the planets do not fall into the Sun. The planets acquired their initial speeds around the Sun when the solar system first formed 4.56 billion years ago. Figure 3-17 shows that a circular orbit is a very special case, so it is no surprise that the orbits of the planets are not precisely circular.

CAUTION Orbiting satellites do sometimes fall out of orbit and crash back to Earth. When this happens, however, the real culprit is not gravity but air resistance. A satellite in a relatively low orbit is actually flying through the tenuous outer wisps of Earth's atmosphere. The resistance of the atmosphere slows the satellite and changes a circular orbit like path E in Figure 3-17 to an elliptical one like path D. As the satellite sinks to lower altitude, it encounters more air resistance and sinks even lower. Eventually, it either strikes Earth or burns up in flight due to air friction. By contrast, the Moon and planets orbit in the near vacuum of interplanetary space. Hence, they are unaffected by this kind of air resistance, and their orbits are much more long lasting.

ConceptCheck 3-20: What keeps the International Space Station from crashing into Earth when it has no rocket engines constantly pushing it around Earth?

VISUAL LITERACY TASK

Kepler's Laws

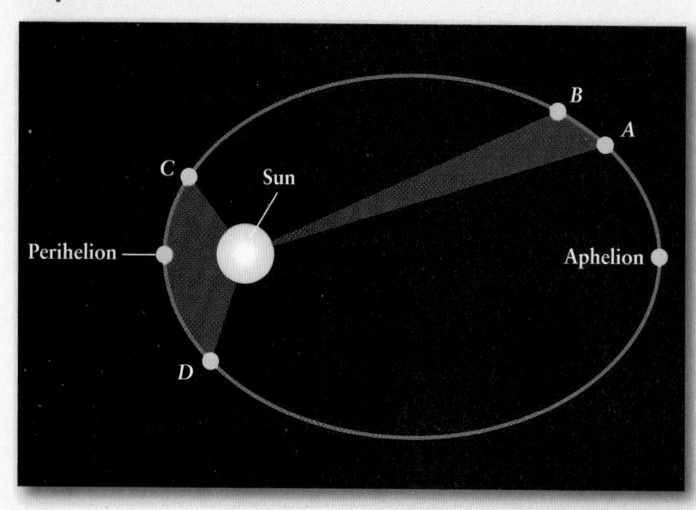

PROMPT: What would you tell a fellow student who said, **"Planets move fastest when farthest from the Sun because there is more gravity at great distances"?**

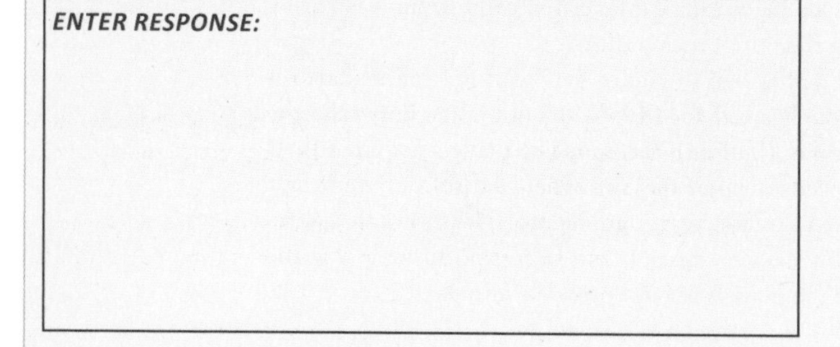

Guiding Questions:

1. Planets move about our Sun
   a. fastest when farthest from our Sun.
   b. fastest when closest to our Sun.
   c. at the same speed no matter their distance.
   d. slower for relatively smaller orbits.

2. The gravitational attraction between an orbiting planet and our Sun is greatest for
   a. longer distances.
   b. shorter distances.
   c. the more rapidly spinning planets.
   d. planets with the thickest atmospheres.

3. For a given amount of time, a line drawn between a planet and the Sun will sweep out
   a. the same size area regardless of distance.
   b. a greater area when close to the Sun.
   c. a smaller area when close to the Sun.
   d. an area that always covers the ellipse's second focus point.

4. With an average distance of about 150 million km (93 million mi), the difference between Earth's closest approach and most distant positions in its elliptical orbit about the Sun is roughly
   a. 5 km (4 mi).
   b. 5 million km (4 million mi).
   c. 50 million km (40 million mi).
   d. 500 million km (400 million mi).

KEY IDEAS AND TERMS

3-1 Astronomers of antiquity used observation and reasoning to develop astonishing advances in the study of astronomy

- Ancient Greeks knew Earth was round because of Earth's curved shadow on the Moon during a lunar eclipse.
- Eratosthenes estimated the size of Earth by comparing noontime shadow lengths at different locations on Earth.
- Aristarchus determined relative distances to the Sun and Moon by carefully measuring the speed of Earth's shadow during an eclipse and the angle between the Sun and a first quarter moon.
- **Powers of ten** are used to express large and small numbers where 10^3 is the same as 1000 and noted using the prefix kilo-.
- An **astronomical unit** is the average distance between Earth and the Sun.
- A **light-year** is the distance light travels in one year in empty space.
- A concept of Earth at the center of the universe with planets, the Sun, and stars orbiting around Earth is known as a **geocentric model.**
- **Retrograde motion** is the apparent westward movement of planets normally traveling from west to east against the apparently stationary background stars.
- Ptolemy imagined a universe where planets could change apparent direction of motion by moving in tiny circular orbits attached to larger and larger circular orbits.

3-2 Nicolaus Copernicus devised the first comprehensive Sun-centered model

- The idea that the most simple and most straightforward explanation of observations in nature is most likely to be correct is called **Occam's razor.**

- Copernicus advocated a conception of the universe with our Sun at the center of the universe and Earth orbiting the Sun, just like the other planets, known as the **heliocentric model.**
- Viewed from Earth, **greatest eastern elongation** is the farthest a planet can be observed from the setting Sun.
- Viewed from Earth, **greatest western elongation** is the farthest a planet can be observed from the rising Sun.
- Planets are at **inferior conjunction** when they are between Earth and the Sun.
- A planet is at **superior conjunction** when the Sun is between Earth and the planet.
- Planets are at **opposition** when Earth is between the planet and the Sun.
- The length of time for a planet to orbit the Sun is known as its **period.**
- The **synodic period** is the time that elapses between two successive identical configurations as seen from Earth—from one opposition to the next, for example, or from one conjunction to the next.
- The **sidereal period** is the true orbital period of a planet, the time it takes the planet to complete one full orbit of the Sun relative to the very distant background stars.

3-3 Galileo's discoveries of moons orbiting Jupiter and the phases of Venus strongly supported a heliocentric model

- Galileo used his telescope to observe moons orbiting Jupiter, demonstrating Earth was not the center of all orbits.
- Galileo's telescope observations of the phases of Venus demonstrated Venus could not be orbiting Earth, but must be orbiting the Sun instead.

3-4 Johannes Kepler proposed that planets orbit the Sun in elliptical paths, moving fastest when closest to the Sun, with the closest planets moving at the highest speeds

- **Kepler's first law** is that the orbit of a planet about the Sun is an **ellipse** with the Sun at one **focus.**
- **Eccentricity** is a measure of the relative roundness of an elliptical shape.
- The **semimajor axis,** *a,* of a planet's orbit is the average distance between the planet and the Sun.
- The closest an orbiting planet gets to its central star is its **perihelion** distance, while the farthest an orbiting planet gets from its central star is its **aphelion** distance.
- **Kepler's second law** is that a line joining a planet and the Sun sweeps out equal areas in equal intervals of time and is also known as the **law of equal areas.**
- **Kepler's third law** says that the larger a planet's orbit—that is, the larger the semimajor axis, or average distance from the planet to the Sun—the longer the sidereal period, which is the time it takes the planet to complete an orbit and is sometimes written as $P^2 = a^3$.

3-5 Isaac Newton formulated three laws relating force and motion to describe fundamental properties of physical reality

- **Newton's first law of motion** is that an object remains at rest, or moves in a straight line at a constant **speed,** unless acted upon by a net outside force.
- **Newton's second law of motion** says that in order to give an object an **acceleration** (that is, to change its **velocity**), a net outside force must act on the object and is often written as $F = ma$.
- **Newton's third law of motion** is the famous statement about action and reaction, which states whenever one object exerts a force on a second object, the second object exerts an equal and opposite force on the first object.

3-6 Newton's description of gravity accounts for Kepler's laws and explains the motions of the planets

- Newton's **law of universal gravitation** is that any two objects attract each other with a force that is directly proportional to the **mass** of each object and inversely proportional to the square of the distance between them.
- Planets do not fall into the Sun during their orbits because they have a sufficiently high velocity moving them forward.

QUESTIONS

Review Questions

1. Write the following numbers using powers-of-ten notation:
   a) 10 million
   b) 60 thousand
   c) four one-thousandths
   d) 38 billion
   e) your age in months
2. How is an astronomical unit (AU) defined? Give an example of a situation in which this unit of measure would be convenient to use.
3. Give the word or phrase that corresponds to the following standard abbreviations:
   a) km
   b) cm
   c) s
   d) km/s
   e) mi/h
   f) m
   g) m/s
   h) h
   i) ly
   j) g
   k) kg

   Which of these are units of speed? (*Hint:* You may have to refer to a dictionary. All of these abbreviations should be part of your working vocabulary.)
4. A reporter once described a light-year as "the time it takes light to reach us traveling at the speed of light." How would you correct this statement?
5. When the *Voyager 2* spacecraft sent back pictures of Neptune during its flyby of that planet in 1989, the spacecraft's radio signals traveled for 4 hours at the speed of light to reach Earth. How far away was the spacecraft? Give your answer in kilometers, using powers-of-ten notation. (*Hint:* See the preceding question.)
6. How did the ancient Greeks explain why the Sun and the Moon slowly change their positions relative to the background stars?
7. In what direction does a planet move relative to the stars when it is in direct motion? When it is in retrograde motion? How do these compare with the direction in which we see the Sun move relative to the stars?
8. (a) In what direction does a planet move relative to the horizon over the course of one night? (b) The answer to (a) is the same whether the planet is in direct motion or retrograde motion. What does this tell you about the speed at which planets move on the celestial sphere?
9. What is the significance of Occam's razor as a tool for analyzing theories?
10. How did the models of Aristarchus and Copernicus explain the retrograde motion of the planets?
11. At what configuration (for example, superior conjunction, greatest eastern elongation, and so on) would it be best to observe Mercury or

Venus with an Earth-based telescope? At what configuration would it be best to observe Mars, Jupiter, or Saturn? Explain your answers.

12. Which planets can never be seen at opposition? Which planets can never be seen at inferior conjunction? Explain your answers.
13. What is the difference between the synodic period and the sidereal period of a planet?
14. What are the foci of an ellipse? If the Sun is at one focus of a planet's orbit, what is at the other focus?
15. What are Kepler's three laws? Why are they important?
16. At what point in a planet's elliptical orbit does it move fastest? At what point does it move slowest? At what point does it sweep out an area at the fastest rate?
17. The orbit of a spacecraft about the Sun has a perihelion distance of 0.1 AU and an aphelion distance of 0.4 AU. What is the semimajor axis of the spacecraft's orbit? What is its orbital period?
18. A comet with a period of 125 years moves in a highly elongated orbit about the Sun. At perihelion, the comet comes very close to the Sun's surface. What is the comet's average distance from the Sun? What is the farthest it can get from the Sun?
19. What observations did Galileo make that reinforced the heliocentric model? Why could these observations not have been made before Galileo's time?
20. Why does Venus have its largest angular diameter when it is new and its smallest angular diameter when it is full?
21. What are Newton's three laws? Give an everyday example of each law.
22. How much force do you have to exert on a 3-kg brick to give it an acceleration of 2 m/s^2? If you double this force, what is the brick's acceleration? Explain your answer.
23. Suppose that Earth were moved to a distance of 3.0 AU from the Sun. How much stronger or weaker would the Sun's gravitational pull be on Earth? Explain your answer.
24. In 2006, Mercury was at greatest western elongation on April 8, August 7, and November 25. It was at greatest eastern elongation on February 24, June 20, and October 17. Does Mercury take longer to go from eastern to western elongation, or vice versa? Explain why, using Figure 3-8.
25. The mass of the Moon is 7.35×10^{22} kg, while that of Earth is 5.98×10^{24} kg. The average distance from the center of the Moon to the center of Earth is 384,400 km. What is the size of the gravitational force that Earth exerts on the Moon?
26. What is the size of the gravitational force that the Moon exerts on Earth? How do your answers compare with the force between the Sun and Earth calculated in the text?

Web Chat Questions

1. Which planet would you expect to exhibit the greatest variation in apparent brightness as seen from Earth? Which planet would you expect to exhibit the greatest variation in angular diameter? Explain your answers.
2. What do you believe to be Galileo's single most important astronomical observation, and why it was most important?

Collaborative Exercises

1. Use two thumbtacks, a loop of string, and a pencil to draw several ellipses. Describe how the shape of an ellipse varies as the distance between the thumbtacks changes.
2. Use data from the appendix to determine how many Martian years old each member of your group would be if they were born on Mars.
3. Considering where your group is sitting right now, how many times dimmer would an imaginary, super-deluxe, ultrabright flashlight be if it were located at the front door of the group member who lives farthest away as compared to if it were at the front door of the group member who lives closest. Explain your reasoning.
4. Galileo's *Dialogue Concerning the Two World Chief Systems* described fictional conversations between three people. Create a short play using this style, describing Kepler's laws of planetary motion using each person in your group.
5. Astronomers use powers of ten to describe the distances to objects. List an object or place that is located at very roughly each of the following distances from you: 10^{-2} m, 100 m, 10^{1} m, 10^{3} m, 10^{7} m, 10^{10} m, and 10^{20} m.

Observing Questions

1. Use the *Starry Night*™ program to observe retrograde motion. Select **Favourites > Explorations > Retrograde** from the menu. The view from Earth is centered upon Mars against the background of stars and the framework of star patterns within the constellations. The **Time Flow Rate** is set to 1 day. Click **Play** and observe Mars as it moves against the background constellations. An orange line traces Mars's path in the sky from night to night. Watch the motion of Mars for at least two years of simulated time. Since the view is centered upon and tracks Mars in the view, the sky appears to move but the relative motion of Mars against this sky is obvious.
   a) For most of the time, does Mars move generally to the left (eastward) or to the right (westward) on the celestial sphere? Select **File > Revert** from the menu to return to the original view. Use the time controls in the toolbar (**Play, Step time forward,** and **Step time backward**) along with the **Zoom** controls (+ and − buttons at the right of the toolbar or the mouse wheel) to determine when Mars's usual *direct* motion ends, when it appears that Mars comes to a momentary halt in the west-east direction, and when *retrograde* motion begins. On what date does retrograde motion end and direct motion resume?
   b) You have been observing the motion of Mars as seen from Earth. You can observe the motion of Earth as seen from Mars by locating yourself on the north pole of Mars. Select **Favourites > Explorations > Retrograde Earth** from the menu. The view is centered upon and will track Earth as seen from Mars, beginning on **June 23, 2010.** Click the **Play** button. As before, watch the motion for two years of simulated time. In which direction does Earth appear to move for most of the time? On what date does its motion change from direct to retrograde? On what date does its motion change from retrograde back to direct? Are these roughly the same dates as you found in part (a)?
   c) To understand the motions of Mars as seen from Earth and vice versa, observe the motion of the planets from a point above the solar system. Select **Favourites > Explorations > Retrograde Overview** from the menu. This view, from a position 5 AU above the plane of the solar system, is centered upon the Sun and the orbits, and positions of Mars and Earth on June 23, 2010, are shown. Click **Play** and watch the motions of the planets for two years of simulated time. Note that Earth catches up with and overtakes Mars as time proceeds. This relative motion of the two planets leads to our observation of retrograde motion. On what date during this two-year period is Earth directly between

Mars and the Sun? How does this date compare to the two dates you recorded in part (a) and the two dates you recorded in part (b)? Explain the significance of this.

2. Use *Starry Night*™ to observe the phases of Venus and of Mars as seen from Earth. Select **Favourites > Explorations > Phases of Venus** and click the **Now** button in the toolbar to see an image of Venus if you were to observe it through a telescope from Earth right at this moment.
   a) Draw the current shape (phase) of Venus. With the **Time Flow Rate** set to **30 days, Step time forward,** drawing Venus to scale at each step. Make a total of 20 time steps and drawings.
   b) From your drawings, determine when the planet is nearer or farther from Earth than is the Sun.
   c) Deduce from your drawings when Venus is coming toward us or is moving away from us.
   d) Explain why Venus goes through this particular cycle of phases. Select **Favourites > Explorations > Phases of Mars** and click the **Now** button in the toolbar. With the **Time Flow Rate** set to **30 days, Step time forward,** and observe the changing phase of Mars as seen from Earth.
   e) Compare this with the phases that you observed for Venus. Why are the cycles of phases as seen from Earth different for the two planets?
3. Use the *Starry Night*™ to observe the moons of Jupiter. Select **Favourites > Explorations > Galilean Moons** from the menu.
   a) With the **Time Flow Rate** set to **2 hours,** use the **Step time forward** button (just to the right of the **Play** button) to observe and draw the positions of the moons relative to Jupiter at 2-hour intervals. From your drawings, which moon orbits closest to Jupiter and which orbits farthest away? Explain your reasoning.
   b) Determine the periods of the orbits of these moons (change the **Time Flow Rate** if necessary).
   c) Are there times when one or more of the satellites are not visible? What happens to the moons at those times?
4. Use *Starry Night*™ to observe the orbits of the planets of the inner solar system. Open **Favourites > Explorations > Kepler.** The view is centered upon the Sun from a position in space 2.486 AU above the plane of the solar system and shows the Sun and the inner planets and their orbits, as well as many asteroids in the asteroid belt beyond the orbit of Mars. Click the **Play** button and observe the motions of the planets from this unique location.
   a) Make a list of the planets visible in the view in the order of increasing distance from the Sun.
   b) Make a list of the planets visible in the view in the order of increasing orbital period.
   c) How do the lists compare?
   d) What might you conclude from this observation?
   e) Which of Kepler's Laws accounts for this observation?

ANSWERS

ConceptChecks

ConceptCheck 3-1: If the noontime position of the Sun on the summer solstice was much closer to being directly overhead, he would assume that Earth was much larger because Earth's surface between the two cities would be much less curved, implying a much larger, circular Earth.

ConceptCheck 3-2: When astronomers say a planet is moving retrograde, a planet is observed night after night to be slowly drifting from east to west compared to the very distant background stars. This is opposite how it typically appears to move. Regardless, all objects always appear to rise in the east and set in the west on a daily basis.

ConceptCheck 3-3: The Greeks' ancient geocentric model used a nonspinning, stationary Earth where the stars, planets, and the Sun all moved around Earth.

ConceptCheck 3-4: In the Ptolemaic model, the planets are continuously orbiting in circles such that they appear to move backward for a brief time. However, the planets never actually stop and change their directions.

ConceptCheck 3-5: No. Planets only appear to move in retrograde motion if seen as two planets moving at different speeds passing one another. An imaginary observer on the stationary Sun would only see planets moving in the same direction as they orbit the Sun.

ConceptCheck 3-6: When Venus is visible in the western sky, it is at its greatest eastern elongation and the faster-moving Venus has not yet caught up with Earth, as Earth is momentarily in front of Venus as they orbit the Sun.

ConceptCheck 3-7: Mars has an orbit around the Sun that is larger than Earth's orbit. As a result, Mars never moves to a position between Earth and the Sun, so Mars never is at inferior conjunction.

ConceptCheck 3-8: In Copernicus's model, the more distant planets are moving slower than the planets closer to the Sun. As a faster-moving Earth moves past a slower-moving Mars, there is a brief time in which Mars appears to move backward through the sky. However, the planets never actually stop and change their directions.

ConceptCheck 3-9: Slowly moving Jupiter does not move very far along its orbit in the length of time it takes for Earth to pass by Jupiter, move around the Sun, and pass by Jupiter again, giving Jupiter a synodic period similar to Earth's orbital period around the Sun. In much the same way, slow-moving Jupiter takes more than a decade to move around the Sun back to its original starting place as measured by the background stars, giving it a large sidereal period.

ConceptCheck 3-10: Copernicus's model was no more accurate at predicting the positions and motions of the planets than Ptolemy's model. However, Copernicus's model turned out to be more closely related to the actual motions of the planets around the Sun than was Ptolemy's model of planets orbiting Earth.

ConceptCheck 3-11: When Venus is on the opposite side of the Sun from Earth, it will be in a full or gibbous phase. The full phase can occasionally be observed because Earth and Venus do not orbit the Sun in the same exact plane.

ConceptCheck 3-12: Galileo was the first person to use and widely share his observations using a telescope, which is necessary in order to observe Jupiter's tiny moons.

ConceptCheck 3-13: An ellipse with an eccentricity of zero is a perfect circle and, compared to Mars's $e = 0.093$, the eccentricity of Venus's orbit is $e = 0.007$. Venus has an eccentricity value closer to zero so its orbit is closer to a perfect circle in shape.

ConceptCheck 3-14: Kepler's second law says that objects are moving slowest when they are farthest from the object they are orbiting, so an Earth-orbiting satellite will move slowest when it is farthest from Earth.

ConceptCheck 3-15: According to Kepler's third law, planets closer to the Sun move faster than planets farther from the Sun. For objects orbiting

Earth, the object closer to Earth's surface is also moving the fastest, which, in this case, is the space shuttle.

ConceptCheck 3-16: No. Kepler's laws of planetary motion apply to any objects in space that orbit around another object, including comets orbiting the Sun, man-made satellites and moons orbiting planets, and even stars orbiting other stars.

ConceptCheck 3-17: According to Newton's first law, if no outside forces are acting on an object, then it will remain in that state of motion. In other words, a spacecraft in outer space will continue at the same speed as it moves far from our Sun's gravitational influence.

ConceptCheck 3-18: Newton's third law states that when one object exerts a force on another object, the second object exerts an equal but opposite force on the first. In other words, if the astronaut touches the door, the door touches the astronaut. The fact that the forces are equal does not mean that the "effects" are equal. If the only force on the door is the one applied by the astronaut, the door will move, following Newton's laws of motion.

ConceptCheck 3-19: According to Newton's universal law of gravitation, the gravitational attraction between two objects depends on the square of the distance between them. In this case, if the asteroids drift 3 times closer together, then the gravitational force of attraction between them increases 3^2 times, or, in other words, becomes 9 times greater.

ConceptCheck 3-20: The International Space Station has an initial forward velocity such that as it falls around Earth, it actually misses Earth because Earth's round surface is curved away.

CalculationChecks

CalculationCheck 3-1: Because the diameters of the Sun and Moon must be in the same proportion as their distances, if Aristarchus assumed the Sun to be 100 times farther away when they appeared to be the same diameter in the sky, he would have proposed that the Sun is 100 times larger than the Moon. Today, we know that the Sun is more than 400 times farther away from Earth, and 400 times larger than the Moon.

CalculationCheck 3-2: According to Kepler's third law, $P^2 = a^3$. So, if $P^2 = (39.5)^3$, then $P = 39.5^{3/2} = 248$ years.

CalculationCheck 3-3: Acceleration is how much the velocity of an object is changing every second. If the space shuttle starts at a velocity of zero on the launch pad and increases its velocity 20 m/s every second, then after 3 s, the space shuttle is moving at roughly 60 m/s (135 mph).

CalculationCheck 3-4: Using Newton's universal law of gravitation, $F_{\text{Mars-astronaut}} = G(m_{\text{Mars}}) \times (m_{\text{astronaut}}) \div (\text{radius})^2 = 6.67 \times 10^{-11} \times 6.4 \times 10^{23} \times 75 \div (3.4 \times 10^6)^2 = 277$ N, which we can covert to pounds because $277 \text{ N} \times 0.255 \text{ lb/N} = 76$ lb.

The *Mars Reconnaissance Orbiter* spacecraft circling Mars (artist's impression). *(NASA/ JPL-Caltech)*

Exploring Our Evolving Solar System

CHAPTER LEARNING OBJECTIVES **By reading the sections of this chapter, you will learn:**

4-1 The solar system has two broad categories of planets orbiting our Sun: terrestrial (Earthlike) and Jovian (Jupiterlike)

4-2 Seven large moons are almost as big as the inner, terrestrial planets

4-3 Spectroscopy reveals the chemical composition of the planets

4-4 Small chunks of rock and ice also orbit the Sun: asteroids, trans-Neptunian objects, and comets

4-5 The Sun and planets formed from a rotating solar nebula

4-6 The planets formed by countless collisions of dust, rocks, and gas in the region surrounding our young Sun

4-7 Understanding how our planets formed around the Sun suggests planets around other stars are common

Fifty years ago, astronomers knew precious little about the worlds that orbit our Sun. Even the best telescopes provided images of the planets that were frustratingly hazy and indistinct. Of smaller objects—asteroids, comets, and the moons orbiting planets—we knew even less.

Today, our knowledge of what makes up the solar system has grown exponentially, due almost entirely to robotic spacecraft. Spacecraft have been sent to fly past all the planets and many of their moons at close range, revealing details unimagined by astronomers of an earlier generation. We have landed spacecraft on the Moon, Venus, and Mars and dropped probes into the atmospheres of Jupiter and Saturn's moon Titan. This is truly the golden age of solar system exploration.

We have two goals in this chapter. First, we paint a broad outline of our present understanding of what makes up the solar system. Then, we will discuss how it came to be—that is, our current best *theory* of the origin of the solar system. We will also consider the nature of nearly invisible planets recently discovered orbiting other stars far beyond our own solar system. Recall from the Chapter 1 that a theory is not merely a set of wild speculations, but a collection of scientific ideas that must be able to be tested and verified by other scientists. Since no humans were present to witness the formation of the planets, scientists must base their theories of solar system origins on their observations of the present-day solar system. (In an analogous way, paleontologists base their understanding of the lives of dinosaurs on the evidence provided by fossils that have survived to the present day.) In so doing, they are following the spirit of the scientific method that we described at the beginning of this text. ■

4-1 The solar system has two broad categories of planets orbiting the Sun: terrestrial (Earthlike) and Jovian (Jupiterlike)

Go to Video 4-1

What do you think of when someone says "solar system"? Do you think of planets, like Earth and Saturn, or of millions of stars and enormous glowing clouds of interstellar gas? When astronomers speak of our solar system, they are specifically talking about the system of planets and other debris that orbit around just a single star, our Sun. Other stars in the night sky also might have planets orbiting them, but those stars are far, far beyond our solar system.

Planetary orbits appear very nearly circular, with the most distant planets moving much slower around the Sun than the innermost planets.

One should pause for a moment and consider, "What is a **planet**?" You probably have heard that there has been considerable debate about whether or not the object called Pluto is a planet. We certainly won't resolve that debate here, but for now, let's consider a planet being an object that has at least two important characteristics. The first is that it is an object that orbits a star. The second is that it is large enough that its own gravity has shaped the object into a sphere. These are defining characteristics nearly all astronomers agree upon, but you can imagine that there are some other characteristics a planet might have that are less agreed upon.

By international agreement, our solar system is defined as currently having eight orbiting planets, with Pluto no longer being officially designated as a planet. Each of the planets orbiting our Sun is unique. Only Earth has liquid water spread widely across its surface and, at the same time, a nitrogen- and oxygen-dominated atmosphere breathable by humans. Only Venus has a perpetual cloud layer made of sulfuric acid droplets. Only Jupiter has immense storm systems that visibly persist for centuries. But there are also striking similarities among the planets. Volcanoes are found not only on Earth but also on Venus, Mars, and several moons orbiting our giant planets; rings encircle Jupiter, Saturn, Uranus, and Neptune; and craters dot the surfaces of Mercury, Venus, Earth, and Mars, showing that all of these planets have been bombarded by interplanetary debris.

How can we make sense of the many similarities and differences among the planets? An important step is to organize our knowledge of the planets in a systematic way. First, we can compare the orbits of different planets around the Sun. Second, we can compare the physical properties of the planets.

ConceptCheck 4-1: How many stars are in our solar system?

Comparing the Planets: Orbits

The planets fall naturally into two classes according to the sizes of their orbits. As **Figure 4-1** shows, the orbits of Mercury, Venus, Earth, and Mars are crowded in quite close to the Sun, and these planets are known as the four inner planets. In contrast, the orbits of the next four planets, Jupiter, Saturn, Uranus, and Neptune, known as the outer planets, are widely spaced from each other and orbit at great distances from the Sun. **Table 4-1** lists the orbital characteristics of these eight planets. If you are looking for Pluto, we haven't forgotten it, but we will need to describe that object later because it has characteristics that make it fundamentally different than these first eight planets.

CAUTION While Figure 4-1 shows the orbits of the planets, it does not show the planets themselves. The reason is straightforward: If Jupiter, the largest of the planets, were to be drawn to the same scale as the rest of this figure, it would be a tiny dot just 0.0002 cm across—about 1/300 of the width of a human hair and far too small to be seen without a microscope. The planets themselves are *very* small compared to the distances between them. Indeed, while light can travel from the Sun to Earth in slightly more than 8 minutes, it takes light more than 10 times that, nearly an hour and a

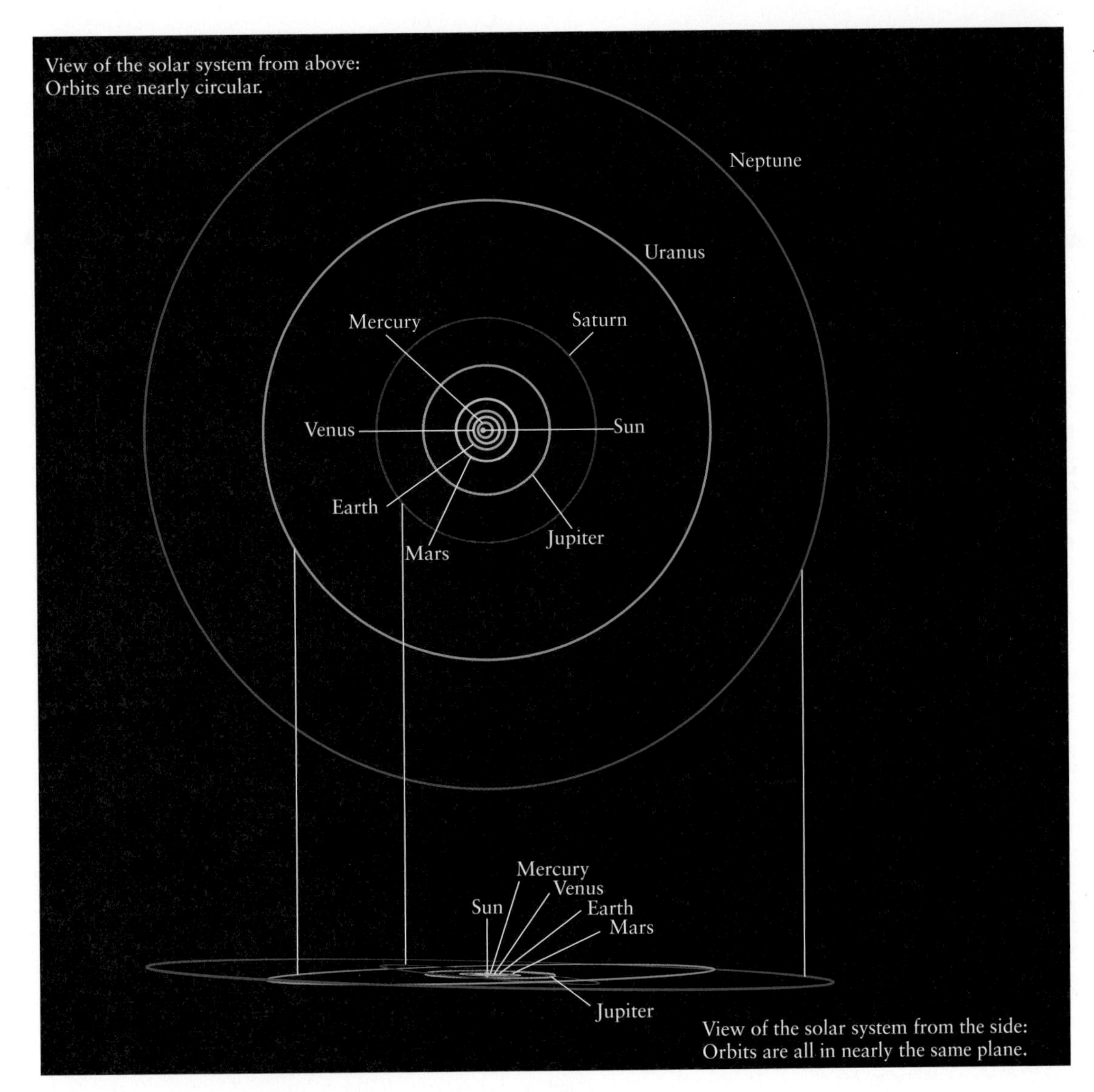

FIGURE 4-1 The Solar System to Scale This scale drawing shows the orbits of the planets around the Sun. The four inner planets are crowded in close to the Sun, while the four outer planets orbit the Sun at much greater distances. On the scale of this drawing, the planets themselves would be much smaller than the diameter of a human hair and too small to see.

half, to travel from the Sun to Saturn. The solar system is a very large and mostly empty place!

Most of the planets have orbits that are nearly circular. As we learned in Chapter 3, Johannes Kepler discovered in the seventeenth century that these orbits are actually ellipses. Astronomers denote the elongation of an ellipse by its *eccentricity* (see Figure 3-14). The eccentricity of a circle is zero, and indeed most of our planets—with the notable exception of Mercury—have orbital eccentricities that appear to be nearly indistinguishable from a perfect circle.

If you could observe the solar system from a point several astronomical units (AU) above Earth's north pole, you would see that all the planets orbit the Sun in the same counterclockwise direction. Furthermore, their orbital paths all lie in nearly the same plane (Figure 4-1). In other words, these orbits are inclined at only slight angles to the plane of the ecliptic, which is the plane of Earth's orbit around the Sun. What is more, the plane of the Sun's equator is very closely aligned with the orbital planes of the planets. As we will see later in this chapter, these near-alignments are not a coincidence. They provide important clues about the origin of the solar system.

ConceptCheck 4-2: **Is Mars classified as an inner planet or an outer planet?**

CalculationCheck 4-1: **Using Table 4-1 on the following page, which of the planets has an orbital path that is most closely a perfect circle in shape?**

Comparing the Planets: Physical Properties

When we compare the physical properties of the planets, we again find that they fall naturally into two classes. All four inner planets have hard, rocky surfaces with mountains,

Table 4-1 Characteristics of the Planets

The Inner (Terrestrial) Planets	Mercury	Venus	Earth	Mars
Average distance from Sun (10^6 km)	57.9	108.2	149.6	227.9
Average distance from Sun (AU)	0.387	0.723	1	1.524
Orbital period (years)	0.241	0.615	1	1.88
Orbital eccentricity	0.206	0.007	0.017	0.093
Inclination of orbit to the ecliptic	7.00°	3.39°	0.00°	1.85°
Equatorial diameter (km)	4880	12,104	12,756	6794
Equatorial diameter (Earth = 1)	0.383	0.949	1	0.533
Mass (kg)	3.302×10^{23}	4.868×10^{24}	5.974×10^{24}	6.418×10^{23}
Mass (Earth = 1)	0.0553	0.815	1	0.1074
Average density (kg/m^3)	5430	5243	5515	3934

The Outer (Jovian) Planets	Jupiter	Saturn	Uranus	Neptune
Average distance from Sun (10^6 km)	778.3	1429	2871	4498
Average distance from Sun (AU)	5.203	9.554	19.194	30.066
Orbital period (years)	11.86	29.46	94.1	164.86
Orbital eccentricity	0.048	0.053	0.043	0.01
Inclination of orbit to the ecliptic	1.30°	2.48°	0.77°	1.77°
Equatorial diameter (km)	142,984	120,536	51,118	49,528
Equatorial diameter (Earth = 1)	11.209	9.449	4.007	3.883
Mass (kg)	1.899×10^{27}	5.685×10^{26}	8.682×10^{25}	1.024×10^{26}
Mass (Earth = 1)	317.8	95.16	14.53	17.15
Average density (kg/m^3)	1326	687	1318	1638

craters, valleys, and volcanoes. These planets are also known as **terrestrial,** or Earthlike, planets. You could stand on the surface of any one of them, although you would need a protective spacesuit on Mercury, Venus, or Mars. The four outer planets resemble Jupiter and are often referred to as **Jovian,** or Jupiterlike, planets (Jove is the mythological name for king of the Roman gods). An attempt to land a spacecraft on the surface of any of the Jovian planets would be futile, because the materials of which these planets are made are mostly gaseous or liquid. The visible "surface" features of a Jovian planet are actually cloud formations in the planet's atmosphere. The photographs in **Figure 4-2** show the distinctive appearances of the two classes of planets.

The most apparent differences between the inner and outer planets are their *diameters*. The diameter of a planet is the distance from one side of the planet to the other. The four Jovian planets are much larger, with much greater diameters, than the terrestrial planets. First place goes to Jupiter, whose diameter across its equator is more than 11 times that of Earth. On the other end of the scale, Mercury's diameter is less than two-fifths that of Earth. Figure 4-2 shows the Sun and the planets drawn to the same scale. The diameters of the planets are given in Table 4-1.

The *masses* of the two categories of planets are also dramatically different. As we saw in Chapter 3, a planet's gravitational attraction to other objects is proportional to its mass. Gravitational attraction between a moon and a planet is greater for the most massive of planets. So, if a planet has a moon like our Moon, astronomers can determine the planet's mass from measurements of how long it takes a moon to orbit the planet for a given distance. In a similar way, astronomers have also measured the mass of each planet by sending a spacecraft to pass near the planet. The planet's gravitational pull deflects the spacecraft's path, and the amount of deflection tells us the planet's mass. Using

FIGURE 4-2 R I V U X G **The Planets to Scale** This figure shows the planets from Mercury to Neptune to the same scale. The four terrestrial planets have orbits nearest the Sun, and, orbiting much farther away, the Jovian planets are the next four planets from the Sun. *(Calvin J. Hamilton and NASA/JPL)*

these techniques, astronomers have found that the four Jovian planets have masses that are tens or hundreds of times greater than the mass of any of the terrestrial planets. Again, first place goes to Jupiter, whose mass is 318 times greater than Earth's mass.

Once we know the diameter and mass of a planet, we can learn something about what that planet is made of on the inside. The trick is to calculate the planet's **average density,** or mass divided by volume, measured in kilograms per cubic meter (kg/m^3). The average density of any substance depends in part on that substance's composition. For example, air near sea level on Earth has an average density of 1.2 kg/m^3, water's average density is 1000 kg/m^3, and a piece of concrete has an average density of 2000 kg/m^3.

The four inner, terrestrial planets have very high average densities (see Table 4-1); the average density of Earth, for example, is 5515 kg/m^3. By contrast, a typical rock found on Earth's surface has a lower average density, about 3000 kg/m^3. Thus, Earth must contain a large amount of material that is denser than rock. The most common material in the solar system denser than Earth's surface rocks is iron. Thus, for Earth to have a density higher than its surface rocks, its core must be dense. This information provides our first clue that terrestrial planets have dense iron cores.

In sharp contrast, the outer, Jovian planets have quite low densities. Saturn has an average density less than that of water. This information strongly suggests that the giant outer planets are composed primarily of light elements such as hydrogen and helium. All four Jupiterlike planets probably have large cores of mixed rock and highly compressed water buried beneath low-density outer layers tens of thousands of kilometers thick.

We can conclude that the following general rule applies to the planets:

The inner planets are made of rocky materials and have dense iron cores, which give these planets high average densities. The outer planets are composed primarily of light elements such as hydrogen and helium, which gives these planets low average densities.

ConceptCheck 4-3: If a planet's density, estimated by measuring how much a planet's gravitation deflects a nearby passing spacecraft's pathway, has the same value as the density of rocks recovered from its surface, what can one infer about the composition of the planet's core?

CalculationCheck 4-2: If Earth's diameter is 12,756 km and Saturn's diameter is 120,536 km, how many Earths could fit across the diameter of Saturn?

4-2 Seven large moons are almost as big as the inner, terrestrial planets

Looking up at Earth's Moon, you might wonder whether all planets have a brilliant, orbiting object visible in the sky above their surfaces. Astronomers formally call these gravitationally bound, orbiting objects natural satellites, but the more casual names for these objects are **moons.** All the planets except Mercury and Venus have moons orbiting them. More than 172 moons are known around the eight planets: Earth has one (named Moon), Mars has two tiny ones, Jupiter has at least 67, Saturn at least 62, Uranus at least 27, and Neptune at least 13. Dozens of other small moons probably remain to be discovered as our telescope technology continues to improve. Like the terrestrial planets, all of the moons of the planets have solid surfaces. More than 500 objects in the solar system are formally classified as moons, at least 76 of which are orbiting tiny objects called asteroids and 84 are orbiting trans-Neptunian objects, which we discuss in Section 4-4.

The many unique moons found in our solar system are often as scientifically intriguing as the planets.

You can see that there is a striking difference between the terrestrial planets, with few or no moons, and the Jovian planets, each of which has so many moons that it resembles a solar system in miniature. Of the known moons, seven are

Table 4-2 The Seven Giant Moons

	Moon	Io	Europa	Ganymede	Callisto	Titan	Triton
Parent planet	Earth	Jupiter	Jupiter	Jupiter	Jupiter	Saturn	Neptune
Diameter (km)	3476	3642	3130	5268	4806	5150	2706
Mass (kg)	7.35×10^{22}	8.93×10^{22}	4.80×10^{22}	1.48×10^{23}	1.08×10^{23}	1.34×10^{23}	2.15×10^{22}
Average density (kg/m^3)	3340	3530	2970	1940	1850	1880	2050
Substantial atmosphere?	No	No	No	No	No	Yes	No

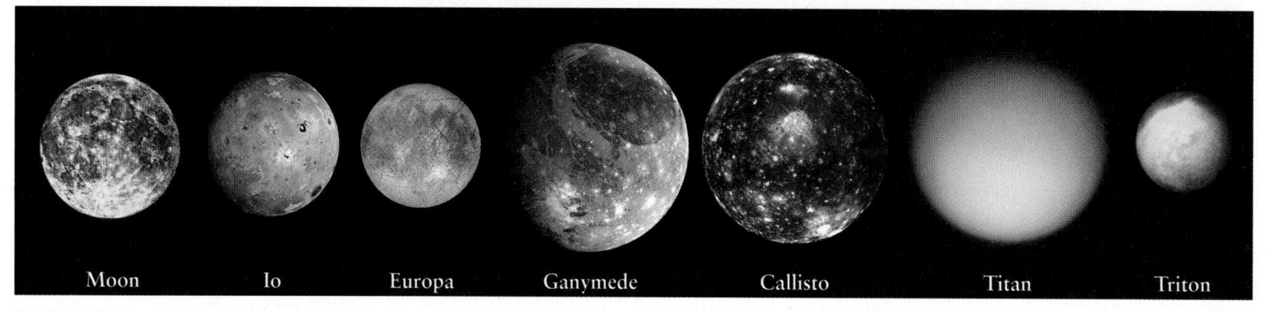

R I V U X G *(NASA/JPL/Space Science Institute)*

roughly as big as the planet Mercury. **Table 4-2** lists these moons and shows them to the same scale. Note that Earth's Moon and Jupiter's moons Io and Europa have relatively high average densities, indicating that these moons are made primarily of rocky materials. By contrast, the average densities of Ganymede, Callisto, Titan, and Triton are all relatively low. Planetary scientists conclude that the interiors of these four moons also contain substantial amounts of water ice, which is less dense than rock.

CAUTION Water ice may seem like a poor material for building a moon, since the ice you find in your freezer can easily be cracked or crushed. But under high pressure, such as is found in the interior of a large moon, water ice becomes as rigid as rock. (It also becomes denser than the ice found in ice cubes, although not as dense as rock.) Note that water ice is an important constituent for moons in the outer solar system, where the Sun is far away and temperatures are very low. For example, the surface temperature of Titan is a frigid 95 K (−178°C = −288°F).

Interplanetary spacecraft have made many surprising and fascinating discoveries about the moons of the solar system. As one example, we now know that Jupiter's moon Io has numerous geyserlike volcanoes that continually belch forth sulfur-rich compounds. The fractured surface of Europa, another of Jupiter's large moons, suggests that a worldwide ocean of liquid water may lie beneath its icy surface. Saturn's largest moon, Titan, is surrounded by a perpetual haze layer that initially gives it a featureless appearance.

ConceptCheck 4-4: How many moons in the solar system are larger than Earth's Moon?

4-3 Spectroscopy reveals the chemical composition of the planets

We have seen that the average densities of the planets and moons give us a crude measure for what they are made of, or their chemical compositions. For example, the low average density of Earth's Moon (3340 kg/m^3) compared with Earth (5515 kg/m^3) tells us that our Moon contains relatively little iron or other dense metals. But to truly understand the nature of the planets and moons, we need to know their chemical compositions in much greater detail than we can learn from average density alone.

The most accurate way to determine a planet's chemical elements would be by directly analyzing samples taken from a planet's atmosphere and soil. Unfortunately, of all the planets and moons, we have such direct information only for Earth and three other worlds on which spacecraft have landed—Venus, Moon, and Mars. However, as we saw in Chapter 2 (Section 2-4), astronomers can use spectroscopy to determine the compositions of even the farthest objects. Spectroscopy reveals the composition of planets both with and without a surrounding envelope of gas, known as the **atmosphere.**

Determining Composition: Planets with Surrounding Atmospheres

If a planet has an atmosphere, then sunlight reflected from that planet must have passed through parts of its atmosphere. During this passage, some of the wavelengths of sunlight will have been absorbed. Hence, the spectrum of this reflected sunlight will have dark absorption lines. Astronomers look at the particular wavelengths absorbed and the amount of light absorbed at those wavelengths. Both of these depend on

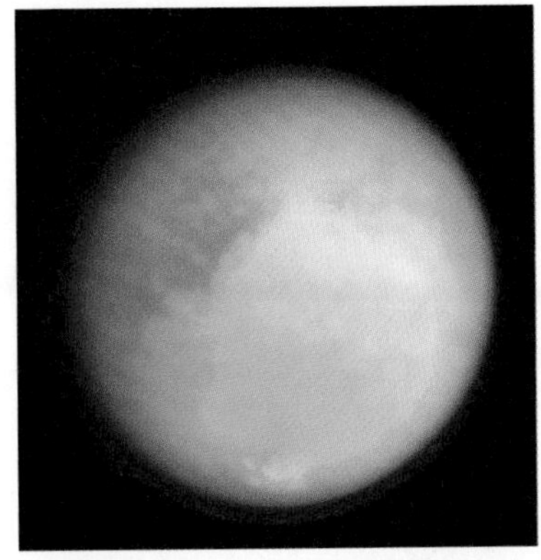

(a) Titan

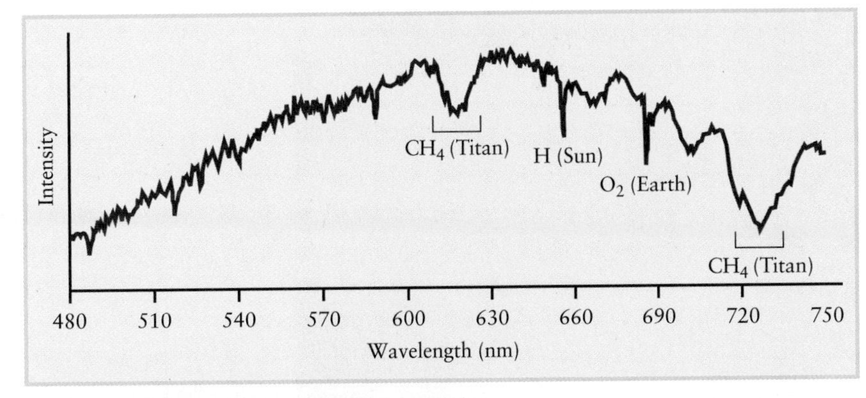

(b) The spectrum of sunlight reflected from Titan

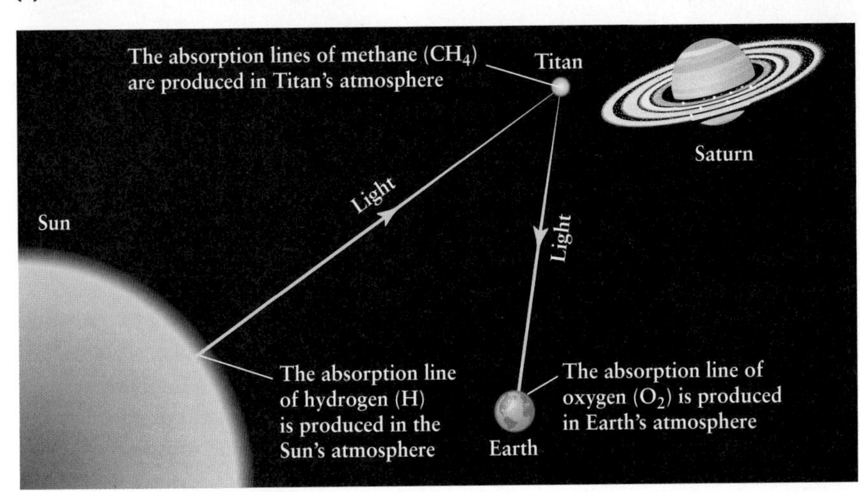

(c) Interpreting Titan's spectrum

FIGURE 4-3 R I V U X G **Analyzing a Moon's Atmosphere through Its Spectrum** (a) Titan is the only moon in the solar system with a substantial atmosphere. (b) The dips in the spectrum of sunlight reflected from Titan are due to absorption by hydrogen atoms (H), oxygen molecules (O_2), and methane molecules (CH_4). Of these, only methane is actually present in Titan's atmosphere. (c) This illustration shows the path of the light that reaches us from Titan. To interpret the spectrum of this light as shown in (b), astronomers must account for the absorption that takes place in the atmospheres of the Sun and Earth. *(a: NASA/JPL/Space Science Institute)*

the kinds of chemicals present in the planet's atmosphere and the abundance of those chemicals. For example, **Figure 4-3** shows how astronomers have used spectroscopy to analyze visible sunlight reflected from Titan, finding the presence of methane (CH_4).

In addition to visible-light measurements, it is sometimes even more useful to study the *infrared* and *ultraviolet* spectra of planetary atmospheres. Many molecules exhibit much more obvious spectral lines in these nonvisible wavelength bands than in the visible regime. As an example, the ultraviolet spectrum of Titan shows that nitrogen molecules (N_2) are the dominant constituent of Titan's atmosphere. Furthermore, Titan's infrared spectrum includes spectral lines of a variety of molecules that contain carbon and hydrogen, indicating that Titan's atmosphere has a very complex chemistry. None of these molecules could have been detected by visible light alone. Since Earth's atmosphere is largely opaque to infrared and ultraviolet wavelengths, telescopes in space are important tools for these spectroscopic studies of the solar system.

ConceptCheck 4-5: If planets reflect the Sun's light rather than emitting light of their own, how can spectroscopy reveal information about a planet's atmosphere?

Determining Composition: Planets without Surrounding Atmospheres

Atmosphere is covered in detail in Chapters 6 and 7.

Spectroscopy can also provide useful information about the *solid surfaces* of planets and moons without atmospheres. When light shines on a solid surface, some wavelengths are absorbed while others are reflected. For example, a plant leaf on Earth absorbs red and violet light but reflects green light—which is why leaves look green. Unlike a gas, a solid object illuminated by sunlight does not produce sharp, definite spectral lines. Instead, only broad absorption features appear in the spectrum. By comparing such a spectrum with the spectra of samples of different substances on Earth, astronomers can infer the chemical composition of the surface of a planet or moon.

As an example, **Figure 4-4*a*** shows Jupiter's moon Europa, and Figure 4-4*b* shows the infrared spectrum of light reflected from the surface of Europa. Because this spectrum is so close to that of water ice—that is, frozen water—astronomers conclude that water ice is the dominant constituent of Europa's surface.

Planetary surfaces are covered in detail in Chapter 6.

Unfortunately, spectroscopy tells us little about what the material is like just below the surface of a moon or planet. For this purpose, there is simply no substitute for sending a spacecraft to a planet and examining its surface directly.

FIGURE 4-4 R I V U X G **Analyzing a Moon's Surface from Its Spectrum** (a) Unlike Titan (Figure 4-3*a*), Jupiter's moon Europa has no atmosphere. (b) Infrared light from the Sun that is reflected from the surface of Europa has almost exactly the same spectrum as sunlight reflected from ordinary water ice. *(a: NASA/JPL-Caltech)*

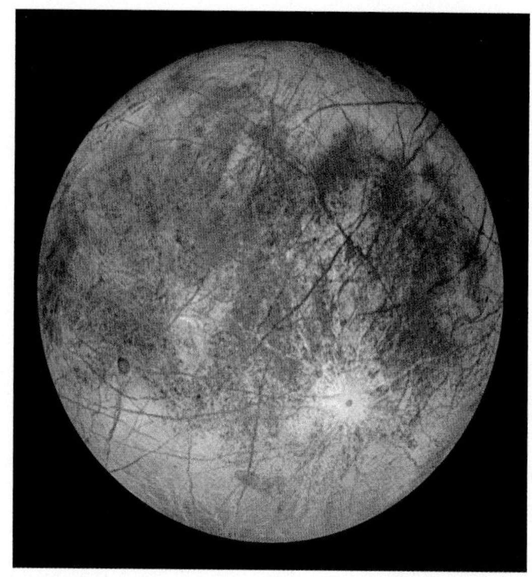

(a) Jupiter's moon Europa

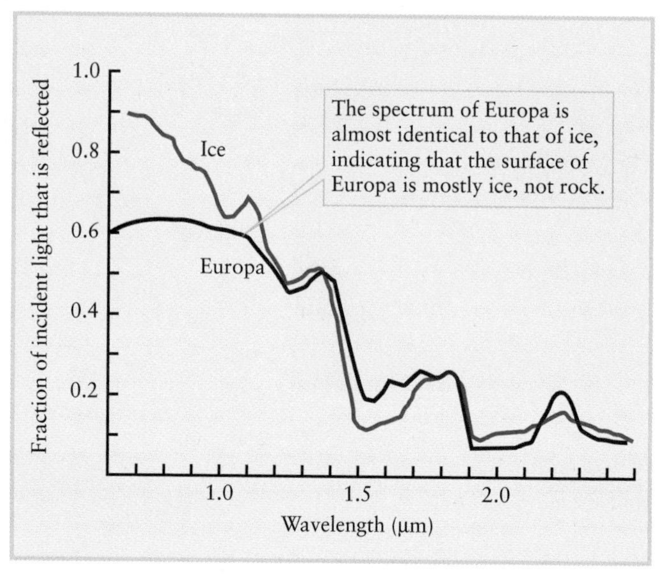

(b) The spectum of light reflected from Europa

ConceptCheck 4-6: **How is the spectrum of light observed from a solid planetary surface different from the spectrum of light observed from a gas composing an atmosphere?**

The Jovian Planets Are Made of Lighter Elements than the Terrestrial Planets

Spectroscopic observations from Earth and spacecraft show that the outer layers of the Jovian planets are comprised primarily of the lightest gases, hydrogen and helium. In contrast, chemical analysis of solid samples from Venus, Earth, and Mars demonstrate that the terrestrial planets are made mostly of heavier elements, such as iron, oxygen, silicon, magnesium, nickel, and sulfur. Spacecraft images such as **Figure 4-5** only hint at these striking differences in chemical composition.

Temperature plays a major role in determining whether the materials of which planets are made exist as solids, liquids, or gases. Hydrogen (H_2) and helium (He) are gaseous except at extremely low temperatures and extraordinarily high pressures. By contrast, rock-forming substances such as iron and silicon are solids except at temperatures well above 1000 K. Between these two extremes are substances such as water (H_2O), carbon dioxide (CO_2), methane (CH_4), and ammonia (NH_3), which solidify at low temperatures (from below 100 K to 300 K) into solids called **ices.** (In astronomy,

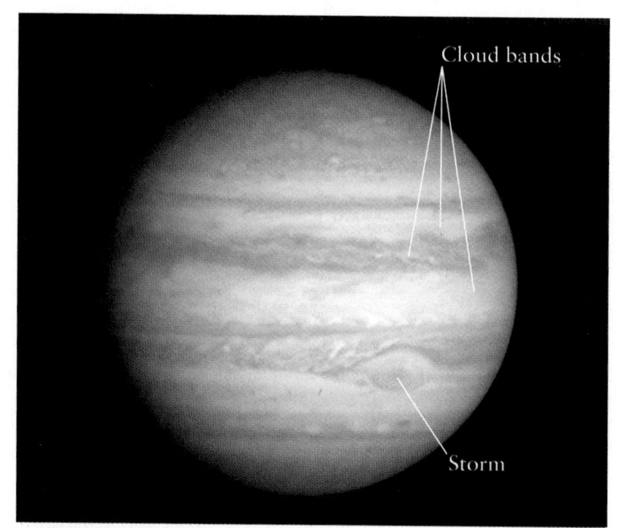

(a) Jupiter

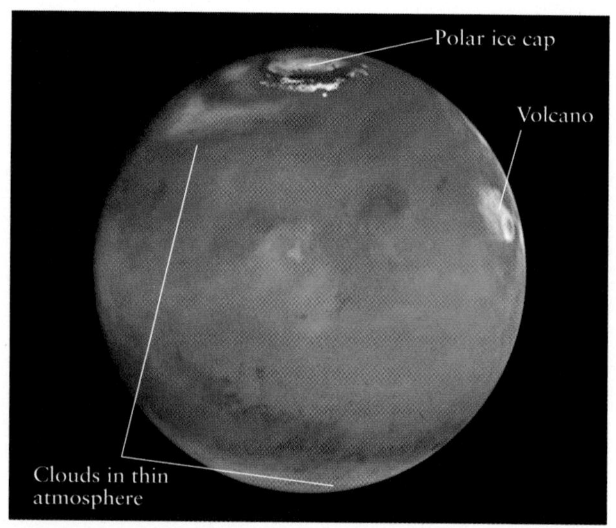

(b) Mars

FIGURE 4-5 R I V U X G **A Jovian Planet and a Terrestrial Planet** (a) This Hubble Space Telescope image gives a detailed view of Jupiter's cloudtops. Jupiter is composed mostly of the lightest elements, hydrogen and helium, which are colorless; the colors in the atmosphere are caused by trace amounts of other substances. The giant storm at lower right, called the Great Red Spot, has been observed raging for more than 300 years. (b) Mars, a terrestrial planet, is composed mostly of heavy elements similar to Earth, including iron, oxygen, silicon, magnesium, nickel, and sulfur. The planet's red surface can be seen clearly in this Hubble Space Telescope image because the Martian atmosphere is thin and nearly cloudless. Olympus Mons, the extinct volcano surrounded by clouds on the right-hand side, is nearly 3 times the height of Mount Everest. *(a: NASA/JPL/University of Arizona; b: © NASA Hubblesite.org)*

frozen water is just one of many kinds of "ice.") At somewhat higher temperatures, they can exist as liquids or gases. For example, clouds of ammonia ice crystals are found in the cold upper atmosphere of Jupiter, but within Jupiter's warmer interior, ammonia exists primarily as a liquid.

CAUTION The Jovian planets are sometimes called "gas giants." It is true that their primary constituents, including hydrogen, helium, ammonia, and methane, are gases under normal conditions on Earth. But in the interiors of these planets, pressures are so high that these substances are *liquids,* not gases. The Jovian planets might be better described as "liquid giants!"

As you might expect, a planet's surface temperature is related to its distance from the Sun. The four inner planets are quite warm. For example, midday temperatures on Mercury may climb to 700 K (= 427°C = 801°F), and during midsummer on Mars, it is sometimes as warm as 290 K (= 17°C = 63°F). The outer planets, which receive much less solar radiation, are cooler. Typical temperatures range from about 125 K (= −148°C = −234°F) in Jupiter's upper atmosphere to about 55 K (= −218°C = −360°F) at the tops of Neptune's clouds.

How might a planet's distance from the Sun, and its resulting surface temperature, explain the chemical elements we find on these planets? As it turns out, the atmospheres of the inner planets, whose orbits keep them close to the Sun, contain virtually *no* hydrogen molecules or helium atoms. Instead, the atmospheres of Venus, Earth, and Mars are composed of heavier molecules such as nitrogen (N_2, 14 times more massive than a hydrogen molecule), oxygen (O_2, 16 times more massive), and carbon dioxide (CO_2, 22 times more massive). To understand the connection between surface temperature and the absence of hydrogen and helium, we need to know a few basic facts about gases.

The temperature of a gas is directly related to the speeds at which the atoms or molecules of the gas move: The higher the gas temperature, the greater the speed of its atoms or molecules. Furthermore, for a given temperature, lightweight atoms and molecules move more rapidly than heavy ones. On the four inner, terrestrial planets, where atmospheric temperatures are high, low-mass hydrogen molecules and helium atoms move so swiftly that they can escape from the relatively weak gravity of these tiny planets. Hence, the atmospheres that surround the inner planets are composed primarily of more massive, slower-moving molecules such as CO_2, N_2, O_2, and water vapor (H_2O). On the four outer, Jovian planets, low temperatures and relatively strong gravitational attraction with molecules prevent even lightweight hydrogen and helium gases from escaping into space, and so their atmospheres are much more extensive. The combined mass of Jupiter's atmosphere, for example, is about a million (10^6) times greater than that of Earth's atmosphere. This is comparable to the mass of the entire Earth!

ConceptCheck 4-7: Why were astronomers surprised when they first discovered giant Jupiterlike planets orbiting very close to other stars?

4-4 Small chunks of rock and ice also orbit the Sun: asteroids, trans-Neptunian objects, and comets

Our solar system is more than a single star orbited by two categories of planets. In addition to the eight planets and their moons, countless smaller objects orbit the Sun. These fall into three broad categories: asteroids, which are rocky objects found in the inner solar system; trans-Neptunian objects, which are found beyond Neptune in the outer solar system and contain both rock and ice; and comets, which are mixtures of rock and ice that originate in the outer solar system but can venture close to the Sun.

Asteroids

Within the orbit of Jupiter are hundreds of thousands of rocky objects called **asteroids.** There is no sharp dividing line between planets and asteroids, which is why asteroids are also called **minor planets.** The largest asteroid, Ceres, has a diameter of about 560 mi (900 km). The next largest, Pallas and Vesta, are each about 300 mi (500 km) in diameter. Still smaller ones, like the asteroids shown in close-up in **Figure 4-6,** are increasingly numerous. Only the largest ones are round. Hundreds of thousands of kilometer-sized asteroids of widely differing shapes are known, and there are probably hundreds of thousands more that are automobile-sized or smaller. All of these objects orbit the Sun in the same clockwise direction as the planets.

Most but not all asteroids orbit the Sun between distances of 2 AU to 3.5 AU. This region of the solar system between the orbits of Mars and Jupiter is called the **asteroid belt.**

FIGURE 4-6 R I V U X G **Asteroid Family Portrait** A collage of asteroids, shown to scale, from largest to smallest: Vesta, 330 mi (530 km) across; Lutetia, 81 mi (130 km); and Mathilde, Ida (and its moon Dactyl), Eros, Gaspra, 2867 Šteins, and Itokawa. Most asteroids are too small to have enough gravity to pull them into spherical shapes. *(NASA/JPL-Caltech/JAXA/ESA)*

CAUTION One common misconception about asteroids is that they are the remnants of an ancient planet that somehow broke apart or exploded, like the fictional planet Krypton in the comic book adventures of Superman. In fact, the combined mass of the asteroids is less than that of the Moon, and they were probably never part of any planet-sized body. The early solar system is thought to have been filled with asteroidlike objects, most of which were incorporated into the planets. The "leftover" objects that missed out on this process make up our present-day population of asteroids.

ConceptCheck 4-8: Is the largest asteroid, Ceres, about the same size as a metropolitan city, a large U.S. state, a large country, an entire continent, or Earth itself?

Trans-Neptunian Objects

While asteroids are the most important small bodies in the inner solar system, the outer solar system is the realm of the **trans-Neptunian objects (TNOs).** As the name suggests, these are small bodies whose orbits lie beyond the orbit of Neptune. The largest of these are known as **dwarf planets.** The first of these to be discovered (1930) was Pluto, with a diameter of only 1413 mi (2274 km). The orbit of Pluto has a greater average orbital radius (39.54 AU), is more steeply inclined to the ecliptic (17.15°), and has a greater eccentricity (0.250) than that of any of the planets (**Figure 4-7**). In fact, Pluto's noncircular orbit sometimes takes it nearer the Sun than Neptune. (Happily, the orbits of Neptune and Pluto are such that they will never collide.) Pluto's density is only 2000 kg/m^3, about the same as Neptune's moon Triton shown in Table 4-2. Hence, its composition is thought to be a mixture of about 70% rock and 30% ice.

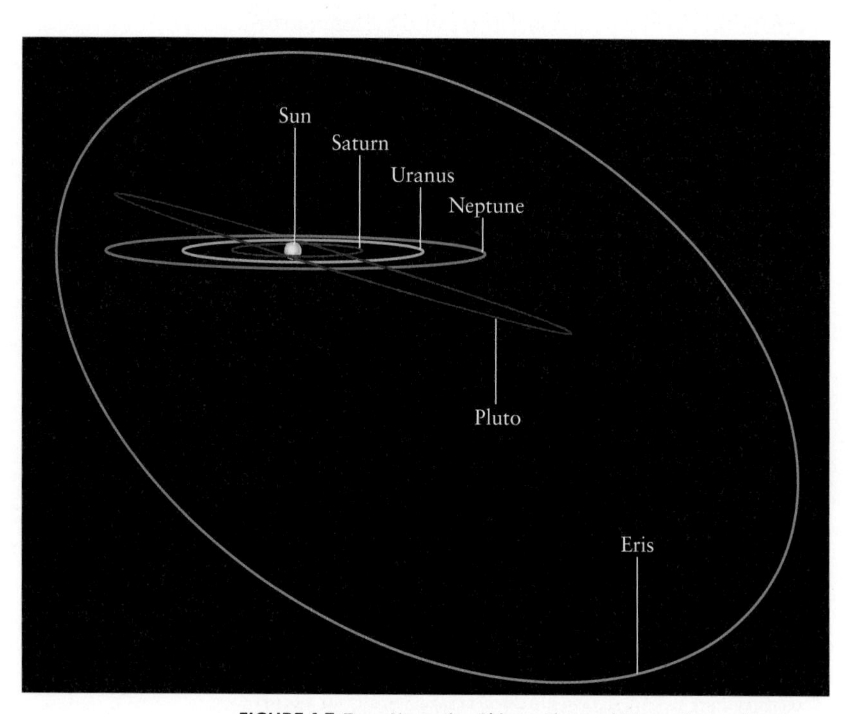

FIGURE 4-7 Trans-Neptunian Objects Pluto and Eris are the two largest trans-Neptunian objects, small worlds of rock and ice that orbit beyond Neptune. Unlike the orbits of the planets, the orbits of these two objects are steeply inclined to the ecliptic: Pluto's orbit is tilted by about 17°, and that of Eris is tilted by 44°.

Since 1992 astronomers have discovered more than 1200 new trans-Neptunian objects, and they are discovering more each year. All trans-Neptunian objects orbit the Sun in the same direction as the planets. The largest of these trans-Neptunian objects are comparable in size to Pluto. At least one dwarf planet, Eris, is even larger than Pluto, as well as being in an orbit that is much larger, more steeply inclined, and more eccentric (Figure 4-7).

Just as most asteroids lie in the asteroid belt, most trans-Neptunian objects orbit within a band called the **Kuiper belt** (pronounced "ki-per") that extends from 30 AU to 50 AU from the Sun and is centered on the plane of the ecliptic. Certainly, many more trans-Neptunian objects remain to be discovered as telescope technology improves. Astronomers estimate that there are 35,000 or more such objects with diameters greater than 100 km. If so, the combined mass of all trans-Neptunian objects is comparable to the mass of Jupiter and is several hundred times greater than the combined mass of all the asteroids found in the inner solar system.

Trans-Neptunian objects are thought to be debris naturally left over from the formation of the solar system. In the inner regions of the solar system, rocky fragments have been able to endure continuous exposure to the Sun's heat, but any ice originally present would have evaporated. Far from the Sun, ice has survived for billions of years.

ConceptCheck 4-9: Is Pluto an asteroid, planet, dwarf planet, Kuiper-belt object, or trans-Neptunian object?

CalculationCheck 4-3: Which has a larger diameter, the asteroid belt or the Kuiper belt?

Comets

Two objects in the Kuiper belt can collide if their orbits cross each other. When this happens, a fragment a few kilometers across can be knocked off one of the colliding objects and be diverted into a new and elongated orbit that brings it close to the Sun. Such small objects, each a combination of rock and ice, get a new name—**comets.** When a comet gets close enough to the Sun, the Sun's radiation vaporizes some of the comet's ices, producing long flowing tails of gas and dust particles (**Figure 4-8**). Astronomers can then deduce the composition of comets by studying the spectra of these tails created by reflected sunlight.

CAUTION Science fiction movies and television programs sometimes show comets tearing across the night sky like a rocket. That would be a pretty impressive sight—but that is not what comets look like. Like the planets, comets orbit the Sun. And like the planets, comets move hardly at all against the background of stars over the course of a single night. If you are lucky enough to see a bright comet, it will not zoom dramatically from horizon to horizon. Instead, it will seem to hang majestically among the stars, so you can admire it at your leisure.

FIGURE 4-8 R I V U X G **A Comet** This photograph shows Comet Hale-Bopp as it appeared in April 1997. The solid part of a comet like this is a chunk of dirty ice a few tens of kilometers in diameter. When a comet passes near the Sun, solar radiation vaporizes some of the icy material, forming a bluish tail of gas and a white tail of dust. Both tails can extend for tens of millions of kilometers. *(NASA Jet Propulsion Laboratory)*

Some comets appear to originate from locations far beyond the Kuiper belt. The source of these is thought to be a swarm of comets that forms a spherical "halo" around the solar system called the **Oort cloud.** This hypothesized "halo" extends to 50,000 AU from the Sun (about one-fifth of the way to the nearest other star). Because the Oort cloud is so distant, it has not yet been possible to detect objects in the Oort cloud directly.

ConceptCheck 4-10: **If a comet is observed to have an orbit that is perpendicular to Earth's orbit about the Sun, did it most likely originate from the Kuiper belt or the Oort cloud?**

4-5 The Sun and planets formed from a rotating solar nebula

How did the Sun and planets form? Where did the solar system come from? These questions have tantalized astronomers for centuries. The first step toward answers lies with an understanding of which elements are most abundant in our solar system and beyond. For a summary of the properties of all the planets, see **Cosmic Connections: Characteristics of the Planets.**

The Origin of the Elements and Cosmic "Recycling"

As we saw earlier in the chapter, one of the key properties of our solar system is that the inner, terrestrial planets are smaller and made of much heavier chemical elements than the significantly larger outer planets, which are made of relatively lightweight elements. One reason for this fundamental difference is that the chemical elements of hydrogen and helium are quite common in the universe, while the heavier elements that make up rocky substances are quite rare.

The dominance of hydrogen and helium is not merely a characteristic of the outer reaches of our solar system. By analyzing the spectra of stars and galaxies outside our solar system, astronomers have found essentially the same pattern of chemical abundances out to the farthest distance attainable by the most powerful telescopes. Hence, the vast majority of the atoms in the universe are hydrogen and helium atoms. The elements that make up the bulk of Earth—mostly iron, oxygen, and silicon—are relatively rare in the universe as a whole, as are the elements of which living organisms are made—carbon, oxygen, nitrogen, and phosphorus, among others.

There is a good reason for this overwhelming abundance of hydrogen and helium. A wealth of evidence has led astronomers to conclude that the universe began some 13.7 billion years ago with a violent event known as the *Big Bang.* Only the lightest elements—hydrogen and helium, as well as tiny amounts of lithium and perhaps beryllium—emerged from the enormously high temperatures following this cosmic event. All the heavier elements were manufactured by stars later, either by the combining of lighter atoms together into heavier atoms deep in their interiors or by the violent explosions that mark the end of massive stars. Were it not for these processes, which take place only in stars, there would be no heavy elements in the universe, no planet like our Earth, and no humans to contemplate the nature of the cosmos.

Because our solar system contains heavy elements, it must be that at least some of its material was once inside other stars. But how did this material become available to help build our solar system? The answer is that near the ends of their lives, stars cast much of their matter back out into space. For most stars this process is a comparatively gentle one, in which a star's outer layers are gradually expelled. **Figure 4-9** shows a star losing material in this fashion. This ejected material appears as the cloudy region that surrounds the star and is illuminated by it. A few stars eject matter much more dramatically at the very end of their lives, in a spectacular detonation called a *supernova,* which blows the star apart.

FIGURE 4-9 R I V U X G **A Mature Star Ejecting Gas and Dust** The star Antares is shedding material from its outer layers, forming a thin cloud around the star. We can see the cloud because some of the ejected material has condensed into tiny grains of dust that reflect the star's light. (Dust particles in the air around you reflect light in the same way, which is why you can see them within a shaft of sunlight in a darkened room.) Antares lies some 600 ly from Earth in the constellation Scorpio. *(Australian Astronomical Observatory/David Malin Images)*

No matter how it escapes, the ejected material contains heavy elements from which to develop a system of planets, moons, comets, and asteroids. Our own solar system must have formed from enriched material in just this way. Thus, our solar system contains "recycled" material that was produced long ago inside a now-dead star. This "recycled" material includes all of the carbon in your body, all of the oxygen that you breathe, and all of the iron and silicon in the soil beneath your feet.

ConceptCheck 4-11: To what does the phrase "we are all made of star stuff" refer?

The Birth of the Sun: The Solar Nebular Hypothesis

Our star, the Sun, did not produce the elements that make up our solar system. These raw materials were first created during the Big Bang and then later in processes within ancient stars that have long since disappeared. But given these ingredients, how might they have combined to make the Sun and planets we see today? Whatever description we devise must account for the two broad categories of planets as well as the shape and rotation of the solar system itself.

The central idea agreed upon by most astronomers dates to the late 1700s, when the German philosopher Immanuel Kant and the French scientist Pierre-Simon de Laplace turned their attention to the manner in which the planets orbit the Sun. Both concluded that the arrangement of the orbits—all in the same direction and in nearly the same plane—could not be mere coincidence. To explain the orbits, Kant and Laplace independently proposed that our entire solar system, including the Sun as well as all of its planets and moons, formed from a vast, rotating cloud of gas and dust called the **solar nebula** (**Figure 4-10**). This explanation is called the **nebular hypothesis.**

Each part of the nebula exerted a gravitational attraction on the other parts, and these mutual gravitational pulls tended to make the nebula contract. As it contracted, the greatest concentration of matter occurred at the center of the nebula, forming a relatively dense region called the **protosun.** As its name suggests, this part of the solar nebula eventually developed into the Sun. The planets formed from the much sparser material in the outer regions of the solar nebula. Indeed, the mass of all the planets together is only 0.1% of the Sun's mass.

When you drop a ball, the gravitational attraction of Earth makes the ball fall faster and faster as it falls; in the same way, material falling inward toward the protosun would have gained speed as it fell. When this fast-moving material slammed into the protosun, the energy of the collision was converted into thermal energy, causing the temperature deep inside the solar nebula to increase.

We will cover the Sun in detail in Chapter 9, but for now there are a few things you need to know about the Sun in the context of the formation of the solar system. The key thing

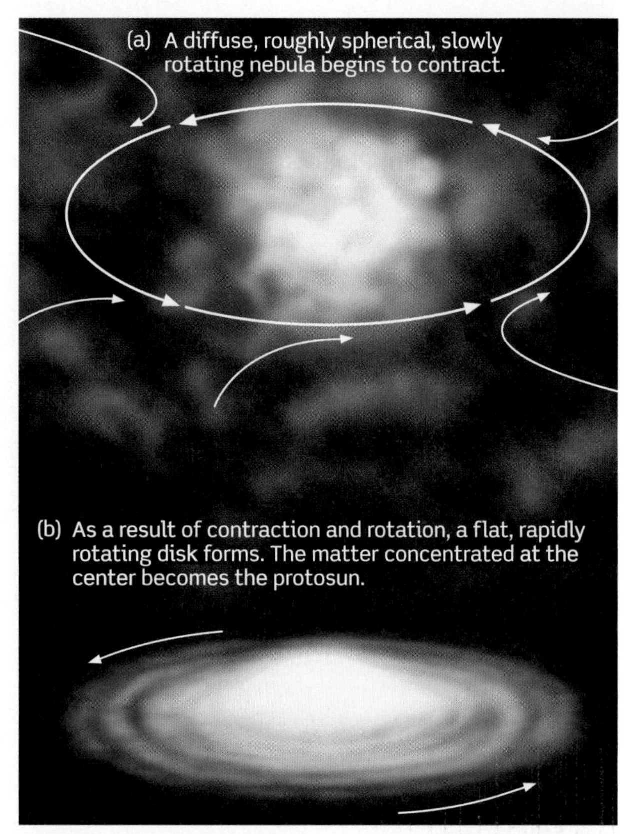

FIGURE 4-10 The Birth of the Solar System (a) A cloud of interstellar gas and dust begins to contract because of its own gravity. (b) As the cloud flattens and spins more rapidly around its rotation axis, a central condensation develops that evolves into a glowing protosun. The planets will form out of the surrounding disk of gas and dust.

to remember is that our solar system formed at nearly the same time as the Sun did. What happens is, as our newly created protosun contracted from its initial solar nebula, its interior temperature dramatically increased as well. About 10^5 (100,000) years after it began to form, the protosun's surface temperature stabilized at about 6000 K, but the temperature in its interior kept increasing to ever higher values as the central regions of the contracting protosun became denser and denser. Eventually, perhaps 10^7 (10 million) years after the solar nebula first began to contract, the gas at the center of the protosun reached a density of about 10^5 kg/m^3 (a hundred times denser than water) and a temperature of a few million kelvins (that is, a few times 10^6 K). Under these extreme conditions, nuclear reactions that convert hydrogen into helium began in the protosun's interior. When this happened, the energy released by these reactions stopped the contraction and a true star was born. Nuclear reactions continue to the present day in the interior of the Sun and are the source of all the energy that the Sun radiates into space.

If the solar nebula had not been rotating at all, everything would have fallen directly into the protosun, leaving nothing behind to form the planets. Instead, the solar nebula must have had an overall slight rotation, which caused its evolution to follow a different path. As the slowly rotating nebula collapsed

(a) (b)

FIGURE 4-11 RIVUXG **Conservation of Angular Momentum** A figure skater who (a) spins slowly with her limbs extended will naturally speed up when (b) she pulls her limbs in. In the same way, the solar nebula spun more rapidly as its material contracted toward the center of the nebula. *(Amy Sancetta/AP Images)*

inward, it would naturally have tended to rotate faster. This relationship between the decreasing size of an object and its increasing rotation speed is an example of a general principle called the **conservation of angular momentum.**

ANALOGY Figure skaters make use of the conservation of angular momentum. When a spinning skater pulls her arms and legs in close to her body, the rate at which she spins automatically increases (**Figure 4-11**). Even if you are not a figure skater, you can demonstrate this by sitting on an office chair. Sit with your arms outstretched and hold a weight, like a brick or a full water bottle, in either hand. Now use your feet to start your body and the chair rotating, lift your feet off the ground, and then pull your arms inward. Your rotation will speed up quite noticeably.

As the solar nebula began to rotate more rapidly, it also tended to flatten out (see Figure 4-10*b*). From the perspective of a particle rotating along with the nebula, it felt as though there were a force pushing the particle away from the nebula's axis of rotation. (Likewise, passengers on a spinning carnival ride seem to feel a force pushing them outward and away from the ride's axis of rotation.) This apparent force was directed opposite to the inward pull of gravity, and so it tended to slow the contraction of material toward the nebula's rotation axis. But there was no such effect opposing contraction in a direction parallel to the rotation axis. Some 10^5 (100,000) years after the solar nebula first began to contract, it had developed the structure shown in Figure 4-10*b*, with a rotating, flattened disk surrounding what became the protosun. This disk is called the **protoplanetary disk,** or **proplyd,** since planets formed from its material. This model explains why their orbits all lie in essentially the same plane and why they all orbit the Sun in the same direction.

There were no humans to observe these processes taking place during the formation of the solar system. But Earth astronomers have seen disks of material surrounding other stars that formed only recently. These, too, are called protoplanetary disks, because it is thought that planets can eventually form from these disks around other stars. Hence, these disks are planetary systems that are still "under construction." By studying these disks around other stars, astronomers are able to examine what our solar nebula may have been like some 4.56×10^9 years ago.

Figure 4-12 shows a number of protoplanetary disks in the Orion Nebula, a region of active star formation. A study of 110 young stars in the Orion Nebula detected protoplanetary disks around 56 of them, which suggests that systems of planets may form around a substantial fraction of stars. Later in this chapter we will see direct evidence for planets that have formed and orbit around stars other than the Sun.

> ConceptCheck 4-12: **If the nebular hypothesis is correct, what must it explain about the planetary orbits?**

(a)

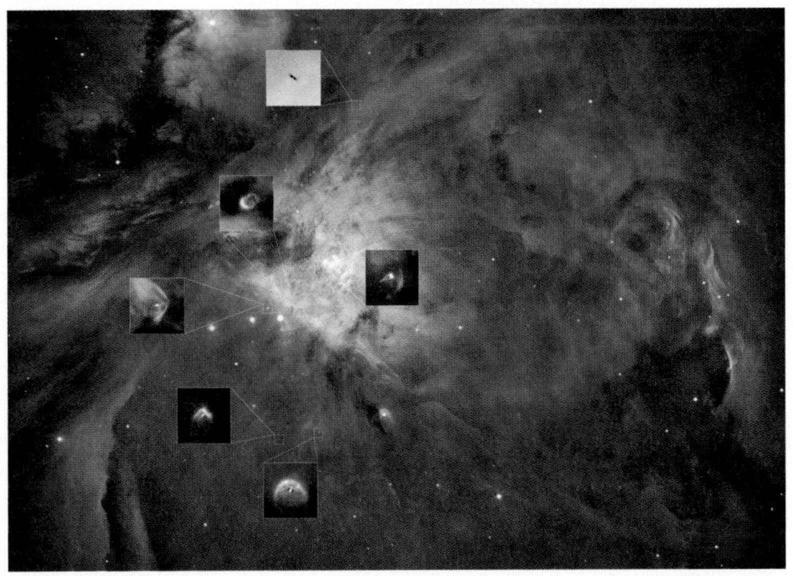

(b)

FIGURE 4-12 RIVUXG **Protoplanetary Disks** (a) The Orion Nebula is a star-forming region located some 1500 ly from Earth. It is the middle "star" in Orion's sword (see Figure 1-3). (b) This view of the Orion Nebula is a mosaic of Hubble Space Telescope images. The insets are close-ups of four protoplanetary disks that lie within the nebula. A young star with orbiting planets is forming at the center of each disk. *(a: Australian Astronomical Observatory/David Malin Images; b: NASA Hubblesite.org)*

4-6 The planets formed by countless collisions of dust, rocks, and gas in the region surrounding our young Sun

We have seen how the solar nebula would have contracted to form a young Sun with a protoplanetary disk rotating around it. But how did the material in this disk form into planets? And why are rocky planets in the inner solar system, while the giant Jupiterlike planets are in the outer solar system? If the nebular hypothesis is to be considered accurate, it needs to be able to successfully provide answers to these overarching questions.

Temperatures in the Solar Nebula

To understand how the planets, asteroids, and comets formed, we must look at the conditions that prevailed within the solar nebula. The density of material in the part of the nebula outside the protosun was rather low, as was the pressure of the nebula's gas. If the pressure is sufficiently low, a substance cannot remain in the liquid state, but must end up as either a solid or a gas. For a given pressure, what determines whether a certain substance is a solid or a gas is its **condensation temperature.** If the temperature of a substance is above its condensation temperature, the substance is a gas; if the temperature is below the condensation temperature, the substance solidifies into tiny specks of dust or snowflakes. You can often see similar behavior on a cold morning. The air temperature can be above the condensation temperature of water, while the cold windows of parked cars may have temperatures below the condensation temperature. Thus, water molecules in the air remain as a gas (water vapor) but form solid ice particles (frost) on the car windows.

Substances such as water (H_2O), methane (CH_4), and ammonia (NH_3) have low condensation temperatures, ranging from 100 K to 300 K. Rock-forming substances have much higher condensation temperatures, in the range from 1300 K to 1600 K. The gas cloud from which the solar system formed had an initial temperature near 50 K, so all of these substances could have existed in solid form. Thus, the solar nebula would have been populated by an abundance of small ice particles and solid dust grains like the one shown in **Figure 4-13.** (Recall that "ice" can refer to frozen carbon dioxide, methane, or ammonia, as well as frozen water.) But hydrogen and helium, the most abundant elements in the solar nebula, have condensation temperatures so near absolute zero that these substances always existed as gases during the creation of the solar system. You can best visualize the initial state of the solar nebula as a thin gas of hydrogen and helium strewn with tiny dust particles.

Planets form as a natural by-product of the formation of stars from the collapse of a giant interstellar cloud of dust and gas.

Go to Video 4-2

Figure 4-14 shows the probable temperature distribution in the solar nebula at this stage in the formation of our solar system. In the inner regions of the solar nebula, water, methane, and ammonia were vaporized by the high temperatures. Only materials with high condensation temperatures could have remained solid. Of these materials, iron, silicon, magnesium, and sulfur were particularly abundant, followed closely by aluminum, calcium, and nickel. (Most of these elements were present in the form of oxides, which are chemical compounds containing oxygen. These compounds also have high condensation temperatures.) In contrast, ice particles and ice-coated dust grains were able to survive in the cooler, outer portions of the solar nebula.

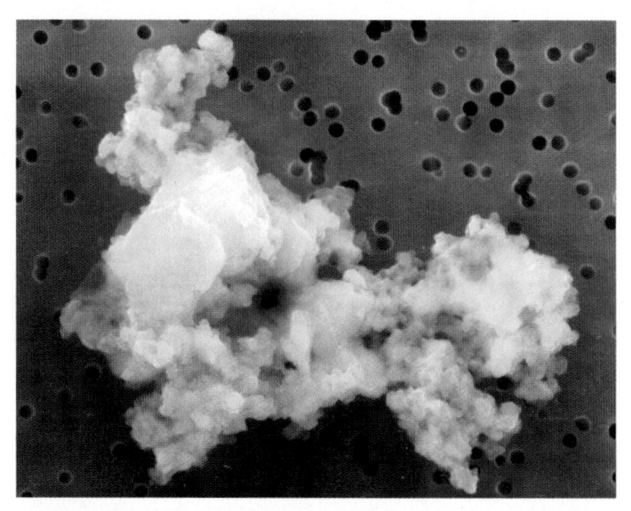

FIGURE 4-13 A Grain of Cosmic Dust This highly magnified image shows a microscopic dust grain that came from interplanetary space. It entered Earth's upper atmosphere and was collected by a high-flying aircraft. Dust grains of this sort are abundant in star-forming regions. These tiny grains were also abundant in the solar nebula and served as the building blocks of the planets. *(NASA)*

ConceptCheck 4-13: How might the solar system be different if the location at which water could freeze in the solar nebula was much more distant than it was in the solar nebula that formed our solar system?

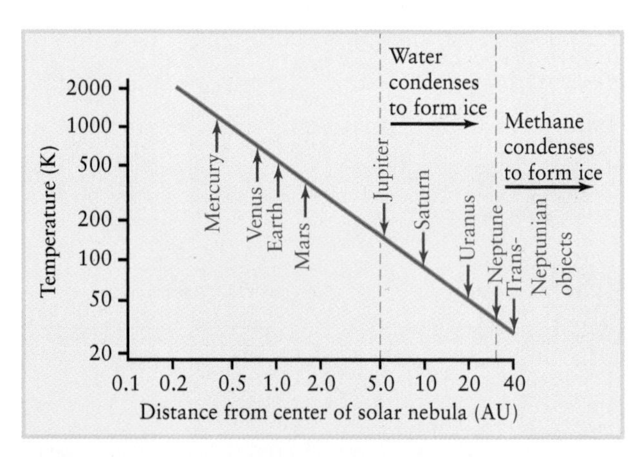

FIGURE 4-14 Temperature Distribution in the Solar Nebula This graph shows how temperatures probably varied across the solar nebula as the planets were forming. Note the general decline in temperature with increasing distance from the center of the nebula. Beyond 5 AU from the center of the nebula, temperatures were low enough for water to condense and form ice; beyond 30 AU, methane (CH_4) could also condense into ice.

Planetesimals Become Protoplanets, Then Rocky Planets

In the inner part of the solar nebula, the grains of high-condensation-temperature materials would have collided and merged into small chunks. Initially, electric forces—that is, chemical bonds—held these chunks together, in the same way that chemical bonds hold an ordinary rock together. Over a few million years, these chunks coalesced into roughly 10^9 asteroidlike objects called **planetesimals,** with diameters of a kilometer or so. These were large enough to be held together by the gravitational attraction of the different parts of the planetesimal for each other. During the next stage, gravitational attraction between the planetesimals caused them to collide and accumulate into still-larger objects called **protoplanets,** which were roughly the size and mass of our Moon. This process of accumulation of material to form larger and larger objects is called **accretion.** During the final stage, these Moon-sized protoplanets collided to form the inner planets. This final episode must have involved some truly spectacular, world-shattering collisions.

In the inner solar nebula only materials with high condensation temperatures could form dust grains and hence protoplanets, so the result was a set of planets made predominantly of materials such as iron, silicon, magnesium, and nickel, and their oxides. This is just the composition of the present-day inner planets. We now understand that the accretion process only requires roughly 10^8 (100 million) years to form four or five rocky planets with orbits between 0.3 AU and 1.6 AU from the Sun.

Figure 4-15 illustrates how planetary systems evolve around a forming star. At first the material that coalesced to form protoplanets in the inner solar nebula remained largely in solid form, despite the high temperatures close to the protosun. But as the protoplanets grew, they were heated by the violent impacts of planetesimals as well as the decay of radioactive elements such as uranium, and this heat caused melting. Thus, the terrestrial planets began their existence as spheres of at least partially molten rocky materials. Material was free to move within these molten spheres, so the denser, iron-rich minerals sank to the centers of the planets while the less dense silicon-rich minerals floated to their surfaces. This process is called **chemical differentiation.** In this way the rocky planets developed their dense iron cores.

Because the materials that went into the terrestrial planets are relatively scarce, these planets ended up having relatively small mass and hence relatively weak ability for gravitational attraction. As a result, the terrestrial protoplanets were unable to capture any of the hydrogen or helium in the solar nebula to form atmospheres. The thin envelopes of atmosphere that encircle Venus, Earth, and Mars evolved much later as trapped gases escaped from the molten interiors of these planets.

ConceptCheck 4-14: How do rocky planets develop a core that has a higher density than rocks on their surfaces?

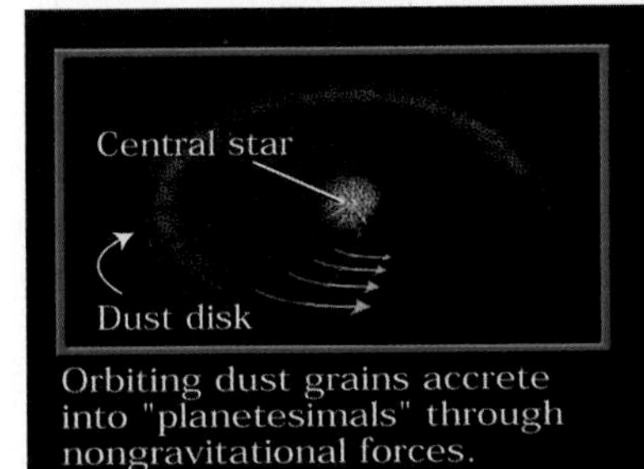

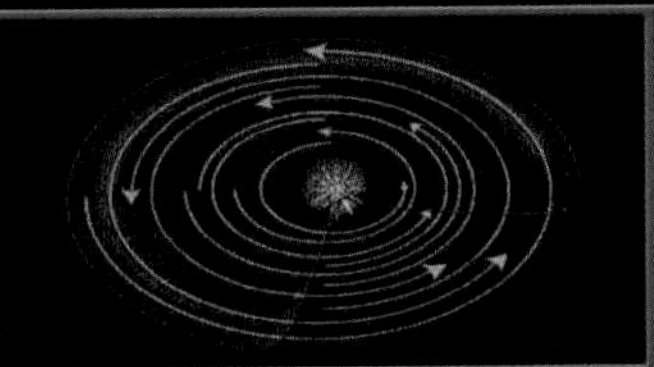

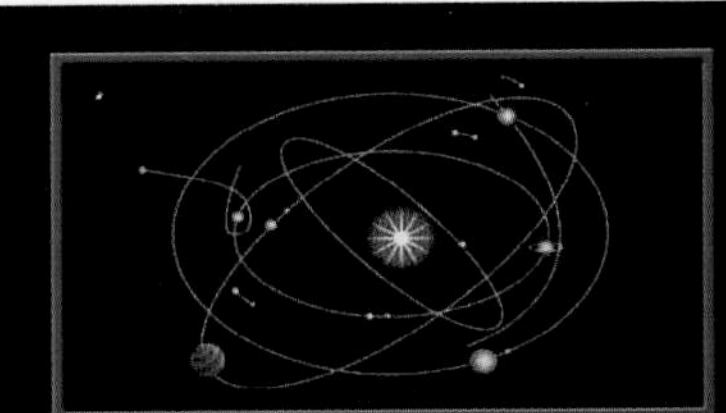

(a)

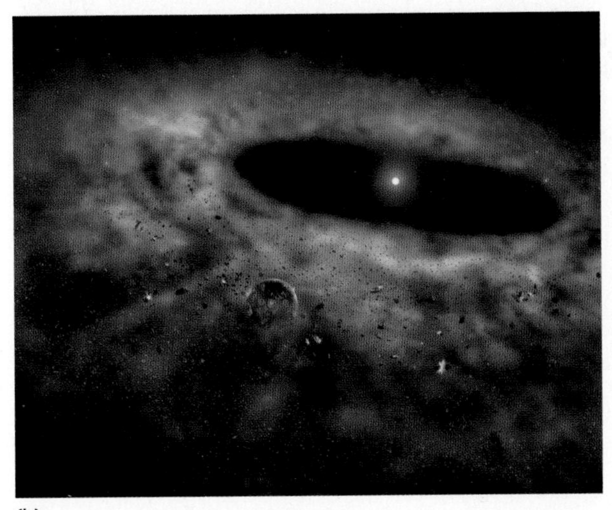

(b)

(c)

FIGURE 4-15 Formation of the Solar System (a) A giant nebula of dust and gas heats, flattens, and spins faster as it collapses under the influence of gravity. (b) and (c) In a process that happens quite quickly, in less than half a billion years, material begins to stick together forming larger and larger objects that eventually clean out most of the solar system becoming Earth and the orbiting planets we see today. *(a: NASA and A. Feild [STScI]; b and c: Gemini Observatory/AURA artwork by Lynette Cook)*

Outer Planet Formation by Core Accretion: Capturing an Envelope of Gas

Like the inner planets, the outer planets may have begun to form by the accretion of planetesimals. The key difference is that ices as well as rocky grains were able to survive in the outer regions of the solar nebula, where temperatures were relatively low (see Figure 4-14). The elements of which ices are made are much more abundant than those that form rocky grains. Thus, much more solid material would have been available to form planetesimals in the outer solar nebula than in the inner part. As a result, solid objects several times larger than any of the terrestrial planets could have formed in the outer solar nebula. Each such object, made up of a mixture of ices and rocky material, could have become the core of a Jovian planet and served as a "seed" around which the rest of the planet eventually grew.

Thanks to the lower temperatures in the outer solar system, gas atoms (principally hydrogen and helium) were moving relatively slowly and so could more easily be captured by the strong gravity of these massive cores. Thus, the core of a Jupiterlike protoplanet could have captured an envelope of gas as it continued to grow by accretion. This picture is called the **core accretion model** for the origin of the outer planets.

Calculations based on the core accretion model suggest that both rocky materials and gas slowly accumulated for several million years, until the masses of the core and the envelope became equal. From that critical moment on, the envelope pulled in all the gas it could get, dramatically increasing the protoplanet's mass and size. This runaway growth of the protoplanet would have continued until all the available gas was used up. The result was a huge planet with an enormously thick, hydrogen-rich envelope surrounding a rocky core with 5 to 10 times the mass of Earth. This scenario could have occurred at four different distances from the Sun, thus creating the four Jovian planets.

In the core accretion model, Uranus and Neptune probably did not form at their present locations, about 19 AU and 30 AU from the Sun, respectively. The solar nebula was too sparse at those distances to allow these planets to have grown to the present-day sizes. Instead, it is thought that Uranus and Neptune formed between 4 AU and 10 AU from the Sun, but were flung into larger orbits by gravitational interactions with Jupiter. **Figure 4-16** summarizes our overall picture of the formation of the terrestrial and Jovian planets.

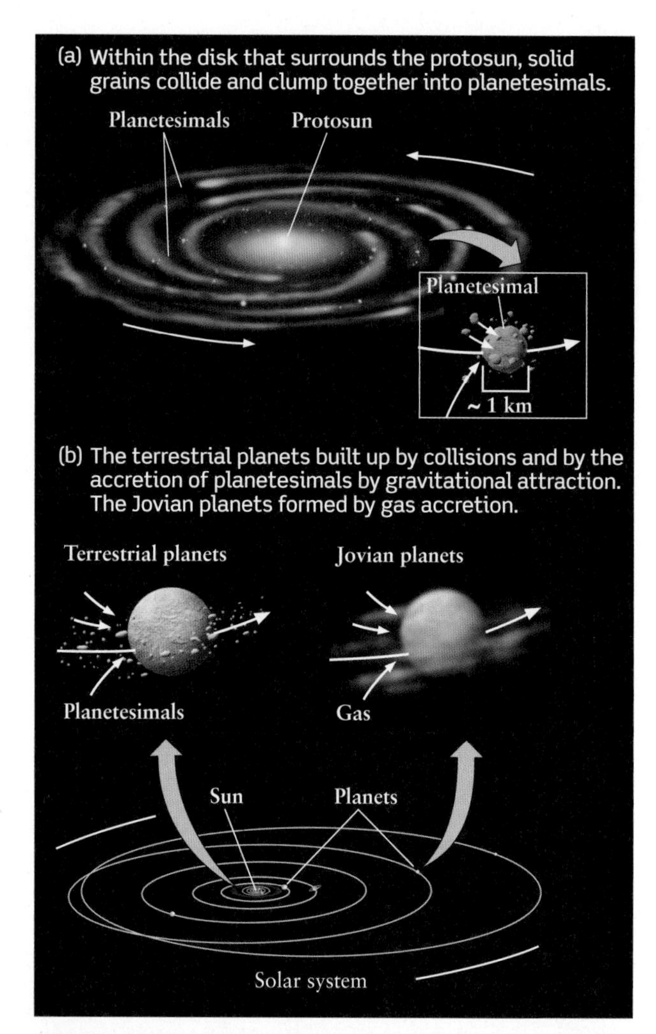

FIGURE 4-16 Terrestrial Versus Jovian Planet Formation The precise details of how planets form are complicated, but terrestrial and Jovian planets might have formed somewhat differently. (a) Planetesimals about 0.5 mi (<1 km) in size most likely form in the solar nebula from small dust grains colliding and sticking together. (b) Whereas planetesimals in the inner solar system grouped together to form the terrestrial planets, the outer Jovian planets could have begun as terrestrial-like planets that accumulated massive envelopes of hydrogen and helium, or, alternatively, the Jovian planets could have formed directly from the gas of the solar nebula.

ConceptCheck 4-15: Why would it be unlikely that the most distant, Jupiterlike planets of Uranus and Neptune formed by core accretion at their current locations?

The Final Stages of Solar System Evolution

While the planets were forming, the protosun was also evolving into a full-fledged star with nuclear reactions occurring in its core. The time required for this to occur was about 10^8 years, roughly the same as that required for the formation of the terrestrial planets. Before the onset of nuclear reactions, however, the young Sun probably expelled a substantial portion of its mass into space (**Figure 4-17**). Magnetic fields within the solar nebula would have funneled a portion of the nebula's mass into oppositely directed jets along the rotation axis of the nebula, such as the jets shown in Figure 4-17*a*.

In addition, observations of our relatively unstable young Sun would have revealed it ejecting its thin outermost layers into space. This brief but intense burst of mass loss, observed in many young stars across the sky, is called a **T Tauri wind**, after the star in the constellation Taurus (the Bull) where it was first identified (an example of a wind-emitting star from the constellation of Orion is shown in

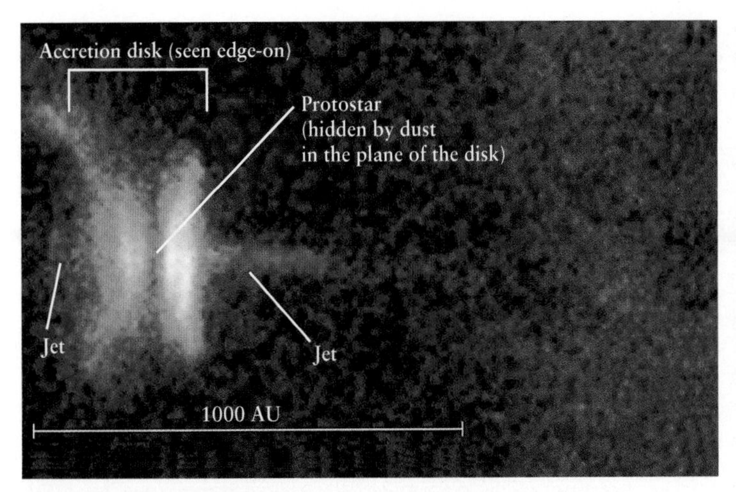

(a) Jets from a young star

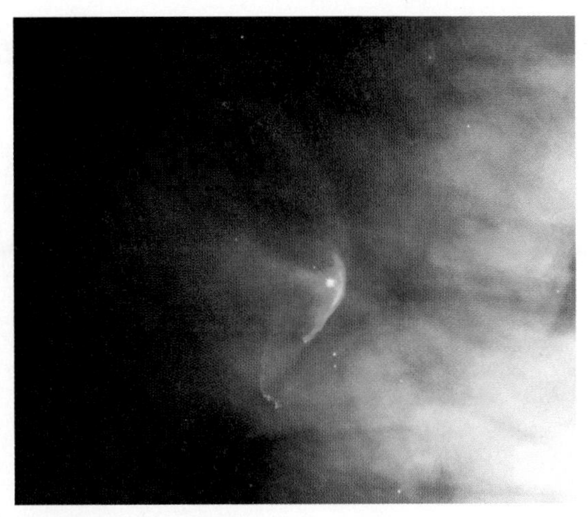

(b) Winds from young stars

FIGURE 4-17 R I V U X G **Jets and Winds from Young Stars** (a) This protostar in the constellation Taurus (the Bull) is ejecting matter in two immense jets directed perpendicular to the plane of the accretion disk. Red denotes light emitted by hot ionized gas in the jet, while green denotes starlight scattered by dust particles in the disk. (b) In this Hubble Space Telescope image an outpouring of particles and radiation from the surfaces of the young star LL Ori can be seen carving out a cavity in the surrounding, slower-moving, dusty material, forming a crescent-shaped bow shock. *(a: C. Burrows, STScI and ESA, the WFPC-2 Investigation Definition Team, and NASA; b: NASA, ESA, R. O'Connell [University of Virginia], F. Paresce [National Institute for Astrophysics, Bologna, Italy], E. Young [Universities Space Research Association/Ames Research Center], the WFC3 Science Oversight Committee, NASA and the Hubble Heritage Team [STScI/AURA])*

Figure 4-17*b*). Each of the protoplanetary disks in Figure 4-12*b* (insets) has a similar wind-emitting T Tauri star at its center. (The present-day Sun also loses mass from its outer layers in the form of high-speed electrons and protons, a flow called the **solar wind.** But this is minuscule compared with a T Tauri wind, which causes a star to lose mass 10^6 to 10^7 times faster than in the solar wind.)

With the passage of time, the combined effects of jets, the T Tauri wind, and accretion onto the planets would have swept the solar system nearly clean of gas and dust. With no more interplanetary material to gather up, the planets would have stabilized at roughly their present-day sizes, and the formation of the solar system would have been complete. The figures in **Cosmic Connections: Formation of the Solar System** summarize our modern-day picture of the origin of our solar system. Prior to 1995 the only fully formed planetary system to which we could apply this model was our own. As we will see in the next section, astronomers can now further test this model on a rapidly growing number of planets known to orbit stars other than our own.

ConceptCheck 4-16: **What are three processes that would have cleaned the early solar system of dust and gas?**

4-7 Understanding how our planets formed around the Sun suggests that planets around other stars are common

If planets formed around our Sun, have they also formed around other stars? That is, are **extrasolar planets** orbiting distant stars just as our planets orbit the Sun? Although just a few years ago astronomers were not entirely sure that planets actually exist around other stars, today we have several lines of evidence convincing us that planets are common around other stars. To appreciate how remarkable these discoveries are, let's consider the process that astronomers go through to search for extrasolar planets.

Directly Photographing Extrasolar Planets

Go to Video 4-3

It is very difficult to make direct observations of planets orbiting other stars. The challenge is that planets are small and dim compared with stars; as seen by an alien astronomer far from our solar system, at visible wavelengths the Sun is 10^9 times brighter than Jupiter and 10^{10} times brighter than Earth. A hypothetical planet orbiting a distant star, even a planet 10 times larger than Jupiter, would be easily lost in the star's glare as seen through even the largest telescope on Earth. Despite these technical challenges, astronomers have been able to fine-tune telescopes to photograph, for the first time, a few planets visible orbiting distant stars. The first planets photographed are shown in **Figure 4-18.** The dramatic "first family" pictures of planets beyond our solar system are shown in Figures 4-18*a* and 4-18*b*, which were taken by the Gemini Observatory near the summit of Mauna Kea in Hawai'i. Figure 4-18*c* was taken by the Hubble Space Telescope and shows the motion of extrasolar planet Fomalhaut b. By 2013, about 30 planets beyond our own solar system have been discovered this way.

We have always suspected that planets might be a common by-product of the formation of stars when giant clouds

COSMIC CONNECTIONS Formation of the Solar System

An overview of our present-day understanding of how the solar system formed.

Vast, rotating cloud of gas and dust (solar nebula)

Increased density, temperature

Protosun forms, begins to grow

Tens of millions of years

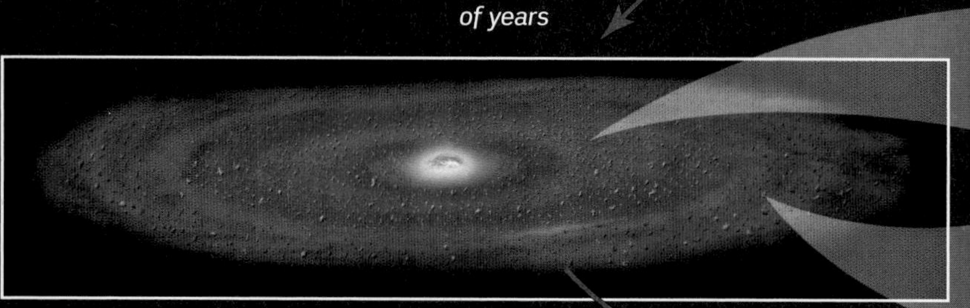
Protoplanetary disk forms, rock and ice particles collide, start to form planetesimals

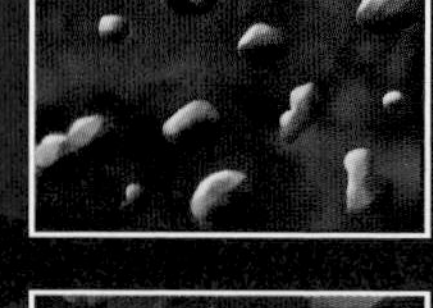
Inner disk: rock particles collide, grow

Outer disk: ice particles collide, attract gas

Several hundred million years

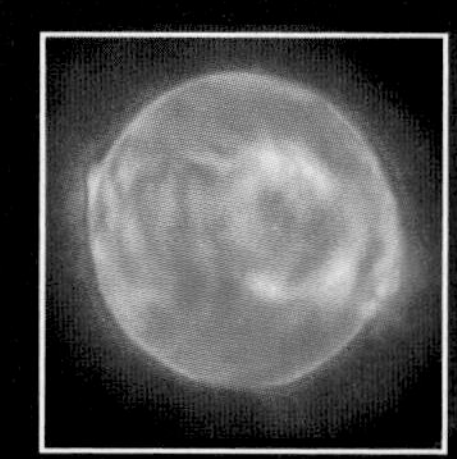
Terrestrial planet in molten state

Accretion of terrestrial planets, protosun becomes hot enough for nuclear fusion to begin

Several hundred million years

Jovian planets accrete from gas in outer disk, terrestrial planets heat up, begin chemical differentiation

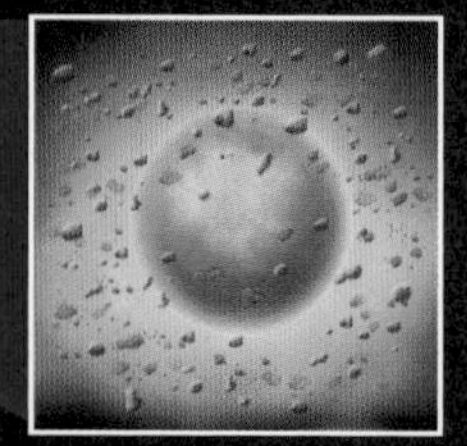
Jovian planet with core of rock, ices

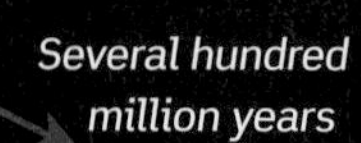

Several hundred million years

T Tauri wind sweeps away gas and dust, leaving planets, moons, asteroids, Kuiper belt, Oort cloud

of dust and gas collapse under gravity into a star and planetary system. The photographs of actual extrasolar planets orbiting other stars are a tremendous vindication of our theory of star and planetary system formation. It would also tell us that our own planetary system is not unique in the universe. Because at least one planet in our solar system—Earth—has the ability to support life, perhaps other planetary systems could also harbor living organisms.

Looking for Transiting Planets

The process of finding exoplanets by directly photographing them is an extremely difficult and time-consuming process. In other words, it isn't very efficient. Instead, the vast majority of planets are found by indirect methods. One of the most surprisingly useful is called the **transit method.** This method looks for the rare situation in which a planet comes between us and its parent star, an event called a transit (**Figure 4-19**). As in a partial solar eclipse (Section 1-6), this causes a small but measurable dimming of the star's light. If a transit is seen, the orbit must be nearly edge-on to our line of sight. As of May 2013, 308 extrasolar planets have been positively identified using ground-based telescopes. But that's not all. Using the NASA Kepler space observatory, a space telescope specifically designed to look for transiting extrasolar planets, another 865 extrasolar planets have been positively identified, and another 2781 highly likely candidates have already been identified since its launch in 2007. Because only a small fraction of exoplanets might have orbits that line up the exoplanet precisely between their host stars and Earth, causing an observable transit, astronomers now estimate that there could be as many as 17 billion Earth-sized exoplanets in our Milky Way Galaxy alone.

In addition to detection, there are at least three benefits of the transit method. One, the amount by which the star is dimmed during the transit depends on how large the planet is, and so tells us the planet's diameter (Figure 4-19*a*). Another benefit is that a star's light passes through the planet's atmosphere while it is transiting and certain wavelengths are absorbed by the atmospheric gases. This absorption affects the spectrum of starlight that we measure and thus allows us to determine the composition of the planet's atmosphere (Figure 4-19*b*). Finally, with an infrared telescope it is possible to detect a slight dimming when the planet goes behind the star (Figure 4-19*c*). That is because the planet emits infrared radiation due to its own temperature, and this radiation is blocked when the planet is behind the star. Measuring the amount of dimming tells us the amount of radiation emitted by the planet, which in turn tells us the planet's surface temperature.

Measuring Stellar Doppler Shifts

The most powerful method available to astronomers for finding and confirming exoplanets is to search for stars that appear to "wobble." If a star has a planet, it is not quite correct to say that the planet orbits the star. Rather, both the planet and the star move in elliptical orbits around a point that is the center of mass. As an example, consider that the Sun and Jupiter both orbit their common center of mass with an orbital period of 11.86 years. (Jupiter has more mass than the other seven planets put together, so it is a reasonable approximation to consider the Sun's wobble as being due to Jupiter alone.) Jupiter's orbit is about 5 times

Planets orbiting other stars appear to be quite common in our Galaxy.

(a)

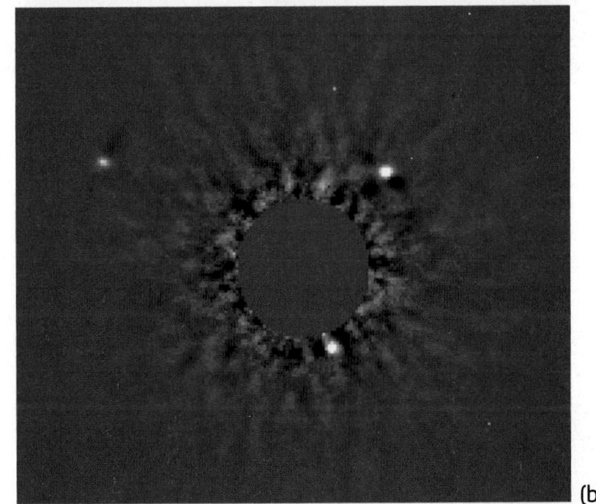
(b)

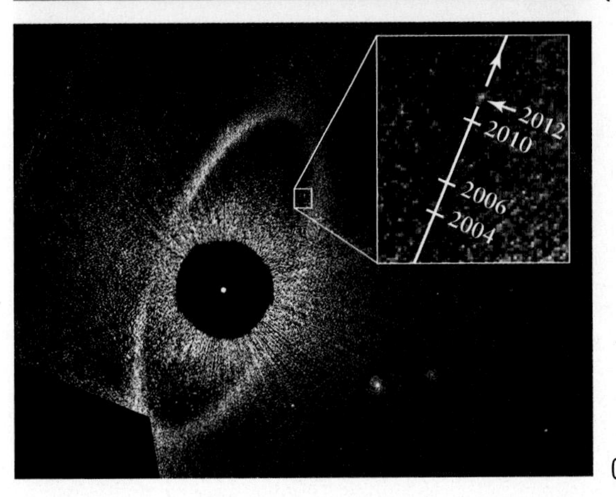

(c)

FIGURE 4-18 R I V U X G **First Direct Images of Extrasolar Planets Orbiting Another Star** Since exoplanets are typically lost in the overbearing glow of their host stars, directly observing exoplanets orbiting other stars is extremely challenging. (a) At a distance of about 500 ly away, the giant exoplanet visible in the upper left orbiting 1RXS is about 8 times more massive than Jupiter and orbits at a great distance—nearly 300 AU—from its star. (b) This image shows the first family of exoplanets ever captured, with three planets visible as tiny white dots with the bright central star blocked out. (c) This false-color composite image, taken from space with the Hubble Space Telescope, reveals the orbital motion of the planet Fomalhaut b following a 2000-year-long, highly elliptical orbit. The black circle at the center of the image blocks out the light from the bright star, allowing reflected light from the debris belt left over from formation of the planetary system and this planet to be photographed. *(a: Gemini Observatory; b: Gemini Observatory/NRC/AURA/Christian Marois et al.; c: NASA, ESA, and P. Kalas [University of California, Berkeley, and SETI Institute])*

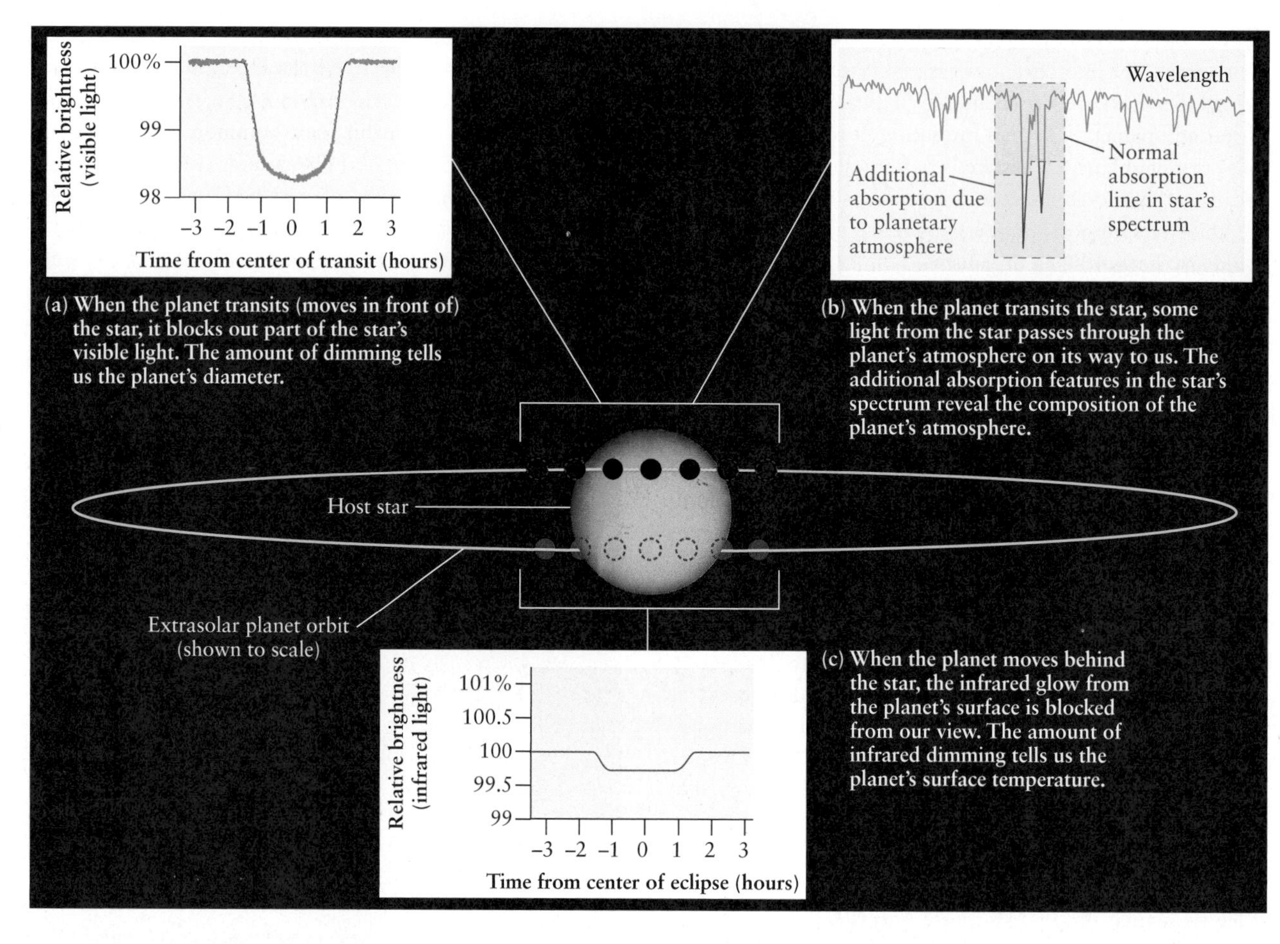

FIGURE 4-19
A Transiting Extrasolar Planet If the orbit of an extrasolar planet is nearly edge-on to our line of sight, we can learn about the planet's (a) diameter, (b) atmospheric composition, and (c) surface temperature. *(S. Seager and C. Reed,* Sky and Telescope*; H. Knutson, D. Charbonneau, R. W. Noyes [Harvard-Smithsonian CfA], T. M. Brown [HAO/NCAR], and R. L. Gilliland [STScI]; A. Feild [STScI]; NASA/JPL-Caltech/ D. Charbonneau, Harvard-Smithsonian CfA)*

the Sun-Earth distance, 7.78×10^8 km, while the Sun's wobble is quite small, just 742,000 km, but detectable. The Sun's radius is 696,000 km, so the Sun slowly wobbles around a point not far outside its surface. If alien astronomers elsewhere in the Galaxy were to detect the Sun's slight wobbling motion in response to its gravitational attraction with Jupiter, they could tell that there was a large planet (Jupiter) orbiting our Sun. They could even determine the planet's mass and the size of its orbit, even though Jupiter is small enough that it would be seemingly invisible.

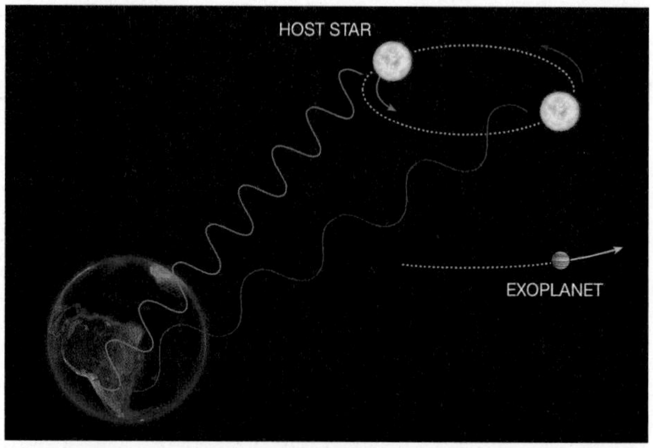

FIGURE 4-20 Detecting an Exoplanet Through Doppler Shift of Its Star
The radial velocity method indirectly detects a tiny and nonvisible exoplanet gravitationally pulling on its host star during its orbit. When the star moves toward us because its planet is moving away from us, its spectrum is blueshifted, while it is alternatively redshifted when the star moves away from us as its planet moves toward us. *(ESO)*

Detecting the wobble of other stars is not an easy task, and we rarely actually see a distant star wobble back and forth because of an orbiting planet pulling on it. Instead, astronomers use the **radial velocity method** (**Figure 4-20**). This is based on the Doppler effect, which we described in Section 2-5. A wobbling star will alternately move away from and toward Earth. This will cause the dark absorption lines in the star's spectrum to change their wavelengths in a periodic fashion. When the star is moving away from us, its spectrum will undergo a redshift to longer wavelengths. When the star is approaching, there will be a blueshift of the spectrum to shorter wavelengths. These wavelength shifts are very small because the star's motion around its orbit is quite slow. As an example, the Sun moves around its small orbit at only 28 mi/h (45 km/h). If the Sun were moving directly toward an observer at this speed, the hydrogen absorption line at a wavelength of 656 nm in the Sun's spectrum would be shifted by only 2.6×10^{-5} nm, or about 1 part in 25 million.

Detecting these tiny shifts requires extraordinarily careful measurements and painstaking data analysis, but it is tremendously fruitful. Since 1995, astronomers have used the radial velocity method to discover more than 500 exoplanets from ground-based observatories, and this same strategy is used to confirm suspected exoplanets identified by the *transit method,* described above.

Microlensing by Extrasolar Planets

In 2004 astronomers began to use a property of space discovered by Albert Einstein as a tool for detecting extrasolar planets. Einstein's general theory of relativity makes the remarkable statement that a star's gravity can deflect the path of a light beam just as it deflects the path of a planet or spacecraft. If a star drifts through the line of sight between Earth and a more distant star, the closer star's gravity acts like a lens that focuses the more distant star's light. Such microlensing causes the distant star's image as seen in a telescope to become brighter. If the closer star has a planet, the planet's gravity will cause a secondary brightening whose magnitude depends on the planet's mass (**Figure 4-21**). Using this technique, astronomers have detected a handful of extrasolar planets at distances up to 21,000 ly and expect to find many more.

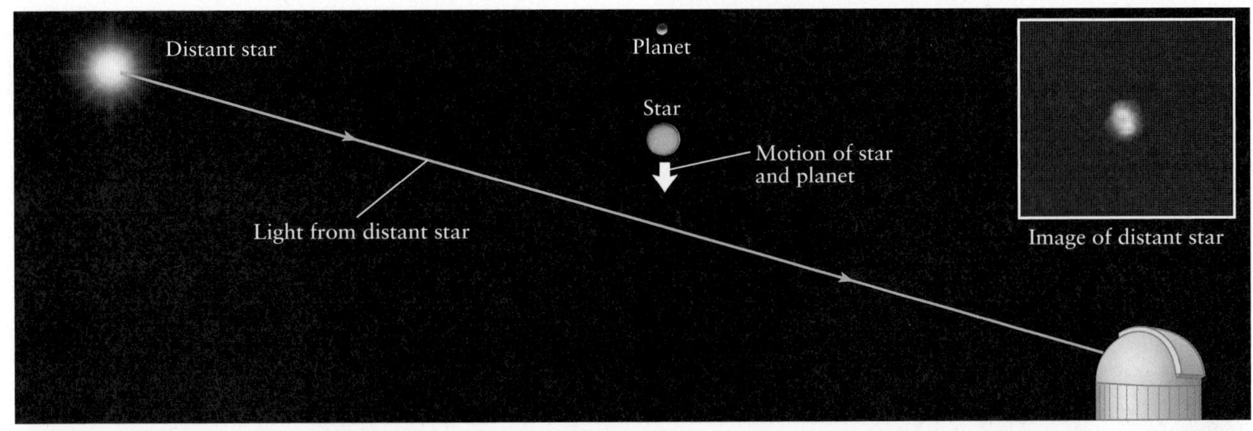

(a) No microlensing

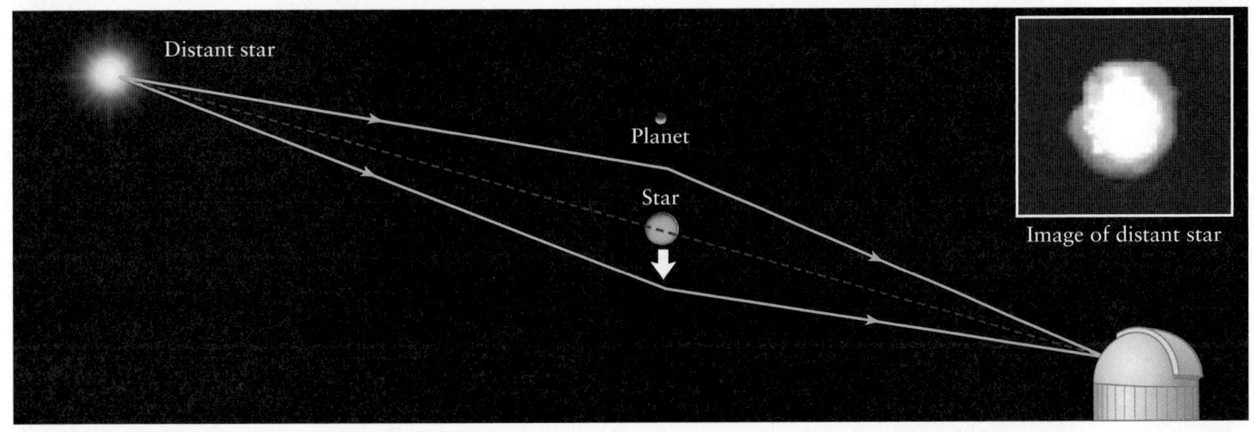

(b) Microlensing by star

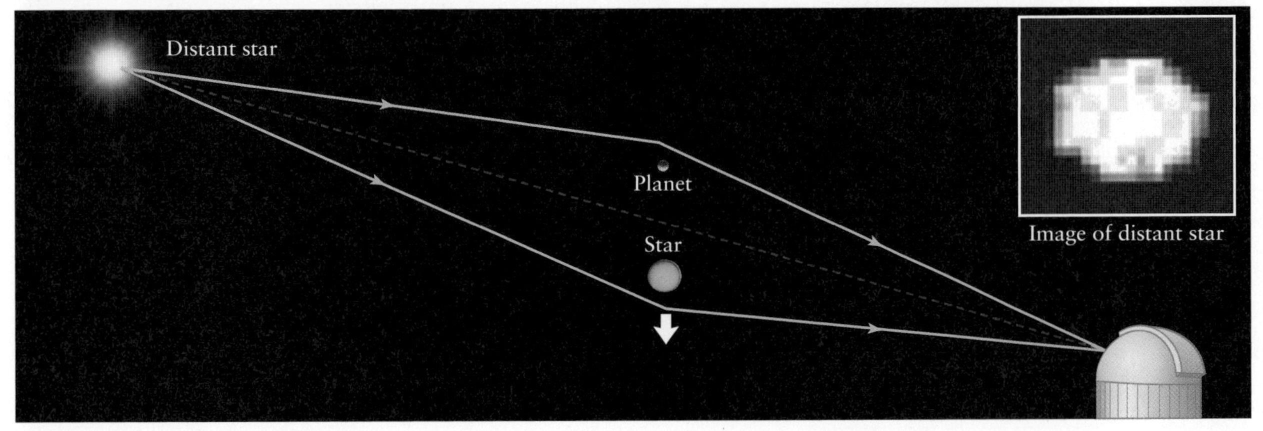

(c) Microlensing by star and planet

FIGURE 4-21
Microlensing Reveals an Extrasolar Planet
(a) A star with a planet drifts across the line of sight between a more distant star and a telescope on Earth. (b) The gravity of the closer star bends the light rays from the distant star, focusing the distant star's light and making it appear brighter. (c) The gravity of the planet causes a second increase in the distant star's brightness.

FIGURE 4-22 **Planets Orbit Most Stars** This artist's illustration shows how common it seems for planets to be orbiting stars throughout the Milky Way. In this illustration, exoplanets, their orbits, and their host stars are all vastly magnified compared to their real separations. *(ESO/M. Kornmesser)*

This *microlensing* technique has only discovered a confirmed 19 exoplanets as of this writing, but this number came from observations covering only a very small portion of the sky. If 19 planets were discovered in just one small region of the sky, astronomers estimate that there could be more than 100 billion planets orbiting stars in our Galaxy alone—an average of about one planet per star, which is the largest estimate ever calculated (**Figure 4-22**).

ConceptCheck 4-17: **Why will the radial velocity method fail to discover an extrasolar planet if the extrasolar planet's orbit is oriented perpendicular to the line between Earth and the wobbling star?**

CalculationCheck 4-4: **How long would the transit last if the star was only half as wide but with the same mass?**

VISUAL LITERACY TASK

Solar System

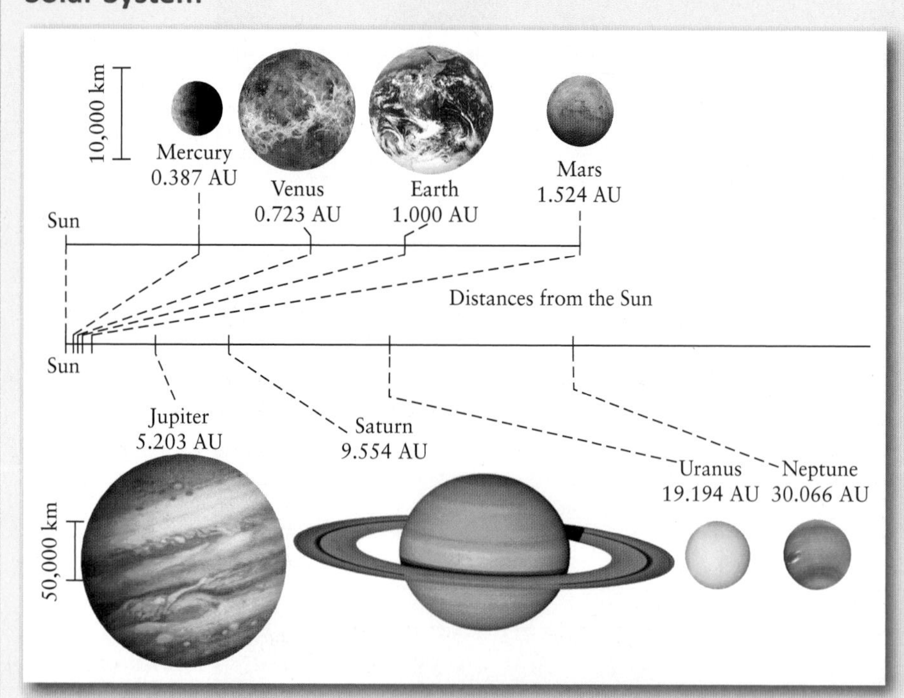

PROMPT: What would you tell a fellow student who said, **"Except for planet Pluto, the largest planets are closest to the Sun, with Mars and Neptune being about the same size and distance"?**

ENTER RESPONSE:

Guiding Questions:

1. Pluto, unlike any of the other planets in size, composition, and orbital shape, is
   a. actually a planet, albeit a small one.
   b. always the most distant object orbiting the Sun.
   c. about the same size as Earth.
   d. classified as a dwarf planet.

2. The largest planets are found orbiting
   a. near the Sun.
   b. at great distances.
   c. in both small and large orbital paths.
   d. relatively close, but just slightly beyond Mars's orbit.

3. Compared to Earth's distance from the Sun, Jupiter's distance is about
   a. 5 times larger.
   b. 50 times larger.
   c. the same, but Jupiter orbits much slower.
   d. the same, but Jupiter orbits a different Sun.

4. If Neptune is 4 times larger than Earth and Mars is half the size of Earth, the size comparison of Mars to Neptune is
   a. Neptune is 2 times larger than Mars.
   b. Neptune is 8 times larger than Mars.
   c. about equal.
   d. about 38 AU.

KEY IDEAS AND TERMS

4-1 The solar system has two broad categories of planets orbiting the Sun: terrestrial (Earthlike) and Jovian (Jupiterlike)

- Two defining characteristics of a **planet** are its orbit around a star and its diameter being large enough to form a spherical shape.
- A planet's **average density** is its mass divided by volume, measured in kilograms per cubic meter (kg/m^3), which characterizes its interior.
- All four inner, **terrestrial planets** have dense iron cores and hard, rocky surfaces with mountains, craters, valleys, and volcanoes.
- All four outer, **Jovian planets** are relatively large with low densities and are Jupiterlike, having thick atmospheres.

4-2 Seven large moons are almost as big as the inner, terrestrial planets

- Naturally forming, rocky moons orbit the Jovian planets much as Earth's Moon orbits Earth.
- The Jovian planets have many natural moons that are almost as large as the terrestrial planets.

4-3 Spectroscopy reveals the chemical composition of the planets

- Spectroscopy reveals the chemical composition of planets both with and without a surrounding envelope of gas, known as its **atmosphere.**
- The inner planets have solid surfaces composed primarily of silicates.
- The outer planets have thick gaseous atmospheres rich in hydrogen, helium, and methane.

4-4 Small chunks of rock and ice also orbit the Sun: asteroids, trans-Neptunian objects, and comets

- Most **asteroids** are rocky objects orbiting our Sun in an **asteroid belt** between the orbits of Mars and Jupiter, the largest of which are known as **minor planets.**
- **Dwarf planets** are the largest of the **trans-Neptunian objects** (TNOs) orbiting our Sun at and beyond Neptune's orbit.
- Icy balls known as **comets** that orbit our Sun just beyond the orbits of the trans-Neptunian objects are found in the **Kuiper belt,** whereas the most distant comets held by our Sun's gravitational attraction are part of the **Oort cloud.**

4-5 The Sun and planets formed from a rotating solar nebula

- A collection of dust and gas illuminated by the star it surrounds is known as a **nebula.**
- The **nebular hypothesis** is that our solar system formed by a flattening and spinning collapse of an initial **solar nebula,** which explains why all planets orbit in the same direction in the same plane.
- This constant relationship between the size of an object and its rotation speed is an example of a general principle called the **conservation of angular momentum.**
- The flattening disk eventually forming the planets of our solar system surrounding our early **protosun** is known as a **protoplanetary disk** or **proplyd.**

4-6 The planets formed by countless collisions of dust, rocks, and gas in the region surrounding our young Sun

- In the early solar system, whether a planet was made of solid or a gas depended on a material's **condensation temperature.**
- Planets formed first as kilometer-sized **planetesimals,** which gravitationally combined through **accretion** to become **protoplanets.**
- In a molten protoplanet, material is free to move, and through **chemical differentiation,** denser, iron-rich minerals sank to the centers of the planets while the less dense, silicon-rich minerals floated to their surfaces.
- Young stars emit a **T Tauri wind** of mass similar to how the present-day Sun loses mass from its outer layers in the form of high-speed electrons and protons known as the **solar wind.**

4-7 Understanding how our planets formed around the Sun suggests planets around other stars are common

- Planets orbiting stars other than our Sun are called **extrasolar planets.**
- Indirect detection using the **transit method,** measuring a star's dimming, and the **radial velocity method,** measuring tiny shifts in a star's position, find more exoplanets than direct imaging.

QUESTIONS

Review Questions

1. What are the two agreed upon defining characteristics of a planet?
2. Compare the characteristics of a terrestrial planet to that of a Jovian planet.
3. In what ways are the largest moons similar to the terrestrial planets? In what ways are they different? Which moons are largest?
4. What is meant by the average density of a planet? Do all the planets orbit the Sun in the same direction? Are all of the orbits circular?
5. What is an asteroid? What is a trans-Neptunian object? In what ways are these minor members of the solar system like or unlike the planets?
6. What are the asteroid belt, the Kuiper belt, and the Oort cloud? Where are they located? How do the objects found in these three regions compare?
7. In what ways is Pluto similar to a terrestrial planet? In what ways is it different?
8. What is the connection between comets and the Kuiper belt? What is the connection between comets and the Oort cloud?
9. Imagine a trans-Neptunian object with roughly the same mass as Earth but located 50 AU from the Sun. (a) What do you think this object would be made of? Explain your reasoning. (b) On the basis of this speculation, assume a reasonable density for this object and calculate its diameter. How many times bigger or smaller than Earth would it be?
10. What is the nebular hypothesis?
11. What is a protosun? What causes it to shine? Into what does it evolve?
12. What are proplyds? What do they tell us about the plausibility of our model of the solar system's origin?
13. (a) What is meant by accretion? (b) Why are the terrestrial planets denser at their centers than at their surfaces?
14. Explain how our current understanding of the formation of the solar system can account for the following characteristics of the solar system: (a) All planetary orbits lie in nearly the same plane. (b) All planetary orbits are nearly circular. (c) The planets orbit the Sun in the same direction in which the Sun itself rotates.

15. Explain why most of the moons of Jupiter orbit that planet in the same direction that Jupiter rotates.
16. What are the differences between radial velocity and the transit method of extrasolar planet detection?

Web Chat Questions

1. Propose an explanation of why the Jovian planets are orbited by terrestrial-like moons.
2. Suppose that a planetary system is now forming around some protostar in the sky. In what ways might this planetary system turn out to be similar to or different from our own solar system? Explain your reasoning.
3. Suppose astronomers discovered a planetary system in which the planets orbit a star along randomly inclined orbits. How might a theory for the formation of that planetary system differ from that for our own?

Collaborative Exercises

1. Imagine that scientists are proposing to send a robotic lander to visit Jupiter's Callisto. Create a 100-word written proposal describing why you would most like to send a robotic lander to another one of the Galilean moons. Explain why your group found it to be the most interesting and why the government should allocate the money for your alternative project. In your proposal, be sure to demonstrate your knowledge of Callisto and at least one other moon.
2. Find objects in the room or among your possessions that can be used to create a reasonably accurate scale model of the planets of our solar system. Try finding a small object to represent Mercury first.

Observing Questions

1. Use *Starry Night*™ to observe the motions of two smaller objects in the solar system, the dwarf planet Ceres and the minor planet Pallas, both members of the asteroid belt, as seen from Earth. Open **Favourites > Explorations > Atlas** to view the sky from the center of a transparent Earth. Select **View > Ecliptic Guides > The Ecliptic** from the menu and set the **Time Flow Rate** to **1 sidereal day.** Open the **Find** pane, type the name **Ceres** in the search box, and press the **Enter** key to center the view on this dwarf planet. Click the **Play** button and observe the motion of Ceres over the course of at least two years of simulated time. (a) Describe how Ceres moves. (b) How can you tell that Ceres orbits the Sun in the same direction as the planets? (c) Return to the Find pane, type the name **Pallas** in the search box, and press the Enter key to center the view upon this minor planet. Watch the motion of Pallas in the sky for at least two years of simulated time. How does the motion of Pallas compare with that of Ceres? (d) Which object's orbit is more steeply inclined to the plane of Earth's orbit? How can you tell?
2. Use *Starry Night*™ to explore some of the dwarf planets of the solar system. Select **Favourites > Explorations > Dwarf Planets** from the menu. This view, from a position in space about 97 AU from the Sun, shows Neptune's orbit as well as the orbits of several dwarf planets. Right-click (Ctrl-click on a Mac) on the Sun and select **Centre** from the contextual menu. (a) Do the dwarf planets revolve about the Sun in the same or opposite direction as the major planets? (b) Use the location scroller to adjust the view so that the plane of Neptune's orbit appears edge-on. How do the orbital planes of the dwarf planets compare to those of the planets?
3. Use *Starry Night*™ to observe a comet. Select **Favourites > Explorations > Hale-Bopp.** (a) From the appearance of this comet, predict the direction of the Sun relative to the comet and explain how you made your prediction. Use the hand tool to find the Sun in the view to verify your hypothesis. (b) Select **File > Revert** from the menu. From the appearance of the comet, can you predict what its motion will be in the future against the background stars? Click the **Play** button to see if your prediction was correct.
4. Use *Starry Night*™ to make observations of the solar system. Select **Favourites > Explorations > Solar System.** The view shows the names and orbits of the major planets of the solar system against the backdrop of the stars of the Milky Way Galaxy, from a location hovering 64 AU from the Sun. You may also see many smaller objects moving in the asteroid belt between the orbits of Mars and Jupiter. (If not, select **View > Solar System** and click on the **Asteroids** box.) (a) Use the location scroller to look at the solar system from different angles, and observe the general distribution and motion of the major planets. Make a list of your observations. (b) How does the nebular hypothesis of solar system formation account for your observations?
5. Use *Starry Night*™ to investigate stars that have planets orbiting them. Click the **Home** button in the toolbar. Open the **Options** pane and use the checkboxes in the Local View layer to turn off **Daylight** and the **Local Horizon.** Expand the **Stars** layer in the Options pane and then expand the Stars item and check the **Mark stars with extrasolar planets** option. Then use the **Find** pane to find and center each of the stars listed below. To do this, click the magnifying glass icon on the side of the edit box at the top of the Find pane and select **Star** from the drop-down menu; then type the name of the star in the edit box and press the Enter or Return key on the keyboard. Click on the **Info** tab for full information about the star. Expand the **Other Data** layer and note the luminosity of each of these stars. (a) Which stars are more luminous than the Sun? (b) Which are less luminous? (c) How do you think these differences would have affected temperatures in the nebula in which each star's planets formed? (i) 47 Ursae Majoris (three known planets); (ii) 51 Pegasi (one known planet); (iii) 70 Virginis (one known planet); (iv) Rho Coronae Borealis (one known planet).

ANSWERS

ConceptChecks

ConceptCheck 4-1: Our solar system contains only one star, the Sun. The other stars are far, far away.

ConceptCheck 4-2: Mars is classified as an inner planet, even though it is farther from the Sun than is Earth. This is because the Earthlike planets all orbit very close to the Sun (within 2 AU), while the outer planets (Jupiter, Saturn, Uranus, and Neptune) all lie much farther (more than 5 AU) from the Sun.

ConceptCheck 4-3: If a planet's overall density is the same as the density of rocks at its surface, one can infer that the rocks deep in its core must be the same density as the rocks found at its surface.

ConceptCheck 4-4: Table 4-2 lists four moons having diameters greater than 3476 km: Io, Ganymede, Callisto, and Titan.

ConceptCheck 4-5: Before sunlight is reflected off a planet, the incoming sunlight must interact with the chemical elements in the planet's atmosphere, altering it in predictable and important ways that allow astronomers to make inferences about what chemical elements are present in the planet's atmosphere.

ConceptCheck 4-6: The spectrum of a solid surface shows broad features whereas the spectrum observed from light passing through a gas shows sharp, definite spectral lines.

ConceptCheck 4-7: Jupiterlike planets are composed of mostly gases and ices of hydrogen and helium, which would normally not seem to be able to combine together to form a planet very close to a hot star. One plausible explanation for recently discovered Jupiterlike planets that are found orbiting nearby stars in the high-temperature region is that these planets formed far from the star and somehow moved to an orbit close to the star after the planet formed.

ConceptCheck 4-8: Ceres has a diameter of about 900 km. This is about the same size as a large U.S. state, such as the length of California.

ConceptCheck 4-9: Because broad categories overlap due to the history of how and when objects were first discovered, Pluto is classified as a dwarf planet, a Kuiper-belt object, and a trans-Neptunian object; it is not, however, an asteroid.

ConceptCheck 4-10: Whereas the Kuiper belt lies in the same plane as Earth's orbit around the Sun, the Oort cloud is a spherical distribution of comets that completely surrounds the solar system. If a comet has an orbit that is considerably different from that of the flat plane of the solar system, it most likely originated in the spherical Oort cloud that exists in all directions around our solar system.

ConceptCheck 4-11: The chemical element of carbon that makes up much of our human bodies and the living plants all around us and the very oxygen atoms we breathe was made inside a star that no longer exists.

ConceptCheck 4-12: To be correct, the nebular hypothesis must explain why all the planets orbit the Sun in the same direction and in nearly the same plane. This arrangement is unlikely to have arisen purely by chance.

ConceptCheck 4-13: The closest a large, Jupiterlike planet composed primarily of hydrogen and helium could form would be much farther away in a much hotter solar nebula.

ConceptCheck 4-14: Called chemical differentiation, denser, iron-rich materials migrated to the centers of the planets while the less dense, silicon-rich minerals float to the outside surface before the initially molten planet solidified in the early solar system.

ConceptCheck 4-15: Neptune and Uranus are more than twice as far from the Sun than Saturn, and in the early solar nebula, there would have been far too little material for them to collect and build a planet.

ConceptCheck 4-16: Jets from the Sun, a T Tauri wind, and accretion onto the planets are the processes that characterize the final stages of solar system evolution.

ConceptCheck 4-17: An extrasolar planet with an orbit that causes a star to wobble side to side will not exhibit any Doppler-shifted spectra as seen from Earth because the star will not be moving alternatively toward and away from Earth, and therefore the radial velocity method will fail to detect such an orbiting extrasolar planet.

CalculationChecks

CalculationCheck 4-1: The shape of a planet's orbit is given by the value of its eccentricity. The closer this value is to zero, the closer the orbit's shape is to that of a perfect circle. According to Table 4-1, the orbit of Venus has the eccentricity closest to zero (0.007), making it the most circlelike of all planetary orbits.

CalculationCheck 4-2: If we divide Saturn's 120,536-km diameter up into Earth's 12,756-km diameter, we find that 120,536 km ÷ 12,756 km = 9.449, so about 9½ Earths would fit across Saturn's diameter.

CalculationCheck 4-3: The asteroid belt is located between Mars and Jupiter at about 3 AU from the Sun, whereas the much larger Kuiper belt is beyond the orbit of Neptune and is located between about 30 AU and 50 AU from the Sun.

CalculationCheck 4-4: The brightness versus time graph in Figure 4-19*a* shows the transit lasting 3 hours. If the star was half the present size, it would take half the time, or 1½ hours.

R I V U X G The two hemispheres of Earth. *(NASA)*

Uncovering Earth's Systems

CHAPTER LEARNING OBJECTIVES **By reading the sections of this chapter, you will learn:**

- **5-1** Most of Earth's surface is covered with flowing water that radically changes the landscape
- **5-2** Earth is surrounded by a thin, multilayered envelope of gas that has changed since life became prominent
- **5-3** Volcanoes and earthquakes reveal energy from a molten interior driving Earth's surface to shift positions
- **5-4** Earth's magnetic field emanates from its spinning, molten interior to create a protective shield from the Sun's harmful radiation
- **5-5** A rapidly growing population is altering our planetary habitat

When you think of astronomy, you might have visions of standing outside looking up into the night sky and trying to recognize constellations or patiently waiting for a shooting star to pass by. Or, you might imagine landscapes of faraway planets orbiting a distant sun, or even of enormous swirling galaxies of stars at the edge of the universe. It might, however, surprise you to know that many astronomers spend much of their careers studying the planet beneath their feet—our home planet Earth. Just like the Sun is one of the most important stars to study because it is so close, Earth is one of the most important planets for astronomers to study because it is the planet we know the best.

Of all the worlds of the solar system, only Earth is uniquely a blue planet, because most of its surface is covered with liquid water. We will learn that water exists in many places throughout the solar system, but water can exist on Earth's surface in all three of its forms: solid ice, liquid water, and gaseous water vapor. The image above shows our water-covered planet, with its distinctive green patches of vegetation dotting the surface and swirling white clouds of water vapor moving just above the surface. Compared with the Moon, whose seemingly dry surface has been ravaged by billions of years of impacts by interplanetary debris, Earth initially seems to be an inviting and tranquil place.

Yet Earth's quiet outer appearance is deceiving. The seemingly solid surface of the planet is in a state of slow but constant motion, driven by energy from flowing molten rock within Earth's interior. At the same time, energy from sunlight drives wind in Earth's atmosphere, the thin layer of gas surrounding our planet, to rapidly move and churn, resulting in sometimes violent weather moving water and energy around our planet.

Unseen in these images is an immense collection of tiny subatomic particles that rapidly wander around the outside of Earth in two giant belts, shaped by Earth's magnetic field. And equally unseen are trace gases in Earth's atmosphere whose delicate abundances may determine the future of Earth's fragile climate, and on which may depend the survival of entire species, humans included.

In this chapter our goal is to learn about Earth's systems—its hot, active interior, its ever-changing surface, and its dynamic atmosphere—and how we interact with these systems as stewards of our planetary home. Understanding Earth from the outside, as an astronomer sees our home planet, makes it easier to understand the similarities and differences of other planets in our solar system, the subjects of the next two chapters. ■

5-1 Most of Earth's surface is covered with flowing water that radically changes the landscape

Upon first glance, Earth is unquestionably a blue planet. Unique among all known planets and moons discovered so far, Earth's surface is mostly covered by liquid water. Other objects in our solar system have surprisingly large amounts of water, as shown in **Figure 5-1**, but rarely so obviously covering much of the surface. This chapter's opening image showing our home planet is famously known as the "blue marble." It was painstakingly created by stitching together months and months of satellite images, giving us an unprecedented view of every square mile of Earth's surface. The water covering our planet has tremendous impacts on how the landscape is shaped, how rocks change, how our air moves, and how the surface temperature is maintained. This same water also influences what life-forms can live here.

FIGURE 5-1 R I V U X G **Earth's Water Amount in Comparison** If you were to collect all of Earth's water, it would be a ball about 860 mi in diameter. If you were to burst this imaginary ball of water, the resulting flow would cover the contiguous United States (lower 48 states) to a depth of about 107 mi. This is similar to how much water is suspected to be hiding on Mars and Saturn's moon Enceladus and far less than exists beneath the surfaces of Jupiter's moon Europa or Saturn's moon Titan. *(Planetary Habitability Laboratory, Univ. Puerto Rico Arecibo Observatory, NASA)*

Nearly all of Earth's water is in the form of liquid salt water like that found in the oceans, amounting to about 97%. The remaining 3% is largely locked up underground in the tiny pore spaces within rocks or as frozen ice near the poles. Only a very small fraction of Earth's water is easily accessible, found as a liquid in rivers and lakes with even a smaller fraction as gaseous water vapor in Earth's atmosphere. All of this water makes us wonder how it might impact the landscape and even how we live on our planet.

Earth's Water Cycle

Go to Video 5-1

One of the most important principles to know about living on a planet where the surface is dominated by liquid water is that water moves in various ways. One way that water moves is by flowing from one place to another, largely due to gravity. Water at the top of a mountain can flow downhill in the forms of streams and rivers, water on the ground can seep deep into the soil and even into and through rocks, and water in the air can fall downward as rain.

Liquid water can also move from one part of the ocean to another. One way this can occur is in the form of water waves driven by the wind or storms. Another way water can move in the ocean is in giant rivers driven by differences in water temperature or by differences in the amount of salt in seawater. Still another way water moves is by Earth's spinning and the gravitational influences with our Sun and Moon.

A quite different, but still important, way that water moves is by changing in and out of its liquid form. When water freezes into a solid, it is called ice. Ice can move around the ocean as icebergs, or even fall from the sky as snow or hail. Alternatively, when water turns into a gas, it is known as water vapor—or steam—and can sometimes be seen as clouds moving across the sky, as seen in Figure 5-2. This water vapor can move around Earth quite quickly. If the water vapor in the sky turns back into a liquid, it can fall to Earth as rain. If this water vapor in the air freezes, it can fall to Earth as snow or hail. There are other ways water moves around our planet, but the important point is that when water changes forms among its liquid, solid, and gaseous states, it can quickly transfer vast amounts of energy around our spinning planet. This energy drives our short-term weather and long-term climate systems.

This movement of water around our planet in different forms through rising evaporation and falling precipitation—including flows in the atmosphere, across the surface, under the ground, and along ocean currents— is known as the **water cycle.** As illustrated in **Figure 5-2,** the water cycle is dominated by an exchange of water between the oceans and the atmosphere. This is because our planet is largely covered by oceans, rather than covered by land. Energy from the Sun evaporates water from the oceans, which rises in the form of water vapor. This water vapor then cools as it moves higher into the sky and, under the right conditions, changes back into liquid and precipitates down into the oceans as rain. Driven by the Sun's energy, the water cycle is the continuous flow of water among the oceans, the atmosphere, and the land.

This process of evaporation of liquid into water vapor and precipitation downward as rain or snow also occurs over land, but in much smaller amounts, mostly because there is more oceanscape than landscape. Some of the rain that falls on land runs off downhill into streams, rivers, lakes, and the oceans. Much rain soaks into the ground and enters rock through cracks or tiny—nearly invisible—pore spaces within rocks. This underground water can then slowly flow toward rivers, lakes, or oceans. Alternatively, this water could evaporate back into the atmosphere. Another portion of this water is taken up by plants and eventually returned to the atmosphere. Snow that falls layer after layer and does not melt could stay locked up in giant ice fields for nearly a million years.

Earth's water cycles through the oceans, the sky, and the land, with the total amount virtually unchanging.

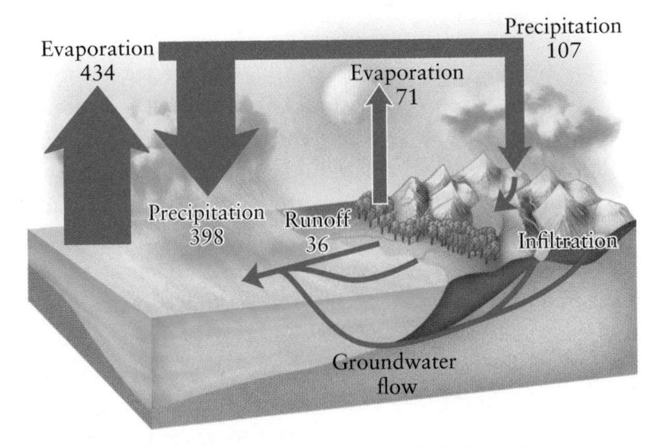

FIGURE 5-2 Earth's Water Cycle The water cycle is the continuous movement of water through Earth's crust, atmosphere, oceans, lakes, and streams. The numbers indicate the amount of water (in thousands of cubic kilometers per year) that flows among these reservoirs annually.

An important point to remember is that water has tremendous influences as it continuously moves through the water cycle. What we are probably most aware of is that this movement of water results in rain that helps crops grow. Perhaps less obvious to us is that flowing water can rapidly change the landscape. As water moves across the surface, whether as waves relentlessly pounding into a cliff or as a river digging deep canyons like the Grand Canyon, it continuously carves away at Earth's surface, reshaping the landscape. Nearly invisible to us, but perhaps the most important, is that the water cycle helps move energy around our planet, impacting our short-term weather system and influencing the nature of our long-term climate.

ConceptCheck 5-1: How does Earth's total amount of water change if precipitation falling on land seeps into deep cracks in rocks?

Water's Unique Properties

You probably remember that water is a molecule composed of two small hydrogen atoms and one larger oxygen atom, called H_2O. What you might not remember is that water has all sorts of fascinating properties. One is that when water gets cold enough to freeze, it expands in size. Perhaps this has unexpectedly happened to you when you put a plastic bottle of water in the freezer. This is a unique property, because most other substances get smaller and take up less space when they get colder. This is important for fish living in lakes that freeze in the winter, because ice takes up more space than the same amount of water. Ice also floats to the top of a lake, rather than sinking to the bottom, leaving space for fish to swim all winter long. This also impacts the landscape because when water seeps into rocks and expands upon freezing it breaks rocks apart.

Another important property of water is that the amount of heat needed to change the temperature of water is larger than many other substances. You might already know that swimming pools and lakes are quite slow to heat up in the summer and equally slow to cool down in the autumn. Water's great resistance to changing temperature compared to other substances makes it a stable substance that life can depend upon even when short-term weather changes the temperature quickly and dramatically. This same ability to hold tremendous amounts of energy allows water vapor to move energy around our planet quickly wherever the wind takes it.

Different sides of an individual water molecule have different characteristics, giving water additional unique properties. One is that water molecules tend to cling to each other, making them flow from one place to another in continuous streams while sticking together. This property accounts for water collecting together and falling from the sky in the form of spherical drops. Water also tends to stubbornly stick to other surfaces. This same property allows it to spread out evenly across a surface, like a lake, or even within a substance, like inside a rock. Another property, compared to other liquids, is that water has a wide ability to break apart and dissolve a large variety of substances, including common substances such as salt, sugar, and ammonia. When mixed with acid, such as pollution-forming acid rain, water can even dissolve solid rock, like marble and limestone. As far as we know, life cannot exist without water. It is these unique properties taken together that make water a highly useful substance for life to thrive on our planet.

FIGURE 5-3 R I V U X G **Earth's Atmosphere** Earth's atmosphere is an extremely thin layer of protective gas covering the entire planet, as seen from a camera mounted on a high-altitude balloon. Today's air is about 78% nitrogen and 21% oxygen, with the remaining mostly composed of water vapor, argon gas, and carbon dioxide. *(Dr. Emrys Hall and Allen Jordan, NOAA/OAR/GMD)*

ConceptCheck 5-2: **According to the water cycle, how would the weight of a cup of liquid water change after it had been placed in a freezer and turned into ice?**

5-2 Earth is surrounded by a thin, multilayered envelope of gas that has changed since life became prominent

As we look around the solar system at our nearest planetary neighbors, Venus and Mars, we notice that these planets have a relatively thin layer of protective gas surrounding them. This gaseous envelope of air surrounding a planet or moon is called an **atmosphere.** While Venus has an incredibly dense atmosphere, Mars has an atmosphere of very low density. Moreover, an analysis of sunlight reflected by the atmospheres of Venus and Mars shows that their air is largely carbon dioxide (CO_2). So, it might be somewhat surprising to learn that we see something quite different when we consider what makes up Earth's air. Our air is about 78% nitrogen and 21% oxygen, with the remaining mostly composed of water vapor, argon gas, and carbon dioxide (**Figure 5-3**). This is quite different than Venus and Mars.

We suspect that most planets' atmospheres initially formed in the same way. When planets like Earth first formed by accretion of planetesimals, as described in Section 4-6, gases were trapped within Earth's interior in the same proportions that they were present in the solar nebula. But since the early Earth was hot enough to be molten throughout its volume, most of these trapped gases eventually leaked into space, easily able to escape Earth's gravity. The atmosphere that remained still contained substantial amounts of hydrogen, but in the form of relatively massive molecules of water vapor (H_2O) made by combining two atoms of hydrogen with one atom of oxygen, the third most common element in our part of the Milky Way Galaxy. In fact, water vapor and carbon dioxide were probably the dominant constituents of the early atmosphere, just like on Venus and Mars today. Somehow our atmosphere changed substantially since the planets were formed about 4.56 billion years ago.

Earth's original CO_2 atmosphere was much like other planets until the emergence of life dramatically changed it.

Emerging Life's Impact on Earth's Atmosphere

Carbon dioxide would have been released into Earth's early atmosphere by erupting volcanoes and much of it would have been trapped within rocks. Carbon dioxide dissolves in rainwater and falls into the oceans, where it can combine with other substances to form a category of rocks known as *carbonates,* which form layers on the ocean floor. (Limestone and marble are examples of carbonate-bearing rock.) The end result is that the amounts of CO_2 that we see in the air today are the result of a natural balance between erupting volcanoes that release CO_2 into the atmosphere and the formation of carbonates that naturally remove CO_2 from the air.

The appearance of life on Earth set into motion a radical transformation of the atmosphere. Early single-cell organisms converted energy from sunlight into chemical energy using **photosynthesis,** a chemical process that consumes CO_2 and water and releases oxygen (O_2). As life proliferated and evolved on Earth, the amount of photosynthesis increased dramatically. Eventually O_2 began to accumulate in the atmosphere. **Figure 5-4** illustrates how the amount of O_2 in the atmosphere has increased over the history of Earth.

About 2 billion (2×10^9) years ago, a new type of life evolved to take advantage of the newly abundant oxygen.

These new organisms produce energy in a process very different from photosynthesis. Instead, these newly abundant life-forms consume oxygen and release carbon dioxide using a process called **respiration.** Respiration is the same process used by all modern animals, including humans. Oxygen-breathing organisms thrived because photosynthetic plants continued to add even more oxygen to the atmosphere.

Several hundred million years ago, the number of oxygen molecules in the atmosphere stabilized at 21% of the total, the same as the present-day value. This value represents a balance between the release of oxygen from plants by photosynthesis and the absorption of oxygen by animals, bacteria, and rocks. The role of life in the production of oxygen in the atmosphere is so important that when they look at planets beyond our own, astronomers look for evidence of oxygen as a hint that life might be present on that planet.

The most numerous molecules in our atmosphere are nitrogen (N_2), which make up 78% of the total. These, too, are a consequence of the presence of life on Earth. Certain bacteria use an energy generation process that extracts oxygen from chemicals called *nitrates,* and in the process nitrogen is released into the atmosphere. Volcanoes outgassing ammonia and ammonia's subsequent dissociation are another source of atmospheric nitrogen. Interestingly, the amount of atmospheric nitrogen is kept in check not by another living entity but by lightning occurring during thunderstorms. It turns out that the sudden burst of energy in a lightning flash causes nitrogen and oxygen in the air to combine into nitrogen oxides, which dissolve in rainwater, fall into the oceans, and form the nitrates that bacteria use to produce energy. What this means is that Earth not only relies on a water cycle, which we saw in the previous section, but there is also a continuous *carbon cycle* and a continuous *nitrogen cycle* running on Earth.

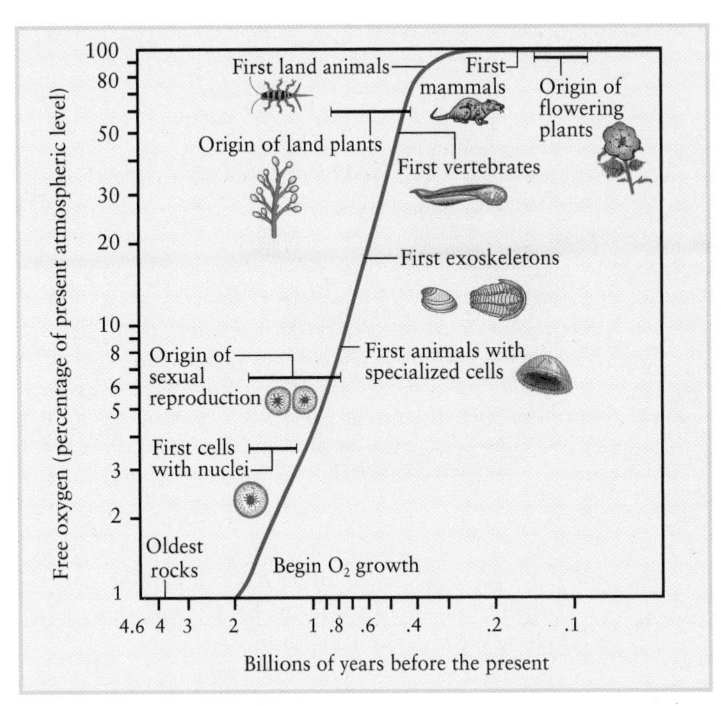

FIGURE 5-4 The Increase in Atmospheric Oxygen This graph shows how the amount of oxygen in the atmosphere (expressed as a percentage of its present-day value) has evolved with time. Note that the atmosphere contained essentially no oxygen until about 2 billion years ago *(Adapted from Preston Cloud, "The Biosphere,"* Scientific American, *September 1983, p. 176)*

ConceptCheck 5-3: What life process converts a predominately carbon dioxide atmosphere to a predominately nitrogen and oxygen atmosphere?

Pressure, Temperature, and Convection in the Atmosphere

We have looked at the contents of Earth's atmosphere, but one wonders how the atmosphere changes as you get farther away from Earth's surface. Imagine smoothly drifting upward in a hot air balloon. As you go up in the atmosphere, there is less and less air pressing down on you from above. Hence, the amount of air above you decreases with increasing altitude. **Atmospheric pressure** at any height in the atmosphere is caused by the weight of all the air above that height. The average atmospheric pressure at sea level is defined to be 1 **atmosphere** (1 **atm**), equal to 1.01×10^5 N/m^2 or 14.7 pounds per square inch.

At the same time, in our imaginary balloon ride, you would notice that the temperature also changes. However, you might be surprised to observe that, unlike atmospheric pressure, temperature varies with altitude in a complex way. **Figure 5-5** shows that temperature decreases with increasing altitude in some layers of the atmosphere, but in other layers actually *increases* with increasing altitude. These differences result from the individual ways in which each layer is heated and cooled.

The lowest layer, called the **troposphere,** extends from the surface to an average altitude of 7.5 mi (roughly 39,000 ft or 12 km). This region closest to Earth's surface is where most of Earth's weather and clouds are found. This part of the atmosphere is heated by energy emitted by Earth's Sun-warmed surface, rather than directly by the Sun itself. Sunlight efficiently warms Earth's surface, which heats the lowest part of our atmosphere. By contrast, the upper part of the troposphere has cooler temperatures because the air there is farther

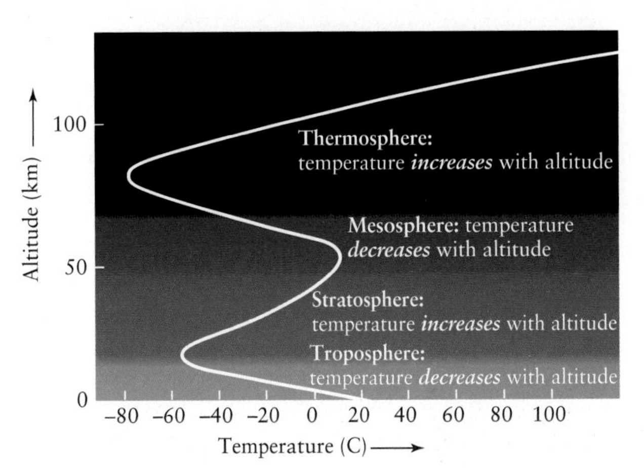

FIGURE 5-5 Temperature Profile of Earth's Atmosphere This graph shows how the temperature in Earth's atmosphere varies with altitude. In the troposphere and mesosphere, temperature decreases with increasing altitude; in the stratosphere and thermosphere, temperature actually increases with increasing altitude.

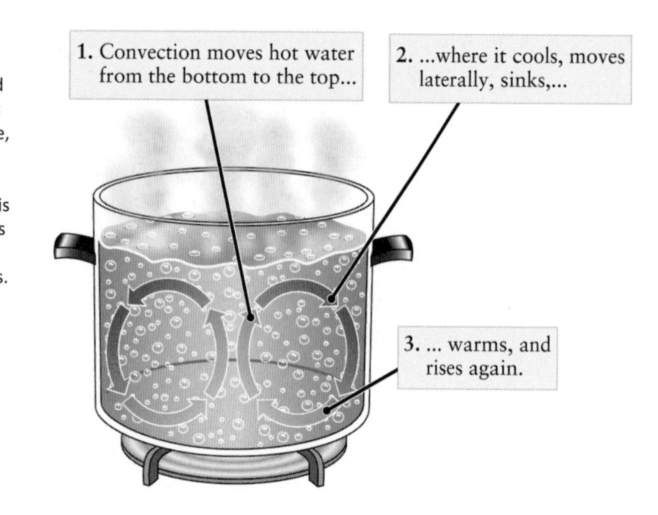

FIGURE 5-6 Convection in the Kitchen Heat supplied to the bottom of a pot warms the water there, making the water expand and lowering its average density. This low-density water rises and transfers heat to its cooler surroundings. The water that began at the bottom thus cools down, becomes denser, and sinks back to the bottom to repeat the process. *(Adapted from F. Press, R. Siever, T. Grotzinger, and T. H. Jordan,* Understanding Earth, *4th ed., W. H. Freeman, 2004)*

from Earth's warming surface. The movement of air and water vapor in this region is what causes much of Earth's weather.

Low-altitude air warmed by Earth's surface expands and becomes less dense. Because this air is now less dense than the cooler air above it, the warm air starts to rise, pushing the cooler air down toward the surface in the process. Of course, that cooler air is now able to be heated by Earth's Sun-warmed surface and it becomes less dense and moves upward. The previously warmed air now cools and contracts because it has moved away from Earth's warming surface and begins to descend, and the process starts all over again. This cycling process of warm air rising and cool air falling over and over again is known as **convection.** It is the same process that helps water to churn over and over in a boiling pot of water, as illustrated in **Figure 5-6.** Convection is a common process throughout the universe that we will encounter again and again: Convection accounts for the movement of energy within Earth and for how some energy generated in the cores of stars makes its way to the surface.

Convection on a grand scale is caused by the temperature difference between Earth's equator and its poles. If Earth did not rotate, heated air near the equator would rise upward and flow at high altitude toward the poles. There it would cool and sink to lower altitudes, at which it would flow back to the equator. However, Earth's rotation breaks up this simple convection pattern into a series of smaller convection circles. In these smaller circles, air flows east and west as well as vertically and in a north-south direction. The structure of these cells explains why the prevailing winds blow in different directions at different latitudes (**Figure 5-7**).

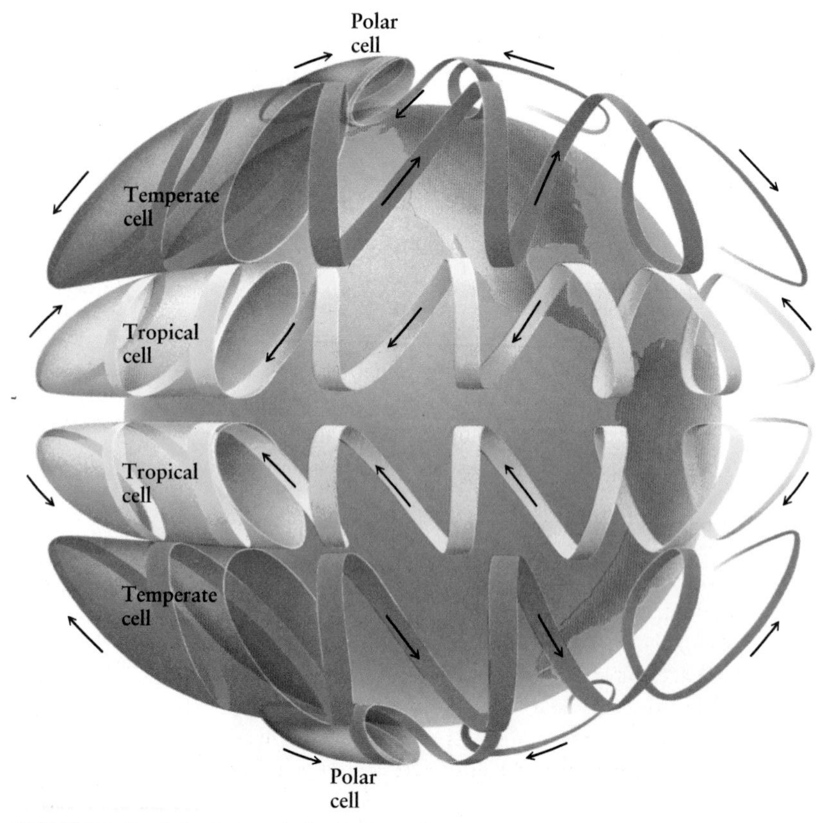

FIGURE 5-7 Circulation Patterns in Earth's Atmosphere The dominant circulation in our atmosphere consists of six convection cells, three in the northern hemisphere and three in the southern hemisphere. In the northern temperate region (including the continental United States), the prevailing winds at the surface are from the southwest toward the northeast. Farther south, within the northern tropical region (for example, Hawaii), the prevailing surface winds are from northeast to southwest.

ConceptCheck 5-4: Is the atmospheric layer called the troposphere heated from above or below?

Upper Layers of the Atmosphere

Almost all the oxygen in the troposphere is in the form of O_2, a molecule made of two oxygen atoms. But in the **stratosphere,** which extends from about 7.5 mi to 31 mi (about 12 km to 50 km) above the surface, some oxygen is in the form of **ozone,** a molecule made of *three* oxygen atoms (O_3). Ozone is very efficient at absorbing potentially harmful ultraviolet radiation from the Sun, which means that the stratosphere can directly absorb solar energy. The result is that the temperature actually increases, rather than decreases, as you move upward in the stratosphere. Convection requires that the temperature must decrease, not increase, with increasing altitude, so there are essentially no convection currents in the stratosphere.

Above the stratosphere lies the **mesosphere.** Very little ozone is found here, so solar ultraviolet radiation is not absorbed within the mesosphere, and atmospheric temperature again declines with increasing altitude as cooling occurs. The temperature of the mesosphere reaches a minimum of about −5°C (= −103°F = 198 K) at an altitude of about 50 mi (80 km). This minimum marks the bottom of the atmosphere's thinnest and uppermost layer, the **thermosphere,** in which temperature once again rises with increasing altitude. This is not due to the presence of ozone, because in this very low–density region oxygen and nitrogen are found mostly as individual atoms rather than in molecules. Instead, the thermosphere is heated because these isolated atoms absorb very-short-wavelength solar ultraviolet radiation, which oxygen and nitrogen molecules are not plentiful enough to absorb.

CAUTION At altitudes near 200 mi above Earth's surface the temperature of the thermosphere is about 1000°C (1800°F). This is near the altitude at which the space shuttle and satellites orbit Earth. Nonetheless, a satellite in orbit does *not* risk being burned up as it moves through the thermosphere. The reason is that the thermosphere is far less dense than the atmosphere at sea level. The high temperature simply means that an average atom in the thermosphere is moving very fast. But because the thermosphere is so thin (only about 10^{-11} as dense as the air at sea level), these fast-moving atoms are few and far between. Hence, the thermosphere contains very little energy. Nearly all the heat that an orbiting satellite receives is from sunlight, not the thermosphere.

ConceptCheck 5-5: What would happen to the temperature in the stratosphere if there was an absence of ozone?

The Sun's Role in the Atmosphere's Energy

The Sun is the principal source of energy for the atmosphere. Earth's surface is warmed by sunlight, which in turn warms the air next to the surface. Solar energy also powers weather systems when water evaporates from the surface. The energy in the water vapor is released when it changes back into water from water droplets, like those that make up clouds. In a typical thunderstorm, some 5×10^8 kg of water vapor is lifted to great heights. The amount of energy released when this water condenses is as much energy as a city of 100,000 people uses in a month!

Solar energy also helps to power the oceans. Warm water from near the equator moves toward the poles, while cold polar water returns toward the equator. In the end, the average surface temperature of Earth depends almost entirely on the amount of energy that reaches us from the Sun in the form of electromagnetic radiation. In an analogous way, whether you feel warm or cool outdoors on a summer day depends on whether you are in the sunlight and receiving lots of solar energy or in the shade and receiving little of this energy.

One might think that if Earth did nothing but *absorb* radiation from the Sun, it would get hotter and hotter until the surface temperature became high enough to melt rock. Happily, there are at least two important reasons why this could not happen. One is that the clouds, snow, ice, and sand *reflect* about 31% of the incoming sunlight back into space. The fraction of incoming sunlight that a planet reflects is called its **albedo** (from the Latin for "whiteness"); thus, Earth's albedo is about 0.31. In other words, only 69% of the incoming solar energy is absorbed by Earth and the rest bounces back into space. A second reason is that Earth also *emits* radiation into space itself because of its own temperature, in accordance with the laws that describe energy emitted by dense objects (see Section 2-3). Earth's average surface temperature is nearly constant, which means that on the whole it is neither gaining nor losing energy.

To better understand this balance between absorbed and emitted radiation, remember that Wien's law tells us that the wavelength at which such an object emits most strongly (λ_{max}) is inversely proportional to its temperature (T) on the Kelvin scale (see Chapter 2). For example, the Sun's surface temperature is about 5800 K, and sunlight has its greatest intensity at a wavelength λ_{max} of 500 nm, in the middle of the visible spectrum. Earth's average surface temperature of 287 K is far lower than the Sun's, so Earth radiates most strongly at longer wavelengths in the infrared portion of the electromagnetic spectrum. The Stefan-Boltzmann law (see Chapter 2) tells us that temperature also determines the *amount* of radiation that Earth emits: The higher the temperature, the more energy it radiates.

Given the amount of energy reaching us from the Sun each second as well as Earth's albedo, we can calculate the amount of solar energy that Earth should *absorb* each second. Since this must equal the amount of electromagnetic energy that Earth *emits* each second, which in turn depends on Earth's average surface temperature, we can calculate what Earth's average surface temperature should be. The result is a very chilly 254 K (= −19°C = −2°F), so cold that oceans and lakes around the world should be frozen over. In fact, Earth's actual average surface temperature is 287 K (= 14°C = 57°F). What is wrong with our model? Why is Earth warmer than we would expect?

The explanation for this discrepancy is called the **greenhouse effect:** Our atmosphere prevents some of the radiation emitted by Earth's surface from escaping into space. This process happens on any planet or moon that has an atmosphere. On Earth, certain gases in our atmosphere, called **greenhouse gases,** among them water vapor and carbon dioxide, are transparent to visible light but not to infrared radiation. Consequently, visible sunlight has no trouble entering our atmosphere and warming the surface. But the infrared radiation coming from the heated surface is partially absorbed by the atmosphere, thus raising the temperatures of both the atmosphere and the surface. As the surface and atmosphere become hotter, they both emit more infrared radiation, part of which is able to escape into space. The temperature levels off when the amount of infrared energy that escapes just balances the amount of solar energy reaching the surface (**Figure 5-8**). The result is that our planet's surface is some 33°C (59°F) warmer than it would be without the greenhouse effect, and water remains unfrozen over most of Earth.

The warming caused by the greenhouse effect helps to give our planet the moderate temperatures needed for the existence of life. For more than a century, however, our technological civilization has been adding greenhouse gases to the atmosphere at an unprecedented rate. In this instance, it is difficult to predict what the effects of this imbalance might be.

ConceptCheck 5-6: What is the positive impact that the greenhouse effect has on our planet?

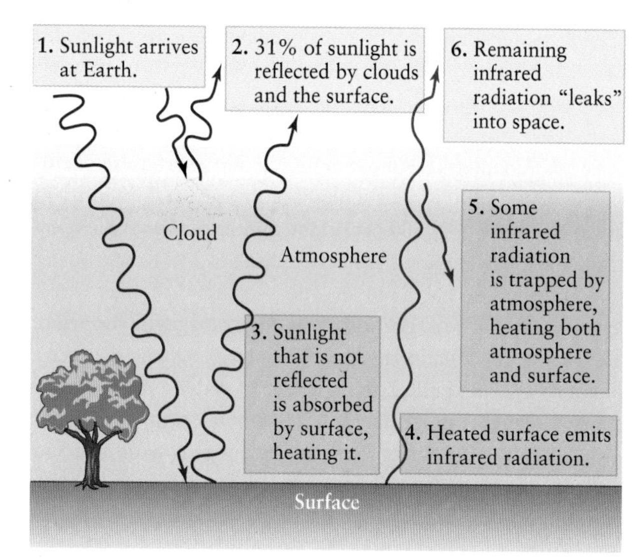

FIGURE 5-8
The Greenhouse Effect Sunlight warms Earth's surface, which due to its temperature emits infrared radiation. Much of this radiation is absorbed by atmospheric water vapor and carbon dioxide, helping to raise the average temperature of the surface. Some infrared radiation does penetrate the atmosphere and leaks into space. In a state of equilibrium, the rate at which Earth loses energy to space in this way is equal to the rate at which it absorbs energy from the Sun.

5-3 Volcanoes and earthquakes reveal energy from a molten interior driving Earth's surface to shift positions

Looking out the window, you might wonder if Earth looks more or less the same today as it always has. Trees and buildings come and go, but what about major landforms such as mountains and oceans? And if something as massive as a mountain could change, what force of nature is strong enough to accomplish such a task? Earth is full of energy. Some of that energy is left over from the force of impacts in the early formation of our planet, a time when the planet was completely molten. But the majority of Earth's internal energy comes from radioactive decay of elements such as uranium, thorium, and potassium deep inside Earth. This energy drives geologic activity that profoundly affects Earth's surface. How do we actually know what is at the center of Earth? Surprisingly, our search starts with the rocks just under our feet.

An Iron-Rich Planet

The specific kinds of rocks found on and near Earth's surface provide important clues to our planet's interior. The densities of typical surface rocks are around 3000 kg/m^3, but the average density of Earth as a whole is 5515 kg/m^3. (As you saw in Chapter 4, average density is total mass divided by total volume.) The interior of Earth must therefore be composed of substances much denser than those found near the surface. But what is this substance? Why is Earth's interior denser than its crust? And is Earth's interior solid like rock or molten like magma?

Iron (chemical symbol Fe) is a good candidate for the substance that makes up most of Earth's interior for two reasons. First, iron atoms are quite massive—a typical iron atom has 56 times the mass of a hydrogen atom—and second, iron is relatively abundant. **Figure 5-9** shows that it is the seventh most abundant element in our part of the Milky Way Galaxy. Other elements such as lead and uranium have more massive atoms, but these elements are quite rare—they would be ranked by their atomic numbers at 82 and 92, respectively, if Figure 5-9 was extended to the right. Hence, the solar nebula could not have had enough of these massive atoms to create Earth's dense interior. Furthermore, iron is common in meteoroids that strike Earth, which suggests that it was abundant in the planetesimals from which Earth formed.

Earth was almost certainly molten soon after its formation, about 4.56×10^9 years ago. Energy released by the violent impacts of numerous meteoroids and asteroids and by the decay of radioactive material likely melted the solid parts

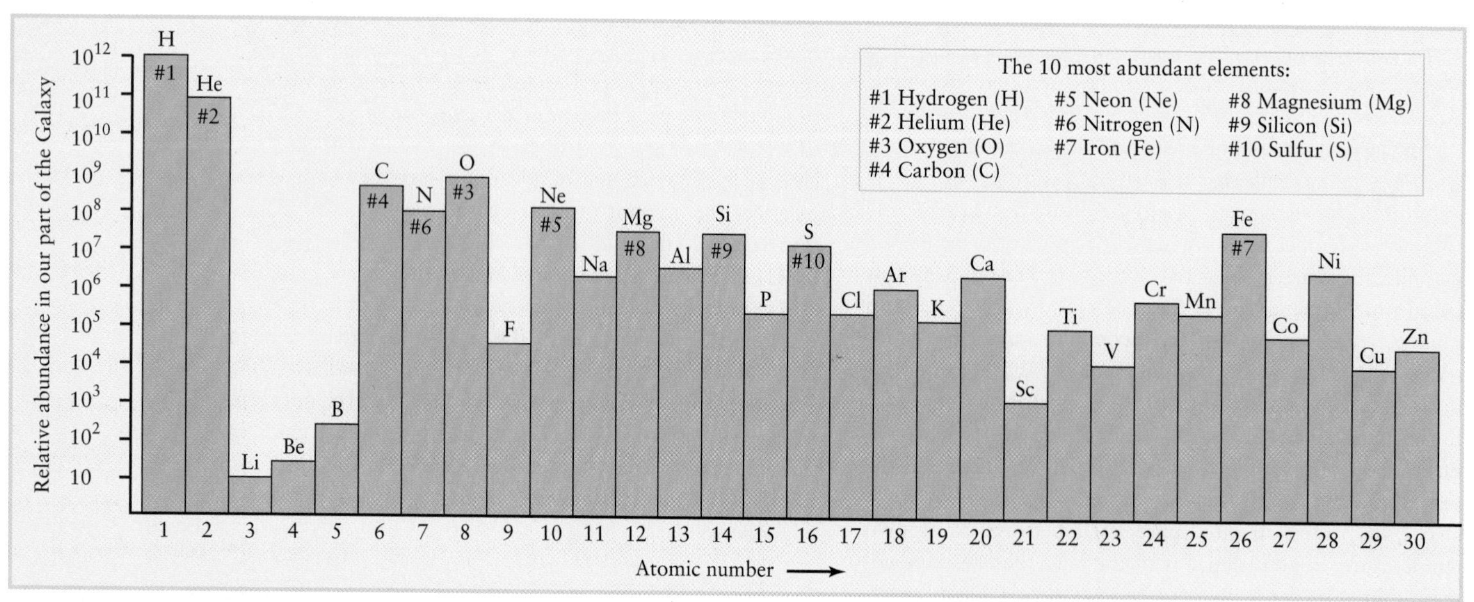

FIGURE 5-9 Abundances of the Lighter Elements This graph shows the abundances in our part of the Galaxy of the 30 lightest elements (listed in order of increasing atomic number) compared to a value of 10^{12} for hydrogen. The inset lists the 10 most abundant of these elements, which are also indicated in the graph. Notice that the vertical scale is not linear; each division on the scale corresponds to a tenfold increase in abundance. All elements heavier than zinc (Zn) have abundances of fewer than 1000 atoms per 10^{12} atoms of hydrogen.

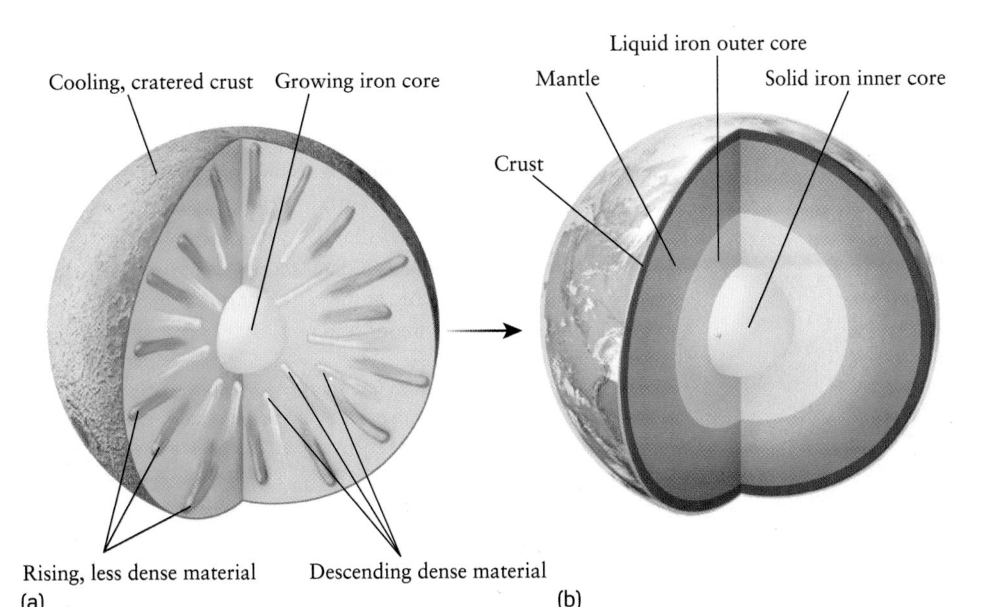

FIGURE 5-10 Chemical Differentiation and Earth's Internal Structure (a) The newly formed Earth was molten throughout its volume. Dense materials such as iron sank toward the center, while low-density materials rose toward the surface. (b) The present-day Earth is no longer completely molten inside. A dense, solid iron core is surrounded by a less dense liquid core and an even less dense mantle. The crust, which includes the continents and ocean floors, is the least dense of all; it floats atop the mantle like the skin that forms on the surface of a cooling cup of cocoa.

collected from the earlier planetesimals. Gravity caused the abundant, dense iron to sink toward Earth's center, forcing less dense material upward to the surface. **Figure 5-10*a*** shows this process of *differentiation*. This is the same process that happens when Italian salad dressing separates into various layers when left undisturbed for a time. In much the same way, we end up with a planet with the layered structure shown in Figure 5-10b—a central **core** composed almost entirely of iron, surrounded by a **mantle** of dense, iron-enriched rocks made from silicon and oxygen. The mantle, in turn, is surrounded by an extremely thin **crust** of relatively light silicon-rich minerals. We live on the surface of this crust. See **Cosmic Connections: Comparing Earth's Atmosphere and Interior** for more information.

> ConceptCheck 5-7: **If a new, Earth-sized planet was discovered that had surface rocks nearly identical to Earth but a much lower average density, what might a scientist assume about iron in its interior?**

Earthquake Waves as Earth Probes

Go to Video 5-2

How might we confirm that Earth does in fact have a layered structure? The challenge in directly testing this notion is that Earth's interior is inaccessible. The deepest wells go down only a few kilometers, barely penetrating the surface of our planet. Despite these difficulties, geologists have learned basic properties of Earth's interior by studying earthquakes and the waves they produce.

Over the centuries, stresses build up in Earth's crust. Occasionally, these stresses are relieved with an energy burst called an **earthquake.** Most earthquakes occur deep within Earth's crust. Earthquakes release energy in the form of at least three different kinds of energy waves, which travel around or through Earth in different ways and at different speeds. The first type of wave, which is analogous to water waves on the ocean, causes the rolling motion that people feel near the source of an earthquake. These are called **surface waves** because they travel only over Earth's surface and can cause a tremendous amount of destruction. The two other kinds of waves, called **P waves** (the P is for "primary") and **S waves** (the capital S is for "secondary," not for the s in "surface" waves, which are sometimes confusingly designated with an L), travel through the interior of Earth. P waves are called *longitudinal* waves because their oscillations are in the same direction the energy wave travels, like a spring that is alternately pushed and pulled. In contrast, S waves are called *transverse* waves because their vibrations are perpendicular to the direction in which the energy waves move. S waves are analogous to waves produced by a person shaking a rope up and down (**Figure 5-11**).

What makes earthquake waves useful for learning about Earth's interior is that they do not travel in straight lines. Instead, the paths that they follow through the body of Earth are bent because of the varying density and composition of Earth's interior. Just as light waves bend when they pass from air into glass or vice versa, earthquake waves bend as they

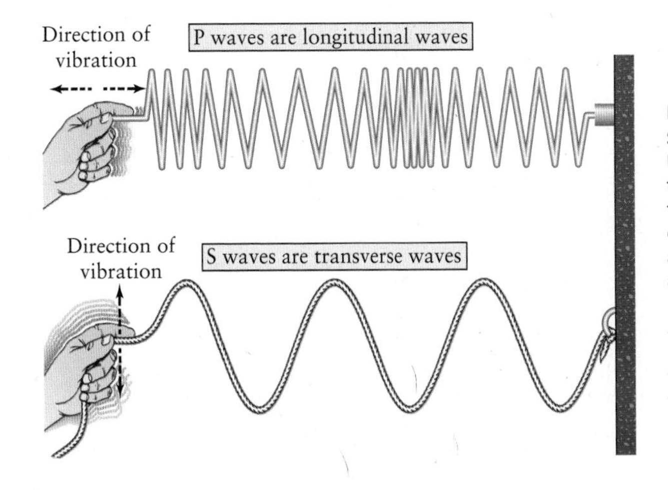

FIGURE 5-11 Seismic Waves Earthquakes produce two kinds of waves that travel through the body of our planet. P waves are longitudinal waves and analogous to those produced by pushing a spring in and out. S waves are transverse waves and analogous to the waves produced by shaking a rope up and down.

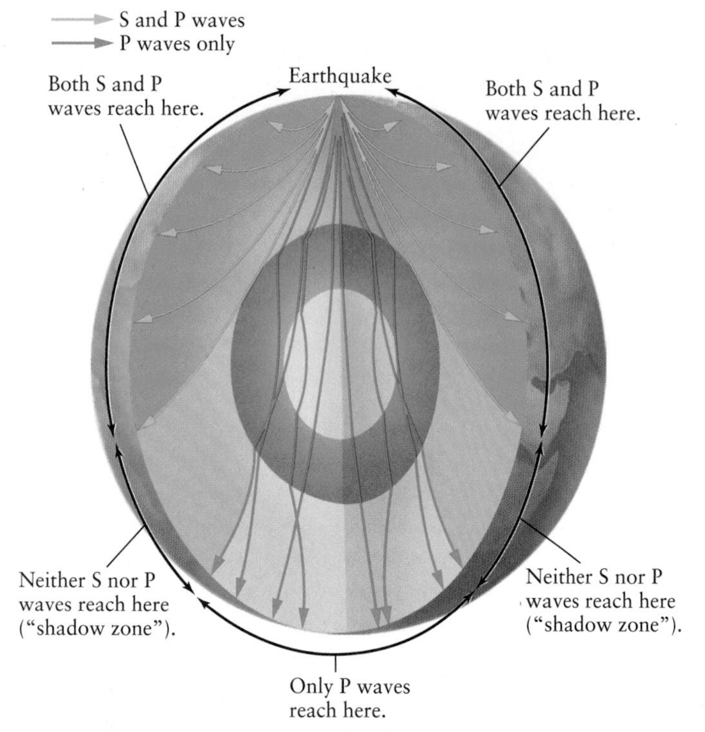

FIGURE 5-12 Earth's Internal Structure and the Paths of Seismic Waves Seismic waves follow curved paths because of differences in the density and composition of the material in Earth's interior. The paths curve gradually where there are gradual changes in density and composition. Sharp bends occur only where there is an abrupt change from one kind of material to another, such as at the boundary between the outer core and the mantle. Only P waves can pass through Earth's liquid outer core.

pass through different parts of Earth's interior. By studying how the paths of these waves bend, geologists can map out the general interior structure of Earth.

One key observation about earthquake waves and how they bend has to do with the differences between S and P waves. When an earthquake occurs and releases energy, both S and P waves are observed near the center of the earthquake, but geologists on the opposite side of Earth observe only P waves. The absence of S waves was first explained in 1906 by the British geologist Richard Dixon Oldham, who noted that transverse vibrations such as S waves cannot travel far through liquids. Oldham therefore concluded that our planet has a molten core. Furthermore, there is a region in which neither S waves nor P waves from an earthquake can be detected (**Figure 5-12**). This "shadow zone" results from the specific way in which P waves are refracted at the boundary between the solid mantle and the molten core. By measuring the size of the shadow zone, geologists have concluded that the radius of the molten core is about 2200 mi (3500 km), about 55% of our planet's overall radius but about double the radius of the Moon (1080 mi =1738 km). In 1936, the Danish geologist Inge Lehmann explained that the rare P wave that did successfully pass through Earth is deflected into the shadow zone by a small, solid **inner core** at the center of our planet. The radius of this inner core is about 800 mi (1300 km).

> ConceptCheck 5-8: **What would have changed about Earth's interior if S waves from earthquakes suddenly started being observed on the side of Earth opposite an earthquake?**

Earth's Major Layers

The earthquake evidence reveals that our planet has a curious internal structure—a liquid **outer core** sandwiched between a solid inner core and a mostly solid mantle. **Table 5-1** summarizes this structure. To understand this arrangement, we must look at how temperature and pressure inside Earth affect the melting point of rock.

Both temperature and pressure increase with increasing depth below Earth's surface. The temperature of Earth's interior rises steadily from about 14°C on the surface to nearly 5000°C at our planet's center (**Figure 5-13**).

Earth's outermost layer, the crust, varies quite a bit in thickness, but overall is only about 3 mi to 20 mi (5 km to 35 km) thick. Given the enormous size of Earth, the crust is an extremely thin layer. It is composed of rocks with a melting temperature far higher than the temperatures actually found in the crust. Thus, the crust is solid.

Earth's mantle, which extends to a depth of about 1800 mi (2900 km), is largely composed of substances enriched by iron and magnesium. On Earth's surface, specimens of these substances will melt at slightly over 1000°C. However, the temperature at which rock melts depends on the pressure to which it is subjected—the higher the pressure, the higher the melting temperature. As Figure 5-13 shows, the actual temperatures throughout most of the mantle are less than the melting temperature of all the substances of which the mantle is made. Hence, the mantle is primarily solid. However, the upper levels of the mantle—called the asthenosphere (from the Greek *asthenia,* meaning "weakness")—are able to flow slowly and are therefore often referred to as being somewhat "plastic."

Table 5-1 Earth's Internal Structure

Region	Depth below surface (km)	Distance from center (km)	Average density (kg/m^2)
Crust (solid)	0–5 (under oceans), 0–35 (under continents)	6343–6378	3500
Mantle (plastic, solid)	from bottom of crust to 2900	3500–6343	3500–5500
Outer core (liquid)	2900–5100	1300–3500	10,000–12,000
Inner core (solid)	5100–6400	0–1300	13,000

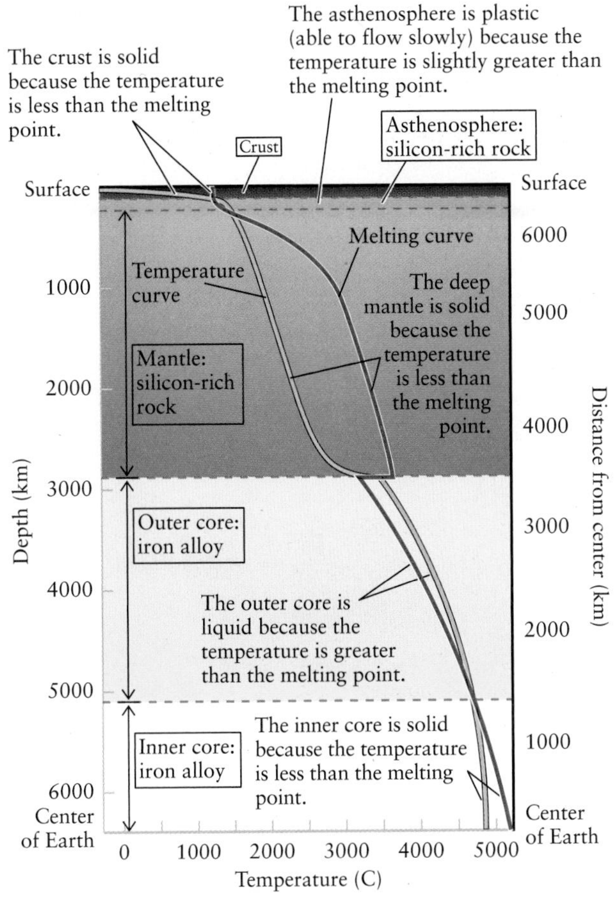

FIGURE 5-13 Temperature and Melting Point of Rock Inside Earth The temperature (yellow curve) rises steadily from Earth's surface to its center. The melting point of Earth's material (red curve) is also shown on this graph. Where the temperature is below the melting point, as in the mantle and inner core, the material is solid; where the temperature is above the melting point, as in the outer core, the material is liquid. *(Adapted from T. Grotzinger, T. H. Jordan, F. Press, and R. Siever,* Understanding Earth, *5th ed., W. H. Freeman, 2007)*

CAUTION It may seem strange to think of a plastic material as one that is able to flow. Normally, we think of plastic objects as being hard and solid, like the plastic out of which compact discs and soft drink bottles are made. But to make these objects, the plastic material is heated so that it can flow into a mold, and then cooled so that it solidifies. Strictly, the material is only "plastic" when it is able to flow. Maple syrup and tree sap are two other substances that flow at warm temperatures but become solid when cooled.

At the boundary between the mantle and the outer core, there is an abrupt change in chemical composition, from silicon-rich materials to almost-pure iron with a small mix of nickel. Because this iron has a lower melting point than the material above it, the melting curve in Figure 5-13 has a "jog" as it crosses from the mantle to the outer core and remains. As a result, the temperature at which rocks melt here is somewhat less than the temperature at depths of about 1800 mi to 3200 mi (2900 km to 5100 km), and the outer core is liquid.

At depths greater than about 3000 mi (5100 km), the pressure is more than 10^6 times ordinary atmospheric pressure, or about 10^4 tons per square inch. Because the melting temperature of the iron-nickel mixture under this pressure is higher than the actual temperature (see Figure 5-13), Earth's inner core is solid.

Heat naturally flows from where the temperature is high to where it is low. Figure 5-13 thus explains why there is a heat flow from the center of Earth outward. We will see in the next section how this heat flow acts as the "engine" that powers our planet's geologic activity.

ConceptCheck 5-9: How would the temperature at which rocks melt be different inside a much larger and denser planet where the internal pressure is higher?

Plate Tectonics

When looking at the shape of the ocean between the continents of Africa and South America (**Figure 5-14**), one naturally wonders if the two continents were closer together in the past. What evidence would you need to be convinced that the two were at one time closer together and that the continents are still moving? What would you think if there were similar fossils on both sides of the Atlantic Ocean? What would you think if you could take GPS satellite measurements and see the two moving apart? Indeed, this is exactly what we observe—the Atlantic Ocean is widening.

Close inspection of Earth's landforms, both above and beneath the oceans, reveal that Earth's crust is divided into

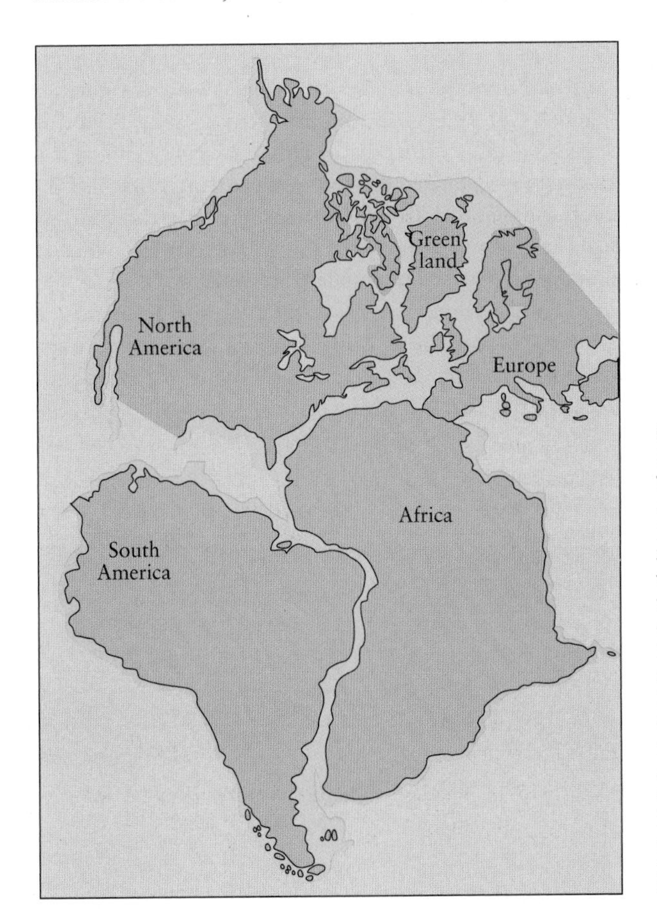

FIGURE 5-14 Fitting the Continents Together Africa, Europe, Greenland, North America, and South America fit together remarkably well. The fit is especially convincing if the edges of the continental shelves (shown in yellow) are used, rather than today's shorelines. This strongly suggests that these continents were in fact joined together at some point in the past. *(Adapted from P. M. Hurley)*

huge plates whose motions produce earthquakes, volcanoes, mountain ranges, and oceanic trenches. This picture of Earth's crust, now known as **plate tectonics** (from the Greek *tekton,* meaning "builder"), has come to be the central unifying theory of geology, much as the theory of evolution has become the centerpiece of modern biology.

Based on the apparent fit of the continents, the German meteorologist Alfred Wegener advocated "continental drift"—the idea that the continents on either side of the Atlantic Ocean have simply drifted apart. In 1915 he published the theory that there had originally been a single gigantic supercontinent, which he called Pangaea (meaning "all lands"). Based on how fast South America and Africa are separating today, this separation probably began roughly 200 million years ago, during what geologists refer to as the early Jurassic period, when dinosaurs dominated the land.

Over time, geologists have refined this theory, arguing that Pangaea must have first split into two smaller supercontinents, which they called Laurasia and Gondwana, separated by what they called the Tethys Ocean. Gondwana later split into Africa and South America, with Laurasia dividing to become North America and Eurasia. According to this theory, the Mediterranean Sea is a surviving remnant of the ancient Tethys Ocean (**Figure 5-15**).

The actual mechanism that was driving the continents' motions was not evident until the mid-1950s, when geologists discovered that material is being forced upward to the crust from deep within Earth. Bruce C. Heezen of Columbia University and his colleagues began discovering long mountain ranges on the ocean floors, such as the Mid-Atlantic Ridge, which stretches all the way from Iceland to Antarctica (**Figure 5-16**). During the 1960s, Harry Hess of Princeton University and Robert Dietz of the University of California carefully examined the floor of the Atlantic Ocean. They concluded that rock from Earth's mantle is being melted and then forced upward along the Mid-Atlantic Ridge, which is in essence a long chain of underwater volcanoes.

The upwelling of new material from the mantle to the crust forces the existing crust to separate, causing **seafloor spreading**. For example, the floor of the Atlantic Ocean to the east of the Mid-Atlantic Ridge is moving eastward and the floor to the west is moving westward. By explaining what fills in the gap between continents as they move apart, seafloor spreading helps to fill out the theories of continental drift. Because of the seafloor spreading from the Mid-Atlantic Ridge, South America and Africa are moving apart at a speed of roughly 3 cm per year. Working backward, these two continents would have been next to each other some 200 million years ago (see Figure 5-14).

FIGURE 5-16 The Mid-Atlantic Ridge This artist's rendition shows the Mid-Atlantic Ridge, an immense mountain ridge that rises up from the floor of the North Atlantic Ocean. It is caused by lava seeping up from Earth's interior along a rift that extends from Iceland to Antarctica. *(Courtesy of M. Tharp and B. C. Heezen)*

ConceptCheck 5-10: Does seafloor spreading result in an underwater mountain range or a deep canyon?

CalculationCheck 5-1: If seafloor spreading was occurring at only 2 cm per year, how long would it have taken for the Atlantic to open to its current width of 600 million (6×10^8) cm?

(a) 237 million years ago: the supercontinent Pangaea

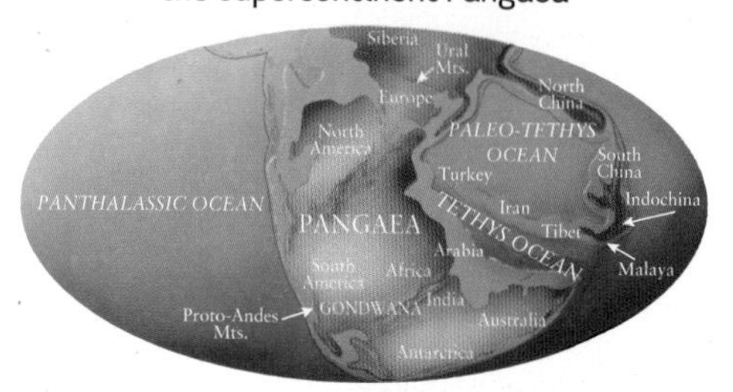

(b) 152 million years ago: the breakup of Pangaea

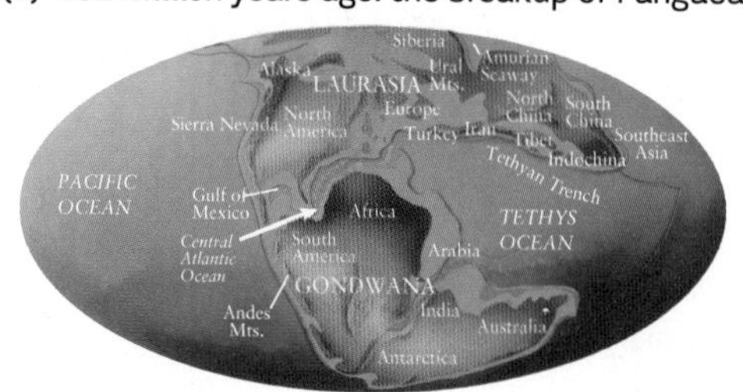

(c) The continents today

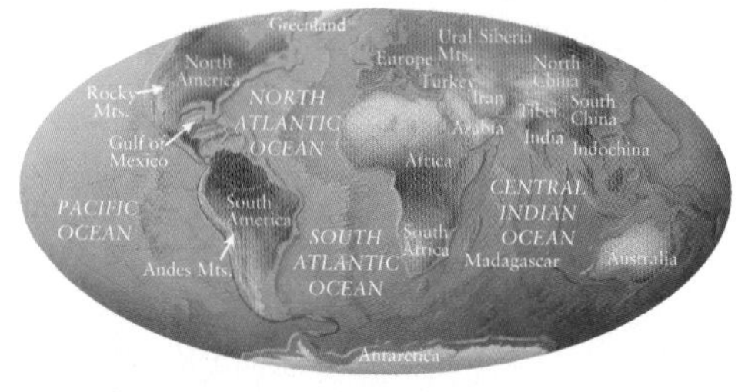

FIGURE 5-15 The Breakup of the Supercontinent Pangaea (a) The shapes of the continents led Alfred Wegener to conclude that more than 200 million (2×10^8) years ago, the continents were merged into a single supercontinent, which he called Pangaea. **(b)** Pangaea first split into two smaller land masses, Laurasia and Gondwana. **(c)** Over millions of years, the continents moved to their present-day locations. Among the evidence confirming this picture are nearly identical rock formations 200 million years in age that today are thousands of kilometers apart but would have been side by side on Pangaea. *(Adapted from F. Press, R. Siever, T. Grotzinger, and T. H. Jordan, Understanding Earth, 4th ed., W. H. Freeman, 2004)*

Convection and Plate Motion

What makes plates move? The high temperatures within Earth cause energy in the form of heat to flow outward from Earth's hot core to its cool crust. Hot material deep in Earth is less dense than cooler material farther away from the core and tends to rise, much the same way heated air in a giant hot air balloon will rise high above Earth's surface. As hot mantle material rises, it transfers heat to its surroundings. As a result, the rising material cools and becomes denser. It then sinks downward to be heated again, and the process starts over. Earlier in this chapter, we considered *convection* cycles that occur in a boiling pot of water (Figure 5-6), which also occur in the atmosphere. Here, too, we find that convection occurs when warmer parts of the partially molten mantle, heated from below, slowly rise. The heat that drives convection in Earth's outer layers actually comes from very far below, at the boundary between the outer and inner cores. As material deep in the liquid core cools and solidifies to join the solid portion of the core, it releases the energy needed to heat the overlying mantle and cause convection.

CAUTION When thinking about the circular pattern that boiling water follows on a kitchen stove, you might be inclined to think that this only occurs with liquids, like water. On the stove, water near the bottom of the heated pan expands and drifts upward because it is now less dense than the cooler water around it. The cooler water on top is pushed out of the way by the rising warmer water, and the cooler water moves to the bottom where it is heated, and the cycle starts over again. It might surprise you, however, to know that convection cycles can happen with any substance that expands when heated. When we talked about atmospheres, we noted that this can happen with gases, just like it can with liquid. It happens when a candle burns because rising air is consumed by the fire and cooler air nearby takes its place. Inside Earth, when heated plastic-like material in the mantle close to the hot core is heated, it expands and begins to rise, pushing the cooler, less dense mantle material out of the way, just like we saw with liquids and with gas.

The upper level of the mantle, the asthenosphere, is hot and soft enough to permit an oozing, plastic flow. Atop the asthenosphere is a rigid layer, called the **lithosphere** (from the Greek word for "rock"). The lithosphere is divided into plates that ride along the convection currents of the asthenosphere. Earth's crust is simply the uppermost layer of the lithosphere, with a somewhat different chemical composition than the lithosphere's lower regions.

Figure 5-17 shows how convection causes plate movement. Molten subsurface rock seeps upward along **oceanic rifts,** where plates are separating. The Mid-Atlantic Ridge, shown in Figure 5-16, is an oceanic rift and seafloor spreading is occurring there. Where plates collide, cool crustal material from one of the plates sinks back down into the mantle along a **subduction zone.** One such subduction zone is found along the west coast of South America, where the oceanic Nazca plate is being subducted into the mantle under the continental South American plate at a relatively speedy 10 cm per year. As the material from the subducted plate sinks, it pulls the rest of its plate along with it, thus helping to keep the plates in motion. New material is added to the crust from the mantle at the oceanic rifts and is "recycled" back into the mantle at the subduction zones. In this way the total amount of crust remains essentially the same.

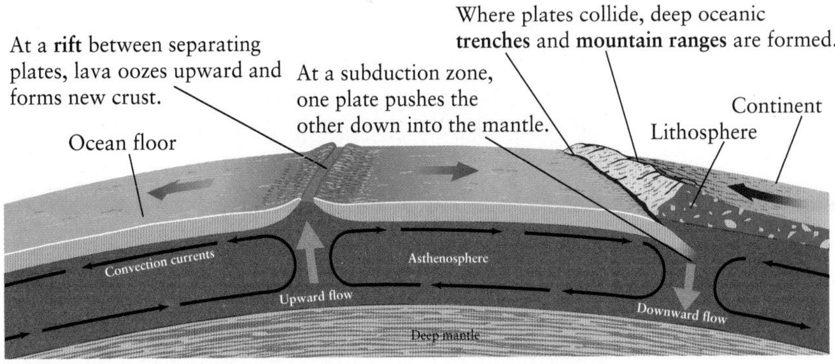

FIGURE 5-17 The Mechanism of Plate Tectonics Convection currents in the asthenosphere, the soft upper layer of the mantle, are responsible for pushing around rigid, low-density crustal plates. New crust forms in oceanic rifts, where lava oozes upward between separating plates. Mountain ranges and deep oceanic trenches are formed where plates collide.

Geologists today realize that earthquakes, and the vast majority of volcanoes, tend to occur at the boundaries of Earth's crustal plates, where the plates are colliding, separating, or rubbing against each other. The boundaries of the plates therefore stand out clearly when earthquake locations are plotted on a map as in **Figure 5-18.**

Convection cycles in Earth's upper mantle drag Earth's plates around so that they separate, collide, and grind against each other.

The boundaries between plates are the sites of some of the most impressive geological activity on our planet. Great mountain ranges, such as the Sierras and Cascades along the western coast of North America and the Andes along South America's west coast, are thrust up by ongoing collisions between continental plates and the plates of the ocean floor. Subduction zones, where old crust is drawn back down into the mantle, are typically the locations of deep oceanic trenches, such as the Peru-Chile Trench off the west coast of South America. Figures 5-18*b*, *c*, and *d* show three well-known geographic features that resulted from tectonic activity at plate boundaries.

ConceptCheck 5-11: **What becomes of the material from the oceanic crust when it is subducted beneath continental crust?**

The Cycle of Supercontinents

Plate tectonic theory offers insight into geology on the largest of scales, that of an entire supercontinent. In recent years, geologists have uncovered evidence that points to a whole succession of supercontinents that once broke apart and then reassembled. Pangaea is only the most recent

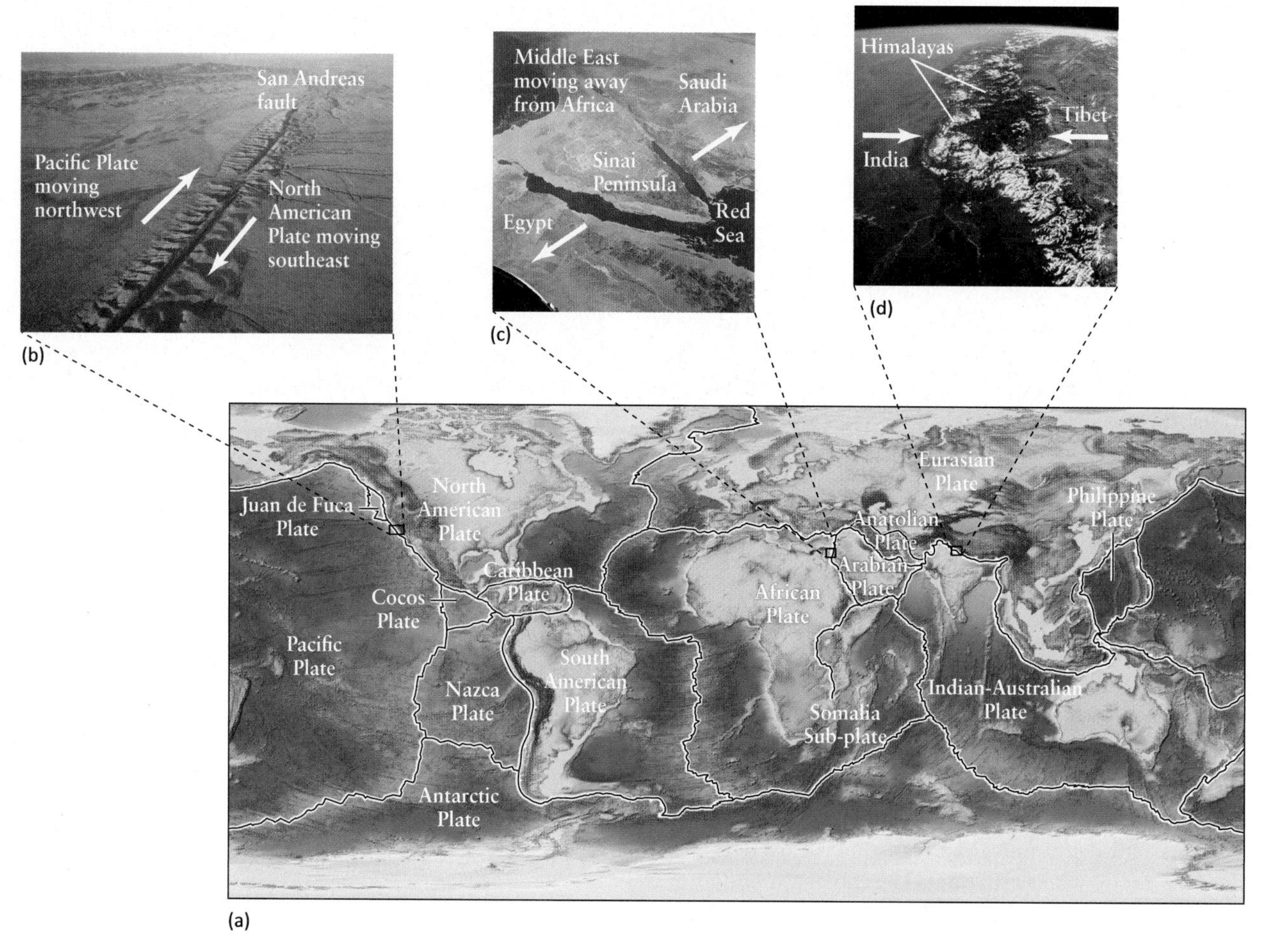

FIGURE 5-18 R I V U X G **Earth's Major Tectonic Plates** (a) Earth's surface is divided into plates that move relative to one another. The plate boundaries are where we find earthquakes, volcanoes, and rising mountain ranges. The arrows indicate whether plates are moving apart (←→), together (→←), or sliding past one another ↑↓. (b) Plates Grinding Past One Another. The San Andreas fault, running up the west coast of North America, formed because the Pacific plate is moving northwest along the North American plate. (c) Plates Pulling Apart. The plates that carry Egypt and Saudi Arabia are moving apart, leaving a space that contains the Red Sea. (d) Plates Colliding. The plates that carry India and China are colliding. As a result, the Himalayas are being thrust upward. In this photograph, taken by astronauts in 1968, Mount Everest is one of the snow-covered peaks near the center. *(a: Smithsonian Institution Volcano Program/ESRI/WorldSat International/U.S. Geological Survey; b: Aurora Photos/Alamy; c: NASA; d: Image courtesy of the Image Science & Analysis Laboratory, NASA Johnson Space Center).*

supercontinent in this cycle, which repeats about every 500 million (5×10^8) years. As a result, intense episodes of mountain building have occurred at roughly 500-million-year intervals.

Apparently, a supercontinent sows the seeds of its own destruction because it blocks the flow of heat from Earth's interior. As soon as a supercontinent forms, temperatures beneath it rise, much as they do under a book lying on an electric blanket. As heat accumulates, the lithosphere domes upward and cracks. Molten rock from the overheated asthenosphere wells up to fill the resulting fractures, which continue to widen as pieces of the fragmenting supercontinent move apart.

It can take a very long time for the heat trapped under a supercontinent to escape. Although Pangaea broke apart some 200 million years ago (see Figure 5-15), the mantle under its former location is still hot and still trying to rise upward. As a result, Africa—which lies close to the center of this mass of rising material—sits several tens of meters higher than the other continents.

The changes wrought by plate tectonics are very slow on the scale of a human lifetime, but they are very rapid in comparison with the age of Earth. For example, the period over which Pangaea broke into Laurasia and Gondwana was only about 0.4% of Earth's age of 4.56×10^9 years. (To put this in perspective, 0.4% of an average human lifetime is about 4 months.) The lesson of plate tectonics is that the seemingly permanent face of Earth is in fact dynamic and ever-changing.

> ConceptCheck 5-12: **What causes a supercontinent to break apart?**
>
> CalculationCheck 5-2: **If supercontinents are formed and break up in a 500-million-year cycle, about how many supercontinents could come and go over Earth's roughly 5-billion-year existence?**

COSMIC CONNECTIONS Comparing Earth's Atmosphere and Interior

Earth's atmosphere (which is a gas) and Earth's interior (which is partly solid, partly liquid) both decrease in density and pressure as you go farther away from Earth's center.

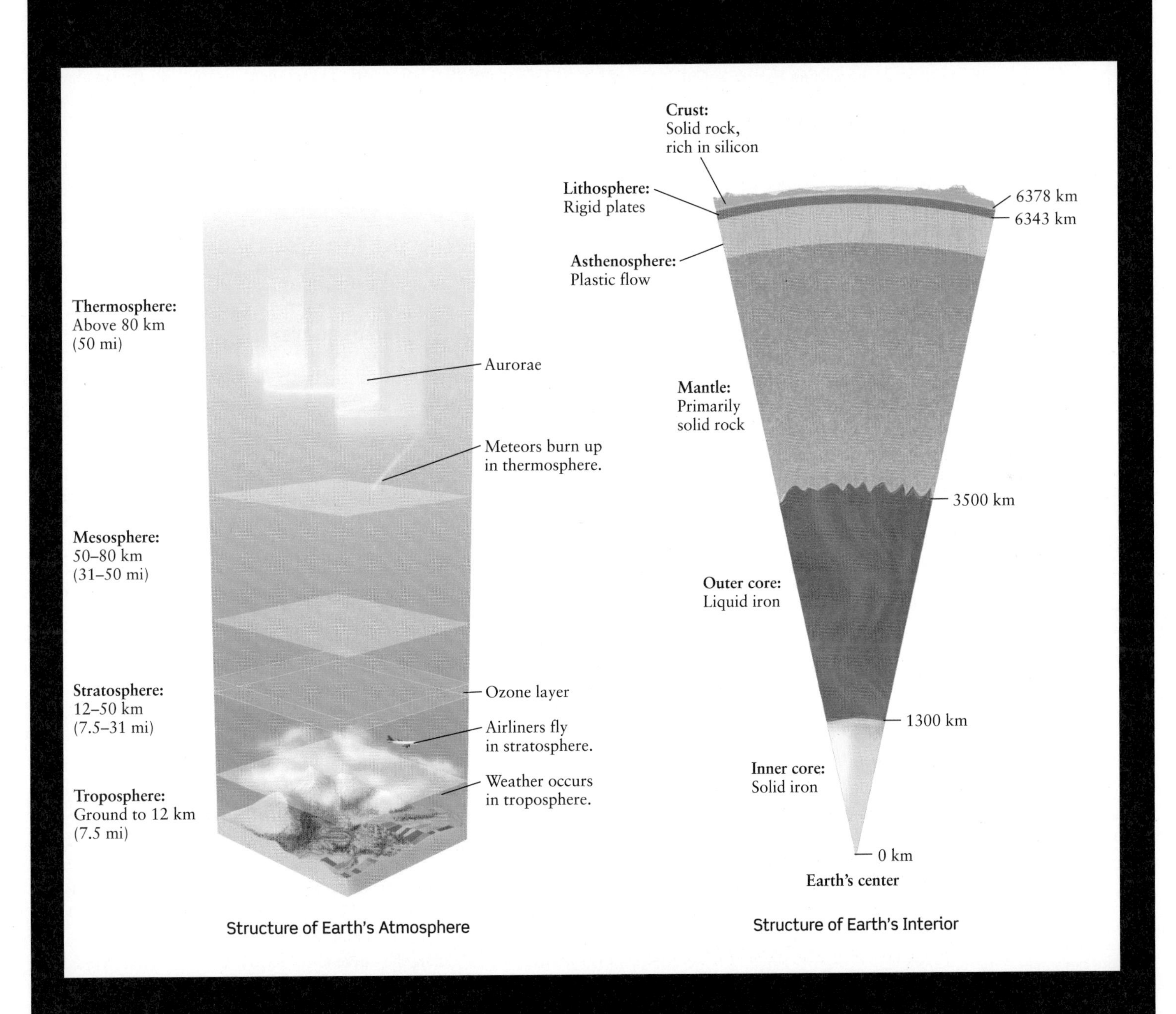

Structure of Earth's Atmosphere

Structure of Earth's Interior

5-4 Earth's magnetic field emanating from its spinning, molten interior creates a protective shield from the Sun's harmful radiation

The extent of a planet's dynamic resurfacing due to plate tectonics provides indirect evidence about whether it still has a molten interior. But another, more direct tool for probing the interior of Earth is an ordinary compass, which senses the magnetic field outside our planet. Magnetic field measurements prove to be an extremely powerful way to investigate the internal structure of a world without having to actually dig into its interior. To illustrate how this works, consider the behavior of a compass on Earth.

Source of Earth's Magnetic Field

The needle of a compass on Earth points north because it aligns with Earth's *magnetic field.* Such fields arise whenever electrically charged particles are in motion. For example, a loop of wire carrying an electric current generates a magnetic field in the space around it. The magnetic field that surrounds an ordinary bar magnet is created by the motions of negatively charged electrons within the iron atoms of which the magnet is made. Earth's magnetic field is similar to that of a bar magnet, as **Figure 5-19** shows. The consensus among geologists is that this magnetic field is caused by the motion of the liquid portions of Earth's interior. Because this molten material—mostly iron—conducts electricity, these motions give rise to electric currents, which in turn produce Earth's magnetic field. Our planet's rotation helps to sustain these motions and hence the magnetic field. This process for producing a magnetic field is called a dynamo.

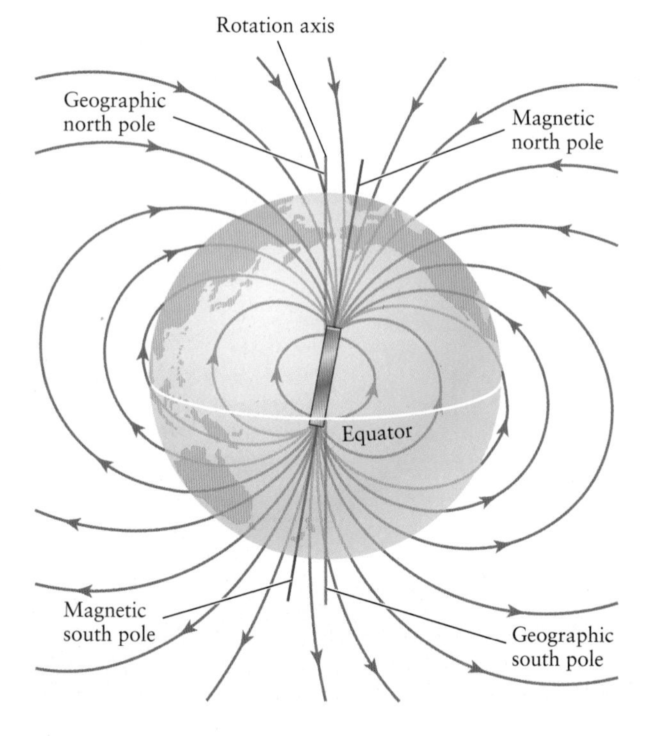

FIGURE 5-19
Earth's Magnetic Field Earth's magnetic field appears to stream from one pole to the other, like a giant bar magnet inside Earth, but Earth's field is produced in a different way—by electric currents in the liquid portion of our planet's interior. This bar magnet is not exactly aligned with Earth's rotation axis, which is why the magnetic north and south poles are not at precisely the same locations as the geographic poles. A compass needle points toward the north magnetic pole, not the geographic north pole.

Studies of ancient rocks reinforce the idea that our planet's magnetism is due to fluid material in motion. When iron-bearing lava cools and solidifies to form igneous rock, it becomes magnetized in the direction of Earth's magnetic field. By analyzing samples of igneous rock of different ages from around the world, geologists have found that Earth's magnetic field actually flips over and reverses direction on an irregular schedule ranging from tens of thousands to hundreds of thousands of years. As an example, lava that solidified 30,000 years ago is magnetized in the opposite direction to lava that has solidified recently. Therefore, 30,000 years ago a compass needle would have pointed south, not north! If Earth were a permanent magnet, like the small magnets used to attach notes to refrigerators, it would be hard to imagine how its magnetic field could spontaneously reverse direction. But computer simulations show that fields produced by moving fluids in Earth's outer core do indeed change direction from time to time.

ConceptCheck 5-13: **If Earth's iron-rich core solidified, would a magnetic compass still point north?**

The Magnetosphere

Is the magnetic field useful for anything other than helping Earth's inhabitants determine direction using a compass? As it turns out, Earth's magnetic field has important effects far above Earth's surface, where it interacts dramatically with charged particles from the Sun, a flow of mostly protons and electrons, known as the *solar wind,* which streams constantly outward from the Sun's upper atmosphere. Near Earth, the particles in the solar wind move at speeds of roughly a million miles per hour or about 450 km/s, considerably faster than sound waves can travel in the very thin gas between the planets, so the solar wind is said to be *supersonic.* (Because the gas between the planets is so thin, interplanetary sound waves carry too little energy to be heard by astronauts.)

If Earth had no magnetic field, it would be continually bombarded by this solar wind of charged particles. But our planet does have a magnetic field, and the forces that this field can exert on charged particles are strong enough to deflect them away from us. The region of space around a planet in which the motion of charged particles is dominated by the planet's magnetic field is called the planet's **magnetosphere. Figure 5-20** is an illustration of Earth's magnetosphere, which was discovered in the late 1950s by the first satellites placed in orbit.

When the supersonic particles in the solar wind first encounter Earth's magnetic field, they abruptly slow to subsonic speeds. Most of the particles of the solar wind are deflected around the magnetosphere, just as water is deflected to either side of the bow of a ship. And, just like ships tend

FIGURE 5-20 **Earth's Protective Magnetosphere** Earth's magnetic field shields our planet from the wind of highly charged particles streaming out of the Sun, shown here in cross section. Most of the solar wind's particles are deflected around Earth, but some do get trapped and follow magnetic field lines to the poles creating the aurora. *(NASA)*

to leak, some charged particles of the solar wind manage to leak into the magnetosphere and make their way to Earth's surface.

> ConceptCheck 5-14: **Does Earth's magnetosphere capture or deflect away particles in the Sun's solar wind?**

Aurorae

Go to Video 5-3

When some particles leak through the magnetic fields at weak points, they gather and cascade down into Earth's upper atmosphere near the poles, usually making an oval-shaped pattern (**Figure 5-21*a***). As these high-speed charged particles collide with atoms in the upper atmosphere, they excite the atoms to high energy levels. The atoms then emit visible light as they drop down to their ground states, like the excited gas atoms in a neon light. The result is a beautiful, shimmering display called an **aurora** (*plural* **aurorae**). These are also called the **northern lights** (**aurora borealis**) or **southern lights** (**aurora australis**), depending on the hemisphere in which the phenomenon is observed. Figure 5-21*b* shows the aurora as seen from Earth's surface. Occasionally, a violent event on the Sun's surface sends a particularly intense burst of protons and electrons toward Earth. The resulting auroral display can be exceptionally bright and can often be seen over a wide range of latitudes. Such events also disturb radio transmissions and can damage commu-

Earth's magnetic field protects Earth from a flow of charged particles from the Sun, which cause aurora when small amounts collide with molecules in Earth's upper atmosphere.

(a)

(b)

FIGURE 5-21 R I V U X G **The Aurora** (a) Captured by the NASA *IMAGE* satellite, this glowing aurora oval is about 3000 mi (4500 km) in diameter and centered over Antarctica just four days just after a record-setting explosion on the Sun. (b) This view from Alaska shows aurorae at their typical altitudes around 100 mi (160 km) above the surface. The green color is from excited oxygen atoms and the red from nitrogen molecules in the upper atmosphere. *(a: NASA; b: NOAA National Climatic Data Center)*

nications satellites and transmission lines. It is remarkable that Earth's magnetosphere and the complex space weather systems it shields us from were entirely unknown until a few decades ago. Such discoveries remind us of how little we truly understand and how much remains to be learned even about our own planet.

ConceptCheck 5-15: **When the aurora is visible, what exactly in the sky is glowing?**

5-5 A rapidly growing population is altering our planetary habitat

One of the distinguishing characteristics of whether or not something is "alive" is whether or not it can change its environment. Indeed, Earth's original atmosphere has been changed from predominantly carbon dioxide–based to nitrogen-oxygen–based because of the biological processes of living things. We can observe the roots of giant trees causing cracks in sidewalks, ant colonies creating giant mounds of soil, and turtles eating away at coral. What makes Earth truly unique among all of the planets is that it is teeming with environment-altering life, from the floors of the oceans to the tops of mountains and from frigid polar caps to blistering deserts. Clearly our planet and its inhabitants interact, but what if things get out of balance?

The Biosphere and Natural Climate Variation

All life on Earth subsists in a relatively thin layer called the **biosphere,** which includes the oceans, the lowest few kilometers of the troposphere, and the crust to a depth of almost 2 mi (3 km). **Figure 5-22** is a portrait of Earth's biosphere based on NASA satellite data. The biosphere, which has taken billions of years to evolve to its present state, is a delicate, highly complex system in which plants and animals depend on each other for their mutual survival.

The state of the biosphere depends crucially on the temperatures of the oceans and atmosphere. Even small temperature changes can have dramatic consequences. An example that recurs every three to seven years is the El Niño phenomenon, in which temperatures at the surface of the equatorial Pacific Ocean rise by 2°C to 3°C. Ordinarily, water from the cold depths of the ocean is able to well upward, bringing with it nutrients that are used by microscopic marine organisms called *phytoplankton* that live near the surface (see Figure 5-22). But during an El Niño, the warm surface water suppresses this upwelling, and the phytoplankton starve. This wreaks havoc on organisms such as mollusks that feed on phytoplankton, on the fish that feed on the mollusks, and on the birds and mammals that eat the fish. During one particularly severe El Niño, one-quarter of the adult sea lions off the Peruvian coast starved, along with all of their pups.

Many different factors can change the surface temperature of our planet. One is that the amount of energy radiated by the Sun can vary up or down by a few tenths of a percent. Reduced solar brightness may explain the period from 1450 to 1850, when European winters were substantially colder than they are today. Another is when volcanoes spew ash into the atmosphere, slightly blocking the Sun's energy for a time.

Other factors are the gravitational influences of the Moon and the other planets. Thanks to these influences, the shape of Earth's orbit varies within a period of 90,000 to 100,000 years, the tilt of its rotation axis varies between

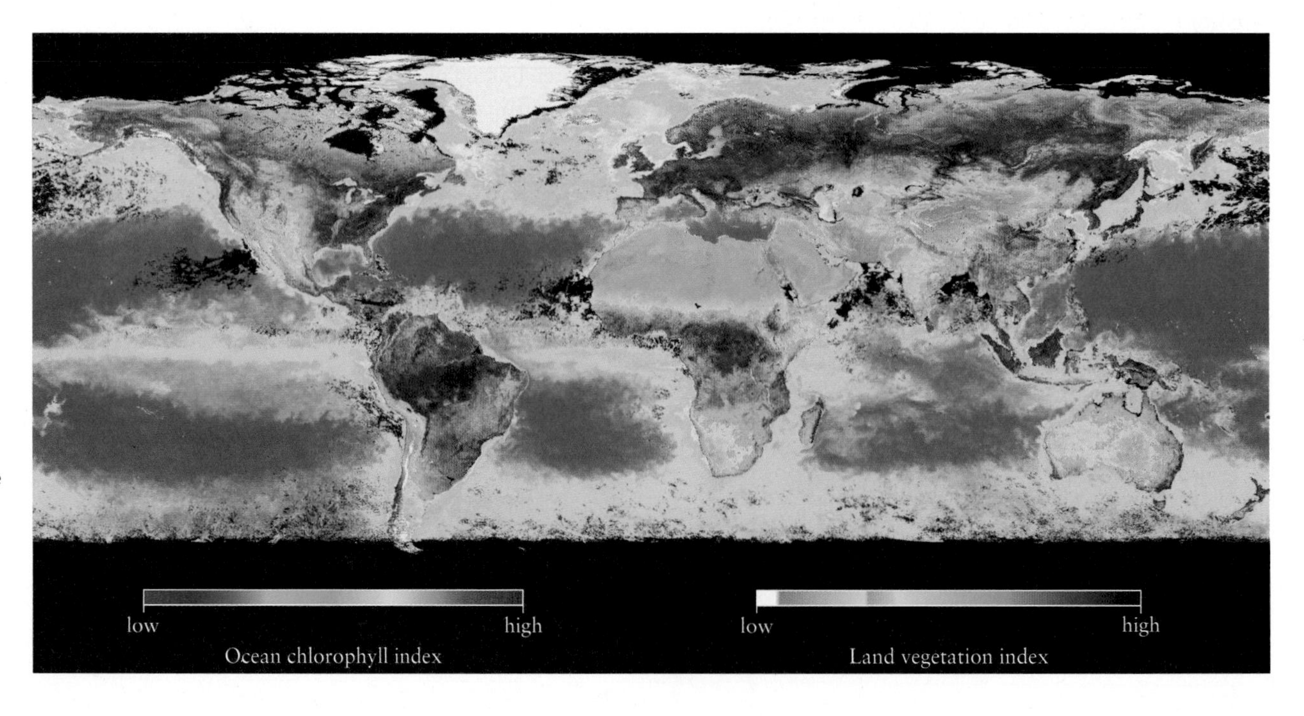

FIGURE 5-22
Earth's Biosphere This image, based on data from the *SeaWIFS* spacecraft, shows the distribution of plant life over Earth's surface. The ocean colors show where free-floating microscopic plants called phytoplankton are found. *(NOAA/Scripps Institution of Oceanography)*

22.1° and 24.5° with a 40,000-year period, and the orientation of its rotation axis changes due to precession with a 26,000-year period. These variations can affect climate by altering the amount of solar energy that heats Earth during different parts of the year. They help explain why Earth periodically undergoes an extended period of low temperatures called an *ice age,* the last of which ended about 11,000 years ago.

One of the most important factors affecting global temperatures is the abundance of greenhouse gases, such as CO_2. Geologic processes can alter this abundance, either by removing CO_2 from the atmosphere (as happens when fresh rock is uplifted and exposed to the air, where it can absorb atmospheric CO_2) or by supplying new CO_2. From time to time in our planet's history, natural events have caused dramatic increases of the amount of greenhouse gases in the atmosphere. One such event may have taken place 251 million (2.51×10^8) years ago, when Siberia went through a period of intense volcanic activity. The tremendous amounts of CO_2 released in this event would have elevated the global temperature by several degrees. Remarkably, the fossil record reveals that 95% of all species on Earth became extinct at this same time. The coincidence of these two events suggests that greenhouse-induced warming can have catastrophic effects on life.

ConceptCheck 5-16: **How can a tiny 3°C increase in ocean surface temperature stifle the food chain?**

Human Effects on the Biosphere: Deforestation

Our species is having an increasing effect on the biosphere because our population is skyrocketing. **Figure 5-23** shows the sharp rise in the human population that began in the late 1700s with the Industrial Revolution and the spread of modern ideas about hygiene. This rise accelerated in the twentieth century thanks to medical and technological advances ranging from antibiotics to high-yield grains. In 1960 there were 3 billion people on Earth; in 1975, 4 billion; and in 1999, 6 billion. We reached 7 billion in 2011. Projections by the United Nations Population Division show that there will be more than 8 billion people on Earth by the year 2030 and more than 9 billion by 2050.

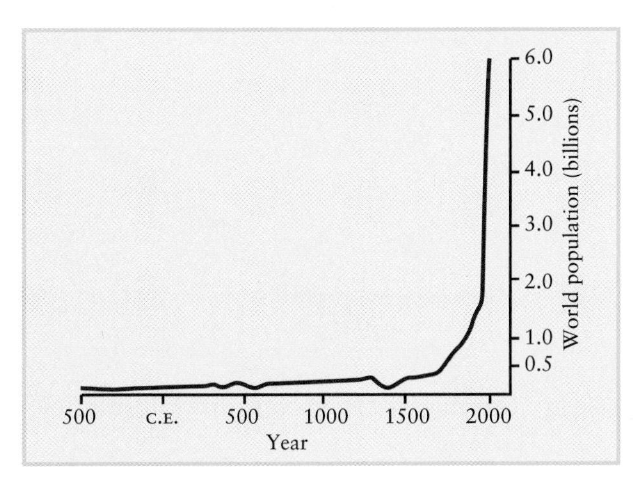

FIGURE 5-23 The Human Population Data and estimates from the U.S. Bureau of the Census, the Population Reference Bureau, and the United Nations Population Fund were combined to produce this graph showing the human population from 500 B.C.E. to 2000 C.E. The population began to rise in the eighteenth century and has been increasing at an astonishing rate since 1900.

Every human being has basic requirements: food, clothing, and housing. We all need fuel for cooking and heating. To meet these demands, we cut down forests, cultivate grasslands, and build sprawling cities. A striking example of this activity is occurring in the Amazon rain forest of Brazil. Tropical rain forests are vital to our planet's ecology because they absorb significant amounts of CO_2 and release O_2 through photosynthesis. Although rain forests occupy only 7% of the world's land areas, they are home to at least 50% of all plant and animal species on Earth. Nevertheless, to make way for farms and grazing land, people simply cut down the trees and set them on fire—a process called *slash-and-burn.* This burning process releases CO_2 once held within the plants back into the atmosphere. Such deforestation, along with extensive lumbering operations, is occurring in Malaysia, Indonesia, and Papua New Guinea. The rain forests that once thrived in Central America, India, and the western coast of Africa are almost gone (**Figure 5-24**).

A rapidly growing population demands more and more energy and causes ever-increasing damage to Earth's fragile environment.

ConceptCheck 5-17: **In what two ways does deforestation through slash-and-burn increase Earth's atmospheric carbon dioxide?**

Human Effects on the Biosphere: The Ozone Layer

Human activity is also having potentially disastrous effects on the upper atmosphere. Certain chemicals released into the air—in particular chlorofluorocarbons (CFCs), which have been used in refrigeration and electronics, and methyl bromide, which is used in fumigation—are destroying the ozone

FIGURE 5-24 R I V U X G **The Deforestation of Amazonia** The Amazon, the world's largest rain forest, is being destroyed at a rate of 20,000 km² per year in order to provide land for grazing and farming and as a source for lumber. About 80% of the logging is being carried out illegally. *(Dr. Morley Read/Shutterstock)*

FIGURE 5-25 The Antarctic Ozone Hole These two false-color images show that there was a net decrease of 50% in stratospheric ozone over Antarctica between October 1979 and September 2003. The amount of ozone at midlatitudes, where most of the human population lives, decreased by 10% to 20% over the same period. *(GSFC/NASA)*

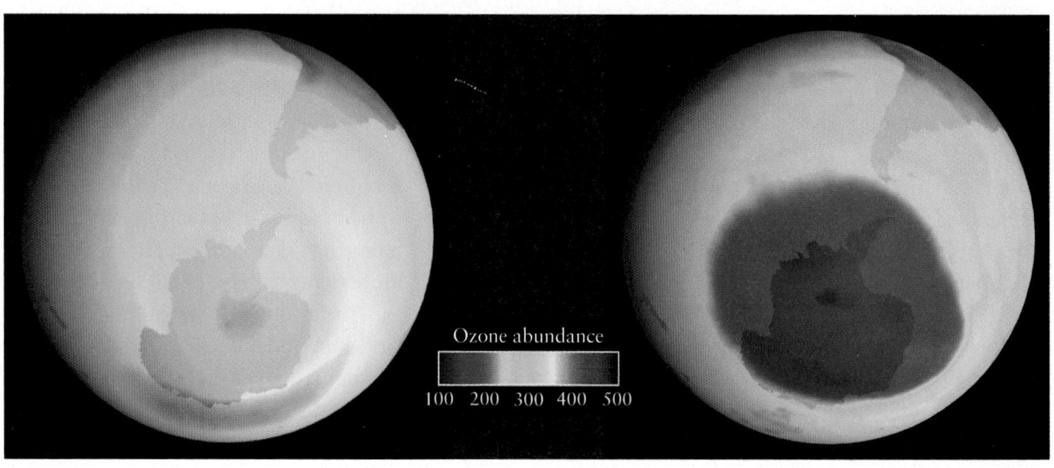

in the stratosphere. Without a high-altitude **ozone layer** to absorb ultraviolet light from the Sun, solar ultraviolet radiation would beat down on Earth's surface with greatly increased intensity. Such radiation breaks apart most of the delicate molecules that form living tissue. Hence, a complete loss of the ozone layer would lead to a catastrophic ecological disaster.

In 1985 scientists discovered a region with an abnormally low concentration of ozone over Antarctica. Since then this **ozone hole** has generally expanded from one year to the next (**Figure 5-25**). Smaller but still serious effects have been observed in the stratosphere above other parts of Earth. As a result, there has been a worldwide increase in the number of deaths due to skin cancer caused by solar ultraviolet radiation. By international agreement, CFCs are being replaced by compounds that do not deplete stratospheric ozone, and sunlight naturally produces more ozone in the stratosphere. In recent years, we have observed a slight slowing of ozone layer damage, but it is not expected to heal for many decades.

ConceptCheck 5-18: If ozone is not actually leaking out of the ozone hole, what is moving through this hole?

Human Effects on the Biosphere: Global Warming and Climate Change

Go to Video 5-4

The most troubling influence of human affairs on the biosphere is a consequence of burning fossil fuels (petroleum and coal) in automobiles, airplanes, and power plants as well as burning forests and brushland for agriculture and cooking. This burning can cause atmospheric pollution, but that is not what concerns scientists the most. Burning fossil fuels rapidly releases carbon dioxide into Earth's atmosphere—extra CO_2 that is predicted to interfere with the normal balancing process of thermal energy being emitted out into space—resulting in an overall increase in average global temperatures. This has been predicted by scientists for some time and we are now starting to see the dramatic effects of excess CO_2 humans have released into our atmosphere. It is not that CO_2 in the atmosphere is unwanted, it is that we are adding CO_2 to the atmosphere much faster than plants and geological processes can extract it, causing Earth's temperature to rapidly rise. **Figure 5-26** shows how the carbon dioxide content of the atmosphere has increased since 1958, when scientists began to measure this quantity on an ongoing basis.

You might wonder what the relationship is between CO_2 and Earth's global temperature. To put the values shown in Figure 5-26 into perspective, we need to know the atmospheric CO_2 concentration in earlier eras. Scientists have learned this by analyzing air bubbles trapped at various depths in the ice that blankets the Antarctic and Greenland. Each winter a new ice layer is deposited, so the depth of the bubble indicates the year in which it was trapped.

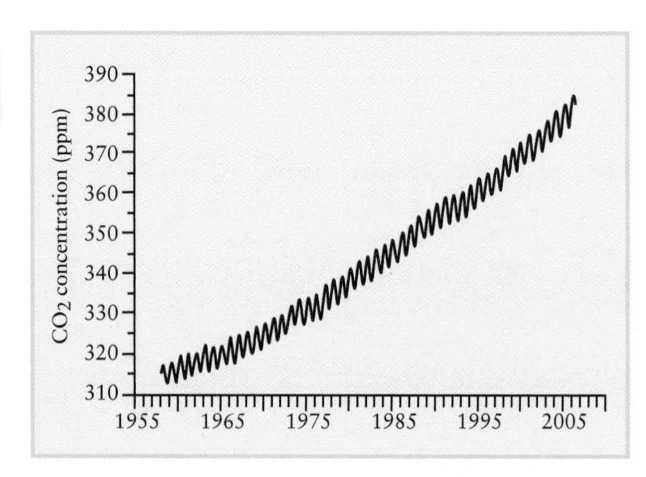

FIGURE 5-26 Atmospheric Carbon Dioxide Is Increasing This graph shows measurements of atmospheric carbon dioxide in parts per million (ppm). The sawtooth pattern results from plants absorbing more carbon dioxide during spring and summer. The CO_2 concentration in the atmosphere has increased by 21% since continuous observations started in 1958. *(NOAA/Scripps Institution of Oceanography)*

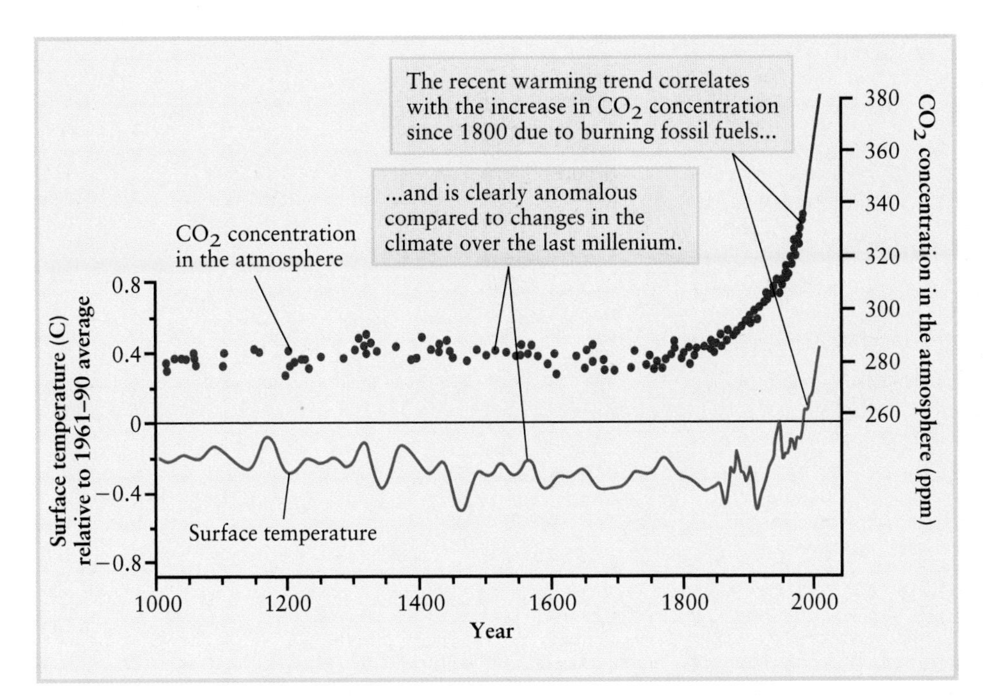

FIGURE 5-27 Atmospheric CO_2 and Changes in Global Temperature This figure shows how the carbon dioxide concentration in our atmosphere (upper curve) and Earth's average surface temperature (lower curve) have changed since 1000 C.E. The increase in CO_2 since 1800 due to burning fossil fuels has strengthened the greenhouse effect and caused a dramatic temperature increase. *(Intergovernmental Panel on Climate Change and Hadley Centre for Climate Prediction and Research, UK Meteorological Office)*

Figure 5-27 uses data obtained in this way to show how the atmospheric CO_2 concentration has varied since 1000 C.E. While there has been some natural variation in the concentration, its value has skyrocketed since the beginning of the Industrial Revolution around 1800. Data from older, deeper bubbles of trapped air show that in the 650,000 years before the Industrial Revolution, the CO_2 concentration was never greater than 300 parts per million. The present-day CO_2 concentration is greater than this by 25% and has grown to its present elevated level in just over half a century. If there are no changes in our energy consumption habits, by 2050 the atmospheric CO_2 concentration will be greater than 600 parts per million.

Increasing atmospheric CO_2 is of concern because this strengthens a greenhouse effect and raises Earth's average surface temperature. Figure 5-27 also shows how this average temperature has varied since 1000 C.E. Like the CO_2 data, the temperature data from past centuries come from analyzing trapped air bubbles. The recent dramatic increase in atmospheric CO_2 concentration has produced an equally dramatic increase in the average surface temperature. This temperature increase is called **global warming.** Other explanations for global warming have been proposed, such as changes in the Sun's brightness, but only greenhouse gases produced by human activity can explain the steep temperature increase shown in Figure 5-27. These gases include methane (CH_4) and nitrous oxide (N_2O), which are released in relatively small amounts by agriculture and industry but which are far more effective greenhouse gases than CO_2.

ConceptCheck 5-19: What is the difference between the greenhouse effect and global warming?

Impact of Global Warming

The effects of global warming can be seen around the world. Each of the last 12 years was one of the warmest on record since 1997, and each of the years since 2000 has seen increasing numbers of droughts, water shortages, and unprecedented heat waves. Glaciers worldwide are receding; the size of the ice cap around the north pole has decreased by nearly 40% since 1979, and portions of the Antarctic ice shelf have broken off (**Figure 5-28**). Earth's most dramatic changes due to global warming are evident in the Arctic Circle. It is generally accepted that the quickly thinning ice at the north pole will thaw and become open sea instead of solid ice within the next decades.

Few scientists disagree that global warming is occurring. Unfortunately, global warming is predicted to intensify in the decades to come. The degree to which this will occur and what

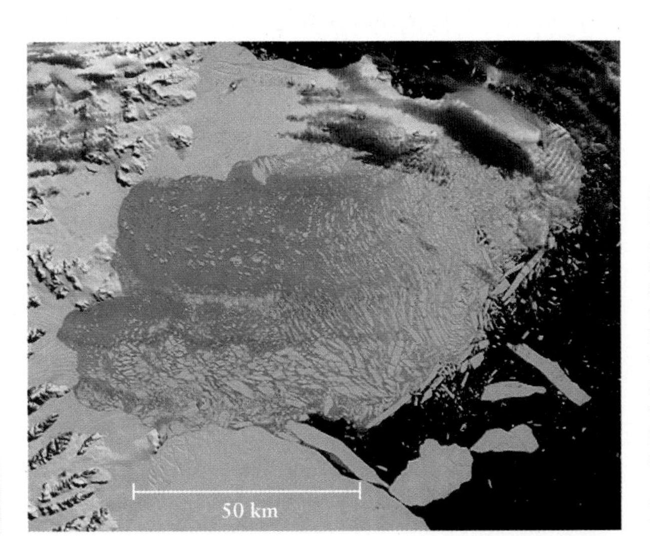

FIGURE 5-28 R I V U X G **A Melting Antarctic Ice Shelf** Global warming caused the Larsen B ice shelf to break up in early 2002. This ice shelf, which was about the size of Rhode Island, is thought to have been part of the Antarctic coast for the past 12,000 years. *(NASA/GSFC/LaRC/JPL, MIST Team)*

(a) 1938

(b) 1981

(c) 1998

(d) 2005

(e) 2009

FIGURE 5-29 R I V U X G **Disappearing Glacier** At an altitude of about 7000 ft (2100 m), Grinnell Glacier is a centerpiece in Montana's Glacier National Park. It is evident in this series of pictures that this glacier has lost about 40% of its size just since 1966 and, at its current melting rate, will disappear by the year 2030, along with all of the other glaciers in the park. *(a: T. J. Hileman/Glacier National Park Archives/USGS; b: Carl Key/USGS; c: Dan Fagre/USGS; d: Blase Reardon/USGS; e: Lindsey Bengtson/USGS)*

might be done about it is under grand debate. The UN Intergovernmental Panel on Climate Change predicts that if nothing is done to decrease the rate at which we add greenhouse gases to our atmosphere, the average surface temperature will continue to rise by an additional 1.4°C to 5.8°C during the twenty-first century. What is worse, the temperature increase is predicted to be greater at the poles than at the equator. The global pattern of atmospheric circulation (Figure 5-7) depends on the temperature difference between the warm equator and cold poles, so this entire pattern will be affected. The same is true for the circulation patterns in the oceans. As a result, temperatures will rise in some regions and decline in others and the patterns of rainfall will be substantially altered. Agriculture depends on rainfall, so these changes in rainfall patterns can cause major disruptions in the world food supply. Studies suggest that the climate changes caused by a 3°C increase in the average surface temperature would cause a worldwide drop in cereal crops of 20 million to 400 million tons, putting 400 million more people at risk of hunger.

The melting of polar ice due to global warming poses an additional risk to our civilization. When floating ice such as that found near the north pole melts, the ocean level remains the same. (You can see this by examining a glass of water with an ice cube floating in it. The water level does not change when the ice melts.) But the ocean level rises when ice on land melts and runs off into the sea. The Greenland ice cap has been melting at an accelerating rate since 2000 and has the potential to raise sea levels by half a meter or more. Low-lying coastal communities such as Boston and New Orleans, as well as river cities such as London, will have a greater risk of catastrophic flooding. Some island nations of the Pacific will disappear completely beneath the waves. Enormous glaciers are melting at an alarming rate (**Figure 5-29**). These observations closely match—if not exceed—the scientific predictions of what happens when atmospheric CO_2 is increased.

While global warming is an unintended consequence of human activity, the solution to global warming will require concerted and thoughtful action. Global warming cannot be stopped completely: Even if we were to immediately halt all production of greenhouse gases, the average surface temperature would increase an additional 2°C by 2100, thanks to the natural inertia of Earth's climate system. Instead, our goal is to minimize the effects of global warming by changing the ways in which we produce energy, making choices about how to decrease our requirements for energy, and finding ways to remove CO_2 from the atmosphere and trap it in the oceans or beneath our planet's surface. Confronting global warming is perhaps the greatest challenge to face our civilization in the twenty-first century.

ConceptCheck 5-20: Is there scientific debate about whether or not global warming is occurring?

KEY IDEAS AND TERMS

5-1 Most of Earth's surface is covered with flowing water that radically changes the landscape

- Seventy-one percent of Earth's surface is covered with liquid water, but water is found in greater amounts elsewhere in the solar system.
- Ninety-seven percent of Earth's water is in the form of seawater, with most of the remaining freshwater locked up as frozen water near the poles or underground. Only a very small portion of Earth's freshwater is easily accessible on Earth's surface.
- The continuous movement of water around our planet in different forms through rising evaporation and falling precipitation, including flows in the atmosphere, across the surface, under the ground, and along ocean currents, is known as the **water cycle.**
- Water is unique in that it expands when frozen, can hold tremendous energy without changing phases, and has a molecular structure that allows it to dissolve substances, flow as a continuous body, and easily attach itself to many surfaces.

VISUAL LITERACY TASK

Earth's Interior and the Paths of Seismic Waves

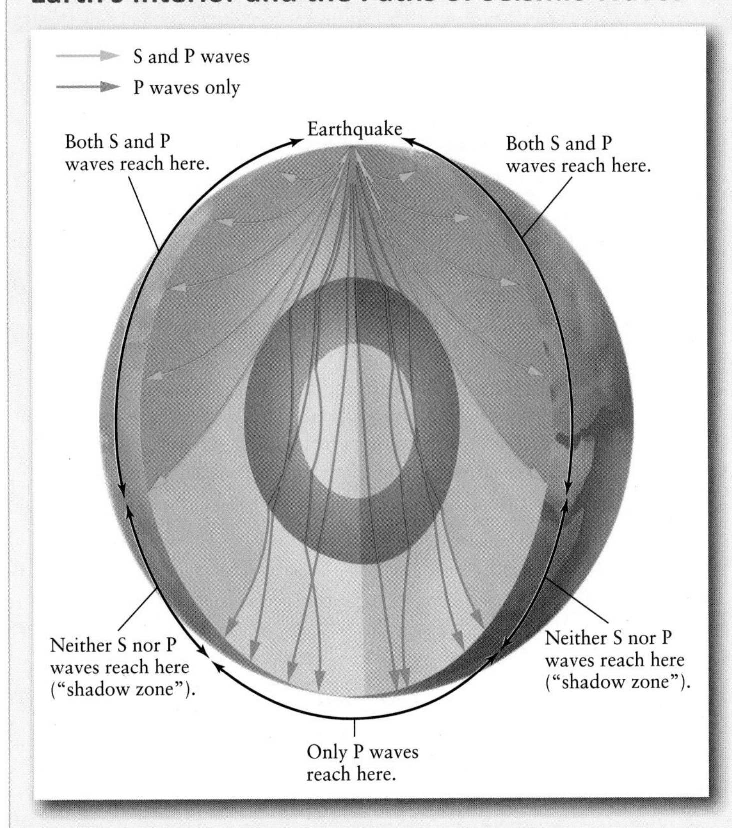

Guiding Questions:

1. P-type seismic waves are observed
   a. everywhere on Earth, except in the shadow zone.
   b. only in places where S waves are not.
   c. only near an earthquake itself.
   d. to be stopped by solid rock.

2. P-type seismic waves will
   a. pass through liquid, just as S waves do.
   b. pass through liquid, whereas S waves will not.
   c. be blocked entirely by Earth's core.
   d. absorb S-type seismic waves.

3. Earth's inner core and outer core take up about
   a. half the size of Earth overall.
   b. one-quarter the size of Earth.
   c. 90% of Earth's volume.
   d. 10% of Earth's volume.

4. Compared to Earth's outer core, Earth's inner core is
   a. liquid, just like the outer core.
   b. the same material as the mantle, only hotter.
   c. solid, whereas the outer core is liquid.
   d. responsible for absorbing P-type seismic waves.

PROMPT: What would you tell a fellow student who said, **"P-type seismic waves are never measured on the side of the planet opposite an earthquake because they will not travel through Earth's liquid inner core"?**

ENTER RESPONSE:

5-2 Earth is surrounded by a thin, multilayered envelope of gas that has changed since life became prominent

- **Photosynthesis** is the chemical energy process by living organisms that consumes carbon dioxide (CO_2) and water and releases oxygen (O_2).
- **Respiration** is the chemical energy process where modern animals consume oxygen and release carbon dioxide.
- A planet's **atmosphere** is the layer of gas surrounding the planet, and the **atmospheric pressure** is the weight of all the air above a particular height.
- Earth's **atmosphere** is divided into layers where the lowest is the **troposphere,** where most weather occurs, surrounded by the **ozone**-containing **stratosphere,** which in turn is surrounded by the **mesosphere** and **thermosphere.**
- The fraction of incoming sunlight that a planet reflects is called its **albedo.**
- The **greenhouse effect** occurs when a planet's **greenhouse gases** in its atmosphere prevent some of the energy emitted by Earth's warm surface from escaping into space.

5-3 Volcanoes and earthquakes reveal energy from a molten interior driving Earth's surface to shift positions

- Earth has a solid **inner core** and a molten **outer core** surrounded by a **mantle** surrounded by a thin **crust.**
- **Earthquakes** release energy in the form of **surface waves,** compressional **P waves,** and transverse **S waves.**
- **Plate tectonics** describes how the process of **convection** drives the movement of Earth's surface plates on the **lithosphere.**
- Where plates *separate*, undersea mountain ranges called oceanic **rifts** are caused by **seafloor spreading.**
- Where plates *collide*, cool crustal material from one of the plates sinks back down into the mantle along a **subduction zone.**

5-4 Earth's magnetic field emanating from its spinning, molten interior creates a protective shield from the Sun's harmful radiation

- Earth's magnetic field is created from electric currents moving in Earth's *outer core.*
- The region of space around a planet in which the motion of charged particles is dominated by the planet's magnetic field is called the planet's **magnetosphere,** which captures the solar wind's charged particles.
- When charged particles excite atoms in Earth's atmosphere and cause them to glow, the **aurora,** also called the **northern lights (aurora borealis)** or **southern lights (aurora australis),** are observed.

5-5 A rapidly growing population is altering our planetary habitat

- All life on Earth subsists in a relatively thin layer called the **biosphere.**
- Without the high-altitude **ozone layer** to absorb ultraviolet light from the Sun, solar ultraviolet light would beat down on Earth's surface with greatly increased intensity. This layer is thinnest over Antarctica and is known as the **ozone hole.**
- The recent dramatic increase in atmospheric CO_2 concentration has produced an equally dramatic increase in Earth's average surface temperature, known as **global warming.**

QUESTIONS

Review Questions

1. How does water falling as snow atop a mountain eventually become a cloud?
2. What is unique about the behavior of water compared to other common substances?
3. How does the greenhouse effect influence the temperature of the atmosphere? Which properties of greenhouse gases in the atmosphere cause this effect?
4. How do we know that Earth was once entirely molten?
5. What are the different types of seismic waves? Why are seismic waves useful for probing Earth's interior structure?
6. Describe the interior structure of Earth.
7. The deepest wells and mines go down only a few kilometers. What, then, is the evidence that iron is abundant in Earth's core? That Earth's outer core is molten but the inner core is solid?
8. Describe the process of plate tectonics. Give specific examples of geographic features created by plate tectonics.
9. Explain how convection in Earth's interior drives the process of plate tectonics.
10. Describe the various ways in which Earth's surface is reshaped over time.
11. Describe Earth's magnetosphere. If Earth did not have a magnetic field, do you think aurorae would be more common or less common than they are today?
12. Ozone and carbon dioxide each make up only a fraction of a percent of our atmosphere. Why, then, should we be concerned about small increases or decreases in the atmospheric abundance of these gases?
13. What is the difference between ozone depletion and global warming?

Web Chat Questions

1. The human population on Earth is currently doubling about every 30 years. Describe the various pressures placed on Earth by uncontrolled human population growth. Can such growth continue indefinitely? If not, what natural and human controls might arise to curb this growth? It has been suggested that overpopulation problems could be solved by colonizing the Moon or Mars. Do you think this is a reasonable solution? Explain your answer.
2. In order to alleviate global warming, it will be necessary to dramatically reduce the amount of carbon dioxide that we release into the atmosphere by burning petroleum. What changes in technology and society do you think will be needed to bring this about?

Collaborative Exercises

1. Using a ruler and self-stick or tape-on labels, create a scale model of Earth on the shortest member of your group. Use the group member's height in inches divided by Earth's diameter (12,800 km) as the scale factor. For example, if the selected group member is 65 in. tall then the 50-km maximum depth of Earth's crust is (65 in. ÷ 12,800 km) × (50-km) = 0.25 in. from the top of the head and 0.25 in. from the bottom of the feet.

Observing Questions

1. Use *Starry Night*™ to view Earth from space. Open the view named **Favourites > Explorations > Earth's Surface.** The view shows the rotating Earth from a point in space about 12,000 km above the surface. The clouds and atmosphere have been removed from the image, and the side of Earth facing away from the Sun is artificially brightened so that you may use the location scroller and **Zoom** controls to inspect the entire surface of our home planet. (a) Can you see any evidence of plate tectonics on Earth? (b) Can you see any evidence of life or of man-made structures or objects?
2. Use *Starry Night*™ to view Earth from space. Select **Favourites > Explorations > Earth** from the menu to place yourself about 12,238 km above Earth's surface. Use the location scroller to rotate Earth beneath you, allowing you to view different parts of its surface. Is there any evidence for the presence of an atmosphere on Earth? Explain.
3. Use *Starry Night*™ to view Earth from space. Click the **Home** button in the toolbar to move your viewing location to your hometown or city. Click the **Increase current elevation** button beneath the **Viewing Location** panel in the toolbar to raise your position to about 11,000 km above your home location. Use the hand tool to move Earth into the center of the view. Locate the position of your home and zoom in on it, using the **Zoom** tool on the right of the toolbar to set the field of view to about 16° × 11°. Use the location scroller to move over various regions of Earth's surface. (a) Can you see any evidence of the presence of life or of man-made objects? (b) Right-click on Earth

(Ctrl-click on a Mac) and select **Google Map** from the contextual menu to examine regions of Earth in great detail on these images from an orbiting satellite. (You will need to have access to the Internet in order to use Google Map.) You can attempt to locate your own home on these maps. As an exercise, find the country of Panama. **Zoom in** progressively upon the Panama Canal and search for the ships traversing this incredible waterway and the massive locks that lift these ships into the upper waterways. What does this suggest about the importance of sending spacecraft to planets to explore their surfaces at close range?

ANSWERS

ConceptChecks

ConceptCheck 5-1: The total amount of water in Earth's water cycle does not change significantly over short periods of time and water soaking through rocks is part of Earth's water cycle because it eventually makes its way back into the atmosphere.

ConceptCheck 5-2: In the water cycle, the total amount of water does not change; rather, water changes forms among solid, liquid, and gas. Because the amount of water would be the same, the cup would have the same weight after freezing as it did before.

ConceptCheck 5-3: In our solar system, Earth is the only planet on which these processes appear to have taken place. Plants using photosynthesis convert carbon dioxide into oxygen, and certain bacteria release nitrogen in the atmosphere.

ConceptCheck5-4: As sunlight warms Earth's surface, energy in the form of heat increases the temperature of the troposphere near Earth's surface such that the troposphere is heated from below.

ConceptCheck 5-5: Ozone in the stratosphere absorbs ultraviolet radiation from the Sun and helps warm the stratosphere. So, a deficit of ozone would make the stratosphere cooler in temperature.

ConceptCheck 5-6: By moderating heat loss into outer space, the greenhouse effect keeps Earth's temperature warmer than it would otherwise be, allowing liquid water and hence life to exist on Earth.

ConceptCheck 5-7: Earth must have significant amounts of iron in its core because Earth's overall density is much higher than the rocks found at its surface, so the new planet must have far less iron if it were to have a lower average density.

ConceptCheck 5-8: Because longitudinal S waves are unable to travel through liquid, the sudden appearance of S waves suggests that the planet's interior would have solidified.

ConceptCheck 5-9: The melting temperature would be higher because regions of high pressure increase the temperature at which rocks melt.

ConceptCheck 5-10: When two oceanic plates move apart, molten material wells up in the gap between them. This piles up and produces an underwater mountain chain like the Mid-Atlantic Ridge.

ConceptCheck 5-11: Where oceanic crust material is subducted beneath continental crust, it is plunged into the high-temperature mantle and melts. Occasionally, this material rises in the form of magma-feeding volcanoes.

ConceptCheck 5-12: A supercontinent traps energy in the form of heat beneath it, causing the material directly beneath it to expand and crack. This process allows fissures to be filled by rapidly upwelling material from underneath, breaking the continent apart.

ConceptCheck 5-13: No. Earth's magnetic field is predominantly created by a dynamo effect caused by the movement of iron-rich material in the liquid core.

ConceptCheck 5-14: The vast majority of charged particles from the Sun's solar wind are deflected away, while a small portion leaks inside the magnetosphere.

ConceptCheck 5-15: What is glowing are atoms in the outer fringes of Earth's atmosphere. When struck by fast-moving charged particles that originated on the Sun, electrons in these atoms are excited and then emit light when they return to their normal state.

ConceptCheck 5-16: A 3°C change in ocean surface temperature is sufficient to starve plankton, the base of the food chain, by preventing cooler, nutrient-rich water from moving to the surface.

ConceptCheck 5-17: Living plants remove CO_2 from the atmosphere through photosynthesis, and burning plants release the CO_2 they have captured back into the atmosphere.

ConceptCheck 5-18: Atmospheric ozone prevents much of the Sun's harmful ultraviolet light from entering Earth's atmosphere. The ozone hole is a region allowing more of this particular energy to enter Earth's atmosphere.

ConceptCheck 5-19: The greenhouse effect is the process by which CO_2 and other gases maintain Earth's temperature at a moderate level by preventing heat loss into outer space. By comparison, global warming is the uncontrolled, rapid increase in Earth's temperature caused by a dramatic increase in atmospheric CO_2 concentration due to human activity, making the effect unusually intense.

ConceptCheck 5-20: No. There is no debate—nearly everyone agrees global warming is occurring. The scientific debate is about how much will occur, why it is occurring, how it might be mitigated, and what will be the extent of damage to the biosphere.

CalculationChecks

CalculationCheck 5-1: Using the relationship that distance equals rate times time $(6 \times 10^8 \text{ cm}) \div 2 \text{ cm/year} = 3 \times 10^8$ years or 300 million years.

CalculationCheck 5-2: 5×10^9 years $\div$ 5×10^8 years/cycle $=$ 10 cycles.

R I V U X G A giant landslide from a collapsing 9-mile (15-km) high cliff partially fills a crater and changes the landscape on Iapetus, a moon orbiting Saturn. *(NASA/JPL/Space Science Institute)*

Exploring Terrestrial Surface Processes and Atmospheres

CHAPTER LEARNING OBJECTIVES **By reading the sections of this chapter, you will learn:**

6-1 Comparing terrestrial planets and moons shows distinct similarities and dramatic differences in appearance

6-2 Many terrestrial world surfaces are dominated by impact craters revealing the age of underlying processes

6-3 Tectonics and volcanism influence surface features

6-4 Atmospheres surrounding terrestrial planets vary considerably

6-5 Evidence exists for water in locations besides Earth

For the past four hundred years, astronomers have gazed at the planets through telescopes and wondered if those worlds are like our own planet Earth. Now, in just the past few decades, advances in spacecraft technology have allowed us to send orbiting probes and robotic rovers to carefully study the planets and their moons. In Chapter 5, we systematically looked at some of the geologic processes that shape Earth, resulting from an energetic molten interior. Using these ideas, we are now in a position to look at other solar system objects and to see how they are similar to or different from our own. This way of thinking leads us to a fascinating question: If planets and moons formed more than 4 billion years ago, how can modern-day astronomers determine the seemingly complex histories of planets and moons beyond Earth? Just as ancient fossils on Earth provide evidence of how life has evolved on our planet, ancient rocks and geological formations help us understand how planets' surfaces and atmospheres have evolved. For example, the oldest rocks on Earth have a chemical structure showing that they formed in a time when there was little oxygen in the atmosphere and, hence, before the appearance of photosynthetic planet life. Planetary scientists who want to learn about the history of Mars are therefore very interested in looking carefully at ancient rocks on that world. Now that we understand the interior, surface, and atmosphere of Earth, we can compare it to the other planets and moons in our solar system, such as the landslide and crater on Iapetus, one of Saturn's moons, in the chapter-opening image. In this chapter we will focus on the terrestrial planets Mercury, Venus, and Mars. We will look first at the forces that have molded them—impacts, volcanism, and tectonics—and then briefly consider their atmospheres. We can also begin to explore where else in the solar system water exists and speculate about whether or not sufficient water is present to sustain living organisms. ■

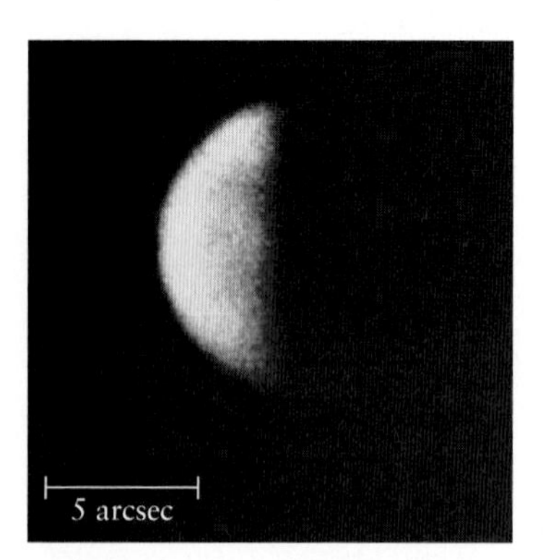

(a) Mercury at greatest elongation

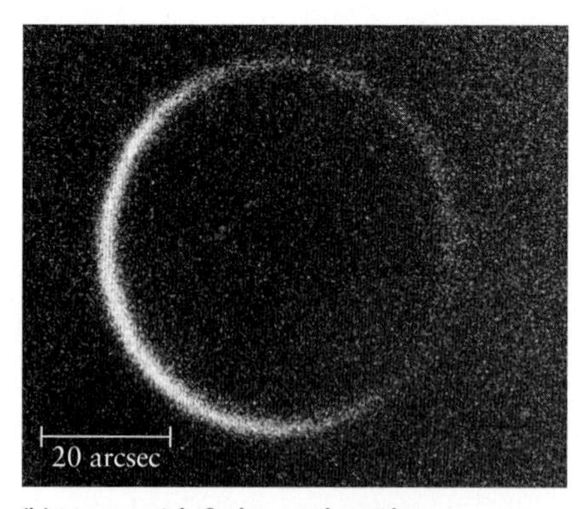

(b) Venus at inferior conjunction

(c) Mars at opposition

FIGURE 6-1 R I V U X G **Earth-Based Views of Mercury, Venus, and Mars** (a) Hazy markings are faintly visible in this photograph of Mercury, one of the best ever produced with an Earth-based telescope. Note that Mercury goes through phases like Venus (see Figure 3-9). (b) Venus at inferior conjunction shows its atmosphere as a faint ring surrounding the planet. This photograph was processed to reveal the ring, which is too faint to show in Figure 3-9. (c) Mars' polar ice cap is clearly visible on this image made by Jamie Cooper using a webcam. *(a: New Mexico State University Observatory; b: Lowell Observatory; c: SSPL via Getty Images)*

6-1 Comparing terrestrial planets and moons shows distinct similarities and dramatic differences in appearance

If some Earth-gazing alien astronomers observe Earth from the scorching surface of Mercury, from above the dense clouds shrouding Venus, or from atop gigantic mountains of Mars with the same telescope technology available to our own astronomers, what might they be able to see? Using today's Earth-based technology from those vantage points, an astronomer would be unable to pick out any specific features or landforms on Earth less than a few hundred miles across. In much the same way, Earth astronomers face the same limitations in studying the other terrestrial planets and moons. To truly understand these other worlds, it is essential to study them using the most modern telescopes or, even better, at close range using spacecraft. Telescopes reveal that the rocky planets near the Sun, and some of the moons orbiting the outer planets, do have surface features similar to, and sometimes quite different from, our home planet.

Tiny planet Mercury is the planet closest to the Sun. Whereas naked-eye observations of Mercury are best made at dusk or dawn when the Sun is blocked by Earth's horizon, you might be surprised to discover that the best telescopic views are obtained in the daytime, when the planet is high in the sky, far above the degrading atmospheric effects occurring near the horizon. The photograph in **Figure 6-1*a*,** which is among the finest Earth-based views of Mercury, was taken during daytime. Only recently has Mercury been visited close up by spacecraft, and, as of this writing, astronomers are still trying to uncover this tiny planet's history.

While an Earth-based telescope reveals some curious surface details on Mercury, Venus appears almost completely featureless, with neither clearly noticeable continents nor mountains. It took astronomers some time to realize that they were not seeing the true surface of the planet. We now understand that Venus is covered by a thick, unbroken layer of clouds that hides the rugged nature of its waterless surface. A heat-capturing cloud layer also explains why Venus reflects such a large fraction of sunlight, making it shine brilliantly in the sky. Direct evidence that Venus has a thick atmosphere came when nineteenth-century astronomers observed Venus near the time of inferior conjunction. This is when Venus lies most nearly between Earth and the Sun, so that we see the planet lit from behind. As Figure 6-1*b* shows, sunlight is scattered by Venus's atmosphere, producing a luminescent ring around the planet that would not otherwise be present. No such ring is seen around Mercury at inferior conjunction, which indicates that Mercury has no appreciable atmosphere.

Unlike cloud-shrouded Venus, Mars has a thin, almost cloudless atmosphere that permits a clear view of the surface. Under the right conditions, even a relatively small telescope can reveal substantial detail on the Martian surface. Figure 6-1*c* shows an image of Mars during 2003, when Mars's orbit brought it closer to Earth than it had been in several thousand years. The thin strip of blue and orange around the right-hand edge of Mars in this image is caused by the Martian atmosphere. This image also shows that different parts of the Martian surface have different colors. Most striking are the whitish ice caps covering the Martian poles, which bear a striking resemblance to the Arctic and Antarctic polar regions on Earth.

What we can easily see is that the planets Mercury, Venus, and Mars have some similarities to Earth and our Moon, but some dramatic differences, too. What can we learn about these objects in terms of comparing them to Earth? There are clues to each terrestrial planet's age and interior found etched into their surfaces, and these clues are the topics of our next two sections: impact craters, tectonics, and volcanism.

ConceptCheck 6-1: If we build bigger visible light telescopes, will astronomers finally be able to see the surface of Venus?

6-2 Many terrestrial world surfaces are dominated by impact craters revealing the age of underlying processes

Probably one of the first things an observer notices about our Moon is that it is pockmarked with circular depressions. The same is true for many of the images of terrestrial planets and other moons. But where do these circular craters come from, and how do they change a planetary landscape?

What Creates an Impact Crater?

Go to Video 6-1

When the Bavarian astronomer Franz von Gruithuisen proposed in 1824 that the craters seen on the Moon could be the result of impacts from flying space debris, a major problem for his theory was the observation that nearly all craters are round. If craters were merely gouged out by high-speed rocks, a rock striking the Moon in any direction except straight downward would have created a noncircular crater, which is different from what we observe. A century after Gruithuisen, it was realized that a high-speed impactor colliding with the Moon generates a shock wave in the lunar surface that spreads out evenly from the point of impact. Perhaps surprisingly, this type of shock wave produces a nearly perfectly circular crater no matter in what direction or impact angle the incoming impactor comes from. Many of the largest lunar craters have a central peak rising up from the crater floor, which is characteristic of a high-speed impact (**Figure 6-2*a***). Craters made by other processes, such as volcanic action, would not have central peaks of this kind. But what object could create such a crater?

The planets orbit the Sun in nearly circular orbits (see Figure 4-1), but many asteroids and comets follow more elongated orbits. These elongated orbits can put these small objects on a collision course with a planet or moon. If an object collides with a Jovian gas-planet, it is swallowed up by the planet's thick atmosphere. (Astronomers actually saw an event of this kind in 1994, when a comet crashed into

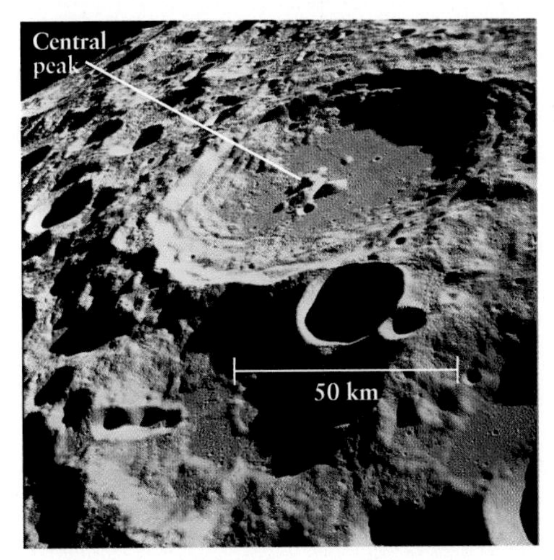

(a) A crater on the Moon

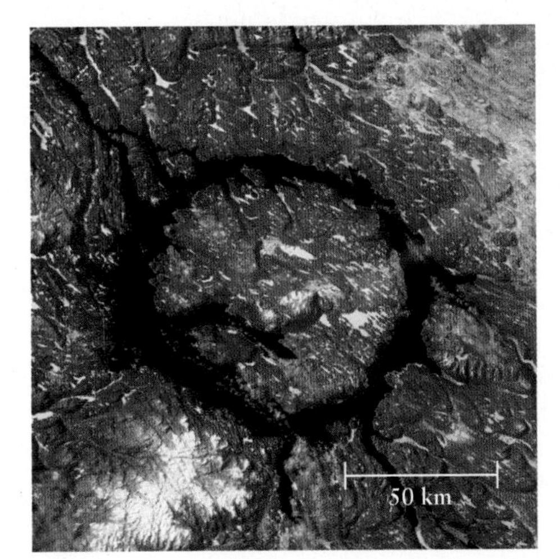

(b) A crater on Earth

(c) A crater on Mars

FIGURE 6-2 R I V U X G **Impact Craters** These images, all taken from spacecraft, show impact craters on three different worlds. (a) The Moon's surface has craters of all sizes. The large crater near the middle of this image is about 50 mi (80 km) in diameter, equal to the length of San Francisco Bay. (b) Manicouagan Reservoir in Quebec is the relic of a crater formed by an impact more than 200 million years ago. The crater was eroded over the ages by the advance and retreat of glaciers, leaving a ring lake 60 mi (100 km) across. (c) Lowell Crater in the southern highlands of Mars is 125 mi (200 km) across. Like the image of the Moon in (a), there are craters on top of craters. Note the light-colored frost formed by condensation of carbon dioxide from the Martian atmosphere. *(a: NASA; b: JSC/NASA; c: NASA/JPL/MSSS)*

and was swallowed up by Jupiter, with all traces of impact disappearing after a few months.) But when an object collides with the solid surface of a terrestrial planet or a moon, the result is an **impact crater** (see Figure 6-2). Such impact craters, found throughout the solar system, offer stark evidence of these violent collisions.

ConceptCheck 6-2: What shape of crater will a potato-shaped impactor, 30 km long and 10 km wide, make if it collides with Earth and impacts at a low angle?

Cratering Measures Geologic Activity

Not all planets and moons show the same amount of cratering. The Moon is heavily cratered over its entire surface, sometimes with craters on top of older craters on top of even older craters, as shown in Figure 6-2*a*. On Earth, by contrast, craters seem to be very rare: Geologists have identified fewer than 200 impact craters on our planet (Figure 6-2*b*). Our understanding is that the Moon formed almost immediately after the formation of Earth, and the Moon and Earth have been bombarded at comparable rates over their history. Why, then, are craters so much rarer on Earth than on the Moon?

The answer is that Earth is a *geologically active* planet. Because of plate tectonics, the continents slowly change their positions over eons. New material flows onto the surface from the interior in volcanic eruptions or from seafloor spreading, and old surface material is pushed back into the interior. These processes, coupled with erosion from wind, water, and ice, cause craters on Earth to be erased over time. The few craters found on Earth today must be relatively recent, since there has not yet been time to erase them.

The Moon, by contrast, is essentially geologically *inactive*. There are no volcanoes and no motion of continental plates (and, indeed, no continents). Furthermore, the Moon has neither oceans nor an appreciable windy and weather-carrying atmosphere, so there is little obvious erosion as we know it on Earth. With none of the processes that rapidly erase evidence of craters on Earth, the Moon's surface remains largely unchanged—pockmarked with the scars of billions of years of impacts.

All terrestrial planets and moons have impact craters, and heavily cratered areas are found on the oldest, least changed surfaces.

As we saw in Chapter 5, in order for a planet to be geologically active, its interior must be at least partially molten, so that continental plates can slide around on the underlying mantle and so that molten lava can come erupting up to the planet's surface (see Figure 5-10). So, we would expect geologically inactive—and hence heavily cratered—worlds like the Moon to have less molten material in their interiors than does Earth. Investigations of these inactive worlds bear this out. But *why* is the Moon's interior less molten than Earth's?

ANALOGY To see one simple answer to this question, notice that a large turkey or roast taken from the oven will stay warm inside for hours, but a single meatball will cool off much more rapidly. The reason is that the meatball has more surface area relative to its volume, so it more easily loses heat to its surroundings. A planet or satellite also tends to cool down as it emits electromagnetic radiation into space; the smaller the planet or satellite, the greater its surface area relative to its volume, and the more readily it radiates away heat (**Figure 6-3**). Both Earth and the Moon were probably completely molten when they first formed, but because the Moon (diameter 2160 mi or 3476 km) is so much smaller than Earth (diameter 7918 mi or 12,742 km), it has lost much of its internal heat and has a much more solid interior.

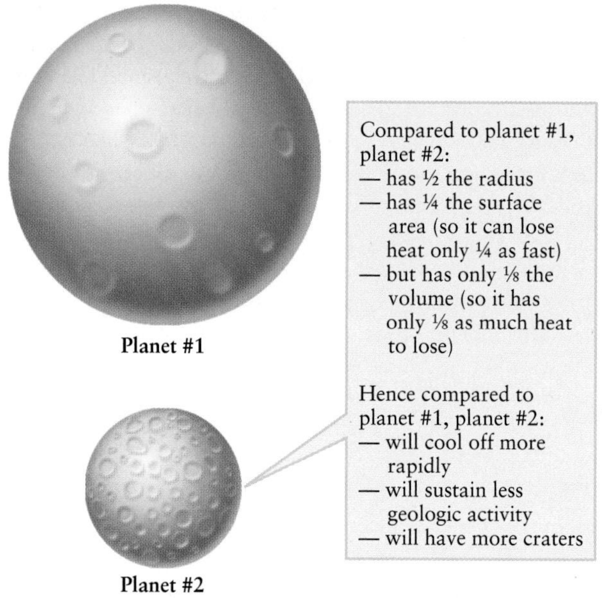

FIGURE 6-3 Planet Size and Cratering Of these two hypothetical planets, the smaller one (#2) has less volume and less internal heat, as well as less surface area from which to radiate heat out into space. But the ratio of surface area to volume is greater for the smaller planet. Hence, the smaller planet will lose heat faster, have a colder interior, and be less geologically active. It will also have a more heavily cratered surface, since it takes active geologic activity to erase craters.

By considering these differences between Earth and the Moon, we have uncovered a general rule for worlds with solid surfaces:

The smaller the terrestrial world, the less internal heat it is likely to have retained from its formation, and, thus, the less geologic activity it will display on its surface. The less geologically active the world, the older and more heavily cratered its surface.

This rule means that we can use the amount of cratering visible on a planet or satellite to estimate the age of its surface and how geologically active it is.

As for all rules, there are limitations and exceptions to the rule relating a world's size to its geologic activity. One limitation is that the four terrestrial planets all have slightly different compositions, which affects the types and extent of geologic activity that can take place on their surfaces.

This also complicates the relationship between the number of craters and the age of the surface. An important exception to our rule is Jupiter's satellite Io, which we will discuss in more detail later in the chapter (see **Box 6-1 The Heavens on Earth: Jupiter's Moon Io Is Covered with Active Volcanoes**). Despite its small size, it is the most volcanic world in the solar system. And in spite of these limitations and exceptions, the relationships between a world's size, internal heat, geologic activity, and amount of cratering are powerful tools for understanding the terrestrial planets and satellites. Let's take a look at cratering on our Moon and the inner planets.

The History of Lunar Cratering By correlating the ages of Moon rocks with the density of craters at the sites where the rocks were collected by NASA astronauts, geologists have found that the rate of impacts on the Moon has changed dramatically over the ages. The ancient, heavily cratered lunar highlands are evidence of an intense bombardment that dominated the Moon's early history. For nearly a billion years, rocky debris left over from the formation of the planets rained down upon the Moon's young surface. As **Figure 6-4** shows, this barrage extended from 4.56 billion years ago, when the Moon's surface solidified, until about 3.8 billion years ago. At its peak, this bombardment from space would have produced a new crater of about 1 mi (1.6 km) radius somewhere on the Moon about once per century. If this seems like a long time interval, remember that we are talking about a bombardment that lasted hundreds of millions of years and produced millions of craters over that time. Many of these impacts could have brought tiny amounts of water to the Moon, adding up over many millions of years to a tremendous amount of water, which could still present today as ice, hidden in the permanently shadowed walls of deep craters near the Moon's poles.

The rate of impacts should have tapered off as small debris left over from the formation of the solar system was swept up by the newly formed planets. In fact, radioactive dating shows that the impact rate spiked upward between 4.0 and 3.8 billion years ago (see Figure 6-4). One proposed explanation is that the much larger orbits of Jupiter and Saturn changed slightly but suddenly, about 4.0 billion years ago during the process of settling into their present-day orbits. This would have changed the gravitational forces that these planets exerted on the asteroid belt, disturbing the orbits of many asteroids and sending them careening toward the inner solar system.

Whatever the explanation for the final epoch of heavy bombardment, major impacts during this period gouged out the mare basins. Between 3.8 and 3.1 billion years ago iron-rich magma oozed up from the Moon's still-molten mantle, flooding the Moon's smooth dark basins with lava that solidified to form the dark rocks we see on the surface. **Figure 6-5** shows rugged terrain created by craters on the Moon, some of which has been smoothed by upwelling lava, whereas other more rugged areas create permanent shadows harboring frozen water ice.

ConceptCheck 6-3: **How would a graph showing the rate of crater formation on the Moon appear if the Moon's impact rate had instead been constant over the past 3 billion years?**

Go to Video 6-2

Mercury's Cratered Surface On first glance, the planet Mercury might look nearly identical to pictures you have seen of the Moon. **Figure 6-6** shows a typical close-up view of Mercury, confirming that it indeed looks

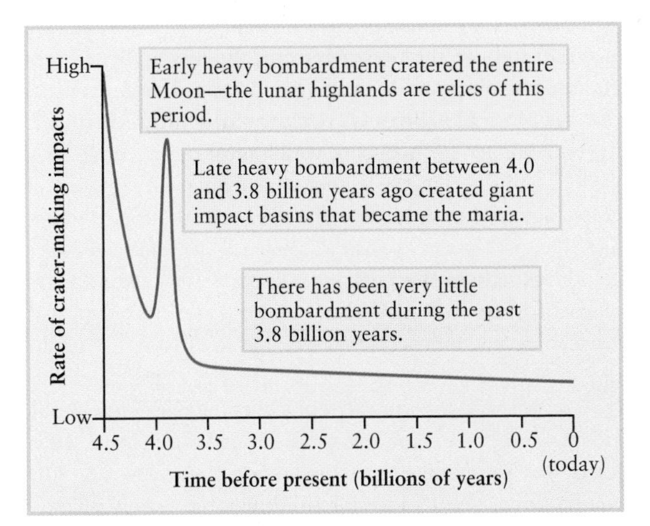

FIGURE 6-4 The Rate of Crater Formation on the Moon This graph shows the rate at which impact craters formed over the Moon's history. The impact rate today is only about 1/10,000 as great as during the most intense bombardment. *(Adapted from T. Grotzinger, T. H. Jordan, F. Press, and R. Siever,* Understanding Earth, *5th ed., W. H. Freeman, 2007)*

FIGURE 6-5 R I V U X G **Craters and Shadows** The complexity of the Moon's surface near its south pole shows light-colored, rugged highlands with ancient craters and smooth younger surfaces covered with maria. The maria formed after the surrounding light-colored terrain, so they have not been exposed to meteoritic bombardment for as long and hence have fewer craters. Some of these shadowed crater walls rarely receive any sunlight and can serve as protected storage places for frozen water ice. *(NASA)*

FIGURE 6-6 R I V U X G **Mercury's Craters, Plains, and Water Ice** Craters dot the smoother intercrater plains in the north polar region of Mercury as seen by NASA's *MESSENGER* probe. The yellow regions indicate the positions of bright radar-reflecting water ice shielded from the Sun along shaded crater walls. *(NASA/Johns Hopkins University Applied Physics Laboratory/ Carnegie Institution of Washington)*

quite a bit like Earth's Moon. Unlike the Moon, Mercury's surface has extensive low-lying plains (examine the left half of Figure 6-6). These large, smooth areas probably have an origin similar to our Moon's smooth basins, where primordial lava flows smoothed the surface. As large impactors punctured the planet's thin, newly formed crust, lava welled up from the molten interior to flood low-lying areas and erase evidence or earlier craters.

At the same time, we find deep crater walls in shadow regions that rarely are bathed in sunlight. These shadow regions can serve as protective areas for frozen water ice to remain on Mercury indefinitely, even with its surface's tremendously high temperatures due to its nearness to the Sun. In recent years, radar observations have detected patches at Mercury's north and south poles that are unusually effective at reflecting radio waves. If these observations are correct, then Mercury is truly a world of extremes, with perpetual ice caps at one or both poles but then midday temperatures at the equator that are high enough to melt lead. Spacecraft have also revealed numerous long cliffs, called scarps or rupes, meandering across Mercury's surface and disrupting craters dotting the landscape (**Figure 6-7**). Some scarps rise as much as 2 mi (3 km) above the surrounding plains and can be hundreds of miles long. These cliffs probably formed as the planet cooled, cracked, and contracted.

FIGURE 6-7 R I V U X G **A Scarp Stretching Across Mercury** Giant cliffs, also known as scarps or rupes, run across this *MESSENGER* image. This towering cliff is nearly a mile (1.6 km) high and extends for several hundred miles across Mercury. Old crater walls have been rearranged due to the more recent scarp's rapid creation as the planet contracted while cooling. *(NASA/Johns Hopkins University Applied Physics Laboratory/Carnegie Institution of Washington)*

ANALOGY The shrinkage of Mercury's crust is probably a result of a general property of most materials: They shrink when cooled and expand when heated. The lid on a jar of jam makes a useful analogy. When the jar is kept cold in the refrigerator, the metal lid contracts more than the glass jar. This makes a tight seal but also makes the lid hard to unscrew. You can loosen the lid by running hot water over it; this makes the metal lid expand more than the glass jar, helping to break the seal.

If there had still been molten material beneath Mercury's surface when this contraction took place, lava would have leaked onto the surface around the scarps. This does not appear to have happened. Therefore, the scarps must have formed relatively late in Mercury's history, after the lava flows had ended and after the planet had solidified to a substantial depth beneath the surface. With the exception of the scarps, there are no features on Mercury that resemble the boundaries of tectonic plates. Thus, we can regard Mercury's crust, like that of the Moon, as a single plate.

ConceptCheck 6-4: What could you infer about the age of the giant and nearly smooth crater on Mercury if instead it was covered with impact craters?

Fewer Craters on Venus and Mars Indicates Geologic Activity

On Venus, by comparison, there are only about a thousand craters larger than a few miles in diameter. This is many more than have been found on Earth, but only a small fraction of the large number on the Moon or Mercury. Venus is only slightly smaller than Earth in diameter, and it should have enough internal heat to power the geologic activity required to erase most of its impact craters.

Mars is an unusual case, in that extensive cratering is found only in the higher terrain; the lowlands of Mars are remarkably smooth and free of craters. Thus, it follows that the Martian highlands are quite old, while the lowlands have a younger surface from which most craters have been erased. Considering the planet as a whole, the amount of cratering on Mars is intermediate between that on Mercury and Earth. This agrees with our general rule, because Mars is intermediate in size between Mercury and Earth. The interior of Mars was once hotter and more molten than it is now, so geologic processes were able to erase some of the impact craters. A key piece of evidence that supports this picture is that Mars has a number of immense volcanoes, like the one shown in **Figure 6-8**. These were active when Mars was young, but as this relatively small planet rapidly cooled down and its interior solidified, the supply of molten material to the volcanoes from the Martian interior was cut off. As a result, all of the volcanoes of Mars are now inactive.

To study Venus and Mars, we need to take a deeper look at volcanism and tectonics on each of these planets.

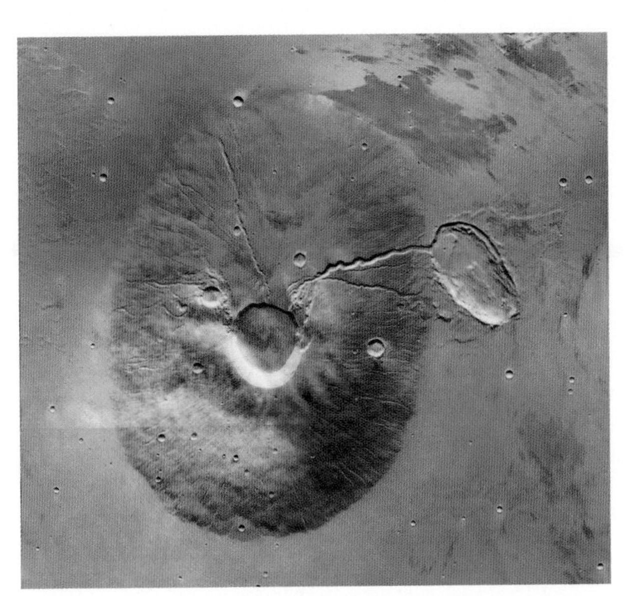

FIGURE 6-8 R I V U X G **Giant Volcano on Mars** Rising 18,000 ft above the smooth Tharsis region of Mars, Ceraunius Tholus is 80 mi (130 km) across and etched with many valleys. This image from *Mars Express* captures icy clouds drifting past the summit. *(© ESA Planetary Science Archive/ Michael Benson/Kinetikon Pictures/Corbis)*

ConceptCheck 6-5: **What will happen to the tire tracks left on the Moon in the 1960s and 1970s from the "Moon-buggy" used by U.S. astronauts to explore the Moon?**

6-3 Tectonics and volcanism influence surface features

Whereas the Moon and Mercury look similar from a distance, Venus and Mars appear radically different, as shown in **Figure 6-9**. Venus is nearly the size of Earth and has a thick atmosphere, while Mars is much smaller and has only a thin atmosphere. Yet spacecraft observations reveal that these two worlds have many surface features in common, including volcanoes and impact craters.

Scientists use orbiting spacecraft to map planets in great detail. To make a detailed study of a planet's surface, a spacecraft that simply flies by the planet and takes a few digital photographs does not provide enough information. Instead, it is necessary to place a spacecraft in long-term orbit around the planet. Since the 1970s, a number of spacecraft have been placed in orbit around Venus and Mars, yielding amazingly detailed information about their surfaces. Once a detailed surface map is made, inferences about the nature of deep planetary interiors can then be proposed.

In order to map the surface of Venus through the perpetual cloud layer, several of the Venus orbiters carried radar devices. A beam of microwave radiation from the orbiter easily penetrated Venus's clouds and reflected off the planet's surface; a receiver on the orbiter then detected the reflected beam. Different types of terrain reflect microwaves more or less efficiently, so astronomers have been able to construct a map of the Venusian surface by analyzing the reflected radiation.

Magellan, the most recent spacecraft to orbit Venus, also carried a radar altimeter that bounced microwaves directly off the ground beneath the spacecraft. By measuring the time delay of the reflected waves, astronomers could determine the height and depth of Venus's terrain.

By contrast, a Mars orbiter can use a more conventional telescope to view the Martian surface through that planet's thin atmosphere. Instead of using a radar altimeter to map surface elevations, the *Mars Global Surveyor* spacecraft (which entered Mars orbit in 1997) used a laser beam for the same purpose.

FIGURE 6-9 R I V U X G **Similarity Among Terrestrial Surfaces** The surfaces of planets orbiting close to the Sun, as well as moons orbiting the outer planets, all experience similar geological processes shaping their landscapes. Earth is unique not in that it has water, but in that liquid water exists widely across its surface, covering 70% of its surface. *(Composition by Mike Malaska; images from NASA/JPL, ESA, ISAS, IKI)*

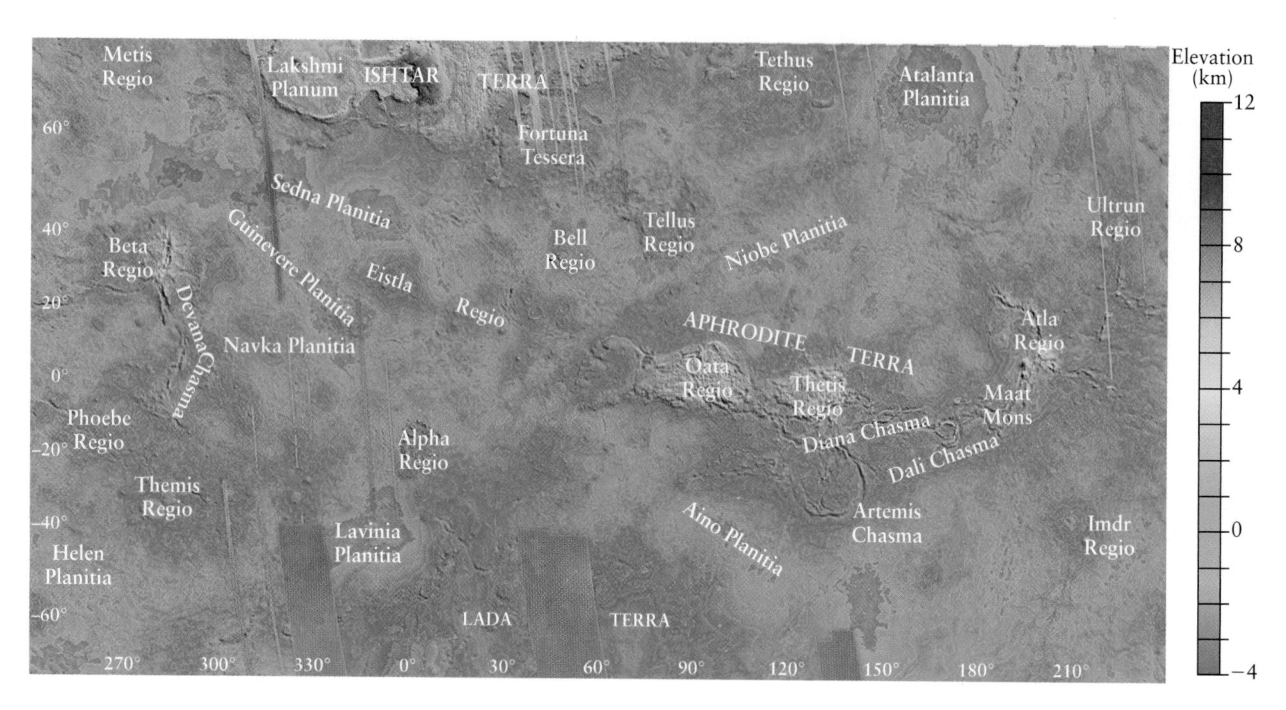

FIGURE 6-10 RIVUXG A **Topographic Map of Venus** Radar altimeter measurements by *Magellan* were used to produce this topographic map of Venus. Color indicates elevations above (positive numbers) or below (negative numbers) the planet's average radius. (The blue areas are *not* oceans!) Gray areas were not mapped by *Magellan.* Flat plains of volcanic origin cover most of the planet's surface, with only a few continentlike highlands. *(Peter Ford, MIT; NASA/JPL)*

Figure 6-10 and **Figure 6-11** are topographic maps of Venus and Mars derived using radar and laser altimeters, respectively. (Note that the scale of these two maps is different: The diameter of Mars is only 56% as large as the diameter of Venus.) The topographies of both worlds differ in important ways from that of Earth. Our planet has two broad classes of terrain: About 71% of the surface is oceanic crust, and about 27% is continental crust that rises above the ocean floors by about 2.5 mi to 4 mi on average. On Venus, by contrast, about 60% of the terrain lies within 600 ft of the average elevation, with only a few localized highlands (shown in yellow and red in Figure 6-10). Mars is different from both Earth and Venus: Rather than having continents scattered among low-lying ocean floors, all of the high terrain on Mars (shown in red and orange in Figure 6-11) is in the southern hemisphere. Hence, planetary scientists refer to Mars as having **northern lowlands** and **southern highlands.**

Tectonics on Venus: A Thin Crust

Whereas Mars seems to have a relatively thick crust, Venus has a relatively thin crust. Before radar maps like Figure 6-10 were available, scientists wondered whether Venus had plate

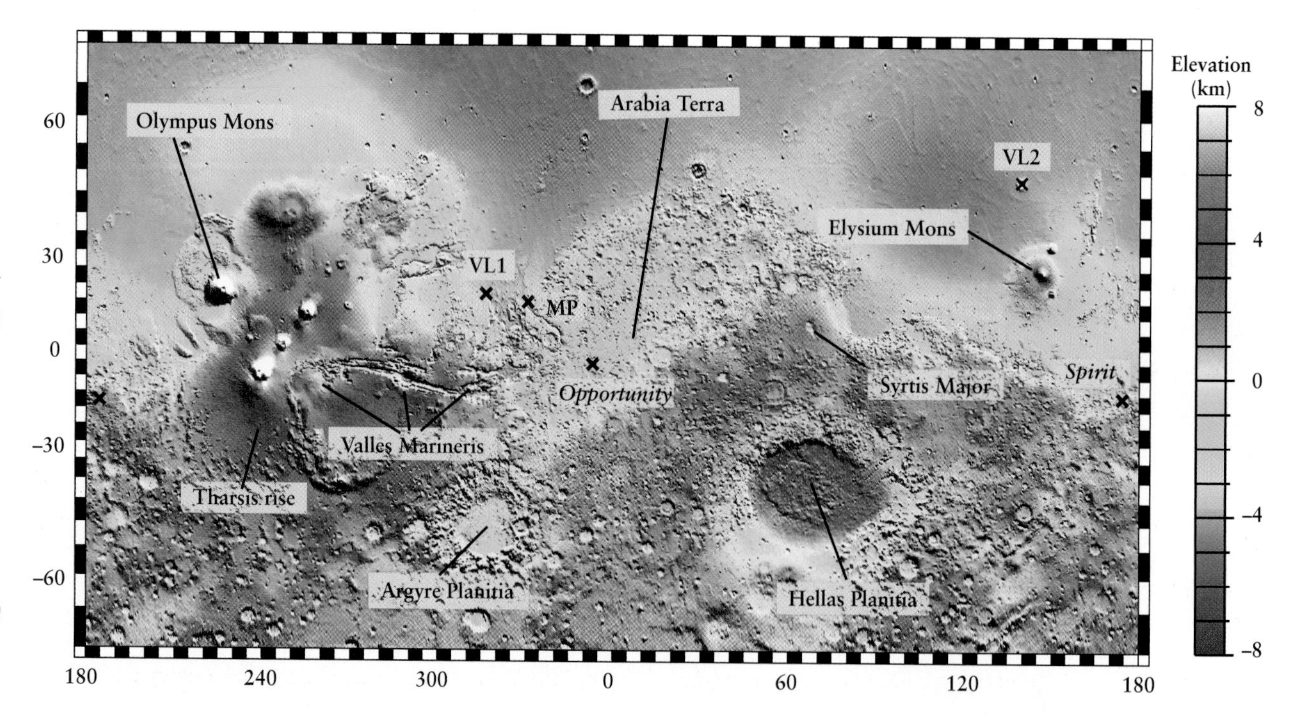

FIGURE 6-11 RIVUXG **A Topographic Map of Mars** This map was generated from measurements made by the laser altimeter on board the *Mars Global Surveyor.* As in Figure 6-10, color indicates elevations above or below the planet's average radius. Most of the southern hemisphere lies several miles above the northern hemisphere, with the exception of the immense impact feature called Hellas Planitia. The landing sites for *Viking Landers 1* and *2* (VL1 and VL2), *Mars Pathfinder* (MP), and the *Mars Exploration Rovers* (*Spirit* and *Opportunity*) are each marked with an X. *(MOLA Science Team, NASA, GSFC)*

tectonics like those that have remolded the face of Earth. Venus is only slightly smaller than Earth and should have retained enough heat to sustain a molten interior and the convection currents that drive tectonic activity on Earth. If this were the case, then the same tectonic effects might also have shaped the surface of Venus. Earth's hard outermost layer is broken into about a dozen large plates that slowly shuffle across the globe (see Figure 5-18).

Radar images from *Magellan* show *no* evidence of Earthlike tectonics on Venus. On Earth, long chains of volcanic mountains (like the Cascades in North America or the Andes in South America) form along plate boundaries where subduction is taking place. Mountainous features on Venus, by contrast, do not appear in chains. There are also no structures like Earth's Mid-Atlantic Ridge (see Figure 5-16), which suggests that there are no places where Venus's surface is tectonically spreading apart, such as occurs on Earth. With no converging or spreading tectonic plates, there can have been only limited horizontal changes of Venus's lithosphere. Thus, like the crusts of the Moon and Mercury, Venus's crust is composed of only one plate.

Unlike the Moon, however, Venus has had some dramatic, but localized, small-scale deformations and reshaping of the surface. One piece of evidence for this is that roughly one-fifth of Venus's surface is covered by folded and faulted ridges. Further evidence comes from close-up *Magellan* images, which show that Venus has about a thousand craters larger than a few kilometers in diameter, many more than have been found on Earth, but as noted, only a small fraction of the number on the Moon or Mercury. The number of craters on Venus indicates that the Venusian surface is roughly 500 million years old. This is about twice the age of Earth's surface but much younger than the surfaces of the Moon or Mercury, each of which is billions of years old. No doubt Venus was more heavily cratered in its youth, but localized activity in its crust has erased the older craters (**Figure 6-12**).

Surprisingly, Venus's craters are uniformly scattered across the planet's surface. We would expect that older regions on the surface—which have been exposed to bombardment for a longer time—would be more heavily cratered, while younger regions would be relatively free of craters. For example, the ancient highlands on the Moon are much more heavily cratered than the younger and smoother dark surfaces. Because such variations are not found on Venus, scientists conclude that the entire surface of the planet has essentially the *same* age. This is very different from Earth, where geological formations of widely different ages are found.

One model that can explain these features suggests that the convection currents in Venus's interior are actually more vigorous than those inside Earth, but that the Venusian crust is much thinner than the continental crust on Earth. Rather than sliding around like the plates of Earth's crust, the thin Venusian crust stays in roughly the same place but undergoes

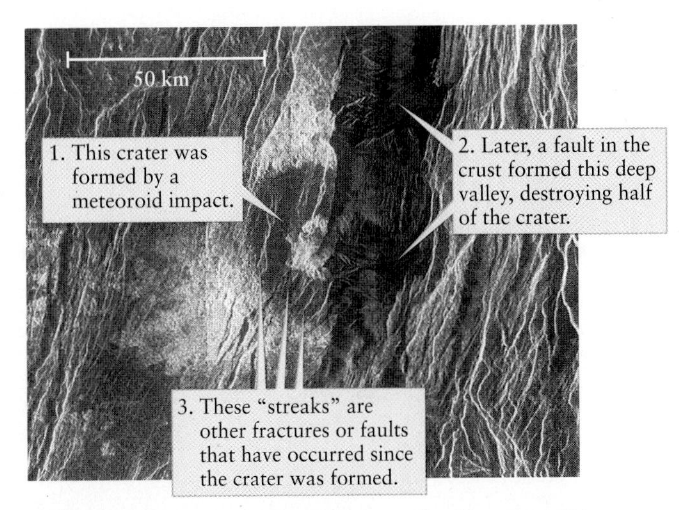

FIGURE 6-12 R I V U X G **A Partially Obliterated Crater on Venus** This *Magellan* image shows how the right half of an old impact crater 20 mi (37 km) in diameter was erased by a fault in the crust. This crater lies in Beta Regio (see the left side of Figure 6-10). Most features on Venus are named for women in history and legend; this crater commemorates Emily Greene Balch, an American economist and sociologist who won the 1946 Nobel Peace Prize. *(NASA/JPL)*

wrinkling and flaking (**Figure 6-13**). Thus, this model is informally known as "flake tectonics." As it turns out, Earth also might have had a short period of flake tectonics billions of years ago when its interior was hotter.

> ConceptCheck 6-6: **What does the absence of mountain chains on Venus's surface suggest about the number of plates that cover its surface?**

Tectonics on Mars: A Thick, Rigid Crust

Like Venus, Mars lacks the global network of ridges and subduction zones that are evidence of active plate tectonics and a molten interior on Earth. Hence, the entire crust of Mars makes up a single tectonic plate, as is the case on the Moon, Mercury, and Venus. The specific explanation, however, is quite different from that for Venus. Because Mars is a much smaller world than Earth or Venus, the outer layers of the red planet have cooled more extensively. Thus, Mars lacks plate tectonics because its crust is too thick for one part of the crust to be subducted beneath another. We see that for a terrestrial planet to have plate tectonics, the crust must not be too thin (like Venus) or too thick (like Mars), but just right (like Earth).

Active tectonic processes continue to resurface the landscape only if the planet or moon still has a hot, molten interior.

Although there is no plate motion on Mars, there are features on the planet's surface that indicate there was once substantial motion of material in the Martian mantle. The region scattered with giant volcanoes, known as the Tharsis rise, visible in Figure 6-11, is a dome-shaped bulge that has been lifted 16,000 ft to 20,000 ft above Mars's average elevation. Apparently, a massive plume of magma once welled upward from a very deep chamber of molten material underlying this region. East of the Tharsis rise, a vast chasm runs roughly parallel to

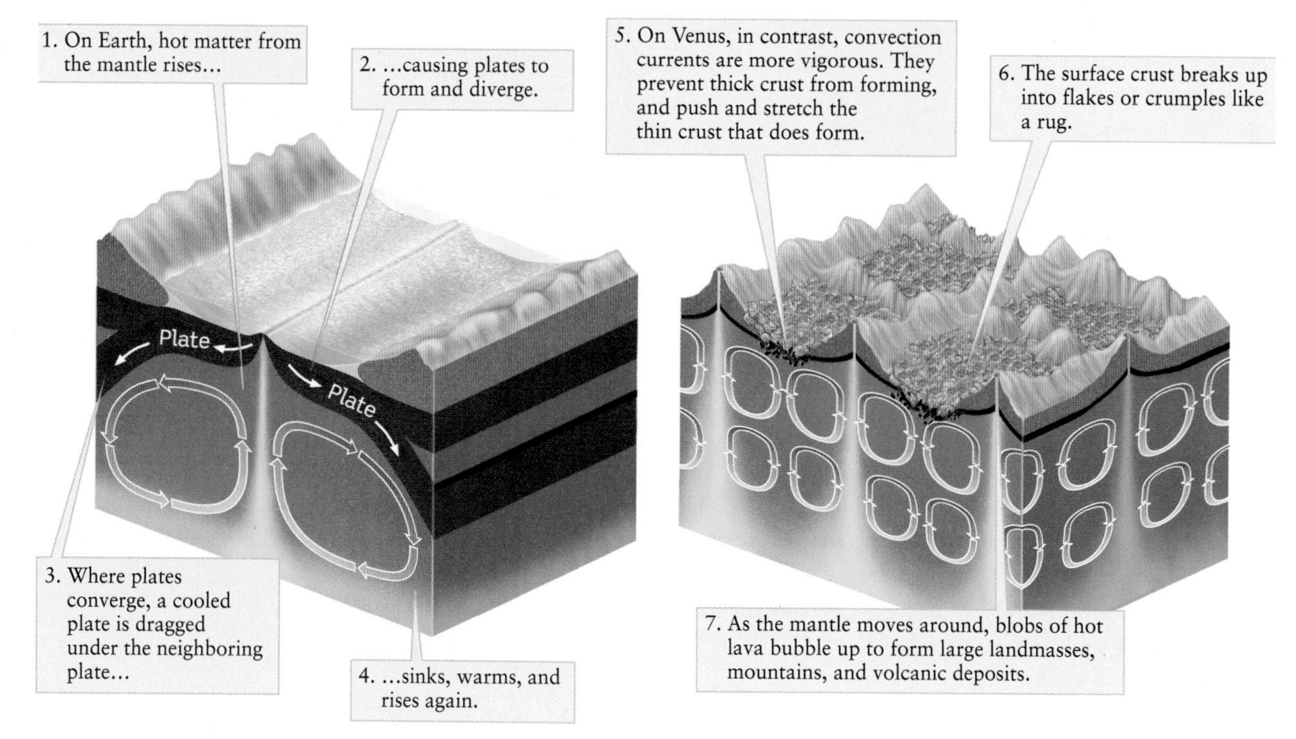

FIGURE 6-13
Plate Tectonics Versus Flake Tectonics This illustration shows the difference between plate tectonics on Earth and the model of flake tectonics on Venus. *(Adapted from T. Grotzinger, T. H. Jordan, F. Press, and R. Siever,* Understanding Earth, *5th ed., W. H. Freeman, 2007)*

the Martian equator (**Figure 6-14**). If this canyon were located on Earth, it would stretch from Los Angeles to New York. In honor of the *Mariner 9* spacecraft that first revealed its presence in 1971, this chasm has been named Valles Marineris.

Valles Marineris has heavily fractured terrain at its western end near the Tharsis rise. At its eastern end, by contrast, it is dominated by ancient cratered terrain. Many geologists suspect that Valles Marineris was caused by the same upwelling of material that formed the Tharsis rise. As the Martian surface bulged upward at Tharsis, there would have been tremendous stresses on the crust, which would have caused extensive fracturing. Thus, Valles Marineris may be a **rift valley,** a feature created when a planet's crust breaks apart along a fault line. Rift valleys are found on Earth; two examples are the Red Sea (see Figure 5-18) and the Rhine River valley in Europe. Other, smaller rifts in the Martian crust are found all around the Tharsis rise. All of these features are very old, however, and it is thought that there has been little geologic activity on Mars for billions of years.

Go to Video 6-3

Additional evidence that may point to ancient geologic activity on Mars is the crustal dichotomy between the northern lowlands and the southern highlands. The

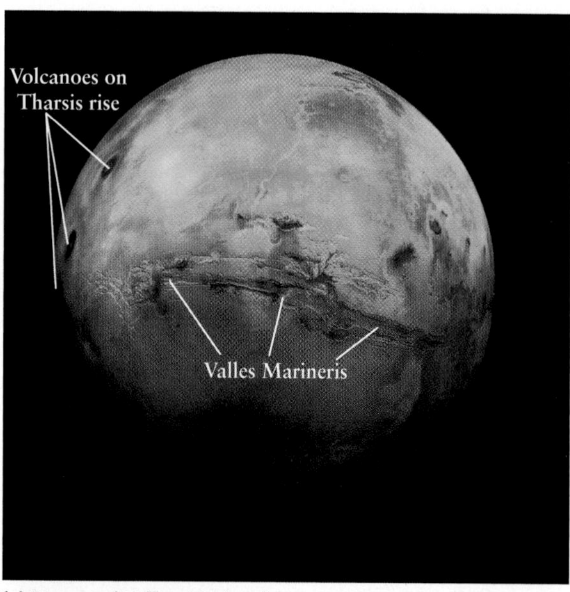

(a) Mars and Valles Marineris

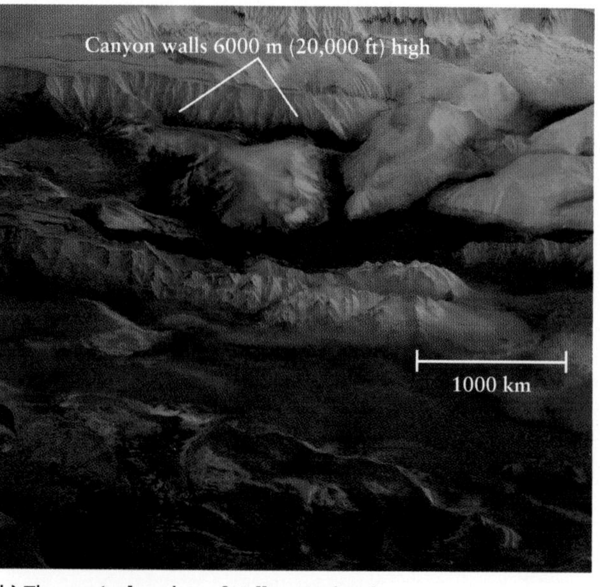

(b) The central region of Valles Marineris

FIGURE 6-14
R I V U X G
Valles Marineris
(a) This mosaic of *Viking Orbiter* images shows the huge rift valley of Valles Marineris, which extends from west to east for more than 2500 mi (4000 km) and is 400 mi (600 km) wide at its center. Its deepest part is 5 mi (8 km) beneath the surrounding plateau. At its western end is the Tharsis rise. Compare this image with Figure 6-11. (b) This perspective image from the *Mars Express* shows what you would see from a point high above the central part of Valles Marineris. *(a: USGS/NASA; b: © European Space Agency/DLR/FU Berlin/G. Neukum/Science Source)*

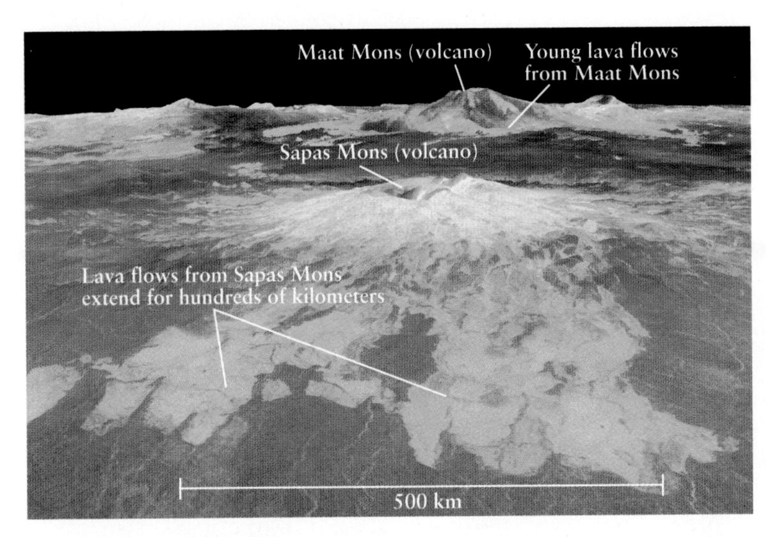

(a) Volcanoes and lava flows on Venus R I V U X G

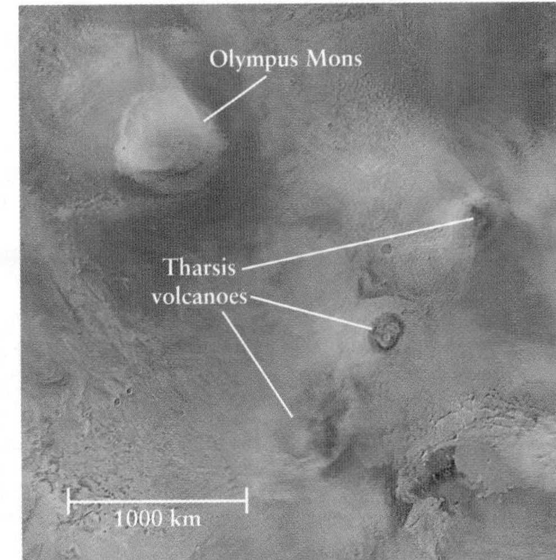

(b) Cloud-topped volcanoes on Mars R I V U X G

FIGURE 6-15 Volcanoes on Venus and Mars (a) The false color in this computer-generated perspective view suggests the actual color of sunlight that penetrates Venus's thick clouds. The brighter color of the extensive lava flows indicates that they reflect radio waves more strongly. To emphasize the gently sloping volcanoes, the vertical scale has been exaggerated 10 times. (b) These volcanoes of Mars, more than 9 mi (15 km) high, also have gently sloping sides and bluish clouds often appear in the afternoons, topping the summits of the volcanoes. These clouds, made of water ice crystals, form on most Martian afternoons as warm air containing water vapor rises up the slopes and freezes. *(a: NASA, JPL Multimission Image Processing Laboratory; b: NASA/JPL/Malin Space Science Systems)*

southern highlands are much more heavily cratered than either Earth or Venus, though less so than the Moon, while the northern lowlands are remarkably smooth and free of craters. (You can see this difference in Figure 6-11.) Since most cratering occurred early in the history of the solar system, this implies that the northern lowlands are relatively young, while the more heavily cratered southern highlands are relatively old. One proposed explanation is that older craters in the northern lowlands were erased by tectonic activity that took place long ago when the Martian crust had not yet cooled to its present thickness.

ConceptCheck 6-7: What is the primary implication of Mars having plates that are significantly thicker than those of Earth or Venus?

Volcanoes on Venus and Mars

Radar images of Venus and visible-light images of Mars show that both planets have a number of large volcanoes (**Figure 6-15**). The Venus-orbiting *Magellan* spacecraft observed more than 1600 major volcanoes and volcanic features on Venus, two of which are shown in Figure 6-15*a*. Both of these volcanoes have gently sloping sides. A volcano with this characteristic is called a **shield volcano,** because in profile it resembles an ancient Greek warrior's shield lying on the ground. Martian volcanoes are less numerous than those on Venus, but they are also shield volcanoes; the largest of these, Olympus Mons, is the largest volcano in the solar system (Figure 6-15*b*). Olympus Mons rises nearly 79,000 ft (15 mi or 24 km) above the surrounding plains. By comparison, the highest volcano on Earth, Mauna Loa in the Hawaiian Islands, has a summit only 33,000 ft (9.750 km) above the surrounding Pacific Ocean floor.

Most volcanoes on Earth are found near the boundaries of tectonic plates, where subducted material becomes molten magma and rises upward to erupt from the surface. This cannot explain the volcanoes of Venus and Mars, since there are no subduction zones on those planets. Instead, Venusian and Martian volcanoes probably formed by magma welling upward from a hot spot in a planet's mantle, elevating the overlying surface and producing a shield volcano.

On Earth, hot-spot volcanism is the origin of the Hawaiian Islands. These islands are part of a long chain of shield volcanoes that formed in the middle of the Pacific tectonic plate as that plate moved over a long-lived hot spot (**Figure 6-16**). On Venus and Mars, by contrast, the absence of plate tectonics means that the crust remains stationary over a hot spot. On Mars, a single hot spot under Olympus Mons probably pumped magma upward through the same vent for millions

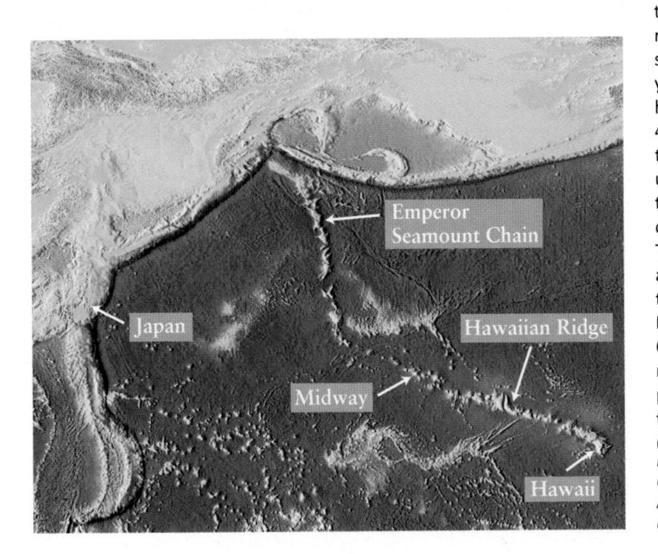

FIGURE 6-16 Hot-Spot Volcanoes on Earth A hot spot under the Pacific plate has remained essentially stationary for 70 million years while the plate has moved almost 4000 mi (6000 km) to the northwest. The upwelling magma has thus produced a long chain of volcanoes. The Hawaiian Islands are the newest of these; the oldest, the Emperor Seamount Chain, has eroded so much that it no longer protrudes above the ocean surface. *(© National Geophysical Data Center/USGS/National Oceanic and Atmospheric Administration/Department of Commerce)*

BOX 6-1 THE HEAVENS ON EARTH

Jupiter's Moon Io Is Covered with Active Volcanoes

If active volcanoes are seemingly absent from the inner planets and their moons, other than Earth, one wonders if there are volcanoes anywhere else in the solar system. As it turns out, there are volcanoes active elsewhere in the solar system at this very moment! One of the most fascinating is not on a planet at all, but on the strangely colored moon closet to planet Jupiter.

Before spacecraft visited the Jovian moons, scientists believed that these natural satellites would be worlds much like our own Moon—geologically dead, with little internal heat available to power tectonic or volcanic activity. They thought that all of Jupiter's moons would be extensively cratered, because there would have been little geologic activity to erase those craters over the satellite's history.

One of Jupiter's moons, Io, proved to be utterly different from these naive predictions. On March 5, 1979, *Voyager 1* came within 21,000 km (13,000 mi) of Io and began sending back a series of bizarre and unexpected pictures of the satellite (**Figure B6-1.1**). These images showed that Io has *no* impact craters at all! Instead, the surface is pockmarked by irregularly shaped pits and is blotched with color.

More recently, the *Galileo* spacecraft, which came as close as 200 km to Io's surface during its eight years in orbit around Jupiter, returned detailed images of several volcanic plumes (**Figure B6-1.2*b***). These plumes rise to astonishing heights of 70 km to 280 km above Io's surface. To reach such altitudes, the material must emerge from volcanic vents with speeds between 300 m/s and 1000 m/s (700 mi/h to 2200 mi/h). Even the most violent terrestrial volcanoes have erup-

FIGURE B6-1.1 R I V U X G **Io** A mosaic of Io's two hemispheres, built up from *Voyager* images. *(NASA/JPL/USGS)*

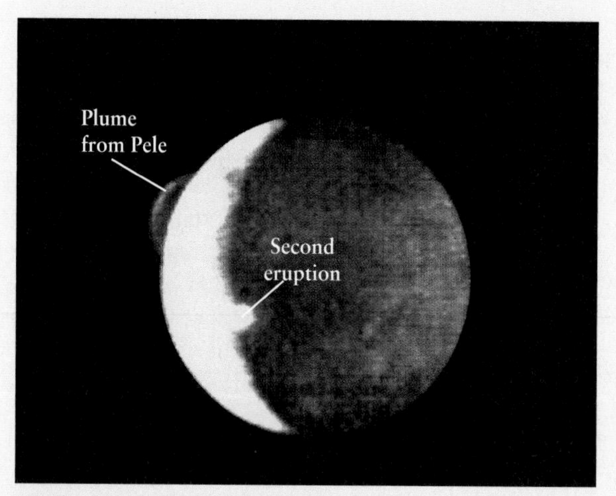

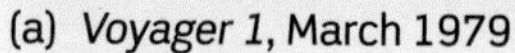

(a) *Voyager 1*, March 1979

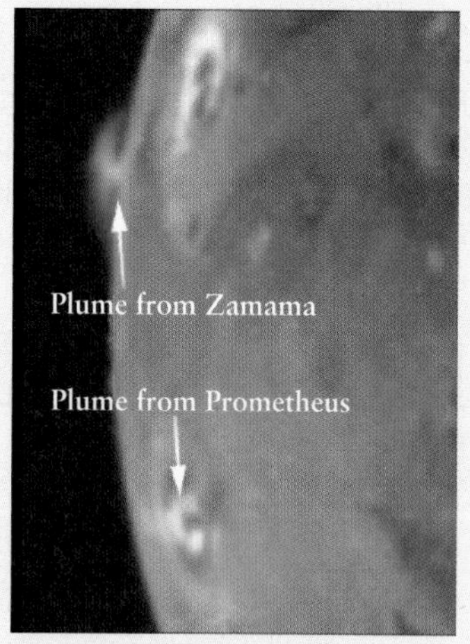

(b) *Galileo*, November 1997

FIGURE B6-1.2 R I V U X G **Volcanic Plumes on Io** (a) Io's volcanic eruptions were first discovered on this *Voyager* image. (b) This *Galileo* image shows plumes from the volcanoes Prometheus and Zamama. *(a: NASA; b: Planetary Image Research Laboratory/University of Arizona/JPL/NASA)*

of years, producing one giant volcano rather than a long chain of smaller ones. The Tharsis rise and its volcanoes (see Figure 6-11, Figure 6-14*a*, and Figure 6-15*b*) may have formed from the same hot spot that gave rise to Olympus Mons. The same process of hot-spot volcanism presumably gave rise to large shield volcanoes on Venus like those shown in Figure 6-15*a*.

About 80% of the surface of Venus is composed of flat plains of volcanic origin. In other words, essentially the entire planet is covered with lava! This observation shows the tremendous importance of volcanic activity in Venusian geology. Most of the volcanoes on Venus are probably inactive at present, just as is the case with most volcanoes on Earth. But *Magellan* found evidence of recent volcanic activity, some of which may be continuing today. The key to estimating the amount of recent volcanic activity is that the radar reflectivity of volcanic materials depends on whether the material is

tion speeds of only around 100 m/s (220 mi/h). Scientists began to suspect that Io's volcanic activity must be fundamentally different from volcanism on Earth.

As it turns out, as Io moves around its orbit, Jupiter and its other moons exert gravitational tugs on it in a regular, rhythmic fashion, alternately squeezing and flexing the innermost moons. Just as a ball of clay or bread dough gets warm as you knead it between your fingers, this squeezing and flexing causes heating of Io's interior. This results in unexpected, yet widespread and recurrent, surface volcanism on Io.

An important clue about how volcanoes on Io are different than volcanoes on Earth came from the infrared spectrometers aboard *Voyager 1,* which detected abundant sulfur and sulfur dioxide in Io's volcanic plumes. This led researchers to suggest that the plumes are actually more like geysers than volcanic eruptions. In geysers on Earth, water seeps down to volcanically heated rocks, where it changes to steam and erupts explosively through a vent. Planetary geologists Susan Kieffer, Eugene Shoemaker, and Bradford Smith suggested that sulfur dioxide, rather than water, could be the principal propulsive agent driving volcanic plumes on Io. Sulfur dioxide is a solid at the frigid temperatures found on most of Io's surface, but it should be molten at depths of a few kilometers. Just as the explosive conversion of water into steam produces a geyser on Earth, the conversion of liquid sulfur dioxide into a high-pressure gas could result in eruption velocities of up to 1000 m/s on Io.

Io's dramatic coloration (see Figure B6-1.1) is probably due to sulfur and sulfur dioxide, which are ejected in volcanic plumes and later fall back to the surface. Sulfur is normally bright yellow, which explains the dominant color of Io's surface. But if sulfur is heated and suddenly cooled, as would happen if it were ejected from a volcanic vent and allowed to fall to the surface, it can change to orange, red, or black. Indeed, these colors are commonly found around active volcanic vents (**Figure B6-1.3**). Whitish surface deposits (examine Figure B6-1.1 and Figure B6-1.3*a*), by contrast, are probably due to sulfur dioxide (SO_2). Volcanic vents on Earth commonly discharge SO_2 in the form of an acrid gas. But on Io, when hot SO_2 gas is released by an eruption into the cold vacuum of space, it crystallizes into white snowflakes. This sulfur dioxide "snow" then falls back onto Io's surface.

What is beneath Io's constantly changing surface? We have never landed a spacecraft on the surface of Io, but precise measurements of Io's gravity taken by the *Galileo* spacecraft demonstrate that Io's interior has layers. Io has a relatively large and dense core surrounded by a partially molten layer enclosed by a very thin crust.

Io may have as many as 300 active volcanoes. Altogether, volcanism on Io could eject enough material to cover the entire surface of Io to a depth of 1 m in a century, or to cover an area of 1000 km^2 in a few weeks (see Figure B6-1.3). Thanks to this continual "repaving" of the surface, there are probably no long-lived features on Io, and any impact craters are quickly obliterated.

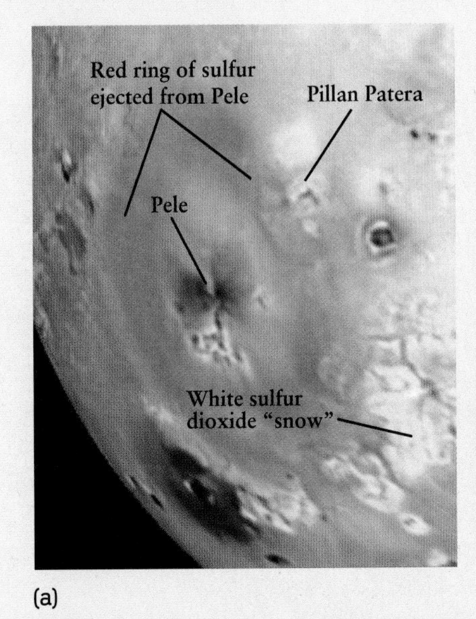

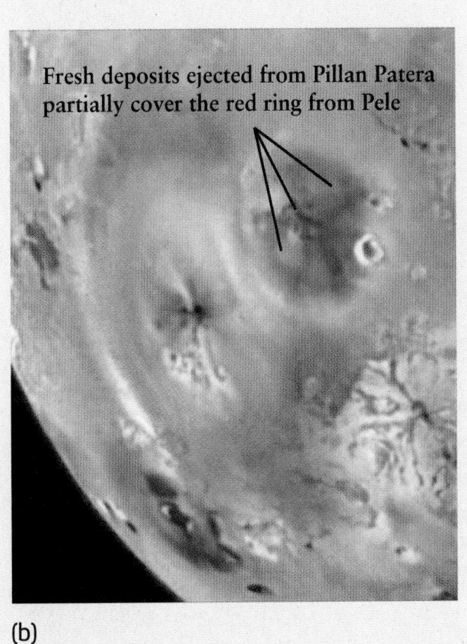

FIGURE B6-1.3 R I V U X G **Colors and Rapid Changes on Io** (a) This *Galileo* image is of Io's southern hemisphere, showing the red ring produced by sulfur ejected from the active volcanic vent Pele. (b) A few months later, a volcanic plume erupted from the Pillan Patera vent. *(NASA/JPL)*

relatively fresh or relatively old. By mapping these reflectivity variations on Venus, *Magellan* found many areas with young lava flows. Some of the youngest material found by *Magellan* caps the volcano Maat Mons (see Figure 6-15*a*). Geologists estimate that the topmost material is no more than 10 million years old and could be much younger. The presence of such young lava flows suggests that Venus, like Earth, has some present-day volcanic activity. (Are you wondering if there are other volcanoes in the universe? See Box 6-1: Jupiter's Moon Io Is Covered with Active Volcanoes)

Another piece of evidence for ongoing volcanic activity on Venus comes from the planet's atmosphere (Section 6-4). An erupting volcano on Earth ejects substantial amounts of sulfur dioxide, sulfuric acid, and other sulfur compounds into the air. Many of these substances are highly reactive and short-lived, forming chemical compounds that become

COSMIC CONNECTIONS Putting It All Together

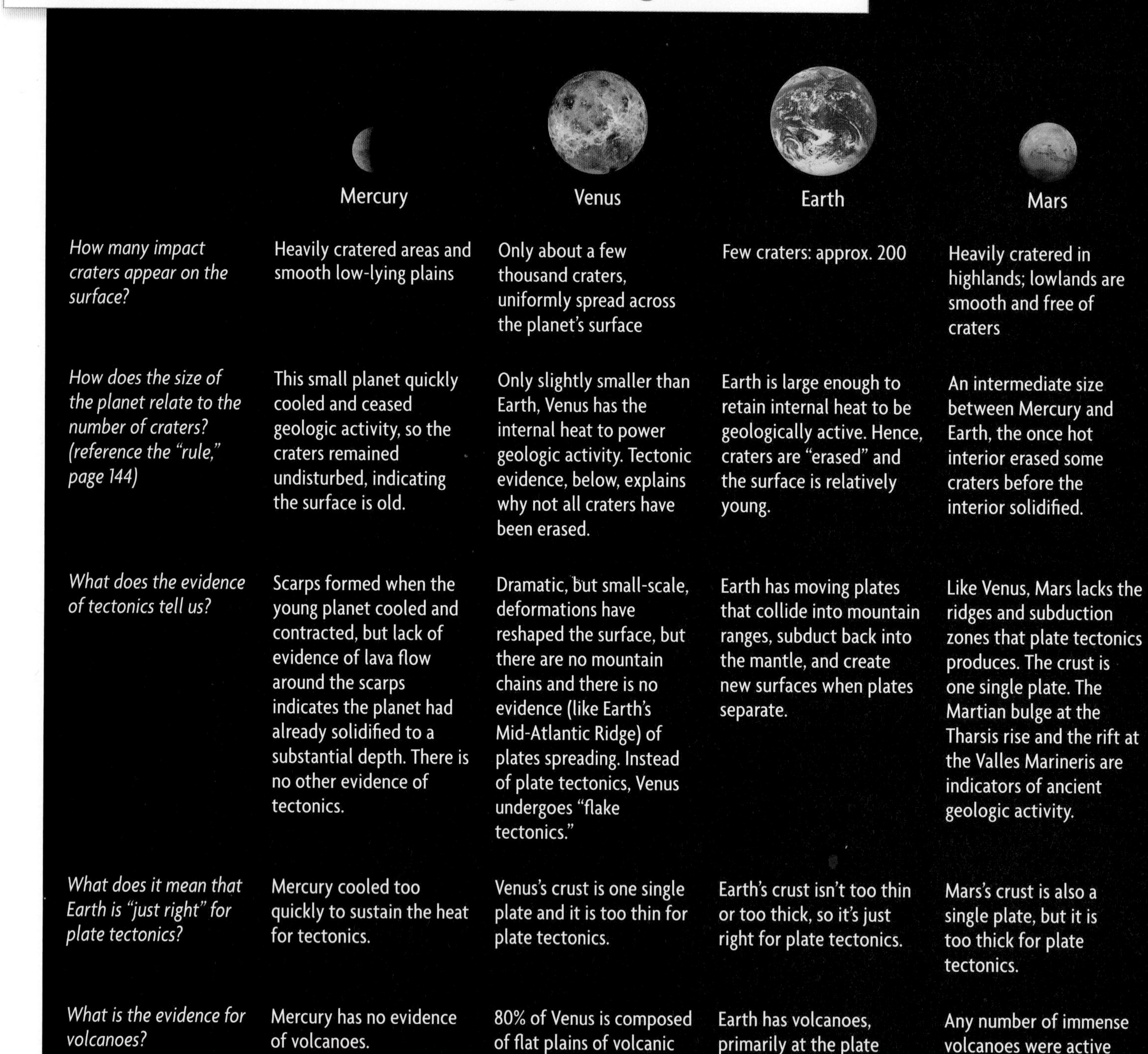

	Mercury	Venus	Earth	Mars
How many impact craters appear on the surface?	Heavily cratered areas and smooth low-lying plains	Only about a few thousand craters, uniformly spread across the planet's surface	Few craters: approx. 200	Heavily cratered in highlands; lowlands are smooth and free of craters
How does the size of the planet relate to the number of craters? (reference the "rule," page 144)	This small planet quickly cooled and ceased geologic activity, so the craters remained undisturbed, indicating the surface is old.	Only slightly smaller than Earth, Venus has the internal heat to power geologic activity. Tectonic evidence, below, explains why not all craters have been erased.	Earth is large enough to retain internal heat to be geologically active. Hence, craters are "erased" and the surface is relatively young.	An intermediate size between Mercury and Earth, the once hot interior erased some craters before the interior solidified.
What does the evidence of tectonics tell us?	Scarps formed when the young planet cooled and contracted, but lack of evidence of lava flow around the scarps indicates the planet had already solidified to a substantial depth. There is no other evidence of tectonics.	Dramatic, but small-scale, deformations have reshaped the surface, but there are no mountain chains and there is no evidence (like Earth's Mid-Atlantic Ridge) of plates spreading. Instead of plate tectonics, Venus undergoes "flake tectonics."	Earth has moving plates that collide into mountain ranges, subduct back into the mantle, and create new surfaces when plates separate.	Like Venus, Mars lacks the ridges and subduction zones that plate tectonics produces. The crust is one single plate. The Martian bulge at the Tharsis rise and the rift at the Valles Marineris are indicators of ancient geologic activity.
What does it mean that Earth is "just right" for plate tectonics?	Mercury cooled too quickly to sustain the heat for tectonics.	Venus's crust is one single plate and it is too thin for plate tectonics.	Earth's crust isn't too thin or too thick, so it's just right for plate tectonics.	Mars's crust is also a single plate, but it is too thick for plate tectonics.
What is the evidence for volcanoes?	Mercury has no evidence of volcanoes.	80% of Venus is composed of flat plains of volcanic origin. Volcanoes would develop from a hot spot, a single place that would produce one giant volcano, rather than a chain. There may be some present-day volcanic activity.	Earth has volcanoes, primarily at the plate boundaries, where subducted material becomes molten magma and rises upward to erupt. Volcanoes may appear in chains as a plate moves over a hot spot.	Any number of immense volcanoes were active when Mars was young. Like on Venus, they formed over a single hot spot. Now all are inactive.

part of the planet's surface rocks. For these substances to be relatively abundant in a planet's atmosphere, they must be constantly replenished by new eruptions. Sulfur compounds make up about 0.015% of the Venusian atmosphere, compared to less than 0.0001% of Earth's atmosphere. This suggests that ongoing volcanic eruptions on Venus are ejecting sulfur compounds into the atmosphere to sustain the high sulfur content.

Unlike lava flows on Venus, however, most of the lava flows on Mars have impact craters on them. This cratering indicates that most Martian lava flows are very old, and that most of the volcanoes on the red planet are no longer active. This is what we would expect from a small planet whose crust has cooled and solidified to a greater depth than that on Earth, making it difficult for magma to travel from the Martian mantle to the surface.

Cosmic Connections: Putting It All Together summarizes the surface features of the terrestrial planets.

ConceptCheck 6-8: Why are there almost no craters on Io when it has no plate tectonic collisions or separations actively resurfacing the landscape?

ConceptCheck 6-9: How can astronomers estimate the age of a lava flow from a volcano?

6-4 Atmospheres surrounding terrestrial planets vary considerably

We have talked about Earth having a special place in the solar system—not too close to the Sun, not too far, but just right. So, what is "just right" about Earth's position as a planet where life can flourish? In short, Earth's position and protective atmosphere of gas surrounding the planet stabilizes the liquid water that life requires to exist. If Earth were any closer to the Sun, then any water would likely boil away; and if Earth were farther away, then any water would likely freeze. If planets nearer and farther away from the Sun have atmospheres so different from Earth, what are they like? Mercury has a daytime temperature too high and gravity too weak to retain any substantial atmosphere, so it will not get our attention here. We have seen that the three large terrestrial planets—Earth, Venus, and Mars—display very different types of geologic activity; as we will discover, these differences help to explain why each of these three worlds has a distinctively different atmosphere. However, there are also similarities.

The original atmospheres of Earth, Venus, and Mars all derived from gases that were emitted, or *outgassed,* from volcanoes (**Figure 6-17**). The gases released by present-day Earth volcanoes are predominantly water vapor (H_2O), carbon dioxide (CO_2), sulfur dioxide (SO_2), and nitrogen (N_2). These gases should therefore have been important parts of the original atmospheres of all three planets. Much of the water on all three planets is thought to have come from impacts from comets, icy bodies from the outer solar system.

In order to better understand the various atmospheres found around the solar system, let's first consider some of the most important aspects of Earth's atmosphere.

- Earth's atmosphere is predominately composed of nitrogen. There are certainly some water molecules in Earth's atmosphere—most easily seen as clouds—but most of Earth's water is found in the oceans.
- Similarly, some carbon dioxide is found in Earth's atmosphere, but much of Earth's carbon dioxide has been captured in carbonate rocks, such as limestone. This occurs because carbon dioxide readily dissolves in water, so rain can remove CO_2 from the atmosphere, which is ultimately deposited on ocean floors where it cements into rocks.
- The oxygen (O_2) in Earth's atmosphere is the result of photosynthesis by plant life, which absorbs CO_2 and releases O_2. The net result is that our planet's atmosphere is predominantly N_2 and O_2, with enough pressure and a high enough temperature (thanks to the greenhouse effect) for water to remain a liquid—and hence for life, which requires liquid water, to exist.

In order to compare the atmospheres of Venus and Mars to Earth's, it is necessary to get information from spacecraft that are able to descend through an atmosphere to take measurements. Both Soviet and U.S. spacecraft have done this for Venus, and five U.S. spacecraft have successfully landed on Mars. The results of these missions are

FIGURE 6-17 R I V U X G **A Volcanic Eruption on Earth** The eruption of Mount St. Helens in Washington State on May 18, 1980, released a plume of ash and gas 25 mi (40 km) high. Eruptions of this kind on Earth, Venus, and Mars probably gave rise to those planets' original atmospheres. *(U.S. Geological Survey, Dept. of the Interior)*

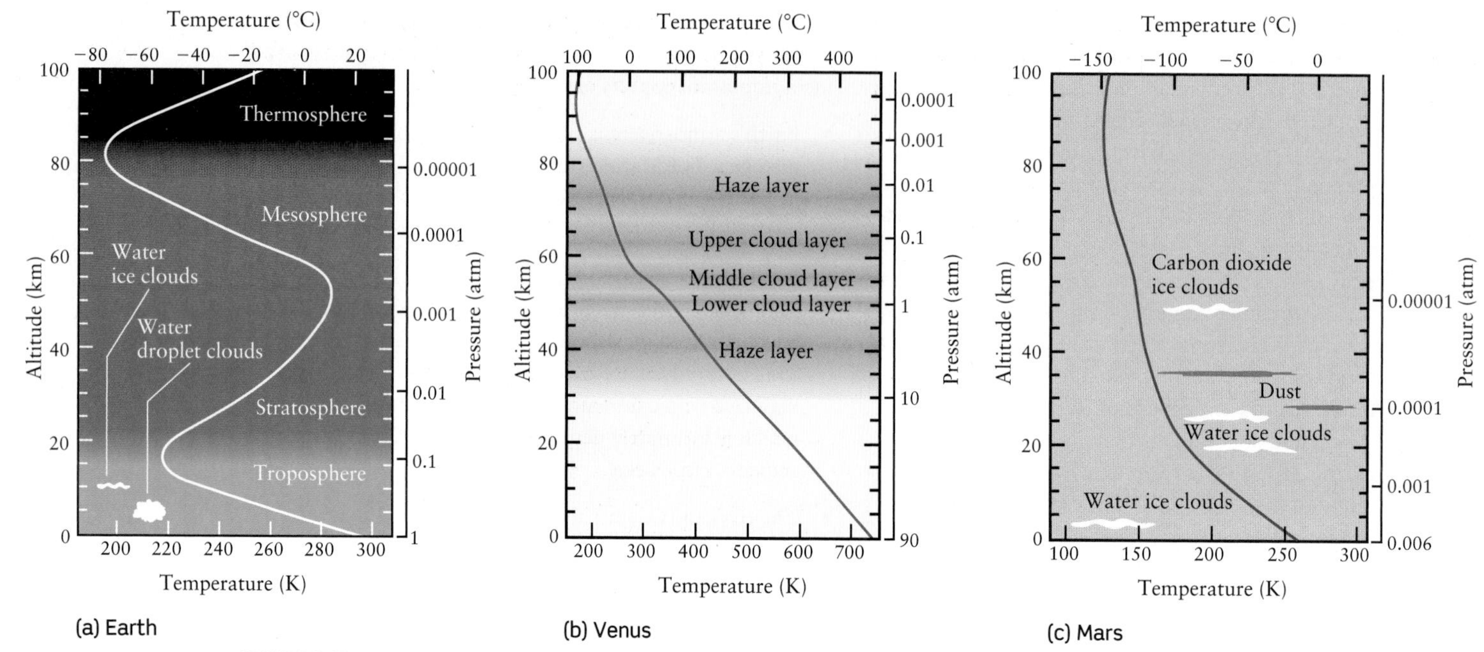

FIGURE 6-18 Atmospheres of the Terrestrial Planets In each graph the curve shows how temperature varies with altitude from 0 km to 100 km above the planet's surface. The scale on the right-hand side of each graph shows how pressure varies with altitude. (a) Clouds in Earth's atmosphere are seldom found above 40,000 ft (12 km). (b) Venus's perpetual cloud layers lie at much higher altitudes. The atmosphere is so dense that the pressure 50 km above the Venusian surface is 1 atm, the same as at sea level on Earth. (c) By contrast, the Martian atmosphere is so thin that the surface pressure is the same as the pressure at an altitude of 20 mi (35 km) on Earth. Wispy clouds can be found at extreme altitudes.

summarized in **Figure 6-18,** which shows how pressure and temperature vary with altitude in the atmospheres of Earth, Venus, and Mars.

Close-up Observations of Venus Reveal a Broiled Planet

Venus is bathed in more intense sunlight than Earth because it orbits closer to the Sun than Earth does. If Venus had no atmosphere, and if its surface reflected sunlight back into space like that of Mercury or the Moon, heating by the Sun would bring its surface temperature to around 113°F (45°C)—comparable to that found in the hottest regions on Earth. However, what is surprising is that Venus is many times hotter than this. Venus's temperatures are well above the boiling point of water and higher even than daytime temperatures on Mercury, which is closest to the Sun and most of whose atmosphere was boiled away billions of years ago.

Most terrestrial planetary atmospheres are dominated by CO_2, unless altered by biological activity or removed by the solar wind.

When Venus was young and had liquid water, the amount of atmospheric CO_2 was kept at small and stable levels, much as occurs on Earth today: CO_2 released by volcanoes quickly dissolved in the oceans and was then bound up in carbonate rocks. Enough of the liquid water would have vaporized to create a thick cover of water vapor clouds. Since water vapor is a greenhouse gas, this humid atmosphere—perhaps denser than Earth's present-day atmosphere, but far less dense than the atmosphere that surrounds Venus today—would have efficiently trapped heat from the Sun. At first, this would have had little effect on the oceans of Venus. Although the temperature would have climbed above 212°F (100°C), the boiling point of water at sea level on Earth, the added atmospheric pressure from water vapor would have kept the water in Venus's oceans in the liquid state.

This hot and humid state of affairs may have persisted for several hundred million years. But as the Sun's energy output slowly increased over time, the temperature at the surface would eventually have risen above 700°F (647 K or 374°C). Above this temperature, no matter what the atmospheric pressure, Venus's oceans would have begun to evaporate. The added water vapor in the atmosphere would have increased the greenhouse effect, making the temperature even higher and causing the oceans to evaporate faster. That would have added even more water vapor to the atmosphere, further intensifying the greenhouse effect and making the temperature climb higher still. This is an example of a **runaway greenhouse effect,** in which an increase in temperature causes a further increase in temperature, and so on.

ANALOGY A runaway greenhouse effect is like a house in which the thermostat has accidentally been connected backward. If the temperature in such a house gets above the value set on the thermostat, the heater comes on and makes the house even hotter.

Today, Venus is dramatically different from what it was in the past, and it isn't like any of the other planets. Venus has a dry, desertlike surface with no oceans, lakes, or rivers. The infrared-absorbing properties of CO_2 have stabilized Venus's surface temperature at its present value of 860°F (460°C). Only minuscule amounts of water vapor—about

30 parts per million, or 0.003%—remain in the atmosphere. As on Earth, volcanic outgassing still adds small amounts of water vapor to the atmosphere, along with carbon dioxide and sulfur dioxide. But on Venus, these water molecules either combine with sulfur dioxide to form sulfuric acid clouds or break apart due to solar ultraviolet radiation. The great irony is that this state of affairs is the direct result of an earlier Venusian atmosphere that was predominantly water vapor.

Figure 6-18*b* depicts just how dense the Venusian atmosphere is. At the surface, the pressure is 90 atm—that is, 90 times greater than the average air pressure at sea level on Earth. This is about the same as the water pressure at a depth of about half a mile below the surface of Earth's oceans. Venus's atmosphere is considerably denser, too. The density of the atmosphere at the surface of Venus is more than 50 times greater than the density of our atmosphere at sea level. The atmosphere is incredibly massive and does not change easily. In fact, once heated by the Sun, the atmosphere retains its heat throughout the long Venusian night. As a result, temperatures on the day and night sides of Venus are almost identical.

Soviet and U.S. spacecraft also discovered that Venus's clouds are primarily confined to three high-altitude layers. An upper cloud layer lies at altitudes between 42 mi and 36 mi (68 km and 58 km), a denser and more opaque cloud layer from 36 mi to 32 mi (58 km to 52 km), and an even denser and more opaque layer between 32 mi and 30 mi (52 km and 48 km). Above and below the clouds are 20-km-thick layers of haze. Below the lowest haze layer, the atmosphere is remarkably clear all the way down to the surface of Venus.

Sulfur plays an important role in the Venusian atmosphere. Sulfur combines with other elements to form gases such as sulfur dioxide (SO_2) and hydrogen sulfide (H_2S), along with sulfuric acid (H_2SO_4), the same acid used in automobile batteries. While Earth's clouds are composed of water droplets, Venusian clouds contain almost no water. Instead, they are composed of droplets of concentrated, corrosive sulfuric acid. Thanks to the high temperatures on Venus, these droplets never rain down on the planet's surface; they simply evaporate at high altitude. (You can see a similar effect on Earth. On a hot day in the desert of the U.S. Southwest, streamers of rain called *virga* appear out of the bottoms of clouds but evaporate before reaching the ground.)

On Earth, friction between the atmosphere and the ground causes wind speeds at the surface to be much less than at high altitude. The same is true on Venus: The greatest wind speed measured by spacecraft on Venus's surface is only about 3 mi/h (5 km/h). Thus, only slight breezes disturb the crushing pressures and infernal temperatures found on Venus's dry, lifeless surface. Whereas Venus has an incredibly thick atmosphere compared to Earth's, observations of Mars show that Mars has an atmosphere far thinner.

Cosmic Connections: Evolution of Terrestrial Atmospheres summarizes the processes that led to the present-day atmospheres of the three large terrestrial planets.

ConceptCheck 6-10: What is "runaway" about Venus's runaway greenhouse?

The Martian Atmosphere: Not a Drop to Drink

Go to Video 6-4

Observations show that Mars has far less atmosphere than either Venus or Earth, but it might not always have been that way. Mars probably had a thicker atmosphere 4 billion years ago. Thanks to Mars's greater distance from the Sun and hence less intense sunlight, temperatures would have been lower than on the young Earth and any water in the atmosphere would more easily have fallen as rain or snow. This would have washed much of the planet's carbon dioxide from its atmosphere, perhaps creating carbonate minerals in which the CO_2 is today chemically bound. Measurements from Mars orbit show only small amounts of carbonate materials on the surface, suggesting that the amount of atmospheric CO_2 that rained out was small. Hence, even the original Martian atmosphere was relatively thin, though thicker than the present-day atmosphere.

Because Mars is so small, it cooled early in its history and volcanic activity came to an end. Hence, any solid carbonates were not recycled through volcanoes as they are on Earth. The depletion of carbon dioxide from the Martian atmosphere into the surface would therefore have been permanent.

As the amount of atmospheric CO_2 declined, the greenhouse effect on Mars would have weakened and temperatures would have fallen. This temperature decrease would have caused more water vapor to condense into rain or snow and fall to the surface, taking even more CO_2 with it and further weakening the greenhouse effect. Thus, a decrease in temperature would have caused a further decrease in temperature—a phenomenon sometimes called a **runaway icehouse effect.** (This is the reverse of the runaway greenhouse effect that has taken place on Venus.) Ultimately, both water vapor and most of the carbon dioxide would have been removed from the Martian atmosphere. With only a very thin CO_2 atmosphere remaining, surface temperatures on Mars eventually would have stabilized at their present frigid values.

The atmosphere of Mars today, as Figure 6-18*c* shows, is very cold and thin compared to the atmosphere of either Earth or Venus—the surface pressure on Mars is a mere 0.006 atm. But its chemical composition is very close to that of Venus—the Martian atmosphere is 95.3% carbon dioxide (versus 96.5% for Venus) and 2.7% nitrogen (versus 3.5% for Venus). The predominance of carbon dioxide means that Mars, like Earth and Venus, is warmed by the greenhouse effect. However, the Martian atmosphere is so thin that the greenhouse effect is very weak and warms the Martian surface by only 40°F

on a planet, much of current planetary research is actually a search for water. Because the presence of water today is not clearly evident on planets besides Earth, astronomers look to rocks on other planets for evidence of past water. We hinted at the beginning of this chapter that water is found as frozen ice hidden in the permanent shadows on the Moon and Mercury. On Venus we cannot look very far into the past, since the planet's volcanic activity and flake tectonics have erased features more than about 500 million years old. On Mars, by contrast, we expect to find rocks that are billions of years old: The thin atmosphere causes very little erosion, there has been little recent volcanic activity to cover ancient rocks, and there is no subduction to drag old surface features back into the planet's interior. (The heavy cratering of the southern highlands bears testament to the great age of much of the Martian surface.) In a quest to find these ancient rocks and learn about the early history of Mars, scientists have sent spacecraft to land on the red planet and explore it at close range.

Water May Have Once Existed on the Surface of Mars

As of 2013, seven spacecraft have successfully landed on the Martian surface. Landing in 2008, NASA's *Phoenix* lander had no capability to move around on the surface, so it could only study objects within reach of the scoop at the end of a mechanical arm (**Figure 6-20*a***). Four more recent unmanned, roving explorers—*Mars Pathfinder* (landed 1997), the two *Mars Exploration Rovers* (landed 2004), and the Mars Science Laboratory's *Curiosity* rover (landed 2012)—were equipped with wheels that allowed travel over the surface (Figure 6-20*b*). Unlike stationary landers, these rovers did not have heavy and expensive rocket engines to lower them gently to the Martian surface. Instead, the spacecraft were surrounded by airbags like those used in automobiles or dropped with speed-slowing rockets and assisted parachutes. Had there been anyone to watch *Mars Pathfinder* or one of the *Mars Exploration Rovers* land, they would have seen an oversized beach ball hit the surface at 50 km/h (30 mi/h), then bounce for a kilometer before finally rolling to a stop. Unharmed by its wild ride, the rover then deflated its airbags and began to study the geology of Mars.

Figure 6-11 shows the landing sites of the *Mars Pathfinder* rover (named *Sojourner*) and the two *Mars Exploration Rovers* (named *Spirit* and *Opportunity*). Following directions from Earth, *Sojourner* spent three months

(a)

(b)

FIGURE 6-20 R I V U X G **Martian Explorers** (a) NASA's *Phoenix Mars Lander* monitors the atmosphere overhead and reaches out to the soil below in this artist's depiction of the spacecraft fully deployed on the surface of Mars. (b) The *Phoenix* lander uncovered ice beneath the Martian surface. This trench, just 8.7 in (22 cm) wide and 13.8 in (35 cm) long and informally called Dodo-Goldilocks, shows white areas that are most likely ice. The image also shows a few white lumps in the middle left of the trench, which disappeared over a few days, in a process similar to evaporation. *(a: NASA/JPL/UA/Lockheed Martin; b: NASA/JPL-Caltech/University of Arizona/Texas A&M University)*

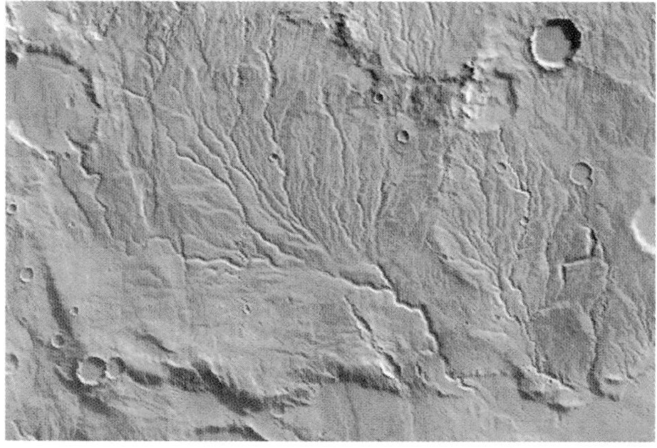

(a)

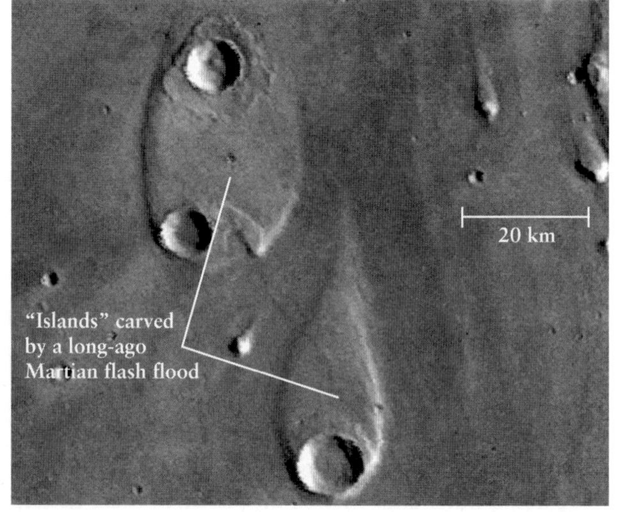

(b)

FIGURE 6-21 R I V U X G **Signs of Ancient Martian Water** (a) This *Viking Orbiter* image shows a network of dry riverbeds extending across the cratered southern highlands. Liquid water would evaporate in the sparse, present-day Martian atmosphere, so these must date from an earlier era when the atmosphere was thicker. (b) These teardrop-shaped islands rise above the floor of Ares Valles. They were carved out by a torrent of water that flowed from the bottom of the image toward the top. Similar flood-carved islands are found on Earth in eastern Washington State. *(a: NASA; b: NASA/USGS)*

exploring the rocks around its landing site. The *Mars Exploration Rovers* were also designed for a three-month lifetime, but *Spirit* functioned for more than seven years after landing and *Opportunity* is still functioning as of this writing in 2013. Over that time *Opportunity* has astonishingly traveled more than 20 mi (35 km) over the Martian surface.

Images made from Mars orbit show a number of features that suggest water flowed there in the past (**Figure 6-21**). To investigate these features, recent rovers have landed in areas that look much like flood plains found on Earth, albeit a much older and ancient flood plain. Figure 6-21*b* shows a portion of this plain. Rushing floodwaters can pick up and move rocks of fairly large sizes great distances from their original locations, so scientists expected that targeted landing sites (**Figure 6-22**) would be rich with numerous different kinds of rocks and show great geologic diversity. They have not been disappointed. Many of the rocks found appear to be rock produced by volcanoes. Other rocks, which resemble jagged rocks found in the ancient highlands of the Moon, appear to have come from areas that were struck by crater-forming impacts. Still others have a layered structure like sedimentary rocks on Earth. Such rocks typically form gradually at the bottom of bodies of water, suggesting that liquid water was stable in at least some regions of Mars for a substantial period of time in the past.

When combined with observations from Mars orbit, data from the rovers show that ancient Mars was a very diverse planet. Most of the planet was probably very dry, but there were isolated areas that had liquid water for a time. Astronomers still do not have a reliable mechanism that explains why liquid water appeared on Mars only in certain places and at certain times.

Whatever liquid water once coursed over parts of the Martian surface is presumably now frozen, either at the polar ice caps or beneath the planet's surface. To check this idea, scientists used a device on board the *Mars Odyssey* orbiter to look for evidence of water beneath the surface. The results

FIGURE 6-22 R I V U X G ***Curiosity* Self-Portrait** Taken at Gale Crater on the surface of Mars, where the mission's first scoop sampling took place, self-portraits like this one allow engineers to monitor how much dust has accumulated on the rover and track how fast the wheels are wearing down. Mount Sharp is visible rising in the background. *(NASA/JPL-Caltech/Malin Space Science Systems)*

from the *Mars Odyssey* measurements suggest that there is abundant water beneath the surface at both Martian poles (**Figure 6-23**). But they also show a surprising amount of subsurface water in two regions near the equator (see Figure 6-11) and another region on the opposite side of the planet. One possible explanation is that about a million years ago, gravitational forces from other planets caused a temporary change in the tilt of Mars's rotation axis. The ice caps were then exposed to more direct sunlight, causing some of the water to evaporate and eventually refreeze elsewhere on the planet.

There are other surprising features on Mars related to the presence of water. A number of *Mars Global Surveyor* images show gullies apparently carved by water flowing down the walls of pits or craters (**Figure 6-24**). These could have formed when water trapped underground—where the pressure is greater and water can remain a liquid—seeped onto the surface. The gullies appear to be geologically young,

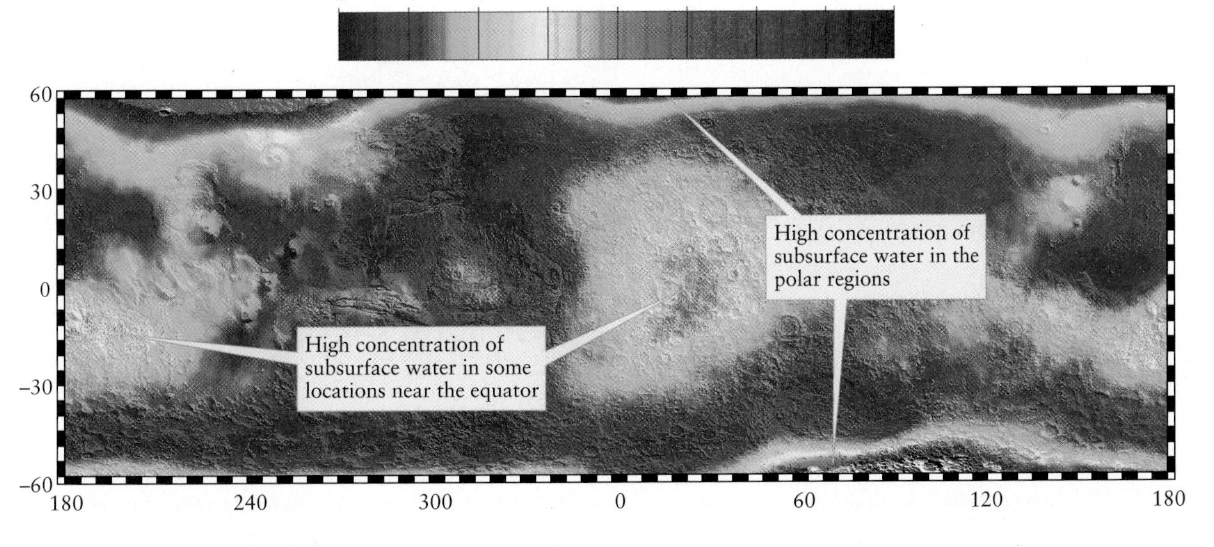

FIGURE 6-23 Water Beneath the Martian Surface This map of the abundance of water (probably in the form of ice) just below the surface of Mars is based on data collected from orbit by the *Mars Odyssey* spacecraft. These measurements can only reveal the presence of water or water ice to a depth of about a meter; there may be much more water at greater depths. *(JPL/NASA/MSSS [LANL])*

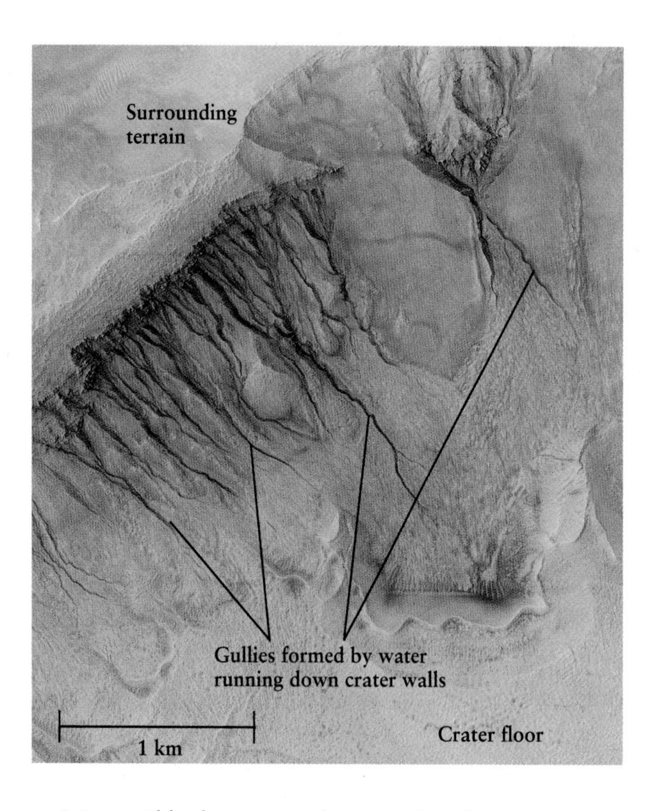

FIGURE 6-24 R I V U X G **Martian Gullies** When the *Mars Global Surveyor* looked straight down into this crater in the Martian southern hemisphere, it saw a series of gullies along the crater wall. These gullies may have been formed by subsurface water seeping out to the surface, or by the melting of snow that fell on crater walls. *(Los Alamos National Laboratory/NASA/JPL/Malin Space Science Systems)*

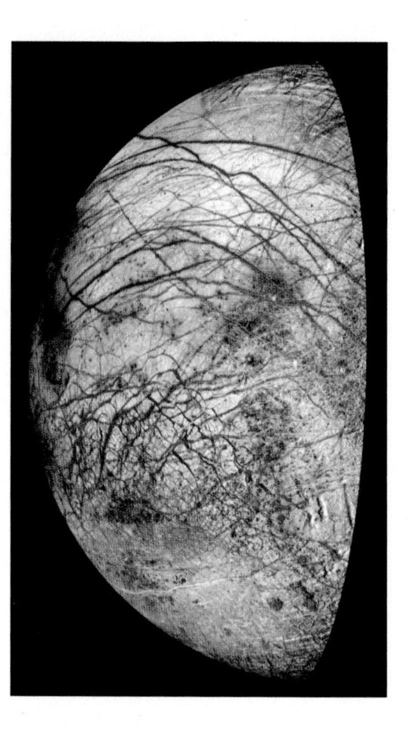

FIGURE 6-25 R I V U X G **Europa** Dark lines crisscross Europa's smooth, icy surface in this false-color composite of visible and infrared images from *Galileo*. These are fractures in Europa's crust that can be as much as 12 mi to 25 mi (20 km to 40 km) wide. Only a few impact craters are visible on Europa, which indicates that this satellite has a very young surface, on which all older craters have been erased. *(NASA/JPL/University of Arizona)*

so it is possible that even today some liquid water survives below the Martian surface.

Alternatively, these gullies may have formed during periods when the Martian climate was colder than usual and the slopes of craters were covered with a layer of dust and snow. Melting snow on the underside of this layer could have carved out gullies, which were later exposed when the climate warmed and the snow evaporated.

Perhaps most exciting, the *Phoenix* lander drilled down into Martian soil to study the history of water on the Martian surface, with a special eye toward determining the potential that life might find Mars to be a habitable environment (see Figure 6-20). By digging trenches, it uncovered water ice that, when exposed to the Martian atmosphere, evaporated, disappearing over just a few days. Before it was covered with ice during the Martian winter, *Phoenix* was the first spacecraft to actually observe water, and even some falling snow, on Mars.

ConceptCheck 6-13: **Why does the presence of a wide diversity of rocks in a single location suggest the presence of water long ago?**

ConceptCheck 6-14: **Why doesn't sunlight evaporate all water on Mercury and the Moon?**

Europa Could Harbor a Subsurface Ocean Beneath Its Icy Crust

Go to Video 6-5

If water abundantly exists on Earth and clear evidence for water recently has appeared on Mars, where else in the solar system might we find water? The next most likely candidates for water are Jupiter's largest moons.

Europa, the second of Jupiter's moons, has the notable distinction of being the smoothest body in the solar system. There are no mountains and no surface features greater than a few hundred meters high (**Figure 6-25**). There are almost no craters, indicating a young surface that has been reprocessed by geologic activity. The dominant surface feature is a worldwide network of stripes and cracks (**Figure 6-26**). Like Io, Europa is an important exception to the general rule that a small world will be cratered and geologically dead.

Confirming earlier speculations, the *Galileo* spacecraft (both Figure 6-25 and Figure 6-26 are *Galileo* images) showed that Europa's infrared spectrum is a close match to

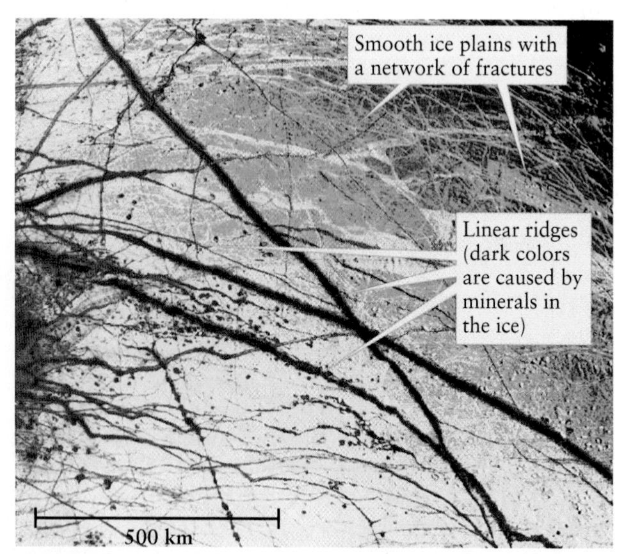

FIGURE 6-26 R I V U X G **Europa's Fractured Crust** False colors in this *Galileo* composite image emphasize the difference between the linear ridges and the surrounding plains. The smooth ice plains (shown in blue) are also the basic terrain found on Io. *(NASA/JPL/University of Arizona)*

that of a thin layer of fine-grained water ice frost on top of a surface of pure water ice. (The brown areas in Figure 6-25 show where the icy surface contains deposits of rocky material from meteoritic impacts, from Europa's interior, or from a combination of these sources.)

The purity of Europa's ice suggests that water is somehow brought upward from the moon's interior to the surface, where it solidifies to make a fresh, smooth layer of ice. Indeed, some *Galileo* images show what appears to be lava flows on Europa's surface, although the "lava" in this case is mostly ice. This idea helps to explain why Europa has very few craters (any old ones have simply been covered up) and why its surface is so smooth. Europa's surface may thus represent a water-and-ice version of plate tectonics.

Although Europa's surface is almost pure water ice, keep in mind that Europa is *not* merely a giant ice ball. The satellite's density shows that rocky material makes up about 85% to 90% of Europa's mass. Hence, only a small fraction of the mass, about 10% to 15%, is water ice. Because the surface is icy, we can conclude that the rocky material is found within Europa's interior.

Europa is too small to have retained much of the internal heat that it had when it first formed. But there must be internal heat nonetheless to power the geologic processes that erase craters and bring fresh water to Europa's surface. What keeps Europa's interior warm? The most likely answer, just as for Io, is heating by being flexed and squeezed gravitationally during its orbit around Jupiter. The rhythmic gravitational tugs exerted by Io and Ganymede on Europa deform its orbit. But because it is farther from Jupiter, tidal effects on Europa are only about one-fourth as strong as those on Io, which may explain why no ongoing volcanic activity has yet been seen on Europa.

Some features on Europa's surface, such as the fracture patterns shown in Figure 6-26, may be the direct result of the crust being stretched and compressed by tidal flexing.

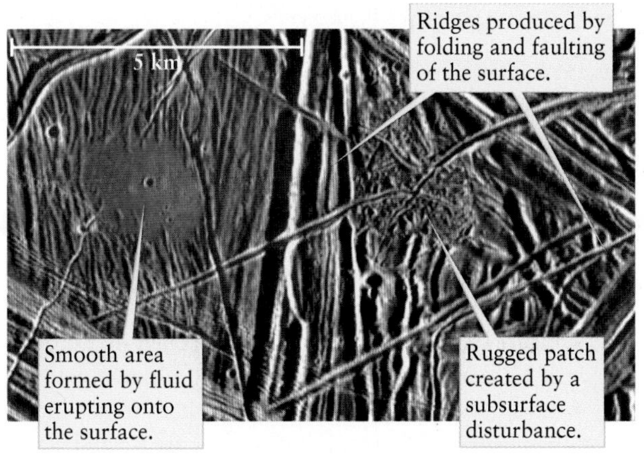

FIGURE 6-27 R I V U X G **Europa in Close-up** This high-resolution *Galileo* image shows a network of overlapping ridges, part of which has been erased to leave a smooth area and part of which has been jumbled into a rugged patch of terrain. Europa's interior must be warm enough to power this complex geologic activity. *(NASA/JPL)*

Figure 6-27 shows other features, such as networks of ridges and a young, very smooth circular area, that were probably caused by the internal heat that tidal flexing generates. The rich variety of terrain depicted in Figure 6-27, with stress ridges going in every direction, shows that Europa has a complex geologic history.

Among the unique structures found on Europa's surface are **ice rafts.** The area shown in **Figure 6-28*a*** was apparently subjected to folding, producing the same kind of linear features as those in Figure 6-27. But a later tectonic disturbance broke the surface into small chunks of crust a few miles across, which then "rafted" into new positions. A similar sort of rafting happens in Earth's Arctic Ocean every spring, when the winter's accumulation of surface ice breaks up into drifting ice floes (Figure 6-28*b*). The existence of such structures on Europa strongly suggests that there is a subsurface layer of liquid water or soft ice over which the ice rafts can slide with little resistance.

Orbiting Jupiter, Europa's interior is kept heated by the continuous stress caused by its intense gravitational interaction with Jupiter.

ConceptCheck 6-15: Why do the long lines of ridges observed on Europa's surface seem to be broken and fractured?

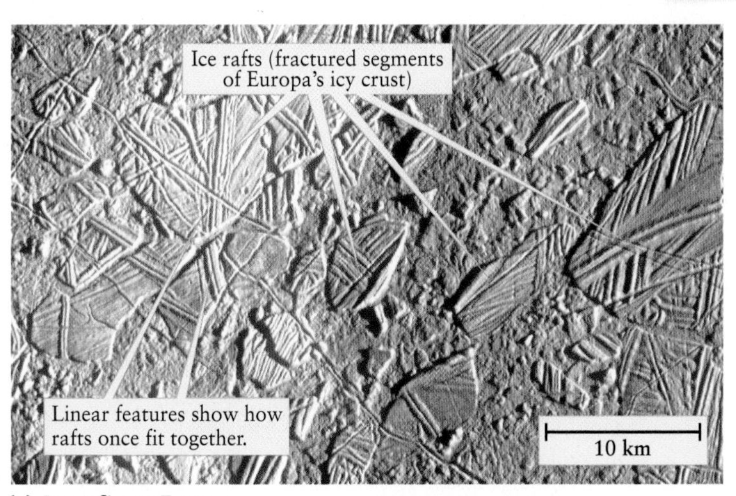

(a) Ice rafts on Europa

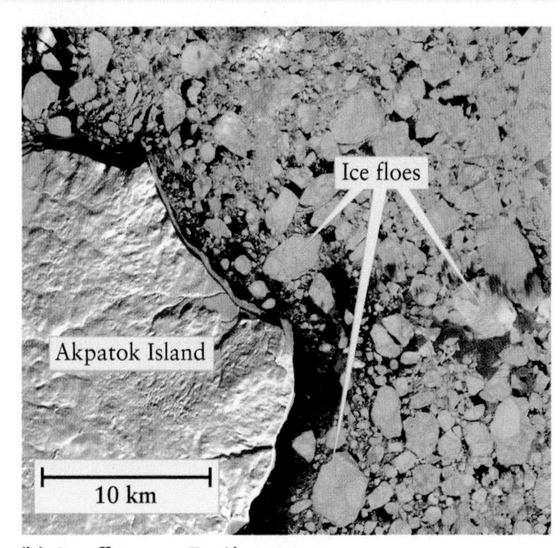

(b) Ice floes on Earth

FIGURE 6-28 R I V U X G **Moving Ice on Europa and Earth** (a) Sometime after a series of ridges formed in this region of Europa's surface, the icy crust broke into "rafts" that were moved around by an underlying liquid or plastic layer. The colors in this *Galileo* image may be due to minerals that were released from beneath the surface after the crust broke apart. (b) Europa's ice rafts are analogous to ice floes on Earth, created when pack ice breaks up, as in this spacecraft view of part of the Canadian Arctic. *(a: NASA/JPL; b: USGS/NASA)*

An Underground Ocean?

The ridges, faults, ice rafts, and other features strongly suggest that Europa has substantial amounts of internal heat. This heat could prevent water from freezing beneath Europa's surface, creating a worldwide ocean beneath the crust. If this idea is correct, geologic processes on Europa may be an exotic version of those on Earth, with the roles of solid rock and molten magma being played by solid ice and liquid water.

A future spacecraft may use radar to penetrate Europa's icy crust to search for definitive proof of liquid water beneath the surface. Scientists conclude that Europa's outermost 60 mi to 120 mi (100 km to 200 km) are ice and water. (It is not clear how much of this is liquid and how much is solid.) Within this outer shell is a rocky mantle surrounding a metallic core some 400 mi (600 km) in radius.

The existence of a warm, subsurface ocean on Europa, if proved, would make Europa the only world in the solar system other than Earth on which there is liquid water. This would have dramatic implications. On Earth, water and warmth are essentials for the existence of life. Perhaps single-celled organisms have evolved in the water beneath Europa's crust, where they would use dissolved minerals and organic compounds as food sources. In light of this possibility, NASA has taken steps to prevent biological contamination of Europa. At the end of the *Galileo* mission in 2003, the spacecraft (which may have carried traces of organisms from Earth) was sent to burn up in Jupiter's atmosphere, rather than remaining in orbit where it might someday crash into Europa. An appropriately sterilized spacecraft may one day visit Europa and search for evidence of life within this exotic moon.

ConceptCheck 6-16: **If Europa has no molten core for an interior heat source, what provides the energy to keep water on Europa from being frozen completely solid?**

VISUAL LITERACY TASK

Mercury and the Moon

PROMPT: What would you tell a fellow student who said, **"Mercury and Earth's Moon are difficult to tell apart because they are about the same size with identical compositions and surface features"?**

ENTER RESPONSE:

Guiding Questions:

1. Compared to Earth's Moon, the planet Mercury is
   a. smaller.
   b. larger.
   c. the same size.
   d. colder.

2. Compared to Earth's Moon, the planet Mercury's surface is
   a. older.
   b. younger.
   c. about the same age.
   d. Astronomers don't know.

3. Compared to the Moon, the interior of Mercury is
   a. more geologically active.
   b. less geologically active.
   c. about the same.
   d. Astronomers don't know.

KEY IDEAS AND TERMS

6-1 Comparing terrestrial planets and moons shows distinct similarities and dramatic differences in appearance

- Comparing planets to Earth helps astronomers understand the solar system.
- Venus is particularly difficult to study because it is covered in a thick layer of clouds.
- Planet interiors are invisible to telescopes.

6-2 Many terrestrial world surfaces are dominated by impact craters revealing the age of underlying processes

- Circular **impact craters** result when one object collides with the solid surface of a terrestrial planet or a satellite.
- Water ice exists in some craters near the poles that have walls permanently shielded from the Sun.
- The smaller the terrestrial world, the less internal heat it is likely to have retained from its formation, and, thus, the less geologic activity it will display on its surface. The less geologically active the world, the older and hence more heavily cratered its surface.
- The rate of impact crater formation has decreased dramatically since the early years of the solar system.

6-3 Tectonics and volcanism influence surface features

- Mars's surface is characterized by relatively smooth **northern lowlands** and jagged **southern highlands.**
- Venus has a relatively thin crust and radar evidence suggests no active plate tectonics.
- Mars has a thick, rigid crust resulting in a **rift valley** stretching across one side of the planet.
- Mars's Olympus Mons is a giant **shield volcano** formed by layer after layer of magma being deposited.
- Jupiter's satellite Io has virtually no impact craters, as it is constantly being smoothed over by active volcano flows.

6-4 Atmospheres surrounding terrestrial planets vary considerably

- Mercury, the Moon, and Mars are easier to study than Venus, which is obscured by its thick atmosphere.
- Venus's atmosphere causes an extremely high surface temperature due to a **runaway greenhouse effect.**
- Mars's atmosphere experiences a **runaway icehouse effect** where CO_2 is pulled from the atmosphere, lowering the temperature even more.

6-5 Evidence exists for water in locations besides Earth

- Frozen water on Mars has been revealed by landers drilling into the surface.
- Frozen water on the Moon has been revealed by studying ejected plumes created by crashing spacecraft into shadowed crater walls.
- Jupiter's satellite Europa may harbor a subsurface ocean that causes **ice rafts** to move across the surface.

QUESTIONS

Review Questions

1. Describe the kinds of features that can be seen on the Moon with a small telescope.
2. Are impact craters on the Moon the same shape as the meteoroids that made the impact? Explain your answer.
3. Why is Earth geologically active while the Moon is not?
4. Rocks found on the Moon are between 3.1 and 4.47 billion years old. By contrast, the majority of Earth's surface is made of oceanic crust that is less than 200 million years old, and the very oldest Earth rocks are about 4 billion years old. If Earth and the Moon are essentially the same age, why is there such a disparity in the ages of rocks on the two worlds?
5. What kind of surface features are found on Mercury? How do they compare to surface features on the Moon? Why are they probably much older than most surface features on Earth?
6. If Mercury is the closest planet to the Sun and has such a high average surface temperature, how is it possible that ice might exist on its surface?
7. What is flake tectonics? Why does Venus exhibit flake tectonics rather than plate tectonics?
8. How was water most recently discovered on Mars?
9. How was water most recently discovered on Earth's Moon?
10. Why do ice rafts indicate the existence of a subsurface ocean on Europa?

Web Chat Questions

1. NASA is planning a new series of manned missions to the Moon. Compare the advantages and disadvantages of exploring the Moon with astronauts as opposed to using mobile, unmanned instrument packages.
2. Describe how you would empirically test the idea that human behavior is related to the phases of the Moon. What problems are inherent in such testing?
3. Imagine that you are planning a lunar landing mission. What type of landing site would you select? Where might you land to search for evidence of recent volcanic activity?
4. If you were planning a new mission to Mercury, what features and observations would be of particular interest to you?
5. The total cost of the *Mars Global Surveyor* mission was about $154 million. (To put this number in perspective, in 2000 the U.S. Mint spent about $40 million to advertise its new $1 coin, which failed to be accepted by the public. Several recent Hollywood movies have had larger budgets than the *Mars Global Surveyor.*) Does this expenditure seem reasonable to you? Why or why not?
6. Is it worthwhile for scientists to actively search for water on planets and satellites?

Collaborative Exercises

1. The image of the Moon in Figure 6-2 reveals numerous craters. Using the idea that the Moon's landscape can only be changed by impacts, make a rough sketch showing 10 of the largest craters and label them from oldest (those that showed up first) to youngest (the most recent ones). Explain your reasoning and any uncertainties.
2. Consider the image of Mars in Figure 6-1. Draw a circle on your paper roughly 5 cm in diameter and, taking turns, have each person in your group sketch a different region of Mars. How is your collaborative sketch different from the other images of Mars found throughout the book?

Observing Questions

1. Use *Starry Night*™ to examine magnified images of the terrestrial major planets Mercury, Venus, Earth, and Mars and the dwarf planet Ceres. Select each of these planets from **Favourites > Explorations** in turn. Use the location scroller to rotate the image to see different views of the planet. (a) Describe each planet's appearance. From what you observe in each case, is there any way of knowing whether you are looking at a planet's surface or at complete cloud cover over the planet? (b) Which planet or planets have clouds? If a planet has clouds, open its contextual menu and choose **Surface Image/Model > Default** and use the location scroller to examine the planet's surface. (c) Which major planet shows the heaviest cratering? (d) Which of these terrestrial planets show evidence of liquid water? (e) What do you notice about Venus's rotation compared to the other planets?
2. Use *Starry Night*™ to examine Mercury. Select **Favourites > Explorations > Mercury** from the menu. Stop the advance of time and use the **Zoom** controls and the location scroller to examine the surface of the planet Mercury. Estimate the diameters of the largest craters on Mercury's surface by measuring their size on the screen with a ruler and comparing these diameters to the diameter of Mercury. (a) What are the diameters (in km) of the largest craters on Mercury? (b) **Zoom in** to examine the surface features of Mercury in more detail and compare these features with those of our Moon. Comment on the similarities and differences between these two planetary objects, neither of which has an atmosphere (i.e., presence of craters, light-ray patterns, maria, peaks within craters).
3. Use *Starry Night*™ to examine Mars. Open **Favourites > Explorations > Mars Surface.** Use the **Zoom** controls and the location scroller to explore this planet's surface features. You will notice that four volcanoes (Mons) and the Valles Marineris have been labeled. (a) Which of the volcanoes appears to be the largest? (b) Right-click on Mars (Ctrl-click on a Mac) and select **Markers and Outlines . . .** from the contextual menu. In the **Mars Markers and Outlines** dialog window, click the lower radio button to the left of the **List** label. Then in the dropdown box to the right of the **List** label, choose **Type.** In the rightmost dropdown box, select **Crater** as the type of feature and then click the **Check all Shown** button. Describe the distribution of craters on the Martian surface. (c) What does the distribution of craters in the region around the volcanoes suggest about the time at which these volcanoes formed on Mars?

ANSWERS

ConceptChecks

ConceptCheck 6-1: No. Venus is perpetually shrouded in a thick layer of clouds and because visible light does not penetrate the clouds, no light escapes for us to observe from Earth.

ConceptCheck 6-2: All large impactors, regardless of shape or angle of impact, will carve out a nearly circular crater with a tall central peak in the middle due to the shock wave created. Earth's thin atmosphere will not protect the surface from large impacts.

ConceptCheck 6-3: The line on the graph would not be horizontal, but instead be sloped upward, showing that more and more impacts have occurred over the duration of the Moon's existence.

ConceptCheck 6-4: Because surfaces that are covered with impact craters have not been smoothed over by lava flows, we would infer that a hypothetical Caloris Basin covered in craters would be very old.

ConceptCheck 6-5: Without wind from a moving atmosphere and without resurfacing from tectonics, those tire tracks will remain until they are obliterated into a crater by an impact.

ConceptCheck 6-6: Mountain chains form where rigid plates collide, and the absence of these features suggests that Venus is a single-plate planet.

ConceptCheck 6-7: When overly thick plates collide, it is nearly impossible for one plate to be subducted beneath another plate, making it impossible for subduction zones and their accompanying mountain chains to form.

ConceptCheck 6-8: The oldest lava flows have craters from ancient impacts that have distorted their appearance, whereas the youngest lava flows are mostly crater free.

ConceptCheck 6-9: Lava flowing from extremely active volcanoes smooths over and covers any craters almost immediately.

ConceptCheck 6-10: The "runaway" term refers to the fact that an increase in temperature causes changes in the atmosphere, which causes the temperature to rise even higher and higher.

ConceptCheck 6-11: The temperature on Mars might be increased through the greenhouse effect if CO_2 could be successfully added to the atmosphere.

ConceptCheck 6-12: Instead of freezing, the water would quickly evaporate into the atmosphere, due to the low atmospheric pressure on Mars.

ConceptCheck 6-13: Rapidly moving water will pick up and move rocks to a flood plain far from where they originated, resulting in a location with highly varying rock types.

ConceptCheck 6-14: Some deposits of frozen water could be shielded from sunlight along darkly shadowed crater walls that never glimpse sunlight.

ConceptCheck 6-15: The seemingly solid, but slushy, ice covering Europa flows like "ice rafts," disrupting the continuous lines formed by the ridges.

ConceptCheck 6-16: The squeezing and flexing of Europa as it moves around Jupiter causes friction-releasing energy in the form of heat, keeping water there from permanently freezing.

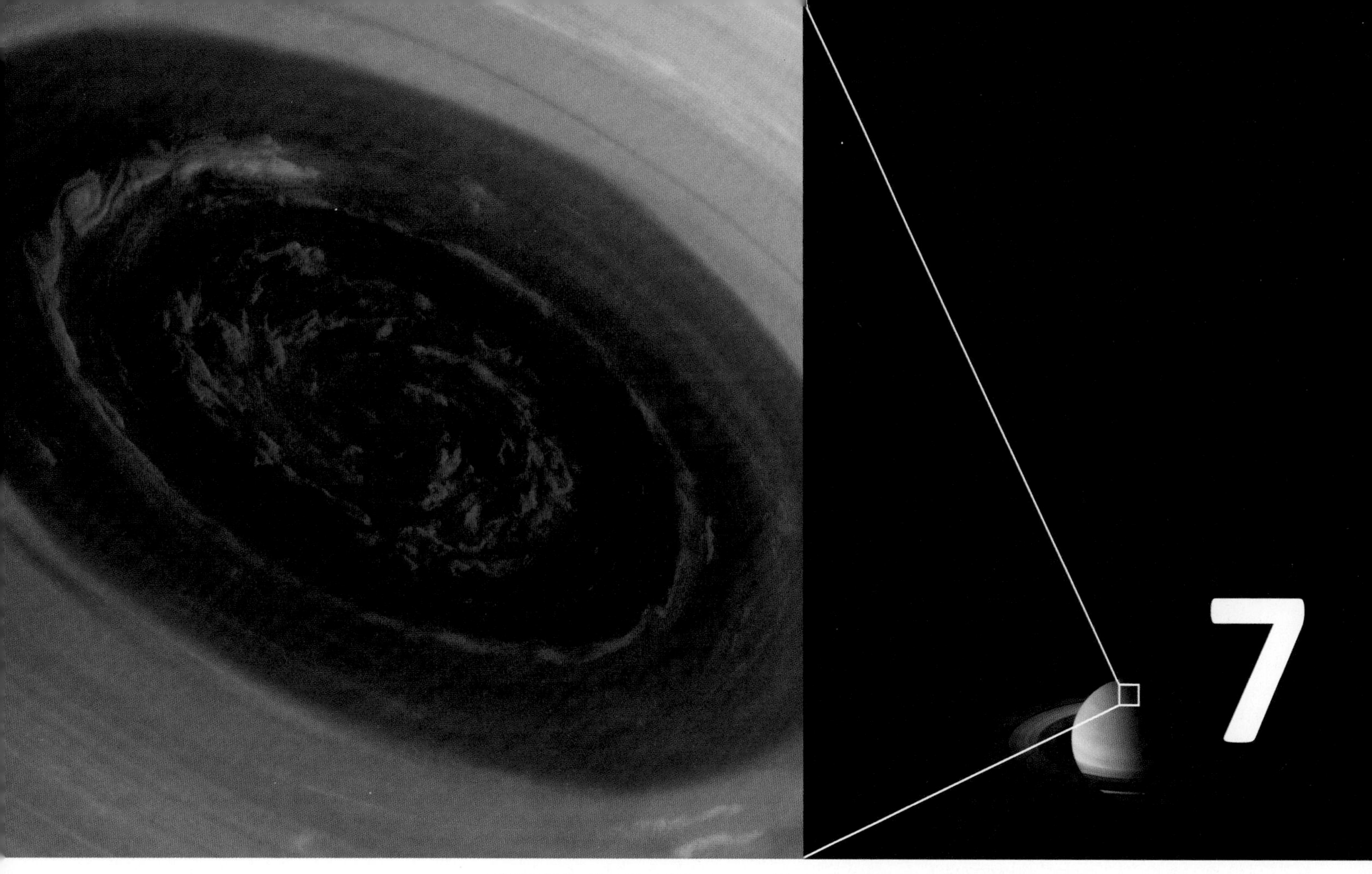

R I V U X G Stretching 1250 mi across with clouds moving nearly 330 mi per hour, this enhanced image of a hurricanelike storm at Saturn's north pole shows low clouds as red and high clouds as green. *(NASA/JPL-Caltech/Space Science Institute)*

Observing the Dynamic Giant Planets

CHAPTER LEARNING OBJECTIVES **By reading the sections of this chapter, you will learn:**

7-1 Dynamic atmospheres of Jupiter and Saturn change rapidly

7-2 Uranus and Neptune have seemingly quiet atmospheres

7-3 Saturn's moon Titan and Neptune's moon Triton exhibit unexpected atmospheres

7-4 All Jovian planet atmospheres are encircled by complex ring systems

Perhaps the most remarkable telescope views in our solar system are of the colorful, turbulent atmosphere of Jupiter and the ethereal beauty of Saturn's rings. Both of these giant worlds dwarf our own planet: More than 1200 Earths would fit inside Jupiter's immense bulk, and more than 700 Earths inside Saturn. Unlike Earth, both Jupiter and Saturn are composed primarily of the lightweight elements hydrogen and helium, in abundances very similar to those in the Sun. The rapid rotations of both Jupiter and Saturn stretch their weather systems into colorful bands that extend completely around each planet. The image on the previous page shows a hurricanelike storm at Saturn's north pole that is 1250 mi across.

Saturn, however, is not merely a miniature version of Jupiter. Its muted colors and more flattened shape are clues that Saturn's atmosphere and interior have important differences from those of Jupiter. The most striking difference is Saturn's elaborate system of rings, composed of countless numbers of icy fragments orbiting in the plane of the planet's equator. The rings display a complex and elegant structure, which is shaped by subtle gravitational influences from Saturn's retinue of moons. Jupiter, too, has rings, but they are made of dark, dustlike particles that reflect little light. These systems of rings, along with the distinctive properties of the planets themselves, make Jupiter and Saturn highlights of any spacecraft's tour of our solar system.

Beyond Saturn, in the cold, dark recesses of the solar system, orbit two gas planets that have long been shrouded in mystery. These planets—Uranus and Neptune—are so distant, so dimly lit by the Sun, and so slow in their motion against the stars that they were unknown to ancient astronomers and were discovered only after the invention of the telescope. Even then, little was known about the seemingly quiet atmospheres of Uranus and Neptune until *Voyager 2* flew past these planets during the 1980s. In this chapter, we will explore the gaseous envelopes that make up the largest planets of our solar system and the complex ring systems that surround them. ■

7-1 Dynamic atmospheres of Jupiter and Saturn change rapidly

Go to Video 7-1

Jupiter and Saturn are, respectively, the largest and second largest of our Sun's planets, and by far the largest objects in the solar system other than the Sun itself. Astronomers have known for centuries about the huge diameters and immense masses of these two planets. Jupiter is about 11 times larger in diameter than Earth, while Saturn's diameter is about 9 times larger than that of Earth. Astronomers also determined that Jupiter and Saturn are, respectively, 318 times and 95 times more massive than Earth. In fact, Jupiter has 2½ times the combined mass of all the other planets, satellites, asteroids, meteoroids, and comets in the solar system. But our Sun is more than 10 times bigger than either of these giants. A visitor from interstellar space might well describe our solar system as mainly one star, the Sun, one big planet, Jupiter, and then some smaller, orbiting debris!

Observing Jupiter and Saturn

As is true for any planet whose orbit lies outside Earth's orbit, the best time to observe Jupiter or Saturn is when they are closest to Earth, and they are most easily observed at midnight. When close, Jupiter can appear nearly 3 times brighter than Sirius, the brightest star in the sky. Only the Moon and Venus can outshine Jupiter then. Saturn's brightness when closest to us is only about one-seventh that of Jupiter, but it still magnificently outshines all of the stars except Sirius and Canopus.

Jupiter takes almost a dozen Earth years to orbit the Sun. Because Saturn is 2 times farther away, it appears to be about half Jupiter's size when viewed through a telescope. But, a view of Saturn presents a more dramatic difference from Jupiter: Saturn is surrounded by a magnificent system of rings. Saturn's orbital period is larger and even longer than Jupiter's, more than 29 Earth years.

As seen through an Earth-based telescope, both Jupiter and Saturn display colorful bands that extend around each planet parallel to its equator. More detailed close-up views of Jupiter from a spacecraft (**Figure 7-1*a***) show alternating dark and light bands parallel to Jupiter's equator in subtle tones of red, orange, brown, and yellow. The dark, reddish bands are called **belts,** and the light-colored bands are called **zones.** The image of Saturn in Figure 7-1*b* shows similar features, but with subdued colors that are much less pronounced.

In addition to these conspicuous stripes, a huge, red-orange oval, called the **Great Red Spot,** is often visible in Jupiter's southern hemisphere. This remarkable feature was seen by the English scientist Robert Hooke in 1664 but it may be much older, having eluded earlier small telescopes. It appears to be an extraordinarily long-lived storm in the planet's dynamic atmosphere. Saturn seems to have no long-lived storm systems of this kind, but it does have numerous shorter-lived "spots," like the large one shown in the chapter opening image. Many careful observers have reported smaller spots and blemishes in the atmospheres of both Jupiter and Saturn that last for only a few weeks or months.

Observations of features like the Great Red Spot and smaller storms allow astronomers to determine how rapidly Jupiter and Saturn rotate. At its equator, Jupiter completes a full rotation in only 9 hours, 50 minutes, and 28 seconds, making it not only the largest and most massive planet in the solar system but also the one with the fastest rotation. Jupiter, however, rotates in a strikingly different way from the solid objects of Earth, the Moon, Mercury, Venus, and Mars.

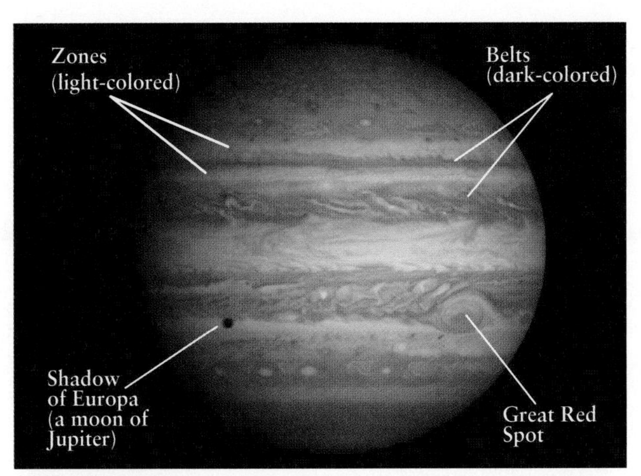

(a) Jupiter

(b) Saturn

FIGURE 7-1 R I V U X G **Jupiter and Saturn as Viewed from Space** (a) This view of Jupiter is a composite of four images made by the *Cassini* spacecraft as it flew past Jupiter in 2000. (b) Images from the *Voyager 2* spacecraft as it flew past Saturn in 1981 were combined to show the planet in approximately natural color. *(NASA/JPL)*

If Jupiter were a solid body like a terrestrial planet (or, for that matter, a billiard ball), all parts of Jupiter's surface would rotate through one complete circle in this same amount of time (**Figure 7-2*a***). But by watching features in Jupiter's cloud cover, Gian Domenico Cassini discovered in 1690 that the polar regions of the planet rotate a little more slowly than do the equatorial regions. Near the poles, the rotation period of Jupiter's atmosphere is about 9 hours, 55 minutes, and 41 seconds. Saturn, too, has a longer rotation period near its poles (10 hours, 39 minutes, and 24 seconds) than at its equator (10 hours, 13 minutes, and 59 seconds).

You can see this kind of rotation, called **differential rotation,** in the kitchen. As you stir the water in a pot, different parts of the liquid take different amounts of time to make one "rotation" around the center of the pot (Figure 7-2*b*). Differential rotation shows that neither Jupiter nor Saturn can be solid throughout their volumes: They must be at least partially fluid, like water in a pot. But, just what are these giant planets made of?

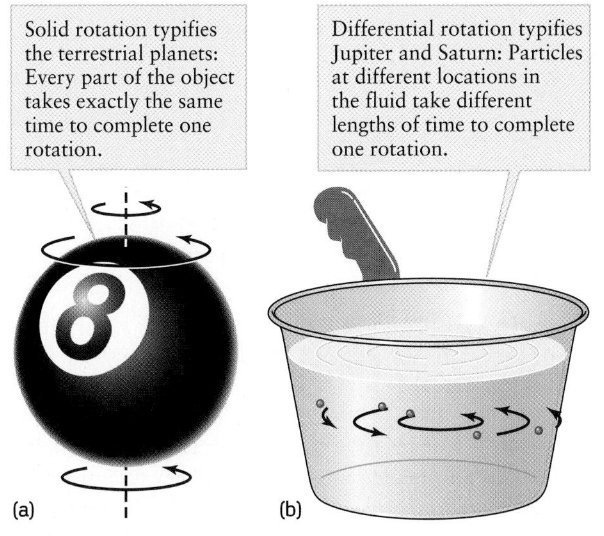

FIGURE 7-2 Solid Rotation Versus Differential Rotation (a) All parts of a solid object rotate together, but (b) a rotating fluid displays differential rotation. To see this, put some grains of sand, bread crumbs, or other small particles in a pot of water. Stir the water with a spoon to start it rotating, and then take out the spoon. The particles near the center of the pot take less time to make a complete rotation than do those away from the center.

ConceptCheck 7-1: Will Jupiter's rocky moons, such as Ganymede, exhibit differential rotation?

The Compositions of Jupiter and Saturn

If Jupiter and Saturn have partially fluid interiors, they cannot be made of the rocky materials that constitute the terrestrial planets. An important clue to the compositions of Jupiter and Saturn is their average densities, which are only 1326 kg/m^3 for Jupiter and 687 kg/m^3 for Saturn. (By comparison, Earth's average density is 5515 kg/m^3.) The only way to account for such giant planets having such low densities is to assume that Jupiter and Saturn are composed mostly of hydrogen and helium atoms—the two lightest elements in the universe—held together by their mutual gravitational attraction to form a planet. This chemical composition is quite similar to the chemical composition of our Sun, and of many distant stars as well.

After centuries of speculation, the presence of helium on Jupiter and Saturn was finally confirmed in the 1970s and 1980s, when spacecraft first flew past these planets and measured their hydrogen spectra in detail. Today we know that the chemical composition of Jupiter's atmosphere is 86.2% hydrogen molecules, 13.6% helium atoms, and 0.2% methane, ammonia, water vapor, and other gases. The percentages in terms of *mass* are somewhat different because a helium atom is twice as massive as a hydrogen molecule. Hence, by mass, Jupiter's atmosphere is approximately 75% hydrogen, 24% helium, and 1% other substances, quite similar to that of the Sun. It is estimated that the breakdown by mass of the planet as a whole (atmosphere plus interior) is approximately 71% hydrogen, 24% helium, and 5% heavier elements.

Data from Earth-based telescopes and spacecraft show that, unlike Jupiter, the atmosphere of Saturn has far less helium than astronomers initially suspected. Its chemical composition is 96.3% hydrogen molecules, 3.3% helium, and 0.4% other substances (by mass, 92% hydrogen, 6% helium, and 2% other substances). This is a puzzle because Jupiter and Saturn are thought to have formed in similar ways from the gases of the solar nebula, and so both planets (and the Sun) should have essentially the same abundances of hydrogen and helium. So where did Saturn's helium go?

The giants Jupiter and Saturn are composed of hydrogen and helium gas surrounding small, Earth-sized rocky cores.

Jupiter and Saturn are thought to have large rocky cores, but they are different. The explanation as to the surprising lack of helium may be simply that Saturn is slightly smaller than Jupiter, and as a result Saturn would have cooled more rapidly. This cooling would have triggered a process analogous to the way rain develops here on Earth. When the air is cool enough, humidity in Earth's atmosphere condenses into raindrops that fall to the ground. On Saturn, however, it is droplets of liquid helium that condense within the planet's cold, hydrogen-rich outer layers. In this scenario, helium is deficient in Saturn's upper atmosphere simply because it has fallen farther down into the planet. By contrast, Jupiter's helium has not rained out because its upper atmosphere is warmer and the helium does not form droplets.

ANALOGY An analogy to helium "rainfall" within Saturn is what happens when you try to sweeten tea by adding sugar. If the tea is cold, the sugar does not dissolve well and tends to sink to the bottom of the glass even if you stir the tea with a spoon. But if the tea is hot, the sugar dissolves with only a little stirring. In the same way, it is thought that the descending helium droplets once again dissolve in hydrogen when they reach the warmer depths of Saturn's interior.

In this scenario, Jupiter and Saturn both have about the same overall chemical composition. But Saturn's smaller mass, less than a third that of Jupiter, means that there is less gravitational force tending to compress its hydrogen and helium. This explains why Saturn's density is only about half that of Jupiter, and is in fact the lowest of any planet in the solar system.

CAUTION Because Jupiter and Saturn are almost entirely hydrogen and helium, it would be impossible to land a spacecraft on either planet (**Figure 7-3**). An astronaut foolish enough to try would notice the hydrogen and helium around the spacecraft becoming denser, the temperature rising, and the pressure increasing as the spacecraft descended. But the hydrogen and helium would never solidify into a surface on which the spacecraft could touch down. Long before reaching the planet's rocky core, the pressure of the hydrogen and helium would reach such unimaginably high levels that any spacecraft, even one made of the strongest known materials, would be crushed.

One might consider that developing this detailed an understanding is nothing short of amazing, considering that Jupiter and Saturn are never closer than 300 million mi (500 million km) and 600 million mi (1 billion km) away from Earth, respectively. We can learn even more when we study them at close range.

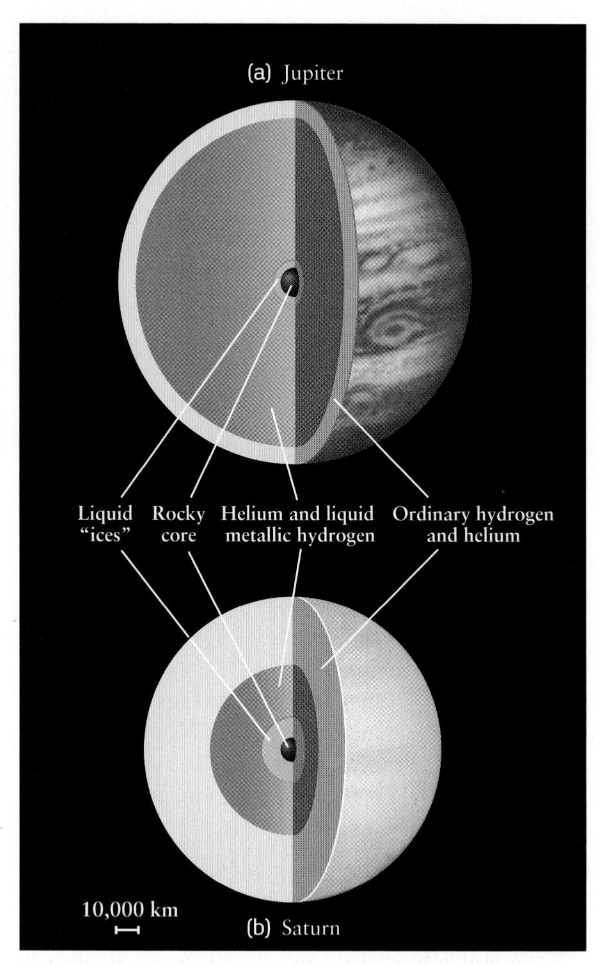

FIGURE 7-3 The Internal Structures of Jupiter and Saturn These diagrams of the interiors of Jupiter and Saturn are drawn to the same scale. Each planet's rocky core is surrounded by an outer core of liquid "ices," a layer of helium and liquid metallic hydrogen, and a layer of helium and ordinary molecular hydrogen (H_2). Saturn's rocky core contains a larger fraction of the planet's mass than Jupiter's core, while Saturn has a smaller relative volume of liquid metallic hydrogen than Jupiter.

ConceptCheck 7-2: If Jupiter and Saturn formed with nearly the same chemical composition, why might Jupiter be observed to have more helium than Saturn?

Robotic Spacecraft Discoveries of Jupiter and Saturn's Atmospheres

Most of our detailed understanding of Jupiter and Saturn comes from a series of robotic spacecraft that have examined these remarkable planets at close range. They found striking evidence of stable, large-scale weather patterns in both planets' atmospheres, as well as evidence of dynamic changes on smaller scales.

The first of several spacecraft to visit Jupiter and Saturn each made a single flyby of the planet. *Pioneer 10* flew past Jupiter in December 1973; it was followed a year later by

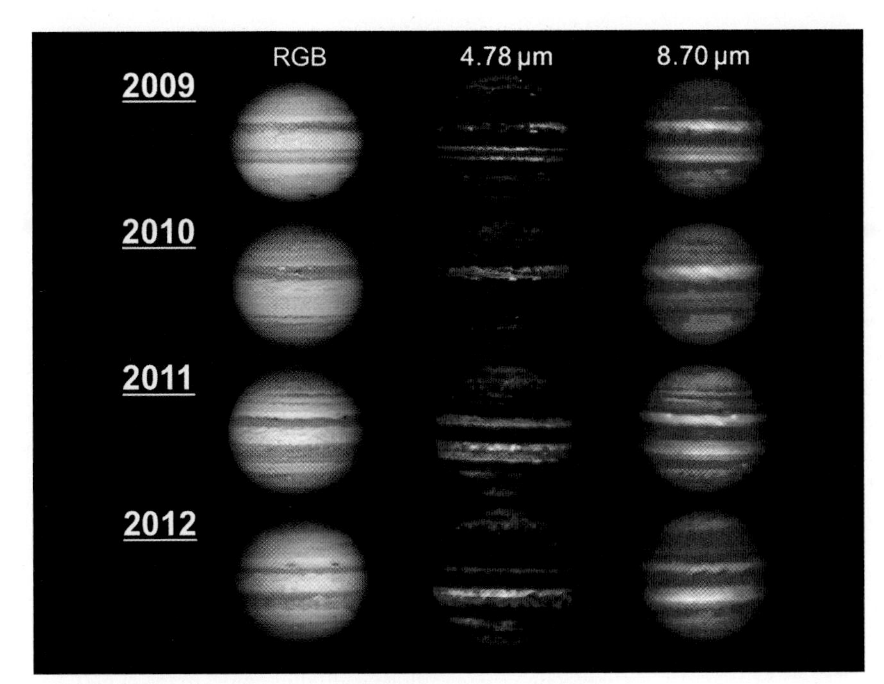

FIGURE 7-4 R I V U X G **Jupiter's Changing Appearance** The images on the left show Jupiter "losing" a southern brown color between 2009 and 2010 that returns in 2011. Just north of the equator, another belt thins and nearly fades from view before returning bright white in 2012. The infrared images in the center and on the right reveal areas that are the clearest and driest regions on the planet, showing up in infrared because they reveal the changing thermal energy emerging from a very deep layer of Jupiter's atmosphere. *(NASA/IRTF/JPL-Caltech/NAOJ/A. Wesley/A. Kazemoto/C. Go)*

the nearly identical *Pioneer 11,* which went on to make the first-ever flyby of Saturn in 1979. Also in 1979, another pair of spacecraft, *Voyager 1* and *Voyager 2,* sailed past Jupiter. These spacecraft sent back spectacular close-up color pictures of Jupiter's dynamic atmosphere. Both *Voyagers* subsequently flew past Saturn.

The first spacecraft to go into Jupiter orbit was *Galileo,* which carried out an extensive program of observations from 1995 to 2003. The *Cassini* spacecraft went into orbit around Saturn in 2004. A cooperative project of NASA, the European Space Agency, and the Italian Space Agency, this spacecraft is named for the astronomer Gian Domenico Cassini. On the way to its destination it viewed Jupiter at close range, recording detailed images such as Figure 7-1*a* and later the images of Saturn that open this chapter.

Over the years, astronomers carefully studying Jupiter from Earth, Earth-orbiting telescopes, and spacecraft have reported many long-term variations in the darkness and size of Jupiter's bands (**Figure 7-4**). Of particular interest are changes in Jupiter's notorious Great Red Spot, which measures 25,000 mi by 8500 mi (40,000 km by 14,000 km)—so large that three Earths could fit side by side across it. At other times (as in 1976 and 1977), the spot almost faded from view. During the *Voyager* flybys of 1979, the Great Red Spot was comparable in size to Earth (**Figure 7-5**).

Other persistent features in Jupiter's atmosphere are the white ovals. Several white ovals are visible in Figure 7-5. As in the Great Red Spot, wind flow in white ovals is counterclockwise. White ovals are also apparently long-lived; Earth-based observers have reported seeing them in the same location since 1938.

Most of the white ovals are observed in Jupiter's southern hemisphere, whereas brown ovals are more common in Jupiter's northern hemisphere. Brown ovals, most easily seen in the 2012 image, appear dark in visible light, but they appear bright in the infrared images shown along the right-hand side of Figure 7-5. For this reason, brown ovals are understood to be holes in Jupiter's cloud cover. They permit us to see into the depths of the Jovian atmosphere, where the temperature is higher and the atmosphere emits infrared light more strongly. White ovals, by contrast, have relatively low temperatures. They are areas with cold, high-altitude clouds that block our view of the lower levels of the atmosphere. What is perhaps even more curious is that these white ovals can merge and form even larger features.

ConceptCheck 7-3: **Besides color, what is the difference between Jupiter's white ovals and brown ovals?**

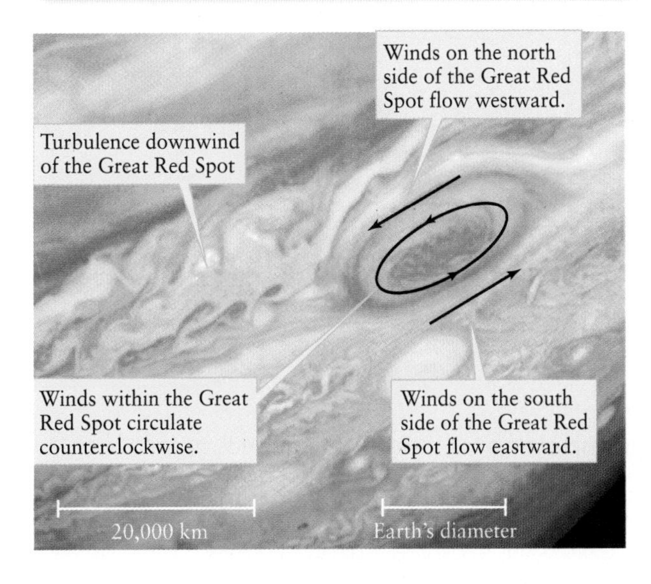

FIGURE 7-5 R I V U X G **Circulation Around the Great Red Spot** This image from *Voyager 2* shows atmospheric turbulence and the direction of winds around the Great Red Spot. Winds within the Great Red Spot itself make it spin counterclockwise, completing a full revolution in about a week. *(NASA/JPL; adapted from A. P. Ingersoll)*

Jupiter and Saturn's Cloud Layers

If white ovals obscure warmer cloud layers below, you might wonder just how many layers exist. From spectroscopic observations and calculations of atmospheric temperature and pressure, scientists conclude that both Jupiter and Saturn have three main cloud layers of differing chemical composition. The uppermost cloud layer is composed of crystals of frozen ammonia. Deeper in the atmosphere, ammonia (NH_3) and hydrogen sulfide (H_2S)—a compound of hydrogen and sulfur—combine to produce ammonium hydrosulfide (NH_4SH) crystals. At even greater depths, the clouds are composed of crystals of frozen water.

Jupiter's strong surface gravity compresses these cloud layers into a region just 50 mi (80 km) deep in the planet's upper atmosphere (**Figure 7-6*a***). But Saturn has a smaller mass and hence weaker surface gravity, so the atmosphere is less compressed and the same three cloud layers are spread out over a range of nearly 175 mi (280 km) (Figure 7-6*b*). The colors of Saturn's clouds are less dramatic than Jupiter's (see Figure 7-1*a*) because deeper cloud layers are partially obscured by the hazy atmosphere above them.

The colors of clouds on Jupiter and Saturn depend on the composition and temperatures of the clouds and, therefore, on the depth of the clouds within the atmosphere. Brown clouds are the warmest and are thus the deepest layers that we can see. Whitish clouds form the next layer up, followed by red clouds in the highest layer. The whitish zones on each planet are therefore somewhat higher than the brownish belts, while the red clouds in Jupiter's Great Red Spot are among the highest found anywhere on that planet.

By observing Jupiter's and Saturn's cloud layers from a distance, we have been able to learn a great deal about the atmospheres beneath the clouds. But there is no substitute for making measurements on site, and for many years scientists planned to send a spacecraft to explore deep into Jupiter's atmosphere. While still 50.66 million mi (81.52 million km) from Jupiter, the *Galileo* spacecraft released the *Galileo Probe,* a cone-shaped body about the size of an office desk. While *Galileo* itself later fired its rockets to place itself into an orbit around Jupiter, the *Galileo Probe* continued on a course that led it, on December 7, 1995, to a point in Jupiter's clouds just north of the planet's equator.

A heat shield protected the *Galileo Probe* as air friction slowed its descent speed from 106,000 mi/h (171,000 km/h) to 25 mi/h (40 km/h) in just 3 minutes. The spacecraft then deployed a parachute and floated down through the atmosphere (**Figure 7-7**). For the next hour, the probe observed its surroundings and radioed its findings back to the main *Galileo* spacecraft, which in turn radioed them to Earth. The mission ceased at a point some 120 mi (200 km) below Jupiter's upper cloud layer, where the tremendous pressure (24 atm)

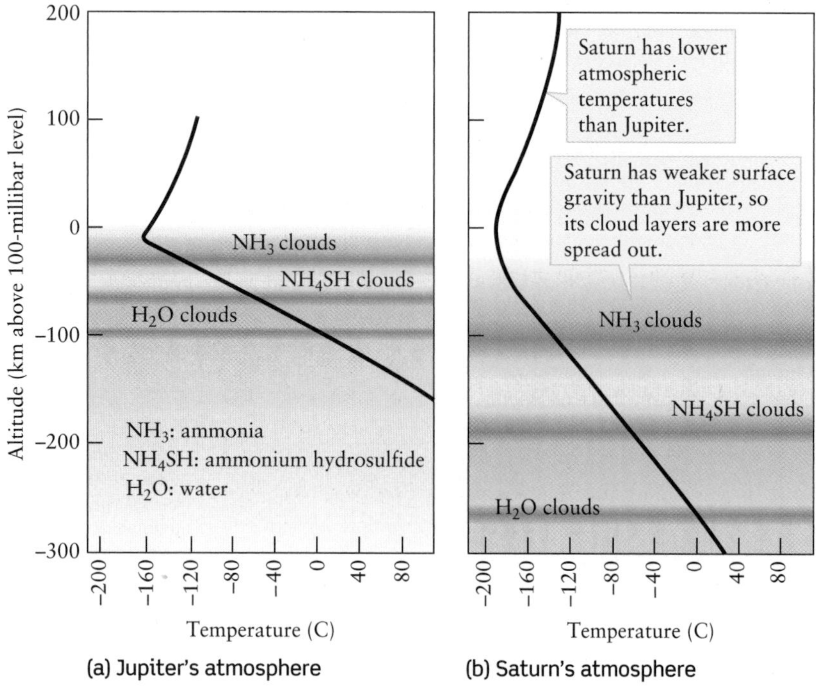

Saturn has lower atmospheric temperatures than Jupiter.

Saturn has weaker surface gravity than Jupiter, so its cloud layers are more spread out.

NH_3: ammonia
NH_4SH: ammonium hydrosulfide
H_2O: water

(a) Jupiter's atmosphere (b) Saturn's atmosphere

FIGURE 7-6 The Upper Atmospheres of Jupiter and Saturn The black curves in these graphs show temperature versus altitude in each atmosphere, as well as the probable arrangements of the cloud layers. Zero altitude in each atmosphere is chosen to be the point where the pressure is 100 millibars, or one-tenth of Earth's atmospheric pressure. Beneath both planets' cloud layers, the atmosphere is composed almost entirely of hydrogen and helium. *(Adapted from A. P. Ingersoll)*

FIGURE 7-7 The *Galileo Probe* Enters Jupiter's Atmosphere This artist's impression shows the *Galileo Probe* descending under a parachute through Jupiter's clouds. The jettisoned heat shield (shown below the probe) protected the probe during its initial high-speed entry into the atmosphere. The *Galileo Probe* returned data for 58 minutes before it was crushed and melted by the pressure and temperature of the atmosphere. *(C. Kallas/Jupiter Mission/NASA)*

and high temperature (305°F = 152°C) finally overwhelmed and crushed the probe's electronics.

Although the *Galileo Probe* did not have a camera, it did carry a variety of instruments that made several new discoveries. A radio-emissions detector found evidence for lightning discharges that, while less frequent than on Earth, are individually much stronger than lightning bolts in our atmosphere. Other measurements showed that Jupiter's winds, which are brisk in the atmosphere above the clouds, are even stronger beneath the clouds: The *Galileo Probe* measured a nearly constant wind speed of 400 mi/h (650 km/h) throughout its descent. This shows convincingly that the energy source for the winds is from Jupiter's own interior heat. If the winds were driven primarily by solar heating, as is the case on Earth, the wind speed would have decreased with increasing depth. A central purpose of the *Galileo Probe* mission was to test a three-layer cloud model depicted in Figure 7-6*a* by making direct measurements of Jupiter's clouds. But the probe saw only traces of the NH_3 and NH_4SH cloud layers and found no sign at all of the low-lying water clouds. One explanation of this surprising result is that the probe by sheer chance may have entered an unusually warm and cloud-free part of Jupiter's atmosphere. Indeed, observations from Earth showed strong infrared emission from the probe entry site, as would be expected where a break in the cloud cover allows a view of Jupiter's warm interior (**Figure 7-8**).

FIGURE 7-8 R I V U X G **Jupiter's Warm Belts and Cool Zones** The larger disk of Jupiter is visible in the background of this Hubble Space Telescope image, showing reddish belts and lighter zones. Pasted over the top is a ground-based infrared image showing bright areas demonstrating that thermal energy is emitted from deep levels of Jupiter's atmosphere. *(NASA/ESA)*

Another surprising and potentially important result from the *Galileo Probe* concerns three elements, argon (Ar), krypton (Kr), and xenon (Xe). These elements are relatively unique in that they do not combine with other atoms to form molecules. All three of these elements appear only in tiny amounts in Earth's atmosphere and in the Sun, and so they must also have been present in the original solar nebula. But the *Galileo Probe* found that argon, krypton, and xenon are about 3 times as abundant in Jupiter's atmosphere as in the Sun's. If these elements were incorporated into Jupiter directly from the gases of the solar nebula, they should be equally as abundant in Jupiter as in the Sun. So, the excess amounts of argon, krypton, and xenon must have entered Jupiter in the form of solid planetesimals, just as did carbon, nitrogen, and sulfur.

Jupiter and Saturn are extremely large planets orbiting far from other planets, but they are not the only gaseous planets in the outer solar system. As we look farther into the distant outer solar system, where temperatures are even lower being so far from our Sun, one wonders if the other gaseous planets are similar or different.

ConceptCheck 7-4: **Why might Jupiter's cloud layers be more compressed than Saturn's?**

7-2 Uranus and Neptune have seemingly quiet atmospheres

Even through a large, backyard telescope, both Uranus and Neptune are dim, mostly uninspiring sights. Each planet appears as a hazy, featureless disk with a faint greenish-blue tinge, with Uranus being a bit greener and Neptune being slightly bluer. Although Uranus and Neptune are both about 4 times larger in diameter than Earth, they are so distant that their angular dimensions as seen from Earth are tiny. To an Earth-based observer, Uranus is roughly the size of a golf ball seen at a distance of 1 mi (1.6 km), and Neptune, being 2 times farther away, appears to be about half that size.

From 2007 through 2011, Uranus and Neptune were less than 40° apart in the sky in the adjacent constellations of Pisces, Aquarius, and Capricornus (the Sea Goat). During these years, the two planets were at opposition in either August or September, which were thus the best months to view them with a telescope. How are these planets similar and how are they different?

Uranus's Atmosphere

Only one space probe has ever visited Uranus. Scientists had hoped that *Voyager 2* would reveal cloud patterns in Uranus's atmosphere when it flew past the planet in January 1986. But even images recorded at close range showed Uranus to be remarkably featureless (**Figure 7-9**). Faint cloud

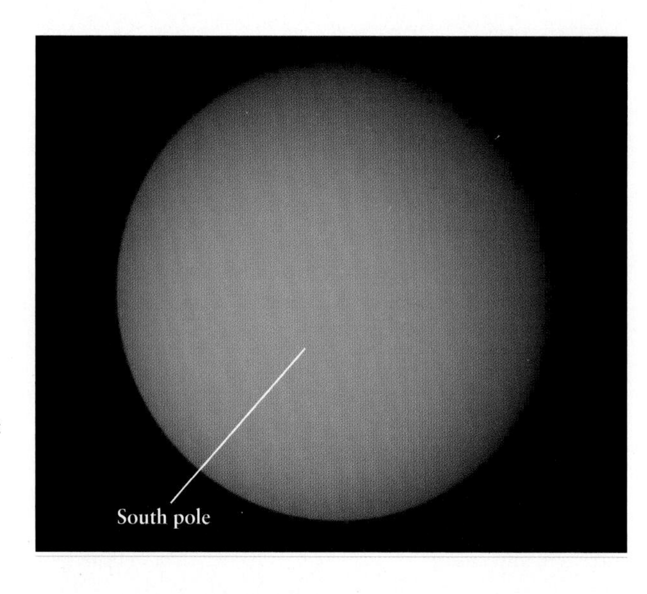

FIGURE 7-9 RIVUXG **Uranus from *Voyager 2*** This image looks nearly straight down onto Uranus's south pole, which was pointed almost directly at the Sun when *Voyager 2* flew past. None of the *Voyager 2* images of Uranus shows any pronounced cloud patterns. The color is due to methane in the planet's atmosphere, which absorbs red light but reflects green and blue. *(JPL/NASA)*

markings became visible in images of Uranus only after extreme computer enhancement (**Figure 7-10**).

Voyager 2 data confirmed that the Uranian atmosphere is dominated by hydrogen (82.5%) and helium (15.2%), similar to the atmospheres of Jupiter and Saturn. Uranus differs, however, in that 2.3% of its atmosphere is methane (CH_4), which is 5 to 10 times the percentages found on Jupiter and Saturn. In fact, Uranus has a higher percentage of all heavy elements—including carbon atoms, which are found in molecules of methane—than do Jupiter and Saturn.

Methane preferentially absorbs the longer wavelengths of visible light, so sunlight reflected from Uranus's upper atmosphere is depleted of its reds and yellows. This gives the planet its distinct greenish-blue appearance. Ultraviolet light from the Sun turns some of the methane gas into a hydrocarbon haze, making it difficult to see the lower levels of the atmosphere.

Uranus and Neptune receive so little of the Sun's energy, astronomers were surprised to see any weather patterns in their atmospheres.

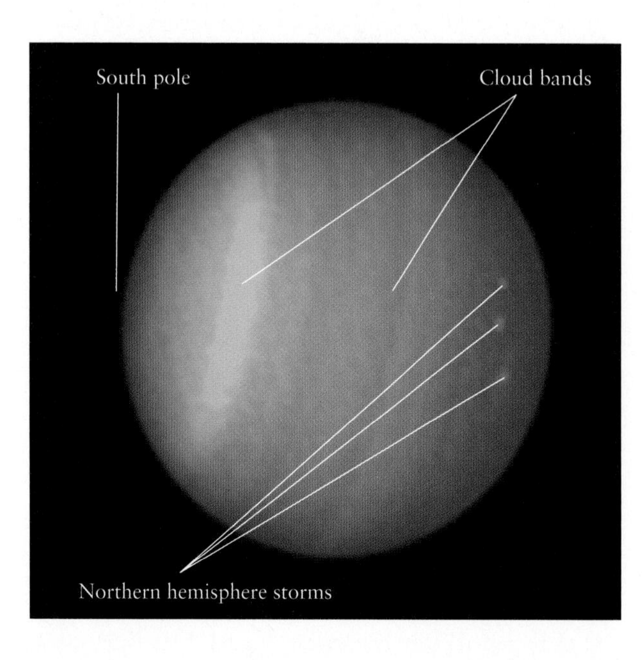

FIGURE 7-10 RIVUXG **Uranus from the Hubble Space Telescope** Images made at ultraviolet, visible, and infrared wavelengths were combined and enhanced to give this false-color view of cloud features on Uranus. These images were captured in August 2004 (18 years after the image in Figure 7-9), during springtime in Uranus's northern hemisphere. *(NASA and Erich Karkoschka, University of Arizona)*

Ammonia (NH_3), which constitutes 0.01% to 0.03% of the atmospheres of Jupiter and Saturn, is almost completely absent from the Uranian atmosphere. The reason is that Uranus is colder than Jupiter or Saturn: The temperature in its upper atmosphere is only −360°F (−218°C = 55 K). Ammonia freezes at these very low temperatures, so any ammonia has long since precipitated out of the atmosphere and into the planet's interior. For the same reason, Uranus's atmosphere is also lacking in water. Hence, the substances that make up the clouds on Jupiter and Saturn—ammonia, ammonium hydrosulfide (NH_4SH), and water—are not available on Uranus. This helps to explain the bland, uniform appearance of the planet shown in Figure 7-9.

The few clouds found on Uranus are made primarily of methane, which condenses into droplets only if the pressure is sufficiently high. Because methane clouds form only at lower levels within the atmosphere, they are difficult for us to see.

By following the motions of clouds and storm systems on Uranus, scientists find that the planet's winds flow to the east—that is, in the same direction as the planet's rotation—at northern and southern latitudes, but to the west near the equator. This is quite unlike the situation on Jupiter and Saturn, where the zonal winds alternate direction many times between the north and south poles. The fastest Uranian winds (about 440 mi/h, or 700 km/h) are found at the equator.

Although Uranus's equatorial region was receiving little sunlight at the time of the *Voyager 2* flyby, the atmospheric temperature there (about −359°F = −218°C = 55 K) was not too different from that at the sunlit pole. Heat must therefore be efficiently transported from the poles to the equator. This north-south heat transport, which is perpendicular to the wind flow, may have mixed and homogenized the atmosphere to make Uranus nearly featureless.

ConceptCheck 7-5: **If methane were absent from Uranus's atmosphere, what color would it appear?**

Neptune Is a Cold, Bluish World with Jupiterlike Atmospheric Features

At first glance, Neptune appears to be the twin of Uranus, but these two planets are by no means identical. While Neptune and Uranus have almost the same diameter, Neptune is 18% more massive. As with Uranus, only one space probe has ever visited Neptune. When *Voyager 2* flew past Neptune in August 1989, it revealed that the planet has a more active and dynamic atmosphere than Uranus. This activity suggests that Neptune, unlike Uranus, has a powerful source of energy in its interior.

The *Voyager 2* data showed that Neptune has essentially the same atmospheric composition as Uranus: 79% hydrogen, 18% helium, 3% methane, and almost no ammonia or water vapor. As for Uranus, the presence of methane gives

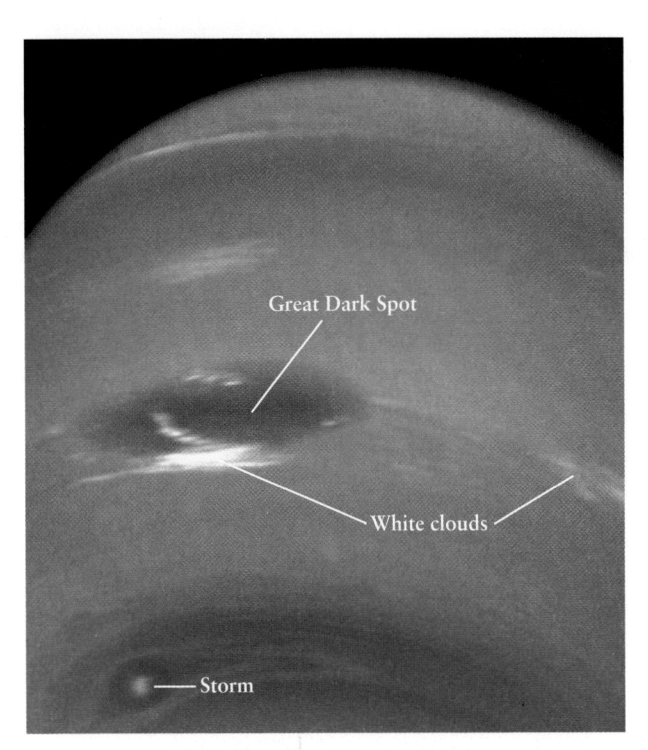

FIGURE 7-11 R I V U X G **Neptune from *Voyager 2*** When this picture of Neptune's southern hemisphere was taken in 1989, the Great Dark Spot measured about 7500 mi by 5000 mi (12,000 km by 8000 km) comparable in size to Earth. (A smaller storm appears at the lower left.) The white clouds are thought to be composed of crystals of methane ice. The color contrast in this image has been exaggerated to emphasize the differences between dark and light areas. *(NASA/JPL)*

FIGURE 7-12 R I V U X G **Cirrus Clouds over Neptune** *Voyager 2* recorded this image of clouds near Neptune's terminator (the border between day and night on the planet). Like wispy, high-altitude cirrus clouds in Earth's atmosphere, these clouds are thought to be made of ice crystals. The difference is that Neptune's cirrus clouds are probably methane ice, not water ice. *(NASA/JPL)*

Neptune a characteristic bluish-green color. The temperature in the upper atmosphere is also the same as on Uranus, about 55 K. That this should be so, even though Neptune is much farther from the Sun, is further evidence that Neptune has a strong internal heat source.

Unlike Uranus, however, Neptune has clearly visible cloud patterns in its atmosphere. At the time that *Voyager 2* flew past Neptune, the most prominent feature in the planet's atmosphere was a giant storm called the Great Dark Spot (**Figure 7-11**). The Great Dark Spot had a number of similarities to Jupiter's Great Red Spot. The storms on both planets were comparable in size to Earth's diameter, both appeared at about the same latitude in the southern hemisphere, and the winds in both storms circulated in a counterclockwise direction (see Figure 7-5). But Neptune's Great Dark Spot appears not to have been as long-lived as the Great Red Spot on Jupiter. When the Hubble Space Telescope first viewed Neptune in 1994, the Great Dark Spot had disappeared. Another dark storm appeared in 1995 in Neptune's northern hemisphere.

Voyager 2 also saw a few conspicuous whitish clouds on Neptune. These clouds are thought to be produced when winds carry methane gas into the cool, upper atmosphere, where it condenses into crystals of methane ice. *Voyager 2* images show these high-altitude clouds casting shadows onto lower levels of Neptune's atmosphere (**Figure 7-12**). Images from the Hubble Space Telescope also show the presence of high-altitude clouds (**Figure 7-13**).

Thanks to its greater distance from the Sun, Neptune receives less than half as much energy from the Sun as Uranus. But with less solar energy available to power atmospheric motions, why are there high-altitude clouds and huge, dark storms on Neptune but not on Uranus? At least part of the answer is probably that Neptune is still slowly contracting, thus converting gravitational energy into thermal energy that heats the planet's core. (The same is true for Jupiter and Saturn.) The evidence for this is that Neptune, like Jupiter and Saturn but unlike Uranus, emits more energy than it receives from the Sun. The combination of a warm interior and a

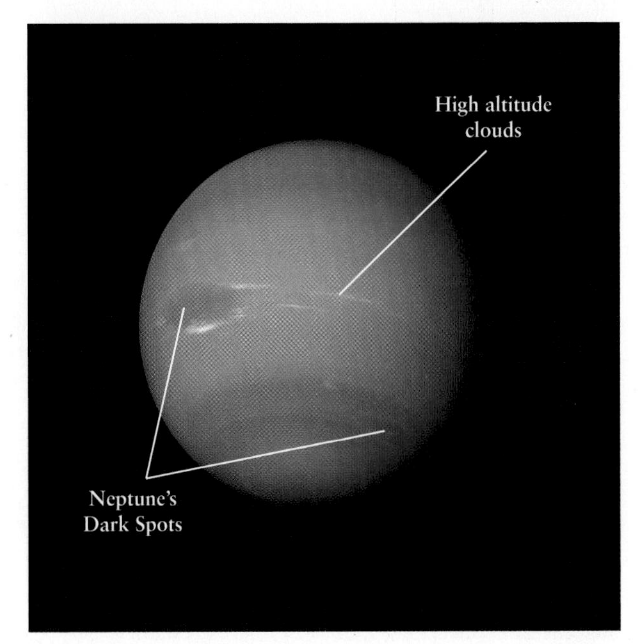

FIGURE 7-13 R I V U X G **Ice Giant Neptune** The ice-giant planets Uranus and Neptune have considerably less hydrogen and more ice than do their much larger counterparts of Jupiter and Saturn. This *Voyager 2* image shows Neptune's two distinctive large dark ovals, dark belts of descending gas, and lighter zones of rising gas. The white areas near the equator are the highest-altitude clouds. *(NASA/JPL)*

cold outer atmosphere can cause convection in Neptune's atmosphere, producing the up-and-down motion of gases that generates clouds and storms. Neptune also resembles Jupiter in having faint belts and zones parallel to the planet's equator (see Figure 7-13).

Like those on Uranus, most of Neptune's clouds are probably made of droplets of liquid methane. Because these droplets form fairly deep within the atmosphere, the clouds are more difficult to see than the ones on Jupiter. Hence, Neptune's belts and zones are less pronounced than Jupiter's, although more so than those on Uranus (thanks to the extra cloud-building energy from Neptune's interior). As described on the previous page, Neptune's high-altitude clouds (see Figure 7-12) are probably made of *frozen* methane.

Go to Video 7-2

Figure 7-14 shows one model for the present-day internal structures of Uranus and Neptune. In this model each planet has a rocky core roughly the size of Earth, although more massive. Each planet's core is surrounded by a mantle of liquid water and ammonia. (This means that the mantle is chemically similar to household window cleaning fluid.) Around the mantle is a layer of liquid molecular hydrogen and liquid helium, with a small percentage of liquid methane. This layer is relatively shallow compared to those on Jupiter and Saturn, and the pressure is not high enough to convert the liquid hydrogen into liquid metallic hydrogen. The figures in **Cosmic Connections: The Outer Planets: A Comparison** summarize the key properties of Uranus and Neptune and how they compare with those of Jupiter and Saturn.

ConceptCheck 7-6: Why would scientists be surprised that Neptune had active belts and zones, unlike Uranus?

7-3 Saturn's moon Titan and Neptune's moon Triton exhibit unexpected atmospheres

Each time we send spacecraft to obtain close-up views of distant planets, we find perhaps even more intriguing and unexpected observations of the unusual moons orbiting our gas giant planets. There are more than 100 of these orbiting objects, each one significantly different from the next (see Table 4.2). We discussed some of the dynamic geological processes shaping the surfaces of these objects in the last chapter. Now that we have discussed the nature of their larger host planets, we are in a position to explore the nature of their atmospheres. We will limit our discussion to focus on the two objects with the most interesting atmospheres.

Saturn's Moon Titan

Unlike Jupiter, Saturn has only one large satellite that is comparable in size to our Moon. This satellite, Titan, has a diameter of 3200 mi (5150 km), which makes it second largest in size among all moons in the solar system. Titan has a low density that suggests it is made of a mixture of ice and rock. But unlike Ganymede or any other satellite in the solar system, Titan has a thick atmosphere with a unique chemical composition.

On Earth water can be a gas (like water vapor in the atmosphere), a liquid (as in the oceans), or a solid (as in the polar ice caps). On Titan, the atmospheric pressure is so high and the surface temperature is so low—about −288°F (−178°C or 95 K)—that any water is frozen solid. But conditions on Titan are just right for methane and ethane to exist as gases, liquids, or solids. This raises the tantalizing possibility that Titan could have methane rain and lakes of liquid ethane.

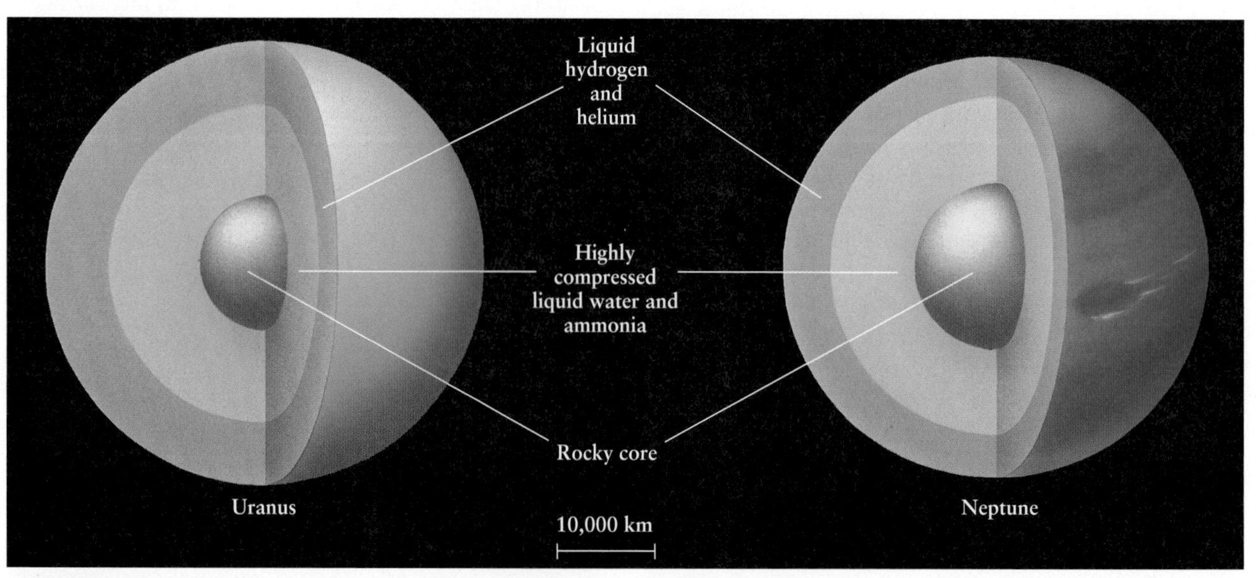

FIGURE 7-14
The Internal Structures of Uranus and Neptune Uranus and Neptune each have a rocky core, resembling a terrestrial planet and probably about the size of planet Earth. A mantle of liquid water with ammonia dissolved in it surrounds the core, and an outer layer of liquid molecular hydrogen and liquid helium surrounds the mantle. Above this, the atmospheres of Uranus and Neptune are very thin shells on top of the hydrogen-helium layers. Uranus and Neptune have about the same diameter, but Neptune is more massive, suggesting it might have a somewhat larger core.

COSMIC CONNECTIONS The Outer Planets: A Comparison

Uranus and Neptune are not simply smaller versions of Jupiter and Saturn. This table summarizes the key differences among the four Jovian planets.

	Interior	Surface	Rings	Atmosphere
Jupiter	Terrestrial core, liquid metallic hydrogen shell, liquid hydrogen mantle	No solid surface, atmosphere gradually thickens to liquid state, belt and zone structure, hurricanelike features	Yes	Primarily H, He
Saturn	Similar to Jupiter, with bigger terrestrial core and less metallic hydrogen	No solid surface, less distinct belt and zone structure than Jupiter	Yes	Primarily H, He
Uranus	Terrestrial core, liquid water shell, liquid hydrogen and helium mantle	No solid surface, weak belt and zone system, hurricanelike features, color from methane absorption of red, orange, yellow	Yes	Primarily H, He, some CH_4
Neptune	Similar to Uranus	Like Uranus	Yes	Primarily H, He, some CH_4

For detailed comparisons between planets, see Appendices 1 and 2.

Nitrogen in Titan's atmosphere can combine with airborne hydrocarbons to produce other compounds, such as hydrogen cyanide (HCN). Hydrogen cyanide, along with other molecules, can join together in long, repeating molecular chains that, when suspended in the atmosphere, are thought to be responsible for the reddish-brown color seen when it was first observed.

The designers of the *Cassini* mission to Saturn made the exploration of Titan a central goal. During its primary mission in Saturn orbit from 2004 to 2008, the *Cassini* spacecraft made 44 close flybys of Titan. Titan's atmosphere is more transparent to infrared wavelengths than to visible light, so *Cassini* used its infrared telescope to obtain the first detailed images of the satellite's surface (**Figure 7-15*a***). Most of the surface is light-colored, but a swath of dark terrain extends around most of Titan's equator.

Cassini has used its on-board radar to map the dark terrain. (Like infrared light, radio waves can pass through Titan's atmosphere.) These images reveal a series of long, parallel lines of sand dunes about a mile (0.6 km) apart. These are aligned west to east, in the same direction that winds blow at Titan's equator, and so presumably formed by wind action. Unlike Earth sand, which is made of small particles of silicate rock, sand on Titan is probably small particles of water ice combined with molecules that have fallen from Titan's atmosphere, which give the dark terrain its color.

The most remarkable images of Titan's surface have come from *Huygens,* a small lander that *Cassini* carried on its journey from Earth and released before entering Saturn orbit. Named for the discoverer of Titan, *Huygens* entered Titan's atmosphere on January 14, 2005. The lander took 2½ hours to descend to the surface under a parachute, during which time it made detailed images of the terrain (Figure 7-15*b*). After touchdown, *Huygens* continued to return data for another 70 minutes (Figure 7-15*c*) before its batteries succumbed to Titan's low temperatures.

Huygens images such as Figure 7-15*b* show that liquids have indeed flowed on Titan like water on Earth. Yet none of the images made by the lander during its descent or after touchdown showed any evidence of standing liquid. There was nonetheless evidence of recent methane rainfall at the *Huygens* landing site. A lander instrument designed to measure atmospheric composition was equipped with a heater for its gas inlet. During the 3 minutes it took to warm this inlet, the measured methane concentration jumped by 30% and then remained steady. The explanation is that the ground was soaked with liquid methane that had fallen as rain and evaporated as it was warmed by the heater. *Huygens* measurements indicate that the average annual methane rainfall on Titan is about 2 inches (5 cm), about the same as the amount of water rain that falls each year at Death Valley in California. The rivers and outflow channels that *Huygens* observed during its descent may date from an earlier period in Titan's history when methane was more abundant and rainfall was more intense. What is curious is that within the next several hundred million years methane outgassing from the surface of Titan will cease for good. In time the hydrocarbon haze will disappear, the skies will clear, and the surface of Titan will be revealed to whichever outside observers may gaze upon it in that distant era.

ConceptCheck 7-7: Prior to the *Huygens* probe, why was so little known about the surface of Titan?

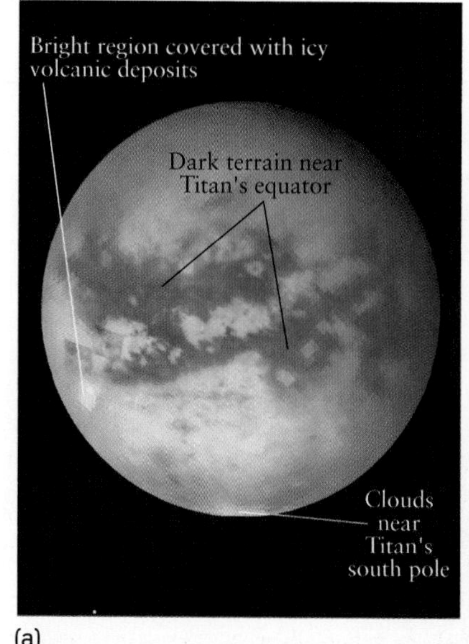

(a)

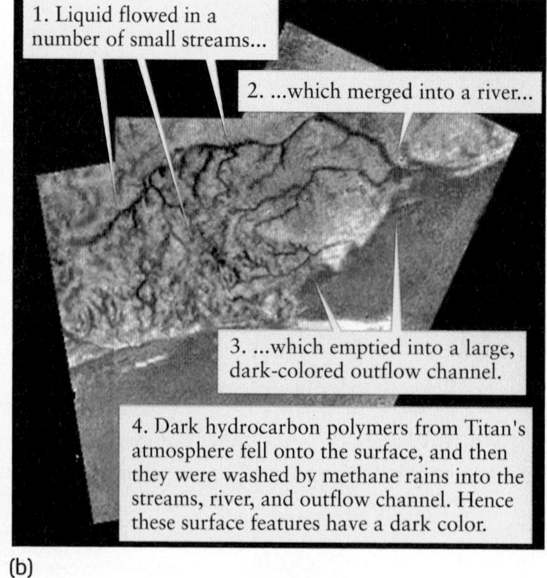

(b)

(c)

FIGURE 7-15 R I V U X G **Beneath Titan's Clouds**
(a) This mosaic of Titan was constructed from *Cassini* images made at visible and infrared wavelengths during a close flyby in December 2005. (b) The *Huygens* lander recorded this view of features on Titan carved by flowing liquid. It is too cold on Titan for water to be a liquid, so these streams and river must have carried liquid methane or ethane. (c) The nearly true-color *Huygens* image shows the view from Titan's surface. The 6 in. (15 cm) wide "rock" is about 33 in. (85 cm) from the camera. Unlike Earth rocks, Titan "rocks" are chunks of water ice. *(NASA/JPL/Space Science Institute)*

Neptune's Moon Triton

Thirteen moons have been discovered so far orbiting the planet Neptune, which is twice as far from the Sun as Uranus and extremely difficult to observe. The moons are named for mythological beings related to bodies of water. (Neptune itself is named for the Roman god of the sea.) Most of these worlds are small, icy bodies, probably similar to the smaller satellites of Uranus. The one striking exception is Triton, Neptune's largest satellite. In many ways Triton is quite unlike any other world in the solar system. One difference is that it harbors a thin atmosphere, making it the only satellite besides Titan to have one.

Figure 7-16 shows the icy, reflective surface of Triton as imaged by *Voyager 2* as it rapidly passed by. There is a conspicuous absence of large craters, which immediately tells us that Triton has a young surface on which the scars of ancient impacts have largely been erased by tectonic activity. There are areas that resemble frozen lakes and may be the calderas of extinct ice volcanoes. In the upper portion of Figure 7-16 you can see dimpled, wrinkled terrain that resembles the skin of a cantaloupe.

There may still be warmth in Triton's interior today. *Voyager 2* observed plumes of dark material being ejected from the surface to a height of 5 mi (8 km). These plumes may have been generated from a hot spot far below Triton's surface, similar to geysers on Earth. Alternatively, the energy source for the plumes may be sunlight that warms the surface, producing subsurface pockets of gas and creating fissures in the icy surface through which the gas can escape.

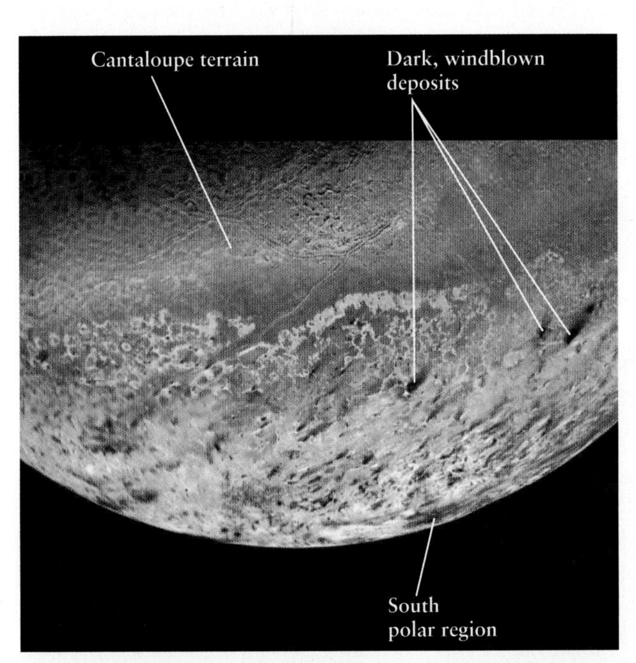

FIGURE 7-16 R I V U X G **Triton** Several high-resolution *Voyager 2* images were combined to create this mosaic. The pinkish material surrounding Triton's south polar region is probably nitrogen frost. Some of this presumably evaporates when summer comes to the south pole; the northward flow of the evaporated gas may cause the dark surface markings. Farther north is a brown area of "cantaloupe terrain" that resembles the skin of a melon. *(NASA/JPL)*

Triton's surface temperature is only −391°F (−235°C = 38 K), the lowest of any world yet visited by spacecraft. This temperature is low enough to solidify nitrogen, and indeed the spectrum of sunlight reflected from Triton's surface shows absorption lines due to nitrogen ice as well as methane ice. But Triton is also warm enough to allow some nitrogen to evaporate from the surface, like the steam that rises from ice cubes when you first take them out of the freezer. *Voyager 2* confirmed that Triton has a very thin nitrogen atmosphere with a surface pressure of only 1.6×10^{-5} atm, about the same as at an altitude of 60 mi (100 km) above Earth's surface. Despite its thinness, Triton's atmosphere has noticeable effects. The first evidence of an atmosphere on Triton came when *Voyager 2* saw streaked areas on Triton's surface where dark material had been blown downwind by a steady breeze (see Figure 7-16). It also observed dark material ejected from the geyserlike plumes being carried as far as 100 mi (160 km) by high-altitude winds.

ConceptCheck 7-8: How was an atmosphere detected on Triton?

7-4 All Jovian planet atmospheres are encircled by complex ring systems

When looking at images of Jovian planets showing swirling atmospheres and at striking spacecraft data that reveal the nature of their moons, it might be easy to forget one of the most fascinating and poorly understood characteristics of these outer planets. Even through a small telescope, Saturn stands out as perhaps the most beautiful of all the worlds of the solar system, thanks to its majestic rings (**Figure 7-17**). In this section, we explore the mysterious ring systems found around the outer planets.

Discovering Saturn's Rings

In 1610, Galileo became the first person to report seeing Saturn through a telescope. He saw few details, but he did notice two puzzling lumps protruding from opposite edges of the planet's disk. Curiously, these lumps disappeared in 1612, only to reappear in 1613. Other observers saw similar appearances and disappearances over the next several decades. In 1655, the Dutch astronomer Christiaan Huygens began to observe Saturn with a better telescope than had been available to any of his predecessors. On the basis of his observations, Huygens suggested that Saturn was surrounded by a thin, flattened ring. At times this ring was edge-on as viewed from Earth, making it almost impossible to see. At other times Earth observers viewed Saturn from an angle either above or below the plane of the ring, and the ring was visible, as in Figure 7-1*b*. (The lumps that Galileo saw were the parts of the ring to either side of Saturn, blurred by the poor resolution of his rather small telescope.) Astronomers confirmed this brilliant deduction over the next several years

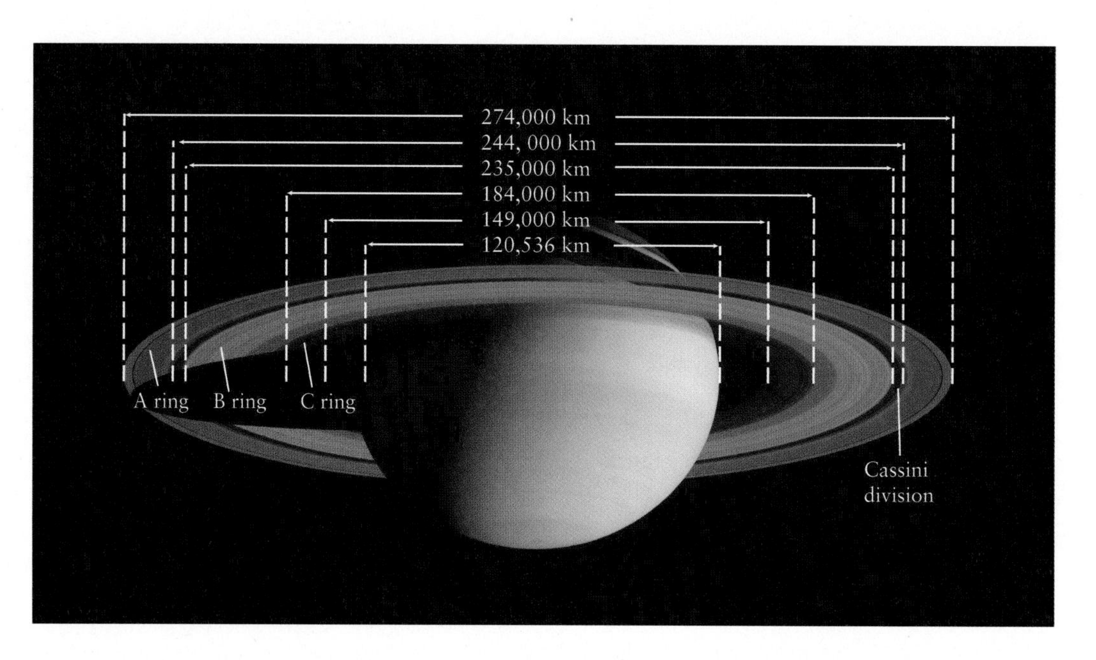

FIGURE 7-17 RIVUXG **Saturn's System of Rings** This *Cassini* image shows many details of Saturn's rings. Saturn's equatorial diameter is labeled, as are the diameters of the inner and outer edges of the rings. The C ring is so faint that it is almost invisible in this view. Saturn is visible through the thinnest C rings, which shows that the rings are not solid. *(NASA/JPL/Space Science Institute)*

as they watched the ring's appearance change, just as Huygens had predicted.

Earth-based views of the Saturnian ring system change as Saturn slowly orbits the Sun. Huygens was the first to understand that this change occurs because the rings lie in the plane of Saturn's equator, and this plane is tilted 27° from the plane of Saturn's orbit. As Saturn orbits the Sun, its rotation axis and the plane of its equator maintain the same orientation in space. Hence, over the course of a Saturnian year, an Earth-based observer views the rings from various angles (**Figure 7-18**). At certain times Saturn's north pole is tilted toward Earth and the observer looks "down" on the "top side" of the rings. Half a Saturnian year later, Saturn's south pole is tilted toward us and the "underside" of the rings is exposed to our Earth-based view.

When our line of sight to Saturn is in the plane of the rings, the rings are viewed edge-on and seem to disappear (see the images at the upper left and lower right corners of Figure 7-18). This disappearance indicates that the rings are very thin. In fact, they are thought to be only a few tens of meters thick. In proportion to their diameter, Saturn's rings are thousands of times thinner than the sheets of paper used to print this book.

As the quality of telescopes improved, astronomers realized that Saturn's "ring" is actually a system of rings, as **Figure 7-19** shows. In 1675, the Italian astronomer Cassini discovered a dark division in the ring. The **Cassini division** is an apparent gap about 4500 kilometers wide that separates the outer A ring from the brighter B ring closer to the planet. In the mid-1800s, astronomers managed to identify the faint C ring, or crepe ring, that lies just inside the B ring.

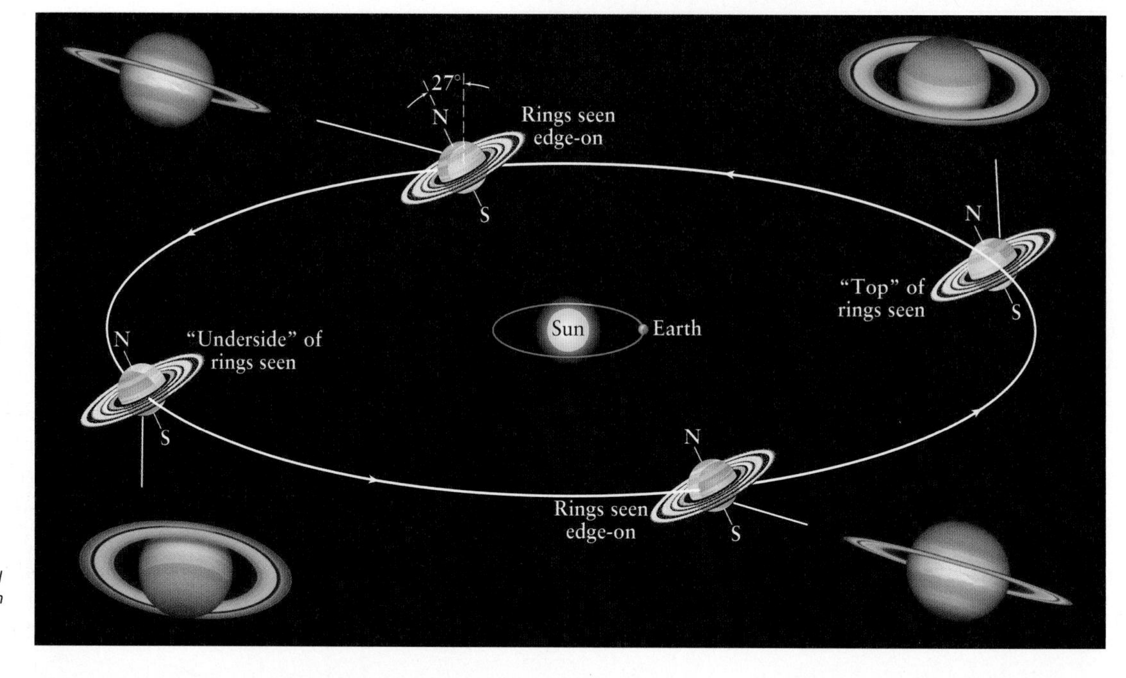

FIGURE 7-18 RIVUXG **The Changing Appearance of Saturn's Rings** Saturn's rings are aligned with its equator, which is tilted 27° from Saturn's orbital plane. As Saturn moves around its orbit, the rings maintain the same orientation in space, so observers on Earth see the rings at various angles. The accompanying Hubble Space Telescope photographs show Saturn at various points in its orbit. Note that the rings seem to disappear entirely when viewed edge-on, which occurs about every 15 years. *(NASA and The Hubble Heritage Team/STScI/AURA/R. G. French [Wellesley College], J. Cuzzi [NASA/Ames], L. Dones [SwRI], and J. Lissauer [NASA/Ames])*

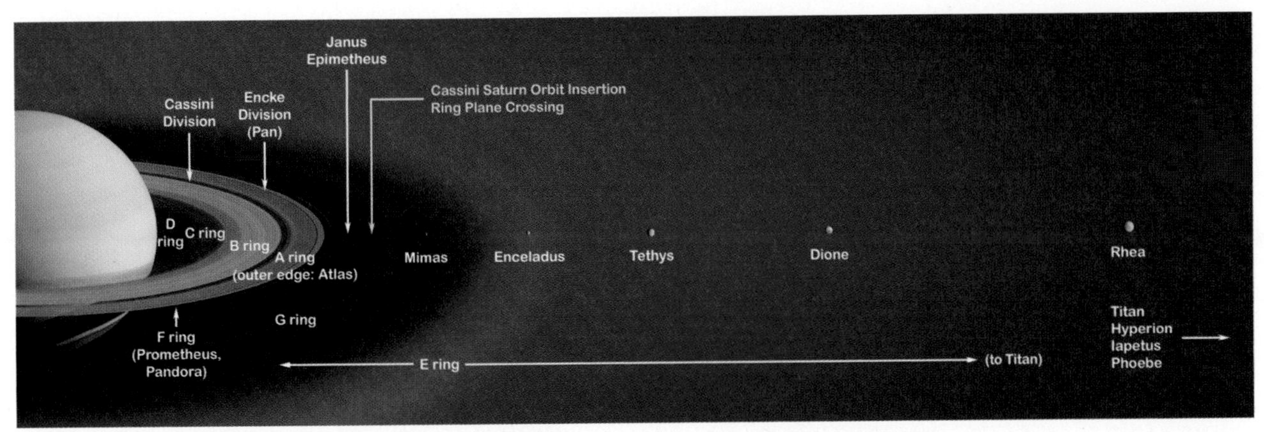

FIGURE 7-19 R I V U X G **Details of Saturn's Rings and Orbiting Moons** From edge to edge, Saturn's enormous ring system would not even fit in the distance between Earth and the Moon. Composed primarily of water ice, the seven main rings are labeled in the order in which they were discovered. From the planet outward, the main rings are labeled, in the order of discovery, D, C, B, A, F, G, and E. The Cassini division is the largest gap in the rings and separates the B ring from the A ring. Just outside the A ring is the narrow F ring, shepherded by the tiny moons Pandora and Prometheus. One moon, Pan, actually orbits inside the A ring in the Encke Gap. Beyond that are two much fainter G and E rings, which extend from Mimas's orbit to Titan's orbit. *(NASA/JPL)*

ConceptCheck 7-9: What causes Saturn's brilliant rings to be sometimes nearly invisible and at other times easily observable with the smallest of telescopes?

Nature of Rings

Saturn's rings are made of an indefinite number of unconnected particles or "moonlets," or **ring particles,** each individually circling Saturn. Saturn's rings are quite bright: They reflect 80% of the sunlight that falls on them. (By comparison, Saturn itself reflects 46% of incoming sunlight.) Astronomers have long suspected that the ring particles are made of ice and ice-coated rock. *Voyager* and *Cassini* spacecraft have made even more detailed infrared measurements, which indicate that the temperature of the rings ranges from −290°F (−180°C) in the sunshine to less than −330°F (−200°C) in Saturn's shadow. Water ice is in no danger of melting or evaporating at these temperatures.

To determine the sizes of the particles that make up Saturn's rings, astronomers analyzed the radio signals received from a spacecraft as it passed behind the rings. How easily radio waves can travel through the rings depends on the relationship between the wavelength and the particle size. The results showed that most of the particles ranged in size from pebble-sized fragments about 0.5 in. (1.25 cm) in diameter to chunks about 5 m across, the size of large boulders. Most abundant were snowball-sized particles about 4 in. (10 cm) in diameter.

All of the material in Saturn's rings may be ancient debris that failed to accrete (fall together) into moons or, more catastrophically, objects that were pulled apart when they came too close to Saturn. The total amount of material in the rings is quite small. If Saturn's entire ring system were compressed together to make a single moon, it would be no more than 60 mi (100 km) in diameter. But, in fact, the ring particles are so close to the gravitational churning of Saturn that they will never be able to form moons.

To see why, imagine a collection of small particles orbiting a planet. Gravitational attraction between neighboring particles tends to pull the particles together. However, because the various particles are at differing distances from the parent planet, they also experience different amounts of gravitational pull from the planet. This difference in gravitational pull is a tidal force that tends to keep the particles separated.

The closer a pair of particles is to the planet, the greater the tidal force that tries to pull the pair apart. At a certain distance from the planet's center, called the **Roche limit,** the disruptive tidal force is just as strong as the gravitational force between the particles. (The concept of this limit was developed in the mid-1800s by the French mathematician Édouard Roche.) Inside the Roche limit, the tidal force overwhelms the gravitational pull between neighboring particles, and these particles cannot accrete to form a larger body. Instead, they tend to spread out into a ring around the planet. Indeed, most of Saturn's system of rings visible in Figure 7-19 lies within the planet's Roche limit. All large planetary satellites are found outside their planet's Roche limit. If any large satellite were to come inside its planet's Roche limit, the planet's tidal forces would cause the satellite to break up into fragments.

CAUTION It may seem that it would be impossible for any object to hold together inside a planet's Roche limit. But the ring particles inside Saturn's Roche limit survive and do not break apart. The reason is that the Roche limit applies only to objects held together by the gravitational attraction of each part of the object for the other parts. By contrast, the forces that hold a rock or a ball of ice together are chemical bonds between the object's atoms and molecules. These chemical forces are much stronger than the disruptive tidal force of a nearby planet, so the rock or ball of ice does not break apart. In the same way, people walking around on Earth's surface (which is inside Earth's Roche limit) are in no danger of coming apart, because we are held together by comparatively strong chemical forces rather than gravity.

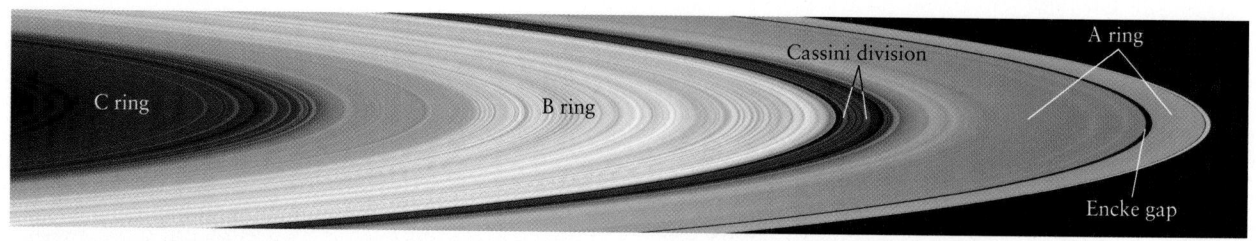

FIGURE 7-20 R I V U X G **Color Variations in Saturn's Rings** This mosaic of *Cassini* images shows the rings in natural color. The color variations are indicative of slight differences in chemical composition among particles in different parts of the rings. *(NASA/JPL/Space Science Institute)*

Pioneer 11, Voyager 1, and *Voyager 2* recorded images of Saturn's rings during their flybys in 1979, 1980, and 1981, respectively. *Pioneer 11* had only a relatively limited capability to make images, but cameras on board the two *Voyager* spacecraft sent back a number of pictures showing the detailed structure of Saturn's rings. Since entering Saturn orbit in 2004, the *Cassini* spacecraft has provided scientists with even higher-resolution images of the rings.

Figure 7-20 shows a *closer view of the rings.* To the amazement of scientists, images like the one in Figure 7-20 revealed that the broad A, B, and C rings are not uniform at all but instead consist of hundreds upon hundreds of closely spaced bands or ringlets.

Spacecraft have done more than show Saturn's rings in greater detail; they have also viewed the rings from perspectives not possible from Earth. Through Earth-based telescopes, we can see only the sunlit side of the rings. From this perspective, the B ring appears very bright, the A ring moderately bright, the C ring dim, and the Cassini division dark (see Figure 7-20). The fraction of sunlight reflected back toward the Sun is directly related to the concentration and size of the particles in the ring. The B ring is bright because it has a high concentration of relatively large, icy, reflective particles, whereas the darker Cassini division has a lower concentration of such particles.

The *Voyagers* and *Cassini* have expanded our knowledge of the ring particles by taking high-resolution images from nearby distances. Some areas of the rings cannot be seen from Earth because they are in the shadow cast by Saturn itself.

Although moons and ring systems are common around gas giants, moons with atmospheres are rare.

One strategy to study the rings is to turn the spacecraft back toward the Sun and look at sunlight passing through Saturn's rings. Using this strategy, we've learned that the Cassini division is not empty but contains a relatively small number of particles. This strategy to study details in the rings is quite similar to how dust particles in the air in your room suddenly become visible when a beam of sunlight passes through them. This strategy has worked well for studying many of the outer planets.

The color variations in Figure 7-20 suggest that the icy particles do not migrate substantially from one ringlet to another. Had such migration taken place, the color differences would have been smeared out over time. The color differences may also indicate that different sorts of material were added to the rings at different times. In this scenario, the rings did not all form at the same time as Saturn but were added to over an extended period. New ring material could have come from small satellites that shattered after being hit by a stray asteroid or comet.

In addition to revealing new details about the A, B, C, and F rings, the *Voyager* cameras also discovered three new ring systems: the D, E, and G rings. Figure 7-19 also shows the layout of all of Saturn's known rings along with the orbits of some of Saturn's satellites. The D ring is Saturn's innermost ring system. It consists of a series of extremely faint ringlets located between the inner edge of the C ring and the Saturnian cloud tops. The E ring and the G ring both lie far from the planet, well beyond the narrow F ring. Both of these outer ring systems are extremely faint, fuzzy, and tenuous. Each lacks the ringlet structure so prominent in the main ring systems. The E ring encloses the orbit of Enceladus, one of Saturn's icy satellites. Some scientists suspect that water geysers on Enceladus are the source of ice particles in the E ring, much as Io's volcanoes supply material to Jupiter's magnetosphere.

ConceptCheck 7-10: If an asteroid entered a low-Earth orbit inside Earth's Roche limit, what would become of it?

Dim Ring Systems Around Jupiter, Neptune, and Uranus

Go to Video 7-3

When the *Voyager 1* spacecraft flew past Jupiter in 1979, it trained its cameras not just on the planet but also on the space around the planet's equator. It discovered that Jupiter, too, has a system of rings that lies within its Roche limit (**Figure 7-21**). These rings differ from Saturn's in two important ways. First, Jupiter's rings are composed

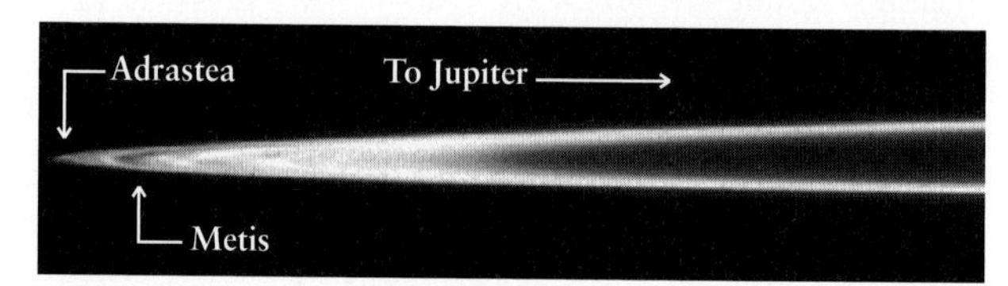

FIGURE 7-21 R I V U X G **Jupiter's Main Ring** This *Galileo* image shows Jupiter's main ring almost edge-on. The ring's outer radius, about 80,000 mi (130,000 km) from the center of Jupiter, is close to the orbit of Adrastea, the second closest (after Metis) of Jupiter's moons. Not shown is an even larger and more tenuous pair of "gossamer rings," one of which extends out past 100,000 mi (160,000 km). *(NASA/JPL; Cornell University)*

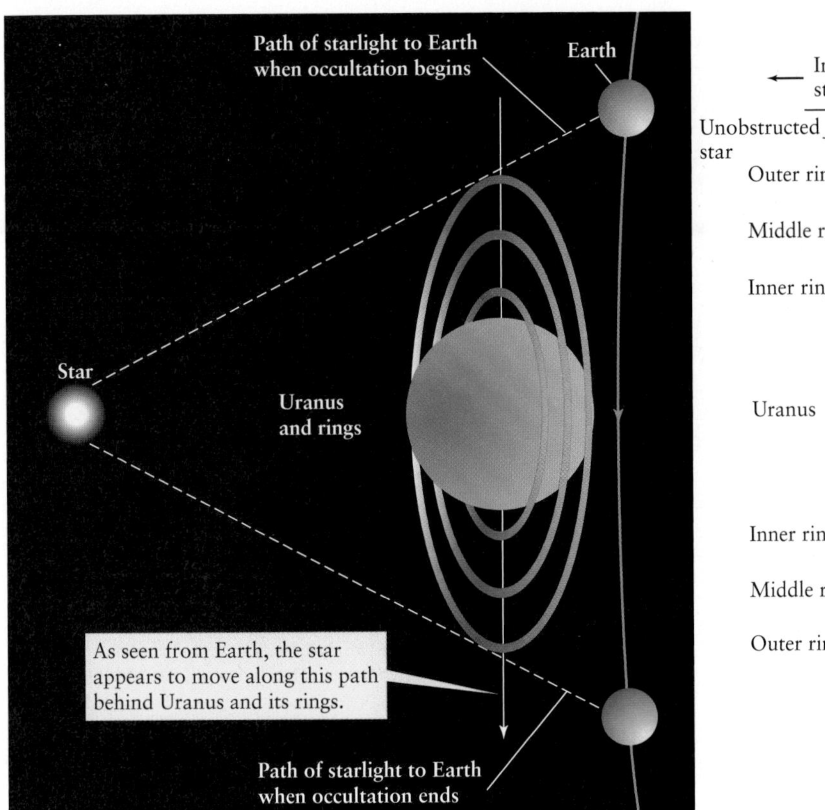

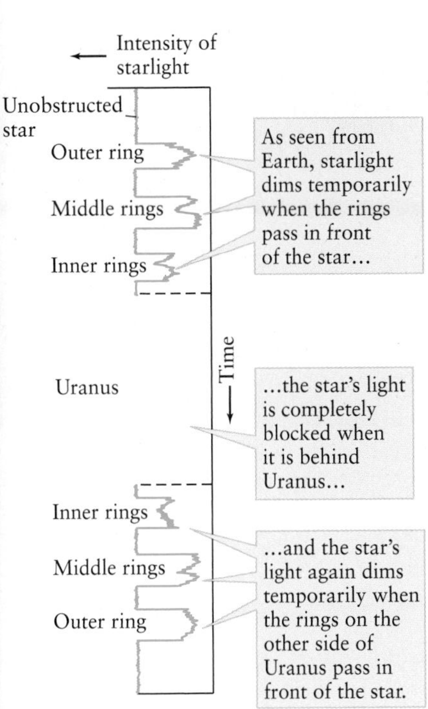

FIGURE 7-22 How the Rings of Uranus Were Discovered As seen from Earth, Uranus occasionally appears to move in front of a distant star. Such an event is called an *occultation*. The star's light is completely blocked when it is behind the planet. But before and after the occultation by Uranus, the starlight dims temporarily as the rings pass in front of the star.

of tiny particles of rock with an average size of only about 1 μm (= 0.001 mm = 10^{-6} m) and that reflect less than 5% of the sunlight that falls on them. Second, there is very little material in the rings of Jupiter, less than 1/100,000 (10^{-5}) the amount of material in Saturn's rings. As a result, Jupiter's rings are extremely faint, which explains why their presence was first revealed by a spacecraft rather than an Earth-based telescope. The ring particles are thought to originate from meteorite impacts on Jupiter's four small, inner satellites, two of which are visible in Figure 7-21.

On March 10, 1977, Uranus was scheduled to move in front of a faint star, as seen from the Indian Ocean. A team of astronomers headed by James L. Elliot of Cornell University observed this event from a NASA airplane equipped with a telescope. They hoped that by measuring how long the star was hidden, they could accurately measure Uranus's size. In addition, by carefully measuring how the starlight faded when Uranus passed in front of the star, they planned to deduce important properties of Uranus's upper atmosphere.

To everyone's surprise, the background star briefly blinked on and off several times just before the star passed behind Uranus and again immediately after (**Figure 7-22**). The astronomers concluded that Uranus must be surrounded by a series of nine narrow rings. In addition to these nine rings, *Voyager 2* discovered two others that lie even closer to the cloud tops of Uranus. Two other extremely faint rings, much larger in diameter than any of the others, were found in 2005 using the Hubble Space Telescope (**Figure 7-23**).

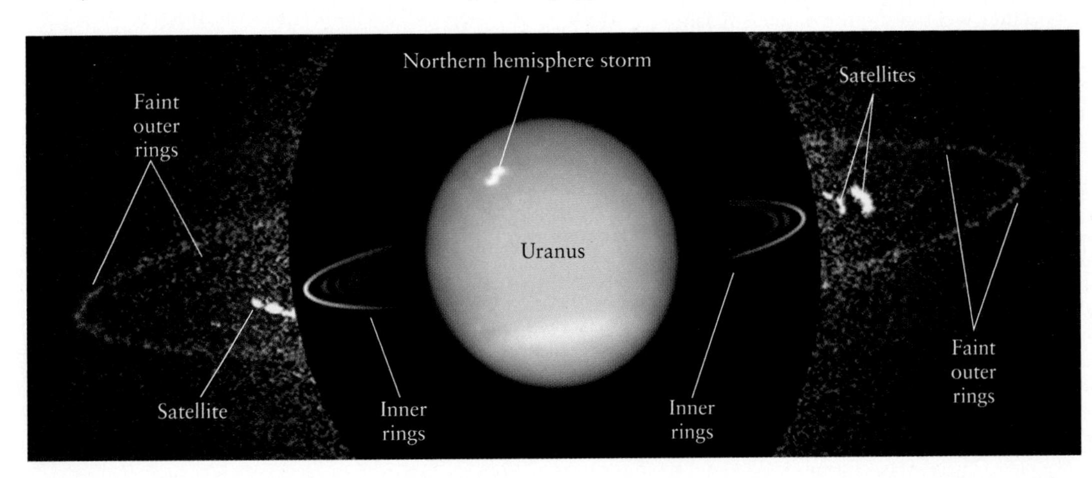

FIGURE 7-23 R I V U X G **Uranus's Rings from the Hubble Space Telescope** Several Hubble Space Telescope images were assembled into this view of Uranus's rings. Relatively short exposure times reveal the planet and the inner rings, while much longer exposure times were needed to reveal the outer rings. During these long exposures several of Uranus's satellites moved noticeably, leaving bright trails on the image. *(NASA, ESA, and M. Showalter, SETI Institute)*

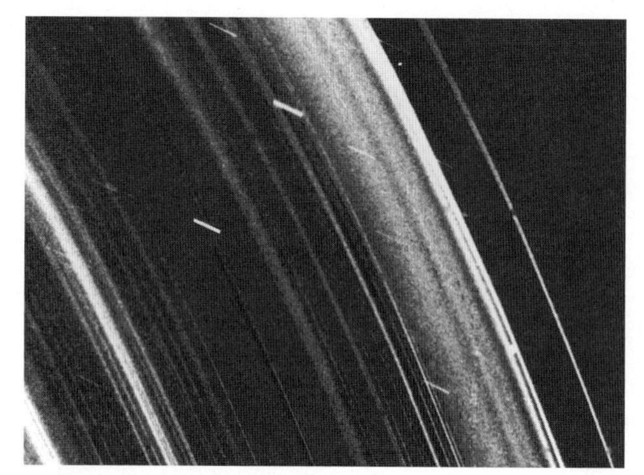

FIGURE 7-24 R I V U X G **The Shaded Side of Uranus's Rings** This view from *Voyager 2,* taken when the spacecraft was in Uranus's shadow, looks back toward the Sun. Numerous fine-grained dust particles gleam in the spaces between the main rings. This dust is probably debris from collisions between larger particles in the main rings. The short horizontal streaks are star images blurred by the spacecraft's motion during the exposure. *(NASA/JPL)*

Unlike Saturn's rings, the rings of Uranus are dark and narrow: Most are less than 6 mi (10 km) wide. Typical particles in Saturn's rings are chunks of reflective ice the size of snowballs (a few centimeters across), but typical particles in Uranus's rings are a few inches to 30 ft (0.1 m to 10 m) in size and are no more reflective than lumps of coal. It is not surprising that these narrow, dark rings escaped detection for so long. **Figure 7-24** is a *Voyager 2* image of the rings from the side of the planet away from the Sun, where light scattering from exceptionally small ring particles makes them more visible. Figure 7-23 shows the Sun-facing side of the rings.

The Uranian rings are so dark because sunlight at Uranus is only one-quarter as intense as at Saturn. As a result, Uranus's ring particles are so cold that they can retain methane ice. Scientists speculate that eons of impacts by electrons trapped in the magnetospheres of the planets have converted this methane ice into dark carbon compounds, accounting for the low reflectivity of the rings.

Uranus's major rings are located less than 2 Uranian radii from the planet's center, well within the planet's Roche limit. Some sort of mechanism, possibly one involving gravitational attraction from nearby moons, efficiently confines particles to their narrow orbits. *Voyager 2* searched for such orbiting satellites but found only two. The others may be so small and dark that they have simply escaped detection.

The two faint rings discovered in 2005 (see Figure 7-23) are different: They lie well *outside* Uranus's Roche limit. The outer of these two rings owes its existence to a miniature satellite called Mab (named for a diminutive fairy in English folklore) that orbits within this ring. Meteorites colliding with Mab knock dust off the moon's surface. Since Mab is so small (just 22 mi (36 km) in diameter) and hence has little gravity, the ejected dust escapes from the satellite and goes into orbit around Uranus, forming a ring. The inner of the two faint rings may have been formed in the same way; however, no small satellite has yet been found within this ring.

FIGURE 7-25 R I V U X G **Neptune's Rings** The two main rings of Neptune and a faint, inner ring can easily be seen in this composite of two *Voyager 2* images. There is also a faint sheet of particles whose outer edge is located between the two main rings and extends inward toward the planet. The overexposed image of Neptune itself has been blocked out. *(NASA)*

Like Uranus, Neptune is surrounded by a system of thin, dark rings that was first detected when the rings passed directly in front of a distant star. **Figure 7-25** is a *Voyager 2* image of Neptune's rings. The ring particles vary in size from a few micrometers (1 μm = 10^{-6} m) to about 30 feet (10 meters). Just as for Uranus's rings, the particles that make up Neptune's rings reflect very little light.

In 2002 and 2003, a team led by Imke de Pater of the University of California, Berkeley, used the 10-m Keck telescope in Hawaii to observe in detail the rings of Neptune. Remarkably, the team found that all of the rings had become fainter since the *Voyager 2* flyby in 1989. Apparently the rings are losing particles faster than new ones are being added. If the rings continue to decay at the same rate, one of them may vanish completely within a century. More research into the nature of Neptune's rings is needed to explain this curious and rapid decay.

ConceptCheck 7-11: **Why are Saturn's rings easier to observe than those of the other planets?**

VISUAL LITERACY TASK

Atmospheric Structure of a Gas Giant

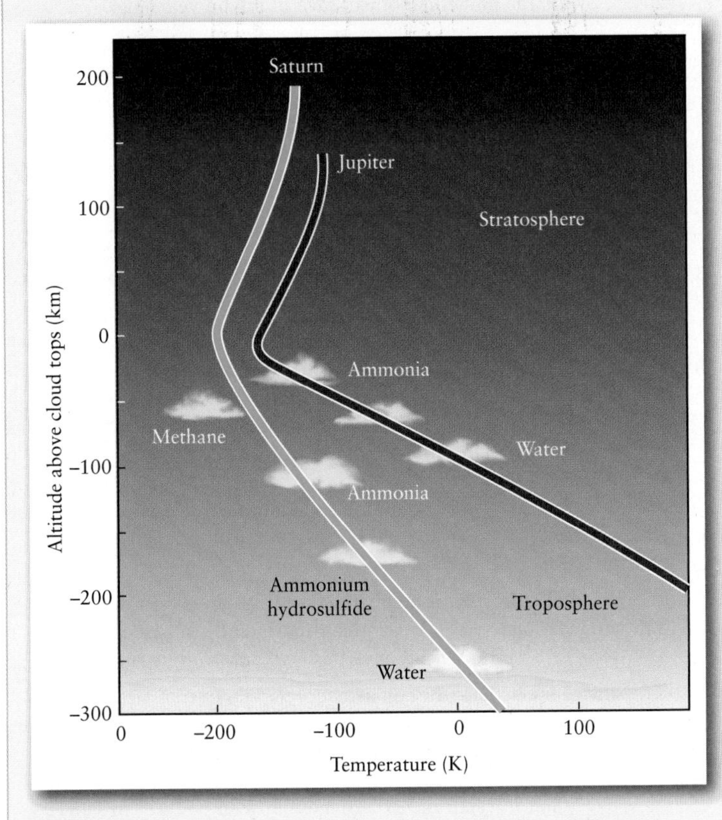

PROMPT: What would you tell a fellow student who said, **"A space probe crashing into Jupiter would encounter continually hotter and hotter temperatures, the atmosphere reaching the boiling point of water at 100 km below the cloud tops"?**

ENTER RESPONSE:

Guiding Questions:

1. At 100 km (60 mi) above Jupiter's cloud tops, the temperature is about
   a. −100°C (−150°F).
   b. the freezing point of water on Earth.
   c. the boiling point of water on Earth.
   d. the same as at 100 km (60 mi) below the cloud tops.

2. Just above the cloud tops, the temperature is
   a. increasing with decreasing height.
   b. decreasing with decreasing height.
   c. stable.
   d. about 0°C (32°F).

3. Beneath the cloud tops, the temperature
   a. decreases with depth.
   b. increases with depth.
   c. is stable.
   d. is about the boiling point of water.

4. Compared to Saturn, the temperature inside Jupiter
   a. increases more greatly with depth.
   b. increases with depth but not as dramatically.
   c. has the same profile.
   d. is much less.

KEY IDEAS AND TERMS

7-1 Dynamic atmospheres of Jupiter and Saturn change rapidly

- Jupiter and Saturn are both much larger than Earth. Each is composed of 71% hydrogen, 24% helium, and 5% all other elements by mass.
- Jupiter probably has a rocky core several times more massive than Earth. The core is surrounded by a layer of liquid "ices" (water, ammonia, methane, and associated compounds). On top of this are a layer of helium and liquid metallic hydrogen and an outermost layer composed primarily of ordinary hydrogen and helium.
- Saturn's internal structure is similar to that of Jupiter, but its core makes up a larger fraction of its volume and its liquid metallic hydrogen mantle is shallower than that of Jupiter.

- The visible "surfaces" of Jupiter and Saturn are actually the tops of their clouds. The rapid rotation of the planets twists the clouds into dark **belts** and light **zones** that run parallel to the equator. The colored ovals visible in the Jovian atmosphere represent gigantic storms. Some, such as the **Great Red Spot,** are quite stable and persist for many years. Storms in Saturn's atmosphere seem to be shorter lived.
- The outer layers of both Jupiter and Saturn's atmospheres show **differential rotation** as the equatorial regions rotate slightly faster than the polar regions.

7-2 Uranus and Neptune have seemingly quiet atmospheres

- Both Uranus and Neptune have atmospheres composed primarily of hydrogen, helium, and a few percent of methane.
- Methane absorbs red light, giving Uranus and Neptune their greenish-blue color.
- Much more cloud activity is seen on Neptune than on Uranus. This is because Uranus lacks a substantial internal heat source.

7-3 Saturn's moon Titan and Neptune's moon Triton exhibit unexpected atmospheres

- Titan, the largest natural satellite of Saturn, is covered with a thick, hazy carbon dioxide atmosphere that may have sufficient energy for methane-based weather and lakes to occur near the surface.
- Triton has a tenuous nitrogen atmosphere and a young, icy surface indicative of tectonic activity.

7-4 All Jovian planet atmospheres are encircled by complex ring systems

- Saturn is circled by a system of thin, broad rings lying in the plane of the planet's equator. This system is tilted away from the plane of Saturn's orbit, which causes the rings to be seen at various angles by an Earth-based observer over the course of a Saturnian year.
- From Earth, Saturn is observed to have three major, broad rings composed of unconnected **ring particles,** ice and ice-coated rock ranging in size from a few micrometers to about 30 feet (10 m) that are kept separate by a **tidal force.** The A ring and the B ring are separated by the **Cassini division.**
- Most of the rings exist inside the **Roche limit** of Saturn, where disruptive tidal forces are stronger than the gravitational forces attracting the ring particles to each other.
- Jupiter's faint rings are composed of a relatively small amount of small, dark, rocky particles that reflect very little light.
- Uranus and Neptune are both surrounded by systems of thin, dark rings. The low reflectivity of the ring particles may be due to radiation-darkened methane ice.

QUESTIONS

Review Questions

1. In what ways are the motions of Jupiter's atmosphere like the motion of water stirred in a pot (see Figure 7-2*b*)? In what ways are they different?
2. How do the swirling atmospheres of Jupiter and Saturn compare?
3. What would happen if you tried to land a spacecraft on the surface of Jupiter?
4. What are the belts and zones in the atmospheres of Jupiter and Saturn? Is the Great Red Spot more like a belt or a zone? Explain your answer.
5. Describe the internal structures of Jupiter and Saturn, and compare them with the internal structure of Earth.
6. Briefly describe the evidence supporting the idea that Uranus was struck by a large planetlike object several billion years ago.
7. Describe the seasons on Uranus. In what ways are the Uranian seasons different from those on Earth?
8. Why are Uranus and Neptune distinctly greenish-blue in color, while Jupiter and Saturn are not?
9. How many rings encircle Saturn? Draw a sketch.
10. If Saturn's rings are not solid, why do they look solid when viewed through a telescope?
11. Compare the rings that surround Jupiter, Saturn, Uranus, and Neptune. Briefly discuss their similarities and differences.

Web Chat Questions

1. The classic science fiction films *2001: A Space Odyssey* and *2010: The Year We Make Contact* both involve manned spacecraft in orbit around Jupiter. What kinds of observations could humans make on such a mission that cannot be made by robotic spacecraft? What would be the risks associated with such a mission? Do you think that a manned Jupiter mission would be as worthwhile as a manned mission to Mars? Explain your answers.
2. Suppose that Saturn were somehow moved to an orbit around the Sun with semimajor axis 1 AU, the same as Earth's. Discuss what long-term effects this would have on the planet and its rings.
3. Sir William Herschel, a British astronomer, discovered Uranus in 1781 and named it Georgium Sidus (Latin for "Georgian Star"), after the reigning monarch, George III. What name might Uranus have been given in 1781 if an astronomer in your country had discovered it? Why? What if it had been discovered in your country in 1881? In 1991?

Collaborative Exercises

1. Using a ruler with millimeter markings on the first image of Jupiter in the text (Figure 7-1*a*), determine the ratio of the longest width of the Great Red Spot to the full diameter of Jupiter. Each group member should measure the image and all values should be averaged.
2. The text provides different years that spacecraft have flown by Jupiter and Saturn. List these dates and create a time line by listing one important event that was occurring on Earth during each of those years.
3. If the largest circle you can draw on a piece of paper represents the largest diameter of Saturn's rings, about how large would Saturn be if scaled appropriately? Which item in a group member's backpack is closest to this size?

Observing Questions

1. Use *Starry Night*™ to examine the Jovian planets Jupiter, Saturn, Uranus, and Neptune. Select each of these planets from **Favourites > Explorations.** Use the location scroller to examine each planet from different views. (a) Describe each planet's appearance. Which has the greatest color contrast in its cloud tops? (b) Which planet has the least color contrast in its cloud tops? (c) What can

you say about the thickness of Saturn's rings compared to their diameter?

2. Use *Starry Night*™ to examine Jupiter. Open **Favourites > Explorations > Jupiter.** Select **Options > Solar System > Planets-Moons . . .** from the menu. In the **Planets-Moons Options** dialog window, click the check box for **Surface guides** and click **OK** to remove the pole sticks and equator from the image. Click the **Stop** button and then use the **Zoom** controls and location scroller to examine Jupiter's atmosphere. (a) Describe the appearance of the atmosphere of the planet. (b) Compare the number of white ovals and brown ovals in Jupiter's southern hemisphere (the hemisphere in which the Great Red Spot is located) with the number of white ovals and brown ovals visible in the northern hemisphere. What general rule can you state about the abundance of these storms in the two hemispheres?
3. Use *Starry Night*™ to observe the changing appearance of Saturn as seen from Earth. Select **Favourites > Explorations > Atlas** from the menu and click the **Now** button in the toolbar. Use the **Find** pane to locate and lock the view on Saturn (double-click the entry for Saturn). **Zoom in** until Saturn and its rings are clearly visible. Set the **Time Flow Rate** to **1 year.** Now use the single-step forward button to observe the changing orientation of the rings as time advances in 1-year steps. (a) Describe qualitatively how the rings change orientation, as seen from Earth. (b) Use the **Time** controls to determine approximately how long it takes for Saturn's rings, as seen from Earth, to go through a complete cycle from edge-on to fully open to edge-on to fully open and then finally to edge-on again. (c) Click the **Now** button to return to the present time. Use the **Time** controls to find when, within the next 30 years, Saturn's rings will appear edge-on as seen from Earth. (d) Why do we see the orientation of Saturn's rings change in the way that you found in part (b)?
4. Use *Starry Night*™ to examine the rings of Saturn. Open **Favourites > Explorations > Saturn** and use the location scroller to view Saturn so that you are looking straight down on the plane of the rings. (a) Draw a copy of what you see and label the different rings and divisions. (b) Using the location scroller, adjust the view so that Saturn's rings appear edge-on and then rotate the image until the Sun comes into view. Which of the following is correct? (i) The Sun is in the same plane as the rings of Saturn. (ii) The rings are in Saturn's equatorial plane. (iii) Neither (i) nor (ii) is correct. (iv) Both (i) and (ii) are correct.

ANSWERS

ConceptChecks

ConceptCheck 7-1: No. All solid objects must have nondifferential rotation or else they would break apart.

ConceptCheck 7-2: If the planets formed with the same chemical composition, then helium in Saturn's colder atmosphere must have condensed and moved deeper inside, resulting in Saturn appearing to be mostly hydrogen.

ConceptCheck 7-3: White ovals are long-lasting, low-temperature, high-altitude features that obscure views of what lies underneath, whereas brown ovals are openings in Jupiter's cloud cover that emit infrared energy to be observed from lower altitudes.

ConceptCheck 7-4: Jupiter has significantly more mass, and as a result the cloud layers are gravitationally compressed much more on Jupiter than on less massive Saturn.

ConceptCheck 7-5: Methane in Uranus's atmosphere absorbs the Sun's longer wavelengths of light, so if it were absent, Uranus would have more reds and, as a result, appear more yellow-white, like Saturn.

ConceptCheck 7-6: We usually think of "weather" being driven in the atmosphere by the Sun's energy; and because Neptune is so far from the Sun, the surprising activity in Neptune's atmosphere must be driven by unexpected energy coming from within Neptune.

ConceptCheck 7-7: Titan is shrouded by a thick atmosphere that does not transmit visible light.

ConceptCheck 7-8: Geyserlike plumes visible in *Voyager 2* images all point in the same direction, suggesting that the ejected material is being carried downwind within a thin atmosphere.

ConceptCheck 7-9: As Saturn travels around the Sun, the tilted and quite thin rings are sometimes seen "face-on" and are easily observed, whereas other times Earth's position makes them "edge-on" and nearly invisible.

ConceptCheck 7-10: An asteroid orbiting Earth within Earth's Roche limit for enough time would be ripped apart by Earth's gravitational attraction differing on the near side and the far side of the asteroid, resulting in a tiny ring around Earth.

ConceptCheck 7-11: Saturn's ring particles are more reflective, being composed of mostly ice, and many times larger than the less reflective and smaller ring particles of the other planets.

R I V U X G Tube worms living in thermal vents deep in Earth's oceans hint that life could exist on other planets in similar environments, such as oceans within Jupiter's moon Europa. *(a: Photo by HOV Alvin © Woods Hole Oceanographic Institution; b: Michael Carrol/NASA-JPL)*

Looking for Life Beyond Earth

CHAPTER LEARNING OBJECTIVES **By reading the sections of this chapter, you will learn:**

8-1 Planets and the chemical building blocks of life are found throughout space

8-2 Europa and Mars have the potential for life to have evolved

8-3 Meteorites from Mars have been scrutinized for life-forms

8-4 The Drake equation helps scientists estimate how many civilizations may inhabit our Galaxy

8-5 Searches with space-based infrared telescopes and Earth-based radio telescopes for Earthlike planets and alien civilizations are under way

One of the most compelling questions in science is also one of the simplest: Are we alone? That is, does life exist beyond Earth? As yet, we have no definitive answer to this question. None of our spacecraft has found life elsewhere in the solar system, and radio telescopes have yet to detect signals of intelligent origin coming from space. Reports of aliens visiting our planet and abducting humans make for compelling science fiction, but none of these reports has ever been verified.

Yet there are strong reasons to suspect that life might indeed exist beyond Earth. For one, we discussed in previous chapters that in just the past few decades, the number of planets known to exist has increased from fewer than 10 to several thousand. This fact makes astronomers consider that there are thousands, if not millions, of more locations where life could possibly reside than previously thought. For another, in the last two chapters we described moons that have liquid oceans heated from below, explosive geysers and volcanoes, and even dynamic weather systems. Life systems need energy and these observations provide evidence that many locations have energy available to support life. We are now ready to consider a third compelling reason to imagine life exists beyond Earth: Earth is home to living and flourishing organisms in some of the most "unearthly" environments on our planet. At the bottom of Earth's oceans, thermal vents feed superheated water into the ocean that serves as the food energy sources for colonies of pink, eyeless, alien-looking worms the size of your thumb. If life can flourish at the bottom of the ocean, might it not also flourish in the equally hostile conditions found on other worlds?

In this chapter, we consider first the basic elements necessary to support life. We look for places in our solar system where life may once have originated, and where it may quietly still exist today. We will see how scientists estimate the chances of finding life beyond our solar system, and how they search for signals from other intelligent species. Then, we will learn how a new generation of telescopes may make it possible to detect the presence of even single-celled organisms on worlds many light-years away. ■

8-1 Planets and the chemical building blocks of life are found throughout space

We are relatively certain that alien life-forms are not quietly observing Earth and occasionally visiting the surface to capture and conduct experiments on unexpecting human beings. But, for a moment, imagine you were an alien from a distant star system sent to explore Earth. If you landed your spacecraft at any random spot on Earth, would you likely see a human being? Remembering that three-quarters of Earth is covered with water, you are much more likely to find fish than humans. But one thing is certain—you would definitely find that tiny bacteria and plankton are the most common forms of life on Earth!

Suppose instead you are assigned the task of being the first person to look for life on another planet beyond Earth. How might you recognize which of the strange objects found around the solar system are living and which are not? Questions such as these are central to **astrobiology,** the study of life in the universe. Most astrobiologists suspect that if we find living organisms on other worlds, they will be "life as we know it"—that is, their biochemistry will be based on the properties of the carbon atom, as is the case for all Earth-based life. Let's first consider the unique nature of carbon atoms.

Organic Molecules in the Universe May Be the Basis for Life

Of all the atoms we know about, carbon seems to be the best candidate around which to build life-forms. The reason is that carbon has the most versatile chemistry of any element. Carbon atoms can form chemical bonds to create especially long and complex molecules (**Figure 8-1**). These carbon-based molecules are the foundation upon which all living organisms are made.

What makes carbon useful for life is that carbon-based molecules can be linked together to form elaborate structures, such as chains, lattices, and fibers. Some of these structures are even capable of complex, self-regulating chemical reactions. These reactions are required for life processes to occur. Furthermore, the primary constituents of the molecules of life found on Earth—carbon, hydrogen, nitrogen, oxygen, sulfur, and phosphorus—are among the most abundant elements across the universe. Taken together, the versatility and abundance of carbon suggest that extraterrestrial life is also likely to be based on organic chemistry.

Carbon molecules exist widely throughout interstellar space and are incorporated into planets during star and planetary system formation. One of the most common carbon-based molecules is carbon monoxide (CO), which is

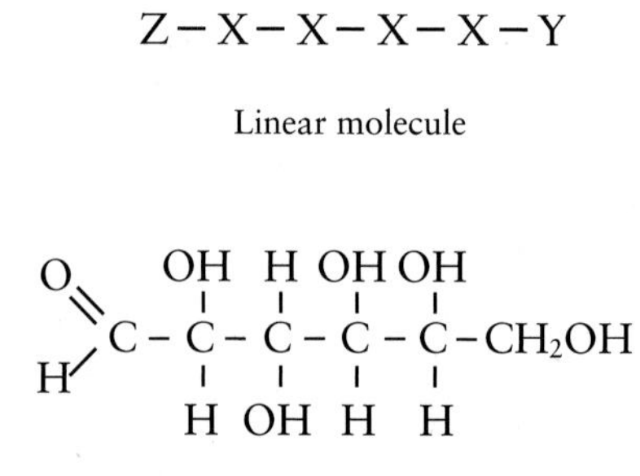

Z – X – X – X – X – Y

(a) Linear molecule

```
O      OH  H  OH OH
 \\     |   |   |  |
  C – C – C – C – C – CH2OH
 /      |   |   |  |
H       H  OH   H  H
```

(b) Glucose

FIGURE 8-1 Complex Molecules and Carbon (a) Atoms that can bond to only two other atoms, like the atoms denoted X shown here, can form a chain of atoms called a linear molecule. The chain stops where we introduce an atom, such as those labeled Y and Z that can bond to only one other atom. (b) A carbon atom (denoted C) can bond with up to *four* other atoms. Hence, carbon atoms can form more complex, nonlinear molecules like glucose. All organic molecules that are found in living organisms have backbones of carbon atoms.

(a) R I V U X G

(b) R I V U X G

FIGURE 8-2 The Orion Nebula—Birthplace of Stars This beautiful nebula is a stellar "nursery" where stars are formed out of the nebula's gas. (a) Intense ultraviolet light from newborn stars excites the surrounding gas and causes it to glow in visible light. (b) Showing the same nebula, infrared light from the young stars embedded in this nebula can be seen passing effortlessly through the clouds. Orion Nebula is some 1500 ly from Earth and is about 30 ly across. *(a: NASA, C. R. O'Dell and S. K. Wong [Rice University]: b: NASA; K. L. Luhman [Harvard-Smithsonian Center for Astrophysics, Cambridge, Mass.]; and G. Schneider, E. Young, G. Rieke, A. Cotera, H. Chen, M. Rieke, R. Thompson [Steward Observatory, University of Arizona])*

made when a carbon atom and an oxygen atom collide and bond together. Carbon monoxide is found in great abundance within giant interstellar clouds like the one shown in **Figure 8-2.** Carbon atoms can also combine with other elements in ways that show up as long wavelength emission lines seen in interstellar clouds.

If life is likely to be based on carbon-based molecules, then these molecules must initially be present on a planet's surface in order for life to arise from nonliving matter. Evidence for this comes from meteorites like the one shown in **Figure 8-3.** These are ancient meteorites that date from the formation of the solar system and that are often found to contain a variety of carbon-based molecules. The spectra of comets—which are also among the oldest objects in the solar system—show that they, too, contain an assortment of molecules needed for life. Comets and meteoroids were much more numerous in the early solar system than they are today, and they were correspondingly more likely to collide with a planet, depositing these materials on the planet's surface in great quantities.

The chemical building blocks for life exist throughout the universe, but we do not yet understand how they came together to form life.

ConceptCheck 8-1: **Why would life-forms throughout the cosmos likely be based on carbon?**

ConceptCheck 8-2: **How could carbon molecules end up on the surfaces of planets?**

The Miller-Urey Experiment Demonstrated Possible Conditions for Life to Exist

Comets and meteorites would not have been the only sources of carbon material on the young planets of our early solar system. Back in 1952, the American chemists Stanley Miller and Harold Urey designed experiments to see if the chemical building blocks of life could be made from simple chemicals, under the right conditions. They created a "soup" of simple chemicals that likely prevailed on the surface of a primitive Earth. In a closed container, they mixed these chemicals into a sample "atmosphere": a mixture of hydrogen (H_2), ammonia (NH_3), methane (CH_4), and water vapor (H_2O), the most common molecules in the solar system. Miller and Urey then exposed this mixture of gases to simulated atmospheric lightning for a week. At the end of this period, the inside of the container was coated with a reddish-brown substance rich in amino acids and other compounds essential to life.

Since Miller and Urey's original experiment, most scientists have come to the conclusion that Earth's primordial atmosphere was composed mostly of carbon dioxide (CO_2), nitrogen (N_2), and water vapor outgassed from volcanoes, along with some hydrogen. Modern versions of the Miller-Urey experiment (**Figure 8-4**) using these common gases have also succeeded in synthesizing a wide variety of carbon-rich compounds.

FIGURE 8-3 R I V U X G **A Meteorite Containing Carbon** Carbonaceous chondrites are primitive meteorites that date back to the very beginning of the solar system. This sample is a piece of the Allende meteorite, a large carbonaceous chondrite that fell in Mexico in 1969. *(Science Source)*

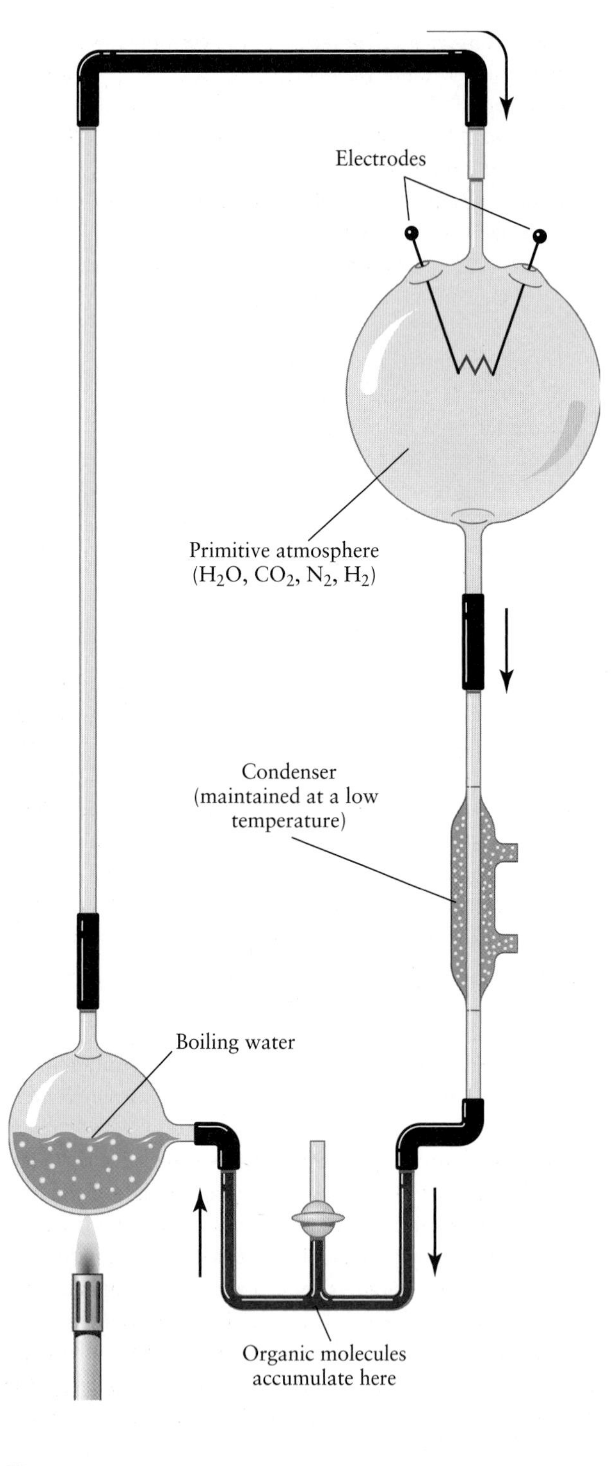

FIGURE 8-4
An Updated Miller-Urey Experiment Modern versions of this classic experiment prove that numerous organic compounds important to life can be synthesized from gases that were present in Earth's primordial atmosphere. This experiment supports the hypothesis that life on Earth arose as a result of ordinary chemical reactions.

CAUTION It is important to emphasize that scientists have *not* created life in a test tube. While organic molecules may have been available on the ancient Earth, biologists have yet to figure out how these molecules gathered themselves into cells and developed systems for self-replication. Nevertheless, because so many chemical components of life are so easily synthesized under conditions that simulate the primordial Earth, it seems reasonable to suppose that life could have originated as the result of chemical processes. Furthermore, because the molecules that combine to form these compounds are rather common, it seems equally reasonable that life could have originated in the same way on other planets. However, having the right chemical building blocks be commonplace throughout the universe does not guarantee that life is equally commonplace.

ConceptCheck 8-3: What did Miller and Urey create when they passed electricity through their sample of "atmosphere" containing a mixture of hydrogen (H_2), ammonia (NH_3), methane (CH_4), and water vapor (H_2O)?

8-2 Europa and Mars have the potential for life to have evolved

If life evolved on Earth from nonliving organic molecules, might the same process have taken place elsewhere in our solar system? Scientists are beginning to carefully scrutinize the planets and moons in an attempt to pursue this question. The first step is to look for places where water is abundant. As it turns out, water exists in many places besides our home planet.

Finding Liquid Water Is Essential

As far as we can tell, liquid water is essential for the survival of life as we know it. Perhaps surprisingly, the water need not be clean, cool, or pleasant by human standards. Earth's living organisms have been found in water that is boiling hot, such as in Wyoming's Yellowstone National Park. At the same time, life has been found to exist happily in environments that are ice cold, such as in Antarctica's ice. Other surprising environments that were once thought to be inhospitable for life have turned out to be perfectly fine for some types of living bacteria—fiercely acidic, amazingly salty, and deep underground environments, and even on the unprotected surface of the Moon. The key is that the water must be liquid at least once in a while in order for life to use the water to survive. In order for water on a planet's surface to remain liquid, the temperature cannot be too hot or too cold. Furthermore, there must be a relatively thick atmosphere to provide enough pressure to keep liquid water from evaporating. Of all the worlds of the present-day solar system, only Earth seems to have the right conditions for water to remain liquid on its surface. But when we expand our search for liquid water that exists within ice or beneath the surface, an entirely new realm of possibilities arises within our solar system. As it turns out, we now know that water is highly common throughout the solar system (Section 5.1), and Earth is not even the largest repository.

Our best example comes from moons orbiting Jupiter. (Chapters 5 and 6 cover evidence for water elsewhere in our solar system.) There is compelling evidence that Europa has an ocean of water *beneath* its icy surface. As this moon orbits Jupiter, Europa is caught in a tug-of-war between

Jupiter's gravitational influence pulling it inward and those of Jupiter's other large moons pulling it outward. This flexes the interior of Europa, and this flexing generates enough heat to keep subsurface water from freezing. There may even be thermal vents on Europa's ocean floor, heating the ocean from underneath, such as those on Earth shown in the chapter opening photos. Chunks of ice on Europa's surface are able to float around on this underground ocean, rearranging themselves into patterns revealing the liquid water beneath. No one knows for sure whether life actually exists in the liquid ocean beneath Europa's thick, icy crust. This might be true on other moons of the outer solar system as well. And, if it is true throughout our solar system, what about the nearly countless other planets orbiting other stars in the night sky?

ConceptCheck 8-4: Why would scientists be quite surprised to find life on the Moon?

Searching for Life on Mars

Another strong possibility for finding evidence of past or present life is Mars. The present-day Martian atmosphere is so thin that water can usually exist only as ice or as a vapor. However, images made from Martian orbit show dried-up streambeds, flash flood channels, and sediment deposits. These features offer tantalizing evidence that the Martian atmosphere was once thicker and more protective, allowing water to flow freely over the planet's surface. Could life have existed on Mars during its "wet" period? If so, could life—even in the form of hard-to-detect microorganisms—have survived just below the surface as the Martian atmosphere thinned and Mars's surface water either froze or evaporated?

How would we know if life exists or existed on Mars? If we consider the history of Mars exploration, we can begin to see how our evolving scientific thinking impacts the way we design experiments to study Mars. One strategy is to recognize that life processes change a planet's environment and go to Mars and look for these signs. Our first direct attempt to look for life on Mars took place in 1976, just after our manned exploration of the Moon was abandoned. Two unmanned spacecraft landed on different parts of Mars in search of answers to these questions. *Viking Lander 1* and *Viking Lander 2* each carried a robotic scoop at the end of a mechanical arm to retrieve surface samples to look for biological activity. These samples were deposited into a compact on-board biological laboratory that carried out three different tests looking not simply for chemical evidence of water but for evidence of Martian microorganisms.

1. The *gas-exchange experiment* was designed to detect any processes that might be broadly considered as respiration. A surface sample was placed in a sealed container along with a controlled amount of gas and nutrients. The gases in the container were then monitored to see if their chemical composition changed.
2. The *labeled-release experiment* was designed to detect metabolic processes. A sample was moistened with nutrients containing radioactive carbon atoms. If any organisms in the sample consumed the nutrients, their waste products would include gases containing the telltale radioactive carbon.
3. The *pyrolytic-release experiment* was designed to detect photosynthesis, the biological process by which terrestrial plants use solar energy to help synthesize organic compounds from carbon dioxide. In the *Viking* experiments, a surface sample was placed in a container along with radioactive carbon dioxide and exposed to artificial sunlight. If plantlike photosynthesis occurred, microorganisms in the sample would take in some of the radioactive carbon from the gas.

The first data returned from these experiments caused great excitement, for in almost every case, rapid and extensive changes were detected inside the sealed containers, suggesting that microbial life had been found on Mars. More in-depth analysis of the data, however, led to the conclusion that these changes were due entirely to nonbiological chemical processes. It appears that the Martian surface is rich in unstable chemicals that react with water to release oxygen gas, a process that is easily mistaken for biological activity. Because the present-day surface of Mars is bone dry, these chemicals had nothing to react with until they were placed inside the moist interior of the *Viking Lander* laboratory.

At best, the results from the *Viking Lander* biological experiments were inconclusive. Perhaps life never existed on Mars at all. Or perhaps it did originate there but failed to survive the thinning of the Martian atmosphere, the unstable chemistry of the planet's surface, and exposure to ultraviolet radiation from the Sun. (Unlike Earth, Mars has no ozone layer to block ultraviolet rays.) Another possibility is that Martian microorganisms have survived only in certain locations that the *Viking Lander*s did not sample, such as isolated spots on the surface or deep beneath the ground. And yet another option is that there is life on Mars, but the experimental apparatus on board the *Viking Lander* spacecraft was not sophisticated enough to detect it.

Mars seems to be the best candidate for finding where life exists beyond Earth because it once had significantly more water on its surface.

As our scientific thinking changed, so did our approaches to searching for life on Mars. What we did not fully understand is the much wider diversity of hospitable living environments that is possible. As we look around own planet, we find life-forms not only just barely surviving, but flourishing and reproducing in environments once considered too extreme to be hospitable for life. As shown in **Figure 8-5**, we have found life-forms living in the superheated water above boiling temperatures in Yellowstone National Park and around geothermal vents near the bottom of

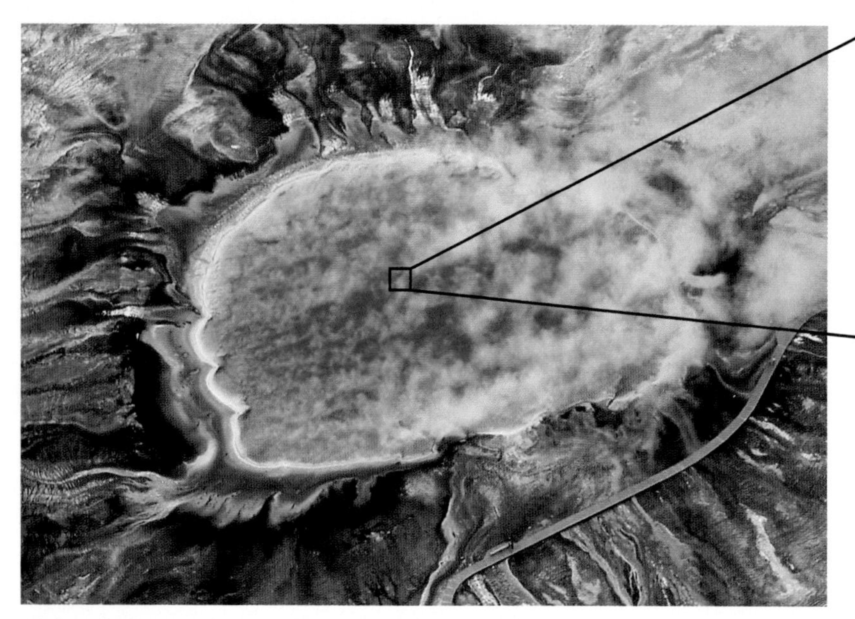

(a) Grand Prismatic Spring, Yellowstone National Park

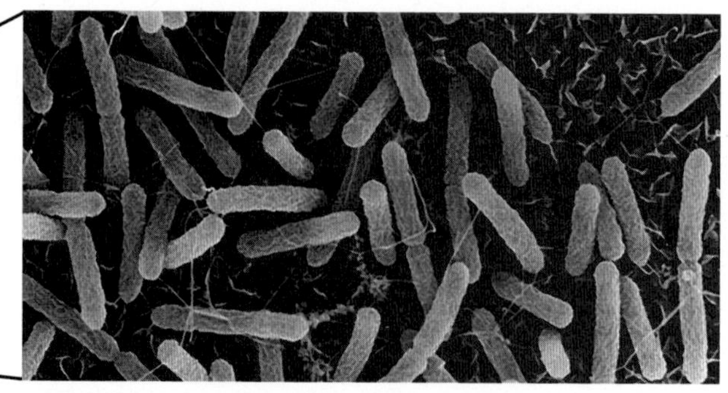

(b) Heat-thriving microscopic organisms

FIGURE 8-5 R I V U X G **Earth's Extreme Life-forms** We often think of boiling water as able to kill most living things, but these microscopic organisms in Yellowstone National Park flourish in boiling water. *(Jim Peaco/ National Park Service Photo; inset:* Applied and Environmental Microbiology, *October 1998, vol. 64, no. 10, pp. 3576–3583/American Society for Microbiology)*

the ocean. They can exist deep under the ice of Antarctica and in both acidic and salty lakes. Perhaps most surprising, scientists have also found extremophiles living as slime inside a nuclear reactor and even some that can survive in the harsh environment of outer space. These discoveries dramatically change how we think about life and help us focus on looking for the only thing common in all these environments—the existence of liquid water, even if for only the briefest amounts of time in the tiniest quantities.

> **ConceptCheck 8-5: What observations were scientists using the *Viking Landers* hoping to make that would allow them to infer that microbes were living in the Martian soil?**

New Evidence for Martian Water

Go to Video 8-1

Scientists searching for evidence of Martian life have been encouraged by other spacecraft that have provided new evidence of water on Mars. The *Mars Reconnaissance Orbiter* began orbiting Mars in 2006 and has been using a host of modern scientific instruments to systematically analyze the landforms, surface composition, weather systems, and ice on Mars. **Figure 8-6** is an image captured by this spacecraft's high-resolution camera and showing channels cut into the walls of a crater, apparently by water. Since that time, we have discovered even more direct evidence for water on Mars. Just a few years later, the *Mars Phoenix Lander* was outfitted to dig for evidence of water near the Martian south pole. Images of a small trench dug by the spacecraft that were taken over several days show that water ice, seen as white in **Figure 8-7**, quickly disappears when hit by warming sunlight.

Scientific discoveries, such as finding evidence for water on Mars, are not conclusive with just a single observation. Instead, scientific discoveries are only taken seriously by the scientific community when there are multiple lines of evidence that all focus on the same idea. As another line of evidence about Martian water, consider that the *Mars Express* spacecraft, which went into orbit around Mars in December 2003, used its infrared cameras to examine the ice caps at the Martian north and south poles (**Figure 8-8**). These cameras allowed scientists to see through the Martian ice cap's surface layer of frozen carbon dioxide and revealed underlying layers that appear to be water ice.

FIGURE 8-6 R I V U X G **Gully Channels** This false-color image of gully channels on Mars, where water has carved into crater walls, was taken by the High Resolution Imaging Science Experiment (HiRISE) camera on the *Mars Reconnaissance Orbiter. (NASA/JPL/University of Arizona)*

In January 2004, NASA successfully landed two robotic rovers named *Spirit* and *Opportunity* at two very different sites on opposite sides of Mars. While the terrain around the *Spirit* landing site appears to have been dry for billions of years, *Opportunity* landed in an area that appears to have been under water for extended periods (**Figure 8-9*a***). Measurements made by *Opportunity* confirm that some of the very dark surface material at its landing site contains an iron-rich mineral called gray hematite (Figure 8-9*b*). On Earth, deposits of gray hematite are commonly found at the bottoms of lakes or mineral hot springs. The presence of gray hematite at the *Opportunity* site reinforces the argument that Mars once had liquid water on its surface and helps hold open the possibility that living organisms could have evolved on Mars.

More recently, the *Curiosity* rover has made significant new observations about the chemical nature of rocks on the

FIGURE 8-7 R I V U X G **Disappearing Ice on Mars** NASA's *Phoenix Mars Lander*'s robotic arm dug into the Martian surface, revealing water ice. Observations of the trench showed that in just four days the previously buried ice quickly started to disappear. *(NASA/JPL-Caltech/University of Arizona/Texas A&M University)*

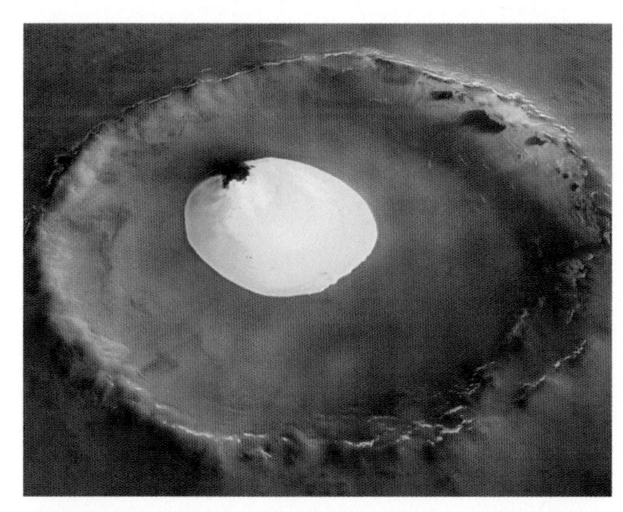

FIGURE 8-8 R I V U X G **Water Ice Hidden in Martian North Pole Crater** The High Resolution Stereo Camera (HRSC) on board the European Space Agency's *Mars Express* spacecraft shows a patch of water ice sitting on the floor of a 21-mi (35-km) wide crater on Mars with residual water ice tucked about 1 mi (1.6 km) beneath the crater's rim. *(© European Space Agency/DLR/FU Berlin/G. Neukum/Science Source)*

(a)

(b)

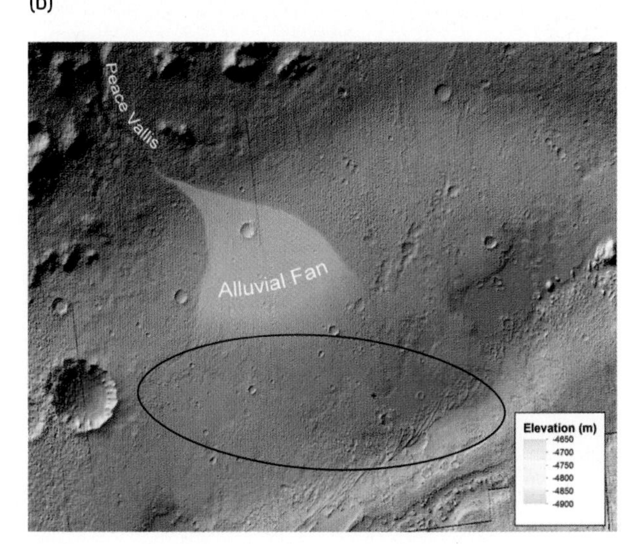

(c)

FIGURE 8-9 R I V U X G **Evidence of Ancient Martian Water** (a) The Mars rover *Opportunity* photographed these sedimentary layers in a region called Meridiani Planum. Some of the layers are made of dust deposited by the Martian winds, but others were laid down by minerals that precipitated out of standing water. (b) In this false-color image from *Opportunity* of Martian sand dunes, the bluish color shows the presence of millimeter-sized spheres of gray hematite. Such spheres naturally form in water-soaked deposits. (c) This false-color map shows highest elevations of Mars in red, revealing a fan-shaped region where water flowing downslope can deposit debris. On the ground, the *Curiosity* rover found rounded pebbles embedded with rocky outcrops characteristic of an ancient river. The black oval shows the targeted landing area for *Curiosity*. *(a: NASA/JPL/Cornell; b: Heka Agence Photo/Alamy; c : NASA/JPL-Caltech/U. of Arizona)*

surface of Mars. *Curiosity's* recent discoveries are described in Chapter 6 and most certainly newer ones exist since the time of this writing.

ConceptCheck 8-6: Why are astrobiologists excited about finding gray hematite rocks on Mars's surface?

8-3 Meteorites from Mars have been scrutinized for life-forms

What would be sufficient evidence for you to believe that life exists on another planet beyond Earth? Would it have to be a message of greetings and salutations from an intelligent being? Is finding fossils of bacteria and plankton on another planet sufficient? While spacecraft can carry biological experiments to other worlds such as Mars, many astrobiologists look forward to the day when a spacecraft will return Martian samples to laboratories on Earth. Until that day arrives, we have the next best thing: A dozen or so meteorites that appear to have formed on Mars have managed to make their way to a variety of locations on Earth.

These meteorites are called **SNC meteorites** after the names given to the first three examples found (Shergotty, Nakhla, and Chassigny). What identifies SNC meteorites as having come from Mars is the chemical composition of trace amounts of gas trapped within them. This composition is very different from Earth's atmosphere but is a nearly perfect match to the composition of the Martian atmosphere found by the *Viking Landers*.

How could a rock have traveled from Mars to Earth? When a large piece of space debris collides with a planet's surface and forms an impact crater, most of the material thrown upward by the impact falls back onto the planet's surface. But some extraordinarily powerful impacts produce large craters on Mars, roughly 100 km in diameter or larger. These tremendous impacts eject some rocks with such speed that they escape the planet's gravitational attraction and fly off into space.

There are numerous large craters on Mars, so a good number of Martian rocks have probably been blasted into space over the planet's history. These ejected rocks then go into elliptical orbits around the Sun. A few such rocks will have orbits that put them on a collision course with Earth, and these are the ones that scientists find as SNC meteorites.

ConceptCheck 8-7: Why are scientists convinced that these SNC meteorites actually came from Mars and not from our Moon?

Martian Meteorites May Show Evidence of the Presence of Liquid Water

Using a radioactive age-dating technique, scientists find that most SNC meteorites are between 200 million and 1.3 billion years old, much younger than the 4.56-billion-year age of the solar system. But one SNC meteorite, denoted by the serial number ALH 84001 and found in Antarctica in 1984, was discovered in 1993 to be 4.5 billion years old (**Figure 8-10*a***). Thus, ALH 84001 is a truly ancient piece of Mars. Analysis of ALH 84001 suggests that it was fractured by an impact between 3.8 and 4.0 billion years ago, was ejected from Mars by another impact 16 million years ago, and landed in Antarctica a mere 13,000 years ago.

Although we have several other meteorites that we are certain came from the surface of Mars, what makes ALH 84001 special is that it is the only known specimen of a rock that was on Mars during the era when liquid water most likely existed on the planet's surface. Scientists have therefore

(a)

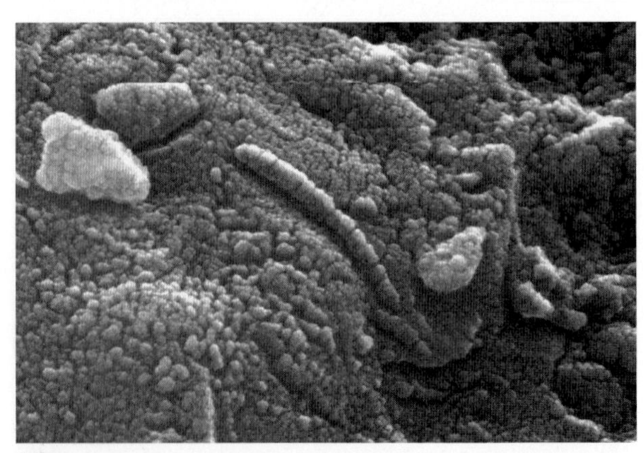

(b)

FIGURE 8-10 R I V U X G **A Meteorite from Mars** (a) This 1.9-kg meteorite, known as ALH 84001, formed on Mars some 4.5 billion years ago. About 16 million years ago a massive impact blasted it into space, where it drifted in orbit around the Sun until landing in Antarctica 13,000 years ago. The small cube at lower right is 0.4 in. (1 cm) across. (b) This electron microscope image, magnified some 100,000 times, shows tubular structures about 100 nanometers (10^{-7} m) in length found within the Martian meteorite ALH 84001. One controversial interpretation is that these are the fossils of microorganisms that lived on Mars billions of years ago. *(a: NASA Johnson Space Center; b: D. McKay [NASA/JSC], K. Thomas-Keprta [Lockheed-Martin], R. Zare [Stanford], NASA/APOD)*

investigated its chemical composition carefully, in the hope that this rock may contain clues to the amount of water that once flowed on the Martian surface. One such clue is the presence of rounded grains of minerals called *carbonates,* which can form only in the presence of water.

In 1996, David McKay and Everett Gibson of the NASA Johnson Space Center and several collaborators reported the results of a two-year study of the carbonate grains in ALH 84001. They made three remarkable findings. First, in and around the carbonate grains were large numbers of elongated, tubelike structures resembling fossilized microorganisms (Figure 8-10*b*). Second, the carbonate grains contain very pure crystals of iron sulfide and magnetite. These two compounds are rarely found together (especially in the presence of carbonates) but can be produced by certain types of bacteria. Indeed, about one-fourth of the magnetite crystals found in ALH 84001 are of a type that on Earth are formed only by bacteria. Third, carbon-based molecules are present—just the sort, in fact, that result from the decay of microorganisms. McKay and Gibson concluded that the structures seen in Figure 8-10*b* are fossilized remains of microorganisms. If so, these organisms lived and died on Mars billions of years ago, during the era when liquid water was abundant.

Are McKay and Gibson's conclusions correct? Their claims of ancient life on Mars are extraordinary, and such claims necessarily require extraordinary proof within the domain of science. With only one rock like ALH 84001 known to science so far and because fully compelling proof is hard to come by, many scientists are skeptical. They argue that the structures found in ALH 84001 could have been formed in other ways that do not require the existence of Martian microorganisms. Future spacecraft may help resolve the controversy by examining rocks on the Martian surface. For now, the existence of microscopic life on Mars in the distant past remains an open question.

Living things could have moved from planet to planet on meteorites.

ConceptCheck 8-8: Why do astrobiologists test meteorites for the presence of magnetite and pure iron sulfide crystals?

8-4 The Drake equation helps scientists estimate how many civilizations may inhabit our Galaxy

We have seen that some notable locations within our solar system might have been suitable for life to exist. But what about life existing on countless other planets orbiting distant stars? Scientists have found and cataloged thousands of such distant planets and can infer that millions, if not billions, more exist, waiting to be discovered. There are, of course, two competing perspectives. One is that life on Earth is unique and only happened one time in the entire universe. The alternative perspective is that the fact that life exists on Earth means that extraterrestrial life, including intelligent species, might evolve on planets around distant stars, given sufficient time and hospitable conditions. How can we learn whether such worlds exist, given the tremendous distances that separate us from them? This is the great challenge facing the **Search for Extraterrestrial Intelligence,** or **SETI.**

Remote Communication Is More Likely than Close Encounters

A tenet of modern folklore is the belief that alien civilizations do exist, and that their spacecraft have visited Earth secretly, mostly avoiding detection. Indeed, public perception surveys show that between one-third and one-half of all Americans believe that unidentified flying objects (UFOs) continue to visit Earth from distant, alien worlds. Despite frequent themes in science-fiction adventure movies and television shows, there is *no* scientifically verifiable evidence of alien visitations. As an example, consider that many UFO proponents believe that the U.S. government is hiding evidence of an alien spacecraft that crashed near Roswell, New Mexico, in 1947. However, the bits of "spacecraft wreckage" found near Roswell turned out to be nothing more than remnants of an unmanned research balloon. To find compelling evidence of the presence or absence of intelligent civilizations on worlds orbiting other stars, we must look elsewhere.

With our present technology, sending even a small unmanned spacecraft to another star requires a flight time of tens of thousands of years. Speculative design studies have been made for unmanned probes that could reach other stars within a century or less, but these are prohibitively expensive. Instead, many astronomers hope to learn about extraterrestrial civilizations by detecting radio transmissions from them. Radio waves are a logical choice for interstellar communication, because they can travel immense distances without being significantly degraded by the interstellar medium, the thin gas and dust found between the stars. They also travel very fast—at light speed of 186,000 mi/s (300,000 km/s).

Over the past several decades, astronomers have proposed various ways to search for alien radio transmissions, and several searches have been undertaken. In 1960, Frank Drake first used a radio telescope at the National Radio Astronomy Observatory in West Virginia to listen to two Sunlike stars, Tau Ceti and Epsilon Eridani, without success. More than 60 more extensive SETI searches have taken place since then, using radio telescopes around the world. Occasionally, a search has detected an unusual or powerful signal. But none has ever repeated, as a signal of intelligent origin might be expected to do. To date, we have no confirmed evidence of radio transmissions from another world.

ConceptCheck 8-9: Why might the use of radio waves for exploration for life in the Galaxy be more fruitful than using unmanned interstellar spaceships?

Are They Out There?

Go to Video 8-2

Should we be discouraged by our failure to make contact? What are the chances that a radio astronomer might someday detect radio signals from an extraterrestrial civilization? The first person to tackle this issue was Frank Drake, who laid out all of the possible characteristics that one would have to consider in order to find another civilization. He proposed that the number of technologically advanced civilizations in the Galaxy could be estimated by combining all of the important variables into a single mathematical sentence. This is now known as the **Drake equation:**

Estimates show that intelligent civilizations could be out there but seem probably too far away to detect easily.

Drake equation

$$N = R^* f_p n_e f_l f_i f_c L$$

N = number of technologically advanced civilizations in the Galaxy whose messages we might be able to detect
R^* = the rate at which solar-type stars form in the Galaxy
f_p = the fraction of stars that have planets
n_e = the number of planets per solar system that are Earthlike (that is, suitable for life)
f_l = the fraction of those Earthlike planets on which life actually arises
f_i = the fraction of those life-forms that evolve into intelligent species
f_c = the fraction of those species that develop adequate technology and then choose to send messages out into space
L = the lifetime of a technologically advanced civilization

Let's consider this one aspect at a time. The first two factors, R^* and f_p, can be determined by observation. In estimating R^*, we should probably exclude stars with masses greater than about 1.5 times that of the Sun. These more massive stars use up the hydrogen in their cores in 3 billion (3×10^9) years or less. On Earth, by contrast, human intelligence developed only within the last million years or so, some 4.56 billion years after the formation of the solar system. If that is typical of the time needed to evolve higher life-forms, then a star of 1.5 solar masses or more probably fades away or explodes into a supernova before creatures as intelligent as us can evolve on any of that star's planets.

Although stars less massive than the Sun have much longer lifetimes, they, too, seem unsuited for life because they are so dim. Only planets very near a low-mass star would be sufficiently warm for life as we know it. More important, a planet sufficiently close to a star is subject to its strong gravitational forces, which impact the speed at which the planet can spin. This is important because a planet that orbits too close to its star would have one hemisphere heated to great temperatures because it would be gravitationally locked toward the star, while the other hemisphere would be in perpetual, frigid darkness (see Figure 1-23).

This leaves us to consider a small range of stars not too different from the Sun. (Like Goldilocks sampling the three bears' porridge, we must have a star that is not too hot and not too cold, but just right.) Based on statistical studies of star formation in the Milky Way, some astronomers estimate that roughly one of these Sunlike stars forms each year in a galactic habitable zone where one might find such stars. This sets R^* at 1 per year.

The planets in our solar system formed as a natural consequence of the birth of the Sun. We have also seen evidence suggesting that planetary formation may be commonplace around single stars. Many astronomers suspect that most Sunlike stars probably have planets, and so they give f_p a value of 1.

Unfortunately, the rest of the terms in the Drake equation are extremely uncertain. Let's consider some hypothetical values. The chances that a planetary system has an Earthlike world suitable for life are not known. Were we to consider our own solar system as representative, we could put n_e at 1. Let's be more conservative, however, and suppose that one in ten solar-type stars is orbited by a habitable planet, making $n_e = 0.1$. The correct number is likely larger than this, but using a smaller number helps keep us from overestimating this unknown value. From what we know about the evolution of life on Earth, we might assume that, given appropriate conditions, the development of life is a certainty, which would make $f_l = 1$. This is an area of intense interest to astrobiologists.

For the sake of argument, we might also assume that evolution might naturally lead to the development of intelligence (a conjecture that is hotly debated) and also make $f_i = 1$. It is anyone's guess as to whether these intelligent extraterrestrial beings would attempt communication with other civilizations in the Galaxy. Even within current political debates, it is an undecided question as to whether we should be trying to find other intelligent civilizations that are perhaps more technologically advanced, because our search might unintentionally alert their attention to our vulnerable existence. If we assume that other civilizations would try to communicate, f_c would be put at 1 also.

The last variable, L, involving the longevity of a civilization, is the most uncertain of all. Looking at our own example, we see a planet whose atmosphere and oceans are increasingly polluted by creatures that possess nuclear weapons. If we are typical, perhaps L is as short as 100 years. If so, there could have been countless civilizations existing elsewhere in our own Galaxy that have long since ceased to exist. Putting all these numbers together, we arrive at

$$N = 1/\text{year} \times 1 \times 0.1 \times 1 \times 1 \times 1 \times 100 \text{ years} = 10$$

In other words, out of the hundreds of billions of stars in the Galaxy, we would estimate that there are only 10 technologically advanced civilizations from which we might receive communications.

A wide range of values has been proposed for the many terms in the Drake equation, and these various guesses produce vastly different estimates of N. Some scientists argue that there is exactly one advanced civilization in the Galaxy and that we are it. Others speculate that there may be hundreds or thousands of planets inhabited by intelligent creatures, and extraterrestrial communication is just too difficult or too expensive in which to participate. If we wish to know whether our Galaxy is devoid of other intelligence, teeming with civilizations, or something in between, we must keep searching the skies.

ConceptCheck 8-10: What makes the longevity of the civilization, *L*, the most difficult to estimate?

CalculationCheck 8-1: If we learned that Sunlike stars are 3 times more frequent than we originally thought, how would that change our estimate of the number of Sunlike stars that form every year in our Galaxy?

8-5 Searches with space-based infrared telescopes and Earth-based radio telescopes for Earthlike planets and alien civilizations are under way

In Chapter 4, we discussed the numerous techniques for finding extrasolar planets orbiting other stars. We also considered the technical challenges that present barriers to finding planets orbiting distant stars. Let's briefly reconsider the most profitable search so far, the search for transiting planets.

Searching for Extrasolar Planets

The most fruitful technique for finding extrasolar planets beyond our own solar system has been the one proven by NASA's Kepler mission, which has been successful in finding thousands of planets. The technique works like this: If a star is orbited by a planet whose orbital plane is oriented edge-on to our line of sight, once per orbit the planet will pass in front of the star in an event called a *transit*. This causes a temporary dimming of the light we see from that star. When using a telescope with a wide field of view, projects like Kepler are able to continuously monitor thousands of stars at once and find hundreds of planets in a very short time.

Much more interesting is to actually take direct images and photographs of extrasolar planets. If we are looking for extraterrestrial life, we sure would like to have an actual picture. Planned for a launch after 2015, Darwin is a European Space Agency telescope that will search for Earthlike planets by detecting their infrared radiation. The rationale is that stars like the Sun emit much less infrared radiation than visible light, while planets are relatively strong emitters of infrared. Hence, observing in the infrared makes it less difficult (although still technically challenging) to detect planets orbiting a star.

The Darwin telescope will also analyze the infrared spectra of any planets that it finds, in the hope of seeing the characteristic absorption of atmospheric gases such as ozone, carbon dioxide, and water vapor (**Figure 8-11**). The relative amounts of these gases, as determined from a planet's spectrum, can reveal whether life is present on that planet.

The Darwin telescope will need to achieve enough resolution to detect individual planets. One proposed mission design makes use of interferometry. By combining the light from three widely spaced dishes, each at least 3 m in diameter, Darwin will make the sharpest infrared images of any telescope in history. NASA has proposed a similar planet-finding interferometry mission, called Terrestrial Planet Finder, but the funding for this mission is uncertain.

A more speculative project is an infrared telescope with sufficient resolution that some detail would be visible in the image of an extrasolar planet. One concept for such a mission would consist of five Darwin-type telescopes flying in a geometrical formation some 6000 km across (equal to the radius of Earth). All five telescopes would collect light from the same extrasolar planet, and then reflect it onto a single mirror. The combined light would go to detectors on board a sixth spacecraft. The technology needed for such an ambitious mission does not yet exist but may become available within a few decades.

Sometime during the twenty-first century, missions such as Darwin may answer the question, "Are there worlds

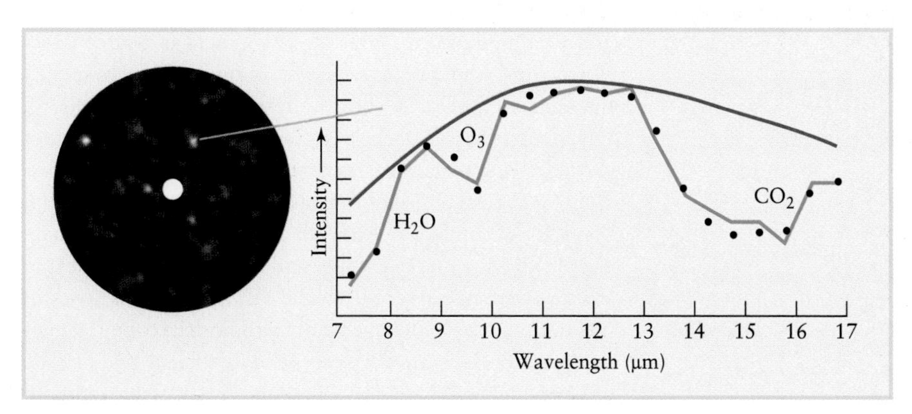

FIGURE 8-11 The Spectrum of a Simulated Planet The image on the left is a simulation of what infrared telescopes might see when looking for habitable planets. The white dot at the center is a nearby Sunlike star, and the smaller dots around it are planets orbiting the star. On the right is the simulated infrared spectrum of one of the planets, showing broad absorption lines of water vapor (H_2O), ozone (O_3), and carbon dioxide (CO_2). While all these molecules can be created by nonbiological processes, the presence of life will change the relative amounts of each molecule in the planet's atmosphere. Thus, the infrared spectrum of such planets will make it possible to identify worlds on which life may have evolved. *(NASA Jet Propulsion Laboratory)*

like Earth orbiting other stars?" If the answer is yes, radio searches for intelligent signals will gain even more impetus.

ConceptCheck 8-11: What exactly is a transit search watching for as it looks for planets around stars?

ConceptCheck 8-12: Which planets will a transit-search telescope not be able to find, even if quite nearby?

ConceptCheck 8-13: How will the Darwin telescope infer whether or not planets it observes harbor life?

Radio Searches for Alien Civilizations

If other civilizations are trying to communicate with us using radio waves, how can we know how to tune our radio telescopes? This is an important question, because if we fail to tune our radio telescopes just right, we might never know whether the aliens are out there.

A reasonable choice would be a frequency that is fairly free of interference from extraneous sources. SETI pioneer Bernard Oliver was the first to draw attention to a range of relatively noise-free frequencies in the neighborhood of the microwave emission lines of hydrogen (H) and hydroxide (OH) (**Figure 8-12**). This region of the microwave spectrum is informally called the **water hole,** not because we are looking for water, but because H and OH together would form the letters H_2O, or water.

In 1989, NASA began work on the High Resolution Microwave Survey (HRMS), an ambitious project to scan the entire sky at frequencies spanning the water hole from 10^3 MHz to 10^4 MHz. HRMS would have observed more than 800 nearby solar-type stars over a narrower frequency range in the hope of detecting signals that were either pulsed (like Morse code) or continuous (like the carrier wave for a TV or radio broadcast). The sophisticated signal-processing technology of HRMS would have been able to sift through tens of millions of individual frequency channels simultaneously. It would even have been able to detect the minute Doppler shifts in a signal coming from an alien planet as that planet spun on its axis and moved around its star.

Sadly, just one year after HRMS began operation in 1992, the U.S. Congress imposed a mandate requiring that NASA no longer support HRMS or any other radio searches for extraterrestrial intelligence. This decision, which was made on budgetary grounds, saved a few million dollars—an entirely negligible amount compared to the total NASA budget. Ironically, the senator who spearheaded this was from the state of Nevada, where tax dollars have been spent to signpost a remote desert road as "The Extraterrestrial Highway."

Even though U.S. government funding is no longer available for this search, several teams of scientists remain actively involved in SETI programs. Funding for these projects has come from nongovernmental organizations such as the Planetary Society and from private individuals. Since 1995 the SETI Institute in California has been carrying out Project Phoenix, the direct successor to HRMS. When complete, this project will have surveyed a thousand Sunlike stars within 200 ly at millions of radio frequencies. At Harvard University, BETA (the *B*illion-channel *E*xtra*T*errestrial *A*ssay) is scanning the sky at even more individual frequencies within the water hole. Other multifrequency searches are being carried out under the auspices of the University of Western Sydney in Australia and the University of California.

A major challenge facing SETI is the tremendous amount of computer time needed to analyze the mountains of data returned by radio searches. To this end, scientists at the University of California, Berkeley, have recruited nearly 5 million personal computer users to participate in a project called SETI@home. Each user receives actual data from a detector called SERENDIP IV (*S*earch for *E*xtraterrestrial *R*adio *E*missions from *N*earby, *D*eveloped, *I*ntelligent *P*opulations) and a data analysis program that also acts as a screensaver. When the computer's screensaver is on, the program runs, the data are analyzed, and the results are reported via the Internet to the researchers at Berkeley. The program then downloads new data to be analyzed. SETI@home has provided as much computer time as a single computer working full time for 3 million years! All current SETI projects make use of existing radio telescopes and must share telescope time with astronomy researchers. The SETI Institute is working to build and put into operation a radio telescope that will be dedicated solely to the search for intelligent signals. This telescope, called the Allen Telescope Array, will eventually

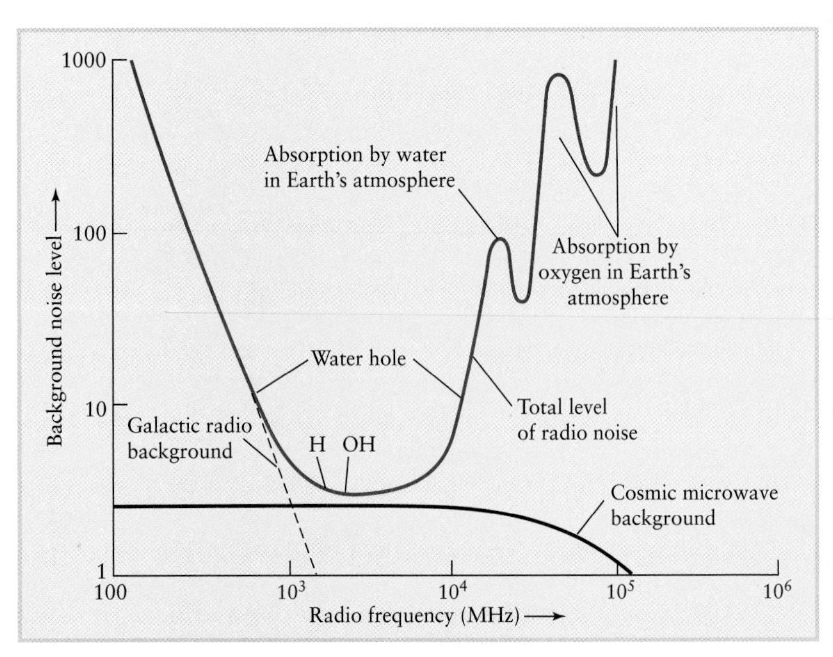

FIGURE 8-12 The Water Hole This graph shows the background noise level from the sky at various radio and microwave frequencies. The so-called water hole is a range of radio frequencies from about 10^3 to 10^4 megahertz (MHz) in which there is little noise and little absorption by Earth's atmosphere. Some scientists suggest that this noise-free region would be well suited for interstellar communication. Within this region dubbed the water hole, the principal source of noise is the afterglow of the Big Bang, called the cosmic microwave background. To put this graph in perspective, a frequency of 100 MHz corresponds to 100 on an FM radio, and 10^3 MHz is a frequency used for various types of radar. *(Adapted from C. Sagan and F. Drake)*

be hundreds of relatively small and inexpensive radio dishes working together. Perhaps this new array will be the first to detect a signal from a distant civilization.

Go to Video 8-3

The potential rewards from such searches are great. Detecting a message from an alien civilization could dramatically change the course of our own civilization, through the sharing of scientific information with another species or an awakening of social or humanistic enlightenment. In only a few years our technology, industry, and social structure might advance the equivalent of centuries into the future. Such changes would touch every person on Earth. Mindful of these profound implications, scientists push ahead with the search for extraterrestrial intelligence.

VISUAL LITERACY TASK

Life in the Universe

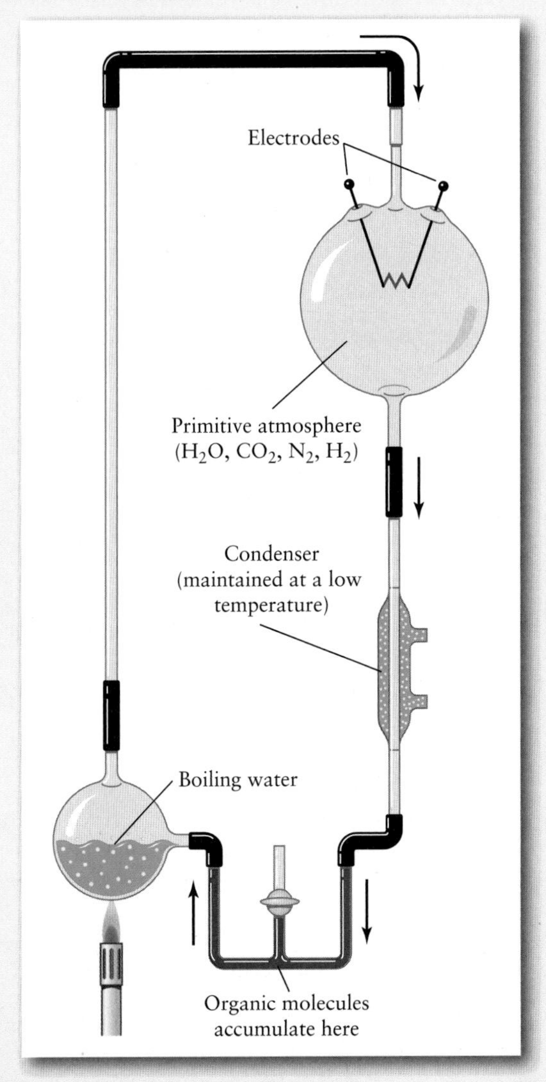

PROMPT: What would you tell a fellow student who said, **"Miller-Urey demonstrated that life can be created by mixing water with soil and energizing it with heat"?**

ENTER RESPONSE:

Guiding Questions:

1. The end result of the Miller-Urey experiment was the creation of
   a. amino acids.
   b. living bacteria.
   c. DNA.
   d. microscopic plants.

2. Miller and Urey utilized substances they believed were
   a. derived from plants.
   b. common to Earth's early atmosphere.
   c. similar to ooze found at the bottom of the ocean.
   d. similar to dirt and mud from riverbanks.

3. Miller and Urey energized the substances in their experiment using
   a. a controlled flame.
   b. electrical shocks, to simulate lightning.
   c. electricity to heat and simmer the concoction.
   d. the steam from boiling water to simulate a volcano.

4. The Miller-Urey experiment was designed to
   a. disprove the existence of a supernatural being.
   b. create living organisms from nonliving materials.
   c. convert plant life into animal life.
   d. recreate the early Earth conditions when life started.

KEY IDEAS AND TERMS

8-1 Planets and the chemical building blocks of life are found throughout space

- **Astrobiology** is the study of how life originates and evolves throughout the universe.
- All life on Earth, and presumably on other worlds, depends on organic (carbon-based) molecules. These molecules occur naturally throughout interstellar space.
- The organic molecules needed for life to originate were probably brought to the young Earth by comets or meteorites.
- The Miller-Urey experiment demonstrated the possible environmental conditions of an early Earth.

8-2 Europa and Mars are promising places for life to have evolved

- Besides Earth, the planet Mars and Jupiter's moon Europa are the most likely candidates to have had the right conditions for the origin of life.
- Mars once had liquid water on its surface, though it has none today. Life may have originated on Mars during the liquid water era.
- The *Viking Lander* spacecraft searched for microorganisms on the Martian surface but found no conclusive sign of their presence. Three decades later, NASA successfully landed two robotic rovers named *Spirit* and *Opportunity* at two very different sites on opposite sides of Mars. More recently, the *Phoenix Lander* and the LCROSS NASA missions found solid evidence of water on Mars and our Moon.
- Europa appears to have extensive liquid water beneath its icy surface. Future missions may search for the presence of life there.

8-3 Meteorites from Mars have been scrutinized for life-forms

- **SNC meteorites** come from Mars and are identified by the chemical composition of trace amounts of Martian atmospheric gas trapped within them.
- ALH 84001 is the only known specimen of a rock that was on Mars during the era when liquid water most likely existed on the planet's surface and provides only circumstantial evidence for Martian life.

8-4 The Drake equation helps scientists estimate how many civilizations inhabit our Galaxy

- The fact that life exists on Earth means that extraterrestrial life, including intelligent species, might evolve on planets around distant stars, given sufficient time and hospitable conditions.
- The collective effort of scientists looking for intelligent life beyond Earth is known as the **Search for Extraterrestrial Intelligence, or SETI.**
- The **Drake equation** is a tool for estimating the number of intelligent, communicative civilizations in our Galaxy.
- As radio waves travel at the speed of light, radio communication is probably more plausible than face-to-face visits with extraterrestrials.

8-5 Searches with space-based infrared telescopes and Earth-based radio telescopes for Earthlike planets and alien civilizations are under way

- Astronomers have discovered many planets by searching for the slight dimming of starlight as a planet "transits" its host star.
- Radio searches focus on a frequency that is fairly free of interference in the neighborhood of the microwave emission lines of hydrogen (H) and hydroxide (OH). This region of the microwave spectrum is called the **water hole,** because the letters H and OH together make H_2O, the chemical symbol for water.
- Astronomers have carried out a number of searches for radio signals from other stars. No signs of intelligent life have yet been detected, but searches are continuing and using increasingly sophisticated techniques.
- A new generation of orbiting telescopes may be able to detect terrestrial planets around nearby stars. If such planets are found, their infrared spectra may reveal the presence or absence of life.

QUESTIONS

Review Questions

1. Why are extreme life-forms on Earth, such as those shown in the photograph that opens this chapter, of interest to astrobiologists?
2. What is meant by "life as we know it"? Why do astrobiologists suspect that extraterrestrial life is likely to be of this form?
3. How have astronomers discovered organic molecules in interstellar space? Does this discovery mean that life of some sort exists in the space between the stars?
4. Mercury and Venus are both considered unlikely places to find life. Suggest why this should be.
5. Many science-fiction stories and movies—including *The War of the Worlds, Invaders from Mars, Mars Attacks!,* and *Martians, Go Home*—involve invasions of Earth by intelligent beings from Mars. Why Mars rather than any of the other planets?
6. Describe how the *Viking Landers* looked for evidence of life on Mars.
7. Explain which variable in the Drake equation is the most difficult to estimate and suggest why this would be.
8. Suppose someone brought you a rock that he claimed was a Martian meteorite. What scientific tests would you recommend be done to test this claim?
9. Why are most searches for extraterrestrial intelligence made using radio telescopes? Why are most of these carried out at frequencies between 10^3 MHz and 10^4 MHz?
10. Explain why planet-hunting infrared telescopes need to be placed in space.

Web Chat Questions

1. Suppose someone told you that the *Viking Landers* failed to detect life on Mars simply because the tests were designed to detect terrestrial life-forms, not Martian life-forms. How would you respond?
2. Science-fiction television shows and movies often depict aliens as looking very much like humans. Discuss the likelihood that intelligent creatures from another world would have (**a**) a biochemistry similar to our own, (**b**) two legs and two arms, and (**c**) about the same dimensions as a human.
3. The late, great science-fiction editor John W. Campbell exhorted his authors to write stories about organisms that think as well as humans but not *like* humans. Discuss the possibility that an intelligent being from another world might be so alien in its thought processes that we could not communicate with it.

4. If a planet always kept the same face toward its star, just as the Moon always keeps the same face toward Earth, most of the planet's surface would be uninhabitable. Discuss why.
5. How do you think our society would respond to the discovery of intelligent messages coming from a civilization on a planet orbiting another star? Explain your reasoning.
6. What do you think will set the limit on the lifetime of our technological civilization? Explain your reasoning.
7. The first of all Earth spacecraft to venture into interstellar space were *Pioneer 10* and *Pioneer 11,* which were launched in 1972 and 1973, respectively. Their missions took them past Jupiter and Saturn and eventually beyond the solar system. Both spacecraft carry a metal plaque with artwork (reproduced below) that shows where the spacecraft is from and what sort of creatures designed it. If an alien civilization were someday to find one of these spacecraft, which of the features on the plaque do you think would be easily understandable to them? Explain.

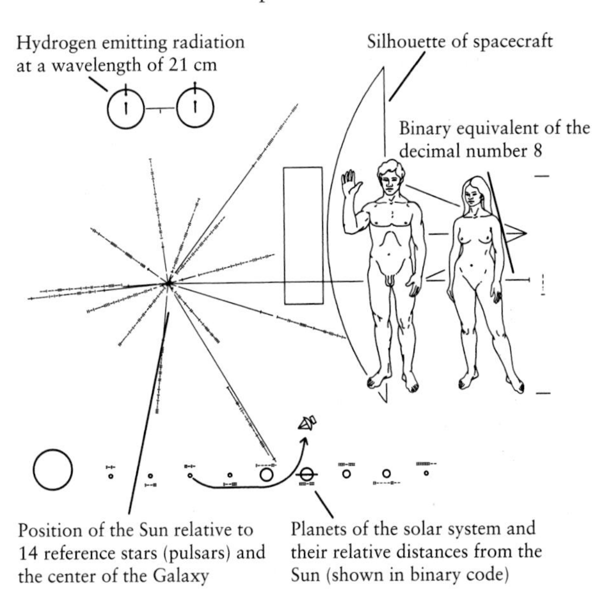

Collaborative Exercises

1. Any living creatures in the subsurface ocean of Europa would have to survive without sunlight. Instead, they might obtain energy from Europa's inner heat. Search the Internet for information about "black smokers," which are associated with high-temperature vents at the bottom of Earth's oceans. What kind of life is found around black smokers? How do these life-forms differ from the more familiar organisms found in the upper levels of the ocean?
2. Like other popular media, the Internet is full of claims of the existence of "extraterrestrial intelligence," namely, UFO sightings and alien abductions. (**a**) Choose a Web site of this kind and analyze its content using the idea of *Occam's razor,* the principle that if there is more than one viable explanation for a phenomenon, one should choose the simplest explanation that fits all the observed facts. (**b**) Read what a skeptical Web site has to say about UFO sightings. A good example is the Web site of the Committee for the Scientific Investigation of Claims of the Paranormal, or CSICOP. After considering what you have read on both sides of the UFO debate, discuss your opinions about whether aliens really have landed on Earth.
3. Imagine that astronomers have discovered intelligent life in a nearby star system. Your group is submitting a proposal for who on Earth should speak for the planet and what 50-word message should be conveyed. Prepare a maximum one-page proposal that states (**a**) who should speak for Earth and why; (**b**) what this person should say in 50 words; and (**c**) why this message is the most important compared to other things that could be said.

Observing Questions

1. Use *Starry Night*™ to examine the planet Mars. Select **Favourites > Explorations > Mars. Zoom** in or out on Mars and use the location scroller to examine the planet's surface. Based on what you observe, where on the Martian surface would you choose to land a spacecraft to search for the presence of life? Explain the reasons for your choice.
2. Use the *Starry Night*™ program to examine Jupiter's moon Europa. Select **Favourites > Explorations > Europa** to view this enigmatic moon of our largest planet. Use the location scroller and **Zoom** controls to examine the surface of this moon. Are there any surface features that suggest that this moon might harbor life? Is there any other evidence, perhaps not visible in this simulation, which suggests the possibility that life exists there?
3. Use *Starry Night*™ to investigate the likelihood of the existence of life in the universe beyond that found on Earth. Select **Favourites > Explorations > Atlas** and then open the **Options** pane. Expand the **Stars** layer and the **Stars** heading and click the checkbox to turn on the **Mark stars with extrasolar planets** option. Then use the hand tool or cursor keys to look around the sky. Each marked star in this view has at least one planet orbiting around it. Click and hold the **Increase current elevation** button in the toolbar until the viewpoint is about 1000 light-years from Earth. Note that the stars with extrasolar planets are contained in a small knot of stars in the region of our Sun, our solar neighborhood. This is because present techniques place a limit upon the distance to which we can find extrasolar planets in space. There is no logical reason why the same proportion of stars in the rest of the Milky Way, or indeed the rest of the universe, should not have companion planets. **Increase current elevation** again to about 90,000 light-years from Earth to see a view of the Milky Way Galaxy. Use the location scroller to view the Milky Way face-on and compare the clump of stars representing the solar neighborhood to the size of the Milky Way Galaxy. Click the **Increase current elevation** button again until the **Viewing Location** panel indicates a distance to Earth of about 0.300 Mly (300,000 light-years). In this view, each point of light represents a separate galaxy containing hundreds of billions of stars. Now, gradually **Increase current elevation** to about 1500 Mly (1.5 billion light-years from Earth) to see the entire database of 28,000 galaxies included in the *Starry Night*™ data bank. The borders of this cube of galaxies reaching out to about 500 million light-years from Earth represent less than 5% of the size of the observable universe and galaxies beyond the limits of this view. To view images of some of these very distant galaxies, open the view named **Hubble Deep Field** from the **Explorations** folder in the **Favourites** pane. This view from the center of Earth is centered upon a seemingly empty patch of sky. **Zoom** in to see this image of more than 1500 very distant galaxies that were found by the Hubble Space Telescope in a small region of our sky. Then open the view named **Hubble Ultra Deep Field** and **Zoom** in on an image of over 10,000 galaxies that extends to the limits of our observable universe. (**a**) Do these views influence your thoughts on the likelihood that life, including intelligent life, might exist elsewhere in the universe other than Earth? Explain your reasoning using the Drake equation as a guide. (**b**) Why is it unlikely that we will be able to communicate with life-forms that might exist on planets in these very distant regions of space?

ANSWERS

ConceptChecks

ConceptCheck 8-1: Carbon atoms can combine with other elements to form an impressively wide array of different molecules, and carbon atoms are plentiful around stars.

ConceptCheck 8-2: Carbon molecules are commonly found in meteorites and comets, which frequently crashed onto planets' surfaces during the early formation of the solar system.

ConceptCheck 8-3: They created amino acids and other compounds that living organisms need—they did not create living entities.

ConceptCheck 8-4: The moon does not have the needed thick atmosphere to keep any liquid water needed for life from evaporating into outer space.

ConceptCheck 8-5: The scientists were looking for bubbles to be given off when water was added, hoping that the microbes would use the water for observable life processes.

ConceptCheck 8-6: On Earth, gray hematite is found in places where water exists or has existed in the past.

ConceptCheck 8-7: The SNC meteorites have trace amounts of gas trapped within them that are nearly identical to the unique Martian atmosphere.

ConceptCheck 8-8: On Earth, magnetite and pure iron sulfide crystals often occur in the presence of certain kinds of living microbes, suggesting that life once existed in this material.

ConceptCheck 8-9: Radio waves travel at more than a hundred thousand kilometers every second whereas our fastest spacecraft can only travel at tens of thousands of kilometers every hour; therefore, the vast distances between stars are simply too far to physically travel.

ConceptCheck 8-10: L is particularly difficult to estimate because at about the same time the technology to communicate beyond one's own planet is developed, the technology of weapons of mass destruction can develop, which could end the civilization just as it starts (alternatively, the civilization can learn to live without war and exist for a very long time).

ConceptCheck 8-11: Transit telescopes are looking for an ever-so-slight dimming of a star as an orbiting planet moves between the star and the observer.

ConceptCheck 8-12: Because a transit-search telescope looks for the dimming of stars, if an orbiting planet never moves directly between the telescope and the star, as if seen from above its orbit, it will not be observed to dim the central star.

ConceptCheck 8-13: The Darwin instruments can measure a planet's spectra, which could have a peculiar look if living organisms are influencing the planet's atmosphere, which is what happens on Earth.

CalculationCheck

CalculationCheck 8-1: Many scientists estimate that about one Sunlike star forms in our Galaxy every year, so if this number increases three times, then the number of Sunlike stars in our Galaxy also increases to about three formed each year.

RIVUXG A composite view of the Sun showing the upheaval on the surface and the dynamic outstretched upper atmosphere of the corona. *(SOHO/LASCO/EIT/ESA/NASA)*

9 Probing the Dynamic Sun

CHAPTER LEARNING OBJECTIVES **By reading the sections of this chapter, you will learn:**

9-1 The Sun's energy is generated by thermonuclear reactions in its core

9-2 Energy slowly moves outward from the solar interior through several processes

9-3 The Sun's outer layers are the photosphere, chromosphere, and corona

9-4 Sunspots are low-temperature regions in the photosphere

9-5 The Sun's magnetic field also produces other forms of solar activity and causes aurorae on Earth

Our Sun is by far the brightest object in the sky. But, the Sun is far more than just an object that illuminates the daytime. Even before the advent of written history, the ancients knew the important role the brilliantly shining Sun has in making plants grow and energizing winds that change the weather. What they did not know was why the Sun shines or how its immense energy impacts Earth in unexpected ways. By earthly standards, the temperature of its glowing surface is remarkably high, reaching thousands of degrees. Yet there are regions of the Sun that reach far higher temperatures of tens of thousands or even millions of degrees. Gases at such temperatures emit high-energy ultraviolet light, which makes their intricate patterns appear prominent with an ultraviolet telescope in space, as shown in the chapter opening image. Some of the hottest and most energetic regions on the Sun spawn immense disturbances that ripple through space. These can eject material from the Sun far enough to reach Earth and other planets.

Just in the past century, by looking carefully at the details of how energy is emitted by the Sun, we have finally learned that the Sun, and in fact most stars, shines because hundreds of millions of tons of hydrogen are squeezed together and converted to helium every second at its core. We have also recently come to understand that the Sun has a surprisingly violent but nearly invisible atmosphere of its own, with a host of dynamic features such as sunspots whose numbers rise and fall on a predictable 11-year cycle. By studying the Sun's vibrations, we have begun to understand new details of the Sun's character far beneath its previously unexplored surface. And, perhaps most important, we are in the beginning phases of investigating how changes in the Sun's activity can affect Earth's fragile environment as well as threaten our technological society. If we can uncover the energetic processes at work inside our closest star, then we can better understand the distant stars far beyond our solar system that dot the night sky. ■

Stars like the Sun shine intensely by converting hydrogen atoms into slightly larger helium atoms releasing tremendous amounts of energy in the process.

9-1 The Sun's energy is generated by thermonuclear reactions in its core

If you were to ask the next five people you meet, "What is the most important object in the sky?" most people would say our Sun. The reasons for the Sun's importance are many, including that it provides light to warm Earth's surface, it provides energy that drives weather, and it underlies the ability of plants to grow through photosynthesis.

Our Sun also plays an important role in the cosmos. The Sun is the largest member of our solar system, hosting the central point our planets orbit around. It has almost a thousand times more mass than all the solar system's planets, moons, asteroids, comets, and meteoroids put together. But the Sun is also a star. In fact, it is a remarkably typical star, with a mass, size, surface temperature, and chemical composition that are roughly midway between the extremes exhibited by the myriad of other stars in the heavens.

Solar Energy

Go to Video 9-1

For most people, what matters most about the Sun is the energy that it radiates into space and lands on Earth. Without the Sun's warming rays, our atmosphere and oceans would freeze into an icy layer coating a desperately cold planet, and life on Earth would be impossible. To understand why we are here, we must understand the nature of the Sun.

Why is the Sun such an important source of energy? One reason is that the Sun has a far higher surface temperature than any of the planets or moons. The Sun's spectrum (see Figure 2-11) is close to that of an idealized blackbody with a temperature of 10,000°F (5800 K). Thanks to this high temperature, each square meter of the Sun's surface emits a tremendous amount of radiation, principally at visible wavelengths. Indeed, the Sun is the only object in the solar system that emits substantial amounts of visible light. The light that we see from the Moon and planets is actually sunlight that struck those worlds and was reflected toward Earth.

The Sun's size also helps us explain its tremendous energy output. Because the Sun is so large, the total number of square feet of radiating surface—that is, its surface area—is immense. Hence, the total amount of energy emitted by the Sun each second, called its **luminosity**, is very large indeed—about 3.9×10^{26} W, or 3.9×10^{26} J of energy emitted every second. Astronomers denote the Sun's luminosity by the symbol $L_\odot$. A circle with a dot in the center is the astronomical symbol for the Sun and was also used by ancient astrologers.

> **ConceptCheck 9-1:** **If the Sun emits light at nearly all possible wavelengths, which range of wavelengths is emitted with the most intensity?**

The Source of the Sun's Energy

What makes the Sun shine so brightly? Albert Einstein discovered the underlying key to the energy source within stars in 1905. According to his *theory of special relativity,* a quantity m of mass can in principle be converted into an amount of energy E, according to a now-famous equation:

Einstein's mass-energy equation

$$E = mc^2$$

E = amount of energy into which the mass can be converted, in joules
m = quantity of mass, in kilograms
c = speed of light = 3×10^8 m/s

The speed of light c is a large number, so c^2 is huge. Therefore, a small amount of matter can release an awesome amount of energy. (*Note that astronomers often find it easier to make these calculations using the metric system of measurement than U.S. standard units of measure.*)

Einstein did not fully appreciate at the time how tremendously his ideas would impact astronomy; it turns out that the temperatures and pressures deep within the core of the Sun are so intense that hydrogen nuclei can combine to produce helium nuclei in a *nuclear reaction* that transforms a tiny amount of mass into a large amount of energy. This process of converting hydrogen into helium is called **thermonuclear fusion.** (It is also sometimes called *thermonuclear burning,* even though nothing is actually burned in the conventional sense. Ordinary burning on Earth involves chemical reactions that rearrange the outer electrons of atoms but have no effect on the atoms' nuclei.) Thermonuclear fusion can take place only at extremely high temperatures. The reason is that all atomic nuclei have a positive electric charge and so tend to repel one another. But in the extreme heat and pressure at the Sun's center, positively charged hydrogen nuclei are moving so fast that they can overcome their electric repulsion and actually touch one another and combine, making new, larger atomic nuclei, and releasing energy in the process. On Earth, the same thermonuclear fusion provides the devastating energy released in a hydrogen bomb.

ANALOGY You can think of protons as tiny electrically charged spheres that are coated with a very powerful glue. If the spheres are not touching, the repulsion between their charges pushes them apart. But if the spheres are forced into contact, the strength of the glue "fuses" them together.

CAUTION Be careful not to confuse thermonuclear fusion with the similar-sounding process of *nuclear fission.* In nuclear fusion, energy is released by joining together nuclei of lightweight atoms such as hydrogen. In nuclear fission, by contrast, the nuclei of very massive atoms such as uranium or plutonium release energy by fragmenting into smaller nuclei. Nuclear power plants produce energy using fission, not fusion. (Generating power using fusion has been a goal of researchers for decades, but no one has yet devised a commercially viable way to do this.)

ConceptCheck 9-2: If hydrogen nuclei are positively charged, under what conditions can two hydrogen nuclei overcome electrical charge repulsion and combine into helium nuclei, thus releasing energy according to Einstein's equation, $E = mc^2$?

CalculationCheck 9-1: How much energy is released when just 5 kg of mass is converted into energy?

Converting Hydrogen to Helium

Without its single electron, the nucleus of a hydrogen atom (H) is the same thing as a single proton. In much the same way, a helium atom (He) nuclei, in the absence of its two electrons, consists of two protons and two neutrons. When they combine, with a concurrent release of energy, we can write the nuclear reaction as:

$$4\ \mathrm{H} \rightarrow \mathrm{He} + \text{energy}$$

In several separate reactions, two of the four protons are changed into neutrons, and eventually combine with the remaining protons to produce a helium nucleus. This sequence of reactions is called the **proton-proton chain** (see **Cosmic Connections: The Proton-Proton Chain**). Each time this process takes place, a small fraction (0.7%) of the combined mass of the hydrogen nuclei does not show up in the mass of the helium nucleus. This "lost" mass is converted into energy.

CAUTION You may have heard the idea that mass is always conserved (that is, it is neither created nor destroyed), or that energy is always conserved in a reaction. Einstein's ideas show that neither of these statements is quite correct, because mass can be converted into energy and vice versa. A more accurate statement is that the total amount of mass *plus* energy is conserved. Hence, the destruction of mass in the Sun does not violate any laws of nature.

For every four hydrogen nuclei converted into a helium nucleus, 4.3×10^{-12} J of energy are released. This may seem like only a tiny amount of energy, but it is about 10^7 times larger than the amount of energy released in a typical chemical reaction, such as occurs in ordinary burning. To produce the Sun's luminosity of 3.9×10^{26} J/s, 6×10^{11} kg (600 million metric tons) of hydrogen must be converted into helium each second. This rate is prodigious, but there is literally an astronomical amount of hydrogen in the Sun. In particular, the Sun's core contains enough hydrogen to have been giving off energy at the present rate for as long as the solar system has existed, about 4.56 billion years, and to continue doing so for more than 6 billion years into the future.

ConceptCheck 9-3: If 1 kg of hydrogen combines to form helium in the proton-proton chain, why is only 0.007 kg (0.7%) available to be converted into energy?

ConceptCheck 9-4: How do astronomers estimate that our Sun has a lifetime of about 10 billion years?

9-2 Energy slowly moves outward from the solar interior through several processes

While thermonuclear fusion is the source of the Sun's energy, this process cannot take place everywhere within the Sun. Extremely high temperatures—in excess of 10^7 K—are required for atomic nuclei to fuse together to form larger nuclei. The

COSMIC CONNECTIONS The Proton-Proton Chain

The most common form of hydrogen fusion in the Sun involves three steps, each of which releases energy.

Hydrogen fusion in the Sun usually takes place in a sequence of steps called the proton-proton chain. Each of these steps releases energy that heats the Sun and gives it its luminosity.

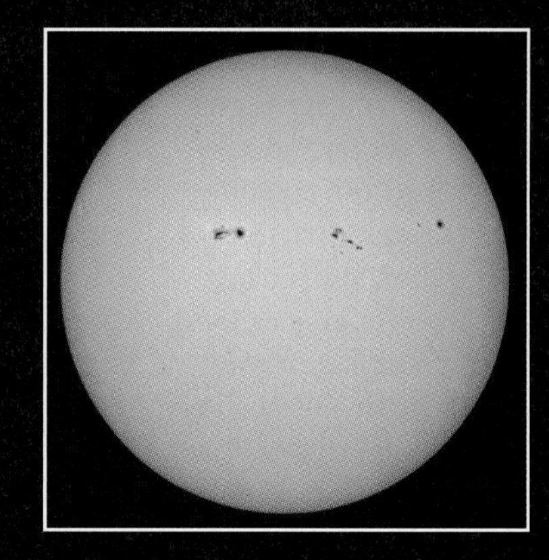

STEP 1

(a) Two protons (hydrogen nuclei, ^{1}H) collide.

(b) One of the protons changes into a neutron (shown in blue). The proton and neutron form a hydrogen isotope (^{2}H).

(c) One by-product of converting a proton to a neutron is a neutral, nearly massless neutrino (ν). This escapes from the Sun.

(d) The other by-product of converting a proton to a neutron is a positively charged electron, or positron (e^+). This encounters an ordinary electron (e^-), annihilating both particles and converting them into gamma-ray photons (γ). The energy of these photons goes into sustaining the Sun's internal heat.

^{1}H ^{1}H ^{2}H e^+ e^- ν γ γ

STEP 2

(a) The ^{2}H nucleus produced in Step 1 collides with a third proton (^{1}H).

(b) The result of the collision is a helium isotope (^{3}He) with two protons and one neutron.

(c) This nuclear reaction releases another gamma-ray photon (γ). Its energy also goes into sustaining the internal heat of the Sun.

^{2}H ^{1}H ^{3}He γ

STEP 3

(a) The ^{3}He nucleus produced in Step 2 collides with another ^{3}He nucleus produced from three other protons.

(b) Two protons and two neutrons from the two ^{3}He nuclei rearrange themselves into a different helium isotope (^{4}He).

(c) The two remaining protons are released. The energy of their motion contributes to the Sun's internal heat.

(d) Six ^{1}H nuclei went into producing the two ^{3}He nuclei, which combine to make one ^{4}He nucleus. Since two of the original ^{1}H nuclei are returned to their original state, we can summarize the three steps as:

$$4\ ^1\text{H} \longrightarrow\ ^4\text{He} + \text{energy}$$

^{3}He ^{3}He ^{4}He + ^{1}H + ^{1}H

Hydrogen fusion also takes place in all of the stars visible to the naked eye. (Fusion follows a different sequence of steps in the most massive stars, but the net result is the same.)

temperature of the Sun's visible surface, about 5800 K, is far too low for these reactions to occur there. Hence, fusion of atoms releasing energy can be taking place only within the Sun's interior. But precisely where does it take place? And how does the energy produced by fusion make its way to the surface, where it is emitted into space in the form of photons?

To answer these questions, we must understand conditions in the Sun's interior. Ideally, we would send an exploratory spacecraft to probe deep into the Sun; in practice, the Sun's intense heat would vaporize even the sturdiest spacecraft. Instead, astronomers use the laws of physics to construct a theoretical model of the Sun.

Hydrostatic and Thermal Equilibrium

There are several competing processes at work here. First, it is important to note that the Sun is neither growing nor shrinking, nor is it quickly becoming either hotter or cooler. To understand the nature of this stability, imagine a slab of material inside the Sun (**Figure 9-1*a***). In equilibrium, the slab on average will move neither up nor down. (In fact, there are upward and downward motions of material inside the Sun, but these motions average out in the long run.) Balance, or equilibrium, is maintained by a balance among three competing forces that act on this imaginary slab:

1. The downward force from the overlying, heavy layers of solar material above the slab due to their weight.
2. The upward forces generated by expanding hot gases pushing the slab upward.
3. The slab's weight itself—that is, the downward gravitational pull it feels from the rest of the Sun.

To be stationary, the upward pushing forces from below must perfectly balance both the slab's weight and the weight of overlying layers from above. Because the slab itself has weight, the upward forces below the slab must be greater than that above the slab for everything to be equal. In other words, force across a slab, better described as *pressure,* has to increase with increasing depth. For the same reason, pressure from the weight of overlying water increases as you dive deeper into the ocean (Figure 9-1*b*) or, in the opposite sense, atmospheric pressure decreases as one climbs up a mountain because of less overlying atmosphere pressing down on you.

In much the same way, we can make inferences about the slab's density. If it is too dense, its weight will be too great and it will sink; if the density is too low, the slab will rise. If things are going to remain suspended with balanced forces, the density of solar material must have a certain value at each depth within the solar interior. The same principle applies to objects that float beneath the surface of the ocean. Scuba divers wear weight belts to increase their average density so that they will neither rise nor sink but will stay submerged at the same level. Astronomers refer to this balanced equilibrium state of a star, such as the Sun, as being in **hydrostatic equilibrium.**

Another consideration is that the Sun's interior is so hot that it is completely gaseous. Gases compress and become more dense when you apply greater pressure to them, so density must increase along with pressure as you go to greater depths within the Sun. Furthermore, when you compress a gas, its temperature tends to rise, so the temperature must also increase as you move toward the Sun's center.

While the temperature in the solar interior is different at different depths, the temperature at each depth remains constant in time. All the energy generated by thermonuclear reactions in the Sun's core must be transported to the Sun's glowing surface, where it can be radiated into space. If too much energy flowed from the core to the surface to be radiated away, the Sun's interior would cool down; the Sun's interior would heat up if too little energy flowed to the surface. This principle describing the Sun is called **thermal equilibrium.**

ConceptCheck 9-5: If our Sun were much less massive and only one-half the diameter, how would the pressure at the Sun's center be different from what it actually is?

ConceptCheck 9-6: If our Sun were not in thermal equilibrium and too little energy successfully made it to the surface, how would the Sun's core be different?

Pressure from gases above the slab

Slab of solar material

Weight of the slab

Pressure from gases below the slab

(a) Material inside the Sun is in hydrostatic equilibrium, so forces balance

(b) A fish floating in water is in hydrostatic equilibrium, so forces balance

FIGURE 9-1 Hydrostatic Equilibrium (a) Material in the Sun's interior tends to move neither up nor down. The upward forces on an imaginary slab of solar material (due to forces from gases pushing up below the slab) must balance the downward forces (due to the slab's weight and the weight of gases above the slab). Hence, the pressure must increase with increasing depth. (b) The same principle applies to a fish floating in water. In equilibrium, the forces balance and the fish neither rises nor sinks. *(b: Sergei25/Shutterstock)*

Transporting Energy Outward from the Sun's Core

But exactly how is energy transported from the Sun's center where it is produced outward to its surface? There are three methods of energy transport: *conduction, convection,* and *radiative diffusion.* Only the last two are important inside the Sun.

If you heat one end of a metal bar with a blowtorch, energy flows to the other end of the bar so that it too becomes warm. The efficiency of this method of energy transport, called **conduction,** varies significantly from one substance to another. For example, copper is a good conductor of heat, but wood is not—copper cooking pots often have wooden handles to keep the chef from being burned. Conduction is not an efficient means of energy transport in substances with low average densities, including the gases inside the Sun.

Inside the Sun, energy moves from center to surface by two other means: convection and radiative diffusion. **Convection** is the circulation of fluids—gases or liquids—between hot and cool regions. Hot gases rise toward a star's surface, while cool gases sink back down toward the star's center. This physical movement of gases transports heat energy outward in a star, just as the physical movement of water boiling in a pot transports energy from the bottom of the pot (where the heat is applied) to the cooler water at the surface (see Figure 5-6).

In **radiative diffusion,** photons emitted from the thermonuclear inferno at a star's center spread outward toward the star's surface. Individual photons are absorbed and reemitted by atoms and electrons inside the star. The overall result is a general outward migration from the hot core, where photons are constantly created, toward the cooler surface, where they escape into space.

ConceptCheck 9-7: Why is the energy transport process of conduction relatively unimportant when studying how energy moves toward the Sun's surface?

Modeling the Sun

To construct a model of a star like the Sun, astronomers express the ideas of hydrostatic equilibrium, thermal equilibrium, and energy transport as a set of equations, most often done using the metric system of measurement. To ensure that the model applies to the particular star under study, they also make use of astronomical observations of the star's surface. (For example, to construct a model of the Sun, they use the data that the Sun's surface temperature is 5800 K, its luminosity is 3.9×10^{26} W, and the gas pressure and density at the surface are almost zero.) The astrophysicists then use a computer to solve their set of equations and calculate conditions layer by layer in toward the star's center. The result is a model of how temperature, pressure, and density increase with increasing depth below the star's surface.

Table 9-1 and **Figure 9-2** show a theoretical model of the Sun that was calculated in just this way. Different models of the Sun use slightly different assumptions, but all models give essentially the same results as those shown here. From such computer models we have learned that at the Sun's center the density is 160,000 kg/m^3 (14 times the density of lead!), the temperature is 1.55×10^7 K, and the pressure is 3.4×10^{11} atm. (One atmosphere, or 1 atm, is the average atmospheric pressure at sea level on Earth.)

Table 9-1 and Figure 9-2 show that the solar luminosity rises to 100% at about one-quarter of the way from the Sun's center to its surface. In other words, the Sun's energy production occurs within a volume that extends out only to 0.25 $R_\odot$. (The symbol $R_\odot$ denotes the solar radius, or radius of the Sun

Table 9-1 A Theoretical Model of the Sun

Distance from the Sun's center (solar radii)	Fraction of luminosity	Fraction of mass	Temperature ($\times 10^{6}$°F)	Density (kg/m^3)	Pressure relative to pressure at center
0.0	0.00	0.00	27.9	160000	1
0.1	0.42	0.07	23.4	90000	0.46
0.2	0.94	0.35	17.1	40000	0.15
0.3	1.00	0.64	12	13000	0.04
0.4	1.00	0.85	8.6	4000	0.007
0.5	1.00	0.94	6.1	1000	0.001
0.6	1.00	0.98	3.9	400	0.0003
0.7	1.00	0.99	2.2	80	4×10^{-5}
0.8	1.00	1.00	1.3	20	4×10^{-6}
0.9	1.00	1.00	0.5	2	3×10^{-7}
1.0	1.00	1.00	0.01	0.00030	4×10^{-13}

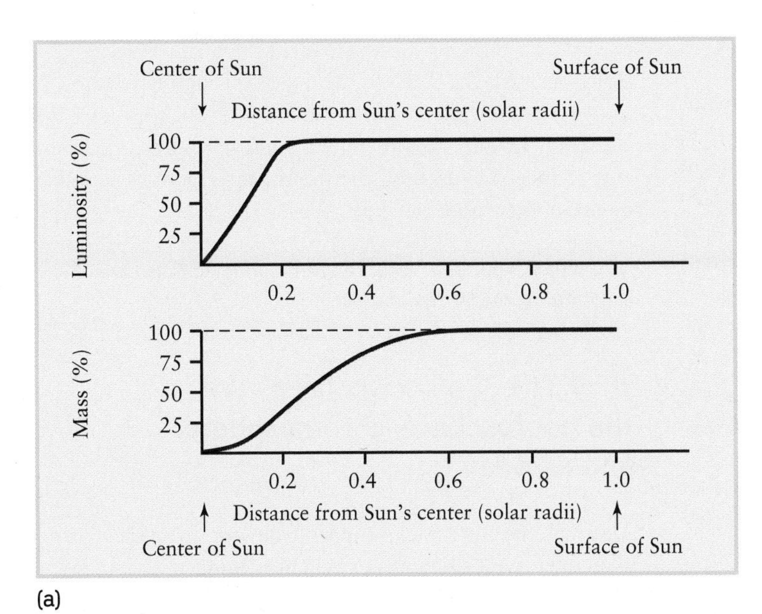

FIGURE 9-2 A Theoretical Model of the Sun's Interior These graphs depict what percentage of the Sun's total luminosity is produced within each distance from the center (upper a), what percentage of the total mass lies within each distance from the center (lower a), the temperature at each distance (upper b), and the density at each distance (lower b). (See Table 9-1 for a numerical version of this model.)

as a whole, equal to 696,000 km.) Outside 0.25 $R_\odot$, the density and temperature are too low for thermonuclear reactions to take place. Also note that 94% of the total mass of the Sun is found within the inner 0.5 $R_\odot$. Hence, the outer 0.5 $R_\odot$ contains only a relatively small amount of material.

Precisely how energy flows from the Sun's center outward depends on how easily photons move through the gas. If the solar gases are comparatively transparent, photons can travel moderate distances before being scattered or absorbed, and energy is thus transported by radiative diffusion. If the gases are comparatively opaque, photons cannot get through the gas easily and heat builds up. Convection then becomes the most efficient means of energy transport. The gases start to churn, with hot gas moving upward and cooler gas sinking downward.

From the center of the Sun out to about 0.71 $R_\odot$, energy is transported by radiative diffusion. Hence, this region is called the **radiative zone.** Beyond about 0.71 $R_\odot$, the temperature is low enough (a mere 2×10^6 K or so) for electrons and hydrogen nuclei to join into hydrogen atoms. These atoms are very effective at absorbing photons, much more so than at absorbing free electrons or nuclei, and this absorption chokes off the outward flow of photons. Therefore, beyond about 0.71 $R_\odot$, radiative diffusion is not an effective way to transport energy. Instead, convection dominates the energy flow in this outer region, which is why it is called the **convective zone.** **Figure 9-3** shows these aspects of the Sun's internal structure.

Although energy travels through the radiative zone in the form of photons, the photons have a difficult time of it. As shown in Table 9-1, the material in this zone is extremely dense, so photons from the Sun's core take a long time to diffuse through the radiative zone. As a result, it takes approximately 170,000 years for energy created at the Sun's center to travel 696,000 km to the solar surface and finally escape as sunlight. The energy flows outward at an average rate of 50 cm per hour, or about 20 times slower than a snail's pace.

Once the energy escapes from the Sun, it travels much faster—at the speed of light. Thus, solar energy that reaches you today took only 8 minutes to travel the 150 million km from the Sun's surface to Earth. But this energy was actually produced by thermonuclear reactions that took place in the Sun's core hundreds of thousands of years ago.

ConceptCheck 9-8: Which of the following decreases when we move from the Sun's central core outward: temperature, mass, or luminosity?

CalculationCheck 9-2: By what percentage does the Sun's temperature drop from its central core temperature moving out to a distance of one-half its radius?

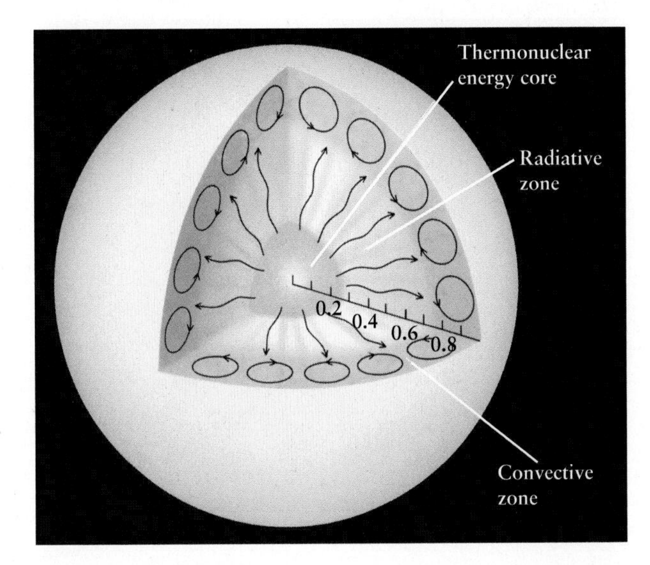

FIGURE 9-3 The Sun's Internal Structure Thermonuclear reactions occur in the Sun's core, which extends out to a distance of 0.25 $R_\odot$ from the center. Energy is transported outward, via radiative diffusion, to a distance of about 0.71 $R_\odot$. In the outer layers between 0.71 $R_\odot$ and 1.00 $R_\odot$, energy flows outward by convection.

Probing the Sun's Interior

If the Sun's interior is not visible from the surface, how might you go about figuring out what is inside? For that matter, how might you determine if a melon is ripe at your local grocery store without cutting it open? Vibrations are a useful tool for examining the hidden interiors of all kinds of objects. Much like food shoppers who tap melons to listen for particular vibrations and much like geologists who determine the structure of Earth's interior by using seismographs to record vibrations during earthquakes, one powerful technique to infer what is going on beneath the Sun's surface involves measuring vibrations of the Sun as a whole. This field of solar research is called **helioseismology.**

The Sun oscillates in millions of ways as a result of waves resonating in its interior. **Figure 9-4** is a computer-generated illustration of one such mode of vibration. Helioseismologists can deduce information about the solar interior from measurements of these oscillations. For example, they have been able to set limits on the amount of helium in the Sun's core and convective zone and to determine the thickness of the transition region between the radiative zone and convective zone. They have also found that the convective zone is thicker than previously thought.

Another approach is to carefully measure everything that comes out of the Sun and then determine how it must have been formed. As part of the process of thermonuclear fusion, protons change into neutrons and release particles called **neutrinos.** Like photons, neutrinos are particles that have no electric charge. Unlike photons, however, neutrinos interact only very weakly with matter and are highly difficult to capture. Even the vast bulk of the Sun offers little impediment to the passage of neutrinos, so neutrinos must be streaming out of the core and into space. Indeed, the conversion of hydrogen into helium at the Sun's center produces 10^{38} neutrinos each second. Every second, about 10^{14} neutrinos created within the Sun must pass through each square meter of Earth. The challenge is that neutrinos are exceedingly difficult to detect. Just as neutrinos pass unimpeded through the Sun, they also pass through Earth almost as if it were not there. When we are careful about how we capture neutrinos, we are able to confirm that the Sun's energy is indeed caused by thermonuclear reactions just like our computer models tell us.

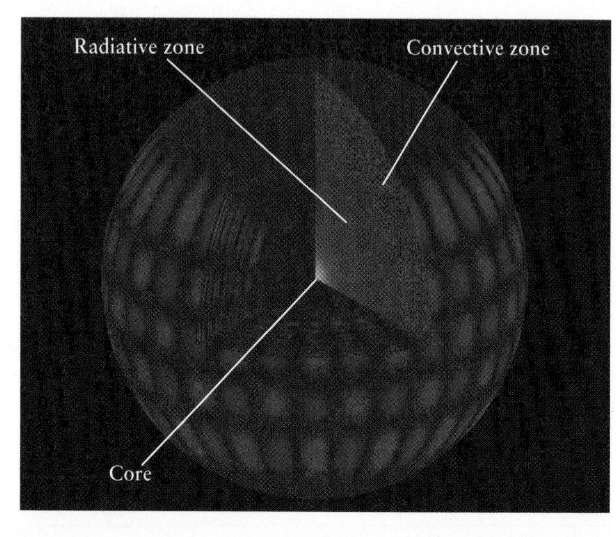

FIGURE 9-4 A Sound Wave Resonating in the Sun This computer-generated image shows one of the millions of ways in which the Sun's interior vibrates. The regions that are moving outward are colored blue; those moving inward are red. As the cutaway shows, these oscillations are thought to extend into the Sun's radiative zone (compare Figure 9-3). *(National Solar Observatory)*

ConceptCheck 9-9: What can be determined from carefully monitoring the Sun's vibrations?

9-3 The Sun's outer layers are the photosphere, chromosphere, and corona

Although the Sun's core is hidden from our direct view using telescopes, we can capture sunlight coming from the high-temperature gases that make up the Sun's atmosphere. These outermost layers of the Sun prove to be the sites of truly dramatic activity, much of which has a direct impact on our planet. By studying these layers, we gain further insight into the character of the Sun as a whole.

Observing the Photosphere

A visible-light photograph like **Figure 9-5** makes it appear that the Sun has a definite surface. This is actually an illusion—the Sun is gaseous throughout its volume because of its high internal temperature, and the gases simply become less and less dense as you move farther away from the Sun's center.

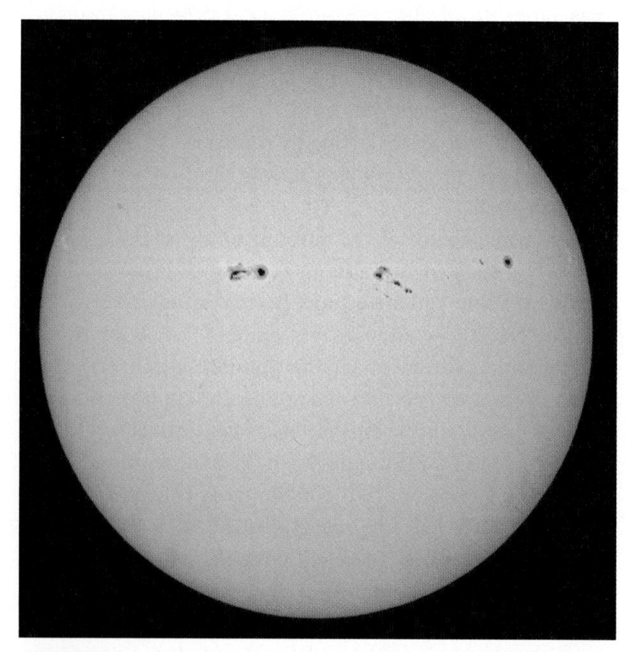

FIGURE 9-5 R I V U X G **The Photosphere** The photosphere is the layer in the solar atmosphere from which the Sun's visible light is emitted. Note that the Sun appears darker around its limb, or edge; here we are seeing the upper photosphere, which is relatively cool and thus glows less brightly. (The dark sunspots, which we discuss in Section 9-4, are also relatively cool regions.) *(Babak Tafreshi/Science Source)*

Why, then, does the Sun appear to have a sharp, well-defined surface? The reason is that essentially all of the Sun's visible light emanates from a single, thin layer of gas called the **photosphere** ("sphere of light"). Just as you can see only a certain distance through Earth's atmosphere before objects vanish in the haze, we can see only about 250 miles (400 km) into the photosphere. This distance is so small compared with the Sun's radius of 865,000 mi (696,000 km) that the photosphere appears to be a definite surface. Although this is an arbitrary line, astronomers usually define everything beneath the photosphere as the Sun's interior and everything above the photosphere as the Sun's atmosphere.

Although the photosphere is a very active place, it actually contains relatively little material. It has a density of only about 10^{-4} kg/m^3, roughly 0.01% the density of Earth's atmosphere at sea level. The photosphere is made primarily of hydrogen and helium, the most abundant elements in the solar system. Despite being such a thin gas, the photosphere is surprisingly opaque to visible light. If it were not so opaque, we could see into the Sun's interior to a depth of hundreds of thousands of miles, instead of a mere 250 mi.

We can learn still more about the photosphere by examining it with a telescope—but only when using special dark filters to prevent eye damage. *Looking directly at the Sun without the correct filter, whether with the naked eye or with a telescope, can cause permanent blindness!* Under good observing conditions, astronomers using such filter-equipped telescopes can often see a blotchy pattern in the photosphere (**Figure 9-6**). Each light-colored **granule** measures about 600 mi (1000 km) across—equal in size to the areas of Texas and Oklahoma combined—and is surrounded by a darkish boundary. The difference in brightness between the center and the edge of a granule corresponds to a temperature drop of about 300 K.

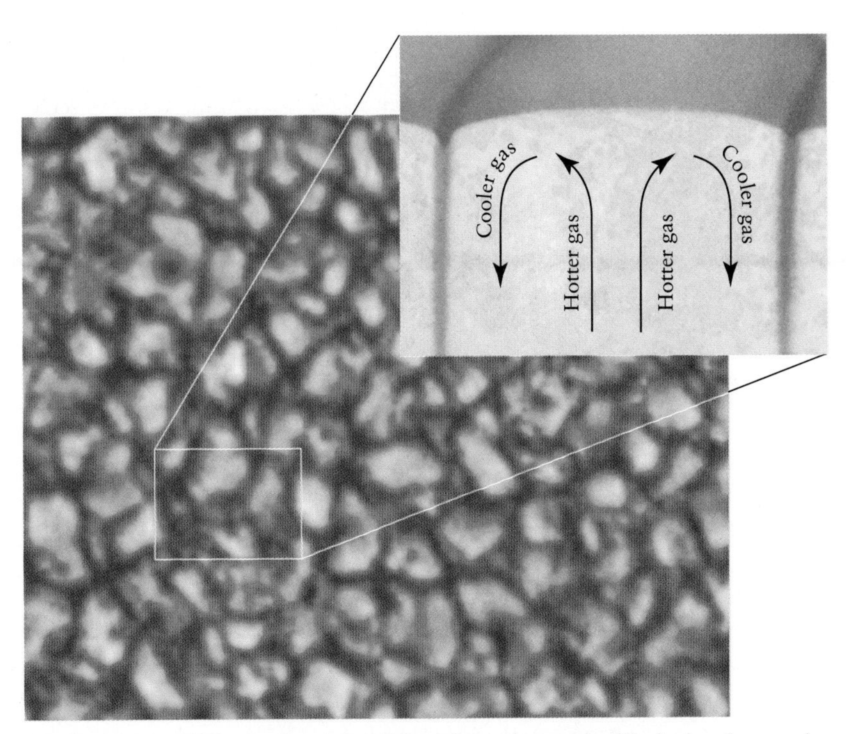

FIGURE 9-6 R I V U X G **Solar Granulation** High-resolution photographs of the Sun's surface reveal a blotchy pattern called granulation. Granules are convection cells about 600 mi (1000 km) wide in the Sun's photosphere. The inset shows rising hot gas coming up making the center of each granule bright. Cooler gas sinks downward along the boundaries between granules; this gas glows less brightly, giving the granule edges their dark appearance. This convective motion transports heat from the Sun's interior outward to the solar atmosphere. *(Hinode JAXA/NASA/PPARC)*

This granulation appearance is caused by convection of the gas in the photosphere. The inset in Figure 9-6 shows how gas from lower levels rises upward in granules, cools off, spills over the edges of the granules, and then plunges back down into the Sun. This can occur only if the gas is heated from below, like a pot of water being heated on a stove. Granules form, disappear, and reform in cycles lasting only a few minutes. At any one time, about 4 million granules cover the solar surface.

Superimposed on the pattern of granulation are even larger cells, or supergranules, that are about 20,000 miles (35,000 km) across, large enough to enclose several hundred granules (**Figure 9-7**). This large-scale convection moves at only about 900 mi/h (1400 km/h), about one-tenth the speed of gases churning in a granule that can last about a day.

ANALOGY Similar patterns of large-scale and small-scale convection can be found in Earth's atmosphere. On the large scale, air rises gradually at a low-pressure area, and then sinks gradually at a high-pressure area, which might be hundreds of kilometers away. Thunderstorms in our atmosphere are small but intense convection cells within which air moves rapidly up and down. Like granules, they last only a relatively short time before they dissipate.

ConceptCheck 9-10: What causes the photosphere to bubble like water boiling on the stove?

FIGURE 9-7 R I V U X G **Supergranules and Large-Scale Convection** Supergranules display relatively little contrast between their center and edges, so they are hard to observe in ordinary images. But they can be seen in a false-color Doppler image like this one. Light from gas that is approaching us (that is, rising) is shifted toward shorter wavelengths, while light from receding gas (that is, descending) is shifted toward longer wavelengths (see Section 2-5). *(NASA/MSFC)*

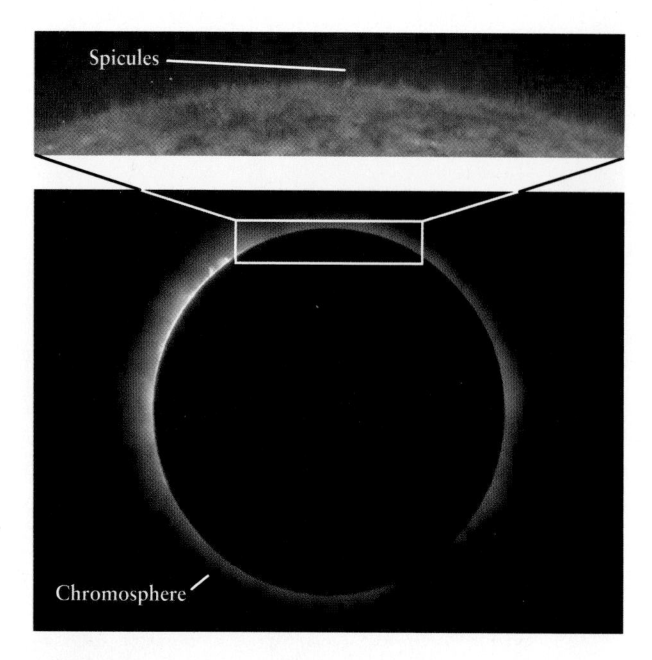

FIGURE 9-8 RIVUXG **The Chromosphere** During a total solar eclipse, the Sun's glowing chromosphere can be seen around the edge of the Moon. It appears pinkish because its hot gases emit light at only certain discrete wavelengths, principally the H_α emission of hydrogen at a red wavelength of 656.3 nm. The expanded area above shows spicules, jets of chromospheric gas that surge upward into the Sun's outer atmosphere. *(top: © Image Courtesy Mr.Eclipse.com; bottom: Science Source)*

The Sun's Chromosphere

Go to Video 9-2

An ordinary visible-light image such as Figure 9-5 gives the impression that the Sun ends at the top of the photosphere. But during a total solar eclipse, the Moon blocks the photosphere from our view, revealing a glowing, pinkish layer of gas above the photosphere (**Figure 9-8**). This is the tenuous **chromosphere** ("sphere of color"), the second of the three major levels in the Sun's atmosphere. The chromosphere is only about one ten-thousandth (10^{-4}) as dense as the photosphere, or about 10^{-8} as dense as our own atmosphere. No wonder it is normally invisible!

Named after the Sun god Helios, the element helium was first discovered in the Sun's chromosphere long before it was found to exist on Earth.

Unlike the photosphere, which has an absorption line spectrum, a spectrum observed from studying gases in the chromosphere is dominated by emission lines. One of the most obvious emission lines in the chromosphere's spectrum is the H_α line at 656.3 nm, which is emitted by a hydrogen atom when its single electron falls from the $n = 3$ level to the $n = 2$ level. This wavelength is in the red part of the spectrum, which gives the chromosphere its characteristic pinkish color. The spectrum also contains emission lines of ionized helium. In fact, helium was originally discovered in the chromospheric spectrum in 1868, almost 30 years before helium gas was first isolated on Earth.

What might be most surprising about the chromosphere is that the temperature *increases* with increasing height in the chromosphere. This is just the opposite of the situation in the photosphere, where temperature decreases with increasing height. This is very surprising, since temperatures should decrease as you move away from the Sun's interior. In fact, though, the temperature is about 10,000°F (5800 K) at the top of the photosphere; just 1250 mi higher, at the top of the chromosphere, the temperature rises incredibly to nearly 50,000°F (28,000 K).

The top photograph in Figure 9-8 is a high-resolution image of the Sun's chromosphere, usually only seen through an H_α filter. This image shows numerous vertical spikes, which are actually jets of rising gas called **spicules.** A typical spicule lasts just 15 minutes or so: It rises at the rate of about 45,000 mi/h (72,000 km/h), can reach a height of several thousand miles, and then rapidly collapses and fades away (**Figure 9-9**).

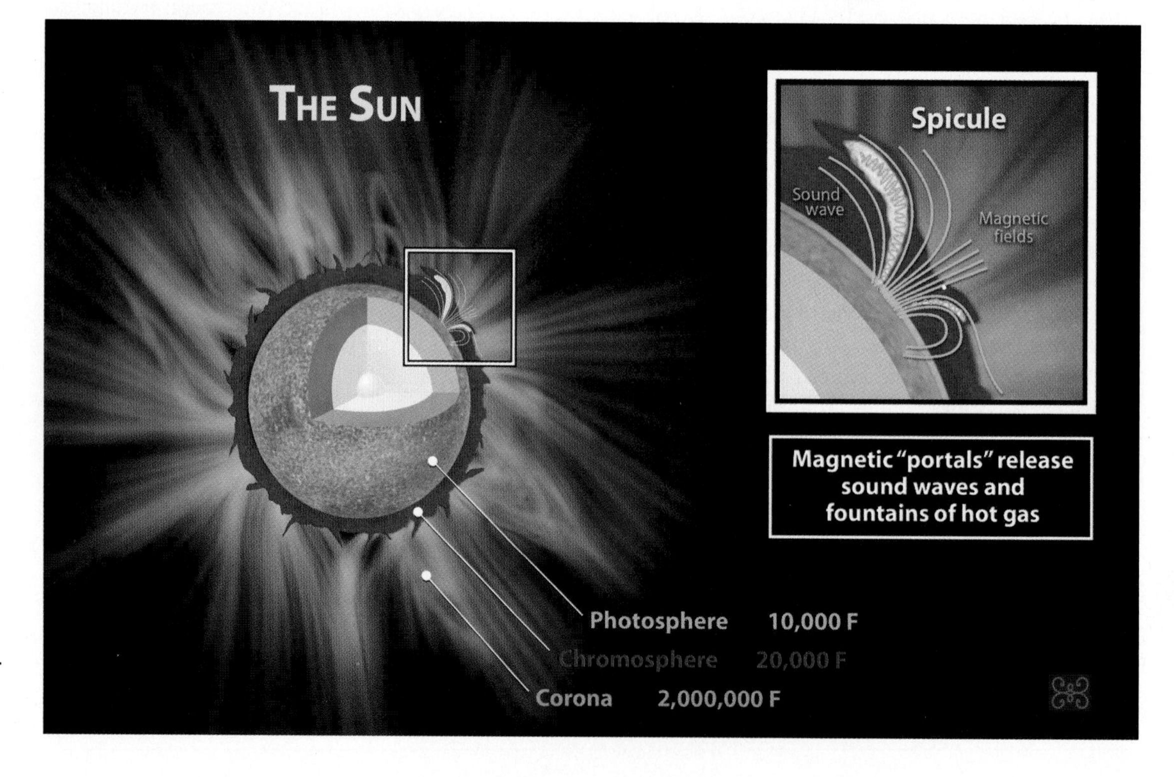

FIGURE 9-9 Spicules in the Chromosphere This schematic diagram shows the three layers of the solar atmosphere. The lowest, the photosphere, is about 250 miles thick. The chromosphere extends about 1250 miles higher, with spicules jutting up to about 6,000 miles above the photosphere. Above a transition region is the Sun's outermost layer, the corona. It extends many millions of miles out into space. *(Zina Deretsky, National Science Foundation)*

Approximately 300,000 spicules exist at any one time, covering about 1% of the Sun's surface.

Spicules are generally located directly above the edges of granule groups. This is a surprising result, because chromospheric gases are rising in a spicule while photospheric gases are *descending* at the edges of granule groups. What, then, is pulling gases upward to form spicules? The answer proves to be the Sun's intense magnetic field, which we discuss in Sections 9-4 and 9-5.

ConceptCheck 9-11: How tall are spicules in the Sun's chromosphere: the height of tall buildings, the distance between large nearby cities, or the distance across the entire United States?

The Corona

The outermost region of the Sun's atmosphere begins at the top of the chromosphere. Called the **corona,** it reaches out to an enormous distance more than several times the diameter of the Sun. Despite its tremendous extent, the corona is only about one-millionth (10^{-6}) as bright as the photosphere—no brighter than the full moon. Hence, the corona can be viewed only when the light from the photosphere is blocked out by the eclipsing Moon during a total eclipse or by the use of a specially designed telescope that can block the majority or the photosphere.

Figure 9-10 is an exceptionally detailed photograph of the Sun's corona taken during a solar eclipse. It shows that the corona is not merely a spherical shell of gas surrounding the Sun. Rather, numerous streamers extend in different directions far above the solar surface. The shapes of these streamers vary on timescales of days or weeks. The temperatures in the corona are extremely high considering how far it is from the Sun's core—temperatures can reach 5 billion degrees Fahrenheit or 3 million Kelvins—far greater than the temperatures in the chromosphere. **Figure 9-11** shows how temperature in both the chromosphere and corona varies with altitude.

FIGURE 9-10 R I V U X G **The Solar Corona** This striking photograph of the corona was taken during the total solar eclipse of March 29, 2006, from the town of Cape Coast in central Ghana. Numerous streamers extend for millions of kilometers above the solar surface. The unearthly light of the corona is one of the most extraordinary aspects of experiencing a solar eclipse. *(Timothy F. Slater)*

CAUTION The corona is actually not very "hot"—that is, it contains very little thermal energy. The reason is that the corona is nearly a vacuum. In the corona there are only about 10^{11} atoms per cubic meter, compared with about 10^{23} atoms per cubic meter in the Sun's photosphere and about 10^{25} atoms per cubic meter in the air that we breathe. Because of the corona's high temperature, the atoms there are moving at very high speeds. But because there are so few atoms in the corona, the total amount of energy in these moving atoms (a measure of how "hot" the gas is) is rather low. If you flew a spaceship into the corona, you would have to worry about becoming overheated by the intense light coming from the photosphere, but you would notice hardly any heating from the corona's ultrathin gas.

ANALOGY The situation in the corona is similar to that inside a conventional oven that is being used for baking. Both the walls of the oven and the air inside the oven are at the same high temperature, but the air contains very few atoms and thus carries little energy. If you put your hand in the oven momentarily, the lion's share of the heat you feel is radiation from the oven walls.

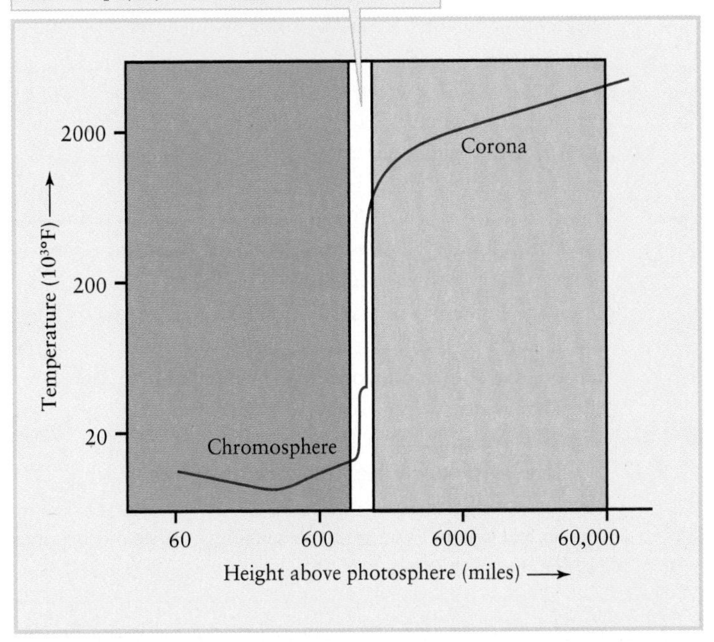

FIGURE 9-11 Temperatures in the Sun's Upper Atmosphere This graph shows how temperature varies with altitude in the Sun's chromosphere and corona and in the narrow transition region between them. In order to show a large range of values, both the vertical and horizontal scales are nonlinear. *(Adapted from A. Gabriel)*

The low density of the corona explains why it is so dim compared with the photosphere. In general, the higher the temperature of a gas, the brighter it glows. But because there are so few atoms in the corona, the net amount of light that it emits is very feeble compared with the light from the much cooler, but also much denser, photosphere.

> ConceptCheck 9-12: **Why is the corona so difficult to see if it is so much hotter than the photosphere?**

The Solar Wind and Coronal Holes

Go to Video 9-3

Earth's gravity keeps our atmosphere from escaping into space. In the same way, the Sun's powerful gravitational attraction keeps most of the gases of the photosphere, chromosphere, and corona from escaping. But the corona's high temperature means that its atoms and ions are moving at very high speeds, nearly a million miles per hour. As a result, some of the coronal gas can and does escape the Sun's gravitational attraction. This outflow of gas, is called the **solar wind.**

Each second the Sun ejects about a million tons of material into the solar wind. But the Sun is so massive that, even over its entire lifetime, it will eject only a few tenths of a percent of its total mass. The solar wind is composed almost entirely of electrons and nuclei of hydrogen and helium. About 0.1% of the solar wind is made up of ions of more massive atoms, such as silicon, sulfur, calcium, chromium, nickel, iron, and argon. The aurorae seen at far northern or southern latitudes on Earth are produced when electrons and ions from the solar wind enter our upper atmosphere.

Figure 9-12 reveals that the corona is not uniform in temperature or density. The densest, highest-temperature regions appear bright, while the thinner, lower-temperature regions are dark. This large dark area, which is not emitting much light, is called a **coronal hole** because it is almost devoid of luminous gas. But this does not mean this region is empty. On the contrary, high-speed particles streaming away from the Sun can most easily flow outward through these particularly thin regions. Therefore, it is thought that coronal holes are the main corridors through which particles of the solar wind rapidly escape from the Sun and flow across the solar system.

The surprisingly high temperatures in the corona and the chromosphere are not at all what we would expect. Just as you feel warm if you stand close to a campfire but become colder the farther away you move, we would naturally expect that the temperature in the corona and chromosphere would *decrease* with increasing distance from the photosphere. Why, then, does the temperature in these regions *increase* with increasing altitude? This has been one of the major unsolved mysteries in astronomy for the past half-century. As astronomers have tried to resolve this dilemma, they have found important clues in one of the Sun's most familiar features—sunspots.

Sunspots appear dark because they are at a lower temperature than the surrounding regions of the Sun.

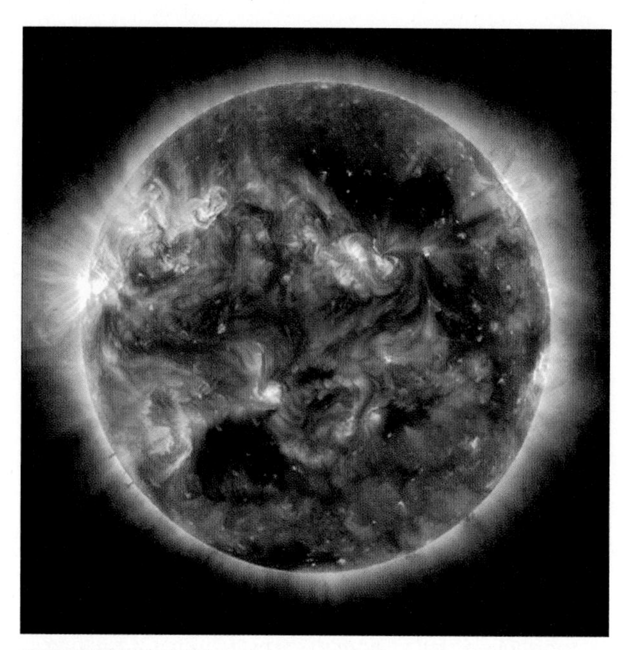

FIGURE 9-12 R I V U X G **The Ultraviolet Corona** The *Solar Dynamics Observatory* spacecraft recorded this false-color ultraviolet view of the solar corona. The dark features are coronal holes, regions where the coronal gases are thinner than elsewhere. Such holes are often the source of strong gusts in the solar wind. *(NASA Solar Dynamics Observatory)*

> ConceptCheck 9-13: **From where on the Sun does the solar wind seem to emanate?**

9-4 Sunspots are low-temperature regions in the photosphere

One might think that the Sun shines pretty much the same, day in and day out. Granules, spicules, and the solar wind occur continuously, and these features are often considered to be aspects of the quiet Sun. But, as it turns out, other, more dramatic features appear periodically, including massive eruptions and regions of concentrated magnetic fields. When these are present, astronomers refer to the active Sun. The features of the active Sun that can most easily be seen with even a small telescope—although only with an appropriate filter attached—are sunspots.

Observing Sunspots

Sunspots are irregularly shaped dark regions in the photosphere. Sometimes sunspots appear in isolation (**Figure 9-13**), but frequently they are found in sunspot groups. Although sunspots vary greatly in size, typical ones measure a few tens of thousands of miles across—comparable to the

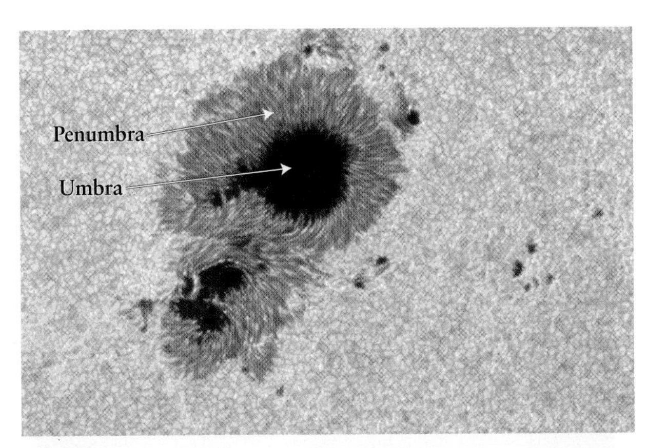

FIGURE 9-13 R I V U X G **Sunspots** This high-resolution photograph of the photosphere shows a mature sunspot pair. The dark center of the spot is called the umbra. It is bordered by the penumbra, which is less dark and has a featherlike appearance. Granulation is visible in the surrounding, undisturbed photosphere. *(NASA)*

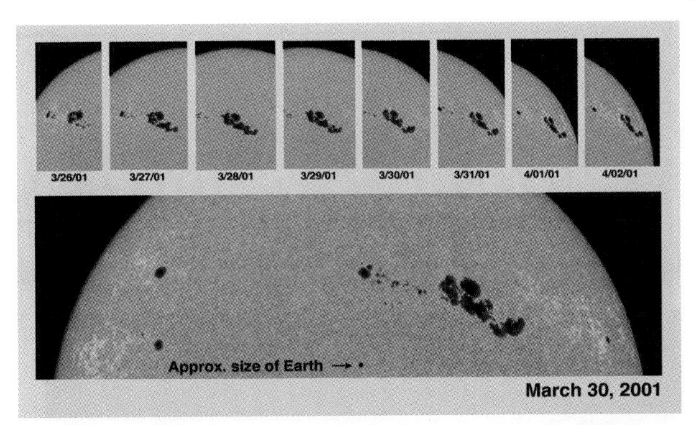

FIGURE 9-14 R I V U X G **Tracking the Sun's Rotation with Sunspots** This series of photographs taken in 2001 by *SOHO* shows the rotation of the Sun. For comparison, a dot the size of Earth is added, showing how large sunspots can be—this sunspot group is more than 13 times larger than Earth. By observing the same group of sunspots from one day to the next, Galileo found 400 years ago that the Sun rotates once in about four weeks. (The equatorial regions of the Sun actually rotate somewhat faster than the polar regions.) Notice how the sunspot group shown here changed its shape. *(SOHO/ESA/NASA)*

diameter of planet Earth. Sunspots are not permanent features of the photosphere but last between a few hours and a few months.

Each sunspot has a dark central core, called the *umbra,* and a brighter border called the *penumbra.* A sunspot is a region in the photosphere where the temperature is relatively low, which makes it appear darker than its surroundings. The colors of a sunspot indicate that the temperature of the umbra is typically 7000°F (4300 K), whereas that of the penumbra is typically 8500°F (5000 K). While high by earthly standards, these temperatures are quite a bit lower than the average temperature of the photosphere of 10,000°F (5800 K). The lower temperature of sunspots explains why these regions appear so dark—they simply emit less light because they are at a cooler temperature.

Occasionally, a sunspot group is large enough to be seen without a telescope. Chinese astronomers recorded such sightings 2000 years ago, and huge sunspot groups visible to the naked eye (with an appropriate filter) were seen in 1989 and 2003. But it was not until Galileo introduced the telescope into astronomy that anyone was able to examine how sunspots move in detail. Galileo discovered that he could determine the Sun's rotation rate by tracking sunspots as they moved across the Sun's face (**Figure 9-14**). He found that the Sun rotates once in about four weeks. A typical sunspot group lasts about two months, so a specific one can be followed for two solar rotations. After more careful study of sunspot movements, it was determined that the equatorial regions rotate more rapidly than the polar regions. This phenomenon is known as **differential rotation.** Thus, while a sunspot near the solar equator takes only 25 days to go once around the Sun, a sunspot at 30° north or south of the equator takes 27½ days. The rotation period at 75° north or south is about 33 days, while near the poles it may be as long as 35 days.

ConceptCheck 9-14: If the center of a sunspot has a temperature of about 7000°F (4300 K), why does it appear dark?

The Sunspot Cycle

Sun observers have been carefully monitoring the Sun's activity since the introduction of the telescope. What is readily observed when looking over sunspot records is that the average number of sunspots on the Sun is not constant, but varies in a predictable **sunspot cycle** (**Figure 9-15**). As Figure 9-15*a* shows, the average number of sunspots varies with a period of about 11 years. A period of exceptionally many sunspots is a **sunspot maximum** (Figure 9-15*b*), as occurred in 1989, 2000, and 2011, and is projected to occur in 2022. Conversely, the Sun is almost devoid of sunspots at a **sunspot minimum** (Figure 9-15*c*), as occurred in 1986, 1996, 2008, and 2016.

The locations of sunspots also vary with the same 11-year sunspot cycle. At the beginning of a cycle, just after a sunspot minimum, sunspots first appear at latitudes around 30° north and south of the solar equator (**Figure 9-16**). Over the succeeding years, the sunspots occur closer and closer to the equator.

Why should the number of sunspots vary with an 11-year cycle? Why should their average latitude vary over the course of a cycle? And why should sunspots exist at all? The first step toward answering these questions came in 1908, when the American astronomer George Ellery Hale discovered that sunspots are associated with intense magnetic fields on the Sun.

When Hale focused a spectroscope on sunlight coming from a sunspot, he found that many spectral lines appear to

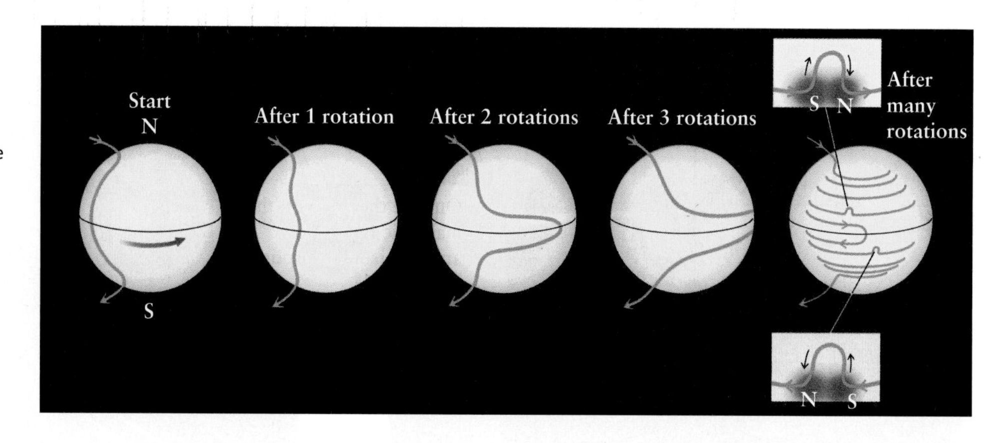

FIGURE 9-18 Babcock's Magnetic-Dynamo Model Magnetic field lines tend to move along with the plasma in the Sun's outer layers. Because the Sun rotates faster at the equator than near the poles, a field line that starts off running from the Sun's north magnetic pole (N) to its south magnetic pole (S) ends up wrapped around the Sun like twine wrapped around a ball. The insets on the far right show how sunspot groups appear where the concentrated magnetic field rises through the photosphere.

fields cancel each other out when they meet at the equator. The following members in each hemisphere have the opposite polarity to the Sun's pole in that hemisphere; hence, when they converge on the pole, the following members first cancel out and then reverse the Sun's overall magnetic field. The fields are now completely relaxed. Once again, differential rotation begins to twist the Sun's magnetic field, but now with all magnetic polarities reversed. In this way, Babcock's model helps to explain the change in field direction every 11 years.

By comparing the speeds of sound waves that travel with and against the Sun's rotation, astronomers now understand that the Sun's rotation rate is different at different depths and latitudes. As shown in **Figure 9-19**, the Sun's surface pattern of differential rotation persists through the convective zone. Farther in, within the radiative zone, the Sun seems to rotate like a rigid object with a period of 27 days at all latitudes. Astronomers suspect that the Sun's magnetic field originates in a relatively thin layer where the radiative and convective zones meet and slide past each other due to their different rotation rates.

Adding to the yet-to-be-fully-understood nature of the Sun, there seem to be times when all traces of sunspots and the sunspot cycle vanish for many years. For example, virtually no sunspots were seen from 1645 through 1715. Curiously, during these same years Europe experienced record low temperatures, often referred to as the Little Ice Age, whereas the western United States was subjected to severe drought. By contrast, there was apparently a period of increased sunspot activity during the eleventh and twelfth centuries, during which Earth was warmer than it is today. Thus, variations in solar activity appear to affect climates on Earth. The origin of this Sun-Earth connection is a topic of ongoing research.

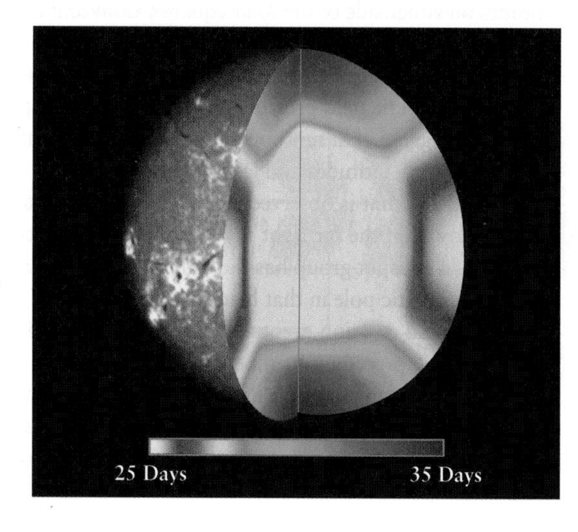

FIGURE 9-19
Rotation of the Solar Interior This cutaway picture of the Sun shows how the solar rotation period (shown by different colors) varies with depth and latitude. The surface and the convective zone have differential rotation (a short period at the equator and longer periods near the poles). Deeper within the Sun, the radiative zone seems to rotate like a rigid sphere. *(Courtesy K. Libbrecht, Big Bear Solar Observatory)*

> ConceptCheck 9-16: **How might the Sun's sunspot cycle change if the Sun were rotating much faster than it is now?**

9-5 The Sun's magnetic field also produces other forms of solar activity and causes aurorae on Earth

If magnetic fields are so powerful on the Sun, what other effects might the Sun's intense magnetic field be able to cause? In fact, the Sun's magnetic field does more than just explain the presence of sunspots.

Magnetic Arches

In a plasma, magnetic field lines and the material of the plasma tend to move together. The tendency of plasma to follow the Sun's magnetic field helps to explain why the temperature of the chromosphere and corona is so high. Spacecraft observations show magnetic field arches extending tens of thousands of kilometers into the corona, with streamers of electrically charged particles moving along each arch (**Figure 9-20*a***). If the magnetic fields of two arches come into proximity, their magnetic fields can rearrange and combine. The tremendous amount of energy stored in the magnetic field is then released into the solar atmosphere. Consider that a single arch contains as much energy as a hydroelectric power plant would generate in a million years on Earth. The amount of energy released in this way appears to be more than enough to maintain the temperatures of the chromosphere and corona.

ANALOGY The idea that a magnetic field can heat gases has applications on Earth as well as on the Sun. In an automobile engine's ignition system an electric current is set up in a coil of wire, which produces a magnetic field. When the current

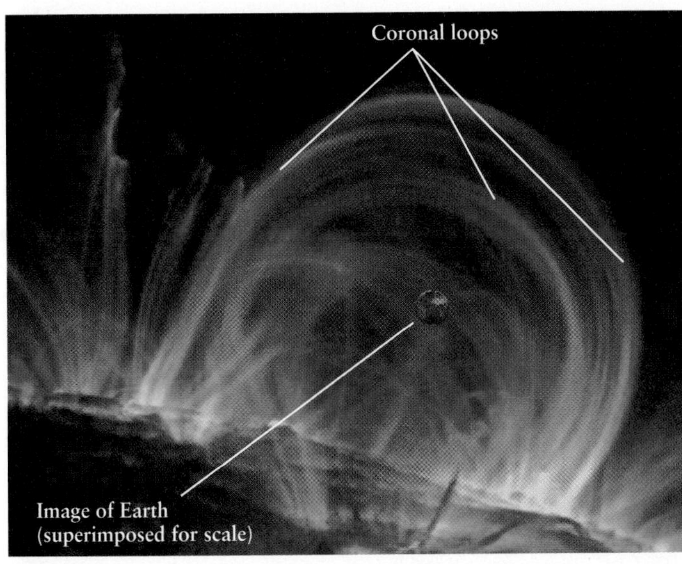

FIGURE 9-20 R I V U X G **Magnetic Arches and Magnetic Reconnection** (a) This false-color ultraviolet image from TRACE (Transition Region and Coronal Explorer) shows magnetic field loops suspended high above the solar surface. The loops are made visible by the glowing gases trapped within them. (b) When the magnetic fields in these loops change their arrangement, a tremendous amount of energy is released and solar material can be ejected upward. *(a: Stanford-Lockheed Institute for Space Research, TRACE, and NASA)*

is shut off, the magnetic field collapses and its energy is directed to a spark plug in one of the engine's cylinders. The released energy heats the mixture of air and gasoline around the plug, causing the mixture to ignite. This drives the piston in that cylinder and makes the automobile go.

Magnetic heating can also explain why the parts of the corona that lie on top of sunspots are often the most prominent in ultraviolet images. (Some examples are the brighter regions in Figure 9-12.) The intense magnetic field of the sunspots helps trap and compress hot coronal gas, giving it such a high temperature that it emits copious amounts of high-energy ultraviolet photons and even more energetic X-ray photons.

ConceptCheck 9-17: Why does glowing plasma on the Sun appear to arch up above the Sun's photosphere?

Prominences, Solar Flares, and Coronal Mass Ejections

Go to Video 9-4

Coronal heating occurs even when the Sun is quiet. But magnetic fields can also push upward from the Sun's interior, compressing and heating a portion of the chromosphere that appears as bright, arching columns of gas called **prominences** (**Figure 9-21**). These can extend for tens of thousands of miles above the photosphere. Some prominences last for only a few hours, while others persist for many months. The most energetic prominences break free of the magnetic fields that confined them and burst into space.

Violent, eruptive events on the Sun, called **solar flares,** occur in complex sunspot groups. Within only a few minutes, temperatures in a compact region may soar to 10 million degrees Fahrenheit (5×10^6 K), and vast quantities of particles and radiation—including as much material as is in the prominence shown in Figure 9-21—are blasted out into space. These eruptions can also cause disturbances that spread outward in the solar atmosphere, like the ripples that appear when you drop a rock into a pond.

The most energetic flares carry as much as 10^{30} J of energy, equivalent to 10^{14} one-megaton nuclear weapons being exploded at once! However, the energy of a solar flare does not come from thermonuclear fusion in the solar atmosphere; instead, it appears to be released from the intense magnetic field around a sunspot group.

As energetic as solar flares are, they are dwarfed by **coronal mass ejections.** One such event is shown in the image

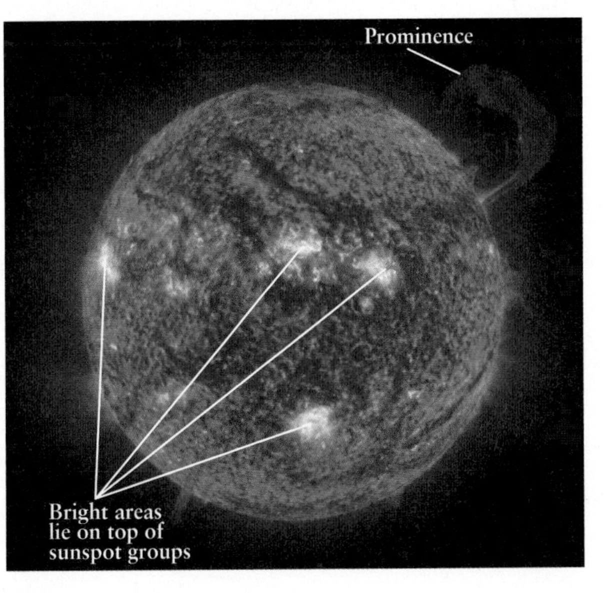

FIGURE 9-21 R I V U X G **A Solar Prominence** A huge prominence arches above the solar surface in this ultraviolet image from the *SOHO* spacecraft. The image was recorded using light at a wavelength of 30.4 nm, emitted by singly ionized helium atoms at a temperature of about 100,000°F. By comparison, the material within the arches in Figure 9-20 reaches temperatures in excess of 3.5×10^{6}°F. *(Joseph B. Gurman, Solar Data Analysis Center, NASA)*

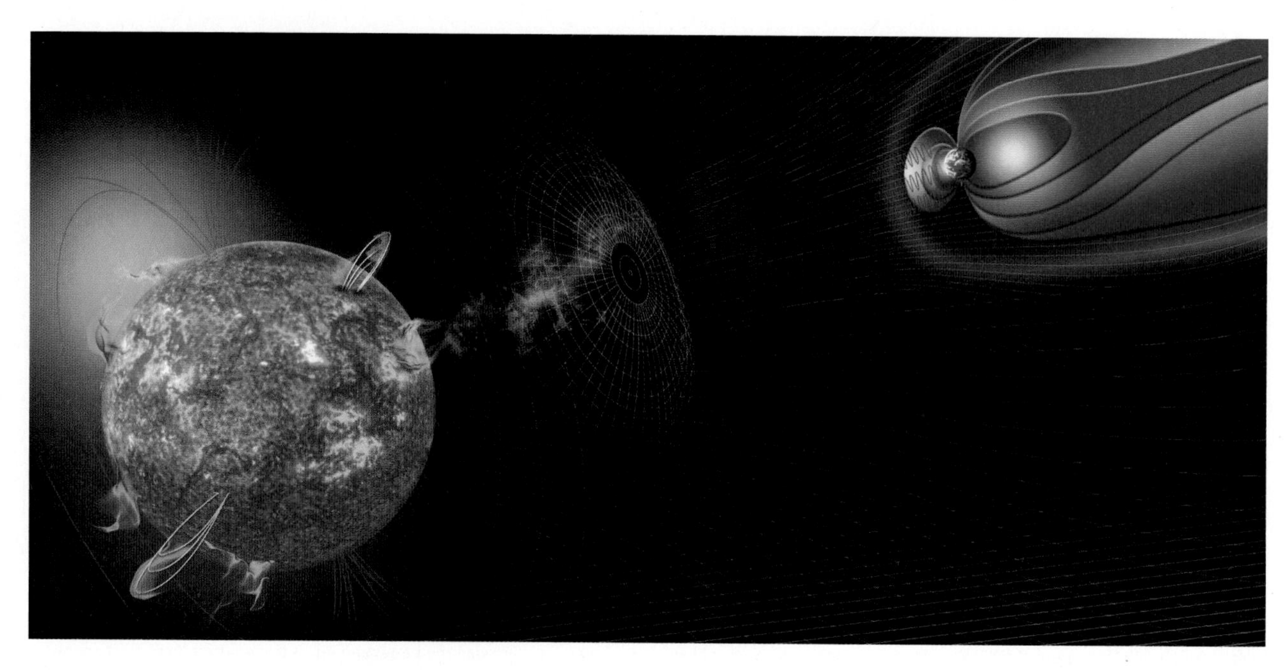

FIGURE 9-22 A Coronal Mass Ejection When a violent coronal mass ejection occurs on the side of the Sun facing Earth, a mountain of material is sent hurtling toward Earth. Within two to four days the fastest-moving ejected material reaches Earth, where most of the particles are deflected by Earth's magnetosphere, but some are able to reach Earth, causing aurorae, as shown in Figure 9-23. (The ejection shown was not aimed toward Earth and did not affect us.) *(NASA)*

that opens this chapter; **Figure 9-22** shows another. In a coronal mass ejection, more than 10^{12} kg (a billion tons) of high-temperature coronal gas is blasted into space at speeds of hundreds of kilometers per second. A typical coronal mass ejection lasts a few hours. These explosive events seem to be related to large-scale alterations in the Sun's magnetic field, like the magnetic reconnection shown in Figure 9-20*b*. Coronal mass ejections occur every few months; smaller eruptions may occur almost daily.

If a solar flare or coronal mass ejection happens to be aimed toward Earth, a stream of high-energy electrons and nuclei reaches us a few days later (on the right of Figure 9-22). When this plasma arrives, it can interfere with satellites, pose a health hazard to astronauts in orbit, and disrupt electrical and communications equipment on Earth's surface. Telescopes on Earth and on board spacecraft now monitor the Sun continuously to provide warnings of dangerous levels of solar particles.

The number of prominences, solar flares, and coronal mass ejections all vary with the same 11-year cycle as sunspots. But unlike sunspots, coronal mass ejections never completely cease, even when the Sun is at its quietest. Astronomers are devoting substantial effort to understanding these and other aspects of our dynamic Sun.

Table 9-2 Sun Data

Distance from Earth:	Average: 1 AU = 93,000,000 mi Maximum: 94,511,923 mi Minimum: 92,955,807 mi
Light travel time to Earth:	8.32 min
Radius:	10^7 Earth radii
Mass:	3.33×10^5 Earth masses
Composition (by mass):	74% hydrogen, 25% helium, 1% other elements
Composition (by number of atoms):	92.1% hydrogen, 7.8% helium, 0.1% other elements
Mean density:	1410 kg/m^3
Mean temperatures:	Surface: 10,000°F (5800 K); Center: 27.9×10^{6}°F (1.55×10^7 K)
Luminosity:	3.90×10^{26} W
Distance from center of Galaxy:	26,000 ly
Orbital period around center of Galaxy:	220 million years
Orbital speed around center of Galaxy:	2000 mph (220 km/s)

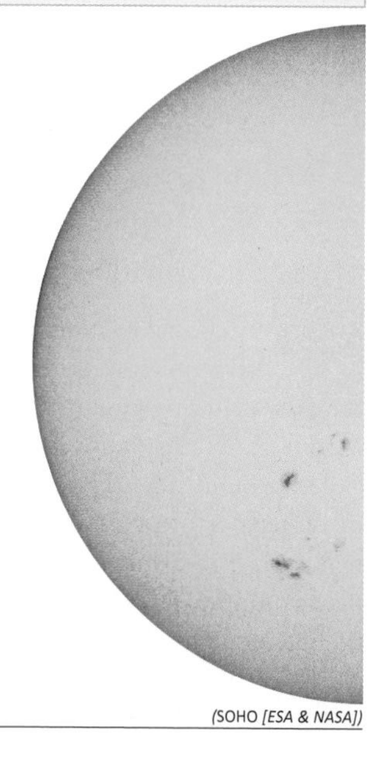

(SOHO [ESA & NASA])

Table 9-2 lists essential data about the Sun, but studies are ongoing to refine the characteristics of our nearest star. One of the most important reasons to understand the changing nature of the Sun is how it impacts Earth. As discussed in Chapter 5, the Sun's energy impacts Earth in many ways. One of the most obvious of these, shown in **Figure 9-23**, is the collision of charged particles from the solar wind with molecules in Earth's upper atmosphere, causing the **northern lights.**

ConceptCheck 9-18: Which of the following are the most energetic: prominences, solar flares, or coronal mass ejections?

FIGURE 9-23 R I V U X G **Aurora in Alaska** These northern lights, or aurora borealis, were photographed dancing above Bear Lake on January 18, 2005. The sky glows as solar particles collide with gases in Earth's upper atmosphere. This usually occurs a few days after a coronal mass ejection collides with Earth. *(United States Air Force photo by Senior Airman Joshua Strang)*

VISUAL LEARNING TASK

The Sun

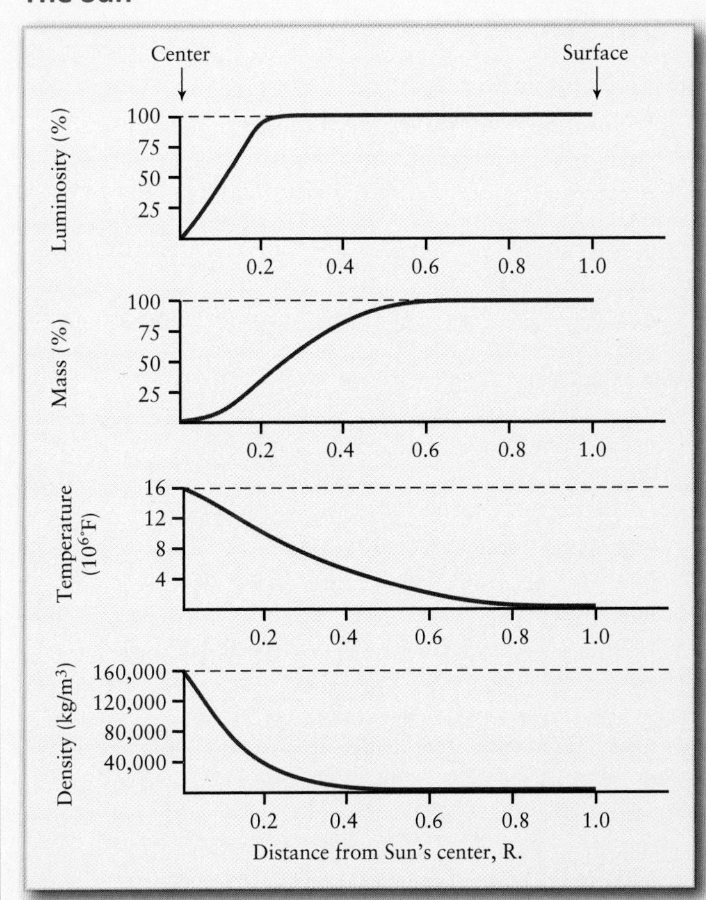

PROMPT: What would you tell a fellow student who said, **"At the halfway point between the Sun's center and its photosphere, it has half the temperature and density of the core, contains half the Sun's total mass, and produces half of the Sun's luminosity"?**

ENTER RESPONSE:

Guiding Questions:

1. At 0.5 of the Sun's radius, the temperature is about
   a. one-fourth of the core temperature.
   b. one-half of the core temperature.
   c. the same as the temperature throughout.
   d. the same temperature as the photosphere.

2. At 0.5 of the Sun's radius, the density is about
   a. one-third of the core density.
   b. one-half of the core density.
   c. the same density as the photosphere.
   d. the same as water.

3. The percentage of mass contained within 0.5 of the Sun's radius is about
   a. 90%.
   b. 50%.
   c. 33%.
   d. 10%.

4. Nearly all of the Sun's luminosity is generated within the inner
   a. one-third of the radius.
   b. one-half of the radius.
   c. 0.8 of the radius.
   d. 0.2 of the radius.

KEY IDEAS AND TERMS

9-1 The Sun's energy is generated by thermonuclear reactions in its core

- The Sun's **luminosity** is the amount of energy emitted each second and is produced by the **proton-proton chain** in which four hydrogen nuclei combine to produce a single helium nucleus.
- The energy released in a nuclear reaction corresponds to a slight reduction of mass, as predicted by Einstein's equation $E = mc^2$.
- **Thermonuclear fusion** occurs only at very high temperatures; for example, hydrogen fusion occurs only at temperatures in excess of about 18,000,000°F (10^7 K). In the Sun, fusion occurs only in the dense, hot core.

9-2 Energy slowly moves outward from the solar interior through several processes

- A theoretical description of a star's interior can be modeled using the laws of physics showing that it is in **hydrostatic equilibrium** where energy moving outward precisely balances its gravitational pull inward.
- The standard model of the Sun suggests that hydrogen fusion takes place in a core extending from the Sun's center to about 0.25 solar radius and that our Sun is in **thermal equilibrium.**
- The core is surrounded by a **radiative zone** extending to about 0.71 solar radius. In this zone, energy travels outward through **radiative diffusion.**
- The radiative zone is surrounded by a rather opaque **convective zone** of gas at relatively low temperature and pressure. In this zone, energy travels outward primarily through **convection.**
- **Neutrinos** emitted in thermonuclear reactions in the Sun's core have been detected, but in smaller numbers than expected. Recent neutrino experiments explain why this is so.
- **Helioseismology** is the study of how the Sun vibrates, which has been used to infer pressures, densities, chemical compositions, and rotation rates within the Sun.

9-3 The Sun's outer layers are the photosphere, chromosphere, and corona

- The visible surface of the Sun, the **photosphere,** is the lowest layer in the solar atmosphere. Its spectrum is similar to that of a blackbody at a temperature of 10,000°F. Convection in the photosphere produces **granules.**
- Above the photosphere is a layer of less dense but higher temperature gases called the **chromosphere. Spicules** extend upward from the photosphere into the chromosphere.
- The outermost layer of the solar atmosphere, the **corona,** is made of very high-temperature gases at extremely low density. A stream of particles making a **solar wind** emanates from thin regions called **coronal holes.**

9-4 Sunspots are low-temperature regions in the photosphere

- **Sunspots** are relatively cool regions produced by local concentrations of the Sun's magnetic field.
- The average number of sunspots increases to a **sunspot maximum** and decreases to a **sunspot minimum** in a regular **sunspot cycle** of approximately 11 years, with reversed magnetic polarities from one 11-year cycle to the next. Two such cycles make up a **22-year solar cycle** in which the surface magnetic field increases, decreases, and then increases again with the opposite polarity.
- The magnetic polarity is measured by observing the **Zeeman effect.**
- The **magnetic-dynamo model** suggests that many features of the solar cycle are due to changes in the Sun's magnetic field. These changes are caused by convection and the Sun's **differential rotation.**

9-5 The Sun's magnetic field also produces other forms of solar activity and causes aurorae on Earth

- **Plasma** on the Sun arranges itself into various observable features, one such being **prominences.**
- A **solar flare** is a brief eruption of hot, ionized gases from a sunspot group. A **coronal mass ejection** is a much larger eruption that involves immense amounts of gas from the corona.
- When charged particles emitted by the Sun interact with Earth's atmosphere, it causes an aurora where the upper atmosphere glows. When observed in the northern hemisphere it is called the **northern lights** or aurora borealis.

QUESTIONS

Review Questions

1. What is meant by the luminosity of the Sun?
2. What is thermonuclear fusion? Why is this fusion fundamentally unlike the burning of a log in a fireplace?
3. Why do thermonuclear reactions occur only in the Sun's core, not in its outer regions?
4. If thermonuclear fusion in the Sun were suddenly to stop, what would eventually happen to the overall radius of the Sun? Justify your answer using the ideas of hydrostatic equilibrium and thermal equilibrium.
5. Give some everyday examples of conduction, convection, and radiative diffusion.
6. What is a neutrino? Why is it useful to study neutrinos coming from the Sun? What do they tell us that cannot be learned from other avenues of research?
7. Briefly describe the three layers that make up the Sun's atmosphere. In what ways do they differ from each other?
8. How do astronomers know when the next sunspot maximum and sunspot minimum will occur?
9. Why do astronomers say that the solar cycle is 22 years long, even though the number of sunspots varies over an 11-year period?
10. Explain how the magnetic-dynamo model accounts for the solar cycle.
11. Why should solar flares and coronal mass ejections be a concern for businesses that use telecommunication satellites?

Web Chat Questions

1. Discuss the extent to which cultures around the world have worshiped the Sun as a deity throughout history. Why do you suppose there has been such widespread veneration?

2. In the movie *Star Trek IV: The Voyage Home,* the starship *Enterprise* flies on a trajectory that passes close to the Sun's surface. What features should a real spaceship have to survive such a flight? Why?
3. Discuss some of the difficulties in correlating solar activity with changes in Earth's climate.
4. Describe some of the advantages and disadvantages of observing the Sun (**a**) from space and (**b**) from Earth's south pole. What kinds of phenomena and issues might solar astronomers want to explore from these locations?

Collaborative Exercises

1. Figure 9-16 shows variations in the average latitude of sunspots. Estimate the average latitude of sunspots in the year you were born and estimate the average latitude on your twenty-first birthday. Make rough sketches of the Sun during those years to illustrate your answers.
2. Create a diagram showing a sketch of how limb darkening on the Sun would look different if the Sun had either a thicker or thinner photosphere. Be sure to include a caption explaining your diagram.
3. Solar granules, shown in Figure 9-6, are about 600 miles (1000 km) across. What city is about that distance away from where you are right now? What city is that distance from the birthplace of each group member?
4. Magnetic arches in the corona are shown in Figure 9-20. How many Earths high are these arches, and how many Earths could fit inside one arch?

Observing Questions

1. Use the *Starry Night*™ program to examine simulations of various features that appear on the surface of the Sun. Select **Favourites > Explorations > Solar Surface** to show a simulated view of the visible surface of the Sun as it might appear from a spacecraft. **Stop Time Flow** and use the location scroller to examine this surface. (**a**) Which layer of the Sun's atmosphere is shown in this part of the simulation? (**b**) List the different features that are visible in this view of the Sun's surface. (**c**) Click and hold the **Decrease current elevation** button in the toolbar to move to a location on the surface of the Sun, from which you can look out into the chromosphere. (The **Viewing Location** panel will indicate the location on the Sun's surface.) This simulated view of the chromosphere is at the color of the Balmer-α wavelength of hydrogen light. The opacity of the gas at this wavelength means that you can see the structure of the hot chromosphere that lies above the visible surface. Use the hand tool or cursor keys to change the gaze direction to view different features of the Sun, zooming in when necessary for a closer look at features on the horizon. (**d**) Provide a detailed description of the various features visible in this simulation of the Sun's surface. You can see current solar images from both ground- and space-based solar telescopes by opening the LiveSky pane if you have an Internet connection on your computer.
2. Use *Starry Night*™ to measure the Sun's rotation. Select **Favourites > Explorations > Solar Rotation** to display the Sun as seen from about 0.008 AU above its surface, well inside the orbit of Mercury. Use the **Time controls** to stop the Sun's rotation at a time when a line of longitude on the Sun makes a straight line between the solar poles, preferably a line crossing a recognizable solar feature. Note the date and time. Run **Time Forward** and adjust the date and time to place this selected meridian in this position again. (**a**) What is the rotation rate of the Sun as shown in *Starry Night*™? This demonstration does not show one important feature of the Sun, namely its differential rotation, where the equator of this fluid body rotates faster than the polar regions. (**b**) To which region of the Sun does your measured rotation rate refer?

ANSWERS

ConceptChecks

ConceptCheck 9-1: The Sun emits most of its energy in the form of visible light.

ConceptCheck 9-2: At the extremely high temperatures and pressures existing in the Sun's core, hydrogen nuclei can move fast enough to overcome the electrical charge repulsion and fuse together into helium nuclei.

ConceptCheck 9-3: When 1 kg of hydrogen combines to form helium, the vast majority of the mass is used as the substance of helium atoms, with only 0.7% of the original mass left over to be converted into energy.

ConceptCheck 9-4: Astronomers use the current energy output of the Sun to estimate how fast the Sun is consuming its usable fuel and how much fuel it has available to continue at its present consumption rate.

ConceptCheck 9-5: Because pressure in the Sun's core is due to the downward pushing weight of the overlying mass of material, having less mass pressing down would result in a lower pressure at the core.

ConceptCheck 9-6: When too little energy flows to the surface, the Sun's core temperature would increase dramatically.

ConceptCheck 9-7: The energy transport process of conduction occurs when energy moves through a relatively dense material by hot material transferring its kinetic energy to nearby cooler material that is in direct contact. The Sun's density is simply too low for conduction to be an important process by which energy moves from one part of the Sun to another part.

ConceptCheck 9-8: Of these three, only the temperature decreases with increasing distance from the Sun's central core.

ConceptCheck 9-9: By carefully monitoring how vibrations within the Sun move through the Sun, astronomers are able to deduce the thickness of various zones and the Sun's internal density at various depths.

ConceptCheck 9-10: The energy transport process of convection causes warmer material to rise to the surface and cooler material to sink back into the photosphere, giving the photosphere a granulelike appearance.

ConceptCheck 9-11: Spicules can have lengths of several thousand kilometers, nearly the distance across the entire United States.

ConceptCheck 9-12: The corona has an extremely low density and can only be observed when the more dominant photosphere is blocked, such as during a solar eclipse.

ConceptCheck 9-13: The solar wind seems to come from places where the corona is most thin, locations called coronal holes.

ConceptCheck 9-14: The photosphere surrounding a sunspot has a temperature of about 5800 K, which far outshines the relatively cooler sunspot region, resulting in the sunspots appearing to be quite dark in comparison.

ConceptCheck 9-15: The length of time between large numbers of sunspots to few numbers of sunspots and back to large numbers of sunspots averages about 11 years. However, the magnetic character of sunspots flips every 11-year cycle, suggesting an overarching 22-year sunspot cycle.

ConceptCheck 9-16: According to the magnetic-dynamo model, the sunspot cycle is a result of twisting magnetic fields. In the event the Sun was turning faster, the twisting would occur more quickly and the length of the sunspot cycle would decrease.

ConceptCheck 9-17: The glowing plasma tends to follow the pathways created by the Sun's invisible magnetic field lines, which form curved arches above the Sun's surface.

ConceptCheck 9-18: Coronal mass ejections are many times more energetic than any other event on the Sun and, when directed at Earth, can cause severe problems with telecommunications, electrical power distribution, and radiation health hazards for astronauts working in space.

CalculationChecks

CalculationCheck 9-1: According to Einstein's equation that $E = mc^2$, a mass of 5 kg is equivalent to 5 kg $\times$ $(3 \times 10^8 \text{ m/s})^2 = 45 \times 10^{16}$ J, which is equivalent to burning about 100,000 metric tons of coal!

CalculationCheck 9-2: According to the graphs, the Sun's core temperature is about 30 million °F or 16 million K. At a distance of 50% of the Sun's radius, the Sun's temperature has dropped to 8 million °F or 4 million K, which is a drop of about 75%.

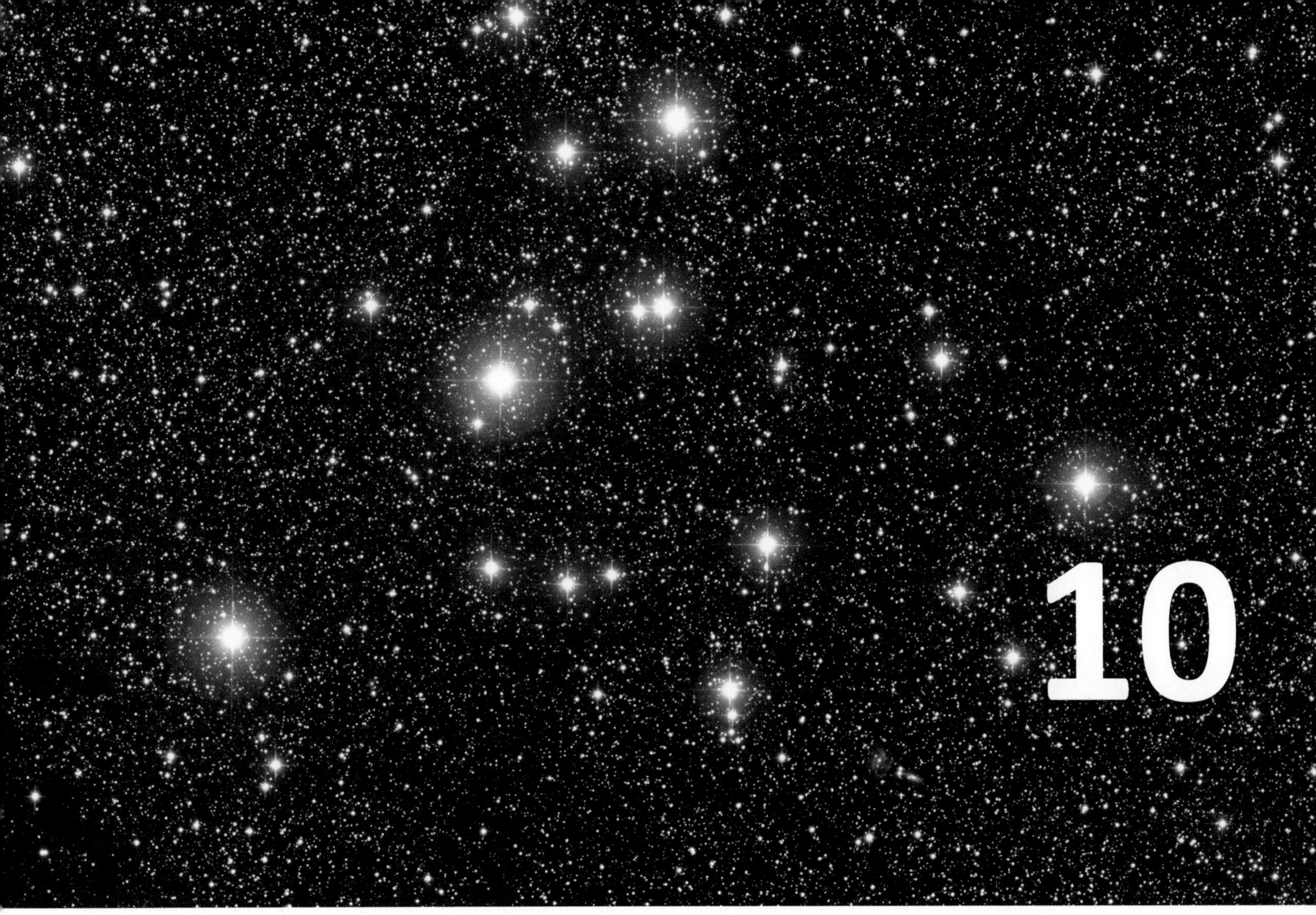

R I V U X G Some stars in this cluster (called M39) are distinctly blue in color, while others are yellow or red. *(Heidi Schweiker/NOAO/AURA/NSF)*

10 Observing Properties of Distant Stars

CHAPTER LEARNING OBJECTIVES By reading the sections of this chapter, you will learn:

10-1 Measuring the distances to nearby stars utilizes an effect called parallax

10-2 A star's brightness can be described in terms of luminosity or magnitude

10-3 A star's distance can be determined by comparing its luminosity and brightness

10-4 A star's color depends on its surface temperature

10-5 The spectra of stars reveal their chemical compositions as well as surface temperatures and sizes

10-6 Stars come in a wide variety of sizes and masses

10-7 Hertzsprung-Russell (H-R) diagrams reveal the different kinds of stars

To the unaided eye, the night sky is spangled with several thousand stars, each appearing as a bright pinpoint of light. With a pair of binoculars, you can see some 10,000 other, fainter stars; with a small backyard telescope, the total rises to more than 2 million. Astronomers now know that there are in excess of 100 billion (10^{11}) stars in our Milky Way Galaxy alone.

But what are these distant pinpoints of light? To the great thinkers of ancient Greece, the stars were bits of light embedded in a vast sphere with Earth at its center. They thought the stars were composed of a mysterious "fifth element," quite unlike anything found on Earth.

Today, we know that the stars are made of the same chemical elements found on Earth. We know their sizes, their temperatures, their masses, and something of their internal structures. We understand, too, why the stars in the chapter-opening image come in a range of beautiful colors: Blue stars have high surface temperatures, while the surface temperatures of red and yellow stars are relatively low.

How have we learned these things? How can we know the nature of the stars, objects so distant that their light takes years or centuries to reach us? Over the next three chapters, we will learn about the measurements astronomers make to determine the properties of stars and infer how they form and change over their life cycles. Our first step is to see how astronomers measure the distances to the stars that are closest to us. We will also take a close look at the Hertzsprung-Russell diagram, an important tool that helps astronomers organize and make sense of the wealth of available information we can learn about the stars. Taken together, we can use these ideas to understand how stars are born, evolve, and eventually die over millions and billions of years. ■

10-1 Measuring the distances to nearby stars utilizes an effect called parallax

The vast majority of stars visible in the night sky are glowing objects very much like the Sun, only much farther away. But if they appear only as tiny pinpoints of light, even in large telescopes, how can we measure the distances to the stars? You might try to estimate distances by comparing the relative brightnesses of different stars, thinking that all stars emit the same amount of energy, with the more distant ones appearing dimmer. Perhaps the brightly shining star Betelgeuse, in the constellation Orion, appears so bright because it is relatively close, while the dimmer and less conspicuous star Polaris (the North Star, in the constellation Ursa Minor) is farther away.

But this line of reasoning is incorrect: Polaris is actually closer to us than the seemingly brighter Betelgeuse! How bright a star appears is *not* a good indicator of its distance. If you see a light on a darkened road, it could be a motorcycle headlight a kilometer away or a person holding a flashlight just a few meters away. In the same way, a bright star might be extremely far away but have an unusually high luminosity, and a dim star might be relatively close but have a rather low luminosity. Astronomers must use other techniques to determine the distances to the stars.

ConceptCheck 10-1: **Which appears brighter, a small handheld flashlight 1 ft (30 cm) away or a large spotlight atop a lighthouse 10 mi (16 km) away?**

Parallax and the Distances to the Stars

Go to Video 10-1

The most straightforward way of measuring stellar distances uses an effect called **parallax.** This is the apparent displacement of an object because of a change in the observer's point of view (**Figure 10-1**). To see how parallax works, hold your arm out straight in front of you. Now look at the hand on your outstretched arm, first with your left eye closed, and then with your right eye closed. When you close one eye and open the other, your hand appears to shift back and forth against the background of more distant objects.

The closer the object you are viewing, the greater the parallax shift. To see this, repeat the experiment with your hand held closer to your face. Your brain analyzes such parallax shifts constantly as it compares the images from your left and right eyes, and in this way determines the distances to objects around you. This is the origin of depth perception.

To measure the distance to a star, astronomers measure the parallax shift of the star using two points of view that are as far apart as possible—at opposite sides of Earth's orbit. The direction from Earth to a nearby star changes as our planet orbits the Sun, and the nearby star appears to move back and forth against the background of more distant stars (**Figure 10-2**). This motion is called **stellar parallax.** The parallax (p) of a star is equal to half the angle through which the star's apparent position shifts as Earth moves from one side of its orbit to the other. The larger the parallax p, the

FIGURE 10-1 Parallax Imagine looking at some nearby object (a tree) against a distant background (mountains). When you move from one location to another, the nearby object appears to shift with respect to the distant background scenery. This familiar phenomenon is called parallax.

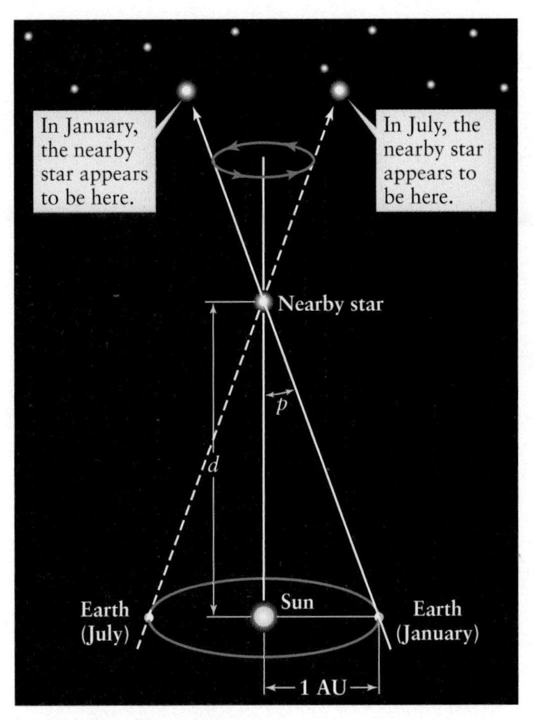

(a) Parallax of a nearby star

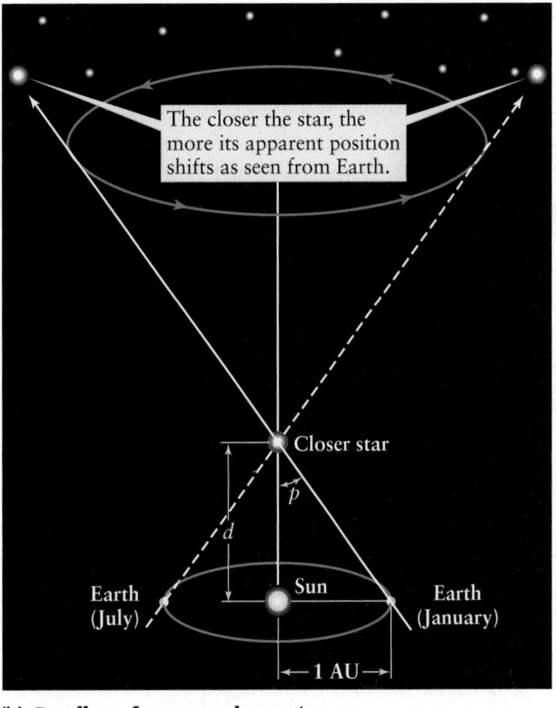

(b) Parallax of an even closer star

FIGURE 10-2 Stellar Parallax (a) As Earth orbits the Sun, a nearby star appears to shift its position against the background of distant stars. The parallax (*p*) of the star is equal to the angular radius of Earth's orbit as seen from the star. (b) The closer the star is to us, the greater the parallax angle *p*. The distance *d* to the star (in parsecs) is equal to the reciprocal of the parallax angle *p* (in arcseconds): $d = 1/p$.

smaller the distance *d* to the star (compare Figure 10-2*a* with Figure 10-2*b*).

How do we change this apparent shift in position into a distance measurement? Imagine taking a journey far into space, beyond the orbits of the outer planets. As you look back toward the Sun, Earth's orbit subtends a smaller angle in the sky the farther you are from the Sun. If we use Earth's distance from the Sun, 1 AU, as our basic measurement, then we can describe distances, *d*, in **parsecs.** As **Figure 10-3** shows, the distance at which 1 AU stretches across an angle of 1 arcsec is defined as 1 parsec (abbreviated pc). A star with a parallax angle of 1 second of arc (p = 1 arcsec) is at a distance of 1 parsec (d = 1 pc). (The word "parsec" is a contraction of the phrase "the distance at which a star has a *par*allax of one arc*sec*ond." One parsec equals 3.26 ly, or 3.09×10^{13} km; see Figure 10-3). If the angle *p* is measured in arcseconds, then the distance *d* to the star in parsecs is given by the following equation:

Relation between a star's distance and its parallax

$$d = \frac{1}{p}$$

d = distance to a star, in parsecs
p = parallax angle of that star, in arcseconds

This simple relationship between parallax and distance reveals that the closest stars have the greatest parallax. For convenience's sake, astronomers like to describe distances in parsecs rather than light-years because they use the observed parallax to measure distance.

All known stars have parallax angles less than 1 arcsecond. In other words, the closest star is more than 1 parsec

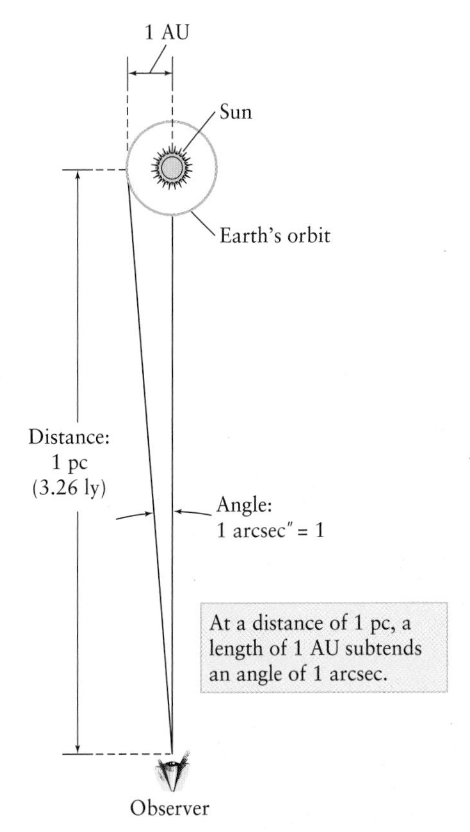

FIGURE 10-3 A Parsec The parsec, a unit of length commonly used by astronomers, is equal to 3.26 ly. The parsec is defined as the distance at which 1 AU perpendicular to the observer's line of sight subtends an angle of 1 arcsec.

away and, in the same way, the closest star is more than 3.26 light-years away. Such small parallax angles are difficult to detect, so it was not until 1838 that the first successful parallax measurements were made by the German astronomer and mathematician Friedrich Wilhelm Bessel. He found the parallax angle of the star 61 Cygni to be just ⅓ arcsec—equal to the angular diameter of a dime at a distance of 7 mi or 11 km. He thus determined that this star is about 3 pc from Earth. (Modern measurements give a slightly smaller parallax angle, which means that 61 Cygni is actually more than 3 pc away.) Using this strategy, he was able to determine the precise positions of more than 50,000 stars. Today we know that the star Proxima Centauri has the largest known parallax angle, 0.772 arcsec, and hence is the closest known star (other than the Sun)—its distance is $1/(0.772) = 1.30$ pc.

Appendix 4 in the back pages of this book lists all the stars within 4 pc of the Sun, as determined by parallax measurements. These are stars very close to our solar system but still very far beyond the orbits of the planets. Perhaps surprisingly, most of these stars are too dim to be seen with the naked eye, which is why their names are probably unfamiliar to you. In our local neighborhood, most stars are very small and dim.

By stark contrast, the majority of the familiar, named bright stars in the nighttime sky (listed in Appendix 5) are so far away that their parallaxes cannot be measured from Earth's surface. They appear bright not because they are close, but because they emit far more energy than the Sun. The brightest stars in the sky are *not* necessarily the nearest stars!

The brightest stars in the night sky are quite far away, but emit tremendous amounts of light.

> ConceptCheck 10-2: **If an astronomer inadvertently measures a parallax angle for a star to be smaller than it actually is, is the star incorrectly assumed to be farther away or closer to Earth than it actually is?**
>
> CalculationCheck 10-1: **How many light-years away is Alpha Centauri if it has a parallax angle of 0.772 arcsec?**

Measuring Parallax from Space

Parallax angles smaller than about 0.01 arcsec are extremely difficult to measure from Earth, in part because of the blurring effects of the atmosphere. Therefore, the parallax method used with ground-based telescopes can give fairly reliable distances only for stars nearer than about $1/0.01 = 100$ pc. But an orbiting observatory in space is unhampered by the atmosphere. Observations made from spacecraft therefore permit astronomers to measure even smaller parallax angles and thus determine the distances to more remote stars.

In 1989 the European Space Agency (ESA) launched the satellite *Hipparcos*, an acronym for *High* Precision *Par*allax *Co*llecting *S*atellite (and a commemoration of the ancient Greek astronomer Hipparchus, who created one of the first star charts). Over more than three years of observations, the telescope aboard *Hipparcos* was used to measure the parallaxes of 118,000 stars with an accuracy of 0.001 arcsec. This has enabled astronomers to determine stellar distances out to several hundred parsecs, and with much greater precision than has been possible with ground-based observations. In the years to come, astronomers will increasingly turn to space-based observations to determine stellar distances. Unfortunately, most of the stars in the galaxy are so far away that their parallax angles are too small to measure even with an orbiting telescope.

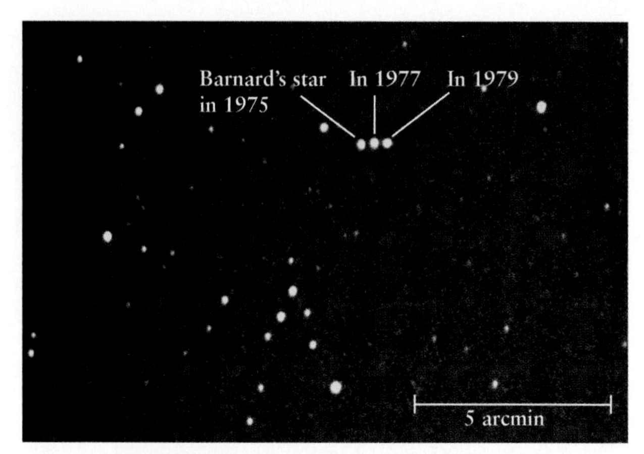

FIGURE 10-4 R I V U X G **The Motion of Barnard's Star** Three photographs taken over a four-year period were combined to show the motion of Barnard's star, which lies 1.82 pc away in the constellation Ophiuchus. Over this time interval, Barnard's star moved more than 41 arcsec on the celestial sphere (about 0.69 arcmin, or 0.012°), more than any other star. *(John Sanford/Science Source)*

Because it can be used only on relatively close stars, stellar parallax might seem to be of limited usefulness. But parallax measurements are the cornerstone for all other methods of finding the distances to more remote objects. Other methods, which we will get to in later chapters as we look beyond the Milky Way Galaxy, require a precise and accurate knowledge of the distances to nearby stars, as determined by stellar parallax. Hence, any inaccuracies in the parallax angles for nearby stars can translate into substantial errors in measurement for the whole universe. For this reason, astronomers are continually trying to perfect their parallax-measuring techniques.

Stellar parallax is an *apparent* motion of stars caused by Earth's orbital motion around the Sun. But stars are not fixed objects and actually do move through space. As a result, stars change their positions on the celestial sphere (**Figure 10-4**). These motions are sufficiently slow, however, that changes in the positions of the stars are hardly noticeable over a human lifetime.

> ConceptCheck 10-3: **Why can't the distances to most stars in our Galaxy be measured using parallax?**

10-2 A star's brightness can be described in terms of luminosity or magnitude

A quick survey of some of the photos in this book suggests that not all stars are the same. Some stars are brilliantly bright, whereas the vast majority is considerably dimmer. Because astronomy is among the most ancient of sciences, some of the words and tools used by modern astronomers to describe the night sky are actually many centuries old. One such tool is the magnitude scale, which astronomers frequently use to denote the brightness of stars. This scale was introduced in the second century B.C.E. by the Greek astronomer Hipparchus, who called the brightest stars first-magnitude stars.

Hipparchus based this scale in part on which stars are the first to be visible in the darkening sky after sunset. In other words, the first group of stars visible as the Sun sets below the horizon are known as magnitude +1 stars. Stars about half as bright as first-magnitude stars, and the second group of stars to become visible as the sky becomes darker, are known as magnitude +2 stars, and so forth, down to sixth-magnitude stars, the dimmest ones that can be seen when the sky is pitch black. After telescopes came into use, astronomers extended Hipparchus's magnitude scale to include even dimmer stars.

This numbering scale that ranks stars, where the brightest objects are "number one," is quite common in our culture. The winner of baseball's World Series is the number one team, the restaurant voted best in your community is the number one restaurant, and the best Olympic performer in gymnastics is the number one gymnast. So, it seems reasonable that when ranking stars by brightness, the brightest stars are magnitude one, the not-as-bright-as-number-one stars are magnitude two, and so on.

CAUTION The magnitude scale can sometimes be confusing because it might seem to work "backward." Keep in mind that the *greater* the apparent magnitude, the *dimmer* the star—just like with rankings of sporting teams. The number five soccer team has fewer wins than the number three soccer team (and likely less talented players). In much the same way, a star that appears to have a magnitude +3 (a third-magnitude star) is dimmer than a star that appears to have a magnitude +2 (a second-magnitude star).

In the nineteenth century, astronomers developed better techniques for measuring the light energy arriving on Earth from a star. These measurements showed that a first-magnitude star is about 100 times brighter than a sixth-magnitude star. In other words, it would take 100 stars of magnitude +6 to provide as much light energy as we receive from a single star of magnitude +1. To make computations easier, the magnitude scale was redefined so that a magnitude difference of 5 corresponds exactly to a factor of 100 in brightness. A magnitude difference of 1 corresponds to a factor of 2.512 in brightness, because

$$2.512 \times 2.512 \times 2.512 \times 2.512 \times 2.512 = (2.512)^5 = 100$$

Thus, it takes 2.512 third-magnitude stars to provide as much light as we receive from a single second-magnitude star.

One might imagine that the objects might have a different magnitude rating if the observer were significantly closer to it or, alternatively, farther away from it. Because Hipparchus's magnitude numbers describe how bright an object *appears* to an Earth-based observer, values in Hipparchus's scale are properly called **apparent magnitudes.**

Figure 10-5 illustrates the modern apparent magnitude scale. The dimmest stars visible through a pair of binoculars

(a) RIVUXG

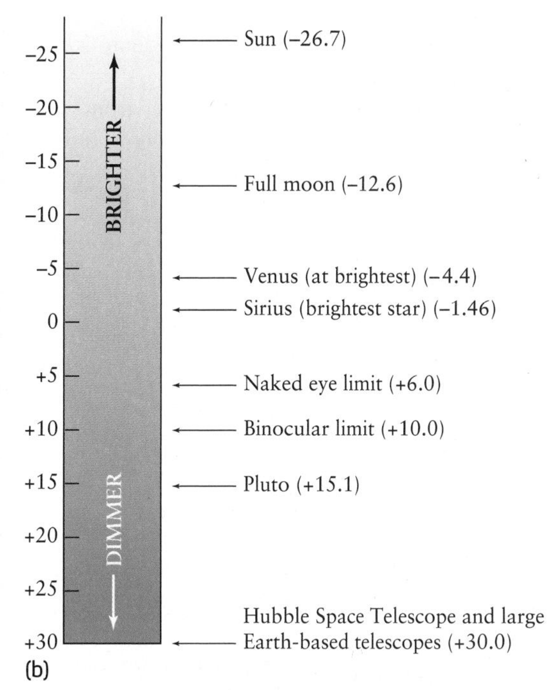

(b)

FIGURE 10-5 The Apparent Magnitude Scale Astronomers denote the apparent brightness of objects in the sky by their apparent magnitudes. The smallest rankings are for the brightest objects. (a) Stars near the constellation Orion, labeled with their names and apparent magnitudes, show a variety of apparent brightnesses. The dimmest stars have the largest apparent magnitude numbers. (b) Stars visible to the naked eye have magnitudes between m = −1.46 (Sirius) and about m = +6.0. However, the Hubble Space Telescope or a specially equipped, large Earth-based telescope can reveal stars and other objects fainter than magnitude m = +30. *(a: Eckhard Slawik/Science Photo Library/Getty Images)*

have an apparent magnitude of +10, and the dimmest stars that can be detected with the Hubble Space Telescope have an apparent magnitude greater than +30.

In contrast to looking for the dimmest objects, we also need to be able to describe objects brighter than typical stars, such as the full moon. Modern astronomers use negative numbers to extend Hipparchus's scale for the brightest of objects. For example, Sirius, the brightest star in the night sky, has a carefully measured apparent magnitude of −1.46. The Sun, the brightest object in the sky, has an apparent magnitude of −26.7.

Using this apparent magnitude scale can be misleading when judging the nature of stars. For example, our star the Sun has about the same energy output as the star Alpha Centauri. However, our Sun is much closer to Earth than Alpha Centauri so it appears many, many times brighter in the sky. In other words, because in fact some stars are relatively close to Earth and some stars are relatively far from Earth, judging all stars only by their apparent magnitudes provides an incomplete picture of the cosmos.

Go to Video 10-2

To solve this problem, astronomers imagine what the brightness or magnitude of a star would be *IF* the star were located exactly 10 pc (32.6 ly) from Earth. The magnitude of a star if it were at a distance of 10 pc is known as the **absolute magnitude,** and it is a quantity that reflects a star's true energy output so it can be compared to other stars.

ANALOGY If you wanted to compare the light output of two different lightbulbs, you would naturally place them side by side so that both bulbs were the same distance from you. In the absolute magnitude scale, we imagine doing the same thing with stars to compare their luminosities.

If the Sun were moved to a distance of 10 pc from Earth, it would have an apparent magnitude of +4.8. The absolute magnitude of the Sun is thus +4.8. The absolute magnitudes of the stars range from approximately +15 for the least luminous to −10 for the most luminous. (*Note:* Like apparent magnitudes, absolute magnitudes work "backward": The *greater* the absolute magnitude, the *less luminous* the star.) The Sun's absolute magnitude is about in the middle of this range.

ConceptCheck 10-4: If we were observing our Sun from a distance of 10 pc, what would be its apparent and absolute magnitudes?

ConceptCheck 10-5: The star Tau Ceti has an apparent magnitude of about +3 and an absolute magnitude of about +6. Is it much closer or much farther from Earth than 10 pc?

10-3 A star's distance can be determined by comparing its luminosity and brightness

Apparent brightness is a measure of how faint a star looks to us, whereas absolute magnitude is a measure of the star's total light output. In addition to absolute magnitude, astronomers have formulated another way to judge the energy output of stars—by comparing how much energy stars emit compared to how much energy the Sun emits. The total amount of energy per second a star emits is called its **luminosity.** As we saw in Chapter 9, the Sun emits about 3.86×10^{26} W (1 watt, or 1 W, is 1 joule per second) and is written as $L_\odot$. Most stars are less luminous than the Sun, but some blaze forth with a million times the Sun's luminosity. For example, a star that emits 5 times more energy per second than our Sun does has a luminosity of 5 $L_\odot$. Knowing a star's luminosity is essential for determining the star's history, present-day internal structure, and future evolution.

Luminosity and the Inverse-Square Law

Go to Video 10-3

To determine the luminosity of a star, we first note that as light energy moves away from its source, it spreads out over increasingly larger regions of space. Imagine a sphere of radius d centered on the light source, as in **Figure 10-6*a*.** The amount of energy that passes each second through a square meter of the sphere's surface area is the total luminosity of the source (L) divided by the total surface area of the sphere (equal to $4\pi d^2$). This quantity is usually known as the apparent brightness of the light (b), because how bright a light source appears depends on how much light energy per second enters through the area of a light detector (such as your eye). Apparent brightness is measured in watts per square meter (W/m^2). In other words, there is a relationship between how bright something appears and how close it is—the closer a luminous object is, the brighter it appears, as illustrated in Figure 10-6*b*. Moreover, this relationship depends not just on the precise distance, but actually on the square of the distance. Written in the form of an equation, the relationship between apparent brightness and luminosity is

Inverse-square law relating apparent brightness and luminosity

$$b = \frac{L}{4\pi d^2}$$

b = apparent brightness of a star's light, in W/m^2
L = star's luminosity, in watts
d = distance to star, in meters

This relationship is called the **inverse-square law,** because the apparent brightness of light that an observer can see or measure is inversely proportional to the square of the observer's distance (d) from the source. This mathematical

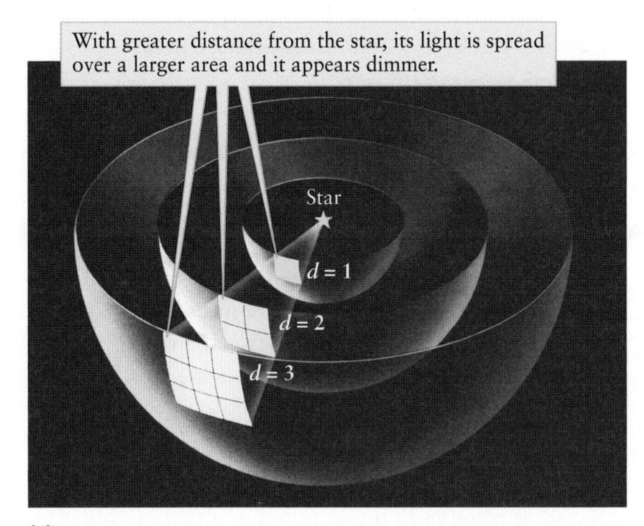

(a)

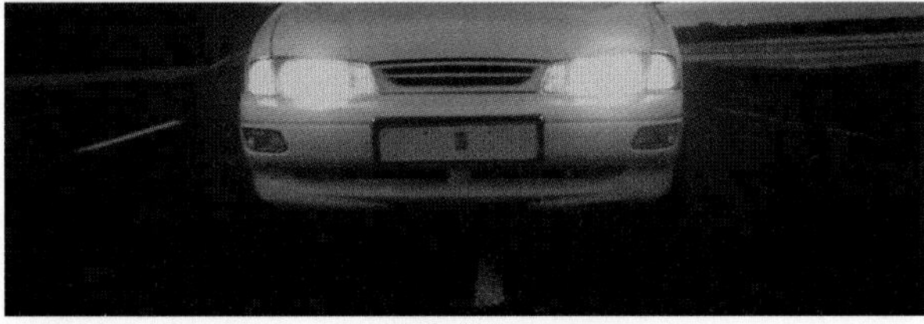

(b) R I V U X G

FIGURE 10-6 The Inverse-Square Law (a) Energy from a light source illuminates an area that increases as the square of the distance from the source. Hence, the apparent brightness decreases as the square of the distance. The brightness at $d = 2$ is $1/(2^2) = \frac{1}{4}$ of the brightness at $d = 1$, and the brightness at $d = 3$ is $1/(3^2) = \frac{1}{9}$ of that at $d = 1$. (b) The change in apparent brightness for a light source with constant luminosity as a result of changing distance is shown using an approaching automobile with headlights. *(b: Cardinal/Corbis)*

equation might look complicated at first glance, but it is not. This equation says simply, if you double your distance from a lightbulb, its light is spread out over an area 4 times larger, so the apparent brightness you see is decreased by a factor of 4. Similarly, at triple the distance, the apparent brightness is one-ninth as great (see Figure 10-6*a*). For example, if we apply the inverse-square law to the Sun, which is 1.50×10^{11} m from Earth, we can calculate the apparent brightness ($b_\odot$) as

$$b_\odot = \frac{3.86 \times 10^{26}\ \text{W}}{4\pi(1.50 \times 10^{11}\ \text{m})^2} = 1370\ \text{W/m}^2$$

Another way of saying this is that a blanket spread out on the ground covering an area of 1 m^2 receives 1370 W of power from the Sun.

Astronomers measure the apparent brightness of a star using a telescope with an attached light-sensitive instrument, similar to the light meter in a camera that determines the proper exposure.

ConceptCheck 10-6: How many times less light falls on a newspaper illuminated by a lightbulb if the newspaper is moved a distance of 3 times farther away?

Calculating a Star's Luminosity

The relationship between the apparent brightness of a shining star and its distance also allows astronomers to calculate the energy output of a star. This is important because a star's appearance is dependent on both how far away it is and how much energy it is actually emitting. As it turns out, we need to know just two things to find a star's luminosity: the distance to a star as compared to the Earth-Sun distance (the ratio $d/d_\odot$), and how that star's apparent brightness compares to the brightness of the Sun (the ratio $b/b_\odot$). This important relationship can be written as a series of ratios as follows:

Determining a star's luminosity from its apparent brightness

$$\frac{L}{L_\odot} = \left(\frac{d}{d_\odot}\right)^2 \frac{b}{b_\odot}$$

$L/L_\odot$ = ratio of the star's luminosity to the Sun's luminosity
$d/d_\odot$ = ratio of the star's distance to the Earth-Sun distance
$b/b_\odot$ = ratio of the star's apparent brightness to the Sun's apparent brightness

In other words, this mathematical relationship gives us a powerful rule relating the luminosity, distance, and apparent brightness of a star. For a specific apparent brightness, the more distant the star, the more luminous it must be to be seen from that distance.

We can determine the luminosity of a star from its distance and apparent brightness. For a given distance, the brighter the star, the more luminous that star must be.

As stars go, our Sun is neither extremely luminous nor extremely dim; it is a rather ordinary, garden-variety star. It is somewhat more luminous than most stars, however. Of the 30 closest stars to the Sun only 3 (α Centauri, Sirius, and Procyon) have a greater luminosity than the Sun.

To better characterize a typical population of stars, astronomers count the stars out to a certain distance from the Sun and plot the number of stars that have different luminosities. **Figure 10-7** shows the distribution of luminosities for stars in our part of the Milky Way Galaxy. The curve

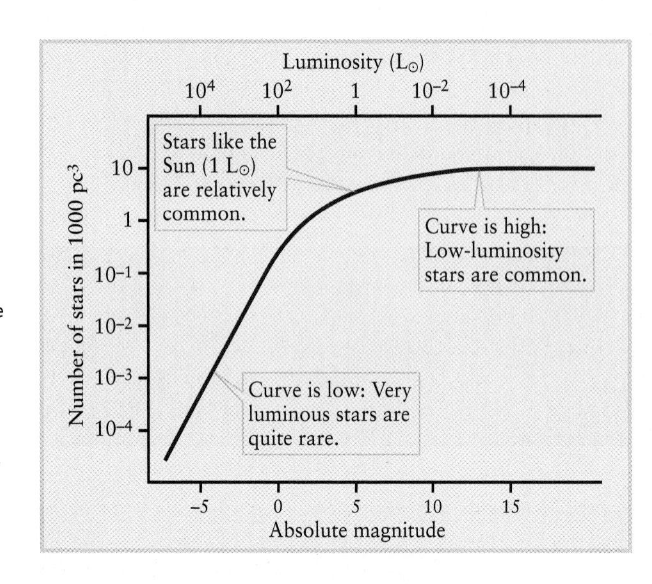

FIGURE 10-7 The Luminosity Function This graph shows how many stars of a given luminosity lie within a nearby representative volume of 1000 pc^3. The scale at the bottom of the graph shows absolute magnitude, an alternative measure of a star's luminosity (described in Section 10-2). *(Adapted from J. Bahcall and R. Soneira)*

Red stars are relatively cold, with low surface temperatures; blue stars are relatively hot, with high surface temperatures.

declines very steeply for the most luminous stars toward the left side of the graph, indicating that they are quite rare. For example, this graph shows that stars like the Sun are about 10,000 times more common than stars like Spica (which has a luminosity of 2100 $L_{\odot}$). The exact shape of the curve in Figure 10-7 applies only to the vicinity of the Sun and similar regions in our Milky Way Galaxy, with other locations having slightly different shaped curves.

CalculationCheck 10-2: The star Pleione in the constellation Taurus is 190 times as luminous as the Sun but appears only 3.19×10^{-13} as bright as the Sun. How far is Pleione from Earth?

10-4 A star's color depends on its surface temperature

As we saw in Chapter 2, stars have different colors, depending on their temperatures. Some people can clearly see these colors, even with the naked eye. For example, some can easily see the red color of Betelgeuse, the star in the "armpit" of the constellation Orion, and the blue tint of Bellatrix at Orion's other "shoulder." Colors are most evident for the brightest stars, because most people's color vision does not work well at low light levels.

CAUTION Remember that the light from a star will appear redshifted if the star is moving away from you and blueshifted if it is moving toward you. But for even the fastest stars, these color shifts are so tiny that it takes sensitive instruments to measure them. The red color of Betelgeuse and the blue color of Bellatrix are not due to their motions; they are the actual colors of the stars.

Like most glowing objects, a star's color is directly related to its surface temperature. The intensity of light from a relatively cool star peaks at long wavelengths, giving off mostly long wavelength light, making the star look red (**Figure 10-8*a***). A hot star's intensity curve peaks at shorter wavelengths, so the star looks blue (Figure 10-8*c*). For a star with an intermediate temperature, such as the Sun, the intensity peak is near the middle of the visible range of the spectrum. This gives the star a yellowish color (Figure 10-8*b*). This leads to an important general rule about star colors and surface temperatures that we will use again and again

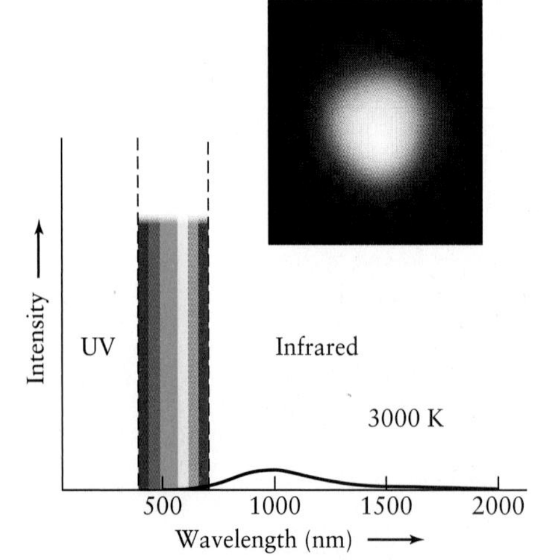

(a) A **cool star** with surface temperature 3000 K emits much more red light than blue light, and so appears red.

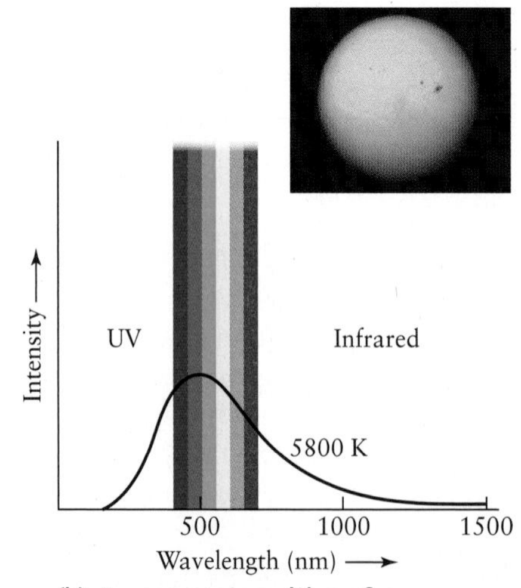

(b) A **warmer star** with surface temperature 5800 K (like the Sun) emits similar amounts of all visible wavelengths, and so appears yellow-white.

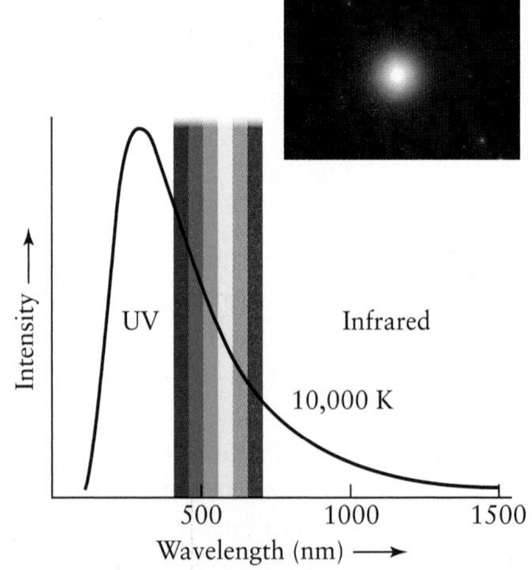

(c) A **hot star** with surface temperature 10,000 K emits much more blue light than red light, and so appears blue.

FIGURE 10-8 Temperature and Color These graphs show the intensity of light emitted by three hypothetical stars plotted against wavelength (compare with Figure 2-10). The rainbow band indicates the range of visible wavelengths. The star's apparent color depends on whether the intensity curve has larger values at the short-wavelength or long-wavelength side of the visible spectrum. The insets show stars of about these surface temperatures. UV stands for ultraviolet, which extends from 10 nm to 400 nm. *(inset a: Andrea Dupree [Harvard-Smithsonian CFA], Ronald Gilliland [STScI], NASA and ESA; inset b: NSO/AURA/NSF; inset c: Alan Dyer)*

throughout our study of astronomy: Red stars are relatively cold, with low surface temperatures; blue stars are relatively hot, with high surface temperatures.

Figure 10-8 is important enough to spend some time studying. It dramatically illustrates how astronomers can accurately determine the surface temperature of a distant star, not by traveling to it, but simply by carefully measuring its color. To do this, the star's light is collected by a telescope and passed through one of a set of differently colored filters. The process is then repeated with each of the filters in the set. The star's image will have a different brightness through each colored filter, and by comparing these brightnesses astronomers can find the wavelength at which the star's intensity curve has its peak—and hence the star's temperature.

CAUTION As we will discuss in Chapter 11, tiny dust particles that pervade interstellar space cause distant stars to appear redder than they really are. In the same way, particles in Earth's atmosphere make the setting Sun look red. Astronomers must take this reddening into account whenever they attempt to determine a star's surface temperature from its color.

ConceptCheck 10-7: **Which is hotter, a green star or a yellow star?**

10-5 The spectra of stars reveal their chemical compositions as well as surface temperatures and sizes

We have seen how careful inspection of a star's color allows astronomers to determine its surface temperature. What happens if we look even more closely by separating starlight into its specific component colors and analyzing each of the colors present in detail?

Classifying Stars: Absorption Line Spectra and Spectral Classes

We observe an absorption line spectrum when a cool gas, lying between us and a hot, glowing object, removes particular wavelengths of light. (This idea is discussed in Chapter 2 and might be worth reviewing.) The light from the hot, glowing object itself has a continuous spectrum. In the case of a star, light with a continuous spectrum is produced at low-lying levels of the star's atmosphere, where the gases are hot and dense. The absorption lines are created when this light flows outward through the upper layers of the star's atmosphere. Atoms in these cooler, less dense layers absorb radiation at specific wavelengths, which depend on the specific kinds of atoms present—hydrogen, helium, or other elements—and on whether or not the atoms are ionized.

Some stars have spectra in which certain hydrogen lines are broad and obvious. But in the spectra of other stars, these hydrogen features are nearly absent and the dominant absorption lines that are obvious are the result of heavier elements such as calcium, iron, and sodium being present in the star. Still other stellar spectra are dominated by broad absorption lines caused by atoms that have combined into molecules, such as titanium oxide. To cope with this diversity, astronomers group similar-appearing stellar spectra into **spectral classes.** In a popular classification scheme that emerged in the late 1890s, a star was assigned a letter from A through O according to the obvious presence of particular hydrogen lines in the star's spectrum, from greatest to least. However, nineteenth-century science could not explain why or how the spectral lines of a particular chemical are affected by the temperature and density of the gas. Since this time, many of the original A-through-O classes were dropped and others were consolidated. The remaining spectral classes were reordered in the sequence **OBAFGKM.** Although there are many strategies to recall this sequence that shows up again and again in this chapter, astronomy students often remember this sequence using the mnemonic "Oh, *Be A Fine Girl* (or *Guy*), *Kiss Me*!"

ConceptCheck 10-8: **What principal characteristic of a star's spectrum most dominates which spectral class letter of the alphabet it is assigned?**

Refining the Classification: Spectral Types

Although stars are nonliving objects far from Earth, astronomy is a human enterprise and its progress is highly dependent on the diversity of people pursuing the scientific questions of astronomy. Working with a team led by E. C. Pickering that included some of the most well-known early twentieth-century astronomers such as Antonia Maury and Williamina Fleming, Annie Jump Cannon refined the original OBAFGKM sequence into even smaller categories called **spectral types.** Each of these spectral type steps are indicated by attaching a number from 0 through 9 to the original letter. For example, the spectral class F includes spectral types F0, F1, F2, . . . , F8, F9, which are followed by the spectral types G0, G1, G2, . . . , G8, G9, and so on. This effort culminated in the *Henry Draper Catalogue,* published between 1918 and 1924. It listed a tremendous number of stars—225,300 stars—each of which Cannon had personally classified herself.

Figure 10-9 shows representative spectra of several spectral types. The strengths of spectral lines change gradually from one spectral type to the next. For example, certain hydrogen absorption lines become increasingly prominent as you go from spectral type B0 to A0. From A0 onward through the F and G classes, the hydrogen lines weaken and almost fade from view. The Sun, whose spectrum is dominated by calcium and iron, is a G2 star.

Based on new understanding of the structure of atoms and thus a new way of thinking about the nature of stars,

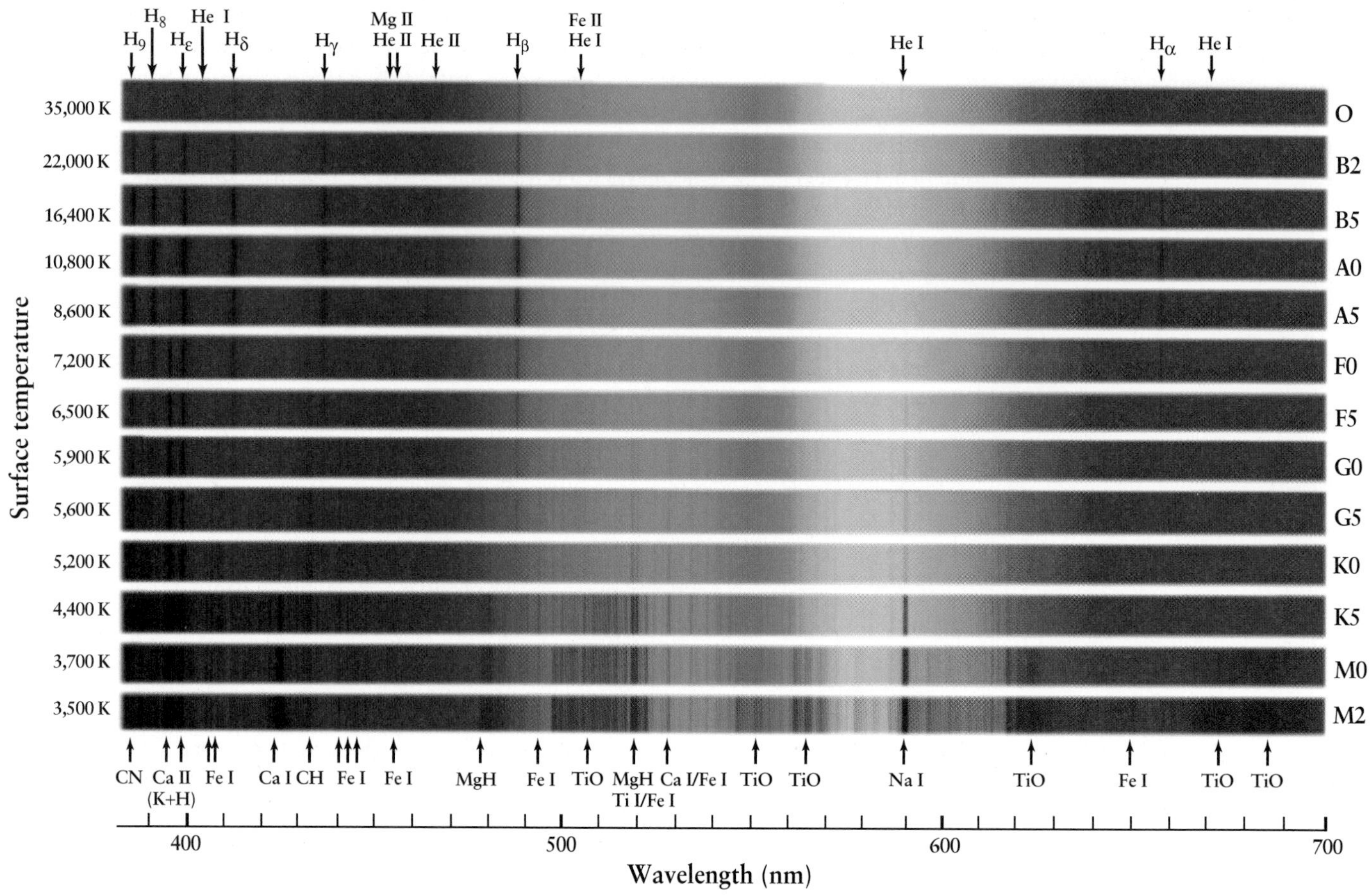

FIGURE 10-9 R I V U X G **Principal Types of Stellar Spectra** Stars of different spectral classes and different surface temperatures have spectra dominated by different absorption lines. Notice how the lines of hydrogen (H_α, H_β, H_γ, and H_δ) are strongest for hot stars of spectral class A, while absorption lines due to calcium (Ca) are strongest in cooler K and M stars. The spectra of M stars also have broad, dark bands caused by molecules of titanium oxide (TiO), which can only exist at relatively low temperatures. A Roman numeral after a chemical symbol shows whether the absorption line is caused by un-ionized atoms (Roman numeral I) or by atoms that have lost one electron (Roman numeral II). *(R. Bell, U. of Maryland, and M. Briley, Appalachian State U.)*

the first tenured, female full professor at Harvard University, Cecilia Payne, working with famed Indian physicist Meghnad Saha, demonstrated in the 1920s that the OBAFGKM spectral sequence is actually a sequence in temperature. The hottest stars are O stars. Their absorption lines can occur only if these stars have enormous surface temperatures, above 25,000 K in this case. M stars are the coolest stars. (Astronomers describe stellar temperatures in terms of Kelvins rather than the more familiar Fahrenheit or Celsius scales—25,000 K is equivalent to about 45,000°F.) The spectral features of M stars are consistent with stellar surface temperatures of about 3000 K (5000°F) much cooler than the surface of our Sun. In other words, the sequence OBAFGKM is also a temperature sequence, from hottest to coldest.

The specific spectral features seen in a star's spectrum are highly dependent on the star's surface temperature.

ConceptCheck 10-9: Is a spectral class F2 star more similar to an A-spectral class star or a G-spectral class star?

ConceptCheck 10-10: As spectral type numbers increase within the G-spectral class of star, do the larger numbers represent higher temperature stars?

Why Surface Temperature Affects Stellar Spectra

What astronomers understand today that they did not 100 years ago is that a star's spectrum is profoundly affected by the star's surface temperature. Hydrogen is by far the most abundant element in the universe, accounting for about three-quarters of the mass of a typical star. Yet these hydrogen lines do not necessarily show up in all stars' spectra. As it turns out, these absorption lines are produced when an electron in the $n = 2$ orbit of hydrogen is lifted into a higher orbit by absorbing a photon with the right amount of energy. If the star is much hotter than 10,000 K, about two times hotter than our Sun's surface, the photons pouring out of the star's interior have such high energy that they easily knock electrons out of hydrogen atoms in the star's surrounding atmosphere. This process ionizes the gas. With its only electron torn away, a hydrogen atom cannot produce any absorption lines. Hence, these lines will be relatively weak in the spectra of such hot stars, such as the hot O and B2 stars in Figure 10-9.

Conversely, if the star's atmosphere is much cooler than 10,000 K, almost all the hydrogen atoms are in the lowest ($n = 1$) energy state. Most of the photons passing through the star's atmosphere possess too little energy to boost electrons up from the $n = 1$ to the $n = 2$ orbit of the hydrogen atoms. Hence, very few of these atoms will have electrons in the $n = 2$ orbit, and only these few can absorb these particular photons. As a result, these lines are nearly absent from the spectrum of a cool star. (You can see this in the spectra of the cool M0 and M2 stars in Figure 10-9.)

For these hydrogen lines to show up in a star's spectrum, the star must be hot enough to excite the electrons out of the ground state but not so hot that all the hydrogen atoms become ionized. A stellar surface temperature of about 9000 K produces the strongest hydrogen lines; this is the case for the stars of spectral types A0 and A5 in Figure 10-9. Every other type of atom or molecule also has a characteristic temperature range in which it produces prominent absorption lines in the observable part of the spectrum. **Figure 10-10** shows the relative strengths of absorption lines produced by different chemicals. By measuring the details of these lines in a given star's spectrum, astronomers can accurately determine that star's surface temperature.

Astronomers use the term metals to refer to all elements other than hydrogen and helium. (This idiosyncratic use of the term "metal" is quite different from the definition used by chemists and other scientists outside of astronomy. To a chemist, sodium and iron are metals but carbon and oxygen are not; to an astronomer, all of these substances are simply referred to as metals.) In this terminology, metals dominate the spectra of stars cooler than 10,000 K, whereas the hydrogen seemingly appears to be absent.

For very low temperature stars, with temperatures below 4000 K, certain atoms in a star's atmosphere can combine to form molecules. (At higher temperatures atoms move so fast that when they collide, they bounce off each other rather than "sticking together" to form molecules.) As these molecules vibrate and rotate, they produce bands of spectral lines that dominate the star's spectrum. Most noticeable are the lines of titanium oxide (TiO), which are strongest for surface temperatures of about 3000 K.

ConceptCheck 10-11: If hydrogen is the most abundant atom in a star, why do the spectra of hot stars show no prominent hydrogen lines?

Spectral Classes for Brown Dwarfs

Since 1995 astronomers have found a number of tiny stars with surface temperatures even lower than those of spectral class M. By the strictest definition these are not stars, but they are not planets either. These "substars" are too small to sustain thermonuclear fusion in their cores. Instead, these objects glow primarily from the heat released as they contract. They are widely known as **brown dwarfs.**

To describe brown dwarf spectra, usually observed with infrared telescopes, astronomers have defined two new spectral classes, L and T. Thus, the modern spectral sequence of stars and brown dwarfs from hottest to coldest surface temperature is OBAFGKMLT. (Can you think of a new mnemonic that includes L and T?) For example, **Figure 10-11** shows a small red star of spectral class M orbited by a significantly smaller brown dwarf of spectral class T at a distance of 200 AU.

By mass, almost all stars (including the Sun) and brown dwarfs are about three-quarters hydrogen, one-quarter helium, and 1% or less metals.

Considering the wide range of star temperatures, astronomers are now in a position to interpret spectra in order to determine the composition of stars. As it turns out, when the effects of temperature are accounted for, astronomers

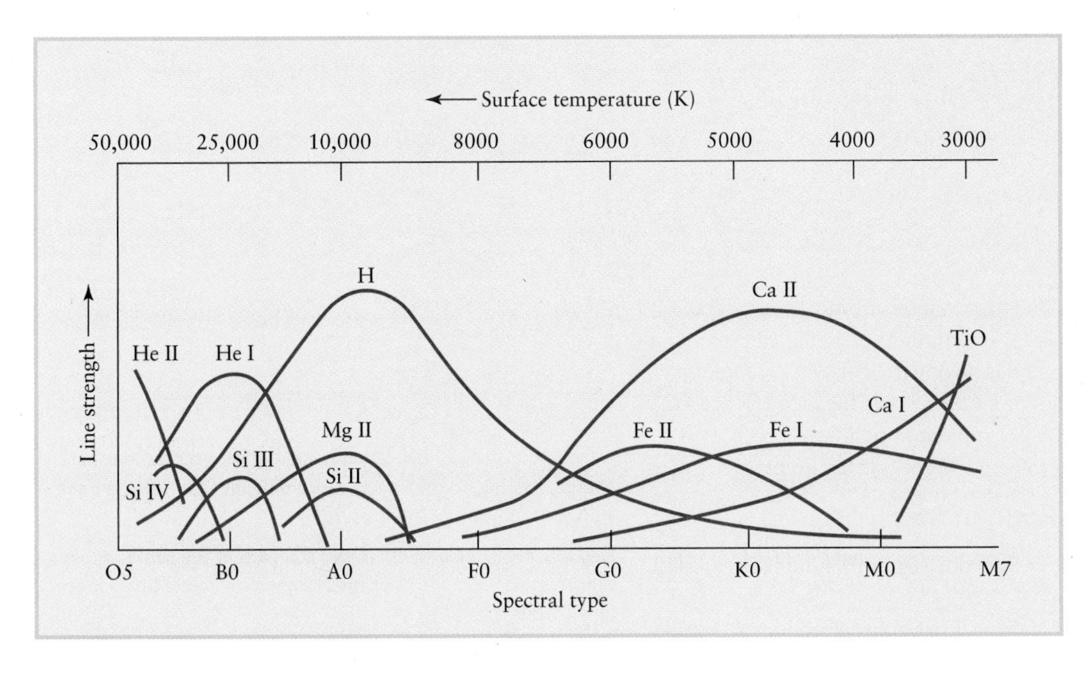

FIGURE 10-10 The Strengths of Absorption Lines Each curve in this graph peaks at the stellar surface temperature for which that chemical's absorption line is strongest. For example, hydrogen (H) absorption lines are strongest in A stars with surface temperatures near 10,000 K. Roman numeral I denotes neutral, un-ionized atoms; II, III, and IV denote atoms that are singly, doubly, or triply ionized (that is, have lost one, two, or three electrons).

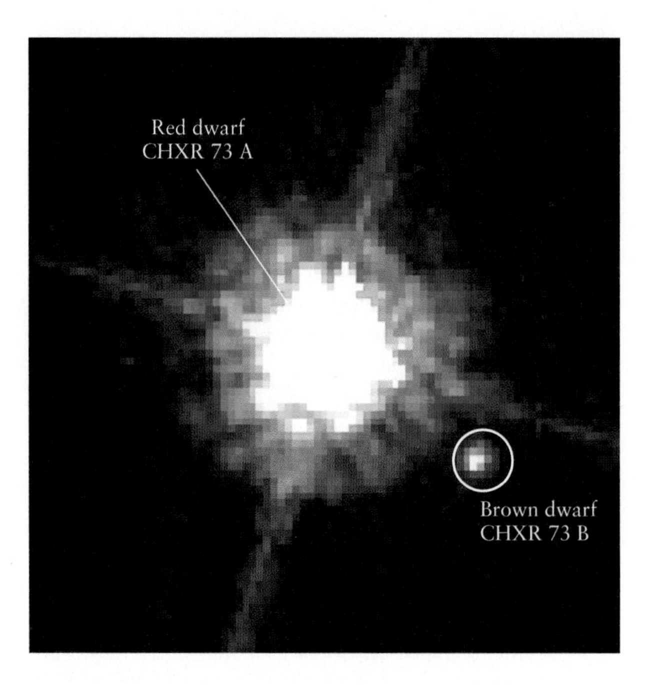

FIGURE 10-11 R I V U X G **Red Dwarf and Brown Dwarf** This false color NASA Hubble Space Telescope image shows a young red dwarf with spectral class M and a much smaller brown dwarf with spectral class T only 200 AU away. The brown dwarf emits most of its light at infrared wavelengths, so an infrared camera was used to record this image. Called CHXR 73 B, this tiny brown dwarf is only 12 times more massive than the planet Jupiter. *(NASA, ESA, and K. Luhman [Penn State University])*

find that *all* stars have essentially the same chemical composition. What we now understand is that by mass, almost all stars (including the Sun) and brown dwarfs are about three-quarters hydrogen, one-quarter helium, and 1% or less metals. Our Sun is about 1% metals by mass, as are most of the stars you can see with the naked eye. But some stars have an even lower percentage of metals.

> **ConceptCheck 10-12: Which brown dwarf is hotter, an L or a T?**
>
> **ConceptCheck 10-13: If our Sun is many times hotter, many times larger, and has more active nuclear reactions than a brown dwarf does, what do they have common?**

We can determine the radius of a star from its luminosity and surface temperature. For a given surface temperature, the greatest luminosity will result from the largest stars.

10-6 Stars come in a wide variety of sizes and masses

Go to Video 10-4

With even the best telescopes, most stars appear as little more than tiny points of light. On a photograph or CCD image, brighter stars appear larger than dim ones (see Figures 10-5*b* and 10-11), but these apparent sizes are due to brightness and give no indication of the star's actual size. So, how could you determine the diameter of an object that is so distant it has no apparent size? One strategy is to combine information about a star's luminosity (determined from its distance and apparent brightness) and its surface temperature (determined from its spectral type).

Calculating the Radii of Stars

The key to finding a star's radius from its luminosity and surface temperature is the Stefan-Boltzmann law. This rule shows up again and again in astronomy and simply says that the amount of energy radiated per second from a square meter of a blackbody's surface—that is, the energy flux (F)—is proportional to the fourth power of the temperature of that surface (T), as given by the equation $F = \sigma T^4$. In other words, how much light a star emits is highly dependent on its temperature, with the hottest stars emitting significantly more energy than cooler stars. (Absorption lines, while important for determining the star's chemical composition and surface temperature, make only relatively small modifications to a star's spectrum.)

A star's luminosity is the amount of energy emitted per second from its entire surface. This equals the energy flux F multiplied by the total number of square meters on the star's surface (that is, the star's surface area). We expect that most stars are nearly spherical, like the Sun, so we can use the formula for the surface area of a sphere. This is $4\pi R^2$, where R is the star's radius (the distance from its center to its surface). Multiplying together the formulas for energy flux and surface area to make a powerful mathematical relationship that can be easily applied, we can write the star's luminosity as follows:

Relationship between a star's luminosity, radius, and surface temperature

$$L = 4\pi R^2 \sigma T^4$$

L = **star's luminosity, in watts**
R = **star's radius, in meters**
σ = **Stefan-Boltzmann constant** = 5.67×10^{-8} **W m^{-2} K^{-4}**
T = **star's surface temperature, in degrees Kelvin**

This equation might look complicated, but it simply says that a relatively cool star (low surface temperature T), for which the energy flux is quite low, can nonetheless be very luminous if it has a large enough radius R. Alternatively, a relatively hot star (large T) can have a very low luminosity if the star has only a little surface area (small R). **Box 10-1 Tools of the Astronomer's Trade: Stellar Radii, Luminosities, and Surface Temperatures** describes and provides examples on how to use the above equation to calculate a star's radius if its luminosity and surface temperature are known.

ANALOGY In a similar way, a roaring campfire can emit more light than a welder's torch. The campfire is at a lower temperature than the torch but has a much larger surface area from which it emits light.

> **ConceptCheck 10-14: What makes lighting a candle using a large, roaring bonfire much more difficult than using a hand lighter of the same temperature?**
>
> **CalculationCheck 10-3: If two stars are at the same temperature, but one is 3 times larger, how many times more luminous is it?**

BOX 10-1 TOOLS OF THE ASTRONOMER'S TRADE

Stellar Radii, Luminosities, and Surface Temperatures

Because stars emit light in almost exactly the same fashion as blackbodies, we can use the Stefan-Boltzmann law to relate a star's luminosity (L), surface temperature (T), and radius (R). The relevant equation is

$$L = 4\pi R^2 \sigma T^4$$

As written, this equation involves the Stefan-Boltzmann constant σ, which is equal to 5.67×10^{-8} W m^{-2} K^{-4}. In many calculations, it is more convenient to relate everything to the Sun, which is a typical star. Specifically, for the Sun we have $L_\odot = 4\pi R_\odot^2 \sigma T_\odot^4$, where $L_\odot$ is the Sun's luminosity, $R_\odot$ is the Sun's radius, and $T_\odot$ is the Sun's surface temperature (equal to 5800 K). Dividing the general equation for L by this specific equation for the Sun, we obtain

$$\frac{L}{L_\odot} = \left(\frac{R}{R_\odot}\right)^2 \left(\frac{T}{T_\odot}\right)^4$$

This is an easier formula to use because the constant σ has cancelled out. We can also rearrange terms to arrive at a useful equation for the radius (R) of a star:

Radius of a star related to its luminosity and surface temperature

$$\frac{R}{R_\odot} = \left(\frac{T_\odot}{T}\right)^2 \sqrt{\frac{L}{L_\odot}}$$

$R/R_\odot$ = ratio of the star's radius to the Sun's radius
$T_\odot/T$ = ratio of the Sun's surface temperature to the star's surface temperature
$L/L_\odot$ = ratio of the star's luminosity to the Sun's luminosity

EXAMPLE: The bright reddish star Betelgeuse in the constellation Orion (see Figure 10-5*a*) is 60,000 times more luminous than the Sun and has a surface temperature of 3500 K. What is its radius?

Situation: We are given the star's luminosity $L = 60{,}000\ L_\odot$ and its surface temperature $T = 3500$ K. Our goal is to find the star's radius R.

Tools: We use the equation for the radius of a star to find the ratio of the star's radius to the radius of the Sun, $R/R_\odot$. Note that we also know the Sun's surface temperature, $T_\odot = 5800$ K.

Answer: Substituting these data into the equation, we get

$$\frac{R}{R_\odot} = \left(\frac{5800\text{ K}}{3500\text{ K}}\right)^2 \sqrt{6 \times 10^4} = 670$$

Review: Our result tells us that Betelgeuse's radius is 670 times larger than that of the Sun. The Sun's radius is 6.96×10^5 km, so we can also express the radius of Betelgeuse as $(670)(6.96 \times 10^5 \text{ km}) = 4.7 \times 10^8$ km. This is more than 3 AU. If Betelgeuse were located at the center of our solar system, it would extend beyond the orbit of Mars!

EXAMPLE: Sirius, the brightest star in the sky, is actually two stars orbiting each other (a binary star). The less luminous star, Sirius B, is a white dwarf that is too dim to see with the naked eye. Its luminosity is $0.0025\ L_\odot$ and its surface temperature is 10,000 K. How large is Sirius B compared to Earth?

Situation: Again we are asked to find a star's radius from its luminosity and surface temperature so we can compare it to Earth's size.

Tools: We use the same equation as in the preceding example.

Answer: The ratio of the radius of Sirius B to the Sun's radius is

$$\frac{R}{R_\odot} = \left(\frac{5800\text{ K}}{10{,}000\text{ K}}\right)^2 \sqrt{0.0025} = 0.017$$

Since the Sun's radius is $R_\odot = 6.96 \times 10^5$ km, the radius of Sirius B is $(0.017)(6.96 \times 10^5 \text{ km}) = 12{,}000$ km. Because Earth's radius (half its diameter) is 6378 km, this star is only about twice the radius of Earth.

Review: Although Sirius B's radius is large compared to a terrestrial planet, it is a minuscule size for a star. The name *dwarf* is well deserved!

Determining a Star's Size from Its Spectrum

You might wonder if the details of a star's size can be deduced from characteristics of its absorption spectrum, and in fact they can be. **Figure 10-12** compares the spectra of two stars of the same spectral type and temperature but different luminosity (and hence different size): a B8 supergiant star and a more common B8 star. Note that the most prominent lines of hydrogen are narrow in the spectrum of the very large, very luminous supergiant but quite broad in the spectrum of the small, less luminous star. In general, for stars of spectral types B through F, the larger and more luminous the star, the narrower is its hydrogen line.

Fundamentally, these differences between stars of different luminosity are due to differences between the stars' atmospheres, where absorption lines are produced. Hydrogen lines in particular are affected by the density and pressure of the gas in a star's atmosphere. The higher the density and

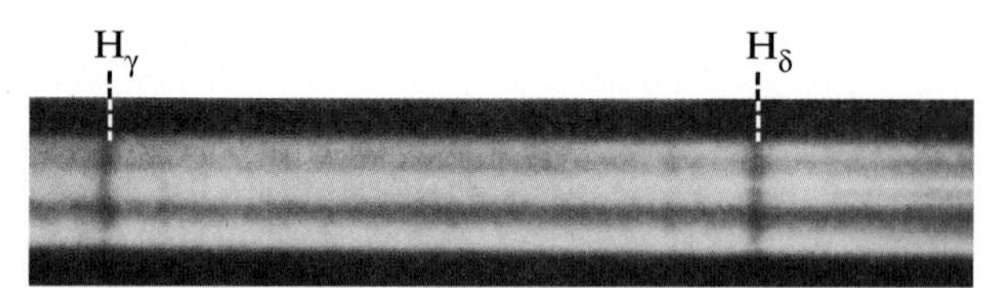

(a) A supergiant star has a low-density, low-pressure atmosphere: Its spectrum has narrow absorption lines.

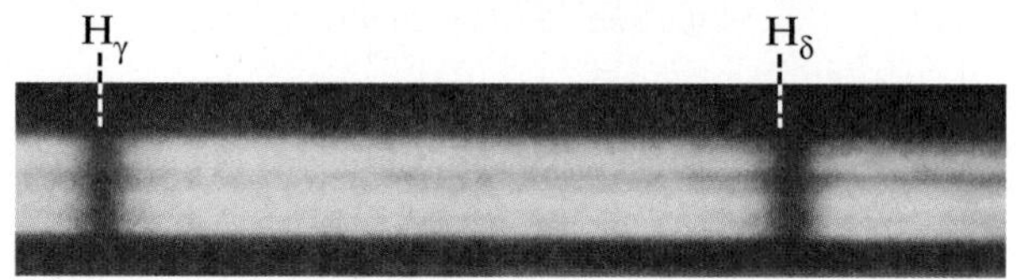

(b) A main-sequence star has a denser, higher-pressure atmosphere: Its spectrum has broad absorption lines.

FIGURE 10-12 RIVUXG **How a Star's Size Affects Its Spectrum** These are the spectra of two stars of the same spectral type (B8) and surface temperature much greater than the Sun (13,400 K) but different radii and luminosities: (a) the B8 supergiant Rigel (luminosity 58,000 $L_\odot$) in Orion, and (b) the B8 main-sequence star Algol (luminosity 100 $L_\odot$) in Perseus. *(From W. W. Morgan, P. C. Keenan, and E. Kellman,* An Atlas of Stellar Spectra *[1943])*

pressure, the more frequently hydrogen atoms collide and interact with other atoms and ions in the atmosphere. These collisions shift the energy levels in the hydrogen atoms and thus broaden the hydrogen spectral lines.

Astronomers find that stars come in a wide range of sizes. The smallest stars visible through ordinary telescopes, called *white dwarfs,* are about the same size as Earth. Although their surface temperatures can be very high (25,000 K or more), white dwarfs have so little surface area that their luminosities are very low (less than 0.01 $L_\odot$). The largest stars, called *supergiants,* are a thousand times larger in radius than the Sun and 10^5 times larger than white dwarfs. If our own Sun were replaced by one of these supergiants, Earth's orbit would lie completely inside the star! (White dwarfs and supergiants are covered in more detail in Section 10-7.)

In the atmosphere of a luminous giant star, the density and pressure are quite low because the star's mass is spread over a huge volume. Atoms and ions in the atmosphere are relatively far apart; hence, collisions between them are sufficiently infrequent that hydrogen atoms can produce narrower lines. A smaller star, however, is much more compact than a giant or supergiant. In the denser atmosphere of a typical, Sunlike star, frequent interatomic collisions perturb the energy levels in the hydrogen atoms, thereby producing broader lines.

ConceptCheck 10-15: **What about the nature of atoms in Sunlike stars makes the notion that the "larger stars have the broadest absorption lines" incorrect?**

Observing Binary Star Systems Reveals the Masses of Stars

We now know something about the sizes, temperatures, and luminosities of stars and how one aspect influences the others. To complete our picture of the physical properties of stars, we need to know how much material these stars are made of—their masses. In this section, we will see that stars come in a wide range of masses. We will also discover an important relationship between the mass and luminosity of unique stars. This relationship is crucial to understanding why some stars are hot and luminous, while others are cool and dim. It will also help us understand what happens to a star as it ages and evolves.

Determining the masses of stars is not an easy task by any means. The problem is that there is no practical, direct way to measure the mass of an isolated star. Fortunately for astronomers, about half of the visible stars in the night sky are not isolated individuals. Instead, they are *multiple-star systems,* in which two or more stars orbit each other. By carefully observing the motions of these stars, astronomers can glean important information about their masses.

A pair of stars located at nearly the same position in the night sky is called a **double star.** The German-born British astronomer William Herschel—who discovered Uranus—made the first organized search for such pairs. Between 1782 and 1821, he published three catalogs listing more than 800 double stars. Late in the nineteenth century, his son, John Herschel, discovered 10,000 more doubles. Some of these double stars are two stars that lie along nearly the same line of sight but are actually at very different distances from us. But many double stars are true **binary stars,** or **binaries**—pairs of stars that actually orbit each other. **Figure 10-13** shows an example of this orbital motion. By observing the binary over an extended period, astronomers can plot the orbit that one star appears to describe around the other, as shown in the center diagram in Figure 10-13.

In fact, *both* stars in a binary system are in motion. They orbit each other because of their mutual gravitational attraction, and their orbital motions can be calculated by adapting Kepler's third law for planet orbits to multiple star systems, as follows:

Kepler's third law for binary star systems

$$M_1 + M_2 = \frac{a^2}{P^2}$$

M_1, M_2 = masses of two stars in binary system, in solar masses

a = semimajor axis of one star's orbit around the other, in AU

P = orbital period, in years

Here a is the semimajor axis of the elliptical orbit that one star appears to describe around the other, plotted as in the center diagram in Figure 10-13. As this equation indicates, if we can measure this semimajor axis (a) and the orbital period (P), we can learn something about the masses of the two stars.

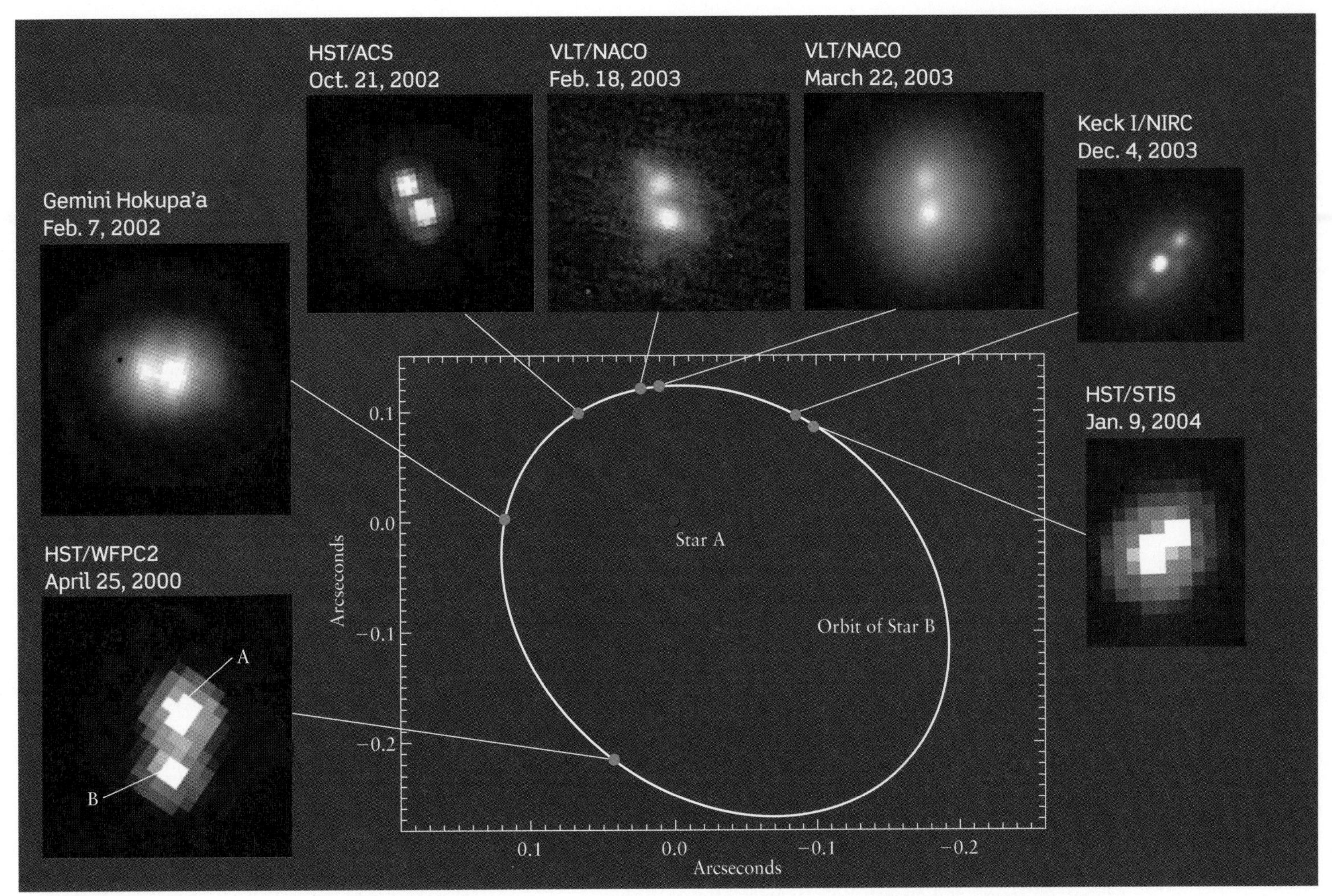

FIGURE 10-13 R I V U X G **A Binary Star System** As seen from Earth, the two stars that make up the binary system called 2MASSW J0746425 + 2000321 are separated by less than ⅓ arcsec. The images surrounding the center diagram show the relative positions of the two stars over a four-year period. These images were made by the Hubble Space Telescope (HST), the European Southern Observatory's Very Large Telescope (VLT), and Keck I and Gemini North in Hawaii (see Figure 2-28). For simplicity, the diagram shows one star as remaining stationary; in reality, both stars move around their common center of mass. *(clockwise from bottom-left: a, c, g: NASA Hubblesite.org; b: Gemini Observatory—Northern Operations Center; d, e: European Southern Observatory; f: Dr. Andrea Ghez)*

In principle, the orbital period of a visual binary is easy to determine. All you have to do is see how long it takes for the two stars to revolve once about each other. The two stars shown in Figure 10-13 are relatively close, about 2.5 AU on average, and their orbital period is only 10 years. Many binary systems have much larger separations, however, and the period may be so long that more than one astronomer's lifetime is needed to complete the observations.

Each of the two stars in a binary system actually moves in an elliptical orbit about the **center of mass** of the system. Imagine two children sitting on opposite ends of a playground seesaw (**Figure 10-14*a***). For the seesaw to balance properly, they must position themselves so that their center of mass—an imaginary point that lies along a line connecting their two bodies—is at the pivot point of the seesaw. If the two children have the same mass, the center of mass lies midway between them, and they should sit equal distances from the center. If their masses are different, the center of mass is closer to the heavier child and the children need to readjust their positions in order to easily balance.

Just as the seesaw naturally balances at its center of mass, the two stars that make up a binary system naturally orbit around their center of mass (Figure 10-14*b*). The center of mass always lies along the line connecting the two stars and is closer to the more massive star.

The center of mass of a visual binary is located by plotting the separate orbits of the two stars, as in Figure 10-14*b*. The center of mass lies at the common focus of the two elliptical orbits. Comparing the relative sizes of the two orbits around the center of mass yields the ratio of the two stars' masses, M_1/M_2. The sum $M_1 + M_2$ is already known

FIGURE 10-14 Center of Mass in a Binary Star System (a) A seesaw balances if the board pivots at the center of mass of the two children. (b) The members of a binary star system orbit around the center of mass of the two stars. Although their elliptical orbits cross each other, the two stars are always on opposite sides of the center of mass and thus never collide.

(a) A "binary system" of two children

The center of mass of the binary star system is nearer to the more massive star.

More massive star

Less massive star

(b) A binary star system

from Kepler's third law, so the individual masses of the two stars can then be determined.

> ConceptCheck 10-16: **If two stars in a binary system were moved farther apart, how would their masses and orbital periods change?**

10-7 Hertzsprung-Russell (H-R) diagrams reveal the different kinds of stars

Astronomers have collected a wealth of data about the stars, but merely having tables of numerical data is not enough. Like all scientists, astronomers want to analyze their data to look for trends and underlying principles. One of the best ways to look for trends in any set of data, whether it comes from astronomy, finance, medicine, or meteorology, is to create a graph showing how one quantity depends on another. For example, investors consult graphs of stock market values versus dates, and weather forecasters make graphs of temperature versus altitude to determine whether thunderstorms will form. Astronomers have found that a particular graph of stellar properties shows that stars fall naturally into just a few categories. This graph, one of the most important in all astronomy, will help us understand how stars form, evolve, and eventually die.

H-R Diagrams

Go to Video 10-5

Which properties of stars should we include in a graph? Most stars have about the same chemical composition, but two properties of stars—their luminosities and surface temperatures—differ substantially from one star to another. What do we learn when we graph the luminosities of stars versus their surface temperatures?

One step toward answering this question was given in 1911 by the Danish astronomer Ejnar Hertzsprung. He pointed out that a regular pattern appears when the absolute magnitudes of stars (which measure their luminosities) are plotted against their colors (which measure their surface temperatures). Two years later, the American astronomer Henry Norris Russell independently discovered a similar regularity in a graph using spectral types (another measure of surface temperature) instead of colors. In recognition of their originators, graphs of this kind are today known as **Hertzsprung-Russell diagrams,** or **H-R diagrams** (**Figure 10-15**).

Figure 10-15*a* is a typical Hertzsprung-Russell diagram and it is a powerful tool for organizing the many characteristics of stars. Each dot represents a single star whose spectral type and luminosity have been determined. The most luminous stars are near the top of the diagram, the least luminous stars near the bottom. Hot stars of spectral classes O and B are toward the left side of the graph and cool stars of spectral class M are toward the right. The Sun is near the center of the various ranges.

CAUTION You are probably accustomed to graphs in which the numbers on the horizontal axis increase as you move to the right. (For example, the business section of a newspaper includes a graph of stock market values versus dates, with later dates to the right of earlier ones.) But on an H-R diagram the temperature scale on the horizontal axis increases toward the *left*. This practice stems from the original diagrams of Hertzsprung and Russell, who placed hot O stars on the left and cool M stars on the right. This arrangement is a tradition that no one has seriously tried to change.

The most striking feature of the H-R diagram is that the data points are not scattered randomly over the graph but are grouped in a few distinct regions. The luminosities and surface temperatures of stars do *not* have random values; instead, these two quantities are related! The band stretching diagonally across the H-R diagram includes about 90% of the stars in the night sky. This band, called the **main sequence,** extends from the hot, luminous, blue stars in the upper left corner of the diagram to the cool, dim, red stars in the lower right corner. A star whose

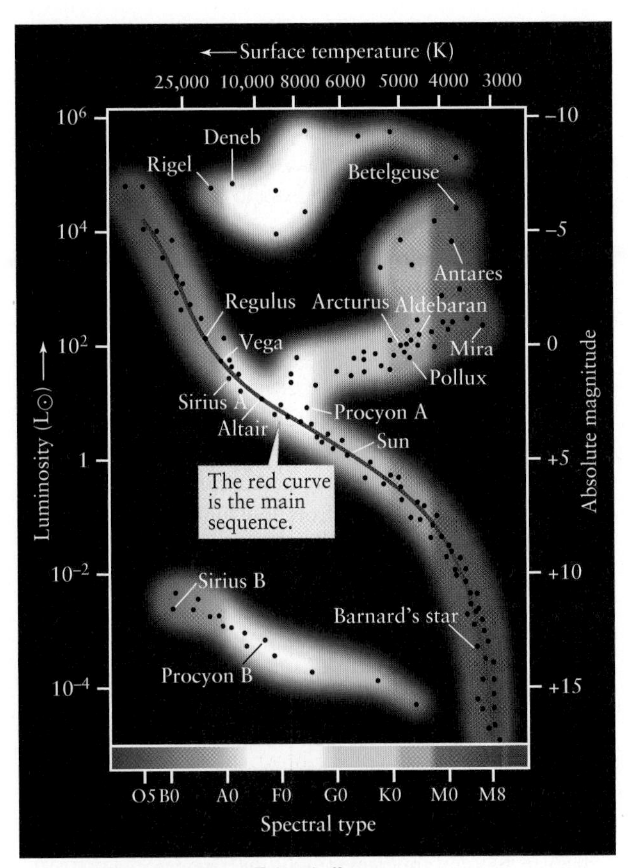

(a) A Hertzsprung-Russell (H-R) diagram

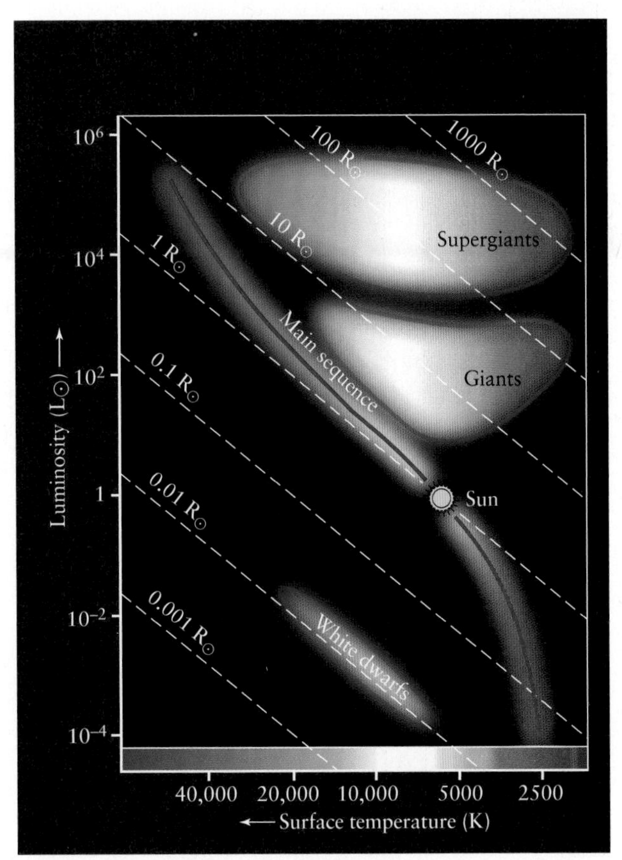

(b) The sizes of stars on an H-R diagram

FIGURE 10-15 Hertzsprung-Russell (H-R) Diagrams On an H-R diagram, the luminosities (or absolute magnitudes) of stars are plotted against their spectral types (or surface temperatures). (a) The data points are grouped in just a few regions on the graph, showing that luminosity and spectral type are correlated. Most stars lie along the red curve called the main sequence. Giants like Arcturus, as well as supergiants like Rigel and Betelgeuse, are above the main sequence, and white dwarfs like Sirius B are below it. (b) The blue curves on this H-R diagram enclose the regions of the diagram in which different types of stars are found. The dashed diagonal lines indicate different stellar radii. For a given stellar radius, as the surface temperature increases (that is, moving from right to left in the diagram), the star glows more intensely and the luminosity increases (that is, moving upward in the diagram). Note that the Sun is intermediate in luminosity, surface temperature, and radius.

properties place it in this region of an H-R diagram is called a main-sequence star. The Sun (spectral type G2, luminosity 1 $L_\odot$, absolute magnitude +4.8) is such a star. We will find that all main-sequence stars are like the Sun in that *hydrogen fusion*—thermonuclear reactions that convert hydrogen into helium—is taking place in their cores.

The upper right side of the H-R diagram shows a second major grouping of data points. Stars represented by these points are both luminous and cool. From the Stefan-Boltzmann law, we know that a cool star radiates much less light per unit of surface area than a hot star. In order for these stars to be as luminous as they are, they must be huge, and so they are called **giants.** These stars are around 10 to 100 times larger than the Sun. You can see this in Figure 10-15*b*, which is an H-R diagram to which dashed lines have been added to represent stellar radii. Most giant stars are around 100 to 1000 times more luminous than the Sun and have surface temperatures less than that of the Sun—only about 3000 K to 6000 K. Cooler members of this class of stars (those with surface temperatures from about 3000 K to 4000 K) are often called **red giants** because they appear reddish. At the same time, they must be enormous to have such high luminosities. A number of red giants can easily be seen with the naked eye, including Aldebaran in the constellation Taurus and Arcturus in Boötes.

A few rare stars are considerably bigger and brighter than typical red giants, with radii up to 1000 $R_\odot$. Appropriately enough, these superluminous stars are called **supergiants.** Betelgeuse in Orion and Antares in Scorpius are two supergiants you can find in the nighttime sky. Together, giants and supergiants make up about 1% of the stars in the sky. Both giants and supergiants have thermonuclear reactions occurring in their interiors, but the character of those reactions and where in the star they occur can be quite different than for a main-sequence star like the Sun.

The remaining 9% of stars form a distinct grouping of data points toward the lower left corner of the Hertzsprung-Russell diagram. Although these stars are hot, their luminosities are quite low; hence, they must be small. They are appropriately called **white dwarfs.** These stars, which are so dim that they can be seen only with a telescope, are approximately the same size as Earth—more than 100 times smaller than the Sun's diameter. No thermonuclear reactions take place within white dwarf stars. Rather, like embers left from a fire, they are the still-glowing remnants of what were once giant stars.

By contrast, *brown dwarfs* are objects that will never become stars. They are comparable in radius to the planet Jupiter. The study of brown dwarfs is still in its infancy, but it appears that there may be twice as many brown dwarfs as there are "real" stars.

ANALOGY You can think of white dwarfs as "has-been" stars whose days of glory have passed. In this analogy, a brown dwarf is a "never-will-be."

The existence of fundamentally different types of stars is the first important lesson to come from the H-R diagram. In Chapters 11 and 12 we will find that these different types represent various stages in the lives of stars. We will use the H-R diagram as an essential tool for understanding how stars evolve.

ConceptCheck 10-17: Where on the H-R diagram are the stars with the greatest luminosity and highest temperatures?

Main-Sequence Masses and the Mass-Luminosity Relation

The greater the mass of a main-sequence star, the greater its luminosity, its surface temperature, and its radius.

Years of careful, patient observations of binaries have slowly yielded the masses of many stars. As the data accumulated, an important trend began to emerge: For main-sequence stars, there is a direct correlation between mass and luminosity. The more massive a main-sequence star, the more luminous it is. **Figure 10-16** depicts this **mass-luminosity relation** as a graph. The range of stellar masses extends from less than 0.1 of a solar mass to more than 50 solar masses. The Sun's mass lies between these extremes.

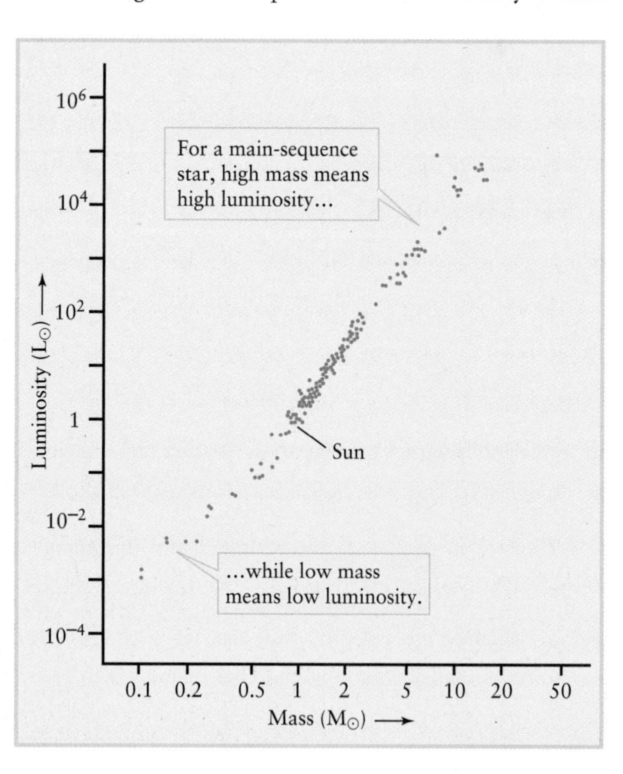

FIGURE 10-16
The Mass-Luminosity Relation For main-sequence stars, there is a direct correlation between mass and luminosity—the more massive a star, the more luminous it is. A main-sequence star of mass 10 $M_\odot$ (that is, 10 times the Sun's mass) has roughly 3000 times the Sun's luminosity (3000 $L_\odot$); one with 0.1 $M_\odot$ has a luminosity of only about 0.001 $L_\odot$.

Cosmic Connections: The Main Sequence and Masses depicts the mass-luminosity relation for main-sequence stars on an H-R diagram. This figure shows that the main sequence on an H-R diagram is a progression in mass as well as in luminosity and surface temperature. The hot, bright, bluish stars in the upper left corner of an H-R diagram are the most massive main-sequence stars. Likewise, the dim, cool, reddish stars in the lower right corner of an H-R diagram are the least massive. Main-sequence stars of intermediate temperature and luminosity also have intermediate masses.

The mass of a main-sequence star also helps determine its radius. Referring back to Figure 10-15*b*, we see that if we go along the main sequence from low luminosity to high luminosity, the radius of the star increases. Thus, we have the following general rule for main-sequence stars: The greater the mass of a main-sequence star, the greater its luminosity, its surface temperature, and its radius.

Why is mass the controlling factor in determining the properties of a main-sequence star? The answer is that all main-sequence stars are objects like the Sun, with essentially the same chemical composition as the Sun but with different masses. Like the Sun, all main-sequence stars shine because thermonuclear reactions at their cores convert hydrogen to helium and release energy. The greater the total mass of the star, the greater the pressure and temperature at the core, the more rapidly thermonuclear reactions take place in the core, and the greater the energy output—that is, the luminosity—of the star. In other words, the greater the mass of a main-sequence star, the greater its luminosity. This statement is just the mass-luminosity relation, which we can now recognize as a natural consequence of the nature of main-sequence stars.

Like the Sun, main-sequence stars are in a state of both hydrostatic equilibrium and thermal equilibrium. Calculations using models of a main-sequence star's interior show that to maintain equilibrium, a more massive star must have a larger radius and a higher surface temperature. This is just what we see when we plot the curve of the main sequence on an H-R diagram (see Figure 10-15*b*). As you move up the main sequence from less massive stars (at the lower right in the H-R diagram) to more massive stars (at the upper left), the radius and surface temperature both increase.

CAUTION The mass-luminosity relation we have discussed applies to main-sequence stars only. There are *no* simple mass-luminosity relations for giant, supergiant, or white dwarf stars. We will find that main-sequence stars evolve into giant and supergiant stars, and that some of these eventually end their lives as white dwarfs.

ConceptCheck 10-18: The brightest of the main-sequence stars have what other relatively large characteristics?

COSMIC CONNECTIONS The Main Sequence and Masses

The main sequence is an arrangement of stars according to their mass. The most massive main-sequence stars have the greatest luminosity, largest radius, and highest surface temperature. This is a consequence of the behavior of thermonuclear reactions at the core of a main-sequence star.

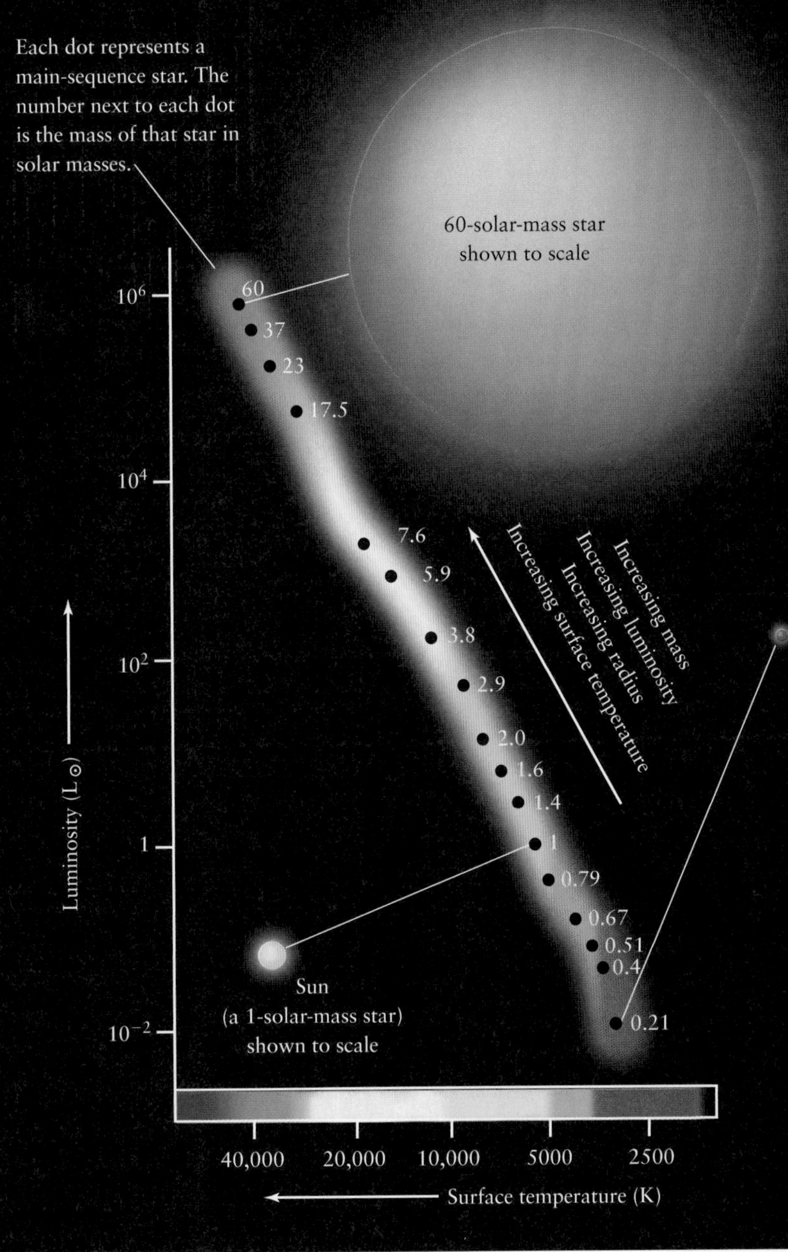

- A star with 60 solar masses has much higher pressure and temperature at its core than does the Sun.
- This causes thermonuclear reactions in the core to occur much more rapidly and release energy at a much faster rate—790,000 times faster than in the Sun.
- Energy is emitted from the star's surface at the same rate that it is released in the core, so the star has 790,000 times the Sun's luminosity.
- The tremendous rate of energy release also heats the star's interior tremendously, increasing the star's internal pressure. This inflates the star to 15 times the Sun's radius.
- The star's surface must be at a high temperature (about 44,500 K) in order for it to emit energy into space at such a rapid rate.

- A star with 0.21 solar mass has much lower pressure and temperature at its core than does the Sun.
- This causes thermonuclear reactions in the core to occur much more slowly and release energy at a much slower rate—0.011 times as fast as in the Sun.
- Energy is emitted from the star's surface at the same rate that it is released in the core, so the star has 0.011 of the Sun's luminosity.
- The low rate of energy release supplies relatively little heat to the star's interior, so the star's internal pressure is low. Hence the star's radius is only 0.33 times the Sun's radius.
- The star's surface need be at only a low temperature (about 3200 K) to emit energy into space at such a relatively slow rate.

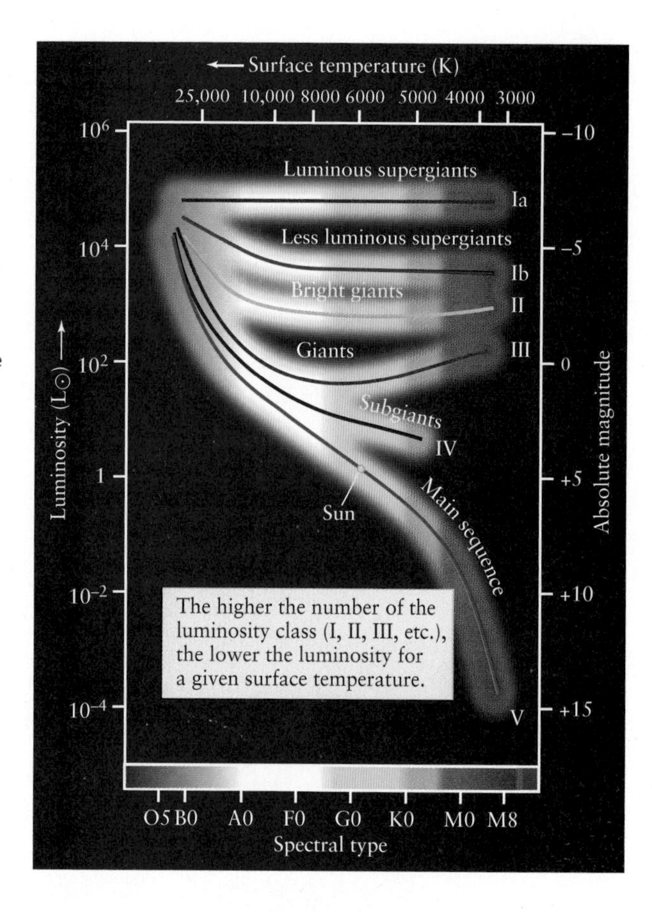

FIGURE 10-17 Luminosity Classes The H-R diagram is divided into regions corresponding to stars of different luminosity classes. (White dwarfs do not have their own luminosity class.) A star's spectrum reveals both its spectral type and its luminosity class; from these, the star's luminosity can be determined.

Luminosity Classes

When carefully looking at small details of spectral lines, astronomers in the 1930s noticed subtle differences that are quite important. This caused astronomers to further classify spectra into **luminosity classes.** When these luminosity classes are plotted on the H-R diagram (**Figure 10-17**), subdivisions of the star types appear in the upper right of the diagram. Luminosity classes use arbitrary letters and Roman numbers such as I, II, and III. Luminosity classes Ia and Ib are composed of supergiants; luminosity class V includes all the main-sequence stars. The intermediate classes distinguish giant stars of various luminosities. Note that for stars of a given surface temperature (that is, a given spectral type), the *higher* the number of the luminosity class, the *lower* the star's luminosity. These different luminosity classes represent different stages in the evolution of a star. White dwarfs are not given a luminosity class of their own, as they represent a final stage in stellar evolution in which no thermonuclear reactions take place.

Astronomers commonly use a shorthand description that combines a star's spectral type and its luminosity class. For example, the Sun is said to be a G2 V star. The spectral type indicates the star's surface temperature, and the luminosity class indicates its luminosity. Thus, an astronomer knows immediately that any G2 V star is a main-sequence star with a luminosity of about 1 $L_\odot$ and a surface temperature of about 5800 K. Similarly, a description of Aldebaran as a K5 III star tells an astronomer that it is a red giant with a luminosity of around 370 $L_\odot$ and a surface temperature of about 4000 K.

ConceptCheck 10-19: How does a K5 V star compare to a K5 II star in terms of temperature, luminosity, and radius?

Spectroscopic Parallax

Now we are in a position to put all of these ideas together to more accurately determine distances to stars. Before, the best strategy we had in determining distance was parallax, but this only works for the nearest stars. Now, using the H-R diagram, we can combine brightness, luminosity, and spectral class together to determine distance for the most distant stars. As an example, consider the star Pleione in the constellation Taurus. Its spectrum reveals Pleione to be a B8 V star (a hot, blue, main-sequence star, like the one in Figure 10-12*b*). An H-R diagram value for luminosity that corresponds to a B8 star would be about 190 $L_\odot$. Given the star's luminosity and its apparent brightness—in the case of Pleione, 3.9×10^{-13} of the apparent brightness of the Sun—we can use the inverse-square law to determine its distance from Earth.

This powerful and relatively straightforward method for determining distance, in which the luminosity of a star is found using spectral analysis, is called **spectroscopic parallax. Figure 10-18** summarizes the method of spectroscopic parallax.

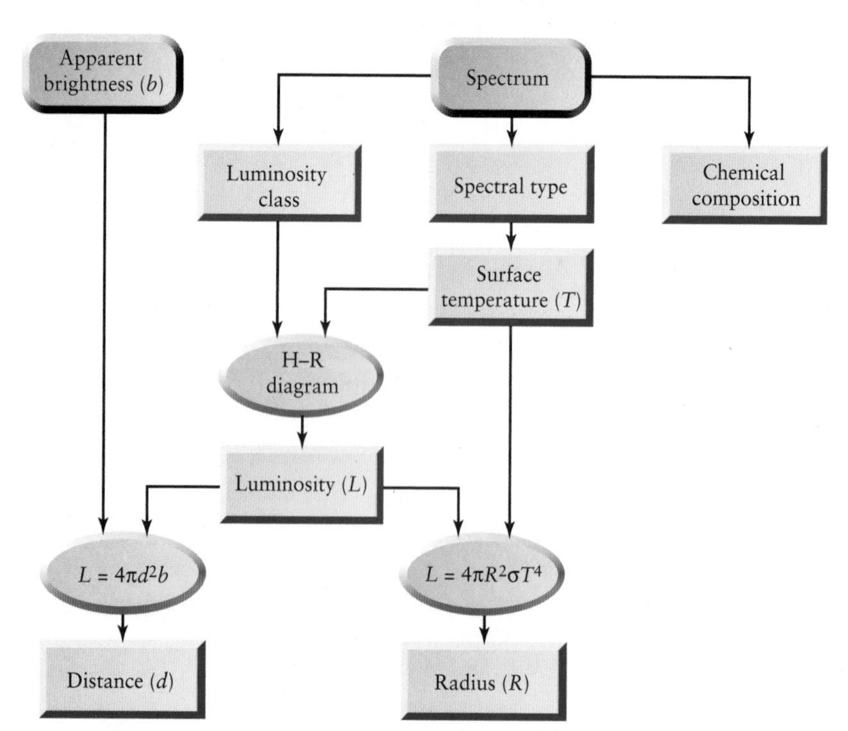

FIGURE 10-18 The Method of Spectroscopic Parallax If a star is too far away, its parallax angle is too small to allow a direct determination of its distance. This flowchart shows how astronomers deduce the properties of such a distant star. Note that the H-R diagram plays a central role in determining the star's luminosity from its spectral type and luminosity class. Just as for nearby stars, the star's chemical composition is determined from its spectrum, and the star's radius is calculated from the luminosity and surface temperature.

CAUTION The name "spectroscopic parallax" is a bit misleading, because no parallax angle is involved! The idea is that measuring the star's spectrum takes the place of measuring its parallax as a way to find the star's distance. A better name for this method, although not the one used by astronomers, would be "spectroscopic distance determination."

Spectroscopic parallax is an incredibly powerful technique. No matter how remote a star is, this technique allows astronomers to determine its distance, provided only that its spectrum and apparent brightness can be measured. Unfortunately, spectroscopic parallax has its limitations; distances to individual stars determined using this method are only accurate to at best 10%. The reason is that the luminosity classes shown in Figure 10-17 are not thin lines on the H-R diagram but are moderately broad bands. Hence, even if a star's spectral type and luminosity class are known, there is still some uncertainty in the luminosity that we read off an H-R diagram. Nonetheless, spectroscopic parallax is often the only means that an astronomer has to estimate the distance to remote stars.

VISUAL LITERACY TASK

The H-R Diagram

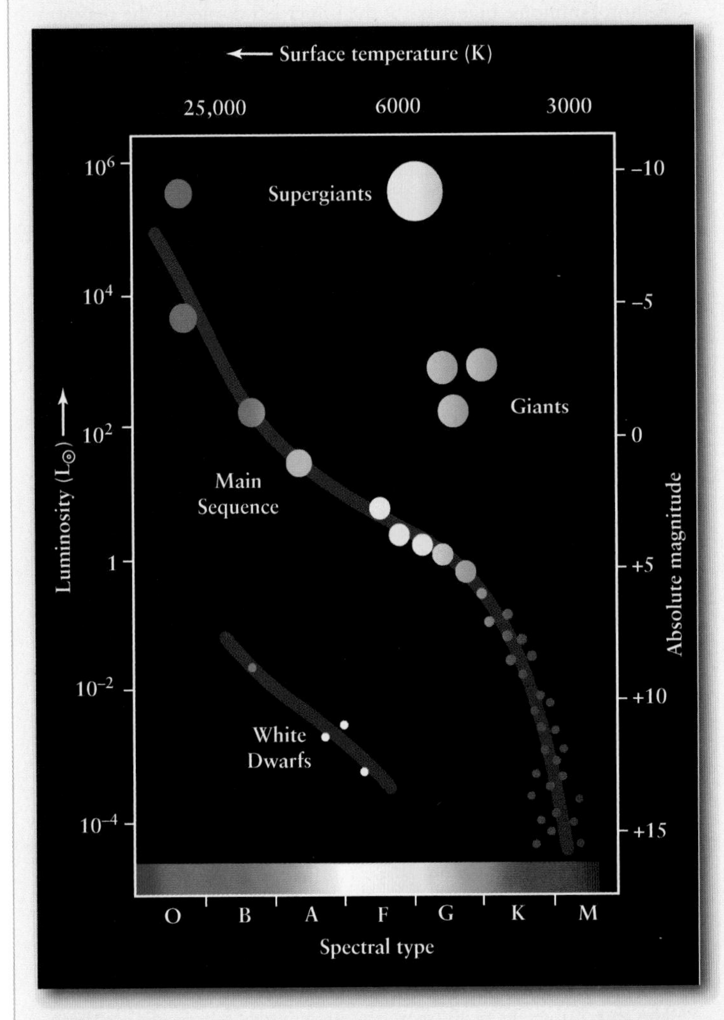

PROMPT: What would you tell a fellow student who said, **"The H-R diagram is a plot of distance versus color and shows that stars of all types are evenly distributed around the Galaxy"?**

ENTER RESPONSE:

Guiding Questions:

1. The horizontal axis of the H-R diagram is
   a. distance.
   b. brightness.
   c. either temperature or spectral class.
   d. either luminosity or absolute magnitude.

2. The farther from the horizontal axis a star plots, the greater its
   a. temperature.
   b. distance.
   c. spectral class.
   d. luminosity or absolute magnitude.

3. The number of red dwarf main-sequence stars is
   a. vastly greater than all other types of stars.
   b. about the same as all other types of stars.
   c. about the same as giant stars.
   d. about the same as white dwarfs.

4. Objects that are relatively hot but still quite dim are
   a. giant stars.
   b. supergiant stars.
   c. white dwarfs.
   d. main-sequence stars.

KEY IDEAS AND TERMS

10-1 Measuring the distances to nearby stars utilizes an effect called parallax

- **Stellar parallax** is the apparent change in position of a star due to Earth's motion around our Sun and is used to measure the distance to nearby stars.
- The **parallax** angle, *p*, is equal to half the angle through which a star's apparent position changes due to the change in position of an observer.
- A star with a parallax angle of 1 second of arc ($p = 1$ arcsec) is defined to have a distance of 1 **parsec** ($d = 1$ pc), which is 3.26 ly.

10-2 A star's brightness can be described in terms of luminosity or magnitude

- How bright a star appears in the sky for an Earth-based observer is called its **apparent magnitude,** with the brightest stars having the lowest numbers.
- Without a telescope, the brightest stars are magnitude 1 stars and the dimmest stars are magnitude 6 stars.
- The magnitude of a star if it were at a distance of 10 pc is its **absolute magnitude,** and it is a quantity that reflects a star's true energy output so it can be compared to other stars.

10-3 A star's distance can be determined by comparing its luminosity and brightness

- A star's **luminosity** is the total amount of energy a star emits each second.
- The **inverse-square law** relationship states that the apparent brightness of light that an observer can see or measure is inversely proportional to the square of the observer's distance (d) from the source.
- For a given distance, the brighter the star, the more luminous that star must be. For a given apparent brightness, the more distant the star, the more luminous it must be to be seen at that distance.

10-4 A star's color depends on its surface temperature

- The primary wavelength of light emitted by a star is dependent on its temperature, with higher temperature stars emitting relatively greater amounts of short wavelength light.
- Red stars are relatively cold, with low surface temperatures.
- Blue stars are relatively hot, with high surface temperatures.

10-5 The spectra of stars reveal their chemical compositions as well as surface temperatures and sizes

- Stars are categorized into **spectral classes,** designated by letters, that are organized by details of the stars' spectra.
- Stars are further subdivided into **spectral types** and numbered 0–9.
- From hottest to coolest, the sequence of spectral classes is **OBAFGKM.**
- Substars too small to sustain thermonuclear fusion in their cores are widely known as **brown dwarfs.**
- By mass, almost all stars (including the Sun) and brown dwarfs are about three-quarters hydrogen, one-quarter helium, and 1% or less metals.

10-6 Stars come in a wide variety of sizes and masses

- The key to finding a star's radius from its luminosity and surface temperature is the Stefan-Boltzmann law.
- We can determine the radius of a star from its luminosity and surface temperature. For a given luminosity, the greater the surface temperature, the smaller the radius must be. For a given surface temperature, the greater the luminosity, the larger the radius must be.
- **Double stars** are two stars that lie along nearly the same line of sight but are actually at very different distances from us.
- Some multiple star systems are true **binary stars,** or **binaries,** which are pairs of stars that actually orbit each other; the details of their orbits allow us to determine sizes and masses.
- Each of the two stars in a binary system actually moves in an elliptical orbit about the **center of mass** of the system.

10-7 Hertzsprung-Russell (H-R) diagrams reveal the different kinds of stars

- Graphs of the luminosity/absolute magnitude versus spectral class/ surface temperature are today known as **Hertzsprung-Russell diagrams** or **H-R diagrams.**
- **Main-sequence** stars are found on the H-R diagram extending from the hot, luminous, blue stars in the upper left corner of the diagram to the cool, dim, red stars in the lower right corner.
- Tremendously luminous and large stars are called **giants.** Cooler members of the largest stars are often called **red giants** because they appear reddish.
- A few rare stars are considerably bigger and brighter than typical giants, with radii up to 1000 $R_{\odot}$; they are called **supergiants.**
- **White dwarfs** are so dim that they can be seen only with a telescope and are approximately the same size as Earth.
- The **mass-luminosity relation** is that the greater the mass of a main-sequence star, the greater its luminosity, its surface temperature, and its radius.
- **Luminosity classes** are based upon the subtle differences in spectral lines, which provide a useful subdivision of the star types in the upper right of the diagram for stars of differing radii.
- **Spectroscopic parallax** is the process of inferring the distance to a main-sequence star based on its apparent magnitude and its spectral classification.

QUESTIONS

Review Questions

1. Explain the difference between a star's apparent brightness and its luminosity.
2. Why does it take at least six months to make a measurement of a star's parallax?
3. What is the inverse-square law? Use it to explain why an ordinary lightbulb can appear brighter than a star, even though the lightbulb emits far less light energy per second.
4. Why is the magnitude scale called a "backward" scale? What is the difference between apparent magnitude and absolute magnitude?
5. The star Zubenelgenubi (from the Arabic for "scorpion's southern claw") has apparent magnitude 2.75, while the star Sulafat (Arabic for "tortoise") has apparent magnitude 3.25. Which star appears brighter? From this information alone, what can you conclude about the luminosities of these stars? Explain your answer.

6. Menkalinan (Arabic for "shoulder of the rein-holder") is an A2 star in the constellation Auriga (the Charioteer). What is its spectral class? What is its spectral type? Which gives a more precise description of the spectrum of Menkalinan?
7. If a red star and a blue star both have the same radius and both are the same distance from Earth, which one looks brighter in the night sky? Explain why.
8. If a red star and a blue star both appear equally bright and both are the same distance from Earth, which one has the larger radius? Explain why.
9. Sketch a Hertzsprung-Russell diagram. Indicate the regions on your diagram occupied by (a) main-sequence stars, (b) red giants, (c) supergiants, (d) white dwarfs, and (e) the Sun.
10. What information about stars do astronomers learn from binary systems that cannot be learned in any other way? What measurements do they make of binary systems to garner this information?
11. Which is more massive, a red main-sequence star or a blue main-sequence star? Which has the greater radius? Explain your answers.

Web Chat Questions

1. In its orbit around Earth, the *Hipparcos* satellite could measure stellar parallax angles with acceptable accuracy only if the angles were larger than about 0.002 arcsec. Discuss the advantages or disadvantages of making parallax measurements from a satellite in a large solar orbit, say at the distance of Jupiter from the Sun. If this satellite can also measure parallax angles of 0.002 arcsec, what is the distance of the most remote stars that can be accurately determined? How much bigger a volume of space would be covered compared to Earth-based observations? How many more stars would you expect to be contained in that volume?
2. As seen from the starship *Enterprise* in the *Star Trek* television series and movies, stars appear to move across the sky due to the starship's motion. How fast would the *Enterprise* have to move in order for a star 1 pc away to appear to move 1° per second? (*Hint:* The speed of the star as seen from the *Enterprise* is the same as the speed of the *Enterprise* relative to the star.) How does this compare with the speed of light? Do you think the stars appear to move as seen from an orbiting space shuttle, which moves at about 8 km/s?

Collaborative Exercises

1. Considering where your group is sitting right now, how many times dimmer would an imaginary, super-deluxe, ultrabright flashlight be if it were located at the front door of the group member who lives farthest away as compared to if it were at the front door of the group member who lives closest. Explain your reasoning.
2. As a group, select any two of the stars in a text appendix listing of the sky's brightest stars and compare the apparent visual magnitudes to determine how many times brighter one is than the other.

Observing Questions

1. Use *Starry Night*™ to examine the 10 brightest stars in Earth's night sky. Select **Favourites > Explorations > Atlas.** Use the **View > Constellations** menu command to display constellation **Boundaries, Labels,** and **Astronomical** stick figures. Use the **File** (Windows) or **Starry Night** (Mac) menu command to open the **Preferences** dialog window. Ensure that the **Cursor Tracking (HUD)** preferences include **Apparent Magnitude, Distance from Observer, Luminosity,** and **Temperature** in the Show list. Before closing the **Preferences** dialog window, it might be helpful to increase the saturation for **Star colour** under the **Brightness/Contrast** preferences. Click on the **Lists** side pane tab, expand the **Observing Lists,** and click the **10 Brightest Stars** option. Then expand the **List Viewer** layer and select **All Targets** from the **Show** dropdown menu to see a list of the 10 brightest stars in Earth's night sky. Double-click on each of the stars in this list in turn to center the star in the view. Use the HUD to compile a table of these stars that includes each star's apparent magnitude, distance, luminosity, and temperature. You may also wish to sketch the star's position within its constellation. Alternatively, you may find it helpful to print out relevant star charts around these stars, using *Starry Night*™. (**a**) Which is the brightest star in Earth's night sky? What features of this star make it so bright in our sky? (**b**) Which of these brightest stars has the highest temperature? What would you expect to be the color of this star compared to others in the list? (**c**) Which of these stars is intrinsically the most luminous? (**d**) Use *Starry Night*™ to determine which of these stars is visible from your location. Click the **Home** button, then the **Stop** button, and finally the **Sunset** button to show the view from your home location today at sunset. Again, it may be helpful to display the constellation **Boundaries, Labels,** and **Astronomical** stick figures in the view. Open the **Lists** side pane and double-click each entry in the list of the 10 Brightest Stars. If the star is visible in your sky, the program will center it in the view or alternately suggest a **Best Time** for observing this star. For those stars in the list that are visible from your home location, go outside if possible and observe them in the real sky. See if you can tell which of these stars has the highest temperature on the basis of your conclusion regarding the star's color and check your estimate against the table you compiled in part a. (*Hint:* The colors of stars are not very distinct and a dark sky background is needed in order to distinguish differences in stellar colors.)
2. Use the *Starry Night*™ program to investigate the Hertzsprung-Russell (H-R) diagram. Select **Favourites > Explorations > Denver.** Open the **Status** pane, expand the **H-R Options** layer, and choose the following options: **Use absolute magnitudes** and **Labels.** In the expanded **Labels** panel, click **On** the **Gridlines, Regions, Main Sequence,** and **Spectral class** options. Now, expand the **Hertzsprung-Russell** layer to show the H-R diagram that plots all of the stars that are currently in the main view. This graphical representation shows the absolute magnitudes of stars as a function of their spectral class. The sequence of spectral class, from O, B, A, F, G, K and M, represents the star's surface temperatures, plotted in an inverse direction, the hottest O-type stars appearing to the left of the diagram. Absolute magnitude is related to the star's luminosity, the smaller the absolute value of absolute magnitude, the larger the luminosity. (**a**) Use the hand tool to scroll around the sky. Watch the H-R diagram change as different stars enter and leave the main window. Right-click (Ctrl-click on a Mac) on a blank part of the sky and select **Hide Horizon** from the contextual menu so that you can survey the entire sky. Does the distribution of stars in the H-R diagram change drastically from one part of the sky to another, or are all types of stars approximately equally represented in all directions from Earth? (**b**) If you place the cursor over a star, a red dot appears in the H-R diagram at the position for this star. Use this facility to estimate and make a note of the position on the H-R diagram of each of the following stars that are labeled in the main window: Altair, Deneb, Enif, 74 Ophiuchi, and 51 Pegasi. Which of these five stars is most similar to the Sun? (**c**) What is the name of the region of the H-R diagram occupied by this star?

When looking into the night sky, one naturally wonders if the stars overhead are the same stars that dotted the skies of our grandparents and that of our great-great-great grandparents. Indeed, the stars that illuminate our nights seem eternal and unchanging. But this permanence is an illusion. Each of the stars visible to the naked eye shines due to thermonuclear reactions and has only a finite amount of fuel available for these reactions. Hence, stars cannot last forever: They form from material in interstellar space, evolve over millions or billions of years, and eventually die. In this chapter, our concern is with how stars form and change over their life cycles.

Observations across the sky reveal that stars form within cold, dark clouds of gas and dust that are scattered abundantly throughout our Galaxy. One such cloud appears as a dark area on the far right-hand side of the photograph opening this chapter. Perhaps a dark cloud like this encounters pressure from one of our Galaxy's spinning arms or an exploding star detonates nearby. From the shock of events like these, the cloud begins to contract under the pull of gravity, forming protostars—the fragments that will one day become stars. The physical process underlying this is that as a protostar develops, its internal pressure builds and its temperature rises. In time, nuclear fusion begins, and a shining star is born. The hottest, bluest, and brightest young stars, like those in the photo, emit ultraviolet light that excites atoms in the surrounding interstellar gas. The result is a beautiful glowing nebula, which typically has the red color characteristic of excited hydrogen (as shown in the photograph). As we will see, stars mature and grow old and their rate of aging is directly related to their mass. In this chapter, we will focus on what happens to the most common of stars—stars like our own Sun. In the following chapter, we will find that the largest of stars even blow themselves apart in death throes that enrich interstellar space with the material for future generations of stars. Thus, like the mythical phoenix, all new stars arise from the ashes of the old. ■

11-1 Stars form from the gravitational collapse of immense clouds of interstellar gas and dust

Because stars consume their own mass as fuel to power their brilliant shining, stars cannot emit light forever. But, because stars are large, they have quite a bit of mass to use as fuel and can shine for a very long time. Stars last much longer than the lifetime of any astronomer—indeed, far longer than the entire history of human civilization. Thus, it is impossible to watch a single star go through its formation, evolution, and eventual demise. Rather, astronomers have to piece together the evolutionary history of stars by studying different stars at different stages in their life cycles.

ANALOGY To gain insight into how an astronomer can build a picture of the life cycle of a star, imagine for a moment that you are a biologist from another planet who sets out to understand the life cycles of human beings. You send a spacecraft to fly above Earth and photograph humans in action. Unfortunately, the spacecraft fails after collecting only 20 s of data, but during that time its sophisticated equipment sends back observations of thousands of different humans. From this brief snapshot of life on Earth—only 10^{-8} (a hundred-millionth) of a typical human lifetime—how would you decide which were the young humans and which were the older ones? Without a look inside our bodies to see the biological processes that shape our lives, could you tell how humans are born and how they die? And how could you deduce the various biological changes that humans undergo as they age?

Astronomers, too, have data spanning only a tiny fraction of any one star's lifetime. A star like the Sun has enough usable fuel to last for about 10 billion years, whereas astronomers have been observing stars in detail for only about a century. Besides the long time frames, astronomers are further frustrated by being unable to see directly into the hidden interiors of stars. For example, we cannot see the thermonuclear reactions that convert hydrogen into helium. But astronomers have an advantage in that there are many, many stars at different stages in their life cycles to observe. Moreover, stars are made of relatively simple substances, primarily hydrogen and helium, that are found almost exclusively in the form of gases. Of the four basic phases of matter—plasma, gas, liquid, and solid—gases are by far the simplest to understand.

Astronomers use our understanding of gases to build theoretical models of the interiors of stars, like the model of the Sun we saw in Chapter 9. In fact, like all great dramas, the story of stellar evolution can be regarded as a struggle between two opposing and unyielding forces: Gravity continually tries to make a star shrink, while the star's internal pressure tends to make the star expand. When these two opposing forces are in balance, the star is in a state of hydrostatic equilibrium. But what happens when changes within the star cause either pressure or gravity to predominate? The star must then either expand or contract until it reaches a new equilibrium. In the process, it will change not only in size but also in luminosity and color.

Interstellar Gas and Dust

Go to Video 11-1

If stars do not exist forever, then somehow they must come into being. Where do stars come from? As we saw in Chapter 4, stars like our Sun condensed from a collection of gas and dust in interstellar space. Observations suggest that other stars originate in a similar way. In order to pursue the origin of stars, we need to first observe the seemingly empty space between stars.

At first glance, the space between the stars seems to be mostly vacant. On closer inspection, we find that it is filled with a thin gas laced with microscopic dust particles. This combination of gas and dust is called the **interstellar medium.**

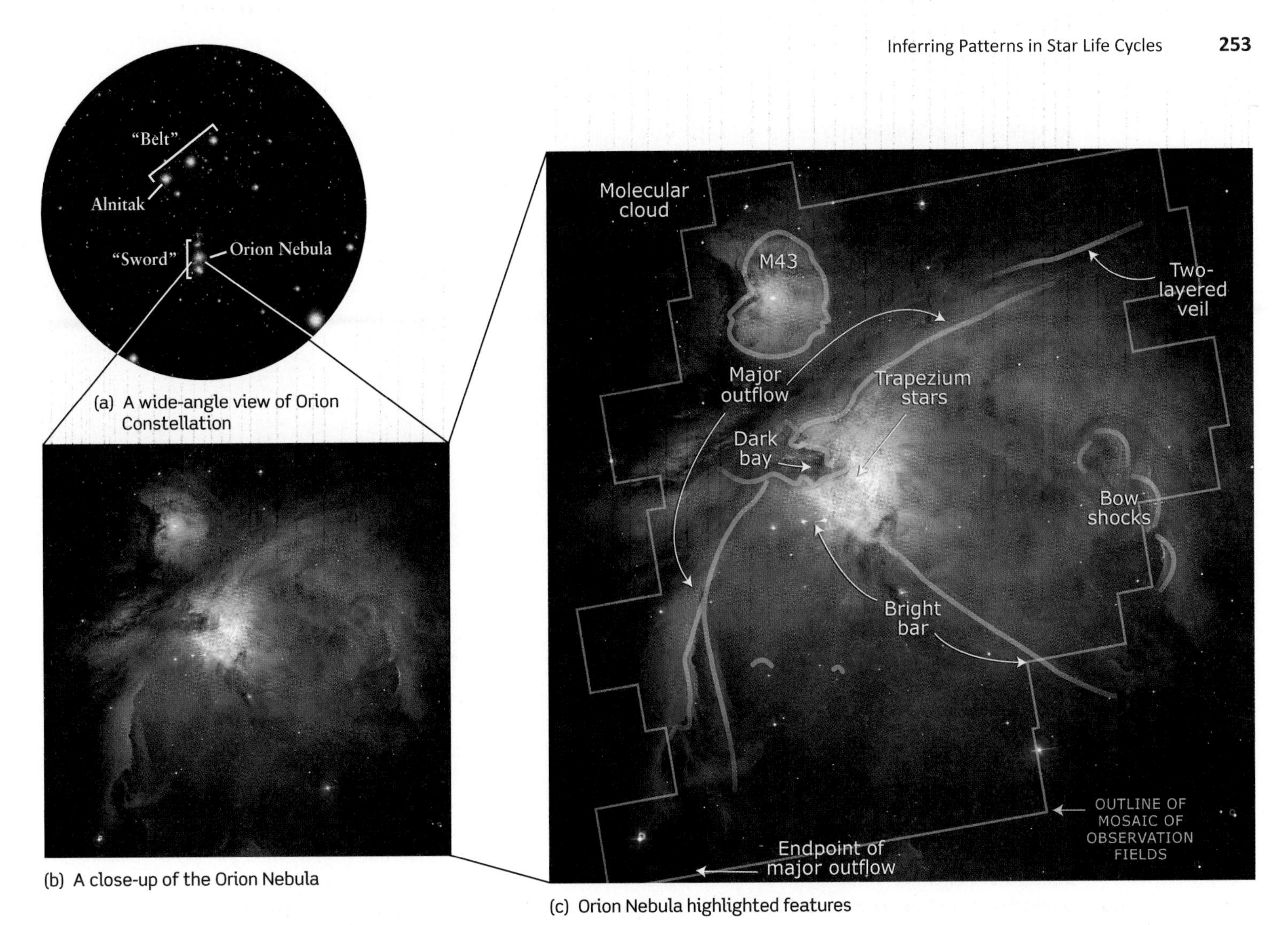

FIGURE 11-1 R I V U X G **The Orion Nebula** (a) The middle "star" of the three that make up Orion's sword is actually an interstellar cloud called the Orion Nebula. (b) A mosaic of HST and ground-based images reveal a bright central region of stars known as the Trapezium. (c) Bow shocks created by young stars emitting streams of material that collide with and heat nearby gas can be seen, as can a "bright bar" emanating from the Trapezium. More outflows of material can be seen to the left and top of the image. About 1500 ly from Earth, the nebula contains thousands of young stars that cause the nebula to glow. *(a: peresanz/Shutterstock; b,c: NASA, ESA, and A. Feild [STScI])*

Evidence for this medium includes interstellar clouds of various types, curious lines in the spectra of binary star systems, and an apparent dimming and reddening of distant stars.

You can see evidence for the interstellar medium with the naked eye. Look carefully at the constellation Orion (**Figure 11-1*a***), easily visible on clear winter nights from the northern hemisphere. While most of the stars in this constellation appear as sharply defined points of light, the middle "star" in Orion's sword has a fuzzy appearance. This becomes more obvious when viewed through binoculars or a telescope. As Figure 11-1*b* shows, this "star" is actually not a star at all, but the Orion Nebula—a cloud in interstellar space. Any interstellar cloud is known to astronomers as a **nebula** (*plural* **nebulae**).

The Orion Nebula emits its own light, with the characteristic emission line spectrum similar to that of a hot, thin gas. For this reason, it is called an **emission nebula.** Typical emission nebulae are large, having masses that range from about 100 solar masses to about 10,000 solar masses. Because this mass is spread over a huge volume that is light-years across, the density is quite low by Earth standards, only a few thousand hydrogen atoms per cubic centimeter (cm^3). (By comparison, the air you are breathing contains more than 10^{19} atoms per cm^3.) Emission nebulae are found near hot, luminous stars of spectral types O and B. Such stars emit copious amounts of ultraviolet radiation. When atoms in the nearby interstellar gas absorb these energetic ultraviolet photons, the atoms become ionized. Indeed, emission nebulae are composed primarily of ionized hydrogen atoms, that is, free protons (hydrogen nuclei) and electrons. These recently ionized atoms are the source of energy emanating from an emission nebula.

In addition to the presence of gas atoms in the interstellar medium, **Figure 11-2*a*** shows evidence for large clouds of dust grains. A **dark nebula** is so opaque that it blocks any visible light coming from stars that lie behind it. Such clouds contain from 10^4 to 10^9 particles (atoms, molecules, and dust grains) per cubic centimeter. Although thin by Earth standards, dark nebulae are large enough—typically many light-years deep—that they block the passage of light. In the same way, a sufficient depth of haze or smoke in our atmosphere can make it impossible to see distant mountains.

Dark nebulae are enormous clouds of dust and gas that do not allow starlight to pass through.

A different observation that serves as evidence for dust is the bluish haze surrounding the stars in Figure 11-2*b*. A haze

(a) Dark Nebula

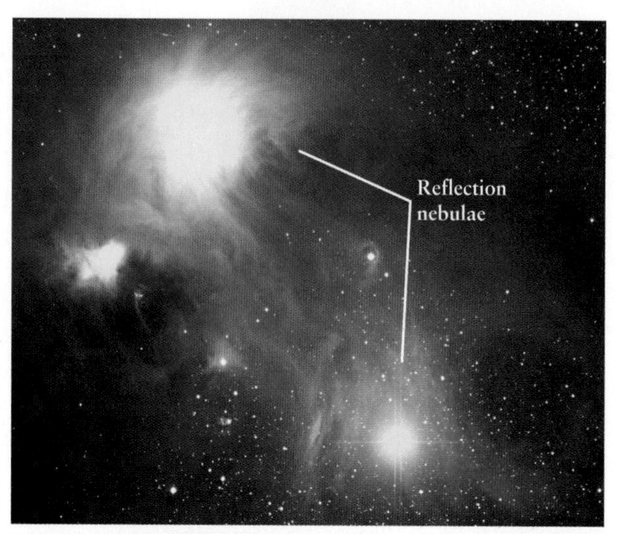

(b) Reflection Nebula

FIGURE 11-2 R I V U X G **A Dark Nebula and a Reflection Nebula**
(a) When first discovered in the late 1700s, dark nebulae were thought to be "holes in the heavens" where very few stars were present. In fact, they are opaque regions of dust grains that block out light from the stars beyond them. The few stars that appear to be within Barnard 86 lie between us and the nebula. Barnard 86 is in the constellation Sagittarius and has an angular diameter of 4 arcmin, about ⅛ the angular diameter of the full moon. (b) A wispy reflection nebula called NGC 6726-27-29 surrounds several stars in the constellation Corona Australis (the Southern Crown). Unlike emission nebulae, reflection nebulae do not emit their own light, but scatter and reflect light from the stars that they surround. This scattered starlight is quite blue in color. *(Australian Astronomical Observatory/David Malin Images)*

of this kind, called a **reflection nebula,** is caused by fine grains of dust in a lower concentration than that found in dark nebulae. The light we see coming from the nebula is starlight that has been scattered and reflected by these dust grains. The grains are submicroscopically small, no larger than a typical wavelength of visible light, and they scatter short-wavelength blue light more efficiently than long-wavelength red light. Hence, reflection nebulae have a characteristic blue color, resulting in an effect similar to what causes Earth's sky to appear blue.

> ConceptCheck 11-1: **What would a reflection nebula become if it could be compressed to 3 times its typical density?**

Evolution of a Protostar

Interstellar gas and dust are the raw materials from which stars are made. The disk of our Galaxy, where most of the matter is concentrated, is therefore the site of ongoing star formation. In order for this interstellar material to condense and form a star, the force of gravity—which tends to draw interstellar material together—must overwhelm the internal pressure pushing the material apart. This means that stars will most easily form in regions where the interstellar material is relatively dense, so that atoms and dust grains are close together and gravitational attraction is enhanced. The only parts of the interstellar medium with high enough density and low enough temperature for stars to form are the dark nebulae. Within these clouds, the densest portions can contract under their own mutual gravitational attraction and form clumps called **protostars.** Each protostar will eventually evolve into a main-sequence star. Because dark nebulae contain many solar masses of material, it is possible for a large number of protostars to form out of a single such nebula. Thus, we can think of dark nebulae as "stellar nurseries."

Many stars are formed from a single, gigantic collapsing cloud.

At first, a protostar is merely a cool blob of gas several times larger than our solar system. The pressure inside the protostar is too low to support all this cool gas against the mutual gravitational attraction of its parts, and so the protostar collapses. As the protostar collapses, gravitational energy is converted into thermal energy, making the gases heat up and start glowing. **Figure 11-3** shows a Hubble Space Telescope observation of a protostar called V838 Monocerotis. This star suddenly brightened for a few weeks, emitting a short burst of light. As that light burst traveled outward, it illuminated each of the surrounding layers of dust for a brief time, one layer at a time. In this picture, we are not actually seeing a movie of previously emitted dust expanding; rather, what we are seeing is light that traveled from the star out to the dust, which was then reflected back to observers here on Earth. Because of the long, indirect path light can take out to where the outermost dust layers are, the reflected starlight arrives at Earth months after light from the star that bounced off of nearby dust layers. This results in a false appearance of expansion, in much the same way medical doctors can look at various layers of a person's brain using a MRI brain scan.

Observing the evolution of a protostar can be quite a challenge. It is quite unlikely that you have ever seen a protostar shining in the night sky because protostars form within clouds that contain substantial amounts of interstellar dust. The dust in a protostar's immediate surroundings, acting somewhat like a protective moth's cocoon, absorbs the vast amounts of visible light emitted by the protostar and makes it very hard to detect using telescopes tuned to visible wavelengths.

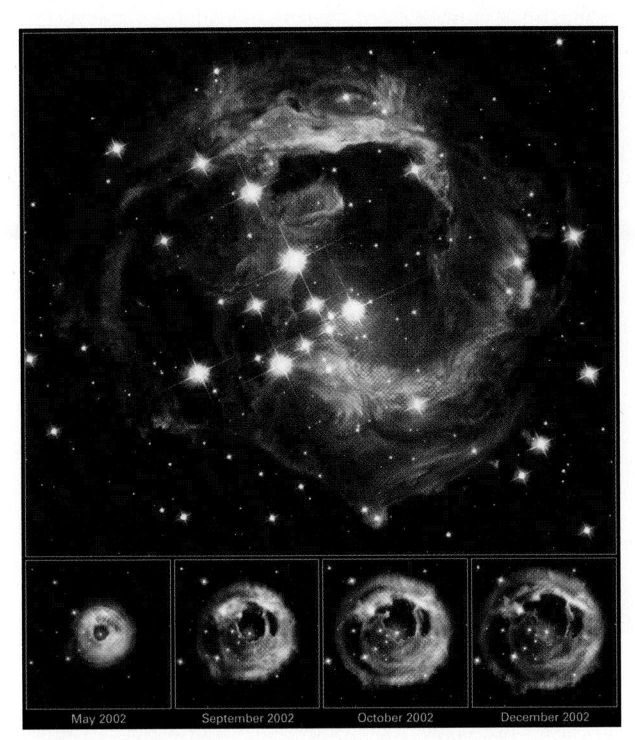

FIGURE 11-3 R I V U X G **Light Echo Reveals Dust Cloud Structures** These Hubble Space Telescope images show various layers of dust in the region around V838 Monocerotis, about 20,000 light-years away. This star unexpectedly brightened for a few weeks, first illuminating the dust nearest the star and then later illuminating dust farther away from the star, revealing the dusty cloud structures. *(NASA, ESA, and The Hubble Heritage Team [STScI/AURA])*

Protostars, however, can be seen hiding inside clouds using infrared telescopes. Because the cloud absorbs so much energy from its central protostar, the dust surrounding a protostar becomes greatly heated to a few hundred kelvins. The warmed dust then reradiates its thermal energy at infrared wavelengths, to which normally opaque dust is relatively transparent. So, by using infrared telescopes, astronomers can see protostars within the "stellar nursery" of a dark nebula.

Figure 11-4 shows visible-light and infrared views of one such stellar nursery, known as the Flame Nebula. The visible-light view (Figure 11-4*a*) shows a dark, dusty nebula that appears completely opaque. The infrared image (Figure 11-4*b*) allows us to easily see through the dust, revealing newly formed protostars within the dark nebula.

It is reasonable to think that a main-sequence star forms simply by collapsing inward. It might be surprising to learn that much of the material of a cold, dark nebula is ejected into space and never incorporated into stars. As it is ejected, this material may help sweep away the dust surrounding a young star, making the star observable at visible wavelengths.

Mass ejection into space is a hallmark of a particular type of star and can be very dramatic, as captured by the Hubble Space Telescope and shown in **Figure 11-5.** Known as **T Tauri stars,** these protostars have emission lines as well as absorption lines in their spectra. Their luminosity can change irregularly on timescales of a few days. The namesake of this class of stars, T Tauri, is a protostar in the constellation Taurus (the Bull).

T Tauri stars have masses less than about 3 times larger than the mass of our Sun (written as 3 $M_{\odot}$) and ages around a million years, so on an H-R diagram, they appear above the right-hand end of the main sequence. The emission lines show that these protostars are surrounded by a thin, hot gas. The Doppler shifts of these emission lines suggest that the protostars eject gas at speeds around 180,000 mi/h (80 km/s).

Protostars are able to slowly add mass to themselves at the same time that they rapidly eject material into space. In fact, the two processes are related. As a protostar's nebula contracts, it spins faster and flattens into a disk with the protostar itself at the center. The same flattening took place in the solar nebula from which the Sun and planets formed. Particles orbiting the protostar within this disk collide with each other, causing them to lose energy, spiral inward onto the protostar, and add to the protostar's mass. The resulting disk of material being added to the protostar in this way is known as

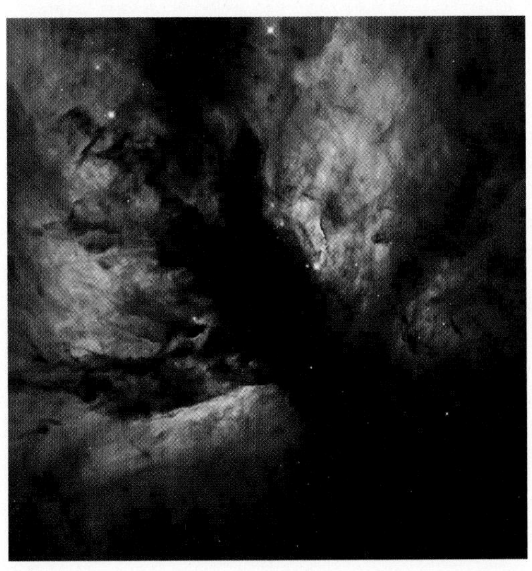
(a)

(b)

FIGURE 11-4 R I V U X G **Hidden Stars Revealed by Infrared Telescope** The Flame Nebula, also known as NGC 2024, hides a cluster of young, blue, massive stars that (a) are invisible when studied with visible light, but (b) are revealed by infrared observations that easily see through the dust. Redder looking stars in the image are more heavily embedded within the nebula than the bluer stars (many of which are in the foreground). The cluster is thought to be less than one million years old. *(a: ESO/IDA/Danish 1.5 m/R. Gendler, J.-E. Ovaldsen, C. Thöne and C. Féron; b: David Thompson)*

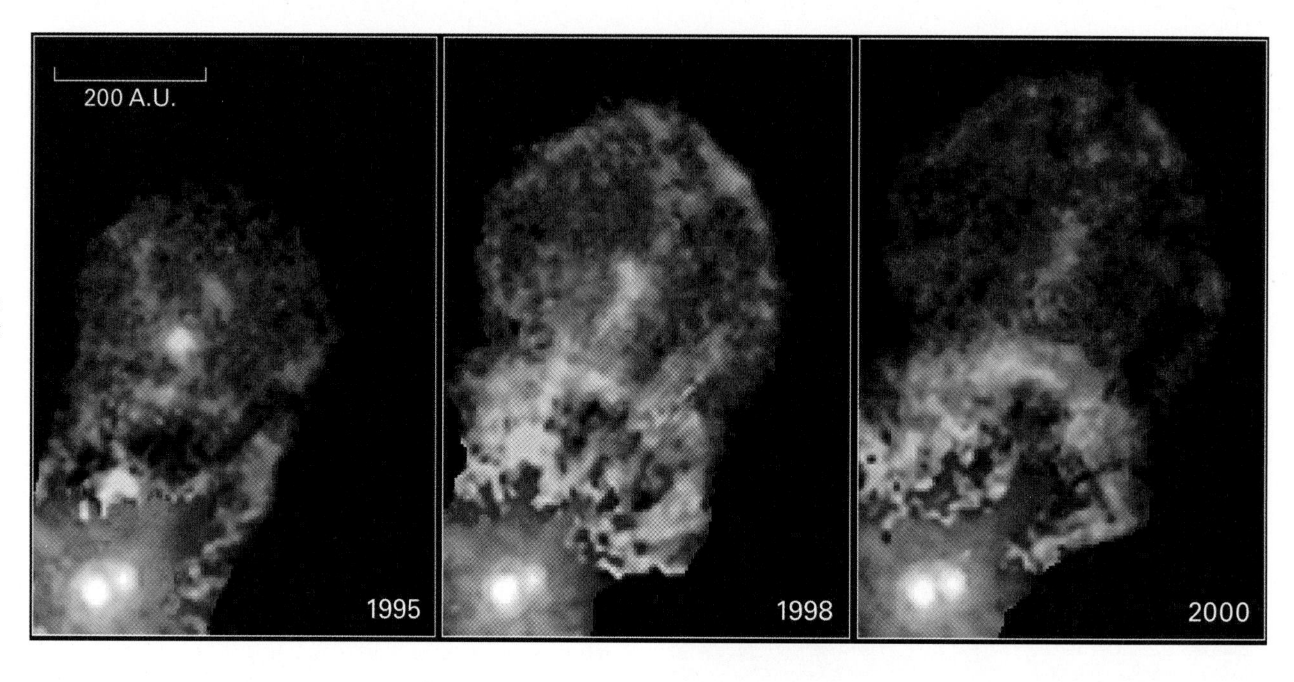

FIGURE 11-5 R I V U X G **T Tauri Star System Ejecting Hot Gas** The Hubble Space Telescope spied this ejection in three dramatic photographs over 5 years. XZ Tau is located about 500 light-years away and has been known to be expanding over the past 30 years. *(NASA, John Krist [Space Telescope Science Institute], Karl Stapelfeldt [Jet Propulsion Laboratory], Jeff Hester [Arizona State University], Chris Burrows [European Space Agency/Space Telescope Science Institute])*

a **circumstellar accretion disk. Figure 11-6** is an edge-on view of one such circumstellar accretion disk, showing one of two oppositely directed jets emanating from a point at or near the center of the disk, where the protostar is located.

In the 1990s, astronomers using the Hubble Space Telescope discovered many examples of disks around newly formed stars in the Orion Nebula (see Figure 11-1 and Figure 4-12*b*), one of the most prominent star-forming regions in the northern sky. If one imagines that planets can form these dusty disks, then they are appropriately called **protoplanetary disks,** or **proplyds,** that surround young stars within the nebula. As the name suggests, protoplanetary disks are thought to contain the material from which planets form around stars.

These protostars can mature into stable main-sequence stars quickly. The speed at which they do so depends entirely on their starting mass. An H-R diagram of protostars, like the one shown in **Figure 11-7,** shows that a protostar 15 times the mass of our Sun becomes a star in about 100,000 years while a much smaller protostar, like one that is about the same mass as our Sun, can take more than 10 million years to form.

ConceptCheck 11-2: Do protostars increasing in brightness gain mass or eject mass during formation?

Supernovae Can Trigger Star Birth

Go to Video 11-2

We will learn in Chapter 14 that a star-filled galaxy's moving spiral arms can cause clouds to compress and initiate star birth, but this is certainly not the only

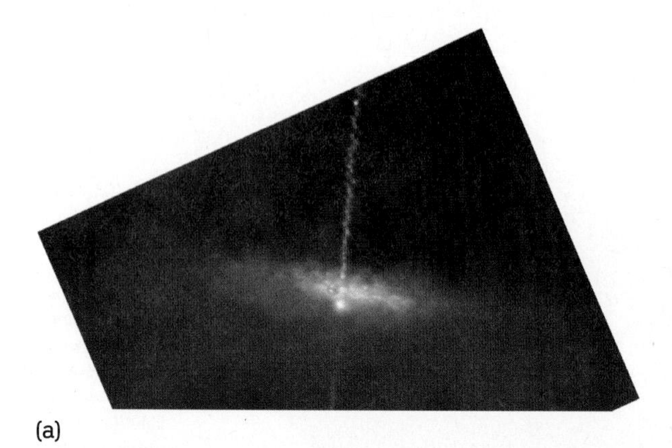

(a)

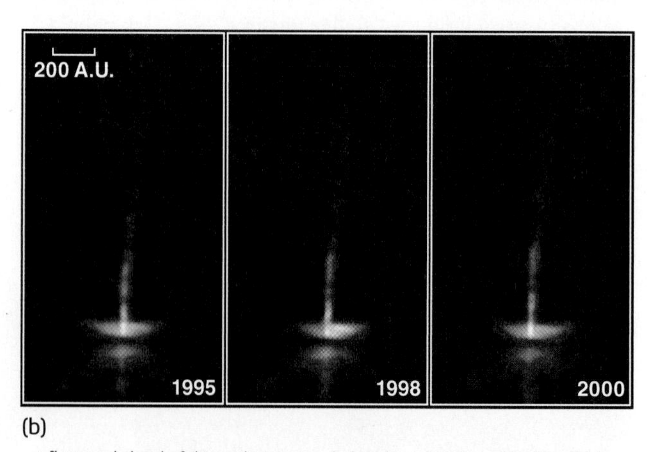

(b)

FIGURE 11-6 R I V U X G **A Circumstellar Accretion Disk and Jets** (a) This is an artist's illustration of excess outflowing material from a gas jet as this forming star is accumulating material from the surrounding disk. Astronomers are interested in this sort of object because it is probably similar to the one from which our solar system formed. (b) These Hubble images of HH 30 show changes over a 5-year period in the disk and jets of this newborn star. There is an edge-on disk (located at the bottom of the images), which appears as a flattened cloud of dust split into two halves by a dark lane, blocking light from the central star. In 1995 and 2000, the left and right sides of the disk were about the same brightness, but in 1998 the right side was brighter. These patterns may be caused by bright spots on the star or variations in the disk near the star. *(a: NASA/IPAC/R. Hurt [SSC]; b: NASA, John Krist [Space Telescope Science Institute], Karl Stapelfeldt [Jet Propulsion Laboratory], Jeff Hester [Arizona State University], Chris Burrows [European Space Agency/Space Telescope Science Institute])*

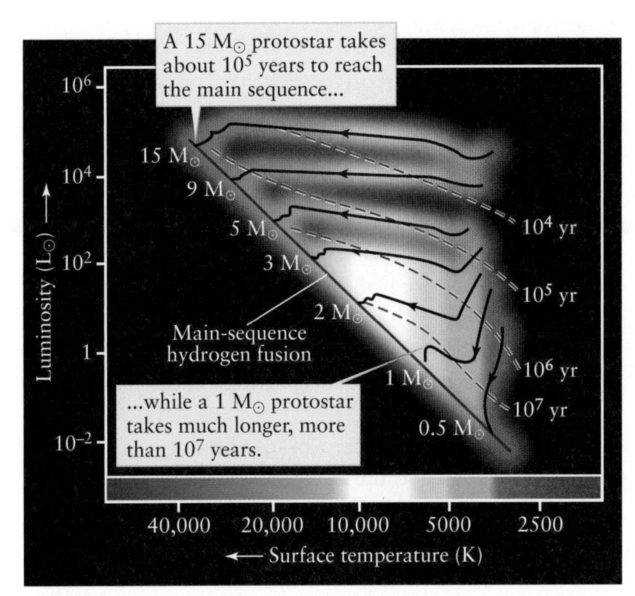

FIGURE 11-7 Pre–Main-Sequence Evolutionary Tracks As a protostar evolves, its luminosity and surface temperature both change. The tracks shown here depict these changes for protostars of seven different masses. Each dashed red line shows the age of a protostar when its evolutionary track crosses that line. The more massive protostars evolve much more quickly than the lower mass protostars.

mechanism for triggering the birth of stars. Presumably, anything that compresses interstellar clouds will do the job. The most dramatic cause is the violent death of a massive star after it has left the main sequence, known as a *supernova*. As we will see in Chapter 12, the core of the doomed large star collapses suddenly, releasing vast quantities of particles and energy that blow the star apart. The star's outer layers are blasted into space at speeds of several thousand kilometers per second.

Astronomers have found many nebulae across the sky that seem to be the shredded remains of these massive dead stars. Such nebulae, like the one shown in **Figure 11-8**, are known as **supernova remnants.** Many supernova remnants have a distinctly circular or arched appearance, as would be expected for an expanding shell of gas. This wall of gas is typically moving away from the dead star faster than sound waves can travel through the interstellar medium. Such supersonic motion produces a shock wave that abruptly compresses the medium through which it passes. When a gas is compressed rapidly, its temperature rises, and this temperature rise causes the gas to glow, as shown in Figure 11-8.

When the expanding shell of a supernova remnant slams into an interstellar cloud, it squeezes the cloud, stimulating star birth, as illustrated in **Figure 11-9.** This kind of star birth is taking place in the stellar nursery seen in Figure 11-9*d*. This stellar nursery is located along an arch of glowing gas that is presumably the remnant of an ancient supernova explosion. There are certainly other compression processes available to trigger star formation. For example, a collision between two interstellar clouds can create new stars. Compression occurs at the interface between the two colliding clouds and vigorous star formation follows. Similarly, intense stellar winds from a group of hot O and B stars

FIGURE 11-8 R I V U X G **A Supernova Remnant** This composite image shows Cassiopeia A, the remnant of a supernova that occurred about 3000 parsecs (10,000 ly) from Earth. In the roughly 300 years since the supernova explosion, a shock wave has expanded about 3 parsecs (10 ly) outward in all directions from the explosion site. The shock wave has warmed interstellar dust to a temperature of about 300 K (Spitzer Space Telescope infrared image in red) and has heated interstellar gases to temperatures that range from 104 K (Hubble Space Telescope visible-light image in yellow) to 107 K (Chandra X-ray Observatory X-ray image in green and blue). *(NASA; JPL-Caltech; and O. Krause, Steward Observatory)*

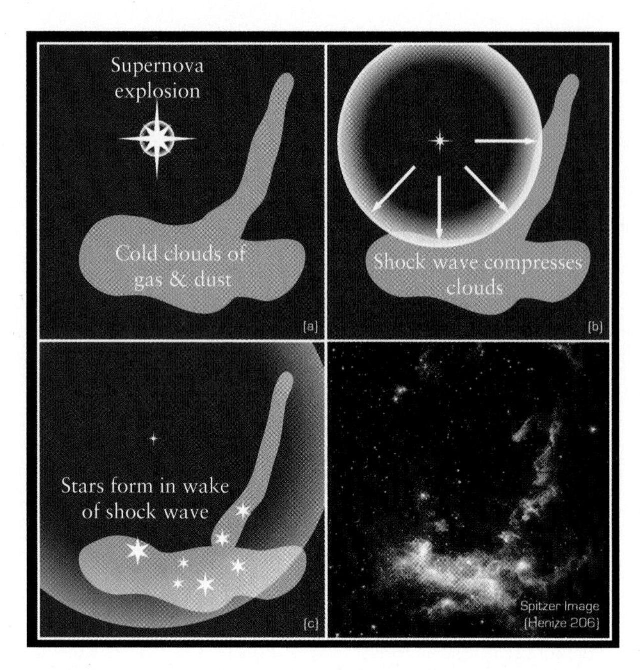

FIGURE 11-9 R I V U X G **Star Formation Caused by an Exploding Star** This three-panel illustration accompanying the photo shows the process of triggered star formation. When a massive, dying star explodes at the end of its life cycle, the shock wave from this explosion passes through clouds of gas and dust. This causes dust and gas to compress, starting star formation within the cloud. The last panel shows a Spitzer Infrared Telescope image of the Henize 206 star-forming region within the Large Magellanic Cloud. *(NASA/JPL-Caltech/ V. Gorjian [JPL])*

may exert strong enough pressure on interstellar clouds to cause compression, followed by star formation.

ConceptCheck 11-3: **What physical process is needed to initiate the formation of a star?**

11-2 Most stars shine throughout their lives by converting hydrogen into helium through nuclear fusion

When a protostar has evolved into a stable main-sequence star, regardless of mass, all stars are fundamentally similar. The fundamental energy source inside stars is the release of energy from the conversion of hydrogen into helium. When hydrogen atoms collide with each other at extremely high temperatures and pressures, they can combine to form larger helium atoms and release tremendous amounts of energy in the process. This process occurs in the very center of Sunlike stars.

The total time that a star will spend fusing hydrogen into helium in its core, and thus the total time that it will spend as a main-sequence star, is called its *main-sequence lifetime*. For our Sun, the main-sequence lifetime is about 12 billion (1.2×10^{10}) years. Hydrogen fusion has been going on in the Sun's core for the past 4.56 billion (4.56×10^{9}) years, so our Sun is less than halfway through its main-sequence lifetime.

Stars Form in Clusters

Dark nebulae contain tens or hundreds of solar masses of gas and dust, enough to form many stars. As a consequence, these nebulae tend to form groups or clusters of young stars. One such cluster is M16, shown in **Figure 11-10.**

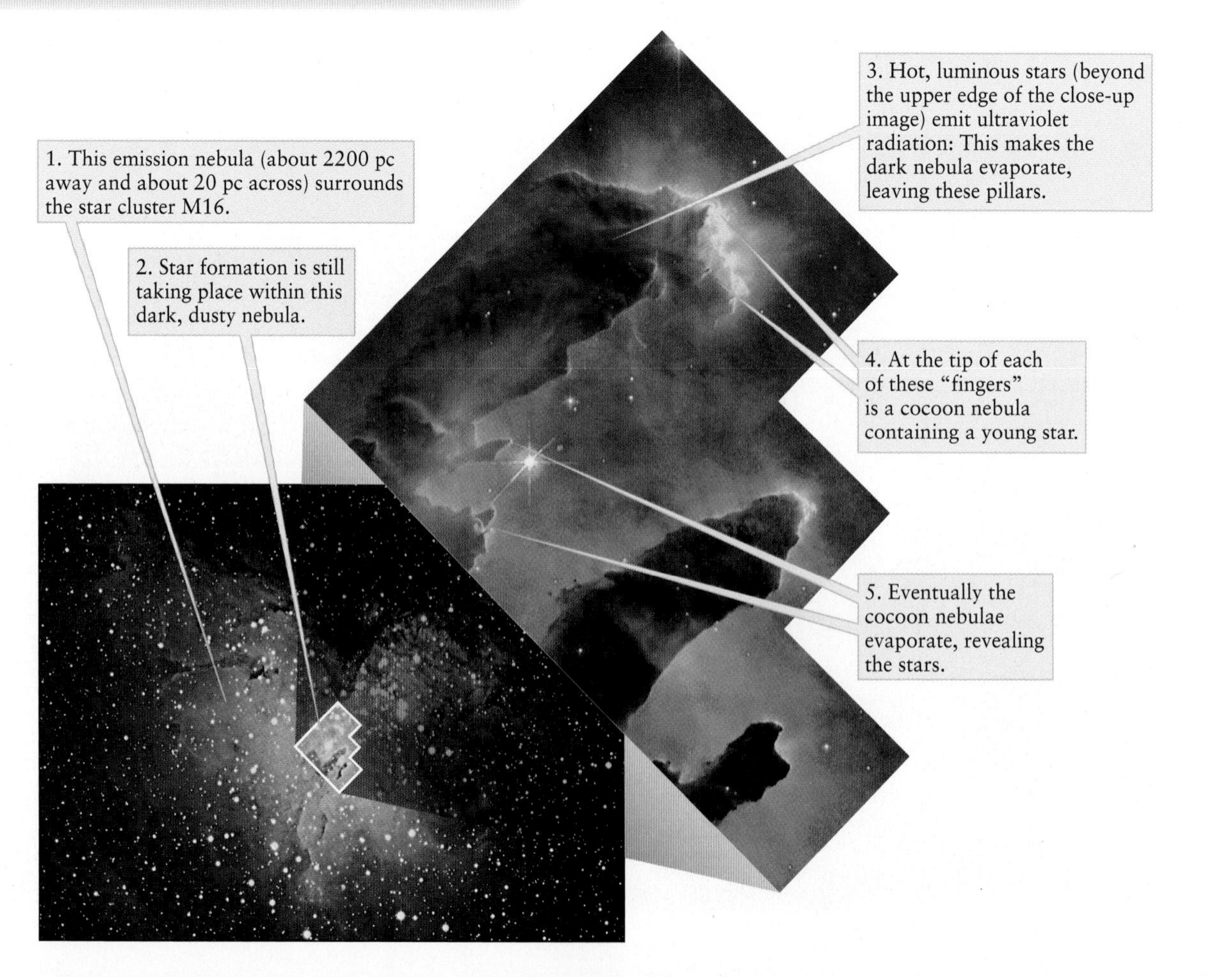

FIGURE 11-10 R I V U X G **A Star Cluster with an H II Region** The star cluster M16 is thought to be no more than 800,000 years old, and star formation is still taking place within adjacent dark, dusty clouds. The inset shows three dense, cold pillars of gas and dust silhouetted against the glowing background of the red emission nebula (called the Eagle Nebula for its shape). The pillar at the upper left extends about 0.3 pc (1 ly) from base to tip, and each of its "fingers" is somewhat broader than our entire solar system. *(Australian Astronomical Observatory/David Malin Images; inset: J. Hester and P. Scowen [Arizona State U.]; NASA)*

In addition to being objects of great natural beauty, star clusters give us a unique way to compare the evolution of stars with differing masses, because clusters typically include stars with a range of different masses, all of which began to form out of the parent nebula at roughly the same time. Thus, they have roughly the same chronological age and similar initial chemical composition.

ANALOGY A foot race is a useful way to compare the performance of sprinters because all the competitors start the race simultaneously. A young star cluster gives us the same kind of opportunity to compare the evolution of stars of different masses that all began to form roughly simultaneously. Unlike a foot race, however, the entire "race" of evolution in a single cluster happens too slowly for us to observe; as Figure 11-7 shows, protostars take many thousands or millions of years to evolve significantly. Instead, we must compare different star clusters at various stages in their evolution to piece together the history of star formation in a cluster.

All the stars in a cluster may begin to form nearly simultaneously, but they do not all become main-sequence stars at the same time. As you can see from their evolutionary tracks (see Figure 11-7), high-mass stars evolve more rapidly than low-mass stars. Consistently, the more massive the protostar, the sooner it develops the central pressures and temperatures needed for steady hydrogen fusion to begin. As we shall see, stars in the same cluster also do not age at the same rate nor do they leave the main sequence at the same time.

Figure 11-11*a* shows a young star cluster called the Pleiades. The photograph shows gas that must have initially surrounded this cluster but has since dissipated into interstellar space, leaving only traces of dusty material that forms nebulae around the cluster's stars. This implies that that the Pleiades must be older than NGC 2264, the cluster shown in **Figure 11-12*a*,** which is still surrounded by a nebula. The H-R diagram for the Pleiades in Figure 11-11*b* bears out this idea. In contrast to the H-R diagram for NGC 2264 (Figure 11-12*b*), nearly all the stars in the Pleiades are on the main sequence. The cluster's age is about 100 million years, which is how long it takes for the least massive stars to finally begin hydrogen fusion in their cores.

CAUTION Note that the data points for the most massive stars in the Pleiades (at the upper left of the H-R diagram in Figure 11-11*b*) are above the main sequence. This is *not* because these stars have yet to arrive at the main sequence. Rather, these stars were the first members of the cluster to arrive at the main sequence some time ago and are now the first members to leave it. They have used up the hydrogen in their cores, so the steady process of core hydrogen fusion that characterizes main-sequence stars cannot continue. A loose collection of stars such as NGC 2264 or the Pleiades possesses barely enough mass to be held together by gravitation. Occasionally, a star moving faster than average will escape, or "evaporate," from the cluster. Indeed, by the time the stars are a few billion years old, they may be so widely separated that a cluster no longer exists.

Go to Video 11-3

What happens to a star like the Sun after the core hydrogen has been used up, so that it is no longer a main-sequence star? As we will see, it expands dramatically to become a red giant. To understand why this happens, it

(a) The Pleiades star cluster RIVUXG

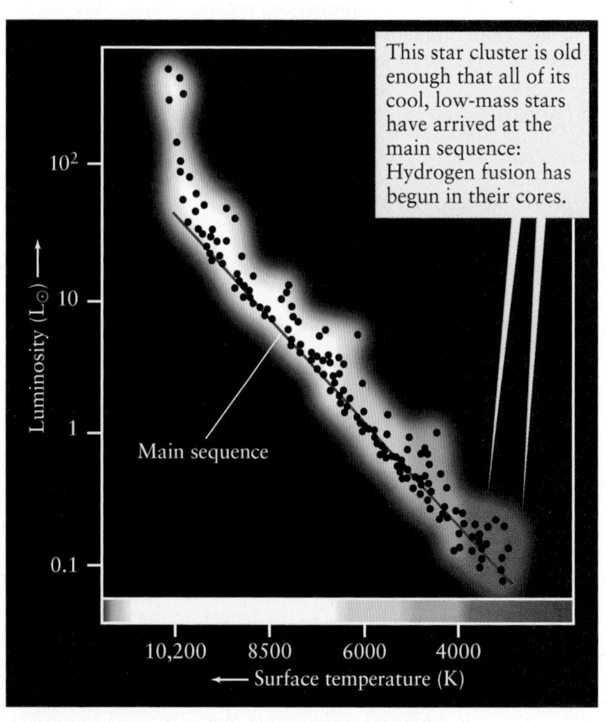

(b) An H-R diagram of the stars in the Pleiades

FIGURE 11-11 The Pleiades and Its H-R Diagram (a) The Pleiades star cluster is 117 pc (380 ly) from Earth in the constellation Taurus and can be seen with the naked eye. (b) Each dot plotted on this H-R diagram represents a star in the Pleiades whose luminosity and surface temperature have been measured. (*Note:* The scales on this H-R diagram are different from those in Figure 11-12*b*.) The Pleiades is about 100 million (10×10^8) years old. *(NASA, ESA and AURA/Caltech)*

(a) The star cluster NGC 2264 RIVUXG

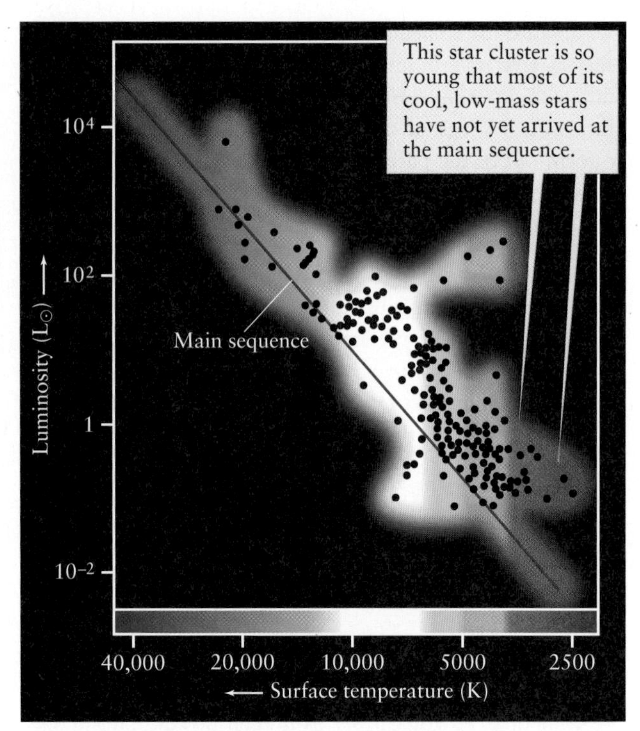

(b) An H-R diagram of the stars in NGC 2264

FIGURE 11-12 A Young Star Cluster and Its H-R Diagram (a) Newborn stars, hidden behind thick dust, are revealed in this image of the NGC 2264 from NASA's Spitzer Space Telescope. These infant stars appear to have formed in regularly spaced intervals along linear structures in a configuration like the pattern of a snowflake. The youngest protostars form straight lines and have yet to scatter from their location of formation. (b) Each dot plotted on this H-R diagram represents a star in NGC 2264, whose luminosity and surface temperature have been determined. This star cluster probably started forming only 2 million years ago. *(NASA/JPL-Caltech/P. S. Teixeira [Harvard-Smithsonian CfA])*

is useful to first look at how a star evolves *during* its main-sequence lifetime. The nature of that evolution depends on whether its mass is less than or greater than about 0.4 $M_\odot$.

ConceptCheck 11-4: Where does the increasing amount of helium come from inside the core of a star?

Main-Sequence Stars of 0.4 $M_\odot$ or Greater: Consuming Core Hydrogen

A protostar becomes a main-sequence star when steady hydrogen fusion begins in its core and it achieves a stable balance between the inward force of gravity and the outward pressure produced by hydrogen fusion. From the point in time a star has stabilized its core hydrogen fusion, a star slowly, but predictably, undergoes noticeable changes in luminosity, surface temperature, and radius during its main-sequence lifetime. These changes are a result of hydrogen fusion in the core, which alters the chemical composition of the core. As an example, when our Sun first formed, its composition was the same at all points throughout its volume: by mass, about 74% hydrogen, 25% helium, and 1% heavy elements. But as **Figure 11-13** shows, the Sun's core now contains a greater mass of helium than of hydrogen. (There is still enough hydrogen in the Sun's core for another 5.4 billion years as a main-sequence star.)

CAUTION Although the outer layers of the Sun are also predominantly hydrogen, there are two reasons why this hydrogen cannot undergo fusion. The first reason is that while the temperature and pressure in the core are high enough for thermonuclear reactions to take place, the temperatures and pressure in the outer layers are not. The second reason is that there is no flow of material between the Sun's core and outer layers, so the hydrogen in the outer layers cannot move into the hot, high-pressure core to undergo fusion. The same is true for main-sequence stars with masses of about 0.4 $M_\odot$ or greater. (We will see below that the outer layers *can* undergo fusion in main-sequence stars with a mass less than about 0.4 $M_\odot$.)

Thanks to hydrogen fusion in the core, the total number of atomic nuclei in a star's core decreases with time: In each reaction, 4 hydrogen nuclei are converted to a single helium nucleus. With fewer particles bouncing around to provide the core's internal pressure, the core contracts slightly under the weight of the star's outer layers. Compression makes the core denser and increases its temperature. As a result of these changes in density and temperature, the pressure in the compressed core is actually higher than before.

As the star's core shrinks, its outer layers expand and shine more brightly. Here's why: As the core's density and temperature increase, hydrogen nuclei in the core collide with one another more frequently, and the rate of hydrogen fusion in the core increases. Hence, the star's luminosity increases. The radius of the star as a whole also increases slightly, because increased core pressure pushes outward on the star's outer layers. The star's surface temperature changes

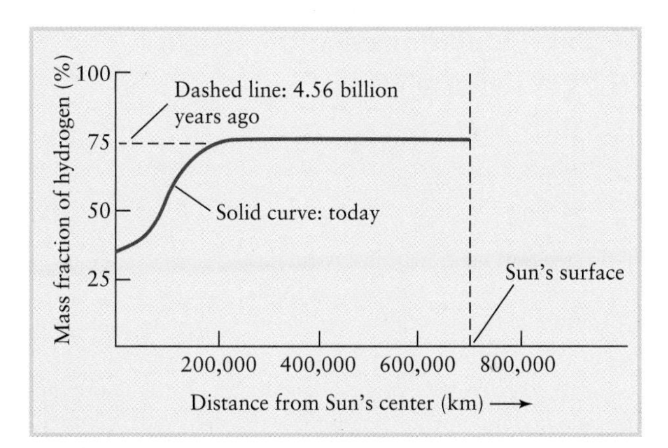

(a) Hydrogen in the Sun's interior

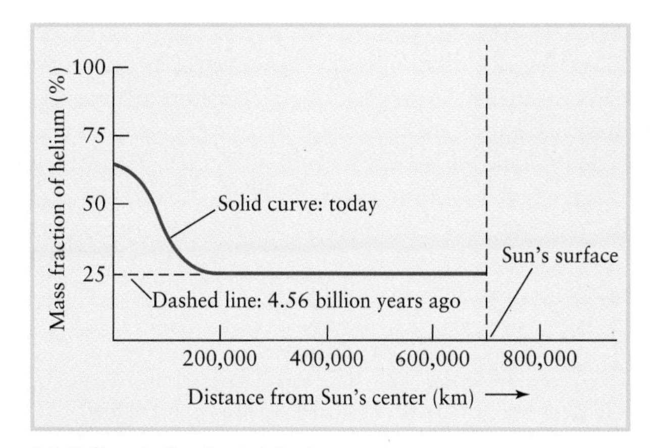

(b) Helium in the Sun's interior

FIGURE 11-13 Changes in the Sun's Chemical Composition These graphs show the percentage by mass of (a) hydrogen and (b) helium at different points within the Sun's interior. The dashed horizontal lines show that these percentages were the same throughout the Sun's volume when it first formed. As the solid curves show, over the past 4.56×10^9 years, thermonuclear reactions at the core have depleted hydrogen in the core and increased the amount of helium in the core.

as well, because it is related to the luminosity and radius. As an example, theoretical calculations indicate that over the past 4.56×10^9 years, our Sun has become 40% more luminous, grown in radius by 6%, and increased in surface temperature by 300 K (**Figure 11-14**).

As a main-sequence star ages and evolves, the increase in energy outflow from its core also heats the material immediately surrounding the core. As a result, hydrogen fusion can begin outside the core in this surrounding material. By tapping this fresh supply of hydrogen, a star manages to eke out a few million more years of main-sequence existence.

ConceptCheck 11-5: **What happens to the size of a star if its core shrinks, increasing in temperature and density?**

Main-Sequence Stars of Less Than 0.4 $M_\odot$: Consuming All Their Hydrogen

The story is somewhat different for the smallest and least massive main-sequence stars, with masses between 0.08 $M_\odot$ (the minimum mass necessary for sustained thermonuclear reactions to take place in a star's core) and about 0.4 $M_\odot$. These stars, of spectral class M, are called red dwarfs because they are small in size and have a red color due to their low surface temperature. They are also very numerous; about 85% of all stars in the Milky Way Galaxy are red dwarfs.

In a red dwarf, helium does *not* accumulate in the core to the same extent as in the core of a larger star, like our Sun. The reason is that in a red dwarf there are convection cells of rising and falling gas that extend throughout the star's volume and penetrate into the core. These convection cells drag helium outward from the core and replace it with hydrogen from the outer layers (**Figure 11-15**). The fresh hydrogen can undergo thermonuclear fusion that releases energy and makes additional helium. This helium is then dragged out of the core by convection and replaced by even more hydrogen from the red dwarf's outer layers.

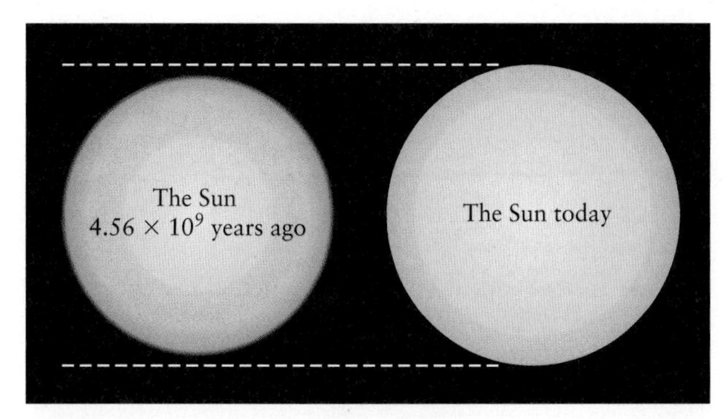

FIGURE 11-14 The Zero-Age Sun and Today's Sun Over the past 4.56×10^9 years, much of the hydrogen in the Sun's core has been converted into helium, the core has contracted a bit, and the Sun's luminosity has gone up by about 40%. These changes in the core have made the Sun's outer layers expand in radius by 6% and increased the surface temperature from 5500 K to 5800 K.

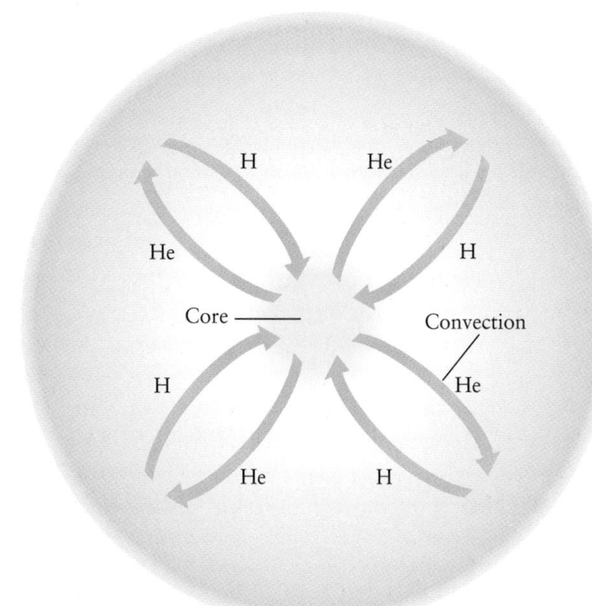

FIGURE 11-15 A Fully Convective Red Dwarf In a red dwarf—a main-sequence star with less than about 0.4 solar masses—helium (He) created in the core by thermonuclear reactions is carried to the star's outer layers by convection. Convection also brings fresh hydrogen (H) from the outer layers into the core. This process continues until the entire star is helium.

As a consequence, over a red dwarf's main-sequence lifetime essentially all of the star's hydrogen can be consumed and converted to helium. The core temperature and pressure in a red dwarf is less than in the Sun, so thermonuclear reactions happen more slowly than in our Sun. Calculations indicate that it takes hundreds of billions of years for a red dwarf to convert all of its hydrogen to helium. The present age of the universe is only 13.7 billion years, so there has not yet been time for any red dwarfs to become pure helium.

> ConceptCheck 11-6: **Why do convection cells in the smallest of main-sequence stars, red dwarfs, result in such long lifetimes when they have such small amounts of fuel?**

11-3 Careful observations of star clusters provide insight into how a star's mass influences how stars change over time

Studies of star clusters reveal a curious difference between the youngest and oldest stars in our Galaxy. Stars in the youngest clusters (those with most of their main sequences still intact) are said to be **metal rich,** because their spectra contain many prominent spectral lines of heavy elements. (Recall that astronomers use the term "metal" to denote any element other than hydrogen and helium, which are the two lightest elements.) Such stars are also called **Population I stars.** Our Sun, as an example, is a relatively young, metal-rich, Population I star.

By contrast, the spectra of stars in the oldest clusters show only weak lines of heavy elements. These ancient stars are thus said to be **metal poor,** because heavy elements are only about 3% as abundant in these stars as in the Sun. They are called **Population II stars. Figure 11-16** shows the difference in spectra between a metal-poor, Population II star and the Sun (a metal-rich, Population I star).

CAUTION Note that "metal rich" and "metal poor" are relative terms. In even the most metal-rich star known, metals make up just a small percentage of the total mass of the star.

> ConceptCheck 11-7: **In which population of stars is Alpha Centauri, a nearby star with a size and chemical composition similar to our Sun?**

Stellar Populations and the Origin of Heavy Elements

To explain why there are two distinct populations of stars, we must go back to the explosive origin of the universe, which took place some 13.7 billion years ago. As we will discuss in Chapter 15, the early universe consisted almost exclusively of hydrogen and helium atoms, with almost no heavy elements (metals). The first stars to form were likewise metal poor. The least massive of these have survived to the present day and are now the ancient stars of Population II.

The more massive of the original stars evolved more rapidly and no longer shine. But as these stars evolved, helium fusion in their cores produced metals—carbon and oxygen. In the most massive stars, as we will learn in the next chapter, further thermonuclear reactions produce even heavier elements. As these massive original stars aged and died, they expelled their metal-enriched gases into space. (The star called HD 65750 creating the drinking mug–shaped nebula IC 2220 shown in **Figure 11-17** is going through such a mass-loss phase late in its life.) This expelled material joined the interstellar medium and was eventually incorporated into a second generation of stars that have a higher concentration of heavy elements. These metal-rich members of the second stellar generation are the Population I stars, of which our Sun is an example.

CAUTION Be careful not to let the designations of the two stellar populations confuse you. *Population I* stars are members of a *second* stellar generation, while *Population II* stars belong to an older *first* generation.

The relatively high concentration of heavy elements in the Sun means that the solar nebula from which both the Sun and planets formed must likewise have been metal rich. Earth is composed almost entirely of heavy elements, as are our bodies. Thus, our very existence is intimately linked to the Sun's being a Population I star. A planet like Earth probably could not have formed from the metal-poor gases that went into making Population II stars.

The concept of two stellar populations provides insight into our own Earthly origins. Helium fusion in red-giant stars produces the same isotopes of carbon (^{12}C) and oxygen (^{16}O) that are found most commonly on Earth. The reason is that Earth's carbon and oxygen atoms, including all of

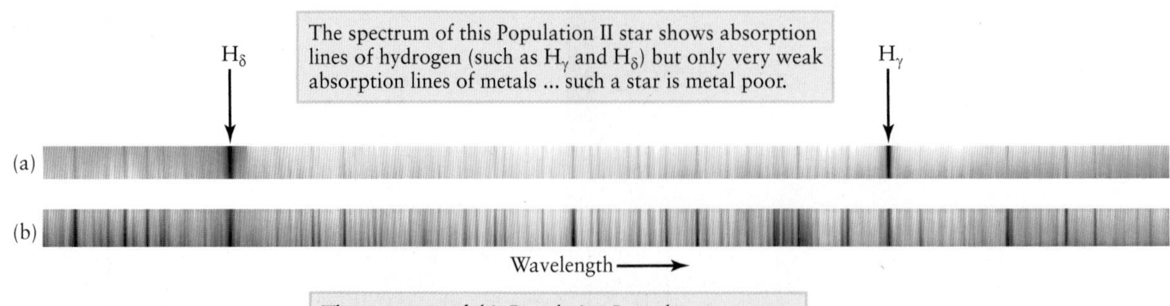

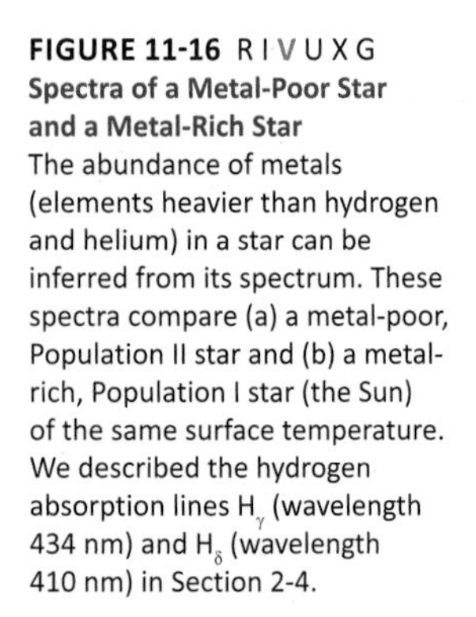
FIGURE 11-16 RIVUXG **Spectra of a Metal-Poor Star and a Metal-Rich Star** The abundance of metals (elements heavier than hydrogen and helium) in a star can be inferred from its spectrum. These spectra compare (a) a metal-poor, Population II star and (b) a metal-rich, Population I star (the Sun) of the same surface temperature. We described the hydrogen absorption lines H_γ (wavelength 434 nm) and H_δ (wavelength 410 nm) in Section 2-4.

FIGURE 11-17 R I V U X G **A Mass-Loss Star** As stars age and become giant stars, they expand tremendously and shed matter into space. This star, HD 65750, is losing matter at a high rate. The "toby jug" nebula shown in this picture is formed by light reflected off dust being ejected by the star. *(Australian Astronomical Observatory/David Malin Images)*

those in your body, actually *were* produced by helium fusion. These reactions occurred billions of years ago within an earlier generation of stars that died and gave up their atoms to the interstellar medium—the same atoms that later became part of our solar system, our planet, and our bodies. We are literally children of the stars.

ConceptCheck 11-8: Was our Sun formed at about the same time that the universe originated?

A Star's Mass Determines Its Main-Sequence Lifetime

Go to Video 11-4

The main-sequence lifetime of a star depends critically on its mass. As **Table 11-1** shows, massive stars have short main-sequence lifetimes because they are also very luminous. To emit energy so rapidly, these stars must be depleting the hydrogen in their cores at a prodigious rate. Hence, even though a massive O or B main-sequence star contains much more hydrogen fuel in its core than is in the entire volume of a red dwarf of spectral class M, the O or B star exhausts its hydrogen much sooner. High-mass O and B stars gobble up the available hydrogen fuel in only a few million years, while red-dwarf stars of very low mass take hundreds of billions of years to use up their hydrogen. Thus, a main-sequence star's mass determines not only its luminosity and spectral type, but also how long it can remain a main-sequence star. In general, the more massive the star, the more rapidly it goes through *all* the phases of its life. Nonetheless, most of the stars we are able to detect are in their main-sequence phase, because this phase lasts so much longer than other luminous phases.

Like so many properties of stars, what happens at the end of a star's main-sequence lifetime depends on its mass. If the star is a red dwarf of less than about 0.4 $M_\odot$, after hundreds of billions of years the star has converted all of its hydrogen to helium. It is possible for helium to undergo thermonuclear fusion, but this requires temperatures and pressures far higher than those found within a red dwarf. Thus, a red dwarf will end its life as an inert ball of helium that still glows due to its internal heat. As it radiates energy into space, it slowly cools and shrinks. This slow, quiet demise is the ultimate fate of the 85% of stars in the Milky Way that are red dwarfs. (As we have seen, there has not yet been time in the history of the universe for any red dwarf to reach this final stage in its evolution.)

ConceptCheck 11-9: If massive O and B stars have the most fuel, why do they have the shortest lifetimes?

11-4 Stars slowly become red giants

What is the fate of stars more massive than about 0.4 $M_\odot$, including the Sun? As we will see, the late stages of their evolution are dramatic. Studying these stages will give us insight into the fate of our solar system and life on Earth.

The most massive stars consume their core fuel rapidly to support the extreme pressure from the overlying weight.

Stars of 0.4 $M_\odot$ or Greater: From Main-Sequence Star to Red Giant

When a star of at least 0.4 solar masses reaches the end of its main-sequence lifetime, all of the hydrogen in its core has been used up and hydrogen fusion ceases there. In this new

Table 11-1 Approximate Main-Sequence Lifetimes

Mass ($M_\odot$)	Surface Temperature (K)	Spectral class	Luminosity ($L_\odot$)	Main-sequence lifetime (10^6 years)
25	35,000	O	80,000	4
15	30,000	B	10,000	15
3	11,000	A	60	800
1.5	7000	F	5	4500
1.0	6000	G	1	12,000
0.75	5000	K	0.5	25,000
0.50	4000	M	0.03	700,000

stage, hydrogen fusion continues only in the surrounding shell of hydrogen-rich material just outside the central core. At first, this process occurs only in the hottest region just outside the core, where the hydrogen fuel has not yet been exhausted. Outside this region, it is not hot enough for any fusion reactions to take place.

Strangely enough, the end of hydrogen fusion in the core *increases* the core's temperature. When thermonuclear reactions first cease in the core, nothing remains to generate heat there. Hence, the core starts to cool and the pressure supporting the core starts to decrease. This pressure decrease allows the star's core to again compress under the weight of the outer layers. As the core contracts, its temperature again increases, and heat begins to flow outward from the core even though no nuclear reactions are taking place there. (Technically, gravitational energy is converted into thermal energy.)

This new flow of heat warms the gases around the core, increasing the rate of hydrogen fusion surrounding the core and making the active shell eat further outward into the surrounding matter. Helium produced by reactions in the shell falls down onto the core, which continues to contract and heat up as it gains mass. Over the course of hundreds of millions of years, the core of a 1-$M_\odot$ star compresses to about one-third of its original radius, while its central temperature increases from about 15 million (1.5×10^7) K to about 100 million (10^8) K.

During this post–main-sequence phase, the star's outer layers expand just as dramatically as the core contracts. As the hydrogen-fusing shell works its way outward, egged on by heat from the contracting core, the star's luminosity increases substantially. This increases the star's internal pressure and makes the star's outer layers expand to many times their original radius. This tremendous expansion causes those layers to cool down, and the star's surface temperature drops. Once the temperature of the star's bloated surface falls, the gases glow with a reddish hue. Based on its new appearance, the star is then appropriately known as a **red giant** (**Figure 11-18**). Thus, we see that red-giant stars are former main-sequence stars that have evolved into a new stage of existence. We can summarize these observations as a general rule:

Stars join the main-sequence group when they begin hydrogen fusion in their cores. They leave the main-sequence group and become giant stars when their core hydrogen is depleted.

Red-giant stars undergo substantial loss of their mass because of their large diameters and correspondingly weak surface gravity. This makes it relatively easy for gases to escape from the red giant into space. Mass loss can be detected in a star's spectrum, because gas escaping from a red giant toward a telescope on Earth produces narrow absorption lines that are slightly blueshifted by the Doppler effect. Typical observed blueshifts correspond to a speed of about 22,000 mi/h (10 km/s). A typical red giant loses roughly 10^{-7} $M_\odot$ of matter per year. For comparison, the Sun's present-day mass-loss rate is only 10^{-14} $M_\odot$ per year. Hence, an evolving star loses a substantial amount of mass as it ages (see Figure 11-17).

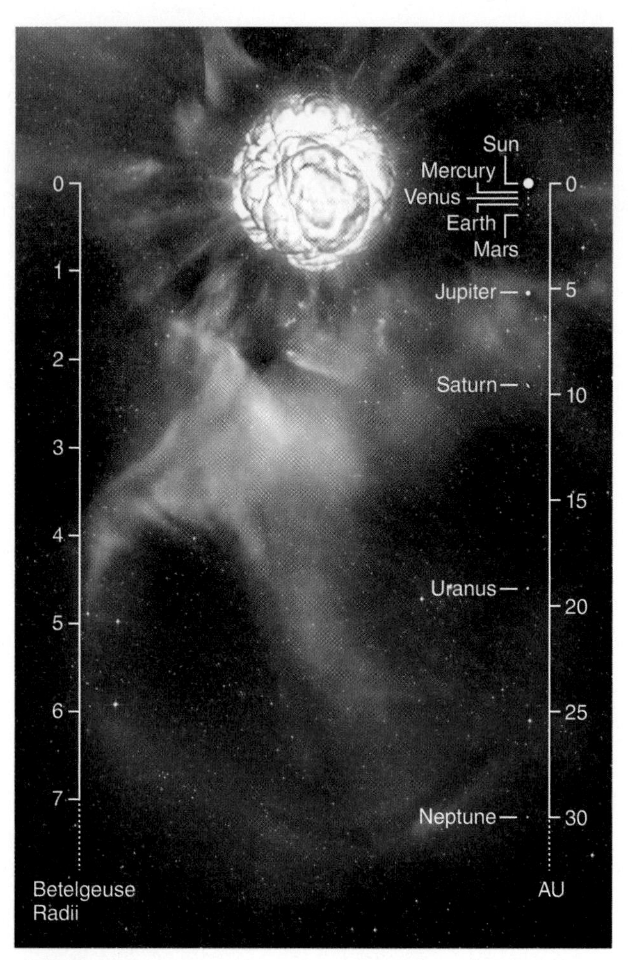

FIGURE 11-18
Red Giant and Plume Main-sequence stars produce energy in a hydrogen-fusing core that will shrink to about ⅓ of its size when it becomes a red giant, causing hydrogen fusion to take place at a furious rate. The Sun's luminosity will be about 2000 times greater than today, and the increased luminosity will make the Sun's outer layers expand to approximately 100 times their present size, swallowing its inner planets and having a hot plume reaching the edge of the solar system. *(ESO/L. Calçada)*

ConceptCheck 11-10: Why does a red giant appear red if it is cool?

Fusion of Helium into Carbon and Oxygen Begins at the Center of a Red Giant

When a star with a mass greater than 0.4 $M_\odot$ first changes from a main-sequence star (**Figure 11-19*a***) to a red giant (Figure 11-19*b*), its hydrogen-fusing shell surrounds a small, compact core of almost pure helium. In a red giant of moderately low mass, which the Sun will become 7 billion years from now, the dense helium core is about twice the diameter of Earth. Most of this core helium was produced by thermonuclear reactions during the star's main-sequence lifetime; during the red-giant era, this helium will undergo thermonuclear reactions.

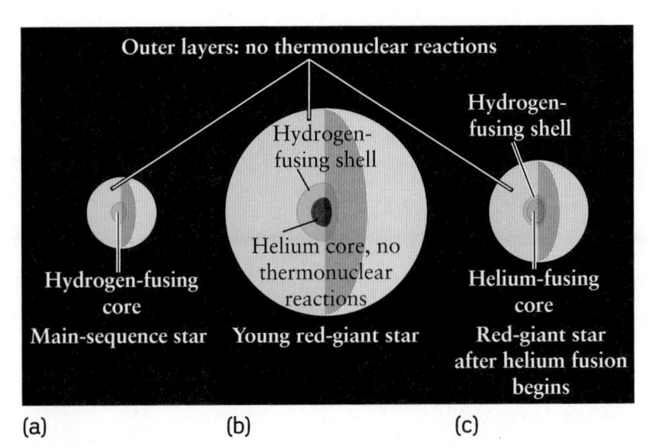

FIGURE 11-19 Stages in the Evolution of a Star with More than 0.4 Solar Masses (a) During the star's main-sequence lifetime, hydrogen is converted into helium in the star's core. (b) When the core hydrogen is exhausted, hydrogen fusion continues in a shell, and the star expands to become a red giant. (c) When the temperature in the red giant's core becomes high enough because of contraction, core helium fusion begins. (These three pictures are not drawn to scale. The star is about 100 times larger in its red-giant phase than in its main-sequence phase, then shrinks somewhat when core helium fusion begins.)

Helium, the "ash" left over from hydrogen fusion, is a potential nuclear fuel, and the thermonuclear fusion of helium nuclei to make even heavier nuclei releases energy in the process. But this reaction cannot take place within the core of our present-day Sun because the temperature there is far too low. Each helium nucleus contains two protons, so it has twice the positive electric charge of a hydrogen nucleus, and there is a much stronger electric repulsion between two helium nuclei than between two hydrogen nuclei. For helium nuclei to overcome this repulsion and get close enough to fuse together, they must be moving at very high speeds, which means that the temperature of the helium gas must be very high.

When a star first becomes a red giant, the temperature of the contracted helium core is still too low for helium nuclei to fuse. But as the hydrogen-fusing shell adds mass to the helium core, the core contracts even more, further increasing the star's central temperature. When the central temperature finally reaches 100 million (10^8) K, thermonuclear fusion of helium in the core begins. As a result, the aging star again has a new central energy source for the first time since leaving the main sequence.

Helium fusion combines helium atoms to form carbon in a new process that releases energy that fuels the star's core. This process uses three helium atoms to form carbon and is called the **triple alpha process,** because helium nuclei (^{4}He) are also called alpha particles. Some of the carbon created in this process can go on to fuse with additional helium to produce a stable isotope of oxygen and release even more energy (see **Cosmic Connections: Helium Fusion in a Red Giant**).

The nuclear fusion processes of helium release tremendous amounts of energy, even more than hydrogen fusion. The onset of these reactions reestablishes stability in the star's core and prevents any further gravitational contraction of the star's core. Curiously, a mature red giant fuses helium in its core for a much shorter time than it spent fusing hydrogen in its core as a main-sequence star. For example, in the distant future the Sun will sustain helium fusion in its core for only about 100 million years. (While this is going on, hydrogen fusion is still continuing in a shell around the core.)

ConceptCheck 11-11: What is different about the core of a red giant that allows helium to fuse when it cannot occur in the core of our present-day Sun?

The Helium Flash and Electron Degeneracy

Precisely how helium fusion begins at a red giant's center depends on the mass of the star. In high-mass red giants (greater than about 2 to 3 $M_\odot$), helium fusion begins gradually as temperatures in the star's core approach 10^8 K. In red giants with a mass less than about 2 to 3 $M_\odot$, helium fusion begins explosively and suddenly, in what is called the **helium flash.** **Table 11-2** summarizes these differences.

The helium flash occurs because of unusual conditions that develop in the core of a moderately low-mass star as it becomes a red giant. To appreciate these conditions we must first understand how an ordinary gas behaves, then we can explore how the densely packed electrons at the star's center alter this behavior.

When a gas is compressed, it usually becomes denser and warmer. In fact, for gas like air on Earth, there is a relatively simple relationship between pressure, temperature, and density. Specifically, the pressure exerted by a gas on its container is directly proportional to both the density and the temperature of the gas. In most circumstances, the gases inside a star behave in this way. If the gas expands, it cools down, and if it is compressed, it heats up. This behavior serves as a safety valve, ensuring that the star remains in a stable thermal equilibrium. For example, if the rate of thermonuclear reactions in the star's core should increase, the additional energy releases heat and expands the core. This cools the core's gases and slows the rate of thermonuclear reactions back to the original value. Conversely, if the rate of thermonuclear reactions should decrease, the core will cool down and compress under the pressure of the overlying layers. The compression of the core will make its temperature increase, thus speeding up the thermonuclear reactions and returning them to their original rate.

In a red giant with a mass between about 0.4 $M_\odot$ and 2–3 $M_\odot$, however, the core behaves very differently than the way just described. In a red giant, the core must be compressed tremendously in order to become hot enough for

Table 11-2 How Helium Core Fusion Begins in Different Red Giants

Mass of star	Onset of helium fusion
More than about 0.4 but less than 2–3 $M_\odot$	Explosive (helium flash)
More than 2–3 $M_\odot$	Gradual

COSMIC CONNECTIONS Helium Fusion in a Red Giant

A star becomes a red giant after the fusion of hydrogen into helium in its core has come to an end. As the red giant's core shrinks and heats up, a new cycle of reactions can occur that create the even heavier elements carbon and oxygen.

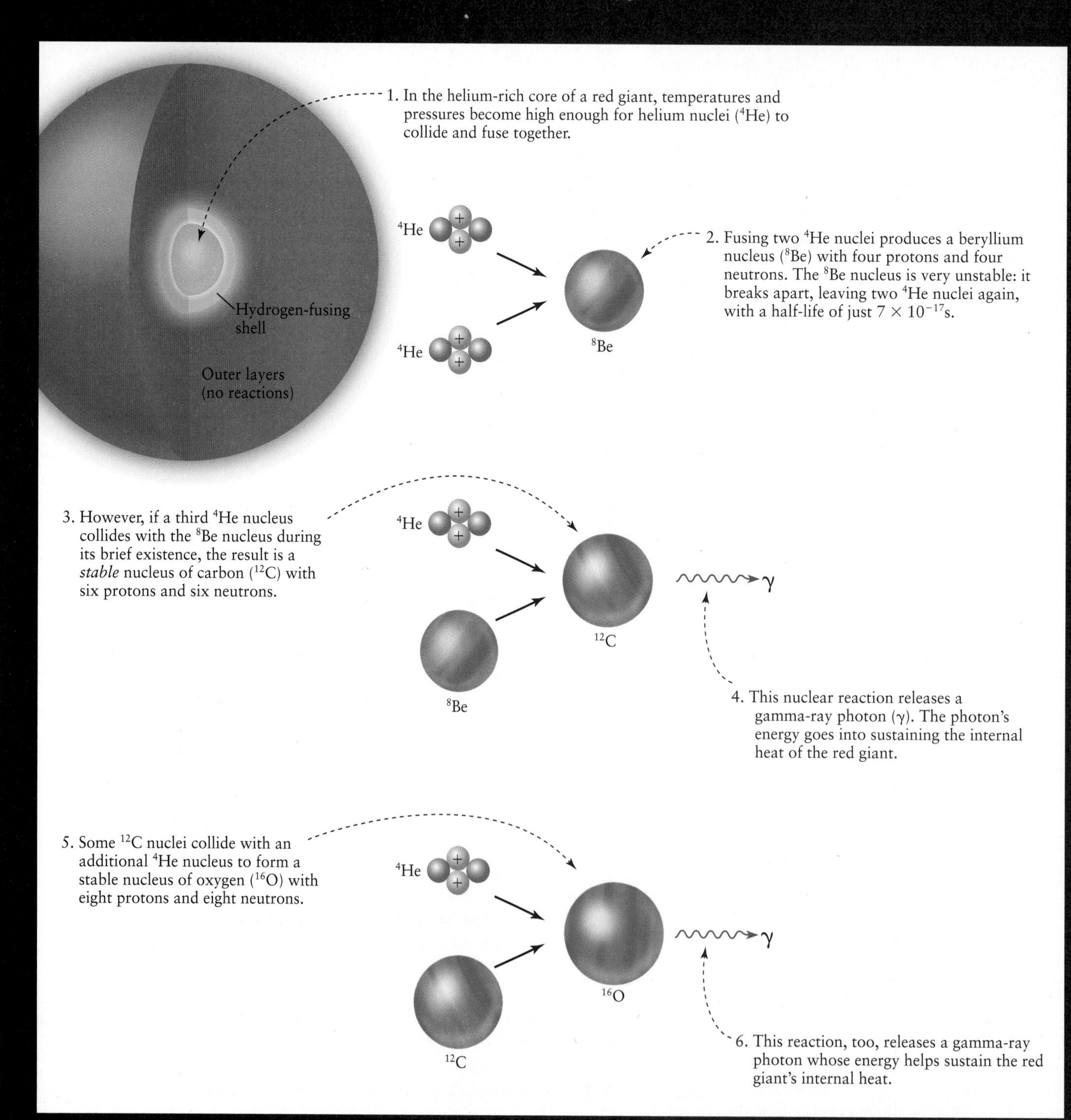

helium fusion to begin. At these extreme pressures and temperatures, the atoms are stripped of their electrons and become completely ionized. As a result, most of the core consists of atomic nuclei and detached electrons. Eventually, the free electrons become so closely crowded that a limit to further compression is reached. Just before the onset of helium fusion, the electrons in the core of a low-mass star are so closely crowded together that they cannot be squeezed any closer together and they produce a powerful pressure that resists any further core contraction.

This phenomenon is called *degeneracy.* Astronomers say that the electrons in the helium-rich core of a low-mass red giant are "degenerate," and that the core is supported by **degenerate-electron pressure.** This degenerate pressure, unlike the more typical gas pressure, does not depend on temperature. When the temperature in the core of a low-mass red giant reaches the high level required for the triple alpha process, energy begins to be released. The helium heats up, which makes the triple alpha process happen even faster. However, the pressure provided by the degenerate electrons is independent of the temperature, so the pressure does not change. Without the "safety valve" of increasing pressure, the star's core cannot expand and cool. The rising temperature causes the helium to fuse at an ever-increasing rate, producing the helium flash.

Eventually, the temperature becomes so high that the electrons in the core are no longer degenerate. Then, when the star's core expands, the helium flash is terminated. These events occur so rapidly that the helium flash is over in seconds, after which the star's core settles down to a steady rate of helium fusion.

CAUTION The term "helium flash" might give the impression that a star emits a sudden flash of light when the helium flash occurs. If this were true, it would be an incredible sight. During the brief time interval when the helium flash occurs, the helium-fusing core is 10^{11} times more luminous than the present-day Sun, comparable to the total luminosity of all the stars in the Milky Way Galaxy! But in fact, the helium flash has no immediately visible consequences—for two reasons. First, much of the energy released during the helium flash goes into heating the core and terminating the degenerate state of the electrons. Second, the energy that does escape the core is largely absorbed by the star's outer layers, which are quite opaque. Therefore, the explosive drama of the helium flash takes place where it cannot be seen directly.

ConceptCheck 11-12: **If a core of degenerate electrons actually depended on temperature, how would it respond to an increase in temperature?**

The Continuing Evolution of a Red Giant

Whether a helium flash occurs or not, the onset of helium fusion in the core actually causes a *decrease* in the luminosity of the star. This is the opposite of what you might expect—after all, turning on a new energy source should make the luminosity greater, not less. What happens is that after the onset of core helium fusion, a star's superheated core expands. Temperatures drop around the expanding core, so the surrounding shell of hydrogen fusion reduces its energy output and the star's overall energy output decreases. This allows the star's outer layers to contract and heat up. Consequently, a post–helium-flash star is less luminous, hotter at the surface, and smaller than a red giant.

Helium fusion occurring in the core lasts for only a relatively short time. Calculations suggest that a 1-$M_\odot$ star like the Sun sustains hydrogen fusion in its core for about 12 billion (1.2×10^{10}) years, followed by about 250 million (2.5×10^8) years of hydrogen fusion in the shell surrounding the core leading up to the helium flash. After the helium flash, such a star can fuse helium in its core (while simultaneously fusing hydrogen in a shell around the core) for only 100 million (10^8) years, a mere 1% of its main-sequence lifetime. Here is the story of post–main-sequence evolution in its briefest form: Before the beginning of helium fusion in the core, the star's core compresses and the outer layers expand, and just after helium fusion begins, the core expands and the outer layers compress. We will see in the next chapter that this behavior, in which the inner and outer regions of the star change in opposite ways, occurs again and again in the final stages of a star's evolution.

ConceptCheck 11-13: **Why is a star fueled by a hotter core of helium fusion less luminous?**

Stars of Between 0.4 $M_\odot$ and 4 $M_\odot$ Go Through Two Distinct Red-Giant Stages

Go to Video 11-5

It is worth emphasizing again that the difference in the final chapters of a star's life cycle primarily depends on one characteristic—the star's mass. Main-sequence stars convert hydrogen to helium in their cores in a series of energy-releasing thermonuclear reactions, but the specific details depend on mass. Let's examine next what happens to a star of moderately low mass, between 0.4 $M_\odot$ and 4 $M_\odot$. One example of such a star is our own Sun, with a mass of 1 $M_\odot$. (Our discussion of the largest stars is in the next chapter.)

We can describe a Sunlike star's post–main-sequence evolution using an evolutionary track on an H-R diagram. **Figure 11-20** shows the track for a 1-$M_\odot$ star like the Sun. When an aging star first becomes a red giant, the luminosity increases and the surface temperature drops, the post–main-sequence star moves up and to the right along a red-giant branch on an H-R diagram (Figure 11-20*a*).

Next, the helium-rich core of the star shrinks and heats until eventually helium fusion begins in the core. This causes the core to expand, which in turn makes the core cool down a bit. The cooling of the core also cools the surrounding hydrogen-fusing shell, so that the shell releases energy more slowly. Hence, the luminosity goes down a bit after core helium fusion begins.

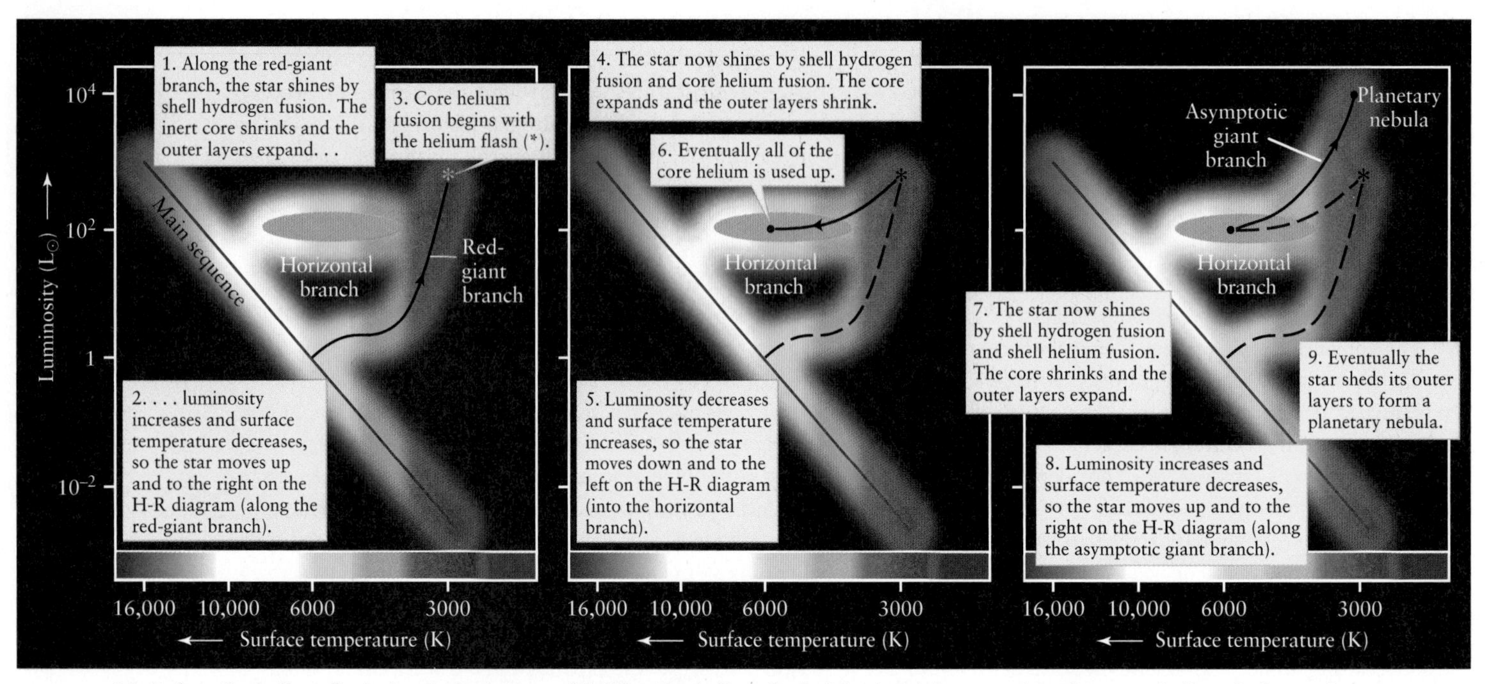

(a) Before the helium flash: A red-giant star (b) After the helium flash: A horizontal-branch star (c) After core helium fusion ends: An AGB star

FIGURE 11-20 The Post–Main-Sequence Evolution of a 1-$M_\odot$ Star These H-R diagrams show the evolutionary track of a star like the Sun as it goes through the stages of being (a) a red-giant star, (b) a horizontal-branch star, and (c) an asymptotic giant branch (AGB) star. The star eventually evolves into a planetary nebula (described in Section 11-5).

The slower rate of energy release also lets the star's outer layers contract. As they contract, they heat up, so the star's surface temperature increases and its evolutionary track moves to the left on the H-R diagram (Figure 11-20*b*). The luminosity changes relatively little during this stage, so the evolutionary track on the H-R diagram moves almost horizontally. These horizontal-branch stars have helium-fusing cores surrounded by hydrogen-fusing shells.

ConceptCheck 11-14: Are horizontal-branch red-giant stars becoming more or less luminous?

AGB Stars: The Second Red-Giant Stage

Helium fusion produces nuclei of carbon and oxygen. After about 100 million (10^8) years of helium fusion in the core, essentially all the core helium of a 1-$M_\odot$ star has been converted into carbon and oxygen, and the fusion of helium in the core ceases. Without thermonuclear reactions to maintain the core's internal pressure, the core again contracts, until it is stopped by degenerate-electron pressure. This contraction releases heat into the surrounding helium-rich gases, and a new stage of helium fusion begins in a thin shell around the core.

History now repeats itself—the star enters a *second* red-giant phase. A star first becomes a red giant at the end of its main-sequence lifetime, when the outpouring of energy from hydrogen fusion in the shell surrounding the core makes the star's outer layers expand and cool. In the same way, the outpouring of energy from shell *helium* fusion causes the outer layers to expand again. The low-mass star ascends into the red-giant region of the H-R diagram for a second time (Figure 11-20*c*), but now with even greater luminosity than during its first red-giant phase.

Stars in this second red-giant phase are commonly called **asymptotic giant branch stars,** or **AGB stars,** and their evolutionary tracks approach the red-giant branch from the left on an H-R diagram. When a low-mass star first becomes an AGB star, it consists of an inert, degenerate carbon-oxygen core and a surrounding shell of helium fusion, both inside a surrounding shell of hydrogen fusion, all within a volume not much larger than Earth. This small, dense central region is surrounded by an enormous hydrogen-rich envelope about as big as Earth's orbit around the Sun. After a while, the expansion of the star's outer layers causes the layers of hydrogen fusion to also expand and cool, and thermonuclear reactions in this shell temporarily cease. This leaves the aging star's structure as shown in **Figure 11-21.**

Remember, the shorter the amount of time it remains on the main sequence, the faster the star will go through these stages. Similarly, the greater the star's mass, the more rapidly it goes through the stages of post–main-sequence evolution. Hence, we can see all of these stages by studying clusters of stars, which contain stars that are all the same age but that have a range of masses. **Figure 11-22** shows an H-R diagram for the globular cluster M55, which is at least 13 billion years old. The least massive stars in this cluster are still on the main sequence. Progressively more massive stars have evolved to the red-giant branch, the horizontal branch, and the asymptotic giant branch.

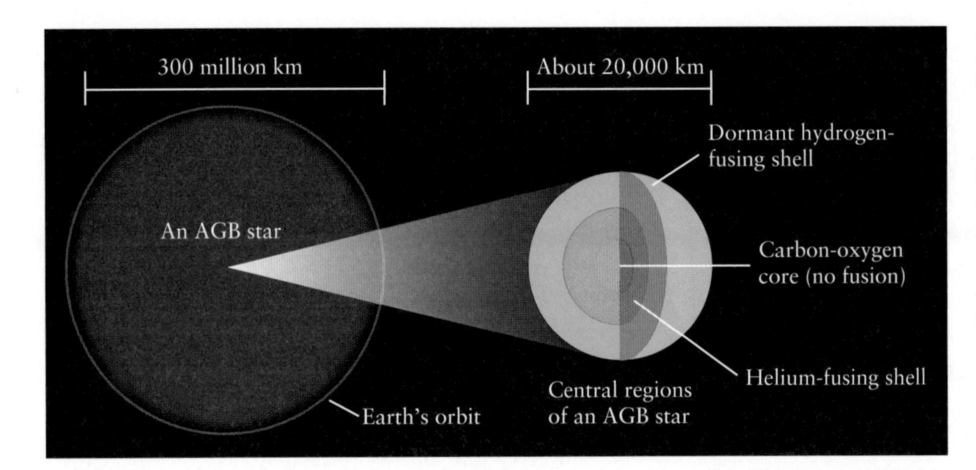

FIGURE 11-21 The Structure of an Old, Moderately Low-Mass AGB Star Near the end of its life, a star like the Sun becomes an immense, red, asymptotic giant branch (AGB) star. The star's inert core, active helium-fusing shell, and dormant hydrogen-fusing shell are all contained within a volume roughly the size of Earth. Thermonuclear reactions in the helium-fusing shell are so rapid that the star's luminosity is thousands of times that of the present-day Sun. (The relative sizes of the shells in the star's interior are not shown to scale.)

A 1-$M_\odot$ AGB star can reach a maximum luminosity of nearly 10^4 $L_\odot$, as compared with approximately 10^3 $L_\odot$ when it reached the helium flash and a relatively paltry 1 $L_\odot$ during its main-sequence lifetime. When the Sun becomes an AGB star some 12.3 billion years from now, this tremendous increase in luminosity will cause the planet Mars and the Jovian planets to largely evaporate away. The Sun's bloated outer layers will reach to Earth's orbit. Mercury and perhaps Venus will simply be swallowed whole.

ConceptCheck 11-15: When the Sun becomes an AGB star, will Earth's orbit be inside the core?

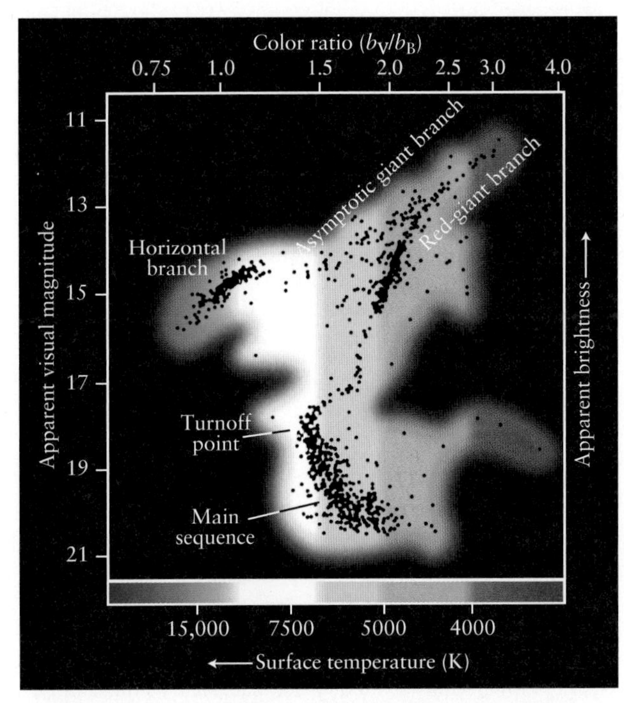

FIGURE 11-22 Stellar Evolution in a Globular Cluster In the old globular cluster M55, stars with masses less than about 0.8 $M_\odot$ are still on the main sequence, converting hydrogen into helium in their cores. Slightly more massive stars have consumed their core hydrogen and are ascending the red-giant branch; even more massive stars have begun helium core fusion and are found on the horizontal branch. The most massive stars (which still have less than 4 $M_\odot$) have consumed all the helium in their cores and are ascending the asymptotic giant branch. *(Adapted from D. Schade, D. VandenBerg, and F. Hartwick)*

11-5 Low-mass stars pulsate and eject planetary nebulae, leaving behind a white dwarf at the end of their life cycles

For a star that began with a moderately low mass (between about 0.4 $M_\odot$ and 4 $M_\odot$), the AGB stage in its evolution is a dramatic turning point. Before this stage, a star loses mass only gradually through steady stellar winds. But as it evolves during its AGB stage, a star divests itself completely of its outer layers. The aging star undergoes a series of bursts in luminosity, and in each burst it ejects a shell of material into space. Eventually, all that remains of a low-mass star is a fiercely hot, exposed core, surrounded by glowing shells of ejected gas. This late stage in the life of a star is called a **planetary nebula. Figure 11-23** shows the wide variety

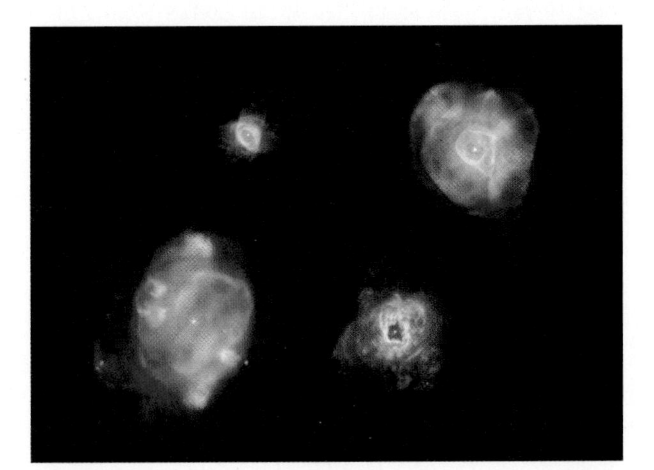

FIGURE 11-23 R I V U X G **Collage of Four Planetary Nebula** These four planetary nebulae imaged by the Hubble Space Telescope are all roughly 7000 ly away. Planetary nebulae show a wide variety of shapes, indicative of the complex processes that occur at the end of stellar life. Henize 2-47 (top left) is dubbed the "starfish," because of its six lobes of gas and dust, which were created by separate ejections of material occurring in different directions. IC 4593 (top right) displays a prominent pair of jets ending in red knots of glowing nitrogen gas. NGC 5307 (bottom left) displays a spiral pattern, which may have been caused by the dying star's wobbling. NGC 5315, the chaotic-looking nebula at bottom right, reveals an X-shaped structure suggesting that the dying star ejected material in two different outbursts in two distinct directions. *(NASA, ESA, and The Hubble Heritage Team [STScI/AURA])*

observed in planetary nebula, often given names reflective of their shapes.

ConceptCheck 11-16: **How many planets make up a planetary nebula?**

Making a Planetary Nebula

To understand how an AGB star can eject its outer layers in a sequence of spherical shells, consider the internal structure of such a star as shown in Figure 11-21. As the helium in the helium-fusing shell is used up, the pressure that holds up the dormant hydrogen-fusing shell decreases. Hence, the dormant hydrogen shell contracts and heats up, and hydrogen fusion begins anew. This revitalized hydrogen fusion creates helium, which rains downward onto the temporarily dormant helium-fusing shell. As the helium shell gains mass, it shrinks and heats up. When the temperature of the helium shell reaches a certain critical value, it reignites. The released energy pushes the hydrogen-fusing shell outward, making it cool off, so that hydrogen fusion ceases and this shell again becomes dormant. The process then starts over again.

When this momentary reignition of the helium shell occurs, the luminosity of an AGB star increases substantially in a relatively short-lived burst called a **thermal pulse.** **Figure 11-24,** which is based on a theoretical calculation of the evolution of a 1-$M_\odot$ star, shows that thermal pulses begin when the star is about 12.365 billion years old. The calculations predict that thermal pulses occur at ever-shorter intervals of about 100,000 years.

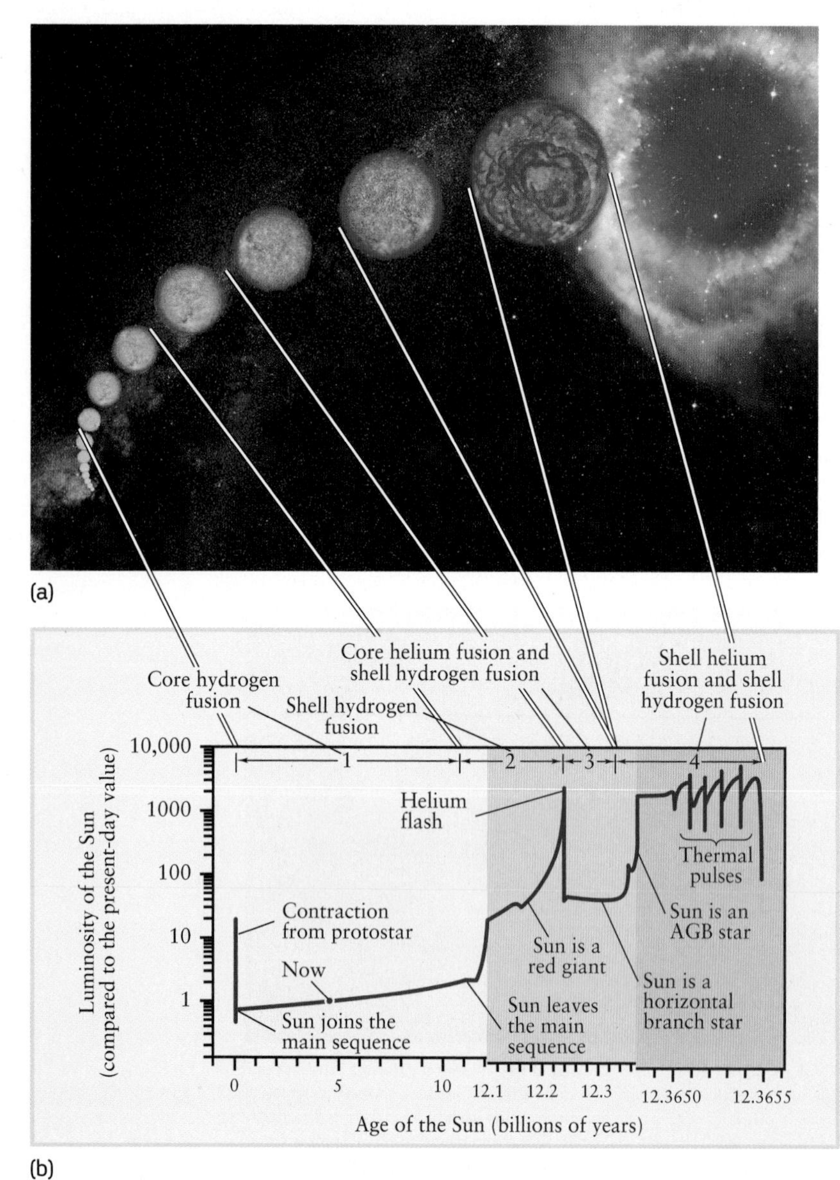

FIGURE 11-24 Further Stages in the Evolution of the Sun This diagram shows how the luminosity of the Sun (a 1-$M_\odot$ star) changes over time. We use different scales for the final stages because the evolution is so rapid. During the AGB stage there are brief periods of runaway helium fusion, causing spikes in luminosity called thermal pulses. All red giants exhibit a slow oscillation in brightness due to their rhythmic "breathing" in and out, and one-third of them are also affected by additional slower and mysterious changes in their luminosity. *(ESO/S. Steinhöfel)*

During these thermal pulses, the dying star's outer layers can separate completely from its carbon-oxygen core. As the ejected material expands into space, dust grains condense out of the cooling gases. Radiation pressure from the star's hot, burned-out core acts on the specks of dust, propelling them further outward, and the star sheds its outer layers altogether. In this way an aging 1-$M_\odot$ star loses as much as 40% of its mass. More massive stars eject even greater fractions of their original masses.

As a dying star ejects its outer layers, the star's hot core becomes exposed. With a surface temperature of about 100,000 K, this exposed core emits ultraviolet radiation intense enough to ionize and excite the expanding shell of ejected gases. These gases therefore glow and emit visible light through the process of fluorescence, producing planetary nebulae like those shown in **Figure 11-25.**

CAUTION Despite their name, planetary nebulae have nothing to do with planets. This misleading term was introduced in the nineteenth century because these glowing objects looked like distant Jovian planets when viewed through the small telescopes then available. The difference between planets and planetary nebulae became obvious with the advent of spectroscopy: Planets have *absorption* line spectra, but the excited gases of planetary nebulae have *emission* line spectra.

ConceptCheck 11-17: **What is the chemical composition of the remaining stellar core after the planetary nebula forms?**

The Properties of Planetary Nebulae

Planetary nebulae are quite common. Astronomers estimate that there are 20,000 to 50,000 planetary nebulae in our Galaxy alone. Many planetary nebulae, such as those in Figure 11-23, are more or less spherical in shape. This is a result of the symmetrical way in which the gases were ejected. But if the rate of expansion is not the same in all directions, the resulting nebula takes on an hourglass or dumbbell appearance (**Figure 11-26**).

(a) RIVUXG

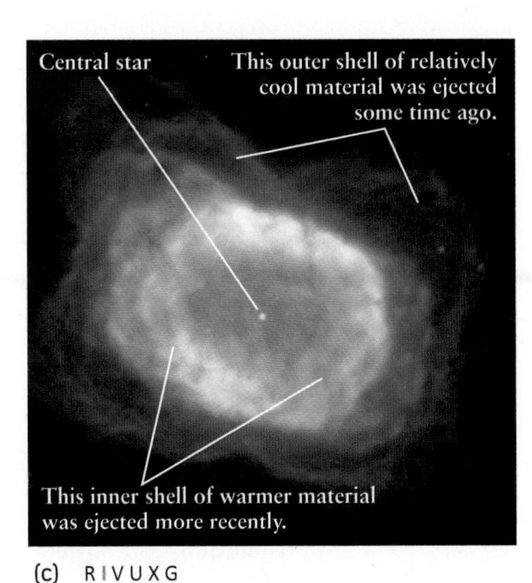

(b) RIVUXG

Central star

Nearly spherical shell of material ejected from the central star

The shell is so thin that we can see stars on the other side...

...but it appears substantial when we look near its rim.

(c) RIVUXG

FIGURE 11-25 Planetary Nebulae (a) The pinkish blob is a planetary nebula surrounding a star in the globular cluster M15, about 33,000 ly from Earth in the constellation Pegasus. (b) The planetary nebula Abell 39 is about 7000 ly from Earth in the constellation Hercules. The almost perfectly spherical shell that constitutes the nebula is about 5 ly in diameter; the thickness of the shell is only about 0.3 ly. (c) This infrared image of the planetary nebula NGC 7027 suggests a more complex evolutionary history than that of Abell 39. NGC 7027 is about 900 pc (3000 ly) from Earth in the constellation Cygnus and is roughly 14,000 AU across. *(a: NASA/The Hubble Heritage Team [STScI/AURA]; b: WIYN/NOAO/NSF, Copyright WIYN Consortium, Inc., all rights reserved; c: William B. Latter [SIRTF Science Center/Caltech], and NASA)*

Spectroscopic observations of planetary nebulae show emission lines of ionized hydrogen, oxygen, and nitrogen. From the Doppler shifts of these lines, astronomers have concluded that the expanding shell of gas moves outward from a dying star at speeds from about 20,000 mi/h to mi/h (10 km/s to 30 km/s). For a shell expanding at such speeds to have attained the typical diameter of a planetary nebula, about 1 light-year, it must have begun expanding about 10,000 years ago. Thus, by astronomical standards, the planetary nebulae we see today were created only very recently.

We do not observe planetary nebulae that are more than about 50,000 years old. After this length of time, the shell has spread out so far from the cooling central star that its gases cease to glow and simply fade from view. The nebula's gases then mix with the surrounding interstellar medium.

Astronomers estimate that all the planetary nebulae in the Galaxy return a total of about 5 $M_\odot$ to the interstellar medium each year. This amounts to 15% of all the matter expelled by all the various sorts of stars in the Galaxy each year. Because this contribution is so significant, and because the ejected material includes heavier elements (metals) manufactured within a nebula's central star, planetary nebulae play an important role in the chemical evolution of the Galaxy as a whole.

ConceptCheck 11-18: For approximately how many years will an isolated planetary nebula continue to expand?

CalculationCheck 11-1: If a planetary nebula is expanding at 20,000 mi/h, how long would it take to become the size of our solar system if Pluto is about 3 billion miles from our Sun?

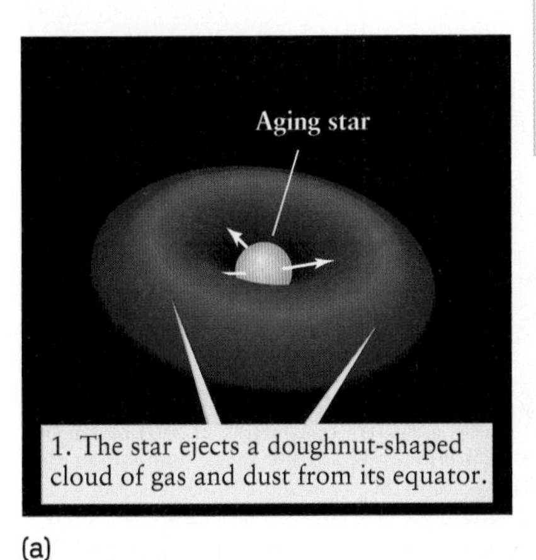

(a)

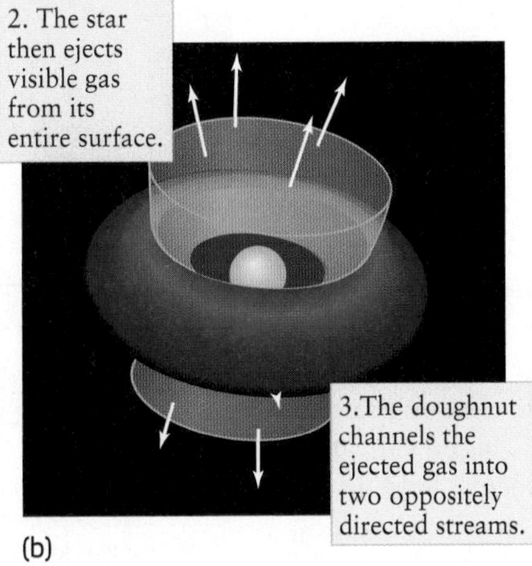

(b)

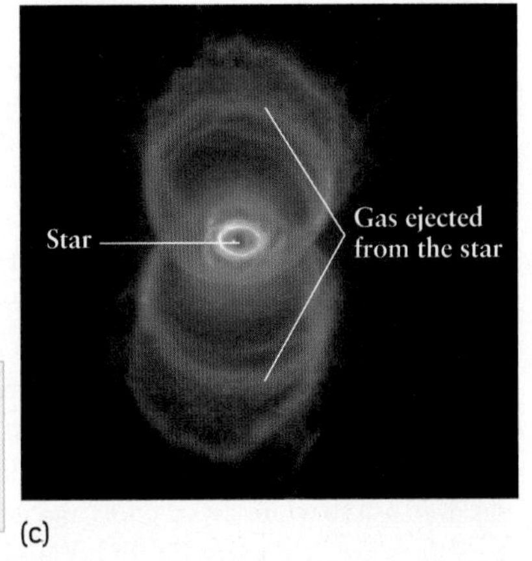

(c)

FIGURE 11-26 Making an Elongated Planetary Nebula (a, b) These illustrations show one proposed explanation for why many planetary nebulae have an elongated shape. (c) The planetary nebula MyCn18, shown here in false color, may have acquired its elongated shape in this way. It lies some 8000 ly from Earth in the constellation Musca (the Fly). *(c: R. Sahai and J. Trauger, Jet Propulsion Laboratory; the WFPC-2 Science Team; NASA)*

Formation of a White Dwarf

We have seen that after a moderately low-mass star (from about 0.4 solar masses to about 4 solar masses) consumes all the hydrogen in its core, it is able to ignite thermonuclear reactions that convert helium to carbon and oxygen. Given sufficiently high temperature and pressure, carbon and oxygen can also undergo fusion reactions that release energy. But for such a moderately low-mass star, the core temperature and pressure never reach the extremely high values needed for these reactions to take place. Instead, as we have seen, the process of mass ejection just strips away the star's outer layers and leaves behind the hot carbon-oxygen core. With no thermonuclear reactions taking place, the core simply cools down like a dying ember. Such a burnt-out relic of a star's former glory is called a **white dwarf.** Such white dwarfs prove to have exotic physical properties that are wholly unlike any object found on Earth.

Our Sun will eventually end its life as a white dwarf.

CAUTION Unfortunately, the word *dwarf* is used in astronomy for several very different kinds of small objects. Here is a review of the three kinds that we have encountered so far in this book. A *white* dwarf is the relic that remains at the very end of the evolution of a star of initial mass between about 0.4 $M_\odot$ and 4 $M_\odot$. Thermonuclear reactions are no longer taking place in its interior; it emits light simply because it is still hot. A *red* dwarf is a cool main-sequence star with a mass between about 0.08 $M_\odot$ and 0.4 $M_\odot$. The energy emitted by a red dwarf in the form of light comes from its core, where fusion reactions convert hydrogen into helium. Finally, a *brown* dwarf is an object like a main-sequence star but with a mass less than about 0.08 $M_\odot$. Because its mass is so small, its internal pressure and temperature are too low to sustain thermonuclear reactions. Instead, a brown dwarf emits light because it is slowly contracting, a process that releases energy. White dwarfs are comparable in size to Earth; by contrast, brown dwarfs are larger than the planet Jupiter, and red dwarfs are even larger.

ConceptCheck 11-19: From largest to smallest, what is the sequence of sizes of a white dwarf, a red dwarf, and a brown dwarf?

Properties of White Dwarfs

Tiny white dwarfs are all that remain from an evaporating planetary nebula.

You might think that without thermonuclear reactions to provide internal heat and pressure, a white dwarf should keep on shrinking under the influence of its own gravity as it cools. Actually, however, a cooling white dwarf maintains its size because the burnt-out stellar core is so dense that most of its electrons are so tightly packed they are classified as being degenerate. Thus, degenerate-electron pressure supports the star against further collapse. This pressure does not depend on temperature, so it continues to hold up the star even as the white dwarf cools and its temperature drops.

Many white dwarfs are found in our local solar neighborhood, but all are too faint to be seen with the naked eye. One of the first white dwarfs to be discovered is a companion to Sirius, the brightest star in the night sky. This companion, designated Sirius B (**Figure 11-27**), was first glimpsed in 1862 by the American astronomer Alvan Clark. Recent Hubble Space Telescope observations at ultraviolet wavelengths, where hot white dwarfs emit most of their light, show that the surface temperature of Sirius B is 25,200 K.

Observations of white dwarfs in binary systems like Sirius allow astronomers to determine the mass, radius, and density of these stars. Such observations show that the density of the degenerate matter in a white dwarf is typically 10^9 kg/m^3 (a million times denser than water). A teaspoonful of white dwarf matter brought to Earth would weigh nearly 5.5 tons—as much as an elephant!

Degenerate matter has a very different relationship between its pressure, density, and temperature than that of ordinary gases. Consequently, white-dwarf stars have an unusual mass-radius relation: The more massive a white-dwarf star, the *smaller* it is.

Figure 11-28 displays the mass-radius relation for white dwarfs. Note that the more degenerate matter you pile onto a white dwarf, the smaller it becomes. However, there is a limit to how much pressure degenerate electrons can produce. As a result, there is an upper limit to the mass that a white dwarf can have. This maximum mass is called the **Chandrasekhar limit,** after the Indian-American scientist Subrahmanyan Chandrasekhar, who pioneered theoretical studies of white dwarfs in the 1930s. The Chandrasekhar limit is equal to 1.4 $M_\odot$, meaning that all white dwarfs must have masses less than 1.4 $M_\odot$.

The material inside a white dwarf consists mostly of ionized carbon and oxygen atoms floating in a sea of degenerate

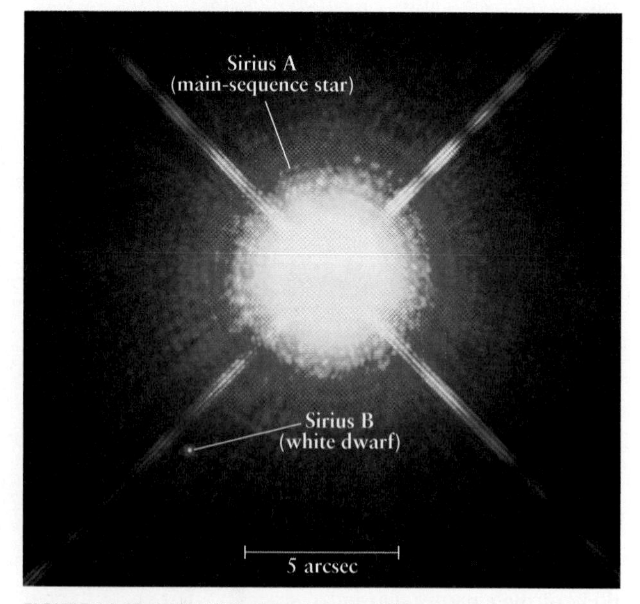

FIGURE 11-27 R I V U X G **Sirius A and Its White-Dwarf Companion** Sirius, the brightest-appearing star in the sky, is actually a binary star. The secondary star, called Sirius B, is a white dwarf. In this Hubble Space Telescope image, Sirius B is almost obscured by the glare of the overexposed primary star, Sirius A, which is about 104 times more luminous than Sirius B. The halo and rays around Sirius A are the result of optical effects within the telescope. *(NASA, H. E. Bond and E. Nelan [Space Telescope Science Institute, Baltimore, Md.]; M. Barstow and M. Burleigh [University of Leicester, U.K.]; and J. B. Holberg [University of Arizona])*

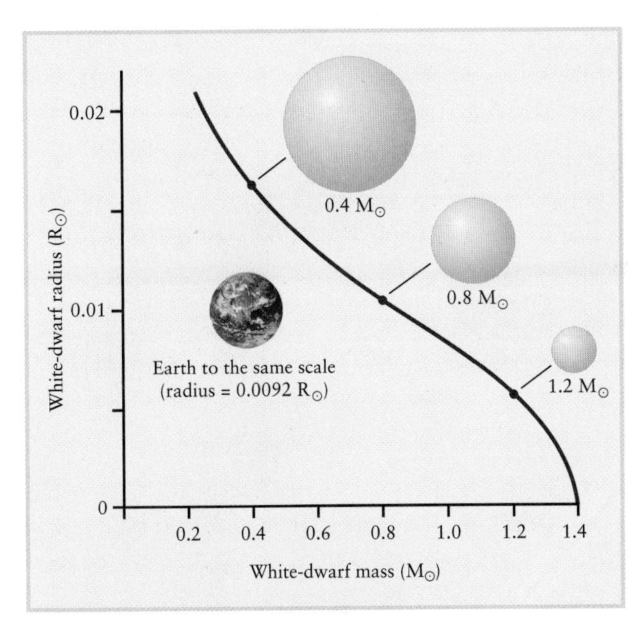

FIGURE 11-28 The Mass-Radius Relationship for White Dwarfs The more massive a white dwarf is, the smaller its radius. (The drawings of white dwarfs of different mass are drawn to the same scale as the image of Earth.) This unusual relationship is a result of the degenerate-electron pressure that supports the star. The maximum mass of a white dwarf, called the Chandrasekhar limit, is 1.4 $M_\odot$.

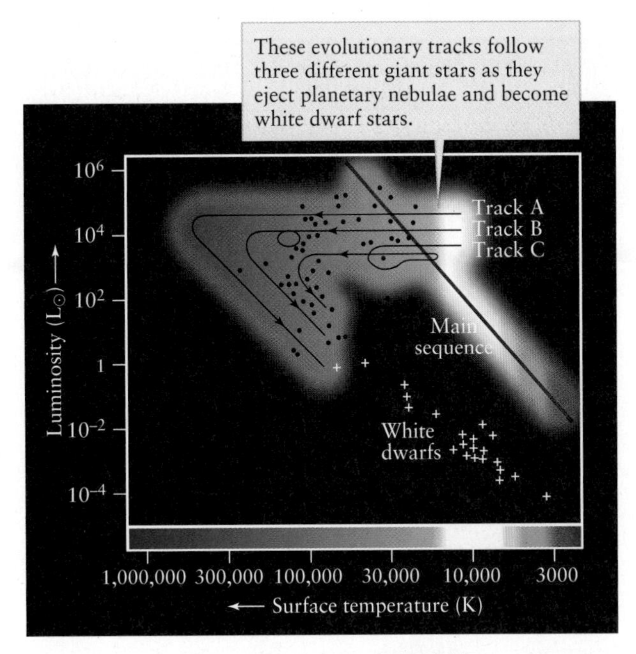

Evolutionary track	Giant star	Ejected nebula	White dwarf
	Mass ($M_\odot$)		
A	3.0	1.8	1.2
B	1.5	0.7	0.8
C	0.8	0.2	0.6

FIGURE 11-29 Evolution from Giants to White Dwarfs This H-R diagram shows the evolutionary tracks of three low-mass giant stars as they eject planetary nebulae. The table gives the extent of mass loss in each case. The dots represent the central stars of planetary nebulae whose surface temperatures and luminosities have been determined; the crosses represent white dwarfs of known temperature and luminosity. *(Adapted from B. Paczynski)*

electrons. As the dead star cools, the carbon and oxygen ions slow down, and electric forces between the ions begin to prevail over the random thermal motions. About 5×10^9 years after the star first becomes a white dwarf, when its luminosity has dropped to about 10^{-4} $L_\odot$ and its surface temperature is a mere 4000 K, the ions no longer move freely. Instead, they arrange themselves in orderly rows, like an immense crystal lattice. From this time on, you could say that the star is "solid." The degenerate electrons move around freely in this crystal material, just as electrons move freely through an electrically conducting metal like copper or silver. A diamond is also crystallized carbon, so a cool carbon-oxygen white dwarf can resemble an immense spherical diamond!

ConceptCheck 11-20: How does the size of a white dwarf change as it cools?

From Red Giant to Planetary Nebula to White Dwarf

Figure 11-29 shows the evolutionary tracks followed by three burned-out stellar cores as they pass through the planetary nebula stage and become white dwarfs. When these three stars were red giants, they had masses of 0.8 $M_\odot$, 1.5 $M_\odot$, and 3.0 $M_\odot$. Mass ejection strips these dying stars of up to 60% of their matter. During their final spasms, the luminosity and surface temperature of these stars change quite rapidly. The points representing these stars on an H-R diagram race along their evolutionary tracks, sometimes executing loops corresponding to thermal pulses (shown as Track B and Track C in Figure 11-29). Finally, as the ejected nebulae fade and the stellar cores cool, the evolutionary tracks of these dying stars take a sharp turn toward the white-dwarf region of the H-R diagram. As the table accompanying Figure 11-29 shows, the final white dwarf has only a fraction of the mass of the giant star from which it evolved.

Although a white dwarf maintains the same size as it cools, its luminosity and surface temperature both decrease with time. Consequently, the evolutionary tracks of aging white dwarfs point toward the lower right corner of the H-R diagram. You can see this in Figure 11-29; **Figure 11-30** shows it in more detail. The energy that the white dwarf radiates into

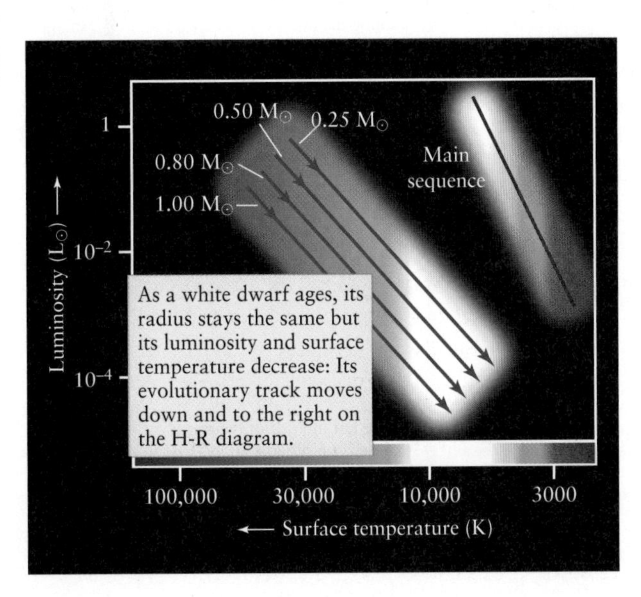

FIGURE 11-30 White-Dwarf "Cooling Curves" As white-dwarf stars radiate their internal energy into space, they become dimmer and cooler. The blue lines show the evolutionary tracks of four white dwarfs of different mass: The more massive a white dwarf, the smaller and hence fainter it is.

COSMIC CONNECTIONS The Sun: The Next 8 Billion Years

The Sun is presently less than halfway through its lifetime as a main-sequence star. The H-R diagram and cross-sections on this page summarize the dramatic changes that will take place when the Sun's main-sequence lifetime comes to an end.

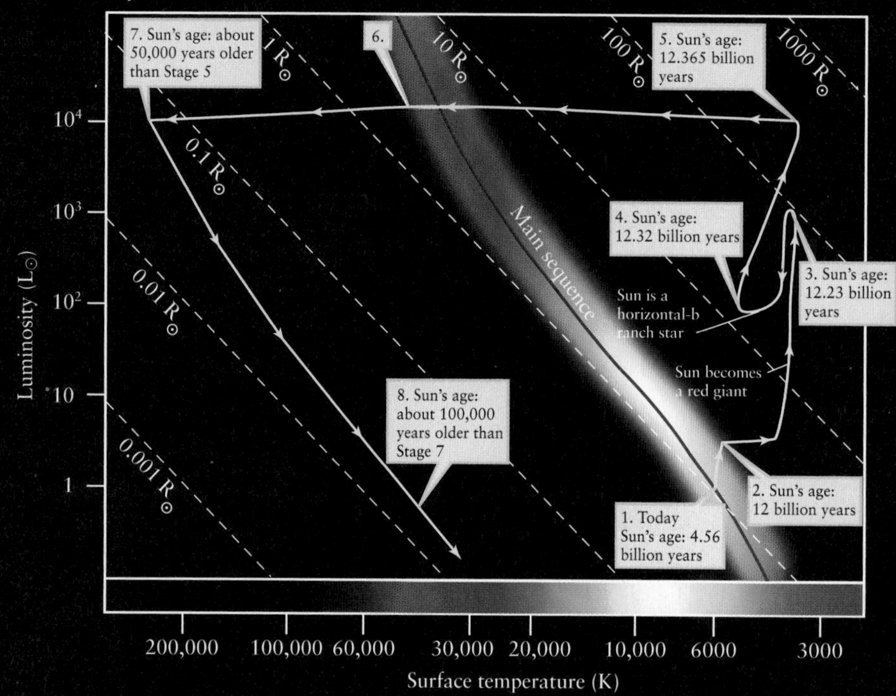

NOTE: The illustrations below do *not* show the dramatic changes in the Sun's radius as it evolves. The sizes of the various layers are not shown to scale.

1. On the main sequence

The present-day Sun is a main-sequence star—in its core, hydrogen fuses to produce helium.

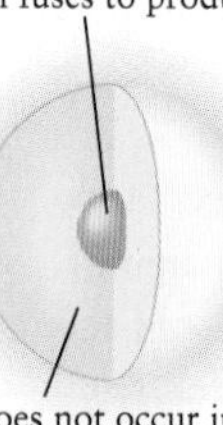

Fusion does not occur in the outer layers (which contain predominantly hydrogen and helium).

2. Becoming a red giant

At the end of the Sun's main-sequence lifetime, fusion stops in the core (which has been converted to helium).

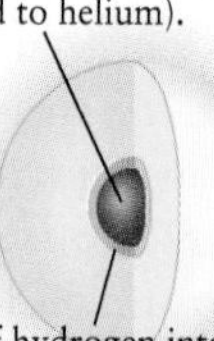

Fusion of hydrogen into helium continues in a shell around the core. The core shrinks, accelerating the fusion reactions in the shell and making the outer layers expand and cool.

3. The helium flash

As the core contracts and heats, the core helium begins to fuse to make carbon and oxygen. The core expands and the rate of energy release slows.

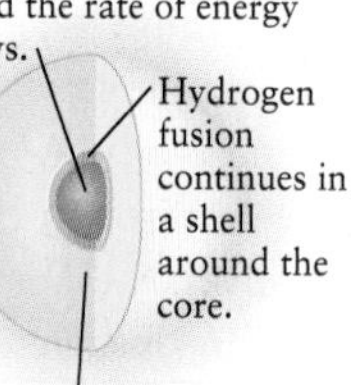

The outer layers (where there are still no fusion reactions) contract and get hotter due to the slower rate of energy release.

4. Beginning the second red-giant phase

Once the core helium is consumed, what remains is an inert core of carbon and oxygen. The core again shrinks and gets hotter.

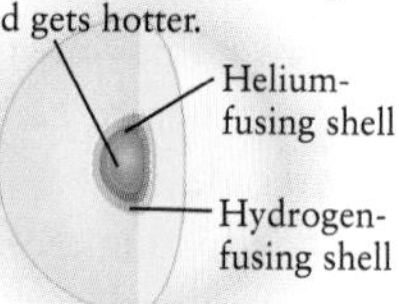

The shrinkage of the core again accelerates fusion reactions in the shells, making the inert outer layers expand and cool.

5. The Sun reaches its maximum size

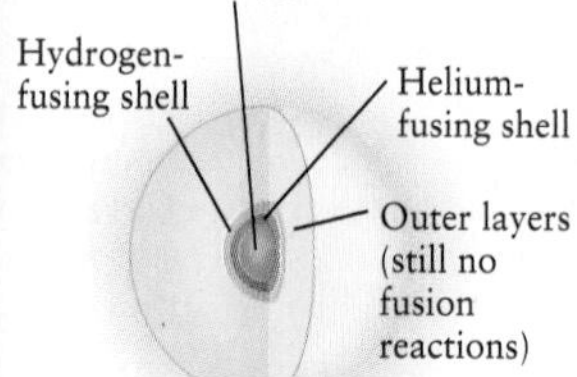

The Sun is more than 100 times larger in radius than when it was a main-sequence star. Part of the outer layers escapes into space in a stellar wind.

6. A planetary nebula

Thermal pulses cause spikes in luminosity that eject the star's outer layers.

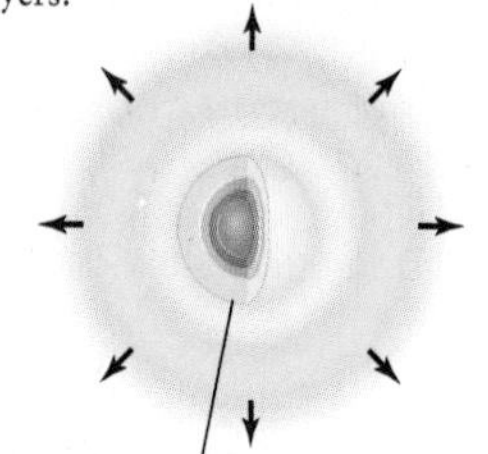

As the hot interior of the star is exposed, we observe an increase in the star's surface temperature.

7. The end of nuclear reactions

With the outer layers gone, the pressure on the shells around the core is too little to sustain nuclear reactions.

The star still glows intensely because of its high temperature. As energy is lost in the form of electromagnetic radiation, the star slowly cools.

8. A white dwarf

The core is now a white dwarf star, and the former shells around the core become its thin atmosphere.

The carbon-oxygen interior of the white dwarf is degenerate, so it does not contract as it cools. Hence the white dwarf's radius no longer changes.

Hydrogen and helium, no fusion | Hydrogen fusion producing helium | Helium, no fusion | Helium fusion producing carbon and oxygen | Carbon and oxygen, no fusion

space comes only from the star's internal heat, which is a relic from the white dwarf's past existence as a stellar core. Over billions of years, white dwarfs grow dimmer and dimmer as their surface temperatures drop toward absolute zero.

After ejecting much of its mass into space, our own Sun will eventually evolve into a white-dwarf star about the size of Earth and with perhaps one-tenth of its present luminosity. It will become even dimmer as it cools. After 5 billion years as a white dwarf, the Sun will radiate with no more than one ten-thousandth of its present brilliance. With the passage of eons, our Sun will simply fade into obscurity. **Cosmic Connections: The Sun: The Next 8 Billion Years** summarizes the full evolutionary cycle of a 1-$M_{\odot}$ star like the Sun, from its birth to becoming a main-sequence star to its demise as a white dwarf.

This process describes the evolutionary process of the vast majority of stars in our Galaxy. Yet, the most dramatic and dynamic changes happen in the rarest of the largest aging stars, which we address in Chapter 12.

ConceptCheck 11-21: As the red giant moves through the planetary nebula stage to the white-dwarf star stage, in which directions on the H-R diagram are the star's graphed positions moving?

VISUAL LITERACY TASK

Star Cluster Ages

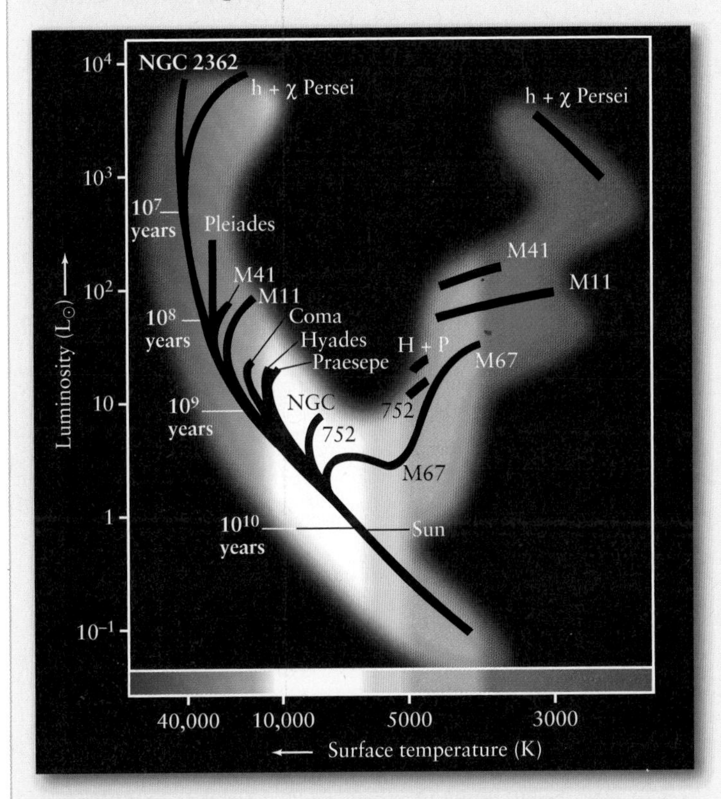

PROMPT: **What would you tell a fellow student who said, "The youngest star clusters have most of their stars in the red-giant phase"?**

ENTER RESPONSE:

Guiding Questions:

1. The age of the Pleiades cluster is about
   a. 10 million years old.
   b. 100 million years old.
   c. 1 billion years old.
   d. 10 billion years old.

2. Compared to the Coma star cluster, the Praesepe star cluster is
   a. older.
   b. younger.
   c. larger.
   d. hotter.

3. The cluster's turnoff point is where the
   a. most massive stars are leaving the main sequence.
   b. least massive stars are leaving the main sequence.
   c. majority of stars form.
   d. black holes form.

4. Compared to the older clusters, the younger clusters have most of their stars
   a. at 5000 K.
   b. in the red-giant phase.
   c. on the main sequence.
   d. with luminosities less than the Sun.

KEY IDEAS AND TERMS

11-1 Stars form from the gravitational collapse of immense clouds of interstellar gas and dust

- Star lifetimes exceed human life spans, so they must be studied by looking at numerous stars at different stages in their life cycles.
- Gas and dust, which make up the **interstellar medium,** are clumped into clouds called **nebulae.**
- **Emission nebulae** are glowing, ionized clouds of gas, powered by ultraviolet light that they absorb from nearby hot stars.
- **Dark nebulae** are so dense that they are opaque. They appear as dark blots against a background of distant stars.
- **Reflection nebulae** are produced when starlight is reflected from dust grains in the interstellar medium, producing a characteristic bluish glow.
- Star formation begins in dense, cold clouds, where gravitational attraction causes a clump of material to condense into a **protostar.**
- **T Tauri stars** are protostars with emission lines as well as absorption lines in their spectra and whose luminosity can change irregularly as they eject material into space.
- The dusty material being added to the protostar is known as a **circumstellar accretion disk,** and the surrounding materials that form planets are called **protoplanetary disks,** or **proplyds.**
- **Supernova remnants** are the remains of large, exploded stars that have run out of usable fuel.

11-2 Most stars shine throughout their lives by converting hydrogen into helium through nuclear fusion

- Nuclear fusion occurs when smaller atoms combine together to make heavier atoms.
- Stars spend the majority of their life cycles consuming hydrogen and forging heavier elements, releasing energy in the process.
- Stars often form in clusters from the same large interstellar cloud, with all stars having a similar chronological age and initial chemical composition.
- A star cluster's age is equal to the age of the main-sequence stars at the upper end of the remaining main sequence. As a cluster ages, the main sequence is "eaten away" from the upper left as stars of progressively smaller mass evolve into red giants.

11-3 Careful observations of star clusters provide insight into how a star's mass influences how stars change over time

- **Population I stars** formed from remains of other stars and are **metal rich.**
- **Population II stars** are first-generation stars formed in the very early galaxy and are **metal poor.**
- The most important characteristic determining a star's main-sequence lifetime is its mass.

11-4 Stars slowly become red giants

- The central core of a star is the location of highest pressure and temperature and where core fusion occurs.
- Spherical shells of material surround a star's core and shell fusion can occur there if hot enough.
- Stars join the main-sequence group when they begin hydrogen fusion in their cores.
- Stars leave the main-sequence group and become brighter and cooler giant stars when their core hydrogen is depleted.
- The **triple alpha process** combines helium atoms to form carbon and release energy.
- In a more massive red giant, fusion of helium begins gradually; in a less massive red giant, it begins suddenly, in a process called the **helium flash.**
- **Degenerate-electron pressure** supports a helium-rich core with atoms packed as closely as negatively charged, electric-repulsing electrons will allow.
- Stars in a second red-giant phase are **asymptotic giant branch stars,** or **AGB stars,** and their evolutionary tracks approach the red-giant branch from the left on an H-R diagram.

11-5 Low-mass stars pulsate and eject planetary nebulae, leaving behind a white dwarf at the end of their life cycles

- Helium shell flashes in an old, moderately low-mass star produce **thermal pulses** during which more than half the star's mass may be ejected into space, exposing the hot, dense, carbon-oxygen core of the star, called a **white dwarf.**
- The maximum mass of a white dwarf is given by the **Chandrasekhar Limit** of 1.4 $M_\odot$.
- Ultraviolet light from the exposed core ionizes and excites the ejected gases, producing a **planetary nebula.**

QUESTIONS

Review Questions

1. If no one has ever seen a star go through the complete formation process, how are we able to understand how stars form?
2. If an interstellar medium fills the space between the stars, how is that we are able to see the stars at all?
3. In Figure 11-2, what makes the nebula dark? What makes the Orion nebula glow?
4. Describe the energy source that causes a protostar to shine. How does this source differ from the energy source inside a main-sequence star?
5. In what ways is the internal structure of a 1-$M_\odot$ main-sequence star different from that of a 5-$M_\odot$ main-sequence star? How is it different from a 0.5-$M_\odot$ main-sequence star? What features are common to all these stars?
6. How does the chemical composition of the present-day Sun's core compare to the core's composition when the Sun formed? What caused the change?
7. On what grounds are astronomers able to say that the Sun has about 5.4×10^9 years remaining in its main-sequence stage?
8. Explain how it is possible for the core of a red giant to contract at the same time that its outer layers expand.
9. Why does helium fusion require much higher temperatures than hydrogen fusion?
10. What does it mean when an astronomer says that a star "moves" from one place to another on an H-R diagram?
11. Explain how and why the turnoff point on the H-R diagram of a cluster is related to the cluster's age.
12. What is the difference between Population I and Population II stars? In what sense can the stars of one population be regarded as the "children" of the other population?

13. What are thermal pulses in AGB stars? What causes them? What effect do they have on the luminosity of the star?
14. How can an astronomer tell the difference between a planetary nebula and a planet?
15. What is a white dwarf? Does it produce light in the same way as a star like the Sun?

Web Chat Questions

1. Some science-fiction movies show stars suddenly becoming dramatically brighter when they are "born" (that is, when thermonuclear fusion reactions begin in their cores). Discuss whether or not this is a reasonable depiction.
2. Eventually the Sun's luminosity will increase to the point where Earth can no longer sustain life. Discuss what measures a future civilization might take to preserve itself from such a calamity.

Collaborative Exercises

1. Imagine that your group walks into a store that specializes in selling antique clothing. Prepare a list of observable characteristics that you would look for to distinguish which items were from the early, middle, and late twentieth century. Also, write a paragraph that specifically describes how this task is similar to how astronomers understand the evolution of stars.
2. Consider advertisement signs visible at night in your community and provide specific examples of ones that are examples of the three different types of nebulae that astronomers observe and study. If an example does not exist in your community, creatively design an advertisement sign that could serve as an example.
3. The inverse relationship between a star's mass and its main-sequence lifetime is sometimes likened to automobiles in that the more massive vehicles, such as commercial semi–tractor-trailer trucks, need to consume significantly more fuel to travel at highway speeds than more lightweight and economical vehicles. As a group, create a table called "Maximum Vehicle Driving Distances," much like Table 11-1: Approximate Main-Sequence Lifetimes, by making estimates for any five vehicles of your group's choosing. The table's column headings should be (1) vehicle make and model; (2) estimated gas tank size; (3) cost to fill tank; (4) estimated mileage (in miles per gallon); and (5) number of miles driven on a single fill-up.

Observing Questions

1. Use the *Starry Night*™ program to observe the characteristics of several nebulae in the Milky Way Galaxy. Open **Favourites > Explorations > Atlas** and then select **View > Galactic Guides > Equator** from the menu to indicate the plane of the disk of the Milky Way Galaxy. Next, open the **Find** pane, click on the search box and select **Nebulae** from the list. For each of the following nebulae, type the name in the search box and press the **Enter** key and then click the icon next to the nebula in the list of found items and select **Magnify** from the menu: (**a**) NGC 7000 (the North America nebula); (**b**) M20 (the Trifid nebula); (**c**) M8 (the Lagoon nebula); and (**d**) M42 (the Orion nebula). **Zoom in** on each of these objects and describe your observations, noting the object's distance from the galactic equator, and whether it is a dark, reflection, or emission nebula or some combination of these.
2. Use *Starry Night*™ to examine the H-R diagram of the Pleiades star cluster. This group of stars was formed relatively recently in astronomical time. Select **Favourites > Explorations > Pleiades** to display this cluster of young stars in the view. Click on the **Status** tab to display the H-R diagram of all stars in this field of view around the Pleiades. Note that this appears to be similar to that of stars in our local neighborhood. However, if you restrict the distance to display only the stars within this localized cluster, a different pattern emerges. Click on the **Distance cut-off** checkbox in the **H-R Options** layer of the **Status** pane to restrict the distance to a range between 320 ly and 420 ly. (**a**) Where on the H-R diagram do you now find the majority of the stars of this cluster? (**b**) In view of the existence within this cluster of very hot stars with high output of energy, what does this tell you about the age of this cluster compared to the general population of stars?
3. Use *Starry Night*™ to look for signs of stellar evolution in M101, the Pinwheel Galaxy. Select **Favorites > Explorations > Atlas** and open the **Find** pane. Click the icon in the search box and select **Messier Objects** from the list. Next, type **M101** in the search box and then click the icon next to M101 in the list box and select **Magnify** from the contextual menu. (**a**) What is the color of the central part of this Galaxy? (**b**) What is the color of the outer regions of this galaxy? (**c**) Based on your observations, what type of stars would you expect to predominate in each of these two regions of this galaxy?
4. Planetary nebulae represent the late stages of the evolution of stars whose masses are similar to that of the Sun. These nebulae are found throughout our Galaxy. You can use *Starry Night*™ to explore the distribution of these objects in our sky and to view several of these spectacular nebulae. Set the **View** for your home location at some time in the evening with a field of view of about 100°. Open the **Options** pane and expand the **Deep Space** panel. Expand the **NGC-IC Database** list, click in its box to activate the display of the objects in this list and click **Off** all entries in the list except **Planetary Nebula.** Use the hand tool to move around the sky. Note that these nebulae are mostly concentrated around the Milky Way in our sky. If you have access to a telescope, try to locate and observe several of these planetary nebulae, if possible on a clear, moonless night. Some of the more notable planetary nebulae include: Little Dumbbell (M76), NGC 1535, Eskimo, Ghost of Jupiter, Owl (M97), Ring (M57), Blinking Planetary, Dumbbell (M27), Saturn Nebula, and NGC 7662. If you do not have access to a telescope, use *Starry Night*™ to examine in detail two of these planetary nebulae, M57 (the Ring Nebula) and M27 (the Dumbbell Nebula), and compare their shapes and sizes. Select **Favourites > Explorations > Atlas.** Open the **Find** pane, click the icon in the search box and select **Messier Objects** from the list. Type the name of the nebula in the search box, then click the icon next to its name in the list box and choose **Magnify** to center upon and magnify the nebula in the view. (**a**) How do you account for the difference in the shape of these two planetary nebulae? (**b**) What is the nature of the central star in each of these nebulae?

ANSWERS

ConceptChecks

ConceptCheck 11-1: It would become a dark nebula because a dark nebula has the same composition as a reflection nebula, but the fine grains of dust comprising it would block out nearly all of the distant starlight if it were compressed to a higher density.

ConceptCheck 11-2: Protostars both gain mass, through dust grain collisions that slow the material causing mass to migrate into the protostar, and simultaneously lose mass, through a T Tauri wind process, during their formation.

ConceptCheck 11-3: Any compressive physical process that will cause an interstellar cloud of dust and gas to collapse can start the stellar formation process.

ConceptCheck 11-4: Helium atoms are created during the nuclear fusion process of combining hydrogen atoms near the star's central core.

ConceptCheck 11-5: A shrinking core will have a higher rate of nuclear reactions causing the outer layers to shine more brightly, heat, and expand.

ConceptCheck 11-6: Red dwarfs have convection cells that move fresh hydrogen from the surface into the core for fuel so that nearly all of the hydrogen in a red dwarf can be used for fuel.

ConceptCheck 11-7: Metal-rich stars, like our Sun, are classified as Population I stars.

ConceptCheck 11-8: No. Our Sun is a metal-rich star, which means it was formed from the remains of previously existing stars.

ConceptCheck 11-9: The inward pull of gravity in the most massive of stars compresses the core, making it hotter and more dense, resulting in dramatically higher rates of thermonuclear reactions occurring in the core.

ConceptCheck 11-10: According to Wien's law, the most common wavelength of light emitted by a glowing object becomes longer, toward the red end of the spectrum, as its temperature decreases.

ConceptCheck 11-11: In a red giant, the core is at a significantly higher temperature and density that allows helium atoms to combine and fuse together, releasing energy.

ConceptCheck 11-12: Normal gases expand when temperature increases, which does not occur in a core of degenerate electrons.

ConceptCheck 11-13: The star is less luminous because a hotter core expands, reducing the temperature of the outer shell, which slows the shell hydrogen fusion reactions, reducing the overall energy output.

ConceptCheck 11-14: Horizontal-branch stars are moving to the left on the H-R diagram and are increasing their temperature, but not changing their luminosity.

ConceptCheck 11-15: No. The core of an AGB star is quite small, about the diameter of planet Earth, not the diameter of Earth's orbit about the Sun.

ConceptCheck 11-16: None. A planetary nebula results from the ejected outer layers of an aging AGB star and does not involve planets.

ConceptCheck 11-17: The remaining core is composed of carbon and oxygen.

ConceptCheck 11-18: Without slowing due to friction with nearby dust, a planetary nebula will continue to expand indefinitely.

ConceptCheck 11-19: A white dwarf is the smallest, about the size of planet Earth; a brown dwarf is in between, being somewhat larger than a gas giant planet like Jupiter; and a red dwarf is largest, being one of the smallest stable stars.

ConceptCheck 11-20: White-dwarf stars do not change their size as they cool because the electrons are in the degenerate state and the white dwarf cannot contract any further, regardless of temperature.

ConceptCheck 11-21: The evolutionary tracks of a red giant becoming a white dwarf move from right to left as the shrouding planetary nebula diffuses and from upper left to the lower right as the star dims.

CalculationCheck

CalculationCheck 11-1: Gases traveling at 2×10^4 mi/h would take 3×10^9 mi $\div$ 2×10^4 mi/h = 1.5×10^5 h to traverse the 3-billion-mile distance from the Sun to Pluto: 1.5×10^5 h $\div$ 24 h/day = 6,250 days, or about 17 years.

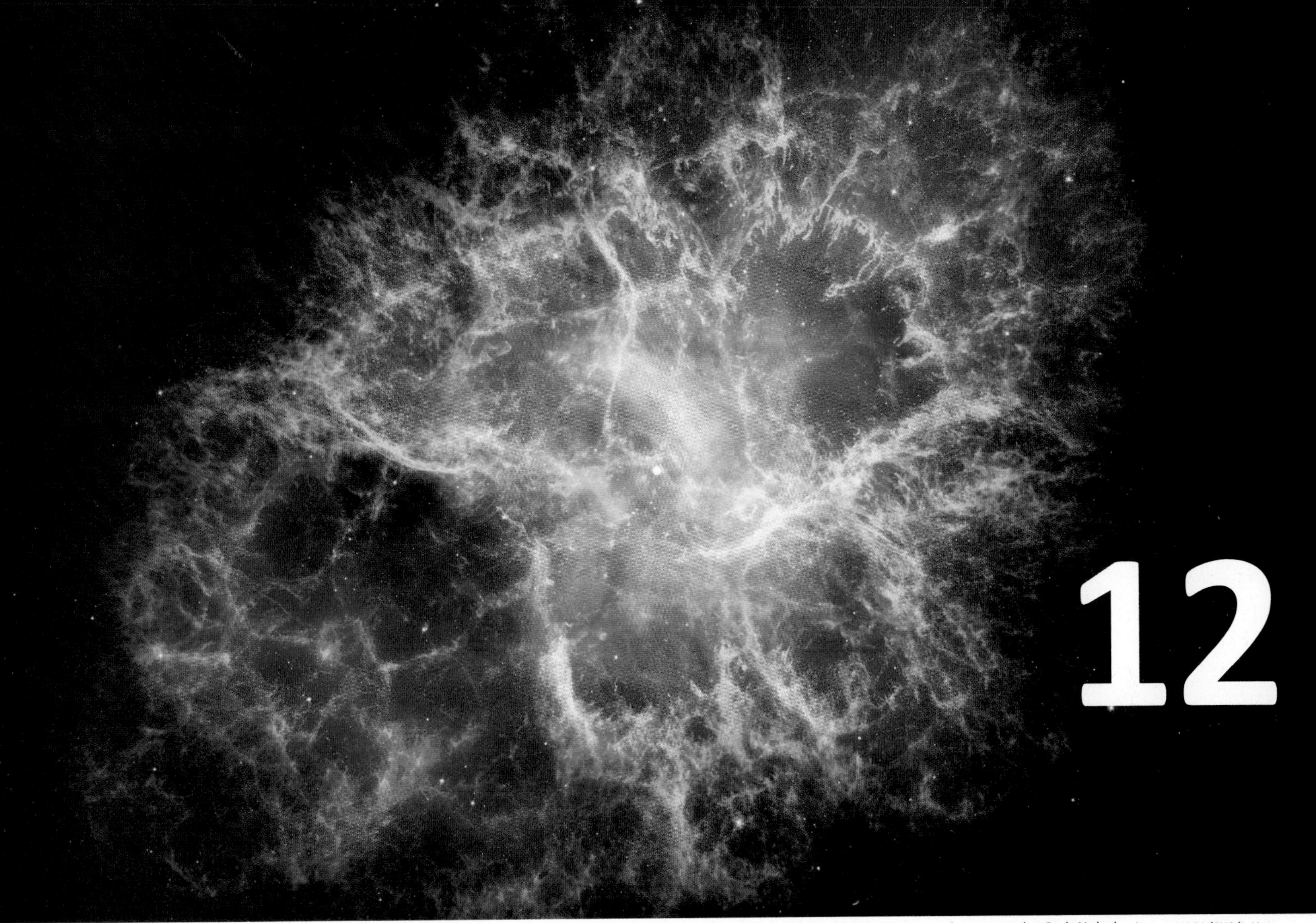

12

RIVUXG A false-color image of the supernova remnant known as the Crab Nebula. *(X-ray: NASA/CXC/J. Hester [ASU]; optical: NASA/ESA/J. Hester & A. Loll [ASU]; infrared: NASA/JPL-Caltech/R. Gehrz [U. of Minnesota])*

Predicting the Violent End of the Largest Stars

CHAPTER LEARNING OBJECTIVES **By reading the sections of this chapter, you will learn:**

- 12-1 High-mass stars create heavy elements in their cores before violently blowing apart in supernova explosions, leaving behind remnants
- 12-2 Core-collapse supernovae can leave behind remnants, neutron stars, and pulsars
- 12-3 Black holes are created in the death throes of the most massive of stars
- 12-4 Black holes cannot be seen directly
- 12-5 White dwarfs and pulsars in close binary systems can become novae, bursters, and supernovae

When a relatively low-mass star like our Sun reaches the end of its main-sequence lifetime and becomes a red giant, it has a hot compressed core and a cooler, bloated atmosphere, ending its evolution by gently expelling its outer layers into space, leaving a burned-out core as remains. In contrast, a high-mass star ends its life in almost inconceivable violence. At the end of its short life, the core of such a star collapses suddenly. This triggers a powerful supernova explosion that can be as luminous as an entire galaxy of stars. The immense force of a supernova combines atoms into heavier elements, which are ejected into interstellar space. Such heavy elements become the essential building blocks for planets, like our Earth. Thus, the deaths of massive stars can provide the seeds for planets orbiting succeeding generations of stars.

In this chapter, we will look at the physical processes that occur when the most massive stars reach the final stages of their life cycles. After a cessation of the nuclear reactions that support a giant stellar core, a star's outer layers can rapidly dissipate, leaving behind a tiny, crushed core that emits very little light as a neutron star or a pulsar. As we will see, it is not only the largest of stars that can end violently. A normally quiet white dwarf can suddenly flare up into an exploding supernova if it gathers explosive gas from a companion star in a close binary system.

But, for the largest of the giant stars, the end can be even more bizarre. Imagine, instead, a swirling disk of gas and dust, orbiting around an object that has more mass than the Sun but is so dark that it cannot be seen. Imagine the material in this disk being compressed and heated as it spirals into the unseen object, reaching temperatures so high that the material emits X-rays. And imagine that the unseen object has such powerful gravity that any material that falls into it simply disappears, never to be seen again. These unseen objects with immensely strong gravity are called black holes. The matter that makes up a black hole has been so greatly compressed that it violently warps space and time. If you get too close to a black hole, the speed you would need to escape it exceeds the speed of light. Because nothing can travel faster than light, nothing—not even light—can escape from a black hole. Even more remarkable is the discovery that black holes of more than a million solar masses lie at the centers of many galaxies, including our own Milky Way. ■

12-1 High-mass stars create heavy elements in their cores before violently blowing apart in supernova explosions, leaving behind remnants

The elements larger than hydrogen and helium up to iron are primarily made within stars.

The swollen red giant and white dwarf end-state of a star like our Sun is stunning indeed. But what about stars many more times massive than the Sun? The life story of a *high-mass* star (with an initial mass greater than about 4 $M_\odot$) begins with the same hydrogen-fusing-into-helium reactions that we observe in smaller stars. But theoretical calculations show that high-mass stars can also go through several additional stages of thermonuclear reactions, involving the fusion of carbon, oxygen, and other heavy nuclei. As it turns out, the failure of internal pressure from thermonuclear reactions to hold up the crushing weight of a giant star's outer layers is much more catastrophic.

Heavy-Element Fusion in Massive Stars

One of the key differences between Sunlike stars and those that are many times more massive is that the largest stars are able to forge the heaviest of chemical elements. Why is fusion of heavy nuclei possible only in a high-mass star? The reason is that heavy nuclei have large electric charges. For example, the nucleus of a carbon atom has 6 positively charged protons and hence 6 times more positive electrical charge than a hydrogen nucleus (which has a single proton). This means that there are strong electric forces that tend to keep these positively charged nuclei apart. Only at the great speeds associated with extremely high temperatures can the nuclei travel fast enough to overcome their mutual electric charge repulsion and fuse together. To produce these very high temperatures at a star's center, the pressure must also be very high. Hence, only the largest stars have enough mass bearing down on the core to produce extremely high temperatures at the center.

As we discussed in Chapter 11, when a main-sequence star with a mass greater than about 0.4 $M_\odot$ uses up its available core hydrogen, it begins shell hydrogen fusion and enters a red-giant phase. But in stars whose overall mass is more than about 4 $M_\odot$, the carbon-oxygen core is more massive than the Chandrasekhar limit of 1.4 $M_\odot$, so degenerate-electron pressure from tightly packed atoms cannot itself prevent the core from contracting and heating. As a result, a high-mass star is able to enter a new phase of core thermonuclear reactions. When the central temperature of such a high-mass star reaches 600 million kelvins (6×10^8 K), the first of the new thermonuclear reactions begins with carbon. The thermonuclear process of carbon fusion consumes carbon nuclei (^{12}C, with 6 protons in each nucleus) and produces oxygen (^{16}O, 8 protons), neon (^{20}Ne, 10 protons), sodium (^{23}Na, 11 protons), and magnesium (^{23}Mg and ^{24}Mg, each with 12 protons).

If a star has an even larger main-sequence mass of about 8 $M_\odot$ or so (before mass ejection), even more thermonuclear reactions can take place. After the cessation of carbon fusion, the core will again contract, and the star's central temperature can rise to 1 billion kelvins (10^9 K). At this temperature, fusion of neon nuclei can begin. This uses up the neon accumulated from carbon fusion and further increases the concentrations of oxygen and magnesium in the star's core.

After neon fusion ends, the core will again contract, and oxygen nuclei will begin to fuse into silicon (^{28}Si, 14 protons) when the central temperature of the star reaches about 1.5 billion kelvins (1.5×10^9 K). Once oxygen fusion is over, the core will contract yet again. If the central temperature reaches

FIGURE 12-1 R I V U X G **Mass Loss from a Supergiant Star** At the heart of this nebulosity, called SMC N76, lies a supergiant star with a mass of at least 18 $M_\odot$. This star is losing mass at a rapid rate in a strong stellar wind. As this wind collides with the surrounding interstellar gas and dust, it creates the "bubble" shown here. SMC N76, which has an angular diameter of 130 arcsec, lies within the Small Magellanic Cloud, a small galaxy that orbits our Milky Way. It is about 60,600 pc (198,000 ly) distant. *(Y. Nazé, G. Rauw, J. Manfroid, and J.-M. Vreux, Liège Institute; Y.-H. Chu, U. of Illinois; and ESO)*

about 2.7 billion kelvins (2.7×10^9 K), silicon will begin to fuse, producing a variety of nuclei from sulfur (^{32}S, 16 protons) to iron (^{56}Fe, 26 protons) and nickel (^{56}Ni, 28 protons). While all of this is going on in the star's interior, at the surface the star is losing mass at a rapid rate (**Figure 12-1**).

As a high-mass star consumes increasingly heavier nuclei, the thermonuclear reactions produce a wider variety of products. For example, oxygen fusion produces not only silicon, but also magnesium (^{24}Mg, with 12 protons), phosphorus (^{31}P, with 15 protons), and sulfur (^{31}S and ^{32}S, each with 16 protons). Some thermonuclear reactions that create heavy elements also release neutrons. A neutron is like a proton except that it carries no electric charge. Therefore, neutrons are not repelled by positively charged nuclei, and so can easily collide and combine with them. This absorption of neutrons by nuclei creates many elements and isotopes that are not produced directly in fusion reactions.

ConceptCheck 12-1: If the core of our Sun was compressed to a much higher density and able to maintain a temperature of 1.5 billion kelvins (1.5×10^9 K) in its core, what would happen to the helium in its very center?

Supergiant Stars and Their Evolution

The increasing density and temperature of the core make each successive thermonuclear reaction more rapid than the one that preceded it. As an example, **Table 12-1** shows a theoretical calculation of the evolutionary stages for a star with a birth mass of 25 $M_\odot$. This calculation indicates that carbon fusion in such a star lasts for 600 years, neon fusion for 1 year, and oxygen fusion for only 6 months. The last, and briefest, stage of nuclear reactions is silicon fusion. The entire core supply of silicon in a 25-$M_\odot$ star is used up in only one day!

Table 12-1 Evolutionary Stages of a 25-$M_\odot$ Star

Stage	Core temperature	Core density (kg/m³)	Duration of stage
Hydrogen fusion	4×10^7	5×10^3	7×10^6 years
Helium fusion	2×10^8	7×10^5	7×10^5 years
Carbon fusion	6×10^8	2×10^8	600 years
Neon fusion	1.2×10^9	4×10^9	1 year
Oxygen fusion	1.5×10^9	10^{10}	6 months
Silicon fusion	2.7×10^9	3×10^{10}	1 day
Core collapse	5.4×10^9	3×10^{12}	¼ second
Core bounce	2.3×10^{10}	4×10^{17}	milliseconds
Explosive supernova	about 10^9	varies	10 seconds

Each stage of core fusion in a high-mass star generates a new shell of material around the core. After several such stages, the internal structure of a truly massive star—say, 25 to 30 $M_\odot$ or greater—resembles that of an onion (**Figure 12-2**). Because thermonuclear reactions can take place simultaneously in several shells, energy is released at such a rapid rate that the star's outer layers expand tremendously. The result is a **supergiant star,** whose luminosity and radius are much larger than those of a giant. Several of the brightest stars in the sky are supergiants, including Betelgeuse in the constellation Orion and Antares in the

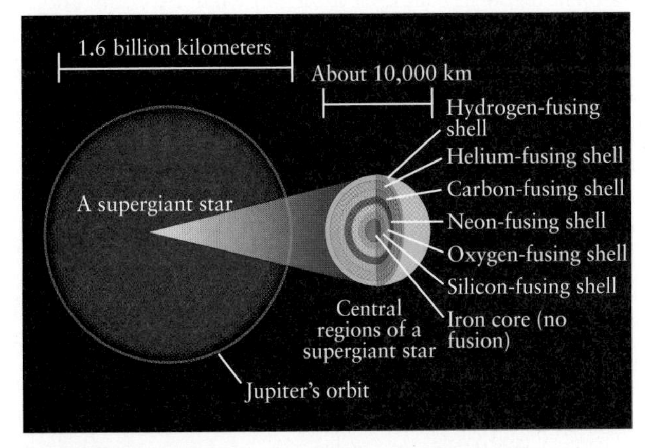

FIGURE 12-2 The Structure of an Old High-Mass Star Near the end of its life, a star with an initial mass greater than about 8 $M_\odot$ becomes a red supergiant. The star's overall size can be as large as Jupiter's orbit around the Sun. The star's energy comes from a series of concentric fusing shells, all combined within a volume roughly the same size as Earth. Thermonuclear reactions do not occur within the iron core, because fusion reactions that involve iron absorb energy rather than release it.

constellation Scorpius. (Figure 10-15 shows the locations of these stars on an H-R diagram.)

A supergiant star cannot keep adding shells to its "onion" structure forever, because the sequence of thermonuclear reactions cannot go on indefinitely. In order for an element to serve as a thermonuclear fuel, energy must be given off when its nuclei collide and fuse. This released energy is a result of the strong nuclear force of attraction that draws nucleons (neutrons and protons) together. However, protons also repel one another by the weaker electric force. As a result of this electric repulsion, adding extra protons to nuclei larger than iron, which has 26 protons, requires an *input* of energy rather than causing energy to be released. Nuclei of this size or larger cannot act as fuel for thermonuclear reactions. Hence, the sequence of fusion stages ends with silicon fusion. One of the products of silicon fusion is iron, and the result is a star with an iron-rich core in which no thermonuclear reactions take place (see Figure 12-2).

Shell fusion in the layers surrounding the iron-rich core consumes the star's remaining reserves of fuel. At this stage the entire energy-producing region of the star is contained in a volume no bigger than Earth, some 10^6 times smaller in radius than the overall size of the star. This state of affairs will soon come to an end, because the buildup of an inert, iron-rich core signals the impending violent death of a massive supergiant star.

ConceptCheck 12-2: How long does the neon fusion stage last in a Sunlike star?

The Violent End of a High-Mass Star

When energy production at the core of a massive star ends, a catastrophic collapse leads to a supernova explosion.

To understand what happens in a core-collapse supernova explosion, we must look deep inside a massive star at the end of its life. Of course, we cannot do this in actuality, because the interiors of stars are opaque. But astronomers have developed theoretical models based on what we know about the behavior of gases and atomic nuclei. The story that follows, while largely theoretical, describes our observations of supernovae fairly well.

The core of an aging, massive star gets progressively hotter as it contracts to ignite successive stages of thermonuclear fusion (Stage 1 in **Figure 12-3**). Remember that as the temperature of an object like a star increases, so does the energy of the photons it emits. When the temperature in the core of a massive star reaches a few hundred million kelvins, the photons are energetic enough to initiate a host of nuclear reactions that create neutrinos that escape, draining the star's energy.

To compensate for the energy drained by the neutrinos, the star must provide energy either by consuming more thermonuclear fuel, by contracting, or both. But when the star's core is converted into iron, no more energy-producing thermonuclear reactions are possible, and the only source of energy is contraction and rapid heating (Stage 2 in Figure 12-3).

Once a star with an original mass of about 8 $M_\odot$ or more develops an iron-rich core, the core contracts very rapidly, so that the core temperature skyrockets to 5×10^9 K within a tenth of a second. The gamma-ray photons emitted by the intensely hot core have so much energy that when they collide with iron nuclei, they begin to break the iron nuclei down into much smaller helium nuclei (^{4}He). As Table 12-1 shows, it takes a high-mass star millions of years and several stages of thermonuclear reactions to build up an iron core; within a fraction of a second, the degradation process undoes the result of those millions of years of reactions.

Within another tenth of a second, the core becomes so dense that the negatively charged electrons within it are forced to combine with the positively charged protons to produce electrically neutral neutrons. This process also releases a flood of neutrinos, denoted by the Greek letter ν (nu):

$$e^- + p \rightarrow n + \nu$$

At about 0.25 second after its rapid contraction begins, the core is less than 12 miles (20 km) in diameter and its enormous density is in excess of 4×10^{17} kg/m^3. This is tremendously dense and is the density with which neutrons and protons are packed together inside the nucleus of an atom. (If Earth were compressed to this density, it would be only 300 meters, or 1000 feet, in diameter.)

Matter at the density where neutrons and protons are packed together is extraordinarily difficult to compress to an even more dense state. Thus, when the density of the neutron-rich core begins to exceed this state, the core suddenly becomes very stiff and rigid. The core's contraction comes to a sudden halt, and the innermost part of the core actually bounces back and expands somewhat. This *core bounce* sends a powerful wave of pressure, like an unimaginably intense sound wave, outward into the outer core (Stage 3 in Figure 12-3).

During this critical stage, the cooling of the core has caused the pressure to decrease profoundly in the regions surrounding the core. Without pressure to hold it up against gravity, the material from these regions plunges inward with great speed and force. When this inward-moving material crashes down onto the rigid core, it encounters the outward-moving pressure wave. In just a fraction of a second, the material that fell onto the core begins to move back out toward the star's surface, propelled in part by the flood of neutrinos trying to escape from the star's core.

In this process, the material surrounding the core behaves somewhat like water boiling furiously in a heated pot. Rising bubbles of superheated gases deliver extra energy to the pressure wave, sustaining it and making it accelerate as it plows outward through the doomed star's outer layers. The wave soon reaches a speed greater than the speed of sound waves in the star's outer layers. When this happens, the wave becomes a *shock wave,* like the sonic boom produced by a supersonic airplane (Stage 4 in Figure 12-3).

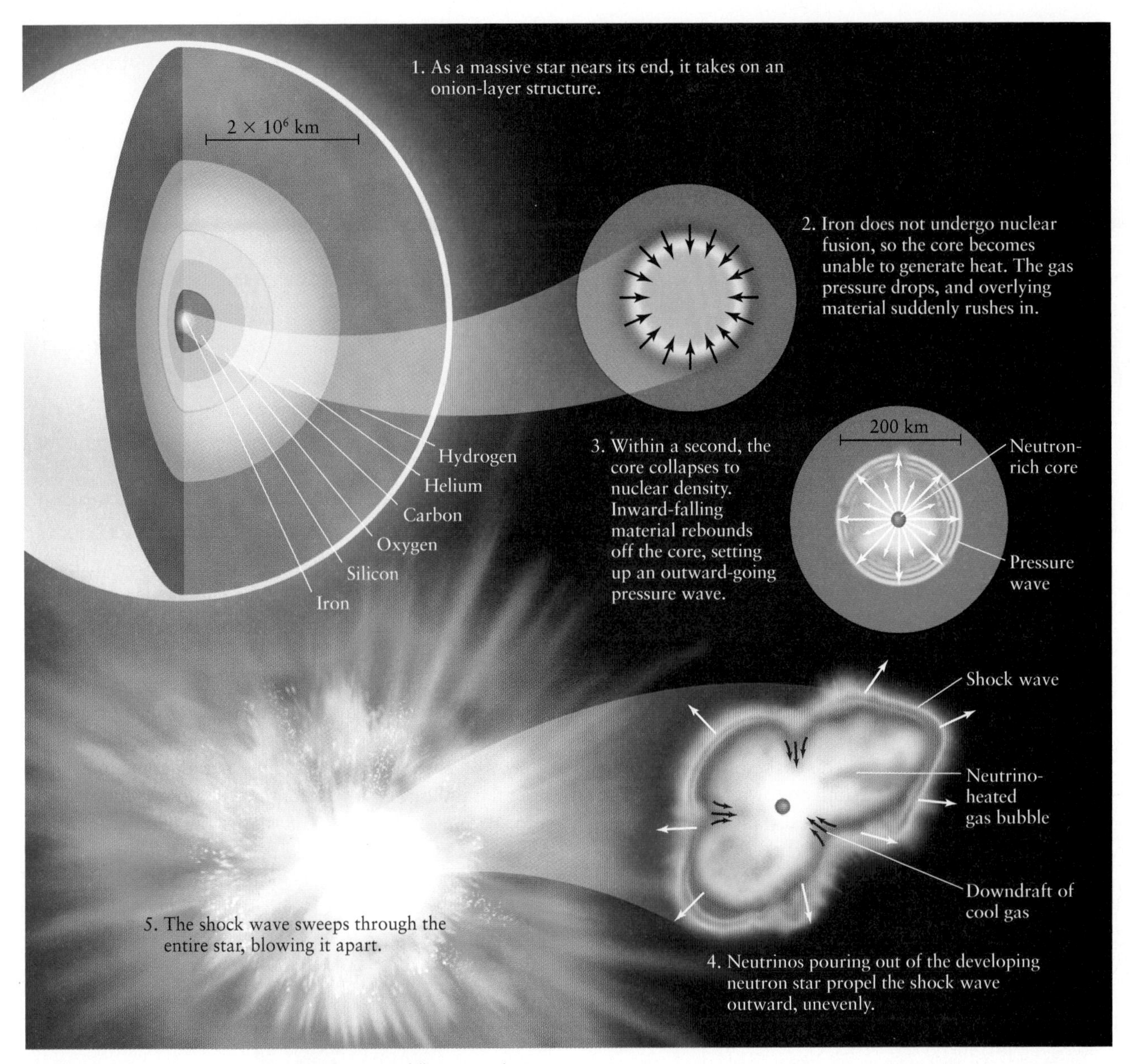

FIGURE 12-3 A Core-Collapse Supernova This series of illustrations depicts our understanding of the last day in the life of a star of more than about 8 $M_{\odot}$. *(Illustration by Don Dixon, adapted from Wolfgang Hillebrandt, Hans-Thomas Janka, and Ewald Müller, "How to Blow Up a Star," Scientific American, October 2006)*

A few hours after the event actually happens, the shock wave eventually reaches the star's surface, by which time the star's outer layers have begun to lift away from the core. When the star's outer layers thin out sufficiently, a portion of this energy escapes in a torrent of light (Stage 5 in Figure 12-3). The star has become a **core-collapse supernova** (*plural* **supernovae**).

CAUTION The energy released in a core-collapse supernova is incomprehensibly large—100 times more energy than the Sun has emitted due to thermonuclear reactions over its entire 4.57-billion-year history. However, it is important to recognize that the source of the supernova's energy release is *not* thermonuclear reactions. Rather, it is the *gravitational* energy released by the collapse of the core and by the inward fall of the star's outer layers. (You release gravitational energy when you fall off a diving board, and this released energy goes into making a big splash in the swimming pool.) The energy released by the collapse of the core reappears in the form of neutrinos; the fall of the outer layers provides the energy to power the nuclear reactions that generate the supernova's electromagnetic radiation. The amount of energy release from the supernova is so great because the star is so massive. Hence, the amount of material that falls inward is immense, it falls a great distance, and it is acted on by a strong gravitational pull as it falls.

ConceptCheck 12-3: Is the supernova resulting from the death of a supermassive star an outwardly directed explosion or an inwardly directed implosion?

FIGURE 12-4 R I V U X G **A Supernova in a Distant Galaxy** (a) Discovered on October 13, 1773, by Charles Messier, the Whirlpool Galaxy, also known as Messier Number 51 (M51), is more than 20 million ly away. (b) A progenitor star that later exploded into supernova SN 2005cs can barely be seen by the Hubble Space Telescope image just before the explosion. (c) The supernova is easily seen in the second image made January 11, 2005, just 12 days after it was first discovered. *(a: NASA, ESA, S. Beckwith [STScI], and The Hubble Heritage Team [STScI/AURA]; b and c: NASA, ESA, W. Li and A. Filippenko [UC Berkeley], S. Beckwith [STScI], and The Hubble Heritage Team [STScI/AURA])*

Supernova 1987A: A Close-up Look at the Death of a Massive Star

Go to Video 12-1

Supernovae can have a maximum luminosity as great as 10^9 $L_\odot$, rivaling the light output of an entire galaxy for a brief period. This makes it possible to see supernovae in galaxies far beyond our own Milky Way Galaxy, and indeed hundreds of these distant supernovae are observed each year (**Figure 12-4**).

On February 23, 1987, a supernova was discovered in the nearby Large Magellanic Cloud (LMC), a companion galaxy to our Milky Way some 168,000 ly from Earth. The supernova, designated SN 1987A because it was the first discovered that year, occurred near a region in the LMC called the Tarantula Nebula (**Figure 12-5**). The supernova was so bright that observers in the southern hemisphere could see it without a telescope. Perhaps surprisingly, the light from a supernova such as SN 1987A does not all come in a single brief flash; the outer layers continue to glow as they expand into space. For the first 20 days after the detonation of SN 1987A, its glow was powered primarily by the tremendous heat that the shock wave deposited in the star's outer layers. As the expanding gases cooled, the light energy began to

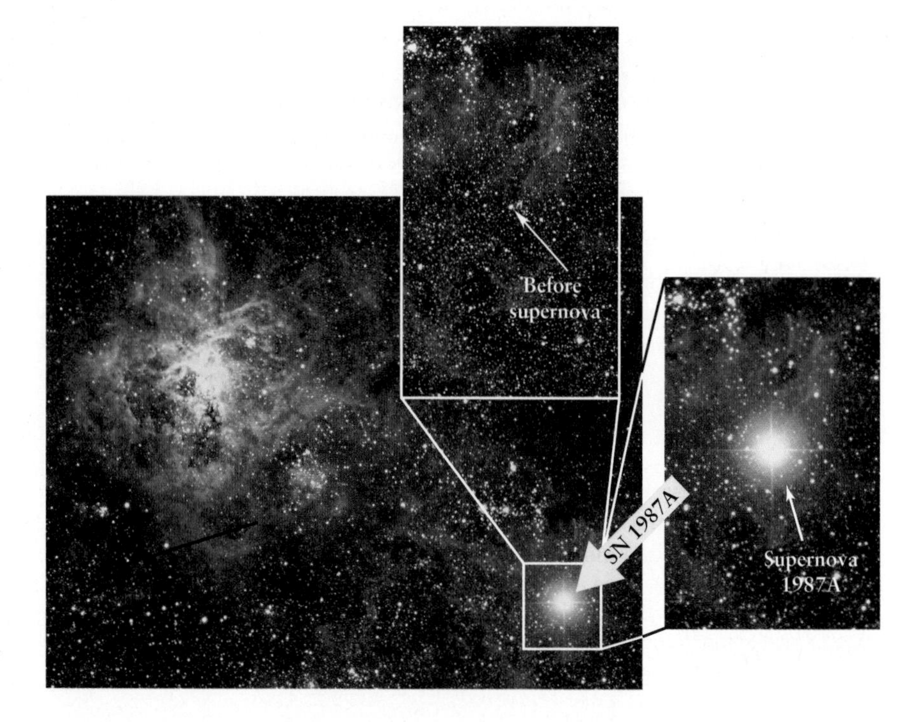

FIGURE 12-5 R I V U X G **Before and After Supernova 1987A Explosion** A supernova was discovered in a nearby galaxy called the Large Magellanic Cloud (LMC) in 1987. This photograph shows part of the Tarantula Nebula, which is the host for SN 1987A. At its maximum brightness, people in the southern hemisphere saw the supernova without a telescope, even though it was 168,000 light-years away. *(Australian Astronomical Observatory/ David Malin Images)*

(a) Supernova 1987A seen in 1996 R I V U X G

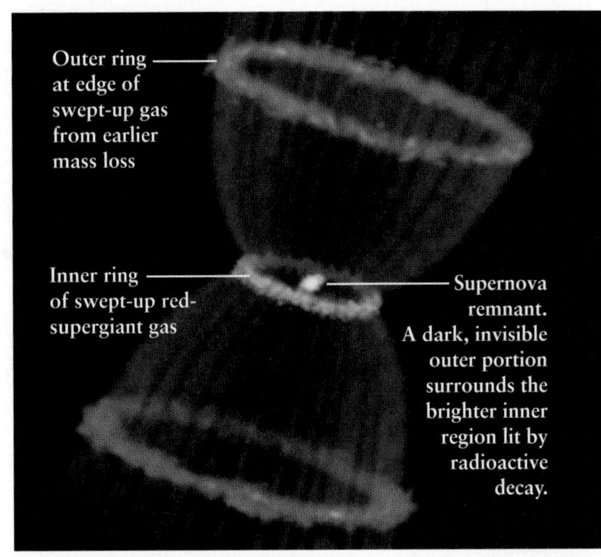

(b) An explanation of the rings

FIGURE 12-6
SN 1987A and Its "Three-Ring Circus"
(a) This true-color view from the Hubble Space Telescope shows three bright rings around SN 1987A.
(b) This drawing shows the probable origin of the rings. A wind from the progenitor star formed an hourglass-shaped shell surrounding the star. (Compare with Figure 11-29.) Ultraviolet light from the supernova explosion ionized ring-shaped regions in the shell, causing them to glow. *(Robert Kirshner and Peter Challis, Harvard-Smithsonian Center for Astrophysics; STScI)*

be provided by a different source—the decay of radioactive isotopes of cobalt, nickel, and titanium produced in the supernova explosion.

Astronomers have been able to pinpoint the specific isotopes involved because different radioactive nuclei emit gamma rays of different wavelengths when they decay. Thanks to the energy released during these radioactive decays, the brightness of SN 1987A actually *increased* for the first 85 days after the detonation, then settled into a slow decline as the radioactive isotopes were used up. The supernova remained visible to the naked eye for several months after the detonation.

Three and a half years after we observed SN 1987A explode, astronomers used the newly launched Hubble Space Telescope to obtain a picture of the supernova. To their surprise, the image showed a set of *three* glowing rings (**Figure 12-6*a***). These rings are relics of a hydrogen-rich outer atmosphere that was ejected by gentle stellar winds from the star when it was a red supergiant, about 20,000 years ago. This diffuse gas expanded in a hourglass shape (Figure 12-6*b*), because it was blocked from expanding around the star's equator either by a preexisting ring of gas or by the orbit of an as yet unseen companion star. The outer rings in Figure 12-6*a* are parts of the hourglass that were ionized by the initial flash of ultraviolet radiation from the supernova.

By the early years of the twenty-first century, the shock wave from the supernova was beginning to be seen to collide with the "waist" of the hourglass shown in Figure 12-6*b*. This collision is making the hourglass glow more brightly in visible wavelengths—though not enough, unfortunately, to make the supernova again visible to the naked eye—and emit copious radiation at X-ray and ultraviolet wavelengths.

ConceptCheck 12-4: How might the appearance of SN 1987A be different if the star had not been encircled by a ring of dust about its equator?

12-2 Core-collapse supernovae can leave behind remnants, neutron stars, and pulsars

There is considerable debris left over from the explosion of a core-collapse supernova. Star parts are scattered in every direction, as *remnants,* and the leftover crushed stellar core forms tiny objects, including *neutron stars* and *pulsars,* each with bizarre properties of their own. Let's consider each of these in turn; then, in the next section, we will turn to the most dramatic remnant of all: black holes.

Supernova Remnants

Astronomers find the remains from exploded massive stars' outer atmospheres, called **supernova remnants,** scattered across the sky. A beautiful example of a supernova remnant is the Veil Nebula, shown in **Figure 12-7.** The doomed star's outer layers were blasted into space with such violence that they are still traveling through the interstellar medium at supersonic speeds 15,000 years later. As this expanding shell of gas plows through space, it collides with atoms in the interstellar medium, exciting the gas and making it glow. We saw in Section 11-1 that the passage of a supernova remnant through the interstellar medium can trigger the formation of new stars, so the death of a single massive star can cause a host of new stars to be born.

An exploding supernova is bright for several months before fading from view.

Many supernova remnants are virtually invisible at the visible wavelengths our human eyes can see. However, when the expanding gases collide with the interstellar medium, they emit energy at a wide range of wavelengths, from X-rays through radio waves. As a rule, radio searches for supernova remnants are more fruitful than visible-light searches. Only two dozen supernova remnants have been found in visible-light images, but more than 100 remnants have been discovered by radio astronomers.

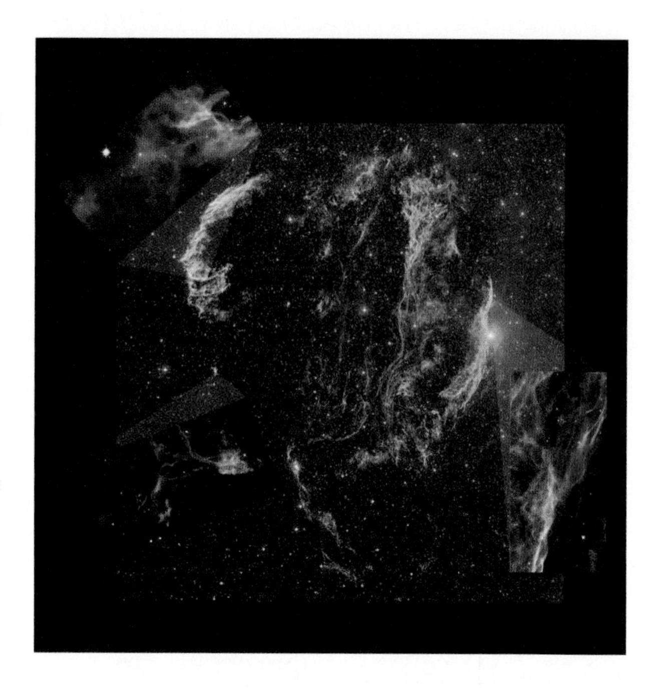

FIGURE 12-7 R I V U X G **The Veil Nebula** The entire shell of the Veil Nebula is too large to be viewed by the Hubble Space Telescope in a single observation. This figure is a collage of images where the background one is a ground-based observation of the Veil Nebula taken by the Digitized Sky Survey Project and the smaller sections are from the Hubble Space Telescope. *(NASA, ESA, the Hubble Heritage [STScI/AURA]-ESA/Hubble Collaboration, and the Digitized Sky Survey 2)*

In general, there are only a few supernova remnants in our local neighborhood. At first glance, this apparent lack of nearby supernovae may seem puzzling. From the frequency with which supernovae occur in distant galaxies, it is reasonable to suppose that a galaxy such as our own should have as many as five supernovae per century. Where have they been?

As we will learn when we study galaxies in Chapter 13, the plane of our Galaxy is where massive stars are born and supernovae explode. This region is so rich in interstellar dust, however, that we simply cannot see very far into space in the directions occupied by the Milky Way. In other words, supernovae probably do in fact erupt every few decades in remote parts of our Galaxy, but their detonations are hidden from our view by intervening interstellar matter.

The cores of the most massive stars are crushed in a supernova leaving behind small but exotic objects.

ConceptCheck 12-5: In which wavelength will an expanding supernova remnant most easily be seen if it is in a region of the galaxy with virtually no interstellar gas or dust around it?

Relics of the Fall: Neutron Stars

In some cases, a supernova remnant may be all that is left after some supernovae explode. In other cases of core-collapse supernovae, some portion of the original stellar core may remain. Depending on the mass of the core and the conditions within it during the collapse, it might leave behind a **neutron star,** an incredibly dense but incredibly small sphere composed primarily of neutrons.

CAUTION Although neutron stars and other relics of a previous stellar core can be part of the debris from a supernova explosion, they are *not* called "supernova remnants." That term is applied exclusively to the gas and dust that spreads away from the site of the supernova explosion.

On the morning of July 4, 1054, Yang Wei-T'e, the imperial astronomer to the Chinese court, made a startling discovery. Just a few minutes before sunrise, a new and dazzling object ascended above the eastern horizon. This "guest star," as Yang called it, was far brighter than Venus and more resplendent than any star he had ever seen.

Yang's records show that the "guest star" was so brilliant that it could easily be seen during broad daylight for the rest of July. Records from Constantinople (now Istanbul, Turkey) also describe this object, and works of art made by the Anasazi culture in the U.S. Southwest suggest that they may have seen it as well (**Figure 12-8**). Over the next 21 months, however, the "guest star" faded to invisibility.

We now know that the "guest star" of 1054 was actually a remarkable stellar transformation: A massive star some 6500 light-years away perished in a supernova explosion, leaving behind both a supernova remnant and a neutron star. Neutrons are a fundamental component of an atom with no positive or negative charge. Perhaps surprisingly, a positively charged proton and a negatively charged electron can combine to form a neutron (as well as a neutrino) in conditions of extraordinarily high temperature and pressure. This is exactly the environment that occurs during a core-collapse supernova.

In much the same way that white-dwarf stars are supported by degenerate-electron pressure due to electrons being packed as closely together as possible, a highly compact ball of neutrons would similarly produce a **degenerate-neutron pressure.** As such, a neutron star must be a rather weird object. If brought to Earth's surface, a single thimbleful of neutron-star matter would weigh 100 million tons! A star compacted to such densities must be very small. A 1.4-$M_\odot$

FIGURE 12-8 R I V U X G **A Supernova Pictograph?** This drawing in an eleventh-century structure in New Mexico shows a 10-pointed star next to a crescent. It may depict the scene on the morning of July 5, 1054, when a "guest star" appeared next to the waning crescent moon. *(Courtesy of National Parks Service)*

neutron star would have a diameter of only 12 miles (20 km), about the size of a moderate-sized city on Earth. Its surface gravity would be so strong that an object would have to travel at one-half the speed of light to escape into space. These conditions seemed outrageous until the late 1960s, when astronomers discovered pulsating radio sources.

ConceptCheck 12-6: Why are neutron stars smaller than white dwarfs?

Pulsars

A discovery in the 1960s stimulated interest in neutron stars. As a young graduate student at Cambridge University, Jocelyn Bell spent many months helping construct an array of radio antennas covering 4½ acres of the English countryside. The instrument was completed by the summer of 1967, and Bell and her colleagues in Anthony Hewish's research group began using it to scrutinize radio emissions from the sky. They were looking for radio sources that "twinkle" like stars; that is, they looked for random small fluctuations in brightness caused by the motion of gas between the source and the observer. What they discovered was something far more exotic.

While searching for random flickering, Bell noticed that the antennas had detected regular pulses of radio noise from one particular location in the sky. These radio pulses were arriving at regular intervals of 1.3373011 seconds—much more rapid than those of any other astronomical object known at that time. Indeed, they were so rapid and regular that the Cambridge team at first suspected that they might not be of natural origin. Instead, it was proposed that these pulses might be signals from an advanced alien civilization.

That possibility had to be discarded within a few months after several more of these pulsating radio sources, which came to be called **pulsars,** were discovered across the sky. In all cases, the rotational periods observed were extremely regular, ranging from about 0.25 second for the fastest to about 1.5 seconds for the slowest (**Figure 12-9**).

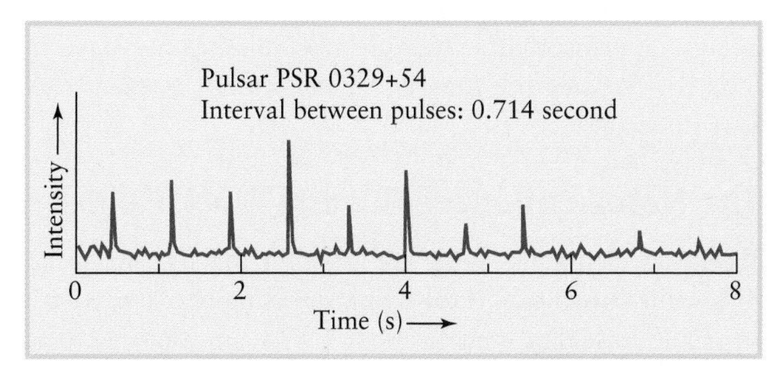

FIGURE 12-9 A Recording of a Pulsar This chart recording shows the intensity of radio emission from one of the first pulsars to be discovered. (The designation PSR 0329+54 means "pulsar at a right ascension of 03 hours 29 minutes and a declination of +54°.") Note that the interval between pulses is very regular, even though some pulses are weak and others are strong. *(Adapted from R. N. Manchester and J. H. Taylor)*

At the time of its discovery, the Crab pulsar was the fastest pulsar known to astronomers. Its period is 0.0333 second, which means that it flashes 1/0.0333 = 30 times each second. Today, the fastest known pulsar spins 716 times every second! An object that changes so quickly must be quite small. But what could these objects be? The only object that could be small enough to spin at such a high rate and not fling apart was a highly compact, high-density, tiny neutron star. Indeed, pulsars are rapidly rotating neutron stars with intense magnetic fields.

Because neutron stars are very small, they should also rotate rapidly. A typical star, such as our Sun, takes nearly a full month to rotate once about its axis. But just as an ice skater doing a pirouette speeds up when she pulls in her arms (see Figure 4-11), a collapsing star also speeds up as its size shrinks, following the principle of *conservation of angular momentum*. Indeed, if our Sun were suddenly compressed to the size of a neutron star, it would spin nearly 1000 times per second! Because neutron stars are so small and dense, they can spin this rapidly without flying apart.

The small size of neutron stars also implies that they have intense magnetic fields. It seems safe to say that every star possesses some magnetic field, but typically the strength of this field is quite low. The magnetic field of a main-sequence star is spread out over billions of square miles of the star's surface. However, if such a star collapses down to a neutron star, its surface area (which is proportional to the square of its radius) shrinks by a factor of about 10^{10}. The magnetic field, which is bonded to the star's ionized gases, becomes concentrated onto an area 10^{10} times smaller than before the collapse, and thus the field strength increases by a factor of 10^{10}. In terms of spectroscopic observations, a star's magnetic field splits one of its spectral lines into two or more lines whose spacing reveals the field's strength. As it turns out, the magnetic fields surrounding typical neutron stars are tens of million times stronger than Earth's magnetic field.

The magnetic field of a neutron star makes it possible for the star to radiate pulses of energy toward our telescopes. As it spins, charged particles are released from the star, not in just any direction, but along preferred directions constrained by the neutron star's magnetic fields. For observers here on Earth, we see a neutron star as flashing on and off as the brightest portions rotate toward and away from us.

A rapidly spinning neutron star is about the same size as a large city on Earth.

ANALOGY A rotating, magnetized neutron star is somewhat like a lighthouse beacon. As the star rotates, the beams of radiation sweep around the sky. If at some point during the rotation one of those beams happens to point toward Earth, as shown in **Figure 12-10*a*,** we will detect a brief flash as the beam sweeps over us. At other points during the rotation, the beam will be pointed away from Earth, and the radiation from the neutron star will appear to have turned off (Figure 12-10*b*). Hence, a radio telescope will detect regular pulses of radiation, with one pulse being received for each rotation of the neutron star.

FIGURE 12-10 A Rotating, Magnetized Neutron Star Charged particles are accelerated near a magnetized neutron star's magnetic poles (labeled N and S), producing two oppositely directed beams of radiation. If the star's magnetic axis (a line that connects the north and south magnetic poles) is tilted at an angle from the axis of rotation, as shown here, the beams sweep around the sky as the star rotates. If Earth happens to lie in the path of one of the beams, we detect radiation that appears to pulse (a) on and (b) off.

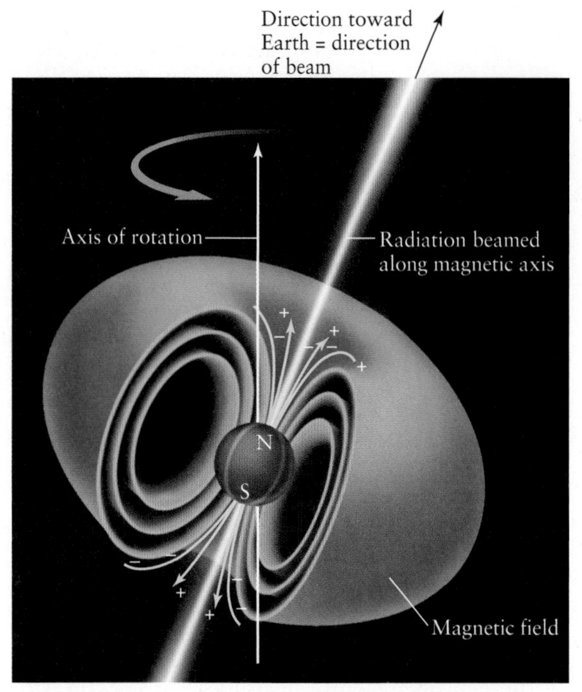

(a) One of the beams from the rotating neutron star is aimed toward Earth: We detect a pulse of radiation.

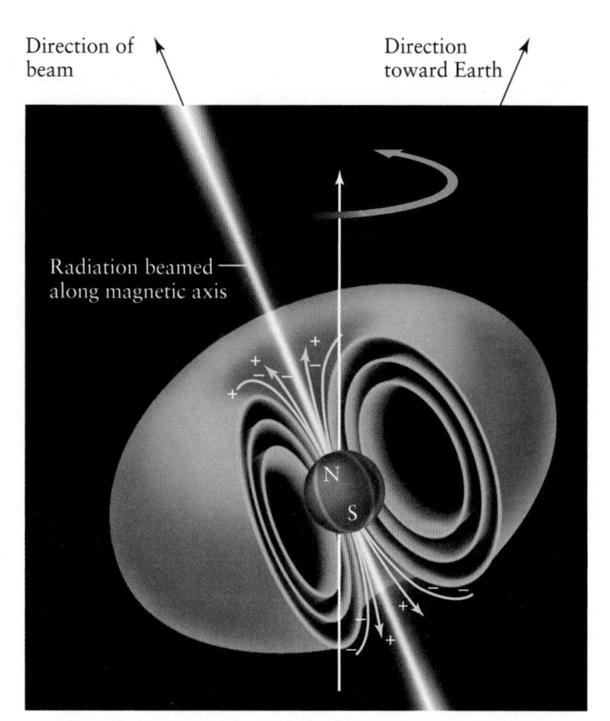

(b) Half a rotation later, neither beam is aimed toward Earth: We detect that the radiation is "off."

CAUTION The name *pulsar* may lead you to think that the source of radio waves is actually pulsing. But in the model just described, this is not the case at all. Instead, beams of radiation are emitted continuously from the magnetic poles of the neutron star. The pulsing that astronomers detect here on Earth is simply a result of the rapid rotation of the neutron star, which brings one of the beams periodically into our line of sight, as Figure 12-10 shows. In this sense, the analogy between a pulsar and a lighthouse beacon is a very close one.

ConceptCheck 12-7: **What is the difference between a pulsar and a neutron star?**

Pulsars Gradually Slow Down as They Radiate Energy into Space

One wonders if it is possible to determine the age of a pulsar. In other words, how long has it been since its original star went supernova? Because a spinning neutron star slows down as it radiates away its rotational energy, it follows that an old neutron star should be spinning more slowly than a young one. This leads us to a general rule:

An isolated pulsar slows down as it ages, so its pulse period increases.

From a pulsar's period (which increases as it ages) and the rate at which its period is increasing (which decreases as it ages), astronomers can estimate the age of a pulsar. This notion tends to work, as long as the pulsar is relatively unaffected by nearby objects. Indeed, in some unusual scenarios, like when pulsars are members of binary star systems, pulsars can be reaccelerated to truly dizzying rotation speeds.

ConceptCheck 12-8: **If two pulsars have differing periods of pulsation, which of the two is the older?**

12-3 Black holes are created in the death throes of the most massive of stars

Go to Video 12-2

If neutron stars result from the crushing collapse of stars with masses of about 4 to 8 times larger than the mass of our Sun, what about even larger stars? To make predictions about tremendous amounts of mass being compressed into the tiniest of volumes, we have to look to the scientific theories of the famous physicist Albert Einstein.

The Nature of Space Around Black Holes

We have seen that if a dying star is not too massive, it ends up as a white-dwarf star. If the dying star is more massive than the Chandrasekhar limit of about 1.4 $M_\odot$, it cannot exist as a stable white-dwarf star and, instead, shrinks down to form a neutron star. But if the dying star has more mass than the maximum permissible for a neutron star, about 2 to 3 $M_\odot$, not even the internal pressure of neutrons can hold the star up against its own gravity, and the star contracts rapidly.

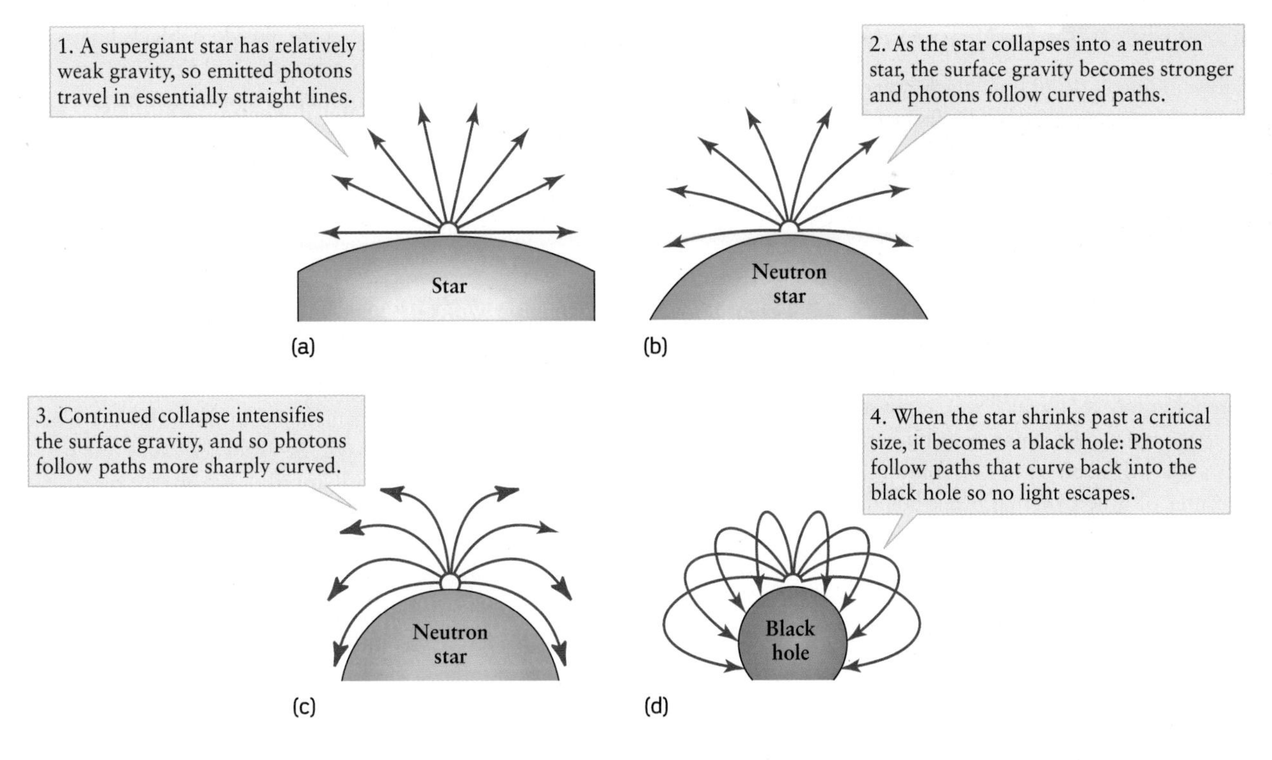

FIGURE 12-11 The Formation of a Black Hole (a, b, c) These illustrations show the steps leading up to the formation of a black hole from a dying star. (d) When the star becomes a black hole, not even photons emitted directly upward from the surface can escape; they undergo an infinite gravitational redshift and disappear.

As the core of a collapsing massive star is crushed by infalling matter, the old core becomes compressed to enormous densities. What is odd here is that in this case, the strength of gravity at the old core's surface of this rapidly shrinking sphere also increases dramatically. According to Einstein's general theory of relativity, the nature of outer space itself is altered. In locations of extremely high gravity, Einstein imagined that the actual fabric of space immediately surrounding the star becomes so highly curved that it closes on itself (**Figure 12-11**). If this were to happen, even massless photons of light flying outward at an angle from the star's surface arc back inward. Perhaps even stranger, photons that fly straight outward undergo such a strong shift in their observed wavelength due to gravity that they lose all their energy and actually cease to exist.

In this condition, strange things indeed start to become normal. Not light nor anything else is able to escape the gravitational attraction of the crushed core. An object from which neither matter nor light (electromagnetic radiation) can escape is called a **black hole.** In a sense, a hole is punched in the fabric of the universe, and the dying star seems to disappears into this cavity (**Figure 12-12**). None of the star's mass is lost when it collapses to form a black hole, and a black hole's gravitational influence can still be felt by other objects.

CAUTION Some low-quality science-fiction movies and books suggest that black holes are evil things that go around gobbling up everything in the universe. Not so! The bizarre effects created by a highly warped location in the universe are limited to a region quite near the hole. For example, the effects of Einstein's general theory of relativity predominate only within 600 miles (1000 km) of a 10-$M_\odot$ black hole. Beyond 600 miles, gravity is weak enough that our everyday experiences with physics can adequately describe everything.

Perhaps the most remarkable aspect of black holes is that they really exist! As we will see, astronomers have located a number of black holes with masses a few times that of the Sun. What is truly amazing is that they have also discovered many truly immense black holes containing millions or billions of solar masses.

ConceptCheck 12-9: If our Sun were to suddenly compress into a black hole of the same mass, what would happen to the planets' orbits?

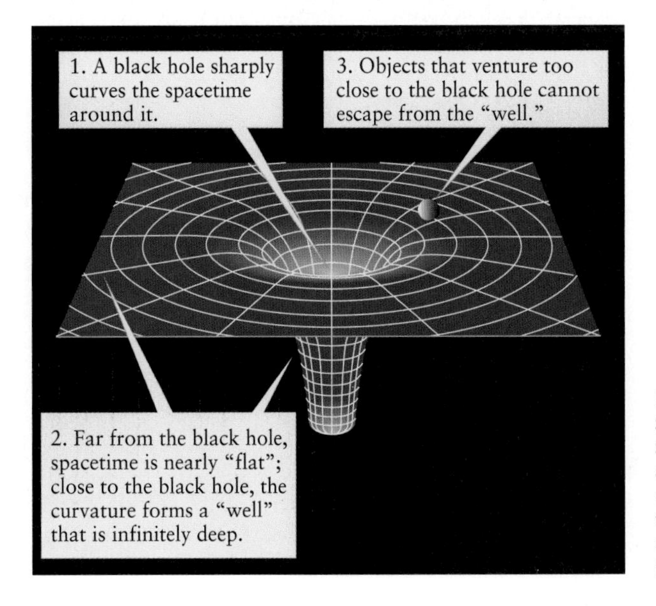

FIGURE 12-12 Space Curves and Time Slows Down Around a Black Hole This diagram suggests how the very fabric of space and time are distorted by a black hole's mass.

FIGURE 12-14 X-rays Emitted by Heated Gas An artist's illustration of a black hole and a normal star shown in blue. Gas is being pulled away from the star and falling onto a red disk spinning around the black hole. Some of this gas is heated as it spirals toward the black hole, generating X-ray amounts of light along the way. *(NASA/CXC/M. Weiss; X-ray spectrum: NASA/CXC/U.Michigan/J. Miller et al.)*

The first sign of such emissions from a binary system with a black hole came shortly after the launch of the *Uhuru* X-ray–detecting satellite in 1971. Astronomers became intrigued with an X-ray source designated Cygnus X-1. The X-rays emitted from Cygnus X-1 are highly variable and irregular; they flicker on timescales as short as one-hundredth of a second. One of the fundamental concepts in physics is that nothing can travel faster than the speed of light. Because of this limitation, an object cannot flicker faster than the time required for light to travel across the object. Because light travels 3000 km in a hundredth of a second, Cygnus X-1 can be no more than 1800 miles (3000 km), or about one-quarter the size of Earth. Yet other observations show that Cygnus X-1 must have a mass of at least 7 solar masses, ruling out either a white dwarf or a neutron star. Only material moving quickly toward a black hole could cause such highly variable X-ray emissions.

It is important to emphasize that the X-rays do not come from the black hole itself. Rather, gas captured from a nearby star goes into orbit about the hole, forming an accretion disk about 2.5 million miles (4×10^6 km) in diameter (**Figure 12-14**). As material in the disk gradually spirals in toward the hole, friction heats the gas to temperatures approaching 2×10^6 K. In the final 125 miles (200 km) above the hole's "surface," these extremely hot gases emit the X-rays that our satellites detect. Presumably the X-ray flickering is caused by small hot spots on the rapidly rotating inner edge of the accretion disk. In this way, the black hole's existence is announced by doomed gases just before they plunge to oblivion.

Black holes can only be "seen" by the effects they have on surrounding matter.

ConceptCheck 12-11: If a gravitationally bound binary system now contains one black hole and one red giant, which of the two original stars was the more massive?

Supermassive Black Holes

Go to Video 12-4

Stellar evolution makes it possible to form black holes with masses several times that of the Sun. But calculations suggest other ways to form black holes that are either extremely large or extraordinarily small.

Galaxies of stars formed in the early universe by the coalescence of gas clouds. During this formation process, some of a galaxy's gas could have plunged straight toward the center of the galaxy, where it collected and compressed together under its own gravity. If this central clump became sufficiently dense, it could have formed a black hole. However, this black hole would be very different from ones created through the core collapse of a dying massive star. Instead, galaxies have masses of 10^{11} $M_\odot$, so even a tiny percentage of the galaxy's gas collected at its center would give rise to a **supermassive black hole** with a truly stupendous mass.

Since the 1970s, astronomers have found evidence suggesting the existence of supermassive black holes in other galaxies. With the advent of a new generation of optical telescopes in the 1990s, it became possible to see the environments of these suspected black holes with unprecedented detail. **Figure 12-15** shows galaxy NGC 3627 as a mosaic of images from a variety of telescopes. Streams of gas jetting

FIGURE 12-15 R I V U X G **A Supermassive Black Hole** The spiral galaxy NGC 3627 is located about 30 million ly from Earth. This composite image is created by combining X-ray data from NASA's Chandra X-ray Observatory (blue), infrared data from the Spitzer Space Telescope (red), visible light from the Hubble Space Telescope and the Very Large Telescope (yellow). The inset shows the central region containing a bright X-ray source powered by material falling onto a supermassive black hole. *(NASA/CXC/Ohio State Univ./C. Grier et al.; optical: NASA/STScI, ESO/WFI; infrared: NASA/JPL-Caltech)*

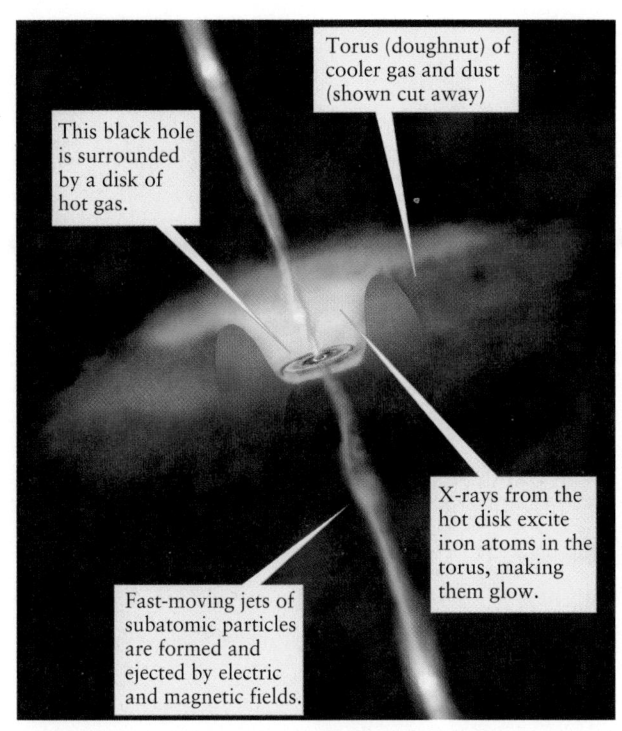

FIGURE 12-16 Environment of an Accreting Black Hole If a black hole is rotating, it can generate strong electric and magnetic fields in its immediate vicinity. These fields draw material from the accretion disk around the black hole and accelerate it into oppositely directed jets along the black hole's rotation axis. This illustration also shows other features of the material surrounding such a black hole. *(CXC/M. Weiss)*

away from the core are visible using X-ray telescopes and are seen moving perpendicular to the plane of the disk, as in **Figure 12-16.**

By measuring the Doppler shifts of light coming from the two sides of the core, astronomers found that the heated material was orbiting around the bright object at the center of the disk at speeds of hundreds of miles per second. Given the size of the material's orbit, and using Newton's form of Kepler's third law, they were able to calculate the mass of the central bright object. The answer is an amazing 1.2 *billion* (1.2×10^9) solar masses! What is more, the observations show that this object can be no larger than our solar system. The only possible explanation is that the object at the center of NGC 3627 is a black hole of super mass.

Dozens of other black holes in the centers of galaxies have been identified by their gravitational effect on surrounding gas and dust. Surveys of galaxies have shown that supermassive black holes are not at all unusual; most large galaxies appear to have them at their centers. As we will see in the next chapter, a black hole with several million solar masses lies at the center of our own Milky Way Galaxy, some 26,000 ly from Earth.

ConceptCheck 12-12: What is the best evidence for supermassive black holes at galactic centers?

12-5 White dwarfs and pulsars in close binary systems can become novae, bursters, and supernovae

Go to Video 12-5

Black holes are indeed fantastic objects. Even stranger, other exotic phenomena can occur when a stellar corpse is part of a gravitationally bound binary star system. One example is a **nova** (*plural* **novae**), in which a faint star suddenly brightens by a factor of 10^4 to 10^8 over a few days or hours, reaching a peak luminosity of about $10^5\ L_\odot$. By contrast, a *supernova* has a peak luminosity of about $10^9\ L_\odot$.

A nova's abrupt rise in brightness is followed by a gradual decline that may stretch out over several months or more. Every year, two or three novae are observed to flare up in our Galaxy, and several dozen more are thought to take place in remote regions of the Galaxy that are obscured from our view by interstellar dust.

Type Ia Supernovae: Detonating a White Dwarf

A supernova is not just a bright nova; they are formed entirely differently.

We need to take a step back from novae for a moment and talk about supernovae. Supernovae that are caused by a core collapse of a massive star of the sort we described earlier are best described as **Type II supernovae.** They are caused by the death of highly evolved massive stars that still have ample hydrogen in their atmospheres when they explode. When the star explodes, the hydrogen atoms are excited and glow prominently, producing hydrogen emission lines. SN 1987A is an example of Type II supernovae.

If there are Type II supernovae, that must mean that there is another category of exploding stars. The other type has no hydrogen lines in the spectrum. This category is called **Type I supernova.** Type I supernovae are further divided into three important subclasses, Type Ia, Ib, and Ic. Although the others are interesting, we are going to focus exclusively on supernovae Type Ia, because it is the most relevant in developing our understanding of the size and structure of our universe.

Type Ia supernovae are thought to result from the thermonuclear explosion of a white-dwarf star. This may sound contradictory, because we saw in Chapter 11 that white-dwarf stars have no thermonuclear reactions going on in their interiors. But these reactions *can* occur if a carbon-oxygen–rich white dwarf is in a close, semidetached binary system with a red-giant star.

Figure 12-17 shows the likely series of events that leads to a Type Ia supernova. Stage 1 in this figure shows a close binary system in which both stars have less than $4\ M_\odot$. The more massive star on the left evolves more rapidly than its less massive companion and eventually becomes a white dwarf. As the companion evolves and its outer layers expand, it

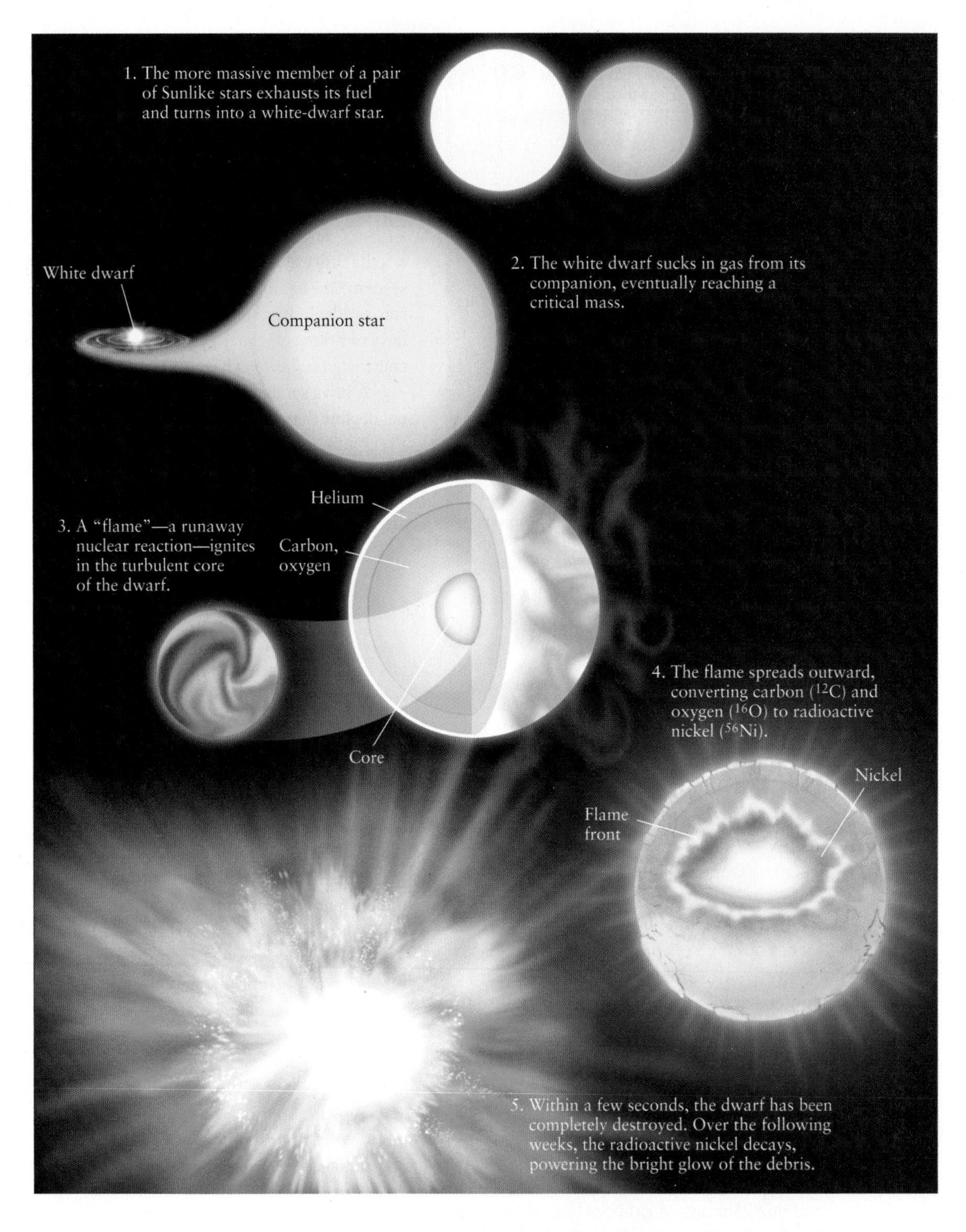

FIGURE 12-17 A Type Ia Supernova This series of illustrations depicts our understanding of how a white dwarf in a close binary system can undergo a sudden nuclear detonation that destroys it completely. Such a cataclysmic event is called a Type Ia supernova or thermonuclear supernova. *(Illustration by Don Dixon, adapted from Wolfgang Hillebrandt, Hans-Thomas Janka, and Ewald Müller, "How to Blow Up a Star," Scientific American, October 2006)*

dumps gas from its outer layers onto the white dwarf (Stage 2 in Figure 12-17). When the total mass of the white dwarf approaches the Chandrasekhar limit, the increased pressure applied to the white dwarf's interior causes carbon fusion to begin there (Stage 3 in Figure 12-17). Hence, the interior temperature of the white dwarf increases.

If the white dwarf were made of ordinary matter, the temperature increase would cause a further increase in pressure, the white dwarf would expand and cool, and the carbon-fusing reactions would abate. But because the white dwarf is composed of degenerate matter, this "safety valve" between temperature and pressure does not operate. Instead, the increased temperature just makes the reactions proceed at an ever-increasing rate, in a catastrophic runaway process reminiscent of the helium flash in low-mass stars (Stage 4 in Figure 12-17). The reaction spreads rapidly outward from the white dwarf's center, with its leading edge (called the *flame front*) being propelled by convection and turbulence in a manner analogous to what happens to the shock wave in a core-collapse supernova. Within seconds the white dwarf blows apart, dispersing 100% of its mass into space (Stage 5 in Figure 12-17).

Before exploding, the white dwarf contained primarily carbon and oxygen and almost no hydrogen or helium, which explains the absence of hydrogen and helium lines in the spectrum of the resulting supernova. Silicon is a byproduct of the carbon-fusing reaction and gives rise to the silicon absorption line characteristic of Type Ia supernovae.

CAUTION Different types of supernovae have fundamentally different energy sources. The Type II core-collapse supernovae are powered by *gravitational* energy released as the star's iron-rich core and outer layers fall inward. Type Ia supernovae, by contrast, are powered by *nuclear* energy released in the explosive thermonuclear fusion of a white-dwarf star. While Type Ia supernovae typically emit more energy in the form of electromagnetic radiation than do supernovae of other types, they do not emit copious numbers of neutrinos because there is no core collapse. If we include the energy emitted in the form of neutrinos, the most luminous supernovae by far are those of Type II.

ConceptCheck 12-13: If a white dwarf has no energy source, where does it get the raw material for fuel to become a Type Ia supernova?

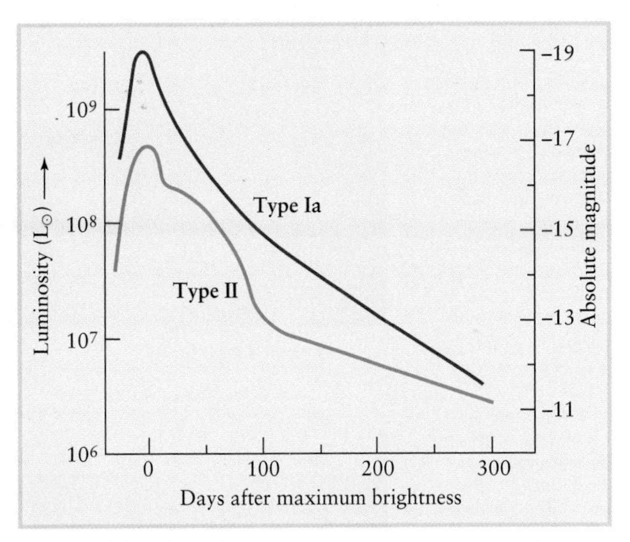

FIGURE 12-18 Supernova Light Curves A Type Ia supernova reaches maximum brightness in about a day, followed by a gradual decline in brightness. A Type II supernova reaches a maximum brightness less than that of a Type Ia supernova and usually has alternating intervals of steep and gradual declines.

The Decay of a Supernova: Light Curves

In addition to the differences in the amount of hydrogen in their spectra, different types of supernovae can be distinguished by differences in how their brightness suddenly increases then slowly decreases (**Figure 12-18**). A graph of the brightness of a supernova versus time is known as a **light curve.** All supernovae begin with a sudden rise in brightness that occurs in less than a day. After reaching peak luminosity, Type Ia supernovae settle into a steady, gradual decline in luminosity. An example is the supernova of 1006, which is thought to have been of Type Ia. This supernova, which at its peak was more than 200 times brighter than any other star in the sky, took three years to fade into invisibility. By contrast, the Type II light curve has a steplike appearance caused by alternating periods of steep and gradual declines in brightness.

For all supernova types, the energy source during the period of declining brightness is the decay of radioactive isotopes produced during the supernova explosion. Because a different set of thermonuclear reactions occurs for each type of supernova, each type produces a unique set of isotopes that decay at different rates. This helps explain the distinctive light curves for different supernova types.

For the same reason, each type of supernova ejects a somewhat different mix of elements into the interstellar medium. As an example, Type Ia supernovae are primarily responsible for the elements near iron in the periodic table, because they generate these elements in more copious quantities than Type II supernovae.

A number of astronomers are now measuring the distances to remote galaxies by looking for Type Ia supernovae in those galaxies. This is possible because there is a simple relationship between the rate at which a Type Ia supernova fades away and its peak luminosity: The slower it fades, the greater its luminosity. Hence, by observing how rapidly a distant Type Ia supernova fades, astronomers can determine its peak luminosity. A measurement of the supernova's peak apparent brightness then tells us (through the inverse-square law) the distance to the supernova and, therefore, the distance to the supernova's host galaxy. The tremendous luminosity of Type Ia supernovae allows this method to be used for galaxies more than 10^9 light-years distant. In a later chapter, we will learn what such studies tell us about the size and evolution of the universe as a whole.

ConceptCheck 12-14: Which is brighter, a Type Ia or a Type II supernova?

ConceptCheck 12-15: Which has a longer-lived brightness, a Type Ia or a Type II supernova?

Novae and White Dwarfs

In the 1950s, painstaking observations of numerous novae at the University of California's Lick Observatory led to the conclusion that all novae are closely orbiting members of

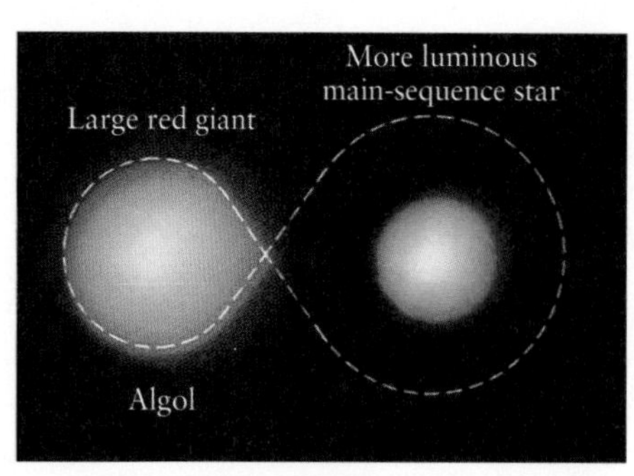

(a) A semidetached binary

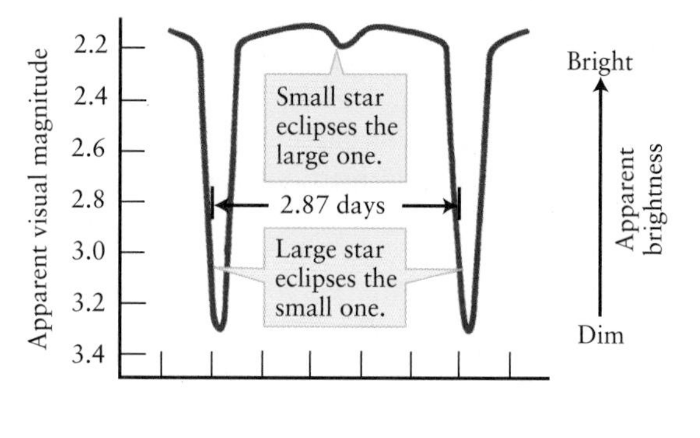

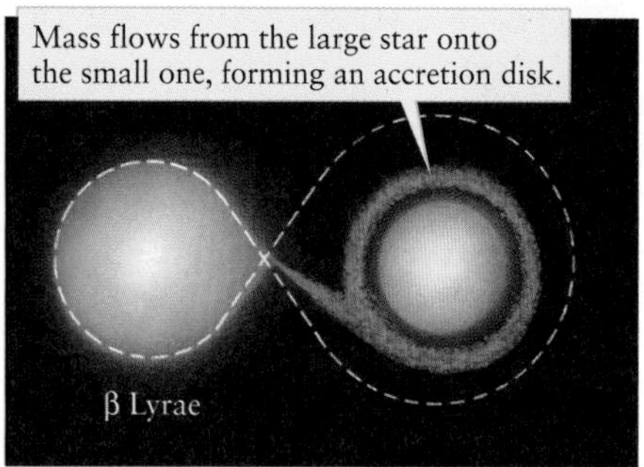

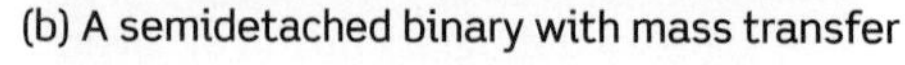

(b) A semidetached binary with mass transfer

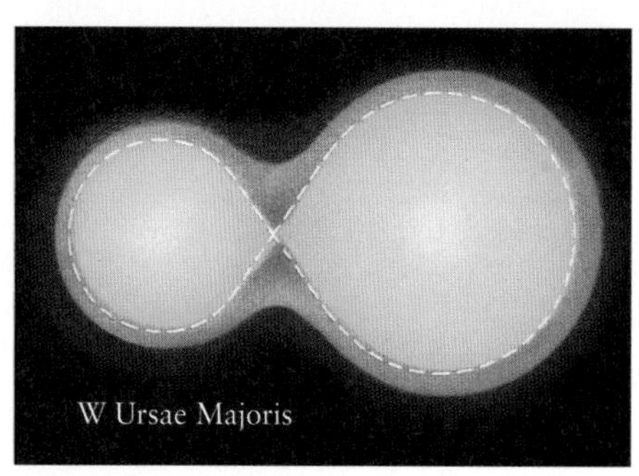

(c) An overcontact binary

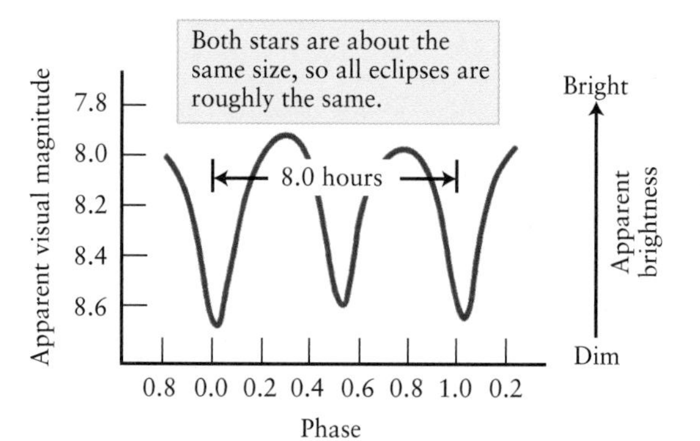

FIGURE 12-19 **Three Eclipsing Binaries** (a) Algol is a semidetached binary. The deep eclipse occurs when the large red-giant star blocks the light from the smaller but more luminous main-sequence star. (b) β Lyrae's light curve is also at its lowest when the larger star completely eclipses the smaller one. Half an orbital period later, the smaller star partially eclipses the larger one, making a shallower dip in the light curve. (c) W Ursae Majoris is an overcontact binary in which both stars overfill their Roche lobes. The extremely short period of this binary indicates that the two stars are very close to each other.

binary star systems containing a white dwarf. Gradual mass transfer from the ordinary companion star deposits fresh hydrogen onto the second star. **Figure 12-19** is a schematic diagram illustrating this sort of mass transfer.

Because of the white dwarf's strong gravity, this hydrogen is compressed into a dense layer covering the hot surface of the white dwarf. As more gas is deposited and compressed, the temperature in the hydrogen layer increases. When the temperature reaches about 10^7 K, hydrogen fusion ignites throughout the gas layer, embroiling the white dwarf's surface in a thermonuclear holocaust that we see as a nova (**Figure 12-20**).

CAUTION It is important to understand the similarities and differences between novae and the Type Ia supernovae that we described earlier. Both kinds of celestial explosions are thought to occur in close binary systems where one of the stars is a white dwarf. But, as befits their name, supernovae are much more energetic. A Type Ia supernova explosion radiates 10^{44} joules of energy into space, while the corresponding figure for a typical nova is 10^{37} joules. (To be fair to novae, this relatively paltry figure is as much energy as our Sun emits in 1000 years.) The difference is thought to be that in a Type Ia supernova, the white dwarf accretes much more mass from its companion. This added mass causes so much compression that nuclear reactions can take place *inside* the white dwarf. Eventually, these reactions blow the white dwarf completely apart. In a nova, by stark contrast, nuclear reactions occur only within the accreted material. The reaction is more sedate because it takes place only on the white dwarf's surface—perhaps because the accretion rate is less or because the white dwarf had less mass in the first place—leaving the white dwarf intact for the process to repeat. As an example, the star RS Ophiuchi erupted as

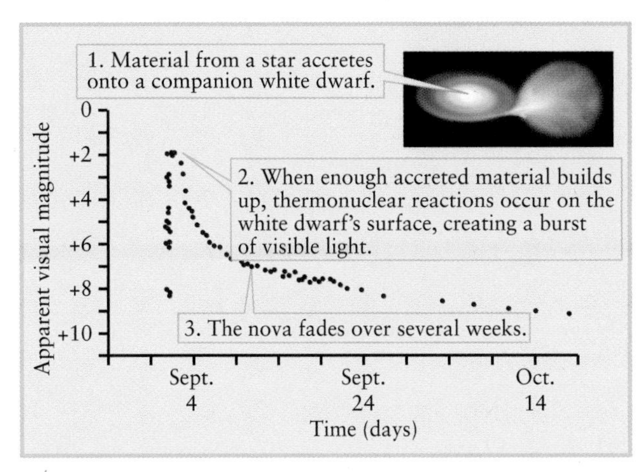

FIGURE 12-20 The Light Curve of a Nova This illustration and graph show the history of Nova Cygni 1975, a typical nova. Its rapid rise and gradual decline in apparent brightness are characteristic of all novae. This nova, also designated V1500 Cyg, was easily visible to the naked eye (that is, was brighter than an apparent magnitude of +6) for nearly a week. *(Illustration courtesy NASA/CXC/M. Weiss)*

a nova in 1898, then put in repeat performances in 1933, 1958, 1967, 1985, and 2006. By contrast, a given supergiant star can only be a Type Ia supernova once, because it is destroyed.

X-ray Bursters and Neutron Stars

A surface explosion similar to a nova also occurs with neutron stars. In 1975, it was discovered that some objects in the sky emit sudden, powerful bursts of X-rays. **Figure 12-21** shows the record of a typical burst. The source emits X-rays at a constant low level until suddenly, without warning, there is an abrupt increase in X-rays, followed by a gradual decline. An entire burst typically lasts for only 20 seconds. Unlike pulsating X-ray sources, there is a fairly long interval of hours or days between bursts. Sources that behave in this fashion are known as **X-ray bursters.** Several dozen X-ray bursters have been discovered in our Galaxy.

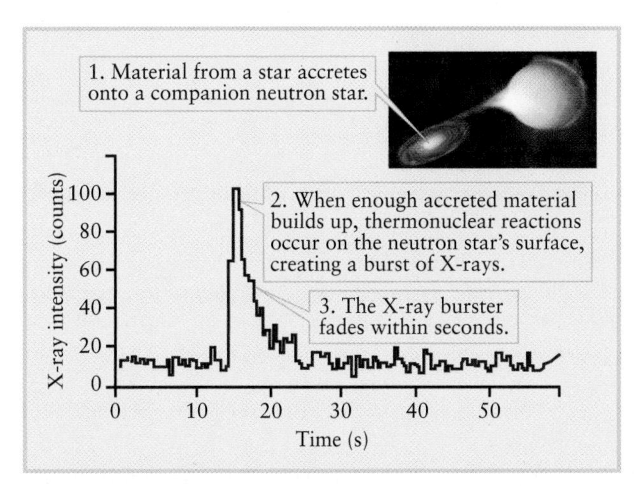

FIGURE 12-21 The Light Curve of an X-ray Burster This illustration and graph show the history of a typical X-ray burster. A burster emits a constant low intensity of X-rays interspersed with occasional powerful X-ray bursts. This burst was recorded on September 28, 1975, by an Earth-orbiting X-ray telescope. Contrast this figure with Figure 12-20, which shows a typical nova. *(Data adapted from W. H. G. Lewin; illustration courtesy NASA/CXC/M. Weiss)*

X-ray bursters, like novae, are thought to involve close binaries whose stars are engaged in mass transfer. With a burster, however, the stellar corpse is a neutron star rather than a white dwarf. Gases escaping from the ordinary companion star fall onto the neutron star. The X-ray burster's magnetic field is probably not strong enough to funnel the falling material toward the magnetic poles, so the gases are distributed more evenly over the surface of the neutron star. The energy released as these gases crash down onto the neutron star's surface produces the low-level X-rays that are continually emitted by the burster.

Most of the gas falling onto the neutron star is hydrogen, which the star's powerful gravity compresses against its hot surface. In fact, temperatures and pressures in this accreting layer become so high that the arriving hydrogen is immediately converted into helium by hydrogen fusion. As a result, the accreted gases develop a layered structure that covers the entire neutron star, with a few tens of centimeters of hydrogen lying atop a similar thickness of helium. The structure is reminiscent of the layers within an evolved giant star (see Figure 11-21), although the layers atop a neutron star are much more compressed, thanks to the star's tremendous surface gravity.

When the helium layer is about 1 m thick, helium fusion ignites explosively and heats the neutron star's surface to about 3×10^7 K. At this temperature the surface predominantly emits X-rays, but the emission ceases within a few seconds as the surface cools. Hence, we observe a sudden burst of X-rays only a few seconds in duration. New hydrogen then flows onto the neutron star, and the whole process starts over. Indeed, X-ray bursters typically emit a burst every few hours or days.

Whereas explosive *hydrogen* fusion on a white dwarf produces a nova, explosive *helium* fusion on a neutron star produces an X-ray burster. In both cases, the process is explosive, because the fuel is compressed so tightly against the star's surface that it becomes degenerate, like the star itself. As with the helium flash inside red giants, the ignition of a degenerate thermonuclear fuel involves a sudden thermal runaway. This is because an increase in temperature does not produce a corresponding increase in pressure that would otherwise relieve compression of the gases and slow the nuclear reactions.

ConceptCheck 12-16: **Why can a nova repeat its brightening but a supernova cannot?**

ConceptCheck 12-17: **What are the fundamental differences between a nova and an X-ray burster?**

VISUAL LITERACY TASK

Supernova Light Curves

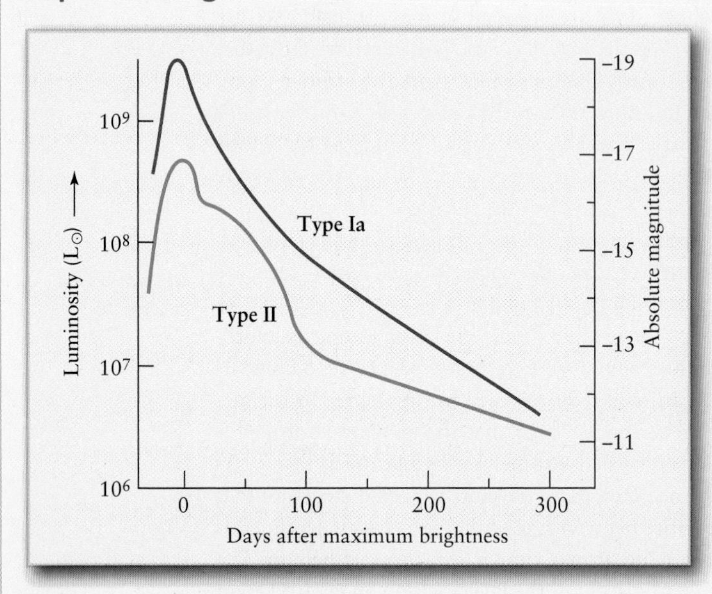

PROMPT: What would you tell a fellow student who said, **"Isolated massive stars that explode by internal collapse are brighter and last longer than other types of supernovae"?**

ENTER RESPONSE:

Guiding Questions:

1. Type Ia supernovae are created by
   a. stars in close binary systems.
   b. isolated supermassive stars.
   c. a wide variety of stellar death scenarios.
   d. collisions between galaxies.

2. The most luminous supernovae are
   a. Type Ia.
   b. Type II.
   c. the same brightness regardless of origin.
   d. highly varying in brightness and origin.

3. One hundred days after the explosion, the most luminous supernovae are
   a. Type Ia.
   b. Type II.
   c. indistinguishable from less luminous supernova.
   d. unpredictable in their luminosity.

4. About a year after the explosion, the most luminous supernovae are
   a. Type Ia.
   b. Type II.
   c. indistinguishable from less luminous supernova.
   d. unpredictable in their luminosity.

KEY IDEAS AND TERMS

12-1 High-mass stars create heavy elements in their cores before violently blowing apart in supernova explosions, leaving behind remnants

- Unlike a moderately low-mass star, a high-mass star larger than 4 $M_\odot$ undergoes an extended sequence of thermonuclear reactions in its core and shells, including processes based on carbon, neon, oxygen, and silicon.
- Because thermonuclear reactions can take place simultaneously in several shells surrounding the core, energy is released at such a rapid rate that the star's outer layers expand tremendously, resulting in a **supergiant star,** whose luminosity and radius are much larger than those of a giant.
- In the last stages of its life, a high-mass star has an iron-rich core surrounded by concentric shells hosting the various thermonuclear reactions.
- A star with an initial mass greater than 8 $M_\odot$ dies in a violent cataclysm in which its core collapses and most of its matter is ejected into space at high speeds, producing a **core-collapse supernova.**

12-2 Core-collapse supernovae can leave behind remnants, neutron stars, and pulsars

- The matter ejected from a core-collapse supernova, moving at supersonic speeds through interstellar gases and dust, glows as **supernova remnants.**
- A **neutron star** is a dense stellar corpse, supported by closely packed neutrons (**neutron degeneracy pressure**) remaining from a core-collapse supernova typically having a diameter of only about 12 miles, a mass less than 3 $M_\odot$, a magnetic field 10^{12} times stronger than that of the Sun, and a rotation period of roughly 1 second.
- Intense beams of radiation emanate from regions of spinning neutron stars near the north and south magnetic poles of a neutron star observed as pulses of radio waves and called a **pulsar,** which slow their rotation as they lose energy with age.

12-3 Black holes are created in the death throes of the most massive of stars

- Surrounding a black hole, where the escape speed from the hole just equals the speed of light, is the **event horizon.**

- An object from which neither matter nor light (electromagnetic radiation) can escape is called a **black hole.**
- The distance from the center of a nonrotating black hole to its event horizon is called the **Schwarzschild radius.**
- The star's entire mass is crushed to zero volume—and hence infinite density—at a single point, known as the **singularity,** at its center.

12-4 Black holes cannot be seen directly

- The evidence for black holes comes from observed regions emanating X-rays where heated material glows as it accelerates toward a black hole, which can occur when a black hole shares a close binary system orbit with a star.
- Gas gravitationally collected at the center of most galaxies gives rise to a central **supermassive black hole** with a truly stupendous mass.

12-5 White dwarfs and pulsars in close binary systems can become novae, bursters, and supernovae

- Explosive hydrogen fusion may occur in the surface layer of a companion white dwarf in a close binary system, producing the sudden increase in luminosity that we call a **nova.**
- Supernovae that are caused by a core-collapse of a massive star are categorized as **Type II supernovae.**
- **Type I supernova** result from an accreting white dwarf in a close binary system, which becomes a supernova when carbon fusion ignites explosively throughout such a degenerate star.
- A graph of the brightness of a supernova versus time is known as a **light curve** showing that supernovae begin with a sudden rise in brightness that occurs in less than a day.
- Rapid and explosive fusion of helium may occur in the surface layer of a companion neutron star producing a sudden increase in X-rays, called an **X-ray burster.**

QUESTIONS

Review Questions

1. What are the important thermonuclear reactions leading up to the formation of iron?
2. Describe the steps leading up to a core-collapse supernova.
3. How do neutron stars form?
4. What are the differences between degenerate-electron pressure and degenerate-neutron pressure?
5. How is a neutron star similar to a coastal lighthouse?
6. What determines if a core-collapse supernova will form a neutron star or a black hole?
7. What is the difference between a black hole's event horizon and its Schwarzschild radius?
8. When we say that the Moon has a radius of 1080 miles (1737 km), we mean that this is the smallest radius that encloses all of the Moon's material. In this sense, is it correct to think of the Schwarzschild radius as the radius of a black hole? Why or why not?
9. Astronomers cannot actually see the black hole candidates in close binary systems. How, then, do they know that these candidates are not white dwarfs or neutron stars?
10. What are the differences between a Type Ia and a Type II supernova?
11. What are the similarities between a nova and a Type Ia supernova? What are the differences?
12. What is the similarity between a nova and an X-ray burster? How are they different?

Web Chat Questions

1. Imagine that our Sun was somehow replaced by a 1-$M_\odot$ white-dwarf star and that Earth continued in an orbit of semimajor axis 1 AU around this star. Discuss what effects this would have on our planet. What would the white dwarf look like as seen from Earth? Could you look at it safely with the unaided eye? Would Earth's surface temperature remain the same as it is now?
2. The similar names *white dwarf, red dwarf,* and *brown dwarf* describe three very different kinds of objects. Suggest better names for these three kinds of objects, and describe how your names more accurately describe the objects' properties.
3. Describe the kinds of observations you might make in order to locate and identify black holes.

Collaborative Exercises

1. Imagine that a supernova originating from a close binary star system, both of whose stars have less than 4 solar masses, began (as seen from Earth) on the most recent birthday of the youngest person in your group. Using the light curves in Figure 12-18, what would its new luminosity be today and how bright would it appear in the sky (apparent magnitude) if it were located 32.6 light-years away? How would your answers change if you were to discover that the supernova actually originated from an isolated star with a mass 15 times greater than our Sun?
2. Consider the graph showing a recording of a pulsar in Figure 12-9. Sketch and label similar graphs that your group estimates for: (1) a rapidly spinning, professional ice skater holding a flashlight; (2) an emergency signal on an ambulance; and (3) a rotating beacon at an airport.
3. As stars go, pulsars are tiny, only about 12 miles (20 km) across. Name three specific things or places that have a size or a separation of about 12 miles.

Observing Questions

1. The red supergiant Betelgeuse in the constellation Orion will explode as a supernova at some time in the future. Use the *Starry Night*™ program to investigate how the supernova might appear if the light from this explosion were to arrive at Earth tonight. Click the **Home** button in the toolbar to show the sky as seen from your location at the present time. Use the **Find** pane to locate Betelgeuse. If Betelgeuse is below the horizon, allow the program to reset the time to show this star.
   a) At what time does Betelgeuse rise on today's date? At what time does it set?
   b) What is the apparent magnitude (m_V) of Betelgeuse? (*Hint:* Use the **HUD** or the **Info** pane to find this information.)
   c) If Betelgeuse became a supernova today, then at peak brightness it would be 11 magnitudes brighter than it is now. (*Note:* Magnitudes increase as brightness decreases for stars.) For comparison, $m_V = -4$ for Venus at its brightest and $m_V = -12.6$ for the full moon. Would Betelgeuse be visible in the daytime? How would it appear at night? Do you think it would cast shadows?
2. Use the *Starry Night*™ program to examine the Veil Nebula, a large supernova remnant. Open **Favourites > Explorations > Veil Nebula** to see a view of the nebula high in the sky of Calgary, Canada, at midnight on August 1, 2013.
   a) What significant feature do you notice about this supernova remnant in this 5-degree field of view?
   b) Use the angular separation tool to measure the approximate angular distance between the components of this nebula. What

angular distance separates these components? What form of optical aid is best suited to observing this object?

3. Use the *Starry Night*™ program to observe the sky in July 1054, when the supernova that spawned the Crab Nebula (M1) would have been visible, probably even in daylight, from North America and may have been recorded as a pictograph at this time by inhabitants of Chaco Canyon, New Mexico. Open **Favourites > Explorations > Crab-Pictograph** to position yourself in Chaco Canyon, at latitude 36°N and longitude 108°W at 5 A.M. on July 5, 1054, looking toward the east, just before sunrise. **Zoom in** to display a field of view of about 10° centered on the Crab Nebula, where you can see the position of the Moon near to the nebula. You may find it helpful to turn daylight on or off (select **Show Daylight** or **Hide Daylight** in the **View** menu).
   a) What is the phase of the Moon?
   b) Investigate how the relative positions of the Moon and the Crab Nebula change when you set the date to July 4, 1054, or July 6, 1054. On which date do the relative positions of the Moon and the Crab Nebula give the best match to the pictograph shown in the textbook? You can now investigate the nebula more closely by zooming in to see a ground-based view of this expanding gas cloud from the violent supernova explosion.
4. Use the *Starry Night*™ program to investigate the X-ray source and probable black hole, Cygnus X-1. This region of space is one of the brightest in the sky at X-ray wavelengths. Click the **Home** button in the toolbar and then use the **Find** pane to center the field of view on Cygnus X-1. If Cygnus X-1 is below the horizon, allow the program to reset the time to when it can best be seen. Click the checkbox to the left of the listing for Cygnus X-1 to apply a label to this object. Use the **Zoom** controls to set the field of view to 100 degrees. Use the **Time** controls in the toolbar to determine when Cygnus X-1 rises and sets on today's date from your location.

ANSWERS

ConceptChecks

ConceptCheck 12-1: The helium would turn into silicon by first fusing into carbon, which would then fuse into neon, which would finally fuse into silicon.

ConceptCheck 12-2: Sunlike stars are far too small for heavy metal fusion to occur at all.

ConceptCheck 12-3: Both. A core-collapse supernova occurs because its central core is no longer able to support its outer layers. These unsupported outer layers first fall inward then bounce back explosively in the outward direction.

ConceptCheck 12-4: A ring of dust might cause part of the expanding envelope to move slower, so if it was not surrounded by a retarding ring of dust, the expansion should have been nearly circular.

ConceptCheck 12-5: Virtually none, if there is no surrounding gas or dust with which to collide. Expanding supernova remnants are most easily seen when they expand and collide with interstellar gas and dust, causing the dust to glow.

ConceptCheck 12-6: Neutrons, which compose a neutron star, can be packed together much more closely than carbon atoms in a white dwarf.

ConceptCheck 12-7: A pulsar can be most easily described as a quickly, rotating neutron star that appears to flash on and off quickly as its brightest portions are alternatively observed and not observed.

ConceptCheck 12-8: The pulsar that is spinning slowest, and thus having longer periods between pulsations, is the oldest, because isolated pulsars tend to slow down over time after their formation.

ConceptCheck 12-9: Nothing would change. A black hole made of the same mass as our Sun would have the same effect on the planets as our Sun currently does because it would have the same gravitational properties at planetary distances.

ConceptCheck 12-10: A nonrotating black hole has infinite density and a zero diameter singularity, independent of its mass.

ConceptCheck 12-11: The more massive of the two stars would have evolved faster, depleted its core fuel supply first, and would have become a black hole, while the second, less massive star would have been evolving more slowly.

ConceptCheck 12-12: Gas near the center of galaxies is observed to be moving at tremendously high speeds, which could only be moved so quickly due to the tremendous gravitational effects of a supermassive black hole.

ConceptCheck 12-13: A white dwarf can only become a Type Ia supernova if it is part of a very close binary star system and it can gravitationally draw matter for fuel from its nearby companion star.

ConceptCheck 12-14: Type Ia supernovae are brighter.

ConceptCheck 12-15: Type II supernovae have a slower brightness decay rate.

ConceptCheck 12-16: The luminosity of a nova occurs as material from a companion is ignited only on the white dwarf's surface for a short time, whereas the luminosity of a Type Ia supernova occurs when the white dwarf completely blows itself apart, leaving nothing behind.

ConceptCheck 12-17: Novae occur when explosive *hydrogen* fusion takes place on a white dwarf, which lasts for weeks, whereas explosive *helium* fusion on a neutron star produces an X-ray burster that lasts for seconds.

13

R I V U X G The Milky Way stretches across the sky in this panorama from telescope observatories high atop mountains in northern Chile. *(Bruno Gilli/ESO)*

Exploring Our Galaxy

CHAPTER LEARNING OBJECTIVES **By reading the sections of this chapter, you will learn:**

- **13-1** The Sun is located in the disk of our Galaxy, about 25,000 light-years from the galactic center
- **13-2** Observations of different types of dust, gas, stars, and star clusters reveal the shape of our Galaxy
- **13-3** Observations of star-forming regions reveal that our Galaxy has spiral arms
- **13-4** Measuring the rotation of our Galaxy reveals the presence of dark matter
- **13-5** Spiral arms are caused by density waves that sweep around the Galaxy
- **13-6** Infrared and radio observations are used to probe the galactic nucleus

On a clear, moonless night, away from the glare of city lights, you can often see a hazy, luminous band stretching across the sky. This band, called the Milky Way, extends all the way around the sky's celestial sphere. Galileo, the first person to write about viewing the Milky Way with a telescope, discovered that this band is composed of countless dim stars. Today, we realize that the Milky Way is actually a disk thousands of light-years across, containing hundreds of billions of stars—one of which is our own Sun—as well as vast quantities of gas and dust. This vast, swirling assemblage of matter gravitationally bound together is collectively called the Milky Way Galaxy.

Just as Galileo's telescope revealed aspects of the Milky Way that the naked eye could not, modern astronomers use telescopes at wavelengths of light invisible to our eyes to peer through our Galaxy's obscuring dust and observe what visible-light telescopes never could. As we will see, radio, infrared, and X-ray observations reveal that the very center of the Galaxy harbors a supermassive black hole with a mass of more than 4 million Suns. Our Sun and planet Earth are not at the galactic center, but rather about 25,000 light-years out along the Galaxy's flat disk, giving us a wondrous view of its center.

Perhaps most surprisingly, astronomers have recently discovered that most of the Milky Way's mass is not in its visible stars, gas, or dust, but in a halo of *dark matter* that emits no measurable light. What the character of this dark matter could be remains one of the greatest open questions in astronomy and physics.

The Milky Way is just one of myriad *galaxies,* or systems of stars and interstellar matter, that are spread across the observable universe. By studying our home Milky Way Galaxy in this chapter, we begin to explore the universe on a grand scale. Instead of focusing on individual stars as we did in earlier chapters, we will now look carefully at the overall arrangement and history of a huge stellar community of which our Sun is but one member. In this way, we gain insights into galaxies, described in the following chapter, and prepare ourselves to ask fundamental questions about the cosmos in the book's final chapter.

13-1 The Sun is located in the disk of our Galaxy, about 25,000 light-years from the galactic center

Peering up into the night sky when it is very dark allows you to see the wispy band of light stretching across the sky, which we call the **Milky Way.** Running from horizon to horizon, the Milky Way is a stunning sight when seen far away from the city lights. As we will see, hundreds of millions of stars collect together in isolated clumps known as **galaxies.** Galaxies are wondrous collections of stars and interstellar matter, far larger than a star cluster. The vast spaces separating galaxies are largely empty.

Perhaps not surprising, the mysterious appearance of the Milky Way has long provided opportunities to share stories, sing songs, and create beautiful pieces of art. One story shared through the ages is an oft-told Cherokee folktale reminding us that a thief stealing cornmeal will always leave a trail behind, like the one scattered through the sky. Plains Indians often described the Milky Way as the pathway of the Wolf Star (Sirius) through the stars. The Māori of Polynesia respect the Milky Way as a canoe, a gift to remind paddlers of the origin and importance of the stars. Others have seen the Milky Way as a reminder to nurture the soul and the body: Greek mythology often refers to the Milky Way as a stream of the goddess Hera's milk, once used to feed Heracles, the son of the god Zeus and the mortal Alcmene, while Roman mythology sometimes attributes this to the milk of Saturn's wife, Opis. Egyptian mythology describes it as cow's milk, emphasizing its importance in feeding the people of the land. Some Hindu stories describe this path through the sky as the Ganges River of the sky. Unquestionably, ancients looked at this same band, knowing it was somehow important and wondering what it might be.

Astronomers, too, have long wondered what the Milky Way could tell us about the nature and structure of the universe. Because the Milky Way completely encircles us, you might be tempted to think Earth and our Sun are located at the center. How could you determine if we are in the center of the Galaxy or off to one side? One strategy is to try to count stars to see if one side has more stars than the other, implying that we are off to one side. As it turns out, when one tries to count the visible stars along the Milky Way, as William Herschel, discoverer of the planet Uranus, described in 1785, it seems that there are a similar number on all sides of us, suggesting that Earth is indeed near the Milky Way's center. But this initial observation is misleading.

Role of Obscuring Interstellar Dust

The reason this seemingly straightforward strategy of counting stars works was discerned in 1930 by Robert J. Trumpler of Lick Observatory. While studying star clusters, Trumpler discovered to his great surprise that the more remote clusters appear unusually dim—more so than would be expected from their distances alone. As a result, Trumpler concluded that interstellar space must not be as empty as everyone had assumed. It must contain dust that absorbs or scatters light from distant stars, causing them to appear dimmer than they otherwise would, so that they seem farther away or that there are even fewer stars than actually exist.

From our vantage point on Earth, we have an edge-on view, which is why the Milky Way appears as a band around the sky (shown in **Figure 13-1*a***). As it turns out, hundreds of millions of stars collect together all over the universe in

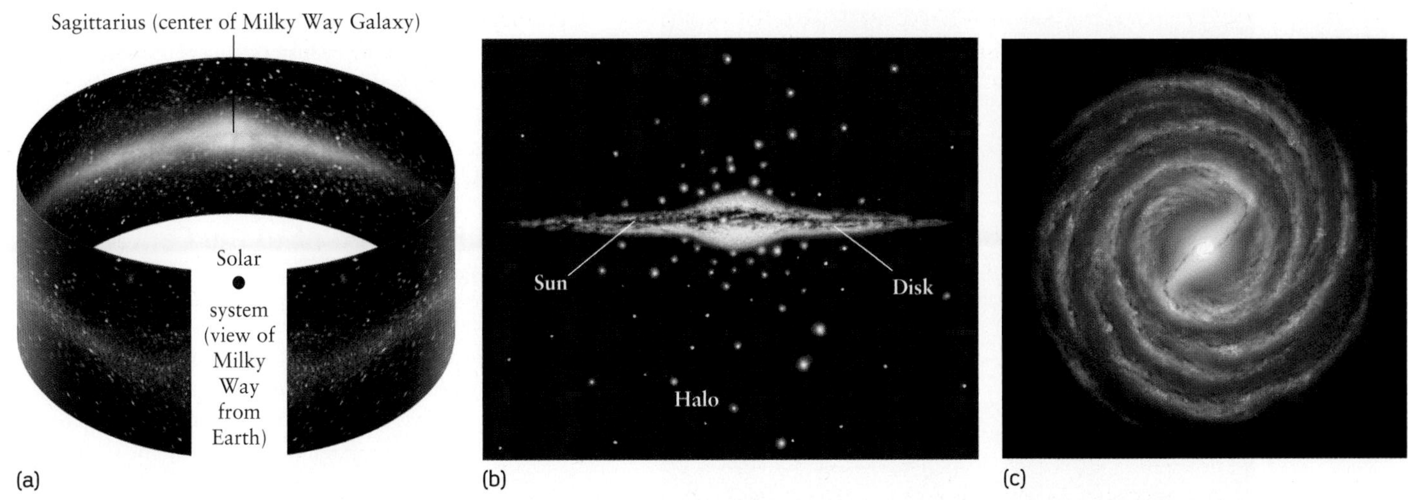

FIGURE 13-1 R I V U X G **Schematic Diagrams of the Milky Way** (a) From our vantage point on Earth sitting inside the Milky Way but not at the center, we see the Milky Way as a thin glowing band stretching across the sky all around us. (b) If we could see it from outside the Galaxy, we could see this edge-on view showing the Milky Way's disk, containing most of its stars, gas, and dust, and its halo, containing many old stars. Individual stars in the halo are too dim to be visible on this scale, so the bright regions in the halo represent clusters of stars. (c) Our Galaxy has two major arms and several shorter segments, all spiraling out from the ends of a bar of stars and gas that passes through the Galaxy's center. *(a: The Cobe Project, DIRBE, NASA; c: NASA/JPL-Caltech/R. Hurt [SSC])*

isolated clumps known as galaxies. As discussed in Chapter 1, astronomers define a *galaxy* as an immense collection of stars and interstellar matter, far larger than a star cluster. The vast spaces separating galaxies are largely empty (Figure 13-1*b* and *c*). Like the stars themselves, this interstellar dust is more concentrated in the disk of our Galaxy. As a result, it obscures our view, making distant objects appear dimmer than they really are, which fools us into thinking those stars are much farther away than they really are. Great patches of interstellar dust are clearly visible in wide-angle photographs such as the one opening this chapter. Because on first glance, we see only the nearest stars in the Galaxy, Herschel had no idea of either the enormous size of our Galaxy or of the vast number of stars concentrated around the galactic center, hidden from view by intervening dust.

ANALOGY Herschel faced much the same dilemma as a lost motorist on a foggy night. Unable to see more than a city block in any direction, a motorist would have a hard time deciding what part of town he was in. If the fog layer were relatively shallow, however, our motorist would be able to see the lights from tall buildings that extend above the fog, and in that way he could determine his location (**Figure 13-2*a***).

The same principle applies to our Galaxy. While interstellar dust in the plane of our Galaxy hides the sky covered by the Milky Way, we have an almost unobscured view out of the plane (that is, to either side of the Milky Way). To find our location in the Galaxy, we need to locate bright objects that are part of the Galaxy but lie outside its plane in unobscured regions of the sky.

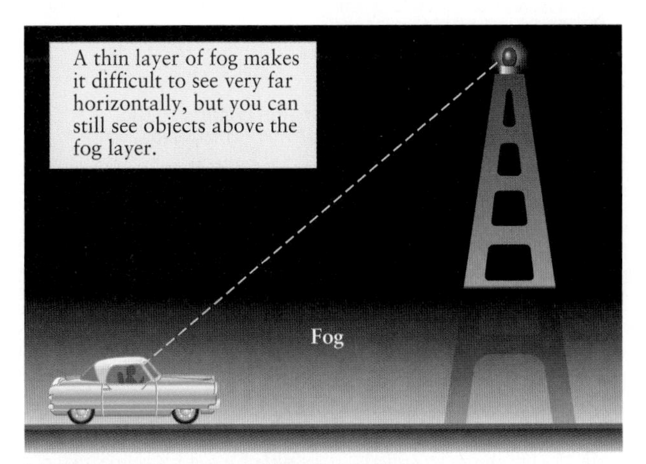

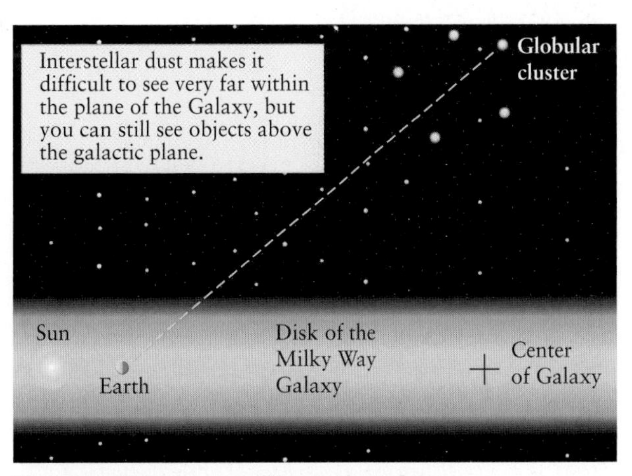

FIGURE 13-2 Finding the Center of the Galaxy (a) A motorist lost on a foggy night can determine his location by looking for tall buildings that extend above the fog. (b) In the same way, astronomers determine our location in the Galaxy by observing globular clusters that are part of the Galaxy but lie outside the obscuring material in the galactic disk. The globular clusters form a spherical halo centered on the center of the Galaxy.

FIGURE 13-3 R I V U X G **Globular Cluster Harboring Variable Stars** M80 is one of the densest of the 147 known globular star clusters in the Milky Way Galaxy, containing hundreds of thousands of stars. From the average apparent brightness (as seen in this photograph) and average luminosity (known to be roughly 100 $L_{\odot}$) of these variable stars, astronomers have deduced that the distance to M80 is about 28,000 ly from Earth. *(Goddard Space Flight Center/Space Telescope Science Institute/NASA)*

Fortunately, bright objects of the sort we need do in fact exist. They are the **globular clusters,** a class of star clusters associated with the Galaxy but which lie outside its plane (Figure 13-2*b*). A typical globular cluster is a spherical distribution of roughly 10^6 stars packed in a volume only a few hundred ly across (**Figure 13-3**).

Globular Clusters Point to the Center of Our Galaxy

To use globular clusters to determine our location in the Galaxy, we must first determine the distances from Earth to these clusters. (Think again of our lost motorist—glimpsing the lights of a skyscraper through the fog may be useful to the motorist, but only if he can tell how far away the skyscraper is.) Pulsating variable stars in globular clusters provide the distances, giving astronomers the key to the dimensions of our Galaxy.

In 1912, the American astronomer Henrietta Leavitt reported her important discovery of the period-luminosity relation for **Cepheid variables.** Cepheid variables are pulsating stars that vary periodically in brightness (**Figure 13-4**). Leavitt found that the longer a Cepheid's period, the greater its luminosity.

(a) The light curve of δ Cephei (a graph of brightness versus time)

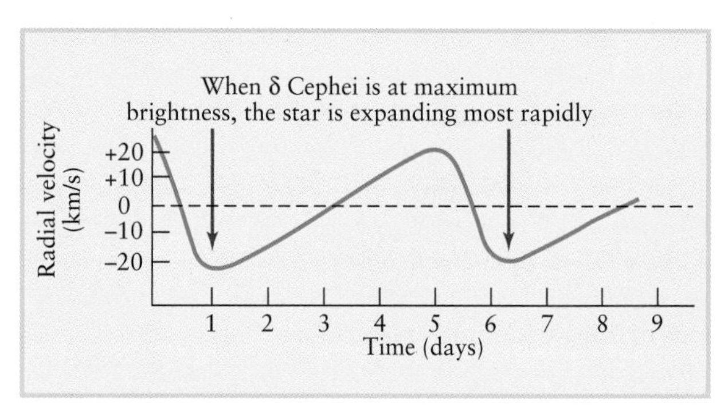

(b) Radial velocity versus time for δ Cephei (positive: star is contracting; negative: star is expanding)

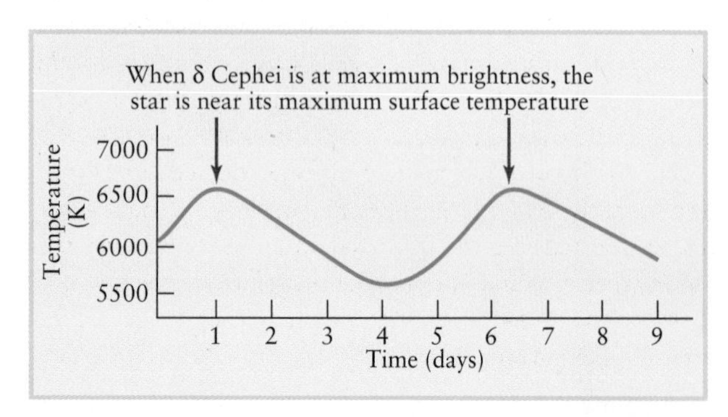

(c) Surface temperature versus time for δ Cephei

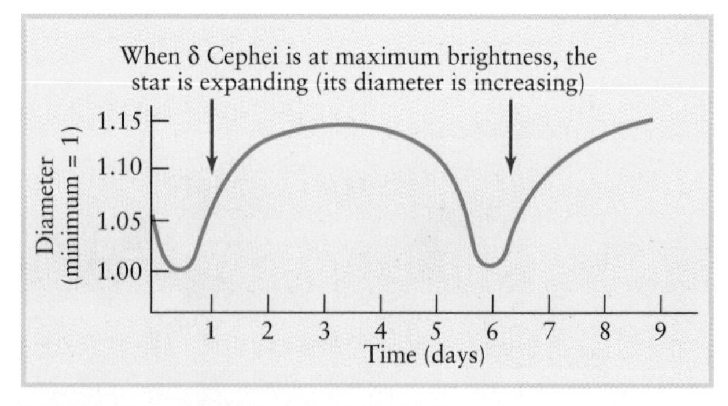

(d) Diameter versus time for δ Cephei

FIGURE 13-4 δ Cephei—A Pulsating Star (a) As δ Cephei pulsates, it brightens quickly (the light curve moves upward sharply) but fades more slowly (the curve declines more gently). The increases and decreases in brightness are nearly in step with variations in (b), the star's radial velocity (positive when the star contracts and the surface moves away from us, negative when the star expands and the surface approaches us), as well as in (c), the star's surface temperature. (d) The star is still expanding when it is at its brightest and hottest (compare with parts a and b).

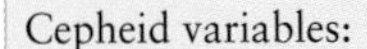
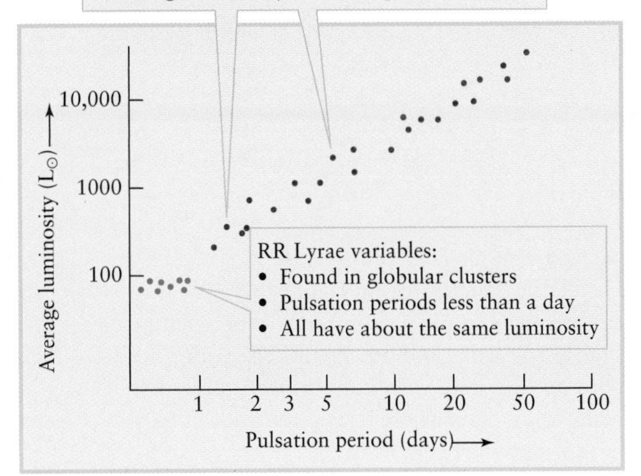

FIGURE 13-5 Period and Luminosity for Cepheid and RR Lyrae Variables This graph shows the relationship between period and luminosity for Cepheid variables and RR Lyrae variables. Cepheids come in a broad range of luminosities: The more luminous the Cepheid, the longer its pulsation period. By contrast, RR Lyrae variables are horizontal-branch stars that all have roughly the same average luminosity of about 100 $L_\odot$.

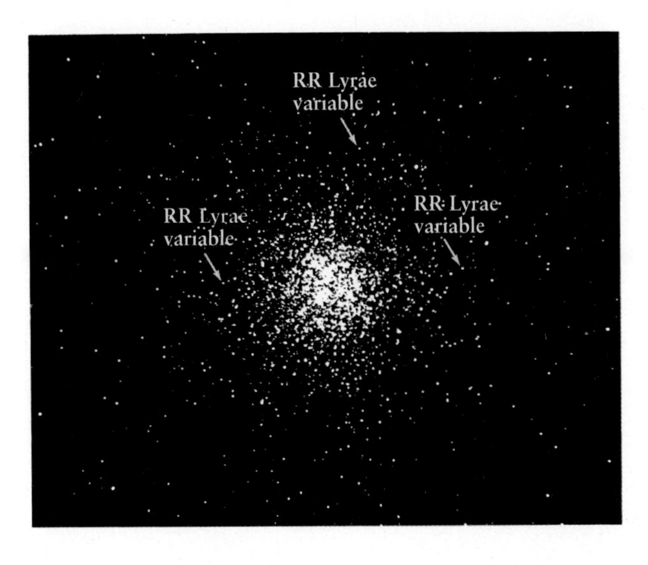

FIGURE 13-6 R I V U X G **RR Lyrae Variables in a Globular Cluster** The arrows point to three RR Lyrae variables in the globular cluster M55, located in the constellation Sagittarius. From the average apparent brightness (as seen in this photograph) and average luminosity (known to be roughly 100 $L_\odot$) of these variable stars, astronomers have deduced that the distance to M55 is 6500 pc (20,000 ly). *(Harvard-Smithsonian Center for Astrophysics)*

The period-luminosity law is an important tool in astronomy because it can be used to determine distances. For example, suppose you find a Cepheid variable in the sky. By measuring its period and using a graph like **Figure 13-5**, you can determine the star's average luminosity. Knowing the star's average luminosity, you can find out how far away the star must be in order to give the observed brightness.

Shortly after Leavitt's discovery of the period-luminosity law, Harlow Shapley, a young astronomer at the Mount Wilson Observatory in California, began studying a family of pulsating stars closely related to Cepheid variables called **RR Lyrae variables.** The light curve of an RR Lyrae variable is similar to that of a Cepheid, but RR Lyrae variables have shorter pulsation periods and lower peak luminosities (see Figure 13-5).

The tremendous importance of RR Lyrae variables is that they are commonly found in globular clusters around the outskirts of galaxies, as mentioned earlier (**Figure 13-6**). By using the period-luminosity relationship for these stars, Shapley was able to determine the distances to the 93 globular clusters then known. He found that some of them were more than 100,000 ly from Earth. The large values of these distances immediately suggested that the Galaxy was much larger than Herschel had thought.

Another striking property of globular clusters is how they are unevenly distributed across the sky. Whereas ordinary stars are rather uniformly spread along the Milky Way, the majority of the 93 globular clusters that Shapley studied are instead located in only one-half of the sky. Most of the globular clusters seem to be clustered around the portion of the Milky Way centered in the general direction of the constellation Sagittarius.

From the directions to the globular clusters and their distances from us, Shapley mapped out the three-dimensional distribution of these clusters in space. In 1920 he concluded that the globular clusters form a huge spherical distribution centered not on Earth but rather about a point in the Milky Way several tens of thousands of light-years away in the direction of Sagittarius (see Figure 13-1*a*). This point, reasoned Shapley, must coincide with the center of our Galaxy, because of gravitational forces between the disk of the Galaxy and the "halo" of globular clusters. Therefore, by locating the center of the distribution of globular clusters, Shapley was in effect measuring the location of the galactic center.

Because globular clusters are distributed around the galactic core, finding their positions shows the direction toward the center of our Galaxy.

Since Shapley's pioneering observations, many astronomers have measured the distance to the center of our Galaxy. Shapley's estimate of this distance was too large by about a factor of 2 because he did not take into account that starlight dims when passing through intervening interstellar dust, which was not well understood at the time. Today, the generally accepted distance to the galactic nucleus is a little more than 25,000 ly. In other words, it would take sunlight about 25,000 years to travel from our Sun to the galactic center. Just as Copernicus and Galileo showed that Earth was not at the center of the solar system, Shapley and his successors showed that the solar system lies nowhere near the center of the Galaxy. Although a great place to live, it seems that Earth itself seems to occupy no particularly special location in the universe.

ConceptCheck 13-1: **If astronomers observed that globular clusters were evenly spread across all parts of the sky, what would astronomers assume about the location of our Sun and Earth in our Galaxy?**

13-2 Observations of different types of dust, gas, stars, and star clusters reveal the shape of our Galaxy

Exploring the Milky Way in the Infrared

Go to Video 13-1

If the Sun is not at the galactic center, what is? When we look toward the center, in the direction of the constellation Sagittarius, our view of the very center is completely blocked by dust. At visible wavelengths, light is so obscured by interstellar dust that the galactic nucleus is blocked from viewing with optical telescopes. Fortunately, longer wavelengths of light can successfully travel through interstellar dust without being scattered or absorbed as much as shorter wavelengths. As a result, we can see farther into the plane of the Milky Way at infrared wavelengths than at visible wavelengths, and radio waves can traverse the Galaxy freely. For this reason, telescopes sensitive to these nonvisible wavelengths are important tools for studying the structure of our Galaxy.

Infrared light is particularly useful for tracing the location of interstellar dust in the Galaxy. Light shining from nearby stars is able to warm the dust grains to temperatures in the chilly range of 10 K to 90 K, and thus the dust emits radiation predominately at wavelengths from about 30 μm to 300 μm. These are called *far-infrared wavelengths,* because they lie in the part of the infrared spectrum most different in wavelength from visible light. At these wavelengths, interstellar dust radiates more strongly than stars, so a far-infrared view of the sky is principally a view of where the dust is. In 1983 the *Infrared Astronomical Satellite* (IRAS) scanned the sky at far-infrared wavelengths, giving the panoramic view of the Milky Way's dust shown in **Figure 13-7*a*.**

In 1990 an instrument on the Cosmic Background Explorer (COBE) satellite scanned the sky at *near-infrared wavelengths,* that is, relatively short wavelengths closer to the visible spectrum. Figure 13-7*b* shows the resulting near-infrared view of the plane of the Milky Way. At near-infrared wavelengths, interstellar dust does not emit very much light. Hence, the light sources in Figure 13-7*b* are stars, which do emit strongly in the near-infrared (especially the cool stars, such as red giants and supergiants). Because interstellar dust does not interfere with light at near-infrared wavelengths, many of the stars whose light is recorded in Figure 13-7*b* lie deep within the Milky Way.

ConceptCheck 13-2: If an astronomer wanted to study the nature of cool stars hiding inside dust clouds, which wavelength would be the best choice, near-infrared or far-infrared?

Structure of Our Galaxy

Observations such as those shown in Figure 13-7, along with the known distance to the center of the Milky Way Galaxy, have helped astronomers to establish the dimensions of the Galaxy. The **disk** of our Galaxy is about 160,000 ly in

(a) Infrared emission from dust at wavelengths of 25, 60, and 100 μm

(b) Infrared emission from dust at wavelengths of 1.2, 2.2, and 3.4 μm

FIGURE 13-7 R I V U X G **The Infrared Milky Way** (a) This view was constructed from observations made at far-infrared wavelengths by the IRAS spacecraft. Interstellar dust, which is mostly confined to the plane of the Galaxy, is the principal source of radiation in this wavelength range. (b) Observing at near-infrared wavelengths, as in this composite of COBE data, allows us to see much farther through interstellar dust than we can at visible wavelengths. Light in this wavelength range comes mostly from stars in the plane of the Galaxy and in the bulge at the Galaxy's center. *(NASA/JPL-Caltech/M. Regan [STScI], and the SINGS Team)*

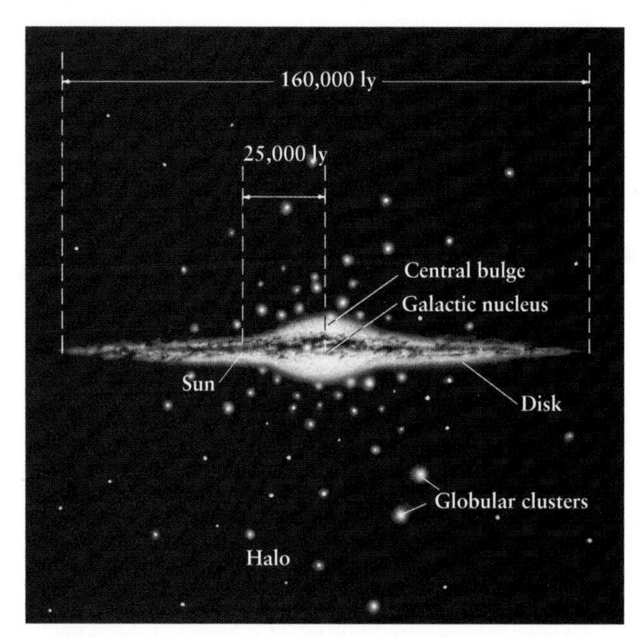

FIGURE 13-8 Our Galaxy (Schematic Edge-on View) There are three major components of our Galaxy: a disk, a central bulge, and a halo. The disk contains gas and dust along with metal-rich (Population I) stars. The halo is composed almost exclusively of old, metal-poor (Population II) stars. The central bulge is a mixture of Population I and Population II stars.

diameter and about 2000 ly thick, as shown in **Figure 13-8.** The center of the Galaxy is surrounded by a distribution of stars, called the **central bulge,** which is about 6500 ly in diameter. This central bulge is clearly visible in Figure 13-7*b*. The spherical distribution of globular clusters traces the **halo** of the Galaxy.

This structure is not unique to our Milky Way Galaxy. **Figure 13-9** shows another galaxy that has dust and stars lying in a disk and that has a central bulge of stars, just like the Milky Way. In the same way that our Sun is a rather ordinary member of the stellar community that makes up the Milky Way, the Milky Way turns out to be a rather common variety of galaxy.

ConceptCheck 13-3: Would an imaginary space traveler have a better view of our Sun and its planets by standing at the galactic center or in the halo?

The Milky Way's Distinct Stellar Populations

Go to Video 13-2

It is estimated that our Galaxy contains about 200 billion (2×10^{11}) stars. Remarkably, different kinds of stars are found in the various components of the Galaxy. The globular clusters in the halo are composed of old, metal-poor, Population II stars. Although these clusters are conspicuous, they contain only about 1% of the total number of stars in the halo. Most halo stars are single Population II stars in isolation. Curiously, these stars move relatively quickly about the galaxies—quickly as compared to the speed at which our Sun moves. These ancient stars orbit the Galaxy along paths tilted at random angles to the disk of the Milky Way, as do the globular clusters. By contrast, stars in the disk travel along orbits that remain in the disk (**Figure 13-10**).

The stars in the disk are mostly young, metal-rich, Population I stars like the Sun. The disk of a galaxy like the

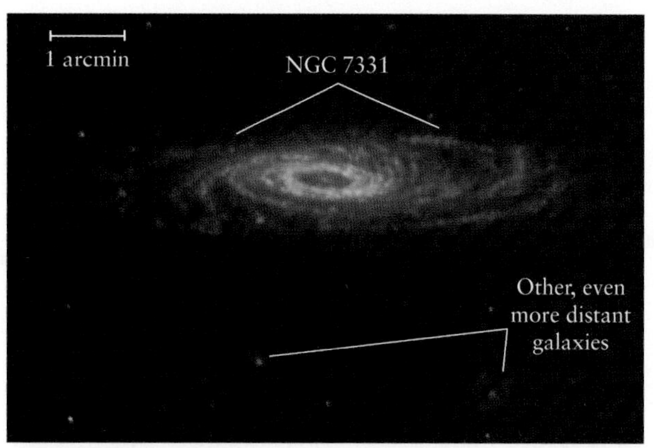

(a) Infrared emission from dust in NGC 7331 at 5.8 and 8.0 μm

(b) Infrared emission from stars in NGC 7331 at 3.6 and 4.5 μm

FIGURE 13-9 R I V U X G **NGC 7331: A Near-Twin of the Milky Way** If we could view our Galaxy from a great distance, it would probably look like this galaxy in the constellation Pegasus. As in Figure 13-7, the far-infrared image (a) reveals the presence of dust in the galaxy's plane, while the near-infrared image (b) shows the distribution of stars. These images of NGC 7331, which is about 50 million ly from Earth, were made with the Spitzer Space Telescope. *(NASA/JPL-Caltech; M. Regan [STScI]; and the SINGS Team)*

FIGURE 13-10 R I V U X G **Star Orbits in the Milky Way** The different populations of stars in our Galaxy travel along different sorts of orbits. Most stars orbit around the galactic center, much as the planets orbit our Sun. Older, Population II (metal-poor) stars follow large looping orbits in the halo, passing through the disk. *(1999–2002, Australian Astronomical Observatory; photograph by S. Lee, C. Tinney, and D. Malin)*

Our Galaxy's oldest stars are found in the halo rather than the disk.

Milky Way appears bluish because its light is dominated by energy emitted from hot O and B main-sequence stars. Such stars have very short main-sequence lifetimes, so they must be quite young by astronomical standards. Hence, their presence shows that there must be active star formation in the galactic disk. By contrast, no O or B stars are present in the halo, which implies that star formation ceased there long ago.

The central bulge contains both Population I stars and metal-poor Population II stars. Since Population II stars are thought to be the first generation of stars to have formed in the history of the universe, some of the stars in the bulge are quite ancient while others were created more recently. From an observational perspective, the central bulge itself looks somewhat yellowish or reddish because it contains many red giants and red supergiants, but does *not* contain many luminous, short-lived, blue O or B stars. Hence, there cannot be much ongoing star formation in the central bulge. The same is true for other galaxies whose structure is similar to that of the Milky Way (**Figure 13-11**).

Why are there such different populations of stars in the halo, disk, and central bulge? Why has star formation stopped in some regions of the Galaxy but continues in other regions? The answers to these questions are related to the way that stars, as well as the gas and dust from which stars form, move within the Galaxy.

FIGURE 13-11 R I V U X G **Stellar Populations: Disk Versus Central Bulge** This illustration shows the Milky Way Galaxy's inner and outer halos. A halo is a spherical cloud of stars surrounding a galaxy and contains similar stars. In contrast, the disk and central bulge of the Milky Way contain different populations of star types. Population I stars, of which the Sun is an example, are found in the disk. The inner and outer halos only contain older Population II stars. The central bulge contains both types, but many more red stars than blue stars. *(NASA, ESA, and A. Feild [STScI])*

ConceptCheck 13-4: If you were looking to find the most recently formed stars in the Galaxy, where would you look?

13-3 Observations of star-forming regions reveal that our Galaxy has spiral arms

The galaxies shown in Figure 13-9 and Figure 13-11 both have **spiral arms,** spiral-shaped concentrations of gas and dust that extend outward from the center in a shape reminiscent of a pinwheel. Although we cannot see the precise shape of the Milky Way because we are inside it, we believe that the Milky Way Galaxy also has spiral arms. Because interstellar dust obscures our visible-light view in the plane of our Galaxy, a detailed understanding of the structure of the galactic disk did not occur until the development of radio astronomy. Thanks to their long wavelengths, radio waves can penetrate the interstellar medium even more easily than infrared light and can travel without being scattered or absorbed. As we will see in this section, both radio and optical observations reveal that our Galaxy does indeed have spiral arms.

Mapping Hydrogen in the Milky Way with Radio Waves

Hydrogen is by far the most abundant element in the universe. Hence, by looking for concentrations of hydrogen gas, we should be able to detect important clues about the distribution of matter in our Galaxy. Unfortunately, ordinary visible-light telescopes are of little use in this quest. This is so because hydrogen atoms can only emit visible light if their electrons are first excited to high energy levels. This is quite unlikely to occur in the cold depths of interstellar space. Furthermore, even if there are some hydrogen atoms that glow strongly at visible wavelengths, interstellar dust would make it impossible to see this glow from distant parts of the Galaxy.

What makes it possible to map out the distribution of hydrogen in our Galaxy is that even cold hydrogen clouds emit *radio* waves. As we mentioned earlier, radio waves can easily penetrate the interstellar medium, so we can detect the radio emission from such cold clouds no matter where they lie in the Galaxy. To understand how hydrogen in these clouds can emit radio waves, we must probe a bit more deeply into the structure of protons and electrons, the particles of which hydrogen atoms are made.

In addition to having mass and charge, particles such as protons and electrons possess a tiny amount of ***spin*** (more formally described as angular momentum). Very roughly, you can visualize each of the electrons around protons in an atom as a tiny, electrically charged sphere that spins on its axis. (This idea also works for protons.) Because electric charges in motion generate magnetic fields, an electron behaves like a tiny magnet with a north pole and a south pole (**Figure 13-12**).

If you have ever played with magnets, you know that two magnets attract when the north pole of one magnet is next to the south pole of the other and repel when two like poles (both north or both south) are next to each other (Figure 13-12*a*). In other words, the energy of the two magnets is least when opposite poles are together and highest when like poles are together. Hence, as shown in Figure 13-12*b*, the energy of a hydrogen atom is slightly different, depending on whether the spins of the proton and electron are in the same direction or opposite directions. According to the laws of physics, these are the only two possibilities; the spins cannot be at random angles.

If the spin of the electron changes its orientation from the higher-energy configuration to the lower-energy one—called a **spin-flip transition**—a photon is emitted. The energy difference between the two spin configurations is very small, only about 10^{-6} times as great as those between different electron orbits. Therefore, the photon emitted in a spin-flip transition between these configurations has only a small amount of energy, and thus its wavelength is a relatively long 21 cm. What is important here is that the photon generated from hydrogen in this scenario has a wavelength of 21 cm and is in the radio wavelengths portion of the electromagnetic spectrum.

The distribution of hydrogen gas in the Milky Way is not uniform but is actually quite frothy. In fact, our Sun lies near the edge of an irregularly shaped region within which the interstellar medium is very thin but at very high temperatures (about 10^6 K). This curious region may have been carved out by a supernova that exploded nearby some 300,000 years ago.

Radio waves of 21 cm are used to map our Galaxy's spiral arms.

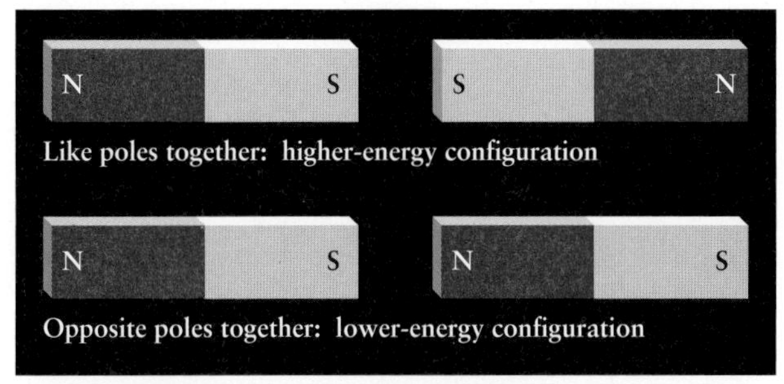

(a) The magnetic energy of two bar magnets depends on their relative orientation.

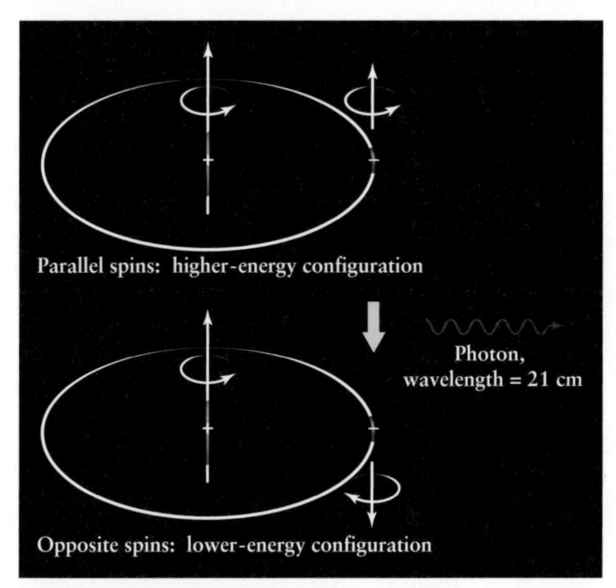

(b) The magnetic energy of a proton and electron depends on their relative spin orientation.

FIGURE 13-12 Magnetic Interactions in the Hydrogen Atom (a) The energy of a pair of magnets is high when their north poles or their south poles are near each other, and low when they have opposite poles near each other. (b) Thanks to their spin, electrons and protons are both tiny magnets. When the electron flips from the higher-energy configuration (with its spin in the same direction as the proton's spin) to the lower-energy configuration (with its spin opposite to the proton's spin), the atom loses a tiny amount of energy and emits a radio wavelength photon at 21 cm.

ConceptCheck 13-5: If a spin-flip transition released substantially more energy, such that a photon was emitted in the far-infrared range, would the Milky Way stretching across the night sky appear brighter or dimmer?

CalculationCheck 13-1: If a 21-cm photon is observed when a single hydrogen atom undergoes a spin-flip transition, what wavelength of photon is observed when 10 hydrogen atoms undergo spin-flip transitions?

Detecting Our Galaxy's Spiral Arms

The detection of 21-cm radio emission was a major breakthrough that permitted astronomers to reveal the presence of spiral arms in the galactic disk. **Figure 13-13** illustrates how this was done. Suppose that you aim a radio telescope along a particular line of sight across the Galaxy. Your radio

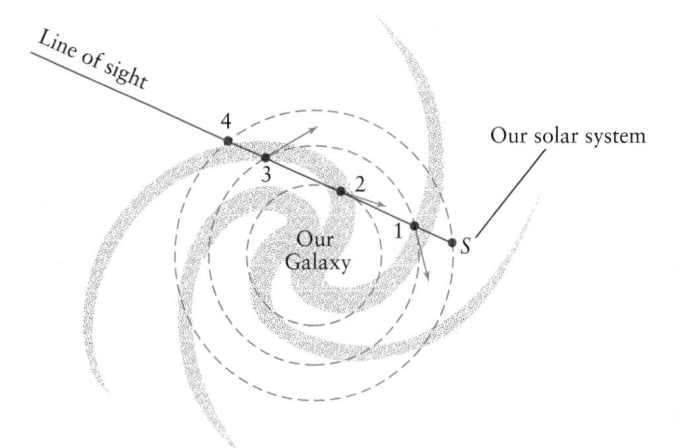

FIGURE 13-13 A Technique for Mapping Our Galaxy If we look within the plane of our Galaxy from our position at *S*, hydrogen clouds at different locations (shown as 1, 2, 3, and 4) along our line of sight are moving at slightly different speeds relative to us. As a result, radio waves from these various gas clouds are subjected to slightly different Doppler shifts. This permits radio astronomers to sort out the gas clouds and thus map the Galaxy.

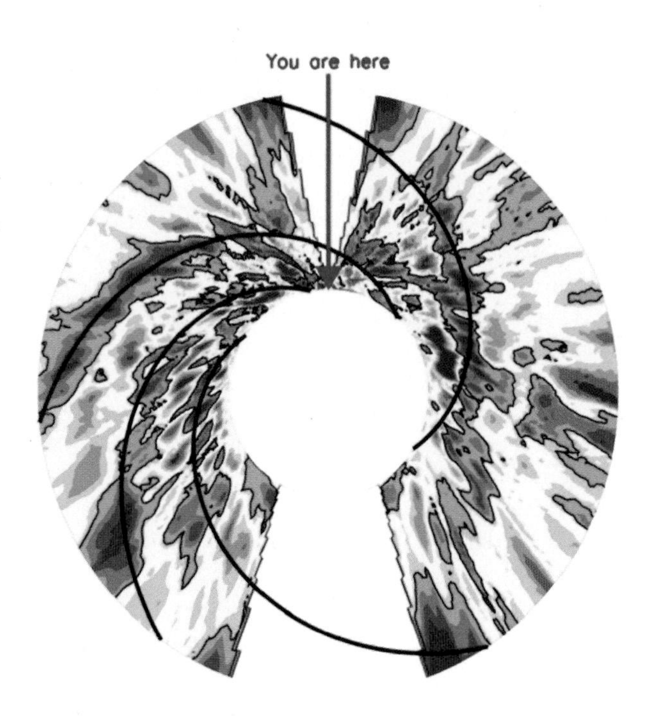

FIGURE 13-14 RIVUXG **A Map of 21-cm Light Emitted by Hydrogen in Our Galaxy** This map, constructed from radio-telescope surveys of 21-cm radiation, shows the distribution of hydrogen gas in a face-on view of our Galaxy. The map suggests a spiral structure. Details in the blank, wedge-shaped region at the bottom of the map are unknown. Gas in this part of the Galaxy is moving perpendicular to our line of sight and thus does not exhibit a detectable Doppler shift. *(Image courtesy Leo Blitz, PhD)*

receiver, located at *S* (the position of our solar system), picks up 21-cm emission from hydrogen clouds at points 1, 2, 3, and 4. However, the radio waves from these various clouds are Doppler shifted by slightly different amounts because the clouds are moving at different speeds as they travel with our rotating Galaxy.

It is important to remember that the Doppler shift reveals only motion along the line of sight. In Figure 13-13, cloud 2 has the highest speed along our line of sight because it is moving directly toward us. Consequently, the radio waves from cloud 2 exhibit a larger Doppler shift than those from the other three clouds along our line of sight. Because clouds 1 and 3 are at the same distance from the galactic center, they have the same orbital speed. The fraction of their velocity parallel to our line of sight is also the same, so their radio waves exhibit the same Doppler shift, which is less than the Doppler shift of cloud 2. Finally, cloud 4 is the same distance from the galactic center as the Sun. This cloud is thus orbiting the Galaxy at the same speed as the Sun, resulting in no net motion along the line of sight. Radio waves from cloud 4, as well as from hydrogen gas near the Sun, are not Doppler shifted at all.

These various Doppler shifts cause radio waves from gases in different parts of the Galaxy to arrive at our radio telescopes with wavelengths slightly different from 21 cm. It is therefore possible to sort out the various gas clouds and thus produce a map of the Galaxy like that shown in **Figure 13-14.**

Figure 13-14 shows that this hydrogen gas is not spread uniformly around the disk of the Galaxy but is concentrated into numerous arched lanes. Similar features are seen in other galaxies beyond the Milky Way. As an example, the galaxy in **Figure 13-15** has prominent spiral arms outlined by hot, luminous, blue main-sequence stars and the red emission nebulae found near many such stars. Stars of this sort are very short-lived, so these features indicate that spiral arms are sites of active, ongoing star formation. A radio telescope tuned to receive 21-cm light shows that spiral arms are also regions where hydrogen gas is concentrated, similar to the structures in our own Galaxy visible in Figure 13-14. This is a strong indication that our Galaxy also has spiral arms.

CAUTION Photographs such as Figure 13-15*a* might lead to the mistaken impression that there are very few stars

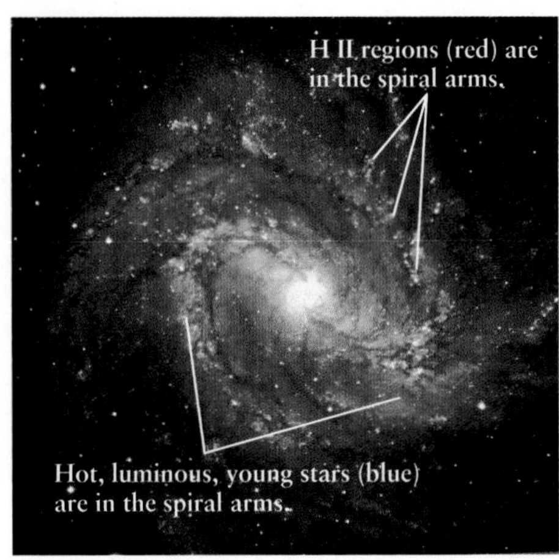

RIVUXG
(a) Visible-light view of M83

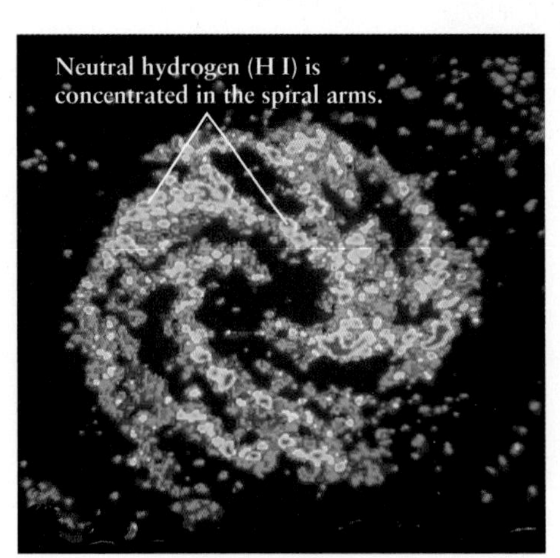

RIVUXG
(b) 21-cm radio view of M83

RIVUXG
(c) Near-infrared view of M83

FIGURE 13-15 A Spiral Galaxy Known as the Southern Pinwheel Galaxy, M83 lies in the southern constellation Hydra about 15 million ly from Earth. (a) This visible-light image clearly shows the spiral arms. The presence of young stars and glowing, energized hydrogen indicates that star formation takes place in spiral arms. (b) This radio view at a wavelength of 21 cm shows the emission from interstellar hydrogen gas (H I). Note that essentially the same pattern of spiral arms is traced out in this image as in the visible-light photograph. (c) M83 has a much smoother appearance in this near-infrared view. This shows that cooler stars, which emit strongly in the infrared, are spread more uniformly across the galaxy's disk. Note the elongated bar shape of the central bulge. *(a: Australian Astronomical Observatory/David Malin Images; b: VLA, National Radio Astronomy Observatory; c: NASA/JPL-Caltech)*

between the spiral arms of a galaxy. Nothing could be further from the truth! In fact, stars are distributed rather uniformly throughout the disk of a galaxy like the one in Figure 13-15*a;* the density of stars in the spiral arms is only about 5% higher than in the rest of the disk. The spiral arms stand out nonetheless because they are where hot, blue O and B stars are found. One such star is about 10^4 times more luminous than an average star in the disk, so the light from O and B stars completely dominates the visible appearance of a spiral galaxy. An infrared image such as Figure 13-15*c* gives a better impression of how stars of all kinds are distributed through a spiral galaxy's disk.

> ConceptCheck 13-6: **When looking at neutral hydrogen gas that is moving away from you, how is its wavelength changed as compared to hydrogen gas that is not moving relative to you?**

Mapping the Spiral Arms and the Central Bulge with Visible-light Observations

Figure 13-15*a* suggests that we can confirm the presence of spiral structure in our own Galaxy by mapping the locations of star-forming regions. Such regions are marked by OB associations, H II regions, and molecular clouds. Unfortunately, the first two of these are best observed using visible light, and interstellar dust limits the range of visual observations in the plane of the Galaxy to less than 10,000 ly from Earth. But there are enough OB associations and H II regions within this range to plot the spiral arms in the vicinity of the Sun.

Molecular clouds are easier to observe at great distances because molecules of carbon monoxide (CO) in these clouds emit radio waves that are relatively unaffected by interstellar dust. Hence, the positions of molecular clouds have been observed even in remote corners of the Galaxy, normally uncharted. Taken together, all of these observations demonstrate that our Galaxy has at least a couple of major spiral arms as well as several shorter arm segments (**Figure 13-16**). The Sun is located just outside a relatively short arm segment called the *Orion Arm,* which includes the Orion Nebula and neighboring sites of vigorous star formation in that constellation.

Two major spiral arms border either side of the Sun's position. The Sagittarius Arm is on the side toward the galactic center. This is the arm you see on June and July nights when you look at the portion of the Milky Way stretching across Scorpius and Sagittarius. In December and January, when our nighttime view is directed away from the galactic center, we see the Perseus Arm. The other major spiral arms cannot be seen at visible wavelengths thanks to the obscuring effects of dust.

Figure 13-16 also shows that the central bulge of the Milky Way is not spherical, but rather is elongated like a bar.

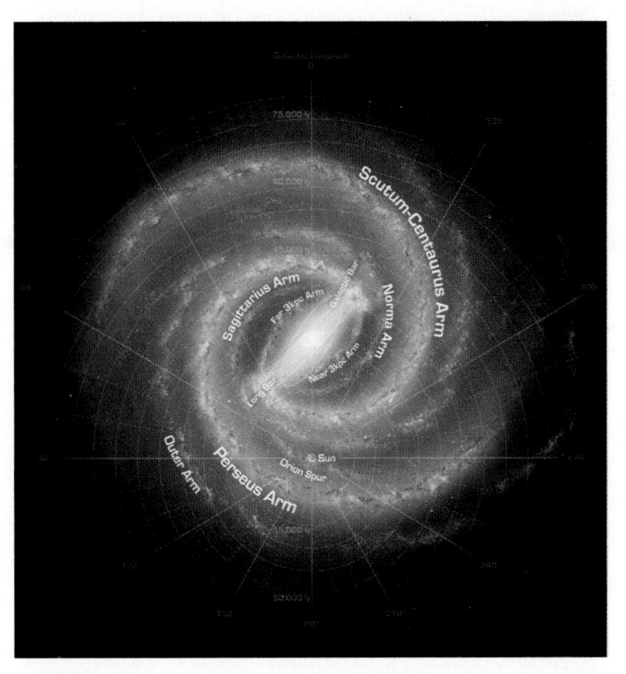

FIGURE 13-16 Our Galaxy Seen Face-on: Artist's Impression The Galaxy's diameter is about 160,000 ly and our solar system is a little more than 25,000 ly from the galactic center. The elongated central bulge is about 27,000 ly long and is oriented at approximately 45° to a line running from the solar system to the galactic center. The Galaxy's two major arms, Scutum-Centaurus and Perseus, can be seen attached to the ends of a thick central bar, while the two smaller arm segments of Norma and Sagittarius are less distinct and located between the Galaxy's two major arms. Our Sun lies near a small, partial arm called the Orion Arm, or Orion Spur, located between the Sagittarius and Perseus arms. *(NASA/JPL-Caltech/R. Hurt [SSC/Caltech/NASA Jet Propulsion Laboratory])*

ED: Yes, art is file posted 2/7

This is unlike the galaxy NGC 7331 shown in Figure 13-9, but similar to the galaxy M83 depicted in Figure 13-15. The elongated shape of the central bulge had been suspected since the 1980s; this was confirmed in 2005 using the Spitzer Space Telescope, which was used to survey the infrared emissions from some 3 million stars in the central bulge. Thus, the artist's impression shown in Figure 13-16 is based on observations using both radio wavelengths (for the spiral arms) and infrared wavelengths (for the central bulge). As it turns out, this elongated shape may play a crucial role in sustaining the Galaxy's spiral structure.

Why are the young stars, star-forming regions, and clouds of neutral hydrogen in our Galaxy all found predominantly in the spiral arms? To answer this question, we must understand why spiral arms exist at all. Spiral arms are essentially cosmic "traffic jams," places where matter piles up as it orbits around the center of the Galaxy. This orbital motion, which is essential to grasping the significance of spiral arms, is our next topic as we continue our exploration of the Galaxy.

> Concept Check 13-7: **When traveling from the edge of the galactic disk toward the galactic bulge passing nearby the Sun, what is the sequence of arms that must be traversed: Orion, Sagittarius, and Perseus?**

13-4 Measuring the rotation of our Galaxy reveals the presence of dark matter

Go to Video 13-3

The spiral arms in the disk of our Galaxy suggest that the disk rotates. This means that the stars, gas, and dust in our Galaxy are all orbiting the galactic center. Indeed, if this were not the case, mutual gravitational attraction would cause the entire Galaxy to collapse into the galactic center. In the same way, the Moon is kept from crashing into Earth and the planets from crashing into the Sun because of their motion around their orbits.

Measuring the rotation of our Galaxy accurately is a difficult business. But such challenging measurements have been made, as we shall see, and the results led to a remarkable conclusion: Most of the mass of the Galaxy is in the form of *dark matter,* a mysterious sort of material that emits no light at all.

Measuring How the Milky Way Rotates

Radio observations of 21-cm light from hydrogen gas across our Galaxy provide important clues about our Galaxy's rotation. Doppler shift measurements of this long-wavelength light indicate that stars and gas all orbit in the same direction around the galactic center, just as the planets all orbit in the same direction around the Sun. Measurements also show that the orbital speed of stars and gas about the galactic center is fairly uniform throughout much of the Galaxy's disk (**Figure 13-17**). As a result, stars inside the Sun's orbit complete a trip around the galactic center more quickly than the Sun because the stars have a shorter distance to travel. Conversely, stars outside the Sun's orbit take longer to go once around the galactic center because they have farther to travel. As seen by Earth-based astronomers moving along with the Sun, stars inside the Sun's orbit overtake and pass us, while we overtake and pass stars outside the Sun's orbit (Figure 13-17*a*).

(a) The orbital speed of stars and gas around the galactic center is nearly uniform throughout most of our Galaxy.

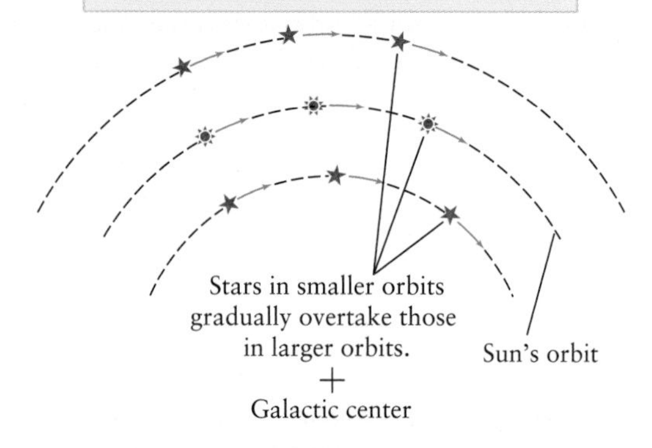

(b) If our Galaxy rotated like a solid disk, the orbital speed would be greater for stars and gas in larger orbits.

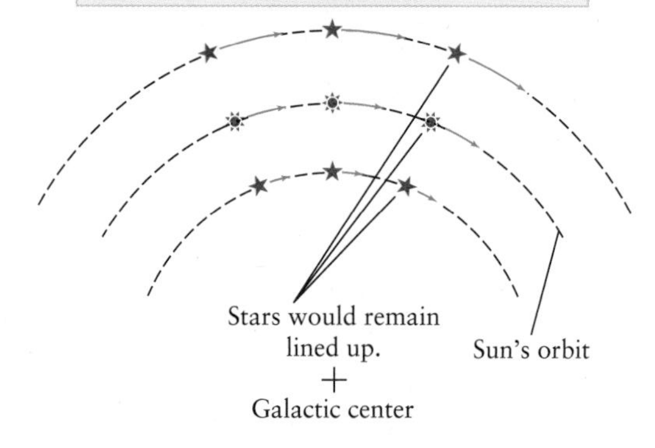

(c) If the Sun and stars obeyed Kepler's third law, the orbital speed would be less for stars and gas in larger orbits.

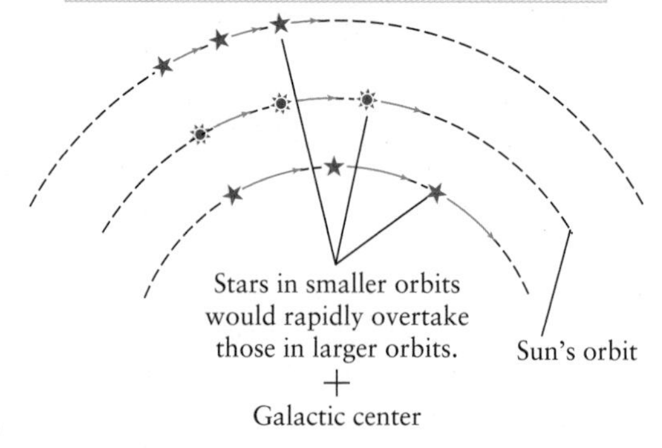

FIGURE 13-17 The Rotation of Our Galaxy (a) This schematic diagram shows three stars (the Sun and two others) orbiting the center of the Galaxy. Although they start off lined up, the stars become increasingly separated as they move along their orbits. Stars inside the Sun's orbit overtake and move ahead of the Sun, while stars far from the galactic center lag behind the Sun. (b) The stars would remain lined up if the Galaxy rotated like a solid disk. This is not what is observed. (c) If stars orbited the galactic center in the same way that planets orbit the Sun, stars inside the Sun's orbit would overtake us faster than they are observed to do.

CAUTION Note that when we say that objects in different parts of the Galaxy orbit at the same speed, we do *not* mean that the Galaxy rotates like a solid disk. All parts of a rotating solid disk—a CD or DVD, for example—take the same time to complete one rotation. Because the outer part of the disk has to travel around a larger circle than the inner part, the speed (distance per time) is greater in the outer part (Figure 13-17*b*). By contrast, the orbital speed of material in our Galaxy is roughly the *same* at all distances from the galactic center.

The most familiar examples of orbital motion are the motions of the planets around the Sun. The farther a planet is from the Sun, the less gravitational force it experiences and the slower the speed it needs to have to remain in orbit. The same would be true for the orbits of stars and gas in the Galaxy *if* they were held in orbit by a single, massive object at the galactic center (Figure 13-17*c*). Hence, the 21-cm observations of our Galaxy, which show that the speed does *not* decrease with increasing distance from the galactic center, demonstrate that there is no such single, massive object holding objects in their galactic orbits.

Instead, what keeps a star in its orbit around the center of the Galaxy is the combined attraction of gravity exerted on it by *all* of the mass, including stars, gas, and dust that lies within the star's orbit. It turns out that the gravitational attraction from the combined matter *outside* a star's orbit has little or no net effect on the star's motion around the galactic center. This gives us a tool for determining the Galaxy's mass and how that mass is distributed.

ConceptCheck 13-8: Which would have a greater impact on changing the length of time it takes the Sun to orbit the galactic center—adding more stars to the galactic center or adding more stars to the galactic halo?

The Sun's Orbital Motion and the Mass of the Galaxy

An important example is the orbital motion of the Sun (and the rest of the solar system along with it) around the center of the Galaxy. If we know the average distance from the galactic center (semimajor axis) and period of the Sun's orbit, we can use Newton's form of Kepler's third law to determine the mass of that portion of the Galaxy that lies within the orbit. We saw earlier that the Sun is about 25,000 light-years from the galactic center; this is the semimajor axis of the Sun's orbit. The orbit is in fact nearly circular, so we can regard 25,000 ly as the radius r of the orbit. In one complete trip around the Galaxy, the Sun travels a distance equal to the circumference of its orbit, which is $2\pi r$. The time required for one orbit, or orbital period P, is equal to the distance traveled divided by the Sun's orbital speed v:

Period of the Sun's orbit around the galactic center

$$P = \frac{2\pi r}{v}$$

P = orbital period of the Sun
r = distance from the Sun to the galactic center
v = orbital speed of the Sun

Unfortunately, we cannot tell the Sun's orbital speed from 21-cm observations, since these reveal only how fast things are moving relative to the Sun. Instead, we need to measure how the Sun is moving relative to a background that is not rotating along with the rest of the Galaxy. Such a background is provided by distant galaxies beyond the Milky Way and by the globular clusters. (Since globular clusters lie outside the plane of the Galaxy, they do not take part in the rotation of the disk, as shown in Figure 13-8.) By measuring the Doppler shifts of these objects and averaging their velocities, astronomers deduce that the Sun is moving along its orbit around the galactic center at about 220 km/s—about 790,000 kilometers per hour or 490,000 miles per hour!

Using this information, we find that the Sun's orbital period is

$$P = \frac{2\pi \times 8000 \text{ pc}}{220 \text{ km/s}} \times \frac{3.09 \times 10^{12} \text{ km}}{1 \text{ pc}} = 7.1 \times 10^{15} \text{ s}$$
$$= 2.2 \times 10^{8} \text{ years}$$

Traveling at 790,000 km/h, it takes the Sun about 220 million years to complete one trip around the Galaxy. As a point of reference, in the 65 million years since the demise of the dinosaurs, our solar system has traveled less than a third of the way around its orbit. The Galaxy is a very large place!

CalculationCheck 13-2: Since it formed 4.57 billion years ago, how many trips around the Galaxy has the Sun made?

Rotation Curves and the Mystery of Dark Matter

Stars in our Galaxy move faster than we expect, suggesting more mass is present than we can see.

In recent years, astronomers have been astonished to discover how much matter may lie outside the Sun's orbit. The clues come from 21-cm light emitted by hydrogen in spiral arms that extend to the outer reaches of the Galaxy. Because we know the true speed of the Sun, we can convert the Doppler shifts of this radiation into actual speeds for the spiral arms. This calculation gives us a **rotation curve,** a graph of the speed of galactic rotation measured outward from the galactic center (**Figure 13-18**). We would expect that for gas clouds beyond the confines of most of the Galaxy's mass, the orbital speed should decrease with increasing distance from the Galaxy's center, just as the orbital speeds of the planets decrease with increasing distance from the Sun (see Figure 13-17c). But as Figure 13-18 shows, the Galaxy's rotation curve is quite flat, indicating roughly uniform orbital speeds well beyond the visible edge of the galactic disk.

To explain these nearly uniform orbital speeds in the outer parts of the Galaxy, astronomers conclude that a large amount of matter must lie outside the Sun's orbit. When this matter is included, the total mass of our Galaxy could exceed 10^{12} $M_{\odot}$ or more, of which about 10% is in the form of stars. This implies that our Galaxy contains roughly 200 billion stars.

These observations lead to a profound mystery. Stars, gas, and dust account for only about 10% of the Galaxy's total mass. What, then, makes up the remaining 90% of

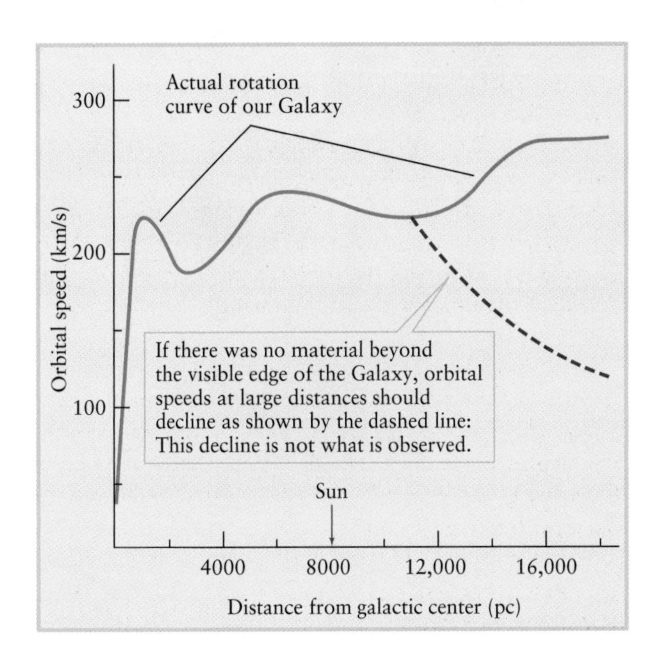

FIGURE 13-18
The Galaxy's Rotation Curve The blue curve shows the orbital speeds of stars and gas in the disk of the Galaxy out to a distance of 60,000 ly (shown as 18,000 pc) from the galactic center. (Very few stars are found beyond this distance.) The dashed purple curve indicates how this orbital speed should decline beyond the confines of most of the Galaxy's visible mass. Because there is no such decline, there must be an abundance of invisible dark matter that extends to great distances from the galactic center.

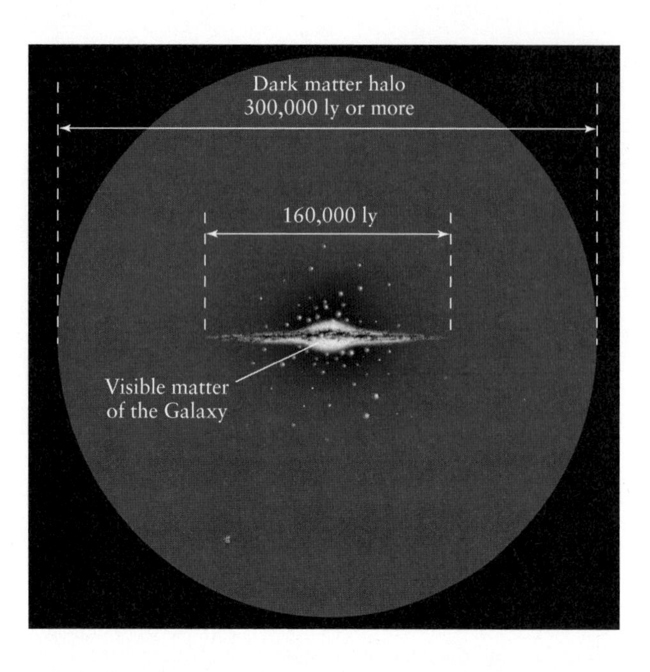

FIGURE 13-19 The Galaxy and Its Dark Matter Halo The dark matter in our Galaxy forms a spherical halo whose center is at the center of the visible Galaxy. The extent of the dark matter halo is unknown, but its diameter is at least 300,000 ly. The total mass of the dark matter halo is at least 10 times the combined mass of all of the stars, dust, gas, and planets in the Milky Way.

the matter in our Galaxy? Whatever it is, it is dark. It does not show up on photographs, nor indeed in images made in any part of the electromagnetic spectrum. This unseen material, which is by far the predominant constituent of our Galaxy, is called **dark matter.** We sense its presence only through its gravitational influence on the orbits of stars and gas clouds.

CAUTION Be careful not to confuse dark *matter* with dark *nebulae*. A dark nebula like the one in Figure 11-2 emits no visible light, but does radiate at longer wavelengths. By contrast, no electromagnetic radiation of any kind has yet been discovered coming from dark matter.

Observations of star groupings outside the Milky Way suggest that our Galaxy's dark matter forms a spherical halo centered on the galactic nucleus, like the halo of globular clusters and high-velocity stars shown in Figure 13-8. However, the dark matter halo is much larger; it may extend to a distance of 326,000 ly to 652,000 ly from the center of our Galaxy, some 2 to 4 times the extent of the visible halo (**Figure 13-19**). Analysis of the rotation curve in Figure 13-18 shows that the density of the dark matter halo decreases with increasing distance from the center of the Galaxy.

ConceptCheck 13-9: How would the motion of stars at the edge of our Galaxy be different if there were no dark matter present?

Dark Matter Speculations

What is the nature of this mysterious dark matter? One proposal is that the dark matter halo is composed, at least in part, of dim objects with masses less than 1 $M_\odot$. These objects, which could include brown dwarfs, white dwarfs, or black holes, are called **massive compact halo objects,** or **MACHOs.** Astronomers have searched for MACHOs by monitoring the light from distant stars. If a MACHO passes between us and the star, its gravity will bend the light coming from the star. As **Figure 13-20** shows, the MACHO's gravity acts like a lens that focuses the light from the star.

Astronomers have indeed detected MACHOs in this way, but not nearly enough to completely solve the dark matter mystery. MACHOs with very low mass (10^{-6} $M_\odot$ to 0.1 $M_\odot$ each) do not appear to be a significant part of the dark matter

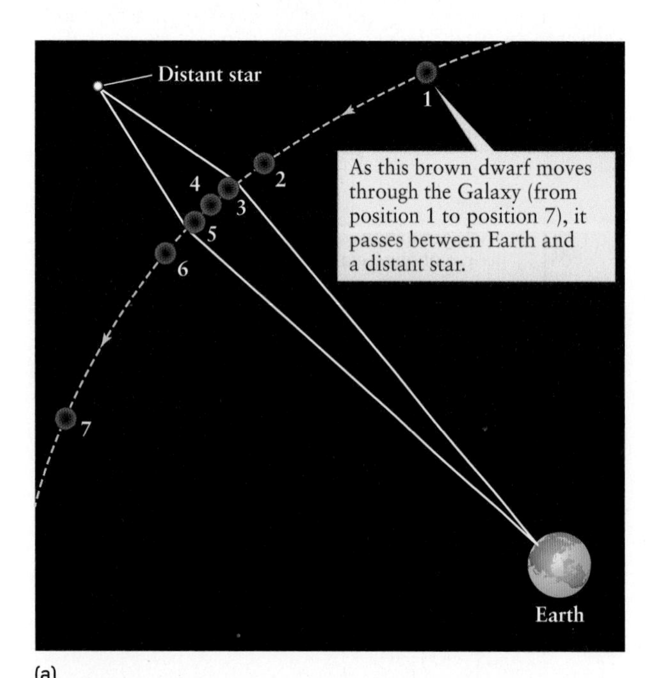

(a)

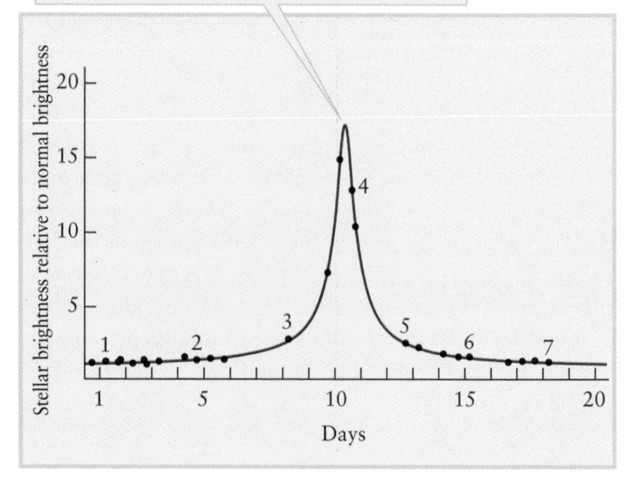

(b)

FIGURE 13-20 Microlensing by Dark Matter in the Galactic Halo (a) If a dense object such as a brown dwarf or black hole passes between Earth and a distant star, the gravitational curvature of space around the dense object deflects the starlight and focuses it in our direction. This effect is called microlensing. (b) This light curve shows the gravitational microlensing of light from a star in the Galaxy's central bulge. Astronomers do not know the nature of the object that passed between Earth and this star to cause the microlensing. *(Data from the MACHO and GMAN Collaborations)*

halo. MACHOs of roughly 0.5 $M_{\odot}$ are more prevalent, but account for only about half of the dark matter halo.

The remainder of the dark matter is thought to be much more exotic. One candidate is a neutrino with a small amount of mass. If these neutrinos are sufficiently massive, and if enough of them are present in the halo of the Galaxy, they might constitute a reasonable fraction of the dark matter. Perhaps surprisingly, one type of neutrino can transform into another. These transformations can take place only if neutrinos have a nonzero amount of mass. Thus, neutrinos must comprise at least part of the dark matter, though it is not known how much.

Another speculative possibility that is gaining support among astronomers is a new class of subatomic particle called **weakly interacting massive particles,** or **WIMPs.** These particles, whose existence is suggested by certain theories but has not yet been confirmed experimentally, would not emit or absorb electromagnetic radiation. Physicists are attempting to detect these curious particles, which would have masses 10 to 10,000 times greater than a proton or neutron, by using a large crystal cooled almost to a temperature of absolute zero. If a WIMP should enter this crystal and collide with one of its atoms, the collision would deposit a tiny but measurable amount of heat in the crystal.

As yet, the true nature of dark matter remains a mystery. Furthermore, this mystery is not confined to our own Galaxy. In the next chapter, we will find that other galaxies have the same sort of rotation curve as in Figure 13-18, indicating that they also contain vast amounts of dark matter. Indeed, dark matter appears to make up most of the mass in our universe. Hence, the quest to understand dark matter is one of the most important in modern astronomy.

ConceptCheck 13-10: Why are astronomers convinced that dark matter exists, even though it has not yet been confirmed?

13-5 Spiral arms are caused by density waves that sweep around our Galaxy

The disk shape of our Galaxy is not difficult to understand. The flattening of spinning matter into a thin disk is what naturally happens when a large number of objects are put into orbit around a common center: Over time the objects tend naturally to orbit in the same plane. This is what happened when our solar system formed from the solar nebula. There a giant cloud of material eventually organized itself into planets, all of which orbit in nearly the same plane. In like fashion, the disk of our Galaxy, which is also made up of a large number of individual objects orbiting a common center, is very flat (see Figure 13-8). Understanding why our Galaxy has spiral arms presents more of a challenge.

The Density-Wave Model

One early explanation for the Galaxy's spiral structure was that the material in the Galaxy somehow condensed into a spiral pattern from the very start. In this view, once stars, gas, and dust had become concentrated within the spiral arms, the pattern would remain fixed. This would be possible only if the Galaxy rotated like a solid disk (see Figure 13-17*b*); the fixed pattern would be like the spokes on a rotating bicycle wheel. But the reality is that the Galaxy is not a solid disk. As we have seen, stars, gas, and dust all orbit the galactic center with approximately the same speed, as shown in Figure 13-17*a*. Let us see why this makes it impossible for a rigid spiral pattern to persist.

Imagine four stars, *A*, *B*, *C*, and *D*, that originally lay on a line extending outward from the galactic center (**Figure 13-21**). In a given amount of time, each of the stars travels the same distance around its orbit. But because the innermost star has a smaller orbit than the others, it takes less

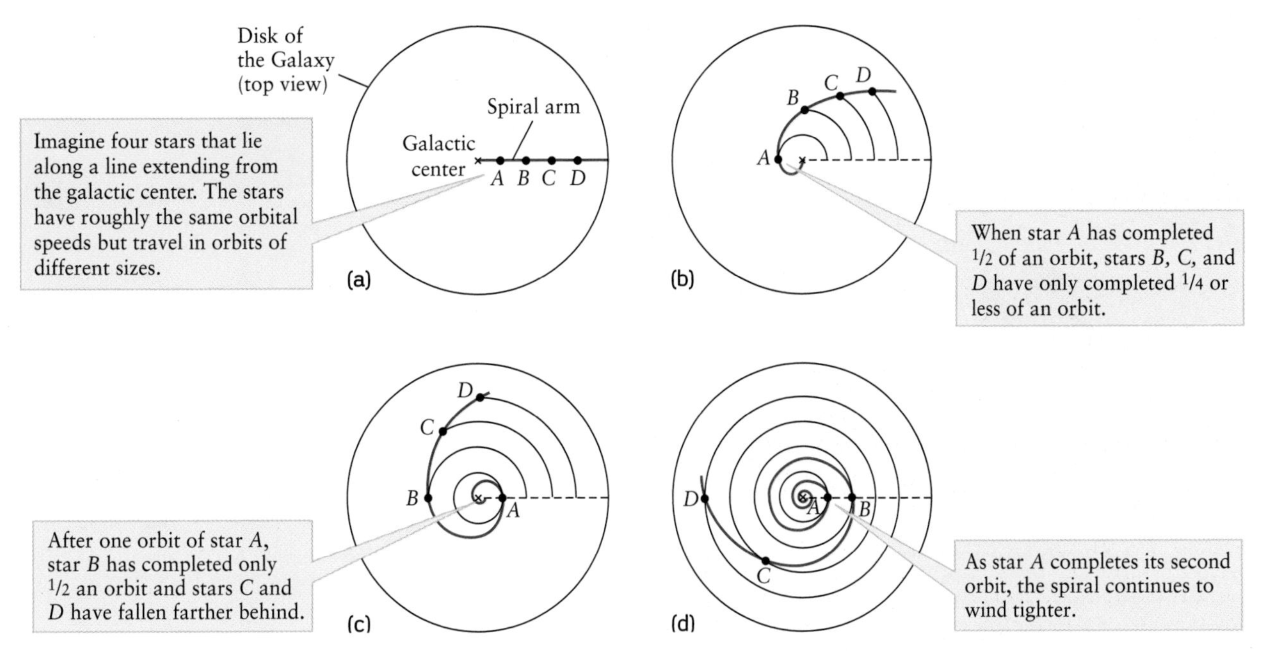

FIGURE 13-21
The Winding Dilemma This series of drawings shows that spiral arms in galaxies like the Milky Way cannot simply be assemblages of stars. If they were, the spiral arms would "wind up" and disappear in just a few hundred million years.

time to complete one orbit. As a result, a line connecting the four stars is soon bent into a spiral (Figure 13-21*b*). Moreover, the spiral becomes tighter and tighter with the passage of time (Figures 13-21*c* and 13-21*d*). This "winding up" of the spiral arms causes the spiral structure to disappear completely after a few hundred million years—a very brief time compared to the age of our Galaxy, thought to be as much as 13.8 billion (1.38×10^{10}) years.

Figure 13-21 suggests that the Milky Way's spiral arms should have disappeared by now. The fact that they have not is puzzling. It shows that the spiral arms cannot simply be assemblages of stars and interstellar matter that travel around the Galaxy together, like a troop of soldiers marching in formation around a flagpole. In other words, the spiral arms cannot be made of anything *material*. What, then, can the spiral arms be?

In the 1940s, the Swedish astronomer Bertil Lindblad proposed that the spiral arms of a galaxy are actually a pattern that moves through the galaxy like ripples on water. This idea was greatly enhanced and embellished in the 1960s by the American astronomers Chia Chiao Lin and Frank Shu. In this picture, spiral arms are a kind of wave, like the waves that move across the surface of a pond when you toss a stone into the water. Water molecules pile up at a crest of the wave but spread out again when the crest passes. By analogy, Lindblad, Lin, and Shu pictured a pattern of **density waves** sweeping around the Galaxy. These waves make matter pile up in the spiral arms, which are the crests of the waves. Individual parts of the Galaxy's material are compressed only temporarily when they pass through a spiral arm. The pattern of spiral arms persists, however, just as the waves made by a stone dropped in the water can persist for some time after the stone has sunk.

To understand better how a density wave operates in a galaxy, think again about a water wave in a pond. If one part of the pond is disturbed by dropping a stone into it, the molecules in that part will be displaced a bit. They will nudge the molecules next to them, causing those molecules to be displaced and to nudge the molecules beyond them. In this way the wave disturbance spreads throughout the pond.

In a galaxy, stars play the role of water molecules. Although stars and interstellar clouds of gas and dust are separated by vast distances, they can nonetheless exert forces on each other because they are affected by each other's gravity. If a region of above-average density should form, its gravitational attraction will draw nearby material into it. The displacement of this material will change the gravitational force that it exerts on other parts of the galaxy, causing additional displacements. In this way a spiral-shaped density wave can travel around the disk of a galaxy.

ANALOGY A key feature of density waves is that they move more slowly around a galaxy than do stars or interstellar matter. To visualize this, imagine workers painting a line down a busy freeway (**Figure 13-22**). The cars normally cruise along the freeway at high speed, but the crew of painters is moving much more slowly. When the cars come upon the painters, they must slow down temporarily to avoid hitting anyone. As seen from the air, cars are jammed together around the painters. An individual car spends only a few

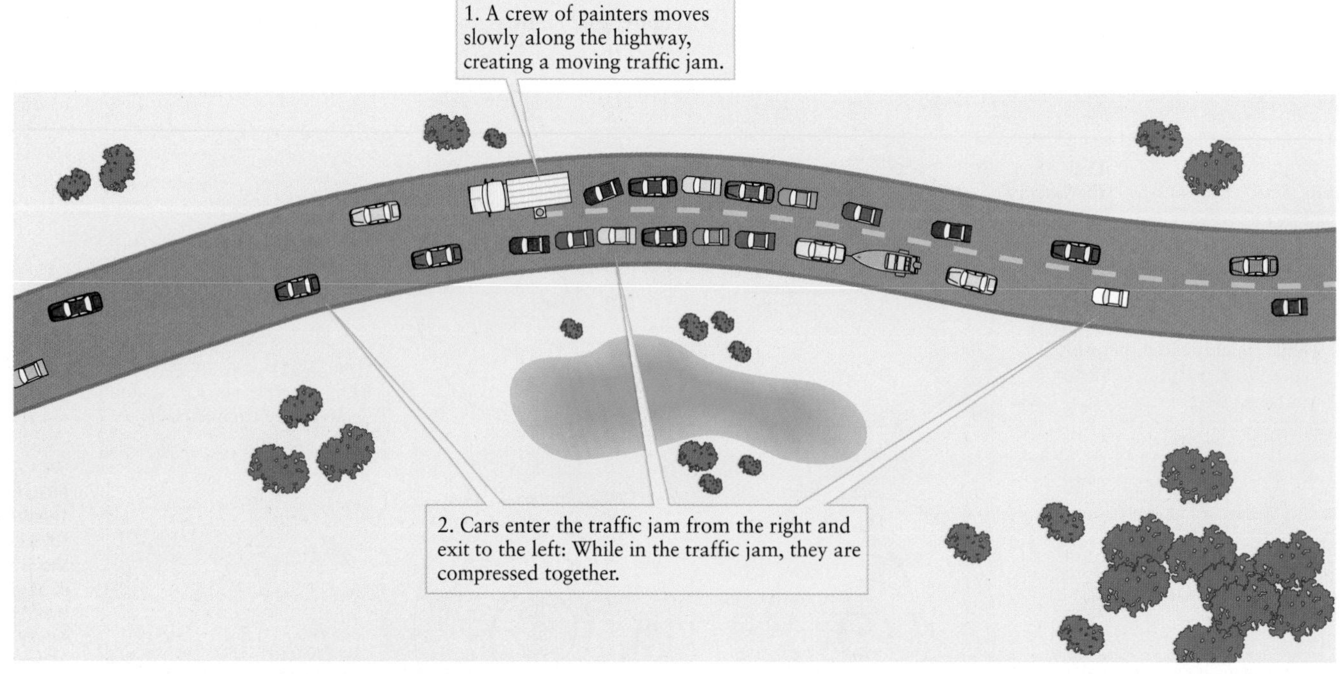

FIGURE 13-22 A Density Wave on the Highway A density wave in a spiral galaxy is analogous to a crew of painters moving slowly along the highway, creating a moving traffic jam. Like such a traffic jam, a density wave in a spiral galaxy is a slow-moving region where stars, gas, and dust are more densely packed than in the rest of the galaxy. As the material of the galaxy passes through the density wave, it is compressed. This triggers star formation, as Figure 13-24 shows.

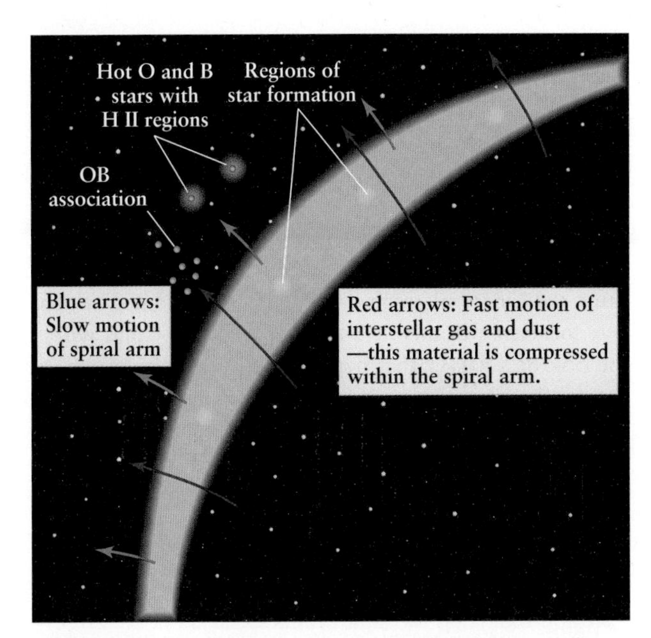

FIGURE 13-23 Star Formation in the Density-Wave Model A spiral arm is a region where the density of material is higher than in the surrounding parts of a galaxy. Interstellar matter moves around the galactic center rapidly (shown by the red arrows) and is compressed as it passes through the slow-moving spiral arms (whose motion is shown by the blue arrows). This compression triggers star formation in the interstellar matter, so that new stars appear on the "downstream" side of the densest part of the spiral arms.

moments in the traffic jam before resuming its usual speed, but the traffic jam itself lasts all day, inching its way along the road as the painters advance.

A similar crowding takes place when interstellar matter enters a spiral arm. This crowding plays a key role in the formation of stars and the recycling of the interstellar medium. As interstellar gas and dust moves through a spiral arm, it is compressed into new nebulae (**Figure 13-23**). This compression begins the very important rebirthing process by which new stars form.

These freshly formed stars continue to orbit around the center of their galaxy, just like the matter from which they formed. The most luminous among these are the hot, massive, blue O and B stars, which may have emission nebulae (H II regions) associated with them. These stars have main-sequence lifetimes of only 3 million to 15 million years, which is very short compared to the 220 million years required for the Sun to make a complete orbit around the Galaxy.

As a result, these luminous O and B stars can travel only a relatively short distance before dying off. Therefore, these stars, and their associated H II regions, are only seen in or slightly "downstream" of the spiral arm in which they formed. **Figure 13-24** illustrates this for the spiral galaxy M51. Less massive stars have much longer main-sequence lifetimes, and

Stars are evenly spread across our Galaxy's disk, but new stars within interstellar clouds clump into spiral arms.

(a) An infrared view of M51 shows the locations of dust

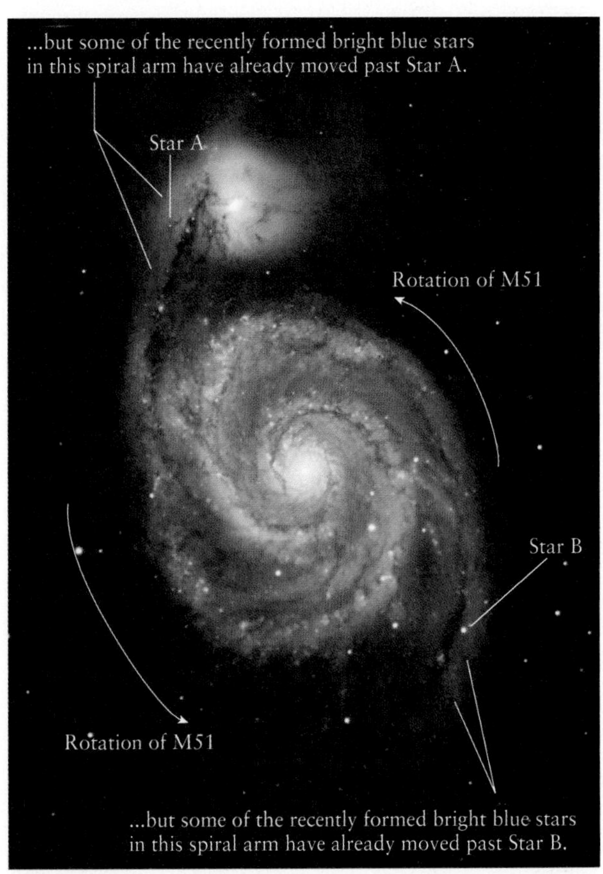

(b) A visible-light view of M51 shows the locations of young stars

FIGURE 13-24 Star Formation in the Whirlpool Galaxy The spiral galaxy M51 (called the Whirlpool) is a real-life example of the density-wave model illustrated in Figure 13-22. (a) This infrared image shows where dust has piled up as the material within M51 passes through its spiral arms. Radio images of M51 show that hydrogen gas also piles up in the same locations, thus beginning the formation of new stars. (b) By the time stars complete their formation process, their motion around the galaxy has swept them "downstream" of the positions of greatest dust density, as depicted in Figure 13-25. *(a: NASA, JPL-Caltech, and R. Kennicutt [U. of Arizona]; b: NASA, ESA, S. Beckwith [STScI], and The Hubble Heritage Team [STScI/AURA])*

thus their orbits are able to carry them all around the galactic disk. These less-luminous stars are found throughout the disk, including between the spiral arms (see Figure 13-15*c*).

The density-wave model of spiral arms explains why the disk of our Galaxy is dominated by metal-rich Population I stars. Because the material left over from the death of ancient stars is rich in heavy elements, new generations of stars formed in spiral arms are likely to be more metal-rich than their ancestors. **Cosmic Connections: Stars in the Milky Way** illustrates this cycle of star birth and death within the disk of our Galaxy.

The density-wave model is still under development. One problem is finding a driving mechanism that keeps density waves going in spiral galaxies. After all, density waves expend an enormous amount of energy to compress the interstellar gas and dust. Hence, we would expect that density waves should eventually die away, just as do ripples on a pond. The American astronomers Debra and Bruce Elmegreen have suggested that gravity can supply that needed energy. As stated earlier, the central bulge of our Galaxy is elongated into a bar shape, much like the central bulge of the galaxy M83 (see Figure 13-15 and Figure 13-16). The asymmetric gravitational field of such a bar pulls on the stars and interstellar matter of a galaxy to generate density waves. Another factor that may help to generate and sustain spiral arms is the gravitational interactions *between* galaxies, which we will discuss more deeply in the next chapter.

ConceptCheck 13-11: If the Galaxy were not experiencing density waves, what shape would it have?

The Self-Propagating Star-Formation Model

Spiral density waves may not be the whole story behind spiral arms in our Galaxy and other galaxies. The reason is that spiral density waves should produce very well defined spiral arms. We do indeed see many so-called **grand-design spiral galaxies** (**Figure 13-25*a***), with thin, graceful, and well-defined spiral arms. But in some galaxies, called **flocculent spiral galaxies** (Figure 13-25*b*), the spiral arms are broad, fuzzy, chaotic, and poorly defined ("flocculent" means "resembling wool").

To explain such flocculent spirals, M. W. Mueller and W. David Arnett in 1976 proposed an alternate theory that arms are a natural result of star formation. Imagine that star formation begins in a dense interstellar cloud within the disk of a galaxy that does not yet have spiral arms. As soon as hot, massive stars form, their radiation and stellar winds compress nearby matter, triggering the formation of additional stars in that gas. When massive stars become supernovae, they produce shock waves that further compress the surrounding interstellar medium, thus encouraging still more star formation.

Although all parts of this broad, star-forming region have approximately the same orbital speed about the galaxy's center, the inner regions have a shorter distance to travel to complete one orbit than the outer regions. As a result, the inner edges of the star-forming region move ahead of the outer edges as the galaxy rotates. The bright O and B stars and their nearby glowing nebulae soon become stretched out in the form of a spiral arm. These spiral arms come and go essentially at random across a galaxy. Bits and pieces of spiral arms appear where star formation has recently begun but fade and disappear at other locations where all the massive stars have died off. This particular genre of star formation therefore tends to produce flocculent spiral galaxies that have a chaotic appearance with poorly defined spiral arms, like the galaxy in Figure 13-25*b*.

The two theories presented here are very different in character. In the density-wave model, star formation is caused by the spiral arms; in the other approach, by contrast, the spiral arms are instead simply caused by star formation.

(a) Grand-design spiral galaxy

(b) Flocculent spiral galaxy

FIGURE 13-25 RIVUXG **Variety in Spiral Arms** The differences from one spiral galaxy to another suggest that more than one process can create spiral arms. (a) NGC 628 is a grand-design spiral galaxy with thin, well-defined spiral arms. (b) NGC 4414 is a flocculent spiral galaxy with fuzzy, poorly defined spiral arms. *(Courtesy Gemini Observatory/Olivier Vallejo)*

COSMIC CONNECTIONS Stars in the Milky Way

Different populations of stars are found in different neighborhoods of our home galaxy. (The galaxy shown here is another spiral galaxy similar to our own.) The variations from one galactic region to another are due to the presence or absence of ongoing star formation.

1. Held together by gravitational attraction, matter in the Milky Way spins counterclockwise.

Orbital motion

2. Dust and gas piles up in arms, providing raw materials and energy for formation of new stars.

3. Giant blue stars form in the dark dusty lanes, then appear fully formed some distance "downstream." These short-lived stars die out before they travel far from their birthplace.

4. Most of the stars produced in the dust lanes are relatively tiny, usually smaller than our Sun. Because they are long-lived, they make many orbits around the Galaxy.

5. Metal-enriched gas and dust are increased in the Galaxy when stars die. When these materials are compressed by a density wave, they are incorporated into a new generation of stars, resulting in metal-rich Population I stars like our Sun.

6. The yellowish bulge contains both old Population I stars and even older Population II stars. No mechanism to trigger star formation exists there and it is too far for hot, blue stars to travel from where they formed.

R I V U X G

7. Also orbiting the Galaxy are the globular clusters containing only very old, first-generation, metal-poor Population II stars. These clusters lack gas and dust to form new stars.

R I V U X G

8. Not visible to any telescope is the Galaxy's far-from-understood dark matter, which emits no light of any kind. The spherical halo of dark matter has at least 10 times the combined mass of all the Galaxy's stars, planets, gas, and dust.

A complete and correct description of spiral arms in our Galaxy remains a topic of active research.

ConceptCheck 13-12: If astronomers learned that the earliest galaxies in the universe were flocculent spiral galaxies, what would this suggest about which came first: stars or spiral arms?

13-6 Infrared and radio observations are used to probe the galactic nucleus

The innermost region of our Galaxy is an active, crowded place. If you lived on a planet near the galactic center, you would see a million stars as bright as Sirius, the brightest single star in our own night sky. The total intensity of starlight from all those nearby stars would be equivalent to 200 of our full moons. In effect, night would never really fall on a planet near the center of the Milky Way. At the center of this empire of light, however, lies the darkest of all objects in the universe—a black hole millions of times more massive than the Sun.

Sagittarius A*: Heart of Darkness

Because of the severe interstellar dust interference blocking visual wavelengths, most of our information about the galactic center comes from infrared and radio observations. **Figure 13-26** shows three infrared views of the center of our Galaxy. Figure 13-26*a* is a wide-angle view covering a 50° segment of the Milky Way from Sagittarius through Scorpius. The prominent reddish band through the center of this false-color infrared image is a layer of dust in the plane of the Galaxy. Figure 13-26*b* is an infrared view of the galactic center. It is surrounded by numerous streamers of dust (shown in blue). The strongest infrared emission (shown in white) comes from a grouping of several powerful sources of radio waves. One of these sources, **Sagittarius A*** (say "A star"), lies at the very center of the Galaxy. (Its position, pinpointed with simultaneous observations by radio telescopes scattered around the world, seems to be very near the gravitational center of the Galaxy.) The high-resolution infrared view in Figure 13-26*c* shows hundreds of stars crowded within about 1 ly of Sagittarius A*. Compare this to our region of the Galaxy, where the average distance between stars is more than a light-year.

Sagittarius A* itself does not appear in infrared images. Nonetheless, astronomers have used infrared observations to make truly startling discoveries about this object. Since the 1990s, two research groups—one headed by Reinhard Genzel of the Max-Planck-Institute for Extraterrestrial Physics in Garching, Germany, and another led by Andrea Ghez at the University of California, Los Angeles—have been using infrared detectors to monitor the motions of stars in the immediate vicinity of Sagittarius A*. They have found a number of stars orbiting around Sagittarius A* (**Figure 13-27**) at speeds in excess of 3 million mi/h (1500 km/s). By comparison, Earth orbits the Sun at a lackadaisical 67,000 mi/h (30 km/s.) In 2000 the UCLA group observed one such star, called SO-16, as its elliptical orbit brought it within a mere 45 AU from Sagittarius A*—1½ times the distance from the Sun to Neptune. At its closest approach, SO-16 was traveling at a breathtaking speed of 27 million mi/h (12,000 km/s), which is about 4% of the speed of light!

(a) A wide-angle infrared view

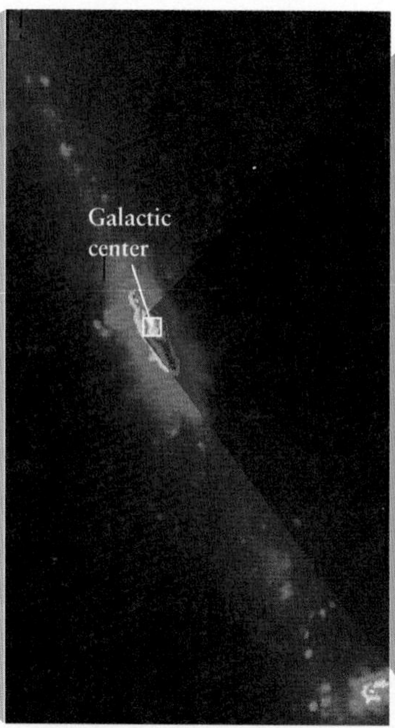

(b) A close-up view shows a more luminous region at the galactic center.

(c) An extreme close-up view centered on Sagittarius A*, a radio source at the very center of the Milky Way Galaxy, shows hundreds of stars within 1 ly (0.3 pc).

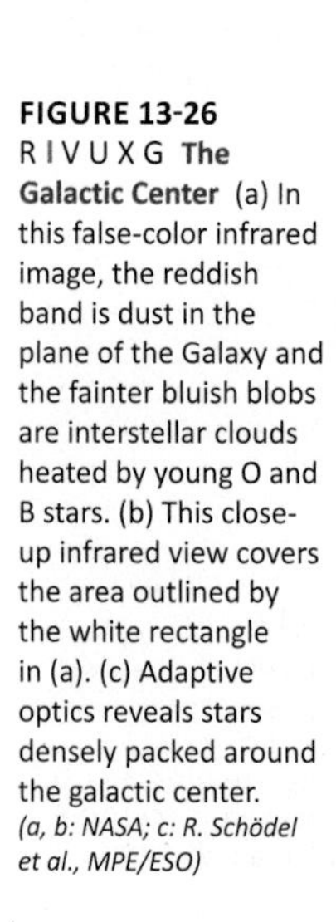

FIGURE 13-26 R I V U X G **The Galactic Center** (a) In this false-color infrared image, the reddish band is dust in the plane of the Galaxy and the fainter bluish blobs are interstellar clouds heated by young O and B stars. (b) This close-up infrared view covers the area outlined by the white rectangle in (a). (c) Adaptive optics reveals stars densely packed around the galactic center. *(a, b: NASA; c: R. Schödel et al., MPE/ESO)*

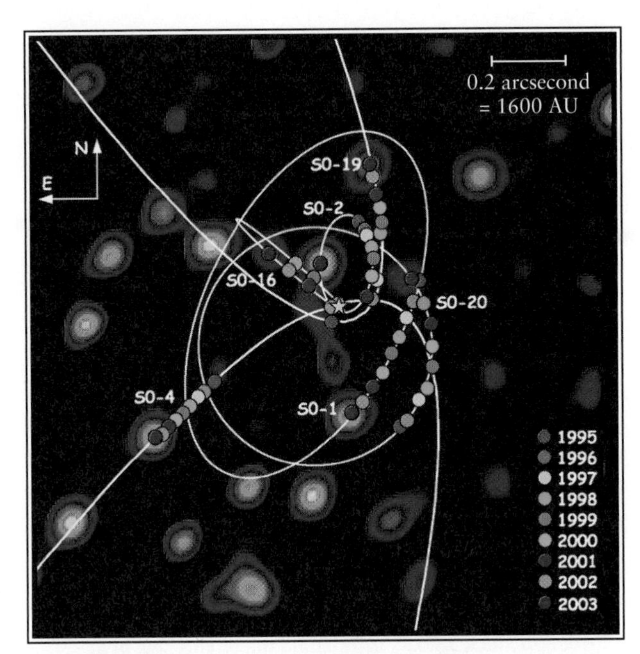

FIGURE 13-27 R I V U X G **Stars Orbiting Sagittarius A*** The colored dots superimposed on this infrared image show the motion of six stars in the vicinity of the unseen massive object (denoted by the yellow five-pointed star) at the position of the radio source Sagittarius A*. The orbits were measured over an 8-year period. This plot indicates that the stars are held in orbit by a black hole of 3.7×10^6 solar masses. *(These images/animations were created by Prof. Andrea Ghez and her research team at UCLA and are from data sets obtained with the W. M. Keck Telescopes. Image creators include Andrea Ghez, Sylvana Yelda, Leo Meyer, Jessica Lu, Seth Hornstein, and Angelle Tanner. UCLA Galactic Center Group)*

In order to keep stars like SO-16 in such small, rapid orbits, Sagittarius A* must exert a powerful gravitational force and hence must be very massive. By applying Newton's form of Kepler's third law to the motions of these stars around Sagittarius A*, the UCLA group calculated the mass of Sagittarius A* to be a remarkable 3.7 *million* solar masses (3.7×10^6 $M_\odot$). Furthermore, the small separation between SO-16 and Sagittarius A* at closest approach shows that Sagittarius A* can be no more than 45 AU in radius. An object this massive and this compact can only be one thing: a supermassive black hole.

ConceptCheck 13-13: How would stars around Sagittarius A* be moving if a supermassive black hole were not found at the galactic center?

X-rays from Around a Supermassive Black Hole

Evidence in favor of this picture comes from the Chandra X-ray Observatory, which has observed X-ray flares coming from Sagittarius A*. The flares brighten dramatically over the space of just 10 minutes, which shows that the size of the flare's source can be no larger than the distance that light travels in 10 minutes. In 10 minutes light travels a distance of about 100 million miles (1.8×10^8 km)—slightly more than the distance between the Sun and Earth—and only a black hole could pack a mass of 3.7×10^6 $M_\odot$ into a volume that size or smaller. The X-ray flares were presumably emitted by blobs of material that were compressed and heated as they fell into the black hole.

The X-ray flares from Sagittarius A* are relatively feeble, which suggests that the supermassive black hole is swallowing only relatively small amounts of material. But the region around Sagittarius A* is nonetheless an active and dynamic place. **Figure 13-28*a*** is a wide-angle radio image of the galactic center covering an area more than 60 pc (200 ly) across. Huge filaments of gas stretch for 65 ly northward

A supermassive black hole exists at the center of the Milky Way as revealed by X-rays and rapid star movements.

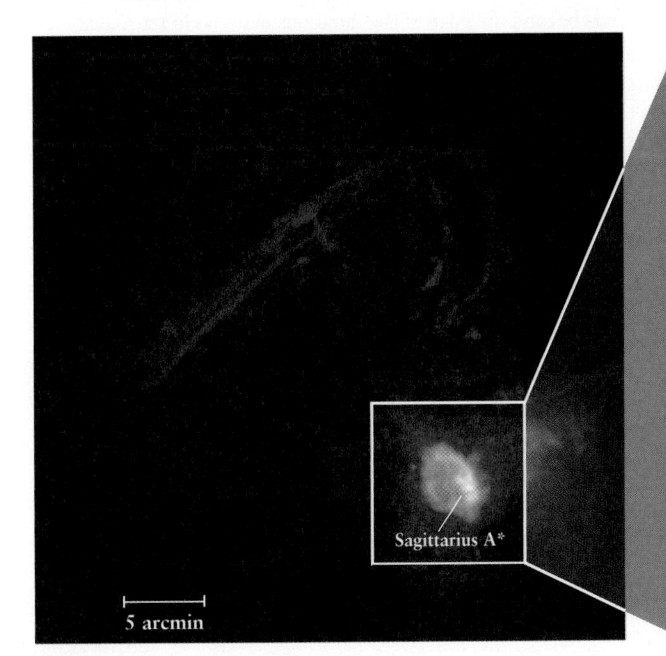

R I V U X G
(a) A radio view of the galactic center

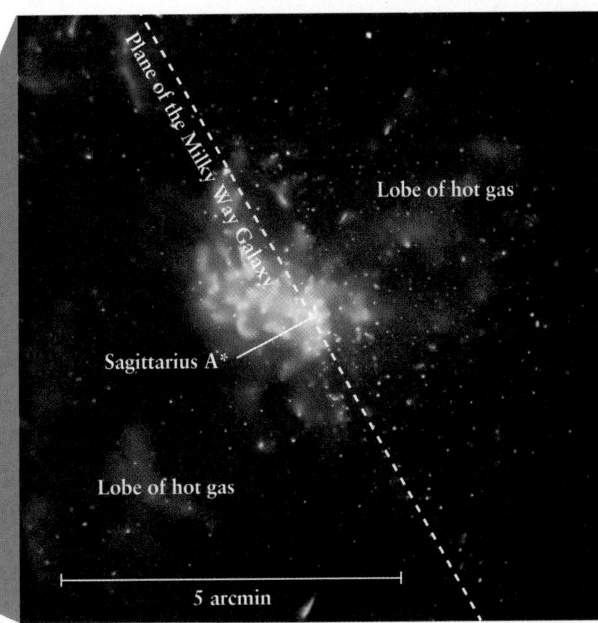

R I V U V G
(b) An X-ray view of the galactic center

FIGURE 13-28 The Energetic Center of the Galaxy (a) The area shown in this radio image has the same angular size as the full moon. Sagittarius A*, at the very center of the Galaxy, is one of the brightest radio sources in the sky. Magnetic fields shape nearby interstellar gas into immense, graceful arches. (b) This composite of images at X-ray wavelengths from 0.16 nm to 0.62 nm shows lobes of gas on either side of Sagittarius A*. The character of the X-ray emission shows that the gas temperature is as high as 2×10^7 K. *(a: F. Yusef-Zadeh et al., AUI/NRAO/NSF; b: NASA/CXC/MIT/F. K. Baganoff et al.)*

of the galactic center (to the right and upward in Figure 13-28*a*) and then they abruptly arch southward (down and to the left in the figure). The orderly arrangement of these filaments is reminiscent of prominences on the Sun. This suggests that, as on the Sun, there is ionized gas at the galactic center that is being controlled by a powerful magnetic field. Indeed, much of the radio emission from the galactic center is radiation produced by high-energy electrons spiraling in a magnetic field.

The false-color X-ray image in Figure 13-28*b* shows the immediate vicinity of Sagittarius A*. The black hole is flanked by lobes of hot, ionized, X-ray–emitting gas that extend for dozens of light-years. These are thought to be the relics of immense explosions that may have taken place over the past several thousand years. Perhaps these past explosions cleared away much of the material around Sagittarius A*, leaving only small amounts to fall into the black hole. This could explain why the X-ray flares from Sagittarius A* are so weak.

There is nonetheless evidence that Sagittarius A* has been more active in the recent past. Between 2002 and 2005, Michael Muno and his colleagues at Caltech observed that certain nebulae that lie about 50 ly from Sagittarius A* suddenly became very bright at X-ray wavelengths. The character of the X-ray light was such that it could not have originated from the nebulae themselves. Muno and colleagues concluded that X-rays emitted from Sagittarius A* about 50 years earlier had struck and excited the nebulae, causing the nebulae to glow intensely at X-ray wavelengths. To produce such an intense X-ray glow, an object the size of the planet Mercury must have fallen into the supermassive black hole. The supermassive black hole at the center of our Galaxy is not unique. Observations show that such titanic black holes are a feature of most large galaxies.

Astronomers are still groping for a better understanding of the galactic center. With new technology, it may be possible to actually obtain a picture of the supermassive black hole lurking there. During the coming years, observations from Earth-orbiting satellites as well as from radio and infrared telescopes on the ground will certainly add to our knowledge of the core of the Milky Way.

ConceptCheck 13-14: Why would the brightening of a nebula near the galactic center suggest the presence of a supermassive black hole?

KEY IDEAS AND TERMS

13-1 The Sun is located in the disk of our Galaxy, about 25,000 light-years from the galactic center

- A **galaxy** is an immense collection of stars and interstellar matter, far larger than a star cluster.
- Our Sun is one of several hundred billion stars in the **Milky Way Galaxy,** which is observed from Earth as a wispy band of light stretching across the sky.
- Our Galaxy is filled with visible-light-obscuring dust and surrounded by a **halo** of **globular clusters** containing **RR Lyrae variables,** which are similar but smaller than **Cepheid variables,** allowing distances to be calculated.

13-2 Observations of different types of dust, gas, stars, and star clusters reveal the shape of our Galaxy

- Infrared light passes more easily through dust and gas in our Galaxy, revealing structures.
- Stars are evenly distributed across the **disk** of our Galaxy, about 160,000 ly (50 kpc) in diameter and about 2000 ly (0.6 kpc) thick.
- The center of the Galaxy is surrounded by a tightly packed distribution of stars, called the **central bulge,** which is about 6500 ly (2 kpc) in diameter.

13-3 Observations of star-forming regions reveal that our Galaxy has spiral arms

- Dust and gas in the plane of the Milky Way organize into **spiral arms,** a shape commonly seen in other galaxies.
- Electrons can undergo a **spin-flip transition** emission of energy, which is seen as a **21-cm radio emission,** allowing the Galaxy's structure to be mapped.

13-4 Measuring the rotation of our Galaxy reveals the presence of dark matter

- From studies of the **rotation curves** of the Galaxy, astronomers estimate that the total mass of the Galaxy is about 10^{12} $M_{\odot}$.
- Only about 10% of this mass is in the form of visible stars, gas, and dust. The remaining 90% is in some nonvisible form, called **dark matter,** that extends beyond the edge of the luminous material in the Galaxy.
- Our Galaxy's dark matter is unknown, but might be a combination of **massive compact halo objects** or **MACHOs** (dim, star-size objects), massive neutrinos, and/or **weakly interacting massive particles** or **WIMPs** (relatively massive subatomic particles).

13-5 Spiral arms are caused by density waves that sweep around the Galaxy

- According to the **density-wave** theory, spiral arms are created by density waves that sweep around the Galaxy. The gravitational field of this spiral pattern compresses the interstellar clouds through which it passes, thereby triggering star formation illuminating the spiral arms.
- According to the theory that arms are caused by star formation, spiral arms are caused by the birth of stars over an extended, stretched region in a galaxy and reveal themselves as **grand-design spiral galaxies** and **flocculent spiral galaxies.**

13-6 Infrared and radio observations are used to probe the galactic nucleus

- The innermost part of the Galaxy, or galactic nucleus, has been studied through its radio, infrared, and X-ray emissions (which are able to pass through interstellar dust).
- A strong radio source called **Sagittarius A*** is located at the galactic center, marking the position of a supermassive black hole with a mass of about 3.7×10^6 $M_{\odot}$.

VISUAL LITERACY TASK

Galaxy Rotation Curve

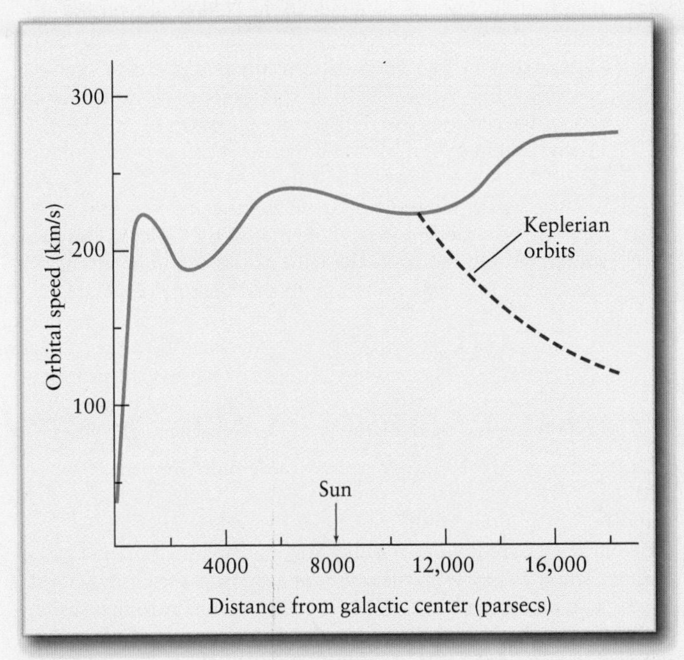

PROMPT: What would you tell a fellow student who said, **"Our Galaxy rotates like a wheel, with the stars farther from the center orbiting fastest, just like our solar system"?**

ENTER RESPONSE:

Guiding Questions:

1. Compared to the Sun's orbital speed, stars orbiting the galactic center at a distance of about 5000 ly orbit
   a. at about the same orbital velocity as the Sun.
   b. at a many-times-lower orbital velocity.
   c. at a many-times-greater orbital velocity.
   d. at highly varying orbital velocities.

2. If the Galaxy were spinning like a solid wheel, stars closer to the center would have
   a. orbital velocities many times greater than the stars in the outer regions.
   b. orbital velocities many times lower than the stars in the outer regions.
   c. about the same orbital velocities.
   d. unpredictable orbital velocities.

3. If the Galaxy were spinning like a solid wheel, stars farthest from the center would
   a. orbit in much greater time than stars closer in.
   b. orbit in much less time than stars closer in.
   c. orbit in the same amount of time as stars closer in.
   d. eventually fracture because of the immense speeds of the outermost stars.

4. If the Galaxy were spinning like our solar system, stars in the outer regions would
   a. have greater orbital velocities than those closer to the center.
   b. have lower orbital velocities than those closer to the center.
   c. have identical orbital velocities to those closer to the center.
   d. orbit the center in the same amount of time as those closer to the center.

5. Our Galaxy has a characteristic rotation that is best described as
   a. much like a solid spinning wheel.
   b. quite similar to our solar system.
   c. stars having similar orbital velocities regardless of their distance from the galactic center.
   d. highly chaotic, with stars having orbital velocities largely unrelated to their distance from the center.

QUESTIONS

Review Questions

1. Why do the stars of the Galaxy appear to form a bright band that extends around the sky?
2. How did observations of globular clusters help astronomers determine our location in the Galaxy?
3. Why are infrared telescopes useful for exploring the structure of the Galaxy? Why is it important to make observations at both near-infrared and far-infrared wavelengths?
4. The galactic halo is dominated by Population II stars, whereas the galactic disk contains predominantly Population I stars. In which of these parts of the Galaxy has star formation taken place recently? Explain your answer.
5. Most interstellar hydrogen atoms emit only radio waves at a wavelength of 21 cm, but some hydrogen clouds emit profuse amounts of visible light (see, for example, Figure 11-1). What causes this difference?
6. The Milky Way map taken at radio wavelengths, shown back in Figure 2-26*b*, has a large gap on the side of the Galaxy opposite to ours. Why is this?

7. In a spiral galaxy, are stars in general concentrated in the spiral arms? Why are spiral arms so prominent in visible-light images of spiral galaxies?
8. How do astronomers determine how fast the Sun moves in its orbit around the Galaxy? How does this speed tell us about the amount of mass inside the Sun's orbit? Does this speed tell us about the amount of mass outside the Sun's orbit?
9. How do astronomers conclude that vast quantities of dark matter surround our Galaxy? How is this dark matter distributed in space?
10. Another student tells you that the Milky Way Galaxy is made up "mostly of stars." Is this statement accurate? Why or why not?
11. What proposals have been made to explain the nature of dark matter?
12. In our Galaxy, why are stars of spectral classes O and B only found in or near the spiral arms? Is the same true for stars of other spectral classes? Explain why or why not.
13. What is the evidence that there is a supermassive black hole at the center of our Galaxy? How is it possible to determine the mass of this black hole?

Web Chat Questions

1. From what you know about stellar evolution, the interstellar medium, and the density-wave theory, explain the appearance and structure of the spiral arms of grand-design spiral galaxies.
2. What observations would you make to determine the nature of the dark matter in our Galaxy's halo?
3. Describe how the appearance of the night sky might change if dark matter were visible to our eyes.
4. Discuss how a supermassive black hole could have formed at the center of our Galaxy.

Collaborative Exercises

1. Student book bags often contain a wide collection of odd-shaped objects. Each person in your group should rummage through her or his own book bag and find one object that is most similar to the Milky Way Galaxy in shape. List the items from each group member's belongings and describe what about the items is similar to the shape of our Galaxy and what about the items is not similar, then indicate which of the items is the closest match.
2. One strategy for identifying a central location is called *triangulation*. In triangulation, a central position can be pinpointed by knowing the distance from each of three different places. First, on a piece of paper, create a rough map showing where each person in your group lives. Second, create a circle around each person's home that has a radius equal to the distance that each home is from your classroom. Label the place where the circles intersect as your classroom. Why can you not identify the position of the classroom with only two people's circles?

Observing Questions

1. Use the *Starry Night*™ program to measure the dimensions of the Milky Way Galaxy. Select **Favourites > Explorations > Milky Way Galaxy** to display a face-on view of a simulation of the Milky Way Galaxy from a distance of 0.128 Mly from Earth. The position of the Sun is labeled, directly below the center of the Galaxy in the view on the screen. Note the spiral arm structure of the galaxy. (Recent research has revealed that our Galaxy appears to have a bar structure surrounding the galactic center.)
    a) Click on the cursor selection tool to the left of the toolbar and activate the angular separation tool to measure the angular separation between the Sun and the center of the Galaxy as seen from this vantage point, and find the corresponding distance in light years (ly). (Both values are shown in the display beside the line drawn by the angular separation tool.) Convert the angular separation to a decimal number in degrees ($1° = 60' = 3600''$), then find the scale of the image of the Galaxy in ly/degree.
    b) Change the cursor to the location scroller and use it to view the Galaxy edge-on and oriented vertically on the screen, with the Sun still below the center. To do this, place the location scroller tool at the center of the right-hand edge of the screen, hold down the mouse button, and move the location scroller directly to the left, toward the center of the Galaxy. Then change to the angular separation tool and use it to measure the angular separation of the Sun from the center of the Galaxy. This value should be approximately the same as you found in part a.
    c) Use the angular separation tool to find the total angular diameter of the Milky Way Galaxy as seen from this viewpoint, measured from one end to the other of this edge-on view (**Zoom** out if necessary). Round off the measurement to the nearest degree and then use the scale that you calculated in part a to find the approximate diameter of the Galaxy in light-years.
2. Use the *Starry Night*™ program to explore the alignment of the Milky Way with respect to our Earth-based coordinate system and its appearance at different electromagnetic wavelengths. Select **Favourites > Explorations > Milky Way** to display a wide-field view of our spiral galaxy from the center of a transparent Earth at 6:30 A.M. on September 1, 2014. The view is similar to that seen by observers on Earth and shows an edge-on view of this Galaxy in a direction toward the galactic center, represented by the marked position of a foreground star, HIP86948.
    a) You can use this view to estimate the alignment of the galactic plane with our reference plane in the sky, the **Celestial Equator,** which is the projection of Earth's equator onto the sky. Display the plane of the Galaxy by clicking on **View > Galactic Guides > Equator** and then show the Celestial Equator by clicking on **View > Celestial Guides > Equator.** Estimate the angle between these two planes.
    b) This visible-light image of the Milky Way shows significant structure, with many dark regions. These dark regions do not represent the absence of material in these directions but the presence of dust and gas clouds obscuring the distant stars. Visible light is scattered and absorbed by these dense clouds. You can examine this galactic plane at other electromagnetic wavelengths. Open the **Options** menu and choose **Stars > Milky Way . . .** to display **Milky Way Options.** Use the brightness slider to show maximum brightness. Click the **Wavelength** box to expand the list of possible wavelengths and display the image of the galactic plane at each wavelength. At which wavelengths does the **Galactic Center** show up most prominently? Why do you think this region shows up brighter at these wavelengths?
3. There are two distinct populations of stars in our Galaxy, the **Population I** young, hot stars, associated with dust and gas clouds in which active star formation continues, and the **Population II** older stars with no associated dust and gas. You can explore the distribution of these populations of stars by examining the distribution of clusters of stars of each population. For example, open clusters are composed of Population I stars while globular clusters consist of older Population II stars. You can use the *Starry Night*™ program to explore the distribution of clusters in and around our Galaxy. Select **Favourites > Explorations > Milky Way** to

view the Galaxy edge-on from the center of a transparent Earth. Before displaying the clusters, ensure that each type of cluster is assigned a different color by opening **Options > Stars > Globular Clusters . . .** and, in the **Globular Clusters Options** panel, change the **Outline Colour** to red and close the panel. Repeat this process to change the color of all clusters by opening **Options > Stars > Star Clusters . . .** and changing the **Outline Color** in the **Star Clusters Options** panel to blue. Open the **View** menu and click on **Stars > Globular Clusters** to display the distribution of these old Population II star clusters around the galactic center. (Early measurements of distances to many of these clusters, using variable stars with known intrinsic brightness as beacons, were used to determine the position of the center of the Galaxy.) You can now display clusters of all types, both the **Open Clusters** of young stars (represented by dotted circles) and the older **Globular Clusters** (represented by solid circles), by selecting **View > Stars > Star Clusters** from the menu. With the above color selections for clusters, the globular clusters will change to purple and will be distinguishable from the remaining open clusters depicted in blue.

a) What do you notice about the distribution of the globular clusters compared to the open clusters? Use the hand tool to move along the Milky Way plane to examine the distribution of each type of cluster, and hence each population of stars, across the sky.
b) Note in particular the two extra concentrations of clusters off the galactic plane. Why do you think that these extra concentrations are in these positions in our sky? (*Hint:* Move the cursor over these regions to identify them.)

ANSWERS

ConceptChecks

ConceptCheck 13-1: Looking at other galaxies, astronomers have observed that globular clusters tend to cluster around the center of galaxies. If globular clusters appeared in every direction equally, then astronomers would assume that we were in the center of the globular clusters and, subsequently, the center of our Galaxy.

ConceptCheck 13-2: While far-infrared wavelengths best reveal the presence of warmed dust grains, near-infrared wavelengths, which can pass through dust, are more predominantly emitted by cool stars, making near-infrared a better choice.

ConceptCheck 13-3: As the view between the galactic center and the Sun is obscured by intervening dust, the better vantage point would be from outside the disk in the halo, where the path of light is largely unobstructed by dust.

ConceptCheck 13-4: There is virtually no star formation occurring in the surrounding halo, so the youngest stars would be found actively forming in the rich dusty regions along the Galaxy's disk.

ConceptCheck 13-5: Because human eyes are most sensitive in the visible range, we would observe virtually no difference in the appearance of the Milky Way.

ConceptCheck 13-6: Neutral hydrogen gas emitting 21-cm photons will appear to be Doppler shifted to longer wavelengths.

ConceptCheck 13-7: Starting at the outside and moving inward, the sequence of galactic arms near the Sun are Perseus Arm, Orion Arm (where the Sun resides), and then the Sagittarius Arm.

ConceptCheck 13-8: The most important factor for how fast a star moves around the galaxy is how much mass is closer to the galactic center than the star.

ConceptCheck 13-9: In the absence of dark matter, the stars most distant from the galactic center should be moving much slower than those stars orbiting closer to the center.

ConceptCheck 13-10: Dark matter is the best explanation, to date, of what sort of unseen mass causes the outermost stars of galaxies to orbit around the galactic center as fast as they do.

ConceptCheck 13-11: Without density waves causing areas of the galaxy to bunch up here and there, the galaxy would have no discernible arms.

ConceptCheck 13-12: Because the arms of flocculent spiral galaxies are likely the result of star formation, this suggests that stars came before spiral arms in the early universe.

ConceptCheck 13-13: The gravitational effects of a supermassive black hole cause stars to move at very high rates of speed; if there was no supermassive black hole, then stars would be moving much, much slower.

ConceptCheck 13-14: The gases in the nebula would need to be energized by an external energy source, and it is proposed that a sudden intense release of X-rays from heated material moving into a supermassive black hole could have externally energized the nebula.

CalculationChecks

CalculationCheck 13-1: No matter how many hydrogen atoms are undergoing spin-flip transitions, only 21-cm photons are released. Fortunately, the more hydrogen atoms present results in a stronger signal received at radio telescopes.

CalculationCheck 13-2: Remembering that it takes 2.2×10^8 years for the Sun to make one orbit around the Galaxy, 4.5×10^9 years $\times$ (1 trip $\div$ 2.2×10^8 years) = 20 orbits.

R I V U X G The two galaxies NGC 1531 and NGC 1532 are so close together that they exert strong gravitational forces on each other. Both galaxies are about 55 million ly from us in the constellation Eridanus. *(Gemini Observatory/Travis Rector, University of Alaska, Anchorage)*

Investigating Other Galaxies

CHAPTER LEARNING OBJECTIVES **By reading the sections of this chapter, you will learn:**

14-1 When galaxies were first discovered, it was not clear that they lie far beyond the Milky Way until their variable stars were carefully observed

14-2 Hubble devised a system for classifying galaxies according to their appearance

14-3 Exploding stars release similar amounts of light and their distance can be inferred by measuring their apparent brightness

14-4 Galaxies are found in clusters and superclusters

14-5 Colliding galaxies produce starbursts, spiral arms, and other spectacular phenomena

14-6 Dark matter can be inferred by observing the motions of galaxy clusters

14-7 Quasars are the ultraluminous centers of the most distant galaxies

14-8 Supermassive black holes may be the "central engines" that power active galaxies

14-9 Galaxies may have formed from the merger of smaller objects

A century ago, most astronomers thought that the entire universe was only a few thousand light-years across and that nothing lay beyond our Milky Way Galaxy. Today, we know that picture was utterly wrong. We now understand that the Milky Way is just one of billions of galaxies strewn across billions of light-years. The image opening this chapter shows two of these galaxies, denoted by rather mundane catalog numbers—NGC 1531 and NGC 1532—that give no hint to these galaxies' magnificence.

Some galaxies are spirals like NGC 1532 or our Milky Way, with arching spiral arms that are active sites of star formation. The bright pink bands we see in NGC 1532 are clouds of excited hydrogen that are set aglow by ultraviolet radiation from freshly formed massive stars. Others, like NGC 1531, are featureless, ellipse-shaped collections of stars, virtually devoid of interstellar gas and dust. Some galaxies are only one-hundredth the size and one ten-thousandth the mass of the Milky Way. Others are giants, with 5 times the size and 50 times the mass of the Milky Way. Only about 10% of a typical galaxy's mass emits light of any kind; the remainder is made up of the mysterious dark matter.

A single galaxy, vast though it may be, is just a tiny part of the entire observable universe. Just as most stars are found within galaxies, most galaxies are located in groups and clusters. In this chapter, we will explore what astronomers know about galaxies and the clusters of galaxies that stretch in huge, lacy patterns across the universe. ■

14-1 When galaxies were first discovered, it was not clear that they lie far beyond the Milky Way until their variable stars were carefully observed

We haven't always known that the universe is filled with countless stars clumped together into galaxies. In fact, as recently as 100 years ago, astronomers were not convinced that any galaxies existed beyond our own. What we were sure of was that there were some objects observed in the night sky besides stars, comets, and planets. Using telescopes, astronomers had cataloged more than a hundred "fuzzy"-looking objects. Originally, all cloudlike-looking objects were described as nebulae. (The word *nebula* is much like the adjective *nebulous,* which means hazy, unclear, or ill-defined.) The object shown in **Figure 14-1** was once known as the "spiral nebula," and the one in **Figure 14-2** is the Great Nebula in Andromeda, and most people assumed that both were relatively nearby objects within the realm of our neighborhood of stars. Comet-hunting astronomers kept catalogs of these diffuse, wispy-looking objects because in small telescopes they often looked like comets. (The objects numbered and cataloged by French astronomer Charles Messier between 1758 and 1782 are often known by their Messier number, or M-number—Figure 14-1 is also known as M51 and Figure 14-2 is also known as M31.)

FIGURE 14-1 R I V U X G **The Spiral Nebula M51** Known today as the Whirlpool Galaxy because of its spinning shape, M51 was initially unknown; a fuzzy sky object such as this was thought to be either nearby gas and dust clouds or very distant, isolated islands of stars separated from us by a vast empty space. *(CFHT Coelum; J-C Cuillandre/G. Ageli)*

A considerable number of nebulae are in fact nearby regions of dust and gas, scattered throughout the Milky Way. (Figures 11-3 and 11-23 show some examples.) It therefore seemed reasonable that "spiral nebulae," even though they are very different in shape from other sorts of nebulae, could also be nearby neighbors. But, other astronomers wondered if these were instead giant objects far beyond our own realm of stars.

Up until about 100 years ago, the astronomical community was increasingly divided over the nature of the spiral nebulae. In April 1920, two opposing ideas were presented before the National Academy of Sciences in Washington, D.C. On one side was Harlow Shapley from the Mount

FIGURE 14-2 R I V U X G **The Andromeda "Nebula"** The Great Nebula in Andromeda, also known as M31, can be seen with even a small telescope. Edwin Hubble was the first to demonstrate that M31 is actually a galaxy that lies far beyond the Milky Way. M32 and M110 are two small satellite galaxies that orbit M31. *(NASA/JPL-Caltech)*

Wilson Observatory. Shapley thought the spiral nebulae were relatively small, nearby objects scattered around our Galaxy like globular clusters. Opposing Shapley was Heber D. Curtis of the University of California's Lick Observatory. Curtis argued that some of these spiral nebulae were an isolated and rotating system of stars much like our own Galaxy.

The Shapley-Curtis "debate" generated much heat but little came to light or was resolved. Nothing was decided because no one could present conclusive evidence to demonstrate exactly the distance to the spiral nebulae. Astronomy desperately needed a definitive determination of the distance to a spiral nebula, which could not be done unless someone could identify an individual star with the nebula. Such a measurement became the first great achievement of a young man who studied astronomy at the Yerkes Observatory, near Chicago. His name was Edwin Hubble.

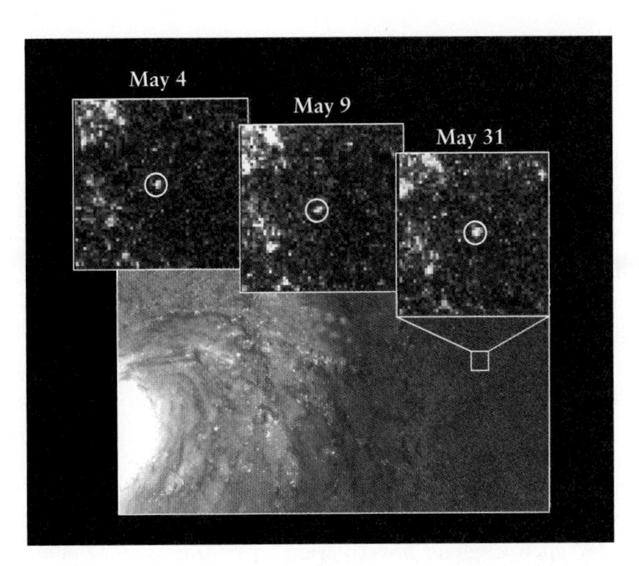

FIGURE 14-3 RIVUXG **Measuring Galaxy Distances with Cepheid Variables** By observing Cepheid variable stars in M100, the galaxy shown here, astronomers have found that it is about 56 million ly from Earth. The insets show one of the Cepheids in M100 at different stages in its brightness cycle, which lasts several weeks. *(Wendy L. Freedman, Carnegie Institution of Washington, and NASA)*

ConceptCheck 14-1: If you were observing one of the "spiral nebulae" and determined that it was closer than some of the stars of our Galaxy, would you be providing evidence in support of Shapley's argument or Curtis's argument?

Hubble Proved that the Spiral Nebulae Are Far Beyond the Milky Way

After completing his studies, Edwin Hubble joined the staff of the Mount Wilson Observatory in Pasadena, California. On October 6, 1923, he took a historic photograph of the Andromeda "Nebula," one of the spiral nebulae around which controversy raged. Figure 14-2 is a modern photograph of this object.

Hubble carefully examined the photo he had taken and discovered what he at first thought to be an exploding nova. Referring to previous photographs of that region, he soon realized that the object was actually a Cepheid variable star. Further scrutiny of additional plates over the next several months revealed several more Cepheids. **Figure 14-3** shows modern observations of a Cepheid in another "spiral nebula."

This discovery was a critical finding because Cepheid variables allow astronomers to determine distances. An astronomer begins by carefully measuring the variations in apparent brightness of a Cepheid variable, then recording the results in the form of a plot of brightness versus time, or light curve (see Figure 13-4*a*). This graph gives the variable star's period and average brightness. Given the star's period, the astronomer then uses the period-luminosity relation shown in **Figure 14-4** to find the Cepheid's average luminosity. (To be more precise, an astronomer must also examine the spectrum of the star to determine whether it is a metal-rich Type I Cepheid or a metal-poor Type II Cepheid. As Figure 14-4 shows, these have somewhat different period-luminosity relations.) Knowing both the apparent brightness and luminosity of the Cepheid, the astronomer can then use the inverse-square law to calculate the distance to the star. **Box 14-1 Tools of the Astronomer's Trade: Cepheids and Supernovae as Indicators of Distance** provides examples of how this is done. Cepheid variables are intrinsically quite luminous, with average luminosities that can exceed $10^4 L_\odot$. Hubble realized that for these luminous stars to appear as dim as they were on his photographs of the Andromeda "Nebula," they must be extremely far away. Straightforward calculations using modern data reveal that M31 is some 2.5 million ly from Earth. Based on its angular size, M31 has a diameter larger than the diameter of our own Milky Way Galaxy!

These results prove that the Andromeda "Nebula" is actually an enormous stellar system, far beyond the confines of the Milky Way with a vast empty space between. Today, this system is properly called the Andromeda Galaxy. (Under good observing conditions, you can actually see this galaxy's central bulge with the naked eye. If you could see the entire Andromeda Galaxy, it would cover an area of the sky roughly 5 times as large as the full moon.)

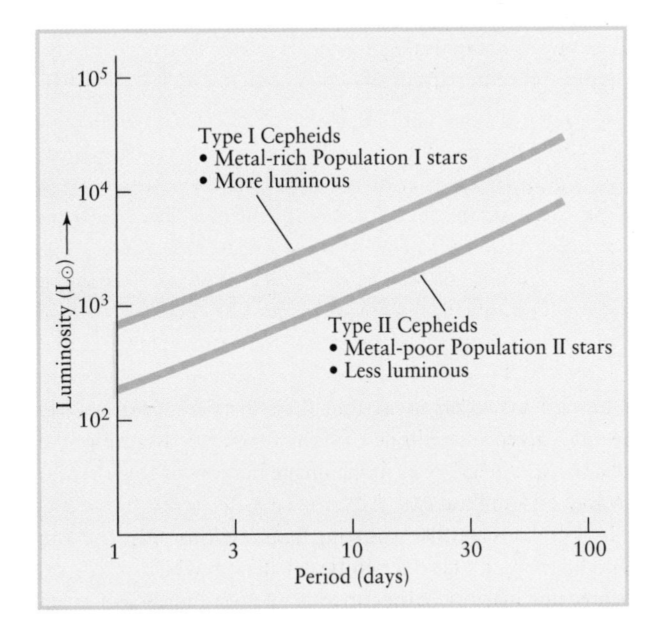

FIGURE 14-4 Period-Luminosity Relations for Cepheids The greater the average luminosity of a Cepheid variable, the longer its period and the slower its pulsations. Note that there are actually two distinct period-luminosity relations—one for Type I Cepheids and one for the less-luminous Type II Cepheids. *(Adapted from H. C. Arp)*

BOX 14-1 **TOOLS OF THE ASTRONOMER'S TRADE**

Cepheids and Supernovae as Indicators of Distance

Because their periods are directly linked to their luminosities, Cepheid variables are one of the most reliable tools astronomers have for determining the distances to galaxies. To this day, astronomers use this link—much as Hubble did back in the 1920s—to measure intergalactic distances. More recently, they have begun to use Type Ia supernovae, which are far more luminous and thus can be seen much farther away, to determine the distances to very remote galaxies.

EXAMPLE: In 1992, a team of astronomers observed Cepheid variables in a galaxy called IC 4182 to deduce this galaxy's distance from Earth. One such Cepheid has a period of 42.0 days and an average apparent magnitude (m) of +22.0. By comparison, the dimmest star you can see with the naked eye has $m = +6$; this Cepheid in IC 4182 appears less than one one-millionth as bright. The star's spectrum shows that it is a metal-rich Type I Cepheid variable.

According to the period-luminosity relation shown in Figure 14-4, such a Type I Cepheid with a period of 42.0 days has an average luminosity of 33,000 $L_\odot$. An equivalent statement is that this Cepheid has an average absolute magnitude (M) of –6.5. (This compares to $M = +4.8$ for the Sun.) Use this information to determine the distance to IC 4182.

Situation: We are given the apparent magnitude $m = +22.0$ and the absolute magnitude $M = -6.5$ of the Cepheid variable star in IC 4182. Our goal is to calculate the distance to this star, and hence the distance to the galaxy of which it is part.

Tools: We use the relationship between apparent magnitude, absolute magnitude, and distance written as a mathematical relationship $m - M = 5 \log d - 5$.

Answer: Given that the apparent magnitude of a star is related to its absolute magnitude and distance in parsecs (d), where one parsec is 3.26 ly, using

$$m - M = 5 \log d - 5$$

This can be rewritten as

$$d = 10^{(m - M + 5)/5} \text{ parsecs}$$

We have $m - M = (+22.0) - (-6.5) = 22.0 + 6.5 = 28.5$. Hence, our equation rearranges to become

$$d = 10^{(28.5 + 5)/5} \text{ parsecs} = 10^{6.7} \text{ parsecs} = 5 \times 10^6 \text{ parsecs}$$

(A calculator is needed to calculate the quantity $10^{6.7}$.)

Review: Our result tells us that the galaxy is 5 million parcsecs, or 5 Mpc, from Earth (1 Mpc = 10^6 pc). This distance can also be expressed as 16 million ly.

EXAMPLE: Astronomers are interested in IC 4182 because a Type Ia supernova was observed there in 1937. All Type Ia supernovae are exploding white dwarfs that reach nearly the same maximum brightness at the peak of their outburst. Once astronomers know the peak absolute magnitude of Type Ia supernovae, they can use these supernovae as distance indicators. Because the distance to IC 4182 is known from its Cepheids, the 1937 observations of the supernova in that galaxy help us calibrate Type Ia supernovae as distance indicators.

At maximum brightness, the 1937 supernova reached an apparent magnitude of $m = +8.6$. What was its absolute magnitude at maximum brightness?

Situation: We are given the supernova's apparent magnitude m, and we know its distance from the previous example. Our goal is to calculate its absolute magnitude M.

Tools: We again use the relationship $m - M = 5 \log d - 5$.

Answer: We could plug in the value of d found in the previous example. But it is simpler to note that the distance modulus $m - M$ has the same value no matter whether it refers to a Cepheid, a supernova, or any other object, just so it is at the same distance d. From the Cepheid example we have $m - M = 28.5$ for IC 4182, so

$$M = m - (m - M) = 8.6 - (28.5) = -19.9$$

This absolute magnitude corresponds to a remarkable peak luminosity of 10^{10} $L_\odot$.

Review: Whenever astronomers find a Type Ia supernova in a remote galaxy, they can combine this absolute magnitude with the observed maximum apparent magnitude to get the galaxy's distance modulus, from which the galaxy's distance can be easily calculated (just as we did above for the Cepheids in IC 4182). This technique has been used to determine the distances to galaxies hundreds of millions of parsecs away.

Galaxies are so far away that their distances from us are usually given in millions of light-years. For example, the distance to the galaxies in the image that opens this chapter is about 55 million ly.

Hubble's results showing stars within Andromeda and allowing its distance to be calculated, which were presented at a meeting of the American Astronomical Society on December 30, 1924, settled the Shapley-Curtis "debate" once and for all. The universe was recognized to be far larger and populated with far bigger objects than anyone had seriously imagined. Hubble had discovered the realm of the galaxies.

ConceptCheck 14-2: **If a nearby galaxy were discovered, why would astronomers immediately look for Cepheids?**

14-2 Hubble devised a system for classifying galaxies according to their appearance

With large telescopes, millions of galaxies are visible across every unobscured part of the sky. Although all galaxies are made up of large numbers of stars, they come in a variety of shapes and sizes. Hubble classified them into four broad categories based on their appearance; this **Hubble classification** is a scheme that is still used today. The four classes of galaxies are the spirals, classified S; barred spirals, or SB; ellipticals, E; and irregulars, Irr. **Table 14-1** summarizes some key properties of each class. These various types of galaxies differ not only in their shapes but also in the kinds of processes that take place within them.

Spiral Galaxies: Stellar Birthplaces

M51 (Figure 14-1), M31 (Figure 14-2), and M100 (Figure 14-3) are examples of **spiral galaxies.** The spiral arms contain young, hot, blue stars and glowing hydrogen gas regions, indicating ongoing star formation.

Thermonuclear reactions within stars create *metals,* that is, elements heavier than hydrogen or helium. These metals are dispersed into space as the stars evolve and die. So, if new stars are being formed from the interstellar matter in spiral galaxies, they will incorporate these metals and be metal-rich Population I stars, similar to our Sun. Indeed, when we measure the visible-light spectrum of the disk of a spiral galaxy, we can see obvious absorption lines from metals. Such a spectrum is a composite of the spectra of many stars and shows that the stars in the disk are principally of Population I. By contrast, there is relatively little star formation in the central bulges of spiral galaxies, and these regions are dominated by old Population II stars that have a low metal content. This also explains why the central bulges of spiral galaxies have a yellowish or reddish color; as a population of stars ages, the massive, luminous blue stars die off first, leaving only the longer-lived, low-mass red stars.

Galaxies are classified by their spiral, elliptical, or irregular appearance.

Edwin Hubble further classified spiral galaxies according to the size of their central bulges and the appearance of their arms. Spirals with the largest and most obvious central bulges and tightly wrapped, smooth, broad spiral arms are called Sa galaxies, for spiral type *a* (**Figure 14-5*a***);

Table 14-1 Some Properties of Galaxies

	Spiral (S) and barred spiral (SB) galaxies	Elliptical galaxies (E)	Irregular galaxies (Irr)
Mass ($M_\odot$)	10^9 to 4×10^{11}	10^5 to 10^{13}	10^8 to 3×10^{10}
Luminosity ($L_\odot$)	10^8 to 2×10^{10}	3×10^5 to 10^{11}	10^7 to 10^9
Diameter (ly)	15,000 to 800,000	3000 to 650,000	3000 to 30,000
Stellar population	Spiral arms: young Population I Nucleus and throughout disk: Population II and the old Population I	Population II and old Population I	mostly Population I
Percentage of observed galaxies	77%	20%*	3%

*This percentage does not include dwarf elliptical galaxies that are as yet too dim and distant to detect. Hence, the actual percentage of galaxies that are ellipticals may be higher than shown here.

(a) Sa (NGC 1357)

(b) Sb (M81)

(c) Sc (NGC 4321)

FIGURE 14-5 R I V U X G **Spiral Galaxies** Edwin Hubble classified spiral galaxies according to the size of their central bulges and the appearance of their arms. At one end of the range, (a) Sa galaxies have the largest and most obvious central bulges and tightly wrapped, smooth, broad spiral arms, while (b) Sb galaxies are more moderate, and (c) Sc galaxies have the smallest central bulges with loosely wrapped arms. *(a: Adam Block/Steve Mandel/Jim Rada and Students/NOAO/AURA/NSF; b: NASA/JPL-Caltech/ESA/Harvard-Smithsonian CfA; c: FORS Team, 8.2-meter VLT, ESO)*

those with moderately well defined spiral arms and a moderate-size central bulge, like M81, are Sb galaxies (Figure 14-5*b*); and galaxies with narrow, well-defined but loosely wrapped spiral arms and a tiny central bulge are Sc galaxies (Figure 14-5*c*).

The differences between Sa, Sb, and Sc galaxies may be related to the relative amounts of gas and dust they contain. Observations with infrared telescopes—which detect the emission from interstellar dust—and radio telescopes—which detect radiation from interstellar gases such as hydrogen and carbon monoxide—show that about 4% of the mass of an Sa galaxy is in the form of gas and dust. This percentage is 8% for Sb galaxies and 25% for Sc galaxies.

Interstellar gas and dust is the material from which new stars are formed, so an Sc galaxy has a greater proportion of its mass involved in star formation than an Sb or Sa galaxy. Hence, an Sc galaxy has a large disk, where star formation occurs, and a small central bulge, where there is little or no star formation. By comparison, an Sa galaxy, which has relatively little gas and dust and thus less material from which to form stars, has a large central bulge and only a small star-forming disk.

ConceptCheck 14-3: Where would you label more and less star formation when making a sketch of a spiral galaxy with considerable star formation?

Barred Spiral Galaxies: Spirals with an Extra Twist

Go to Video 14-1

In **barred spiral galaxies,** such as those shown in **Figure 14-6,** the spiral arms originate at the ends of a bar-shaped region running through the galaxy's nucleus rather than from the nucleus itself. As with ordinary spirals, Hubble subdivided barred spirals according to the relative size of their central bulge and the character of their spiral arms. An SBa galaxy has a large central bulge and thin, tightly wound spiral arms (Figure 14-6*a*). Likewise, an SBb galaxy is a barred spiral with a moderate central bulge and moderately wound spiral arms (Figure 14-6*b*), while an SBc galaxy has lumpy, loosely wound spiral arms and a tiny central bulge (Figure 14-6*c*). As for ordinary spiral galaxies, the difference between SBa, SBb, and SBc galaxies may be related to the amount of gas and dust in the galaxy.

Bars appear to form naturally in many spiral galaxies. This conclusion comes from computer simulations of galaxies, which set hundreds of thousands of simulated "stars" into orbit around a common center. As the "stars" orbit and exert gravitational forces on one another, a bar structure forms in most cases. Indeed, barred spiral galaxies outnumber ordinary spirals by about two to one.

ConceptCheck 14-4: How are galaxies with loosely wrapped arms categorized using Hubble's scheme?

Elliptical and Irregular Galaxies

Elliptical galaxies, so named because of their distinctly elliptical shapes, have no spiral arms. Hubble subdivided these galaxies according to how round or flattened they look. The roundest elliptical galaxies are called E0 galaxies and the flattest, E7 galaxies. Elliptical galaxies with intermediate amounts of flattening are given designations between these extremes (**Figure 14-7**).

CAUTION Unlike the designations for spirals and barred spirals, the classifications E0 through E7 may not reflect the true shape of elliptical galaxies. An E1 or E2 galaxy might actually be a very flattened disk of stars that we just happen to view face-on, and a cigar-shaped E7 galaxy might look spherical if seen end-on. The Hubble scheme classifies galaxies entirely by how they *appear* to us on Earth.

(a) SBa (NGC 4650)

(b) SBb (M83)

(c) SBc (NGC 1365)

FIGURE 14-6 R I V U X G **Barred Spiral Galaxies** As with spiral galaxies, Hubble classified barred spirals according to the texture of their spiral arms (which correlates to the sizes of their central bulges). SBa galaxies have the smoothest spiral arms and the largest central bulges, while SBc galaxies have narrow, well-defined arms and the smallest central bulges. *(a: AURA/NOAO/NSF; b: FORS Team, 8.2-meter VLT, ESO; c: FORS Team, 8.2-meter VLT Antu, ESO)*

(a) E0 (M105)

(b) E3 (NGC 4406)

(c) E6 (NGC 3377)

FIGURE 14-7 R I V U X G **Elliptical Galaxies** Hubble classified elliptical galaxies according to how round or flattened they look. A galaxy that appears round is labeled E0, and the flattest-appearing elliptical galaxies are designated E7. *(a: Karl Gebhardt (Univ. of Michigan), Tod Lauer (NOAO) and NASA; b: CFHT/Coelum; J-C Cuillandre/G. Ageli; c: Karl Gebhardt (Univ. of Michigan), Tod Lauer [NOAO], and NASA)*

Elliptical galaxies look far less dramatic than their spiral and barred spiral cousins. The reason is that ellipticals have virtually no interstellar gas and dust. Consequently, there is little material from which stars could have recently formed, and indeed there is no evidence of young stars in most elliptical galaxies. For the most part, star formation in elliptical galaxies ended long ago. Hence, these galaxies are composed of old, red, Population II stars with only trace amounts of any metals.

Elliptical galaxies come in a wide range of sizes and masses. Both the largest and the smallest galaxies in the known universe are elliptical. **Figure 14-8** shows two **giant elliptical galaxies** that are about 20 times larger than an average galaxy. These giant ellipticals are located near the middle of a large cluster of galaxies in the constellation Virgo.

FIGURE 14-8 R I V U X G **Giant Elliptical Galaxies** The Virgo cluster is a rich, sprawling collection of more than 2000 galaxies about 56 million ly from Earth. Only the center of this huge cluster appears in this photograph. The two largest members of this cluster are the giant elliptical galaxies M84 and M86. *(CFHT/Coelum; J-C Cuillandre/G. Ageli)*

Giant ellipticals are rather rare, but **dwarf elliptical galaxies** are quite common. Dwarf ellipticals are only a fraction of the size of their normal counterparts and contain so few stars—only a few million, compared to more than 100 billion (10^{11}) stars in our Milky Way Galaxy—that these galaxies are completely transparent. You can actually see straight through the center of a dwarf galaxy and out the other side, as **Figure 14-9** shows.

Hubble also identified galaxies that are midway in appearance between ellipticals and the two kinds of spirals. These are denoted as S0 and SB0 galaxies, and are also called **lenticular galaxies.** Although they look somewhat elliptical, lenticular ("lens-shaped") galaxies have both a central bulge and a disk like spiral galaxies, but no clear spiral arms. They are therefore sometimes referred to as "armless spirals," mostly because we are seeing them edge-on.

FIGURE 14-9 R I V U X G **A Dwarf Elliptical Galaxy** This diffuse cloud of stars is a nearby E4 dwarf elliptical called Leo I. It actually orbits the Milky Way at a distance of about 600,000 ly. Leo I is about 3000 ly in diameter but contains so few stars that you can see through the galaxy's center. *(Australian Astronomical Observatory/David Malin Images)*

FIGURE 14-10 RIVUXG **A Lenticular Galaxy** NGC 5866 is tilted nearly edge-on to our line of sight. Close inspection reveals a crisp dust lane dividing the galaxy into two halves centered on a subtle, reddish bulge surrounding a bright nucleus, a blue disk of stars running parallel to the dust lane, and a transparent outer halo. Viewed face on, it would look like a smooth, flat disk with little spiral structure. At a distance of 44 million ly, it has a diameter of roughly 60,000 ly—only two-thirds the diameter of the Milky Way, although its mass is similar to our Galaxy. *(NASA, ESA, and the Hubble Heritage Team [STScI/AURA])*

Figure 14-10 shows an example of an S0 (pronounced "ess-zero") lenticular galaxy.

Hubble summarized his classification scheme for spiral, barred spiral, and elliptical galaxies in a diagram, now called the **tuning fork diagram** for its shape (**Figure 14-11**).

CAUTION When Hubble first drew his tuning fork diagram, he was thinking that it might represent an evolutionary sequence of galaxy formation and evolution. He thought that galaxies evolved over time from the left to the right of the diagram, beginning as ellipticals and eventually becoming either spiral or barred spiral galaxies. We now understand that this is not the case at all! For one thing, elliptical galaxies have little or no overall rotation, while spiral and barred spiral galaxies have a substantial amount of overall rotation. There is no way that an elliptical galaxy could suddenly start rotating, which means that it could not evolve into a spiral galaxy.

Finally, galaxies that do not fit into the scheme of spirals, barred spirals, and ellipticals are conveniently referred to as **irregular galaxies.** They are generally rich in interstellar gas and dust, and have both young and old stars. For lack of any better scheme, the irregular galaxies are sometimes placed between the ends of the tines of the Hubble tuning fork diagram, as in Figure 14-11.

ConceptCheck 14-5: Which type of galaxy has almost no ongoing active star formation?

14-3 Exploding stars release similar amounts of light and their distance can be inferred by measuring their apparent brightness

A key question that astronomers ask about galaxies is, "How far away are they?" Unfortunately, many of the techniques that are used to measure distances within our Milky Way Galaxy cannot be used for the far greater distances to other galaxies. The extremely accurate parallax method that we described earlier can be used only for stars within about 1500 ly. Beyond that distance, parallax angles become too small to measure.

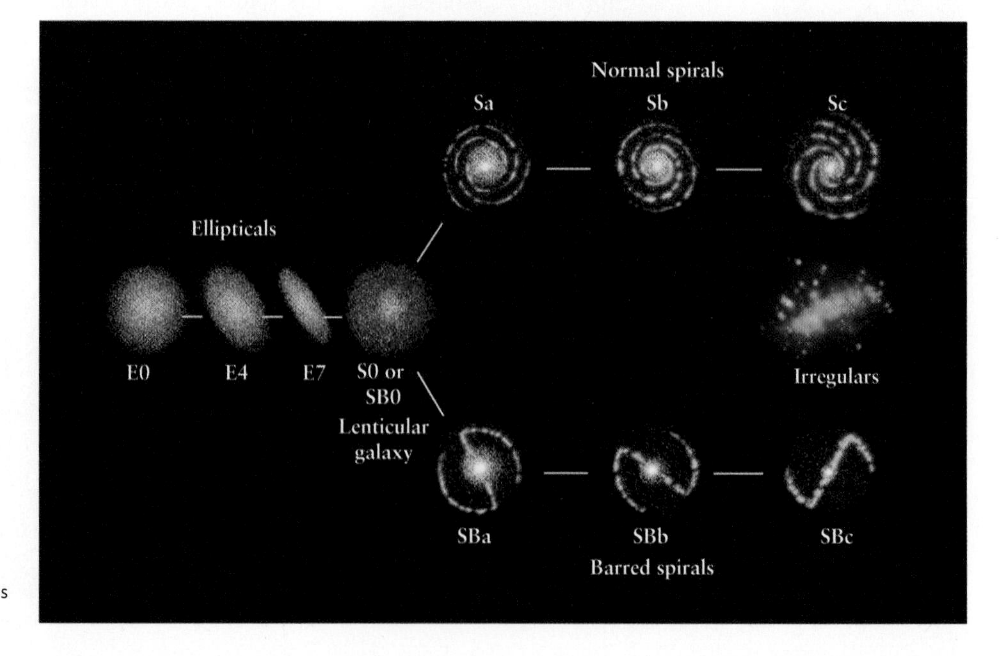

FIGURE 14-11 Hubble's Tuning Fork Diagram Edwin Hubble's classification of regular galaxies is shown in his tuning fork diagram. An elliptical galaxy is classified by how flattened it appears. A spiral or barred spiral galaxy is classified by the texture of its spiral arms and the size of its central bulge. A lenticular galaxy is intermediate between ellipticals and spirals. Irregular galaxies do not fit into this simple classification scheme.

Standard Candles: Variable Stars and Type Ia Supernovae

To determine the distance to a remote galaxy, astronomers look instead for a **standard candle**—an object, such as a star, that lies within that galaxy and for which we know its luminosity. By measuring how bright the standard candle appears, astronomers can calculate its distance—and hence the distance to the galaxy of which it is part—using the inverse-square law.

The challenge is to find standard candles that are luminous enough to be seen across the tremendous distances to galaxies. To be useful, standard candles should have four properties:

1. They should be luminous, so we can see them out to great distances.
2. We should be fairly certain about their luminosities, so we can be equally certain of any distance calculated from a standard candle's apparent brightness and luminosity.
3. They should be easily identifiable—for example, by the shape of the light curve of a variable star.
4. They should be relatively common so that astronomers can use them to determine the distances to many different galaxies.

For nearby galaxies, Cepheid variable stars make reliable standard candles. These variables can be seen out to about 100 million ly using the Hubble Space Telescope, and their luminosity can be determined from their period through the period-luminosity relation shown in Figure 14-4. Beyond about 100 million ly, even the brightest Cepheid variables, which have luminosities of about 2×10^4 $L_\odot$, fade from view.

One class of standard candles that astronomers have used beyond 100 million ly are Type Ia supernovae. Type Ia supernovae are stars that explode when they capture enough material from a very nearby star to ignite in a tremendous explosion and blow themselves apart. A Type Ia supernova can reach a maximum luminosity of about 3×10^9 $L_\odot$ (**Figure 14-12**). If a Type Ia supernova is seen in a distant galaxy and its maximum apparent brightness measured, the inverse-square law can be used to find the galaxy's distance.

One complication is that not all Type Ia supernovae are equally luminous. Fortunately, there is a simple relationship between the peak luminosity of a Type Ia supernova and the rate at which the luminosity decreases after the peak: The more slowly the brightness decreases, the more luminous the supernova. Using this relationship, astronomers have measured distances to supernovae more than 3 billion ly from Earth.

Unfortunately, this technique can be used only for galaxies in which we happen to observe a Type Ia supernova. But telescopic surveys now identify many dozens of these supernovae every year, so the number of galaxies whose distances can be measured in this way is continually increasing.

ConceptCheck 14-6: If a thin cloud of intergalactic dust reduced the observed brightness of a Type Ia supernova found in a distant galaxy, would astronomers who did not know the dust was there mistakenly assume the galaxy is farther or closer than it actually is?

(a) Before a Type Ia supernova explosion

(b) After a Type Ia supernova explosion

FIGURE 14-12 RIVUXG **A Supernova in a Spiral Galaxy** These images from NASA's SWIFT orbiting telescope show the spiral galaxy M100 (a) before and (b) after a Type Ia supernova exploded within the galaxy in 2006. On the left is an image taken on November 11, 2005, before the explosion, while the image on the right was taken on February 8, 2006, after the explosion. The position of the supernova is indicated by the arrow. Such luminous supernovae, which can be seen at extreme distances, are important standard candles used to determine the distances to faraway galaxies. *(NASA Goddard Space Flight Center)*

Distance Determination Without Standard Candles

Go to Video 14-2

Other methods for determining the distances to galaxies do not make use of standard candles. One was discovered in the 1970s by the astronomers Brent Tully and Richard Fisher. They found that the variability in the light emitted by glowing hydrogen in a certain galaxy is related to the galaxy's overall luminosity. This correlation is the **Tully-Fisher relation**—the more variability in this particular light emitted by glowing hydrogen, the more luminous the galaxy.

Distances to most galaxies are measured by identifying their brightest stars, such as supernovae.

Such a relationship exists because radiation from the approaching side of a rotating galaxy is changed to shorter wavelengths of light, while that from the galaxy's receding side is changed to longer wavelengths of light. Thus, the variability in the light is increased by an amount directly related to how fast a galaxy is rotating. Furthermore, astronomers know that the more stars a galaxy contains, the more luminous it is. Consequently, the variability in this particular amount of light emitted by hydrogen is directly related to a galaxy's luminosity. By combining this information with measurements of apparent brightness, they can calculate the distance to the galaxy. This technique can be used to measure distances of 300 million ly or more.

ConceptCheck 14-7: **Is a galaxy that appears to be quite bright but has almost no variability in the hydrogen light emitted relatively close to or very distant from our own Galaxy?**

The Distance Ladder

Figure 14-13 shows the ranges of applicability of several important means of determining astronomical distances. Because these ranges overlap, one technique can be used to calibrate another. As an example, astronomers have studied Cepheids in nearby galaxies that have been host to Type Ia supernovae. The Cepheids provide the distances to these nearby galaxies, making it possible to determine the peak luminosity of each supernova using its maximum apparent brightness and the inverse-square law. Once the peak luminosity is known, it can be used to determine the distance to Type Ia supernovae in more distant galaxies. Because one measuring technique leads us to the next one like rungs on a ladder, the techniques shown in Figure 14-13 (along with others) are referred to collectively as the **distance ladder.**

ANALOGY If you give a slight shake to the bottom of a tall ladder, the top can wobble back and forth alarmingly. A change in distance-measuring techniques used for nearby objects can also have substantial effects on the distances to remote galaxies. For example, if astronomers discovered that the distances to nearby Cepheids were in error, distance measurements using any technique that is calibrated by Cepheids would be affected as well. (As an example, the distance to the galaxy M100 shown in Figure 14-3 is determined using Cepheids. A Type Ia supernova has been seen in M100, as Figure 14-12 shows, and its luminosity is determined using the Cepheid-derived distance to M100. Any change in the calculated distance to M100 would change the calculated luminosity of the Type Ia supernova, and so would have an effect on all distances derived from observations of how bright these supernovae appear in other galaxies.) For this reason, astronomers go to great lengths to check the accuracy and reliability of their standard candles.

ConceptCheck 14-8: **Which of the rungs on the distance ladder depend on an accurate measurement of parallax?**

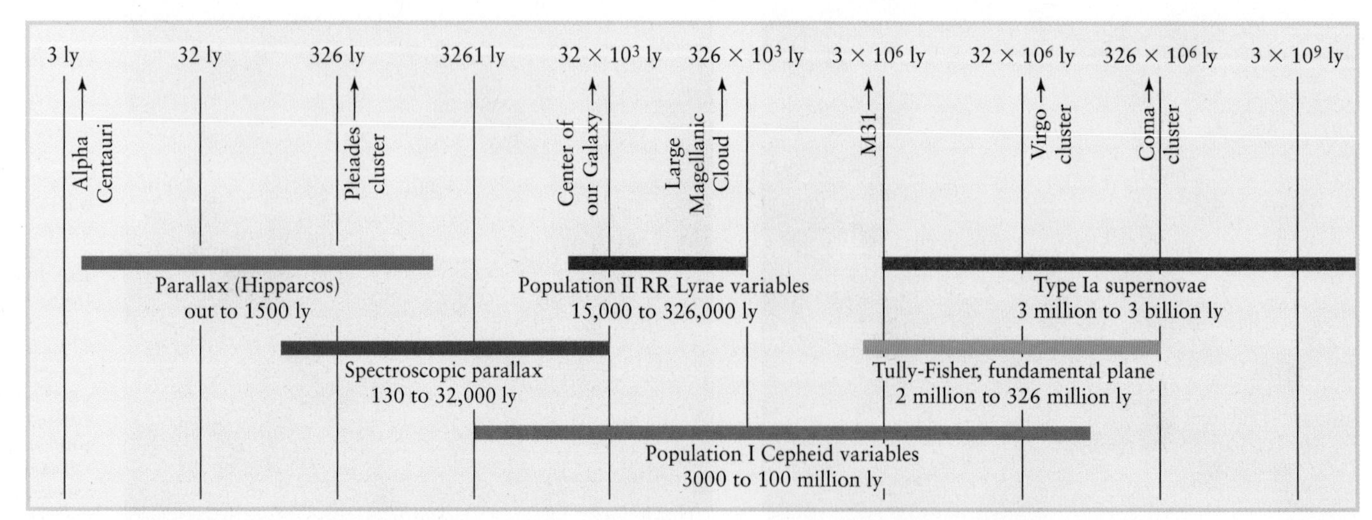

FIGURE 14-13 The Distance Ladder Astronomers employ a variety of techniques for determining the distances to objects beyond the solar system. Because their ranges of applicability overlap, one technique can be used to calibrate another. The arrows indicate distances to several important objects.

FIGURE 14-14 R I V U X G **The Hercules Cluster** This irregular cluster of galaxies is about 650 million ly from Earth. The Hercules cluster contains many spiral galaxies, often associated in pairs and small groups. *(NOAO)*

14-4 Galaxies are found in clusters and superclusters

Galaxies are not scattered randomly throughout the universe but are found in **clusters. Figure 14-14** shows one such cluster. Like stars within a star cluster, the members of a cluster of galaxies are in continual motion around one another.

Clusters of Galaxies: Regular and Irregular

Go to Video 14-3

Galaxies tend to be gravitationally bound into **groups.** For example, the Milky Way Galaxy, the Andromeda Galaxy (M31), and the Large and Small Magellanic Clouds of our neighborhood are part of a cluster called the **Local Group.** The Local Group contains more than 54 galaxies, most of which are quite small. **Figure 14-15** is a map of a portion of the Local Group.

In recent years, astronomers have discovered several smaller, and previously unknown, galaxies in the Local Group. As of 2014, the nearest and most recently discovered is the Canis Major Dwarf, so named for the constellation in whose direction it lies as seen from Earth. It is about 42,000 ly from the center of the Milky Way Galaxy and a mere 25,000 ly from Earth (about the same as the distance from Earth to the center of the Milky Way). Tidal forces exerted by the Milky Way on the Canis Major Dwarf are causing this dwarf galaxy to gradually disintegrate and leave a trail of debris behind it (**Figure 14-16**).

We may never know the total number of galaxies in our Local Group, because dust in the plane of the Milky Way

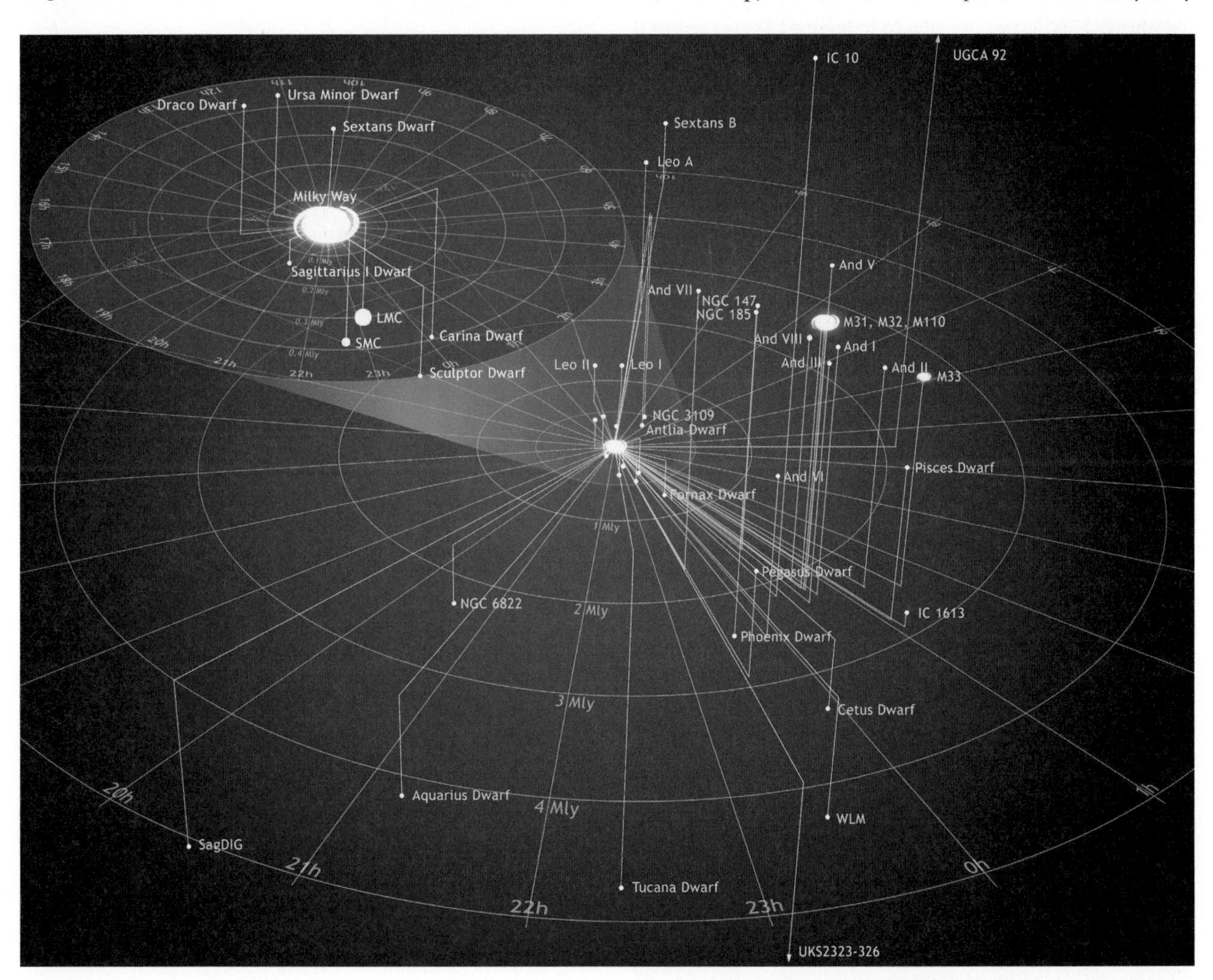

FIGURE 14-15 **The Local Group** This illustration shows the relative positions of the galaxies that constitute the Local Group, a poor, irregular cluster of which our Galaxy is part. (The blue rings represent the plane of the Milky Way's disk; 0h is the direction from Earth toward the Milky Way's center. Solid lines point to galaxies above and below the plane.) The largest and most massive galaxy in the Local Group is M31, the Andromeda Galaxy; in second place is the Milky Way, followed by the spiral galaxy M33. Both the Milky Way and M31 are surrounded by a number of small satellite galaxies. *(Mark Garlick/Science Source)*

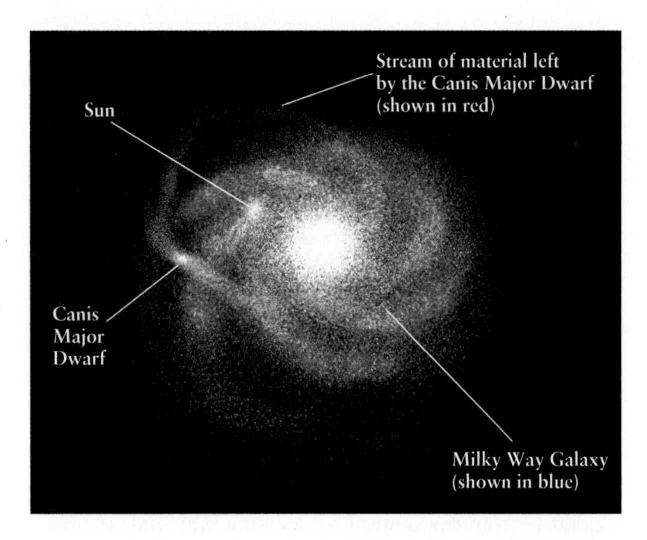

FIGURE 14-16 **The Canis Major Dwarf** Discovered in 2003, this dwarf elliptical galaxy is actually slightly closer to Earth than is the center of the Milky Way Galaxy. This illustration shows the stream of material left behind by the Canis Major Dwarf as it orbits the Milky Way. *(R. Ibata et al., Observatoire de Strasbourg/Université Louis Pasteur; 2MASS; and NASA)*

obscures our view over a considerable region of the sky. Nevertheless, we can be certain that no additional large spiral galaxies are hidden by the Milky Way because at least some of their light would not be completely absorbed by interstellar dust.

The nearest fairly rich cluster is the Virgo cluster, a collection of more than 2000 galaxies covering a 10° × 12° area of the sky. Figure 14-8 shows a portion of this cluster. One member of this cluster not shown in Figure 14-8 is the spiral galaxy M100; measurements of Cepheid variables in M100 (see Figure 14-3) give a distance of about 56 million ly. The Tully-Fisher relation and observations of Type Ia supernovae (see Figure 14-12) give similar distances to this cluster. The overall diameter of the cluster is about 9 million ly.

The center of the Virgo cluster is dominated by three giant galaxies. You can see two of these, M84 and M86, in Figure 14-8. The diameter of each of these enormous galaxies is comparable to the distance between the Milky Way and M31. In other words, one giant galaxy is approximately the same size as the entire Local Group!

Galaxies are clumped together in gravitationally connected groups, with vast empty spaces between galaxy clusters.

Astronomers categorize clusters of galaxies as regular or irregular, depending on the overall shape of the cluster. The Virgo cluster, for example, is called an **irregular cluster**, because its galaxies are scattered throughout a sprawling region of the sky. Our own Local Group is also an irregular cluster. In contrast, a **regular cluster** has a distinctly spherical appearance, with a marked concentration of galaxies at its center.

The nearest example of a regular cluster is the Coma cluster, located about 300 million ly from us toward the constellation Coma Berenices (Berenice's Hair) (**Figure 14-17**). Despite its great distance, telescopic images of this cluster show more than 1000 galaxies, many of which are faint, dwarf galaxies. The Coma cluster is dominated by two large elliptical galaxies, but it almost certainly contains many thousands of tiny galaxies as yet unseen, and the total membership of the cluster may be as many as 10,000 galaxies. The overall shape of a cluster is related to the dominant types of galaxies it contains. Regular clusters contain galaxies that lack an apparent spiral structure. For example, about 80% of the brightest galaxies in the Coma cluster (see Figure 14-17) are of this type. Alternatively, irregular clusters, such as the Virgo cluster and the Hercules cluster shown in Figure 14-14, have a more even mixture of galaxy types, some with spiral structures and others without.

FIGURE 14-17 R I V U X G **The Coma Cluster** This rich, regular cluster is about 300 million ly from Earth. Almost all of the spots of light in this image, and there are thousands of them, are individual galaxies of the cluster. Two giant elliptical galaxies, NGC 4889 and NGC 4874, dominate the center of the cluster. *(NASA, JPL-Caltech, SDSS, Leigh Jenkins, Ann Hornschemeier [Goddard Space Flight Center] et al.)*

ConceptCheck 14-9: **How would a photograph of a regular cluster's appearance compare to that of an irregular cluster?**

Superclusters and Voids: Clusters of Clusters of Galaxies

Go to Video 14-4

Clusters of galaxies are themselves grouped together in huge associations called **superclusters**. A typical supercluster contains dozens of individual clusters spread over a region of space up to 150 million ly across. **Figure 14-18** shows the distribution of clusters in our part of

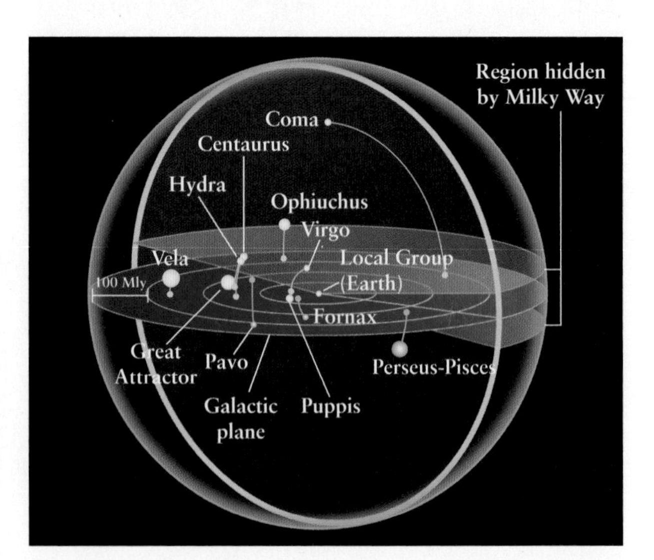

FIGURE 14-18 **Nearby Clusters of Galaxies** This illustration shows a sphere of space 800 million ly in diameter centered on Earth in the Local Group. Each spherical dot represents a cluster of galaxies. To better see the three-dimensionality of this figure, colored arcs are drawn from each cluster to the green plane, which is an extension of the plane of the Milky Way outward into the universe. Note that clusters of galaxies are unevenly distributed here, as they are elsewhere in the universe.

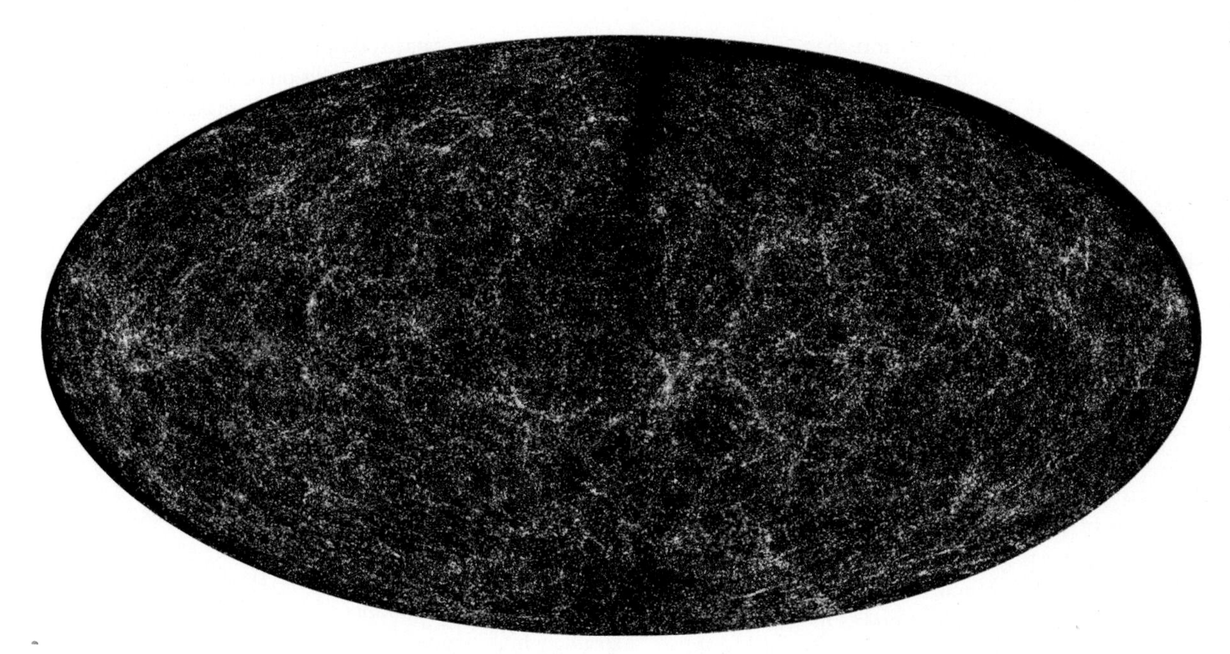

FIGURE 14-19 R I V U X G **Structure in the Nearby Universe** This composite infrared image from the 2MASS (Two-Micron All-Sky Survey) project shows the light from 1.6 million galaxies. In this image, the entire sky is projected onto an oval; the blue band running vertically across the center of the image is light from the plane of the Milky Way. Note that galaxies form a lacy, filamentary structure. Note also the large, dark voids that contain few galaxies. *(2MASS/T. H. Jarrett, J. Carpenter, & R. Hurt)*

the universe. The nearer ones out to the Virgo cluster, including our own Local Group, are members of the *Local Supercluster.* The other clusters shown in Figure 14-18 belong to other superclusters. The most massive cluster in the local universe is called the *Great Attractor.*

Observations indicate that, unlike clusters, superclusters are not bound together by gravity. That is, most clusters in each supercluster are drifting away from the other clusters in that same supercluster. Furthermore, the superclusters are all moving away from one another.

Since the 1980s, astronomers have been working to understand how superclusters are distributed in space. Some of this structure is revealed by maps such as **Figure 14-19,** which displays the positions on the sky of 1.6 million galaxies. Such maps reveal that superclusters are not randomly distributed, but seem to lie along filaments. But to comprehend more fully the distribution of superclusters, it is necessary to map their positions in three dimensions. This is done by measuring both the position of a galaxy on the sky as well as the galaxy's distance and thus its position in three-dimensional space.

The most extensive galaxy maps available at this writing are those from the Sloan Digital Sky Survey, a joint project of astronomers from the United States, Japan, and Germany, and from the Two Degree Field Galactic Redshift Survey (2dFGRS), a collaboration between Australian, British, and U.S. astronomers. (The name refers to the 2° field of view of the telescope used for the observations, which is unusually wide for a research telescope.) **Figure 14-20*a*** shows a map

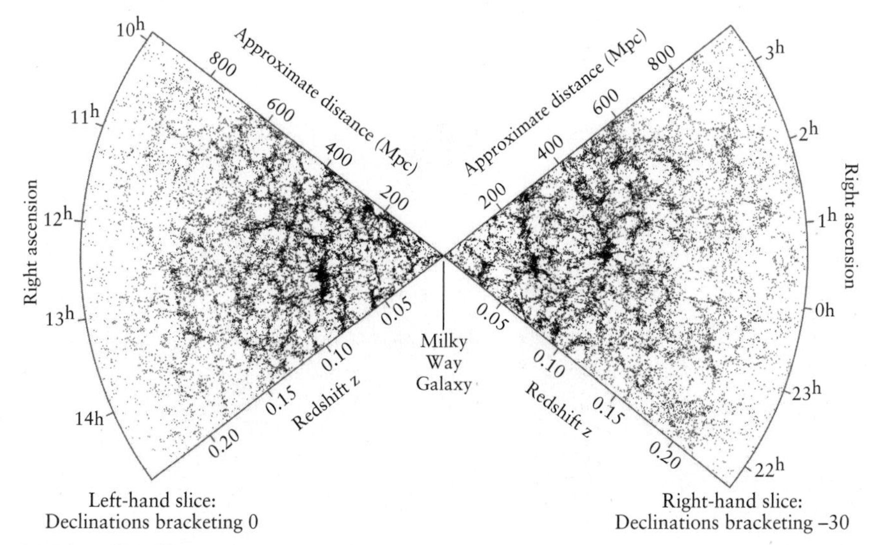

(a) The 2dF galaxy survey

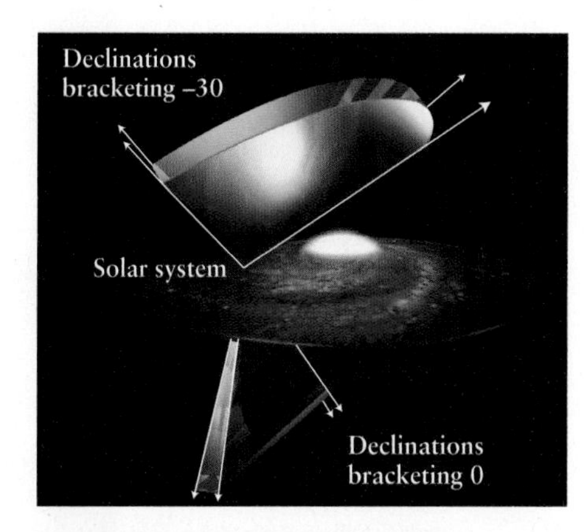

(b) Fields of view in the 2dF survey

FIGURE 14-20 The Large-Scale Distribution of Galaxies (a) This map shows the distribution of 62,559 galaxies in two wedges extending out to about 5 billion ly. Note the prominent voids surrounded by thin regions full of galaxies. (b) The two wedges shown in (a) lie roughly perpendicular to the plane of the Milky Way. These were chosen to avoid the obscuring dust that lies in our Galaxy's plane. *(2dF Galaxy Redshift Survey Team/Australian Astronomical Observatory)*

made from 2dFGRS measurements of more than 60,000 galaxies. This particular map encompasses two wedge-shaped slices of the universe, one on either side of the plane of the Milky Way (Figure 14-20*b*). Earth (in the Milky Way) is at the apex of the wedge-shaped map; each dot represents a galaxy and extends out to a distance of nearly 1000 Mpc, or 3 *billion* ly from Earth.

Maps such as that shown in Figure 14-20*a* reveal enormous **voids** where exceptionally few galaxies are found. These voids are roughly spherical and measure 100 million to 400 million ly in diameter. They are not entirely empty, however. There is evidence for hydrogen clouds in some voids, while others may be subdivided by strings of dim galaxies.

Figure 14-20*a* shows that most galaxies are concentrated in sheets on the surfaces between voids. This pattern is similar to that of soapsuds in a kitchen sink, with sheets of soap film (analogous to galaxies) surrounding air bubbles (analogous to voids). These titanic sheets of galaxies are the largest structures known in the universe: On scales much larger than 300 million ly, the distribution of galaxies in the universe appears to be roughly uniform.

ConceptCheck 14-10: **Are superclusters evenly distributed throughout the universe?**

ConceptCheck 14-11: **What is the most common type of galaxy in our Local Group?**

14-5 Colliding galaxies produce starbursts, spiral arms, and other spectacular phenomena

Go to Video 14-5

Occasionally, two galaxies within a cluster or from adjacent clusters can collide with each other. Past collisions have hurled vast numbers of stars into intergalactic space. In some cases, we can even observe a collision in progress, a cosmic catastrophe that gives birth to new stars. And astronomers can predict collisions that will not take place for billions of years, such as the collision that is fated to occur between the galaxy M31 and our own Milky Way Galaxy.

High-Speed Galaxy Collisions: Shredding Gas and Dust

When two galaxies collide at high speed, the huge clouds of interstellar gas and dust in the galaxies slam into each other and can be completely stopped in their tracks. In this way, two colliding galaxies can be stripped of their interstellar gas and dust.

The best evidence that such collisions take place is that many clusters of galaxies are strong sources of X-rays (**Figure 14-21**). This emission reveals the presence of substantial amounts of hot **intracluster gas** (that is, gas within the cluster) at temperatures between 10^7 and 10^8 K. The only way that such large amounts of gas could be heated to such extremely high temperatures is in violent collisions between galaxies.

CAUTION Although galaxies can and do collide, it is highly unlikely that the *stars* from two colliding galaxies actually run into each other. The reason is that the stars within a galaxy are very widely separated from one another, with a tremendous amount of space between them.

In a less-violent collision or a near miss between two galaxies, the compressed interstellar gas may have more time to cool, allowing many protostars to form. Such collisions may account for **starburst galaxies** such as M82 (**Figure 14-22**), which blaze with the light of numerous newborn stars. These galaxies have bright centers surrounded by clouds of warm interstellar dust, indicating recent, vigorous star birth. Their warm dust is so abundant that starburst galaxies are among the most luminous objects in the universe at infrared wavelengths.

FIGURE 14-21 X-ray Emission from a Cluster of Galaxies (a) An X-ray image of this cluster of galaxies shows emission from hot gas between the galaxies. The gas was heated by collisions between galaxies within the cluster. (b) The galaxies themselves are too dim at X-ray wavelengths to be seen in (a), but are apparent at visible wavelengths. This cluster, one of many cataloged by the UCLA astronomer George O. Abell, is about 1 billion ly from Earth in the constellation Serpens. *(a: NASA, CXC, and University of California, Irvine/A. Lewis et al.; b: Palomar Observatory DSS)*

(a) An X-ray image of Abell 2029 shows emission from hot gas.

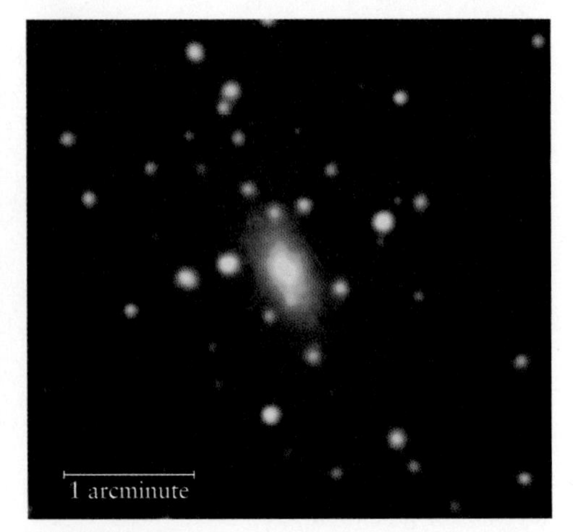

(b) A visible-light image of Abell 2029 shows the cluster's galaxies.

FIGURE 14-22 R I V U X G **A Starburst Galaxy** Prolific star formation is occurring at the center of the irregular galaxy M82, which lies about 12 million ly from Earth in the constellation Ursa Major. *(NASA; ESA; and the Hubble Heritage Team [STScI/AURA])*

The starburst galaxy M82 shown in Figure 14-22 also shows the effects of strong winds from young, luminous stars. It also contains a number of luminous globular clusters. Unlike the globular clusters in our Galaxy, whose stars are about 12.5 billion years old, those in M82 are no more than 600 million years old. These young globular clusters are another sign of recent star formation.

M82 is one member of a nearby cluster of galaxies that includes the beautiful spiral galaxy M81 and a fainter elliptical companion called NGC 3077 (**Figure 14-23*a***). Radio

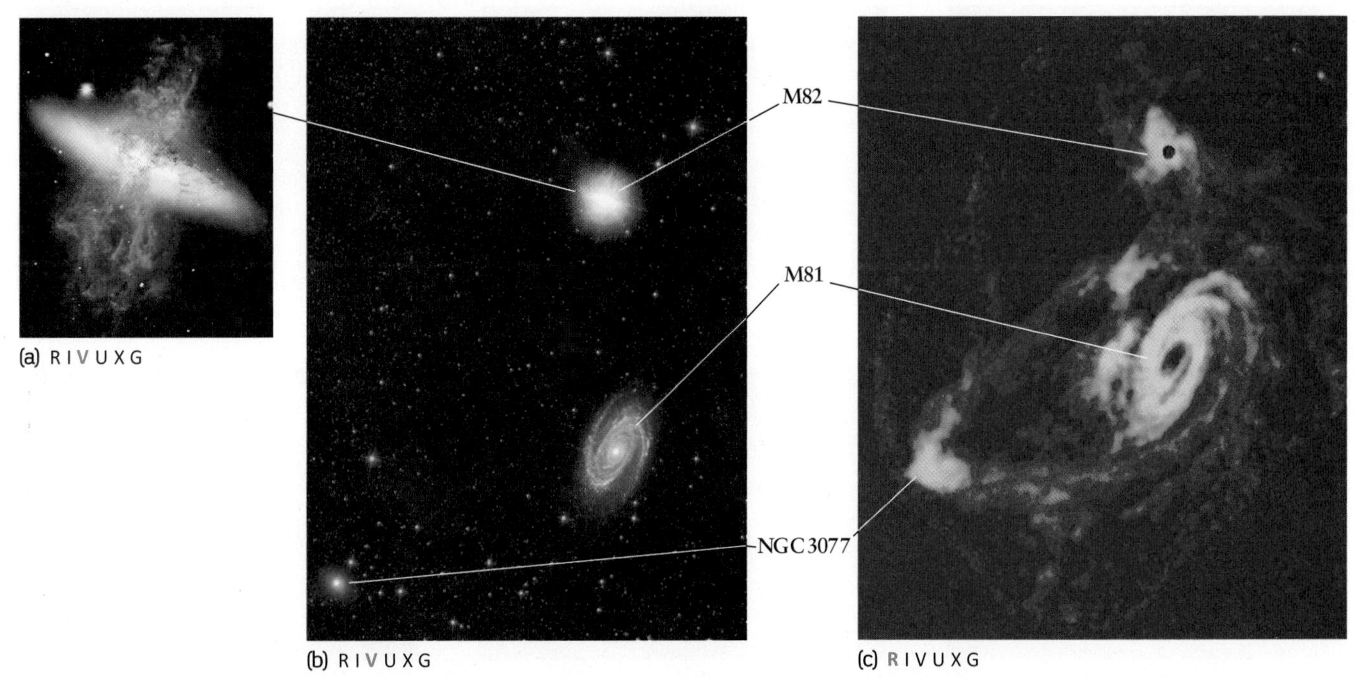

FIGURE 14-23 The M81 Group (a) The irregular starburst galaxy M82 is shown close up in a nearby cluster of about a dozen galaxies, dominated by the spectacular spiral M81. (b) Several of the galaxies in this cluster are connected by streamers of hydrogen gas. This wide-angle visible-light photograph shows the three brightest galaxies of the cluster, dominated by M81. The area shown is about 1° across. (c) This false-color radio image of the same region, created from data taken by the Very Large Array, shows streamers of hydrogen gas that connect the three bright galaxies as well as several dim ones. *(a: NASA/JPL-Caltech/WISE Team; b: M. S. Yun, VLA and Harvard; c: NASA/JPL-Caltech/WISE Team)*

surveys of that region of the sky reveal enormous streams of hydrogen gas connecting the three galaxies (Figure 14-23*b*). The loops and twists in these streamers suggest that the three galaxies have had several close encounters over the ages. A similar stream of hydrogen gas connects our Galaxy with its second-nearest neighbor, the Large Magellanic Cloud, suggesting a history of close encounters between our Galaxy and the LMC.

> **ConceptCheck 14-12:** What is the energy source accounting for intergalactic gas between galaxies that is often quite hot?

Tidal Forces and Galaxy Mergers

Tidal forces between colliding galaxies can deform the galaxies from their original shapes. The galactic deformation is so great that thousands of stars can be hurled into intergalactic space along huge, arching streams. (This same effect has stripped material away from the Canis Major Dwarf as it orbits the Milky Way, as shown in Figure 14-16.) Supercomputer simulations of such collisions show that while some of the stars are flung far and wide, other stars slow down and the galaxies may merge.

Figure 14-24 shows one such simulation. As the two galaxies pass through each other, they are severely distorted by gravitational interactions and throw out a pair of extended tails. The interaction also prevents the galaxies in the simulation from continuing on their original paths. Instead, they fall back together for a second encounter (at 625 million years). The simulated galaxies merge soon thereafter, leaving a single object. **Cosmic Connections: When Galaxies Collide** explores a real-life example of two galaxies that are colliding in just this manner.

Our own Milky Way Galaxy is expected to undergo a galactic collision like that shown in Figure 14-24. The Milky Way and the Andromeda Galaxy, shown in Figure 14-2, are actually approaching each other and should collide in another 6 billion years or so. (Recall that our solar system is only 4.57 billion years old.) When this happens, the sky will light up with a plethora of newly formed stars, followed in rapid succession by a string of supernovae, as the most massive of these stars complete their life spans. Any inhabitants of either galaxy will see a night sky far more dramatic and tempestuous than our present one.

Go to Video 14-6

When two galaxies merge, the result is a bigger galaxy. If this new galaxy is located in a rich cluster, it may capture and devour additional galaxies, growing to enormous dimensions by a sort of galactic cannibalism. Cannibalism, in this sense, differs from mergers in that the galaxy that does the devouring is bigger than its "meal," whereas merging galaxies are about the same size.

Many astronomers suspect that galactic cannibalism is the reason that giant ellipticals are so huge. As we have seen, giant galaxies typically occupy the centers of rich clusters. In many cases, smaller galaxies are located around

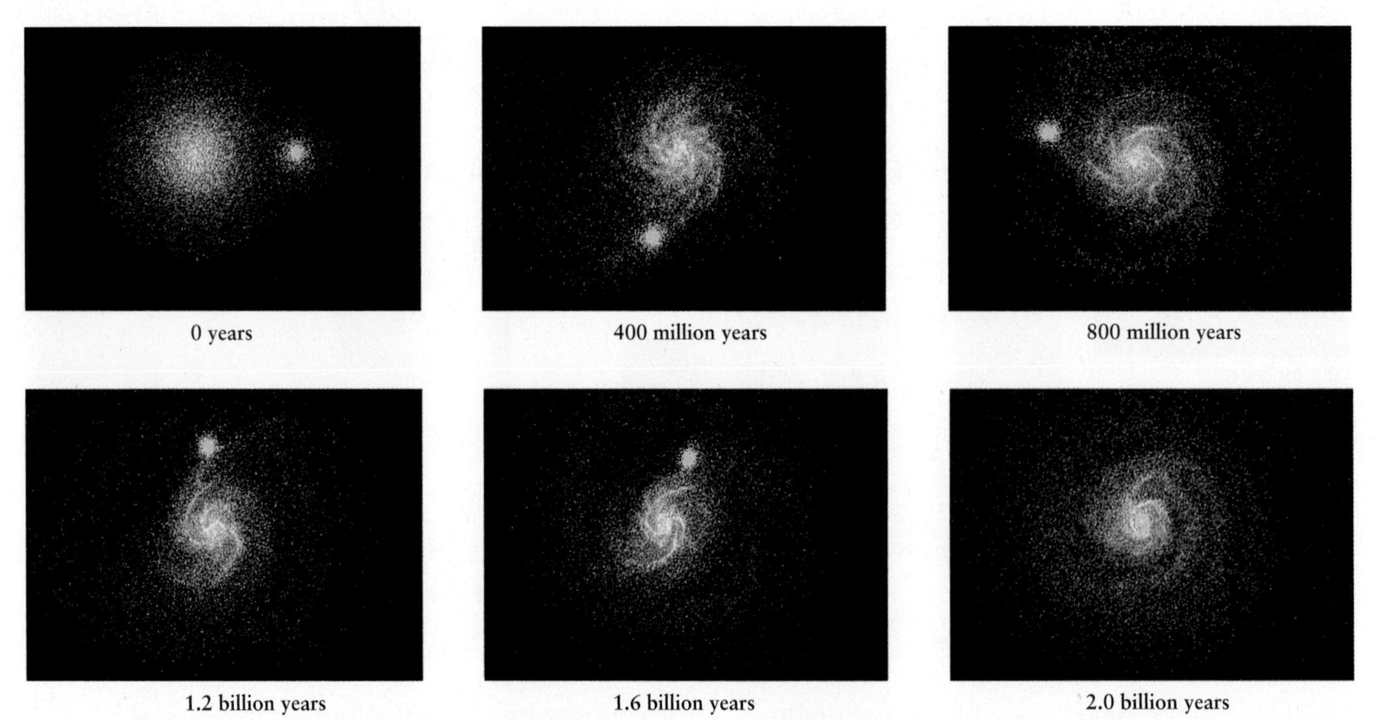

FIGURE 14-24 A Simulated Collision Between Two Galaxies These frames from a supercomputer simulation show the collision and merger of two galaxies accompanied by an ejection of stars into intergalactic space. It shows a small galaxy (yellow stars) being devoured by a larger, disk-shaped one (blue stars, white gas). Note how spiral arms are generated in the disk galaxy by its interaction with the satellite galaxy. *(Lars Hernquist, Institute for Advanced Study, with simulations performed at the Pittsburgh Supercomputing Center)*

COSMIC CONNECTIONS When Galaxies Collide

Although galaxies can collide at very high speeds by Earth standards, they are so vast that a collision can last hundreds of millions of years. Understanding what happens during a galactic collision requires ideas about tidal forces, star formation, and stellar evolution.

1. One example of a galactic collision is the pair of galaxies called the Antennae, which lies 19 Mpc (16 million ly) from Earth in the constellation Corvus (the Crow). They probably began to interact several hundred million years ago.

Tidal forces between the galaxies pulled out these long "tidal tails" 200 to 300 million years ago.

RIVUXG

2. As the gas and dust clouds of the two galaxies collide with each other, they are greatly compressed. This causes stars to form in tremendous numbers.

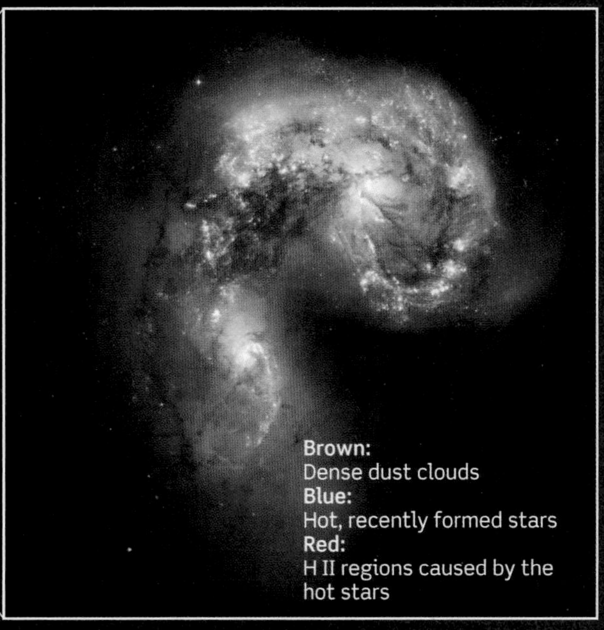

RIVUXG

RIVUXG

3. This composite infrared and visible-light image of the Antennae allows us to see inside the two galaxies and reveals clouds of dust warmed by the light of hot young stars.

4. The globular clusters that orbit our Milky Way Galaxy contain only old stars; all of the short-lived blue stars have long since died. But some of the globular clusters that orbit the Antennae galaxies *do* have hot blue stars. Hence these clusters must be young. These, too, are a result of the compression of gas and dust that takes place in a collision between galaxies.

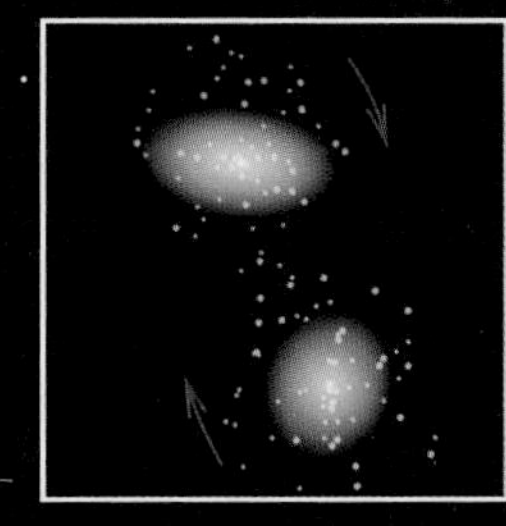

a. Two galaxies, each with old globular clusters (yellow), begin to interact.

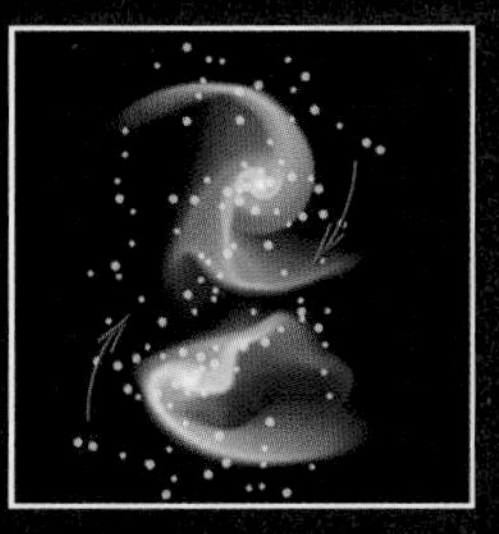

b. The two galaxies swoop past each other before finally settling down as a single merged galaxy. Gas and dust is compressed in both galaxies in the process, creating new star clusters.

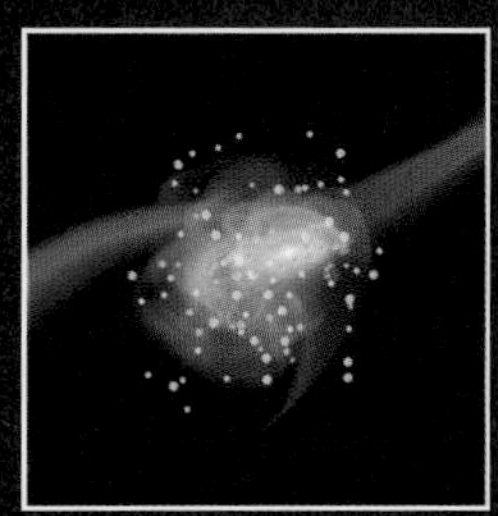

c. The combined galaxy has the original globular clusters (shown in yellow) as well as new ones (shown in blue).

these giants (see Figure 14-8 and Figure 14-17). As they pass through the extended halo of a giant elliptical, these smaller galaxies slow down and are eventually devoured by the larger galaxy.

ConceptCheck 14-13: When the Milky Way Galaxy and the Andromeda Galaxy finish colliding with each other, what will be left over?

14-6 Dark matter can be inferred by observing the motions of galaxy clusters

A cluster of galaxies must be held together by gravity. In other words, there must be enough matter in the cluster to prevent the galaxies from wandering away. Nevertheless, careful examination of a rich cluster, like the Coma cluster shown in Figure 14-17, reveals that the mass of the visually luminous matter (principally the stars in the galaxies) is not at all sufficient to bind the cluster gravitationally. The observed line-of-sight speeds of the galaxies, measured by Doppler shifts, are so large that the cluster should have broken apart long ago. Considerably more mass than is visible is needed to keep the galaxies bound in orbit about the center of the cluster.

We encountered a similar situation in studying our Milky Way Galaxy in the previous chapter: The total mass of our Galaxy is more than the amount of visible mass. As for our Galaxy, we conclude that clusters of galaxies must contain significant amounts of nonluminous *dark matter.* If this dark matter were not there, the galaxies would have long ago dispersed in random directions and the cluster would no longer exist today. Analyses demonstrate that the total mass needed to bind a typical densely packed galaxy cluster is about 10 times greater than the mass of material that shows up on visible-light images.

Evidence from Rotation Curves and Gravitational Lensing

As for our Galaxy, the problem is to determine what form the invisible mass takes. A partial solution to this **dark-matter problem,** which dates from the 1930s, was provided by the discovery in the late 1970s of hot, X-ray–emitting gas within clusters of galaxies (see Figure 14-21*a*). By measuring the amount of X-ray emission, astronomers find that the total mass of intracluster gas in a typical rich cluster can be greater than the combined mass of all the stars in all the cluster's galaxies. But this is sufficient to account for only about 10% of the invisible mass. The remainder is dark matter of unknown composition.

Although we do not know what dark matter is made of, it is possible to investigate how dark matter is distributed in galaxies and clusters of galaxies. It appears that dark matter

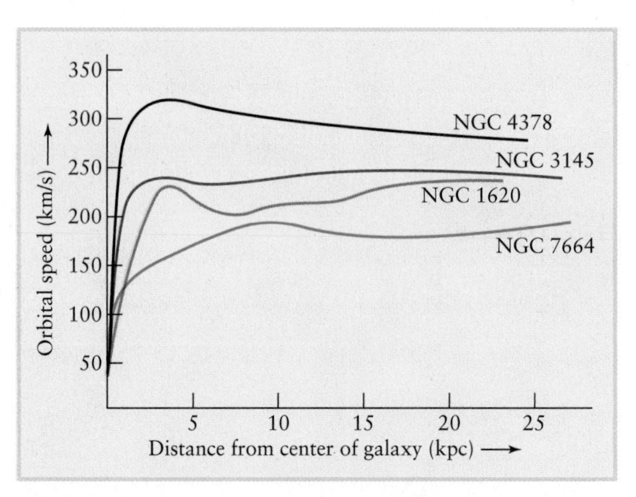

FIGURE 14-25 The Rotation Curves of Four Spiral Galaxies This graph shows how the orbital speed of material in the disks of four spiral galaxies varies with the distance from the center of each galaxy. If most of each galaxy's mass were concentrated near its center, these curves would fall off at large distances. But these and many other galaxies have flat rotation curves that do not fall off. This indicates the presence of extended halos of dark matter. *(Adapted from V. Rubin and K. Ford)*

lies within and immediately surrounding galaxies, but is not in the vast spaces between galaxies. The evidence for this comes principally from observations of the rotation curves of galaxies and of the gravitational bending of light by clusters of galaxies.

As we saw in the last chapter, a rotation curve is a graph that shows how fast stars in a galaxy are moving at different distances from that galaxy's center. As **Figure 14-25** illustrates, many other spiral galaxies have similar rotation curves that remain remarkably flat out to surprisingly great distances from each galaxy's center. In other words, the orbital speed of the stars remains roughly constant out to the visible edges of these galaxies.

This observation tells us that we still have not detected the *true* edges of these galaxies (and many similar ones). Near the true edge of a galaxy, we should see a decline in orbital speed. Because this decline has not been observed, astronomers conclude that there must be a considerable amount of dark material that extends well beyond the visible portion of the disk.

Further evidence about how dark matter is distributed comes from the gravitational bending of light rays. As **Figure 14-26** shows, the gravity of a single star like the Sun can deflect light by only a few arcseconds. But a more massive object such as a galaxy can produce much greater deflections, and the amount of this deflection can be used to determine the galaxy's mass. For example, suppose that Earth, a massive galaxy, and a background light source (such as a more distant galaxy) are in nearly perfect alignment, as sketched in **Figure 14-27*a*.** Because of the warped space around the massive galaxy, light from the background source curves around the galaxy as it heads toward us. As a result, light rays can travel along two paths from the background

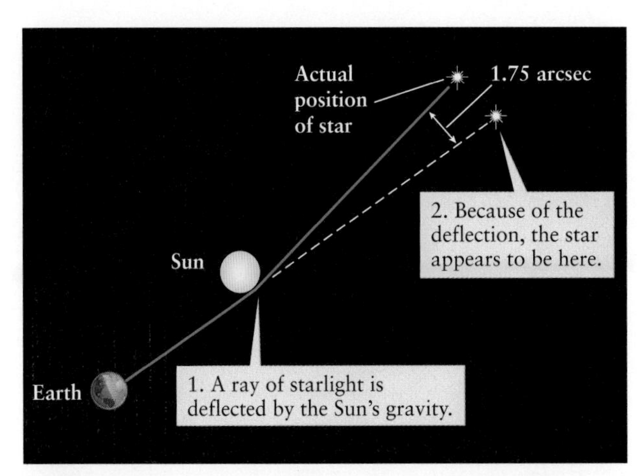

FIGURE 14-26 The Gravitational Deflection of Light Light rays are deflected by the curved spacetime around a massive object like the Sun. The maximum deflection is very small, only 1.75 arcsec for a light ray grazing the Sun's surface. By contrast, Newton's theory of gravity predicts *no* deflection at all. The deflection of starlight by the Sun was confirmed during a solar eclipse in 1919.

source to us here on Earth. Thus, we should see two images of the background source.

A powerful source of gravity that distorts background images is called a **gravitational lens.** For gravitational lensing to work, the alignment between Earth, the massive galaxy, and a remote background light source must be almost perfect. Without nearly perfect alignment, the second image of the background star is too faint to be noticeable.

Most of the mass within a galaxy emits no light and is of unknown composition.

Beginning in 1979, astronomers have discovered a great number of examples of gravitational lensing. The example shown in Figure 14-27*b* is almost exactly like the ideal situation depicted in Figure 14-27*a*. If the alignment is very slightly off, the image of the distant galaxy is distorted into an arc, as shown in Figure 14-27*c*. Figure 14-27*d* shows a more complicated example of lensing that results when the gravitational lens is not one but several massive galaxies.

By measuring the distortion of the images of such background galaxies, J. Anthony Tyson of Bell Laboratories and his colleagues have determined that dark matter, which constitutes about 90% of the cluster's mass, is distributed much like the visible matter in the cluster. In other words, the overall arrangement of visible galaxies seems to trace the location of dark matter.

Many proposals have been made to explain the nature of dark matter. One reasonable suggestion was that clusters might contain a large number of faint, red, low-mass (0.2 $M_\odot$ or less) stars. These faint stars could be located in extended halos surrounding individual galaxies or scattered throughout the spaces between the galaxies of a cluster. They would have escaped detection because their luminosity and hence

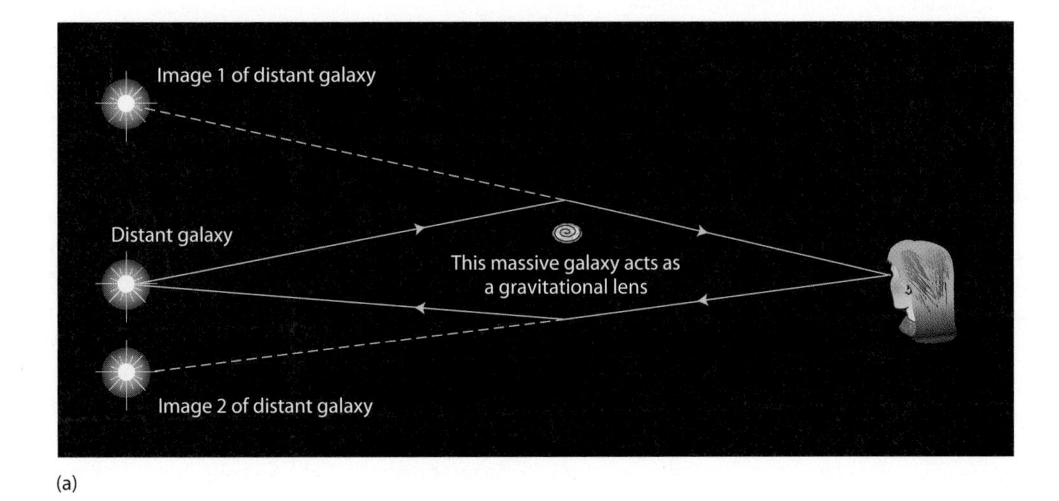

(a)

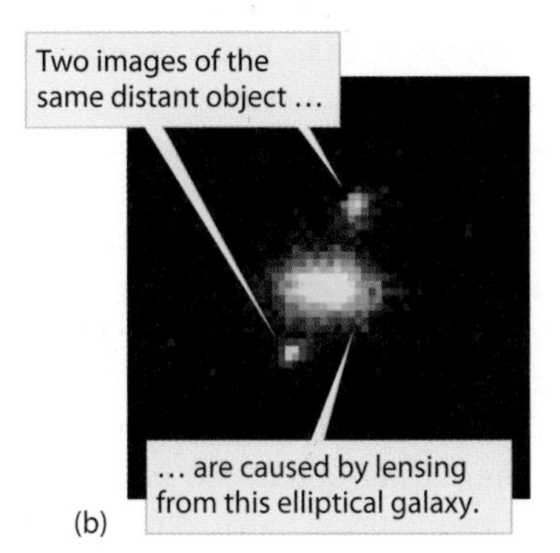

(b)

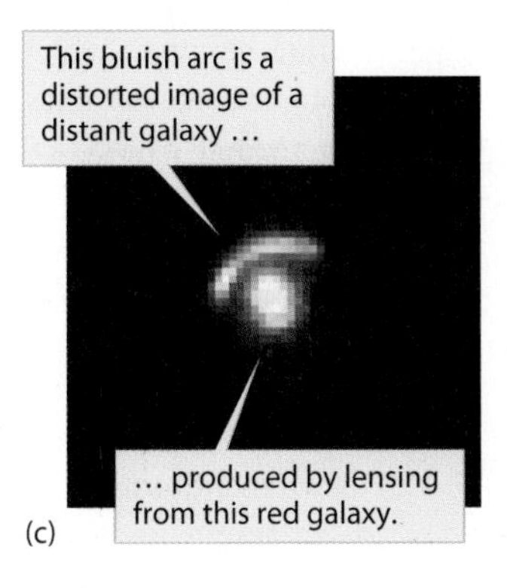

(c)

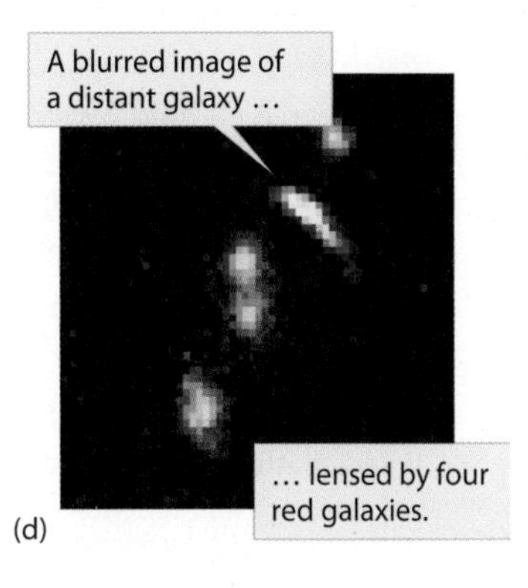

(d)

FIGURE 14-27 R I V U X G **Gravitational Lensing** (a) A massive object such as a galaxy can deflect light rays like a lens so that an observer sees more than one image of a more distant galaxy. (b, c, d) Three examples of gravitational lensing. In each case a single, distant blue galaxy is "lensed" by a closer red galaxy or galaxies. *(b, c, d: Kavan Ratnatunga, Carnegie Mellon University; ESA; and NASA)*

(a) Composite image of galaxy cluster 1E0657-56 showing visible galaxies, X-ray-emitting gas (red) and dark matter (blue)

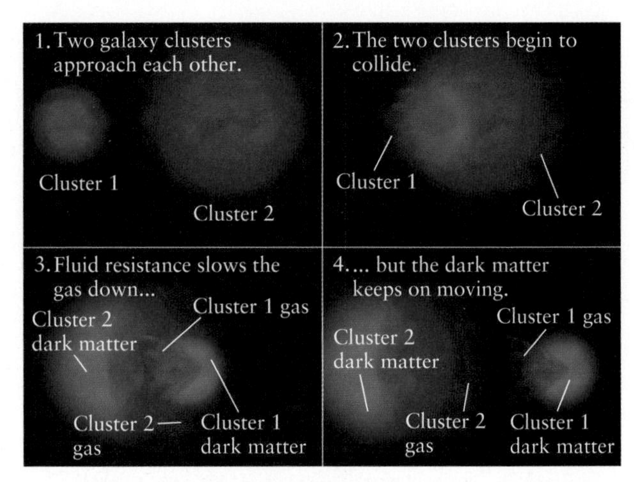

(b) A model of how the gas and dark matter in 1E0657-56 could have become separated

FIGURE 14-28 R I V U X G **Isolated Dark Matter in a Cluster of Galaxies** (a) This visible-light image of the galaxy cluster 1E0657-56 shows more than a thousand galaxies. The superimposed image in red shows the distribution of the cluster's hot, X-ray-emitting gas, and the blue image shows the distribution of dark matter as determined by gravitational lensing (see Figure 14-27). (b) We can understand the separation of dark matter and gas in this cluster if we assume that dark matter does not feel any force of fluid resistance. This is what we would expect if dark matter were to respond to gravitational forces only. *(a: NASA/STScI; Magellan/U. of Arizona/D. Clowe et al.; b: NASA/CXC/M. Weiss)*

apparent brightness would be very low. Searches for these stars around other galaxies as well as around the Milky Way have been carried out using the Hubble Space Telescope. None has yet been detected, so it is thought that faint stars are unlikely to constitute the majority of the dark matter in the universe.

An important clue about the nature of dark matter was discovered in 2006 by examining a rich cluster of galaxies called 1E0657-56. Remarkably, the visible matter and dark matter in this cluster do *not* have the same distribution (**Figure 14-28*a***). The best explanation for how this could have come about is that 1E0657-56 is the result of a collision between two galaxy clusters, one larger than the other (Figure 14-28*b*). During the collision, the gas from one cluster slams into the gas from the other cluster and slows down due to fluid resistance. (You feel the force of fluid resistance pushing against you whenever you try to move through a liquid or gas—for example, when you swim in a lake or put your hand outside the window of a fast-moving car.) Fluid resistance is a consequence of the electric forces between adjacent atoms and molecules in a fluid. But if dark matter is made up of some curious material that responds only to *gravitational* forces, it is unaffected by fluid resistance. As a result, during the collision sketched in Figure 14-28*b* the gas is slowed by fluid resistance but the dark matter is not. The agreement of the simulation in Figure 14-28*b* with the observations in Figure 14-28*a* strongly reinforces the idea that dark matter, though mysterious, is quite real.

ConceptCheck 14-14: If dark matter cannot be seen, what are the two primary lines of evidence that it exists?

14-7 Quasars are the ultraluminous centers of the most distant galaxies

Grote Reber, a radio engineer and ham radio enthusiast, built the first true radio telescope in 1936 in his backyard in Illinois. By 1944, he had detected three strong radio emissions coming from the general directions of the constellations Cassiopeia, Sagittarius, and Cygnus.

Two of these heavenly sources of radio waves, named Cassiopeia A and Sagittarius A, happen to lie in our own Galaxy; these two different sources are the remains of an exploded star and, much farther away, a mysterious source at the center of the Galaxy. The nature of the third source, called Cygnus A (**Figure 14-29*a***), proved more elusive. The mystery only deepened in 1951, when Walter Baade and Rudolph Minkowski used the 200-in. (5-m) optical telescope on Palomar Mountain to discover a dim, strange-looking galaxy at the position of the Cygnus A radio wave source (Figure 14-29*b*).

When Baade and Minkowski photographed the spectrum of Cygnus A, they were surprised to find a number of bright *emission* lines. This was wholly unexpected because a normal galaxy shows *absorption* lines in its spectrum. This typically observed absorption takes place in the atmospheres of the individual stars that make up the galaxy. In order for Cygnus A to have emission lines, something must be exciting and ionizing its atoms. Furthermore, the wavelengths of Cygnus A's emission lines are all shifted by 5.6% toward the red end of the spectrum. This corresponds to a tremendous distance from Earth—about 740 million ly. Yet despite its tremendous distance from Earth, radio waves

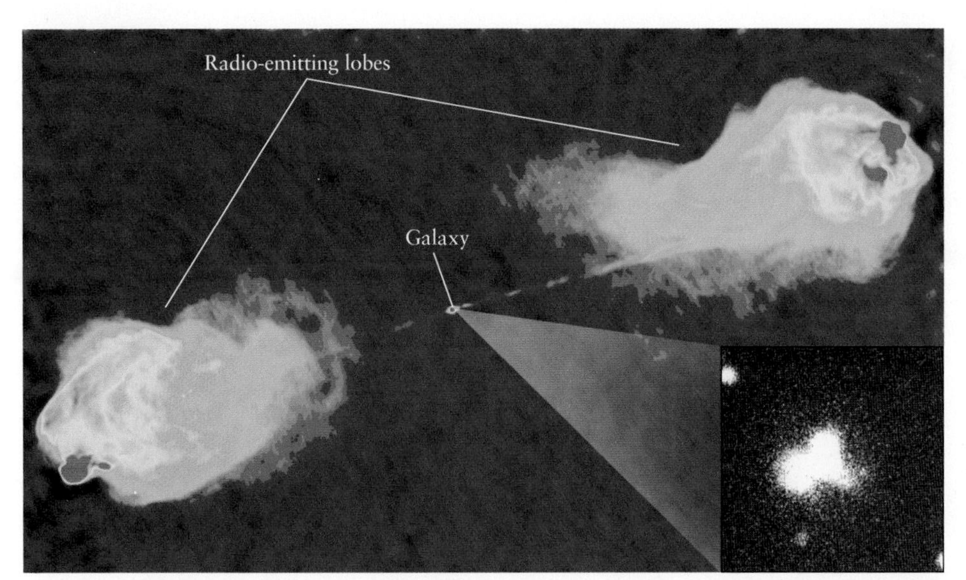

FIGURE 14-29 Cygnus A (3C 405) (a) This false-color radio image from the Very Large Array shows that most of the emission from Cygnus A comes from luminous radio lobes located on either side of a peculiar galaxy. (Red indicates the strongest radio emission, while blue indicates the faintest.) Each lobe extends about 230,000 ly from the galaxy. (b) The galaxy at the heart of Cygnus A has a substantial redshift, so it must be extremely far from Earth, about 740 million ly. To be so distant and yet be one of the brightest radio sources in the sky, Cygnus A must have an enormous energy output. *(a: R. A. Perley, J. W. Dreher, and J. J. Cowan, NRAO; b: NRAO/AUI)*

from Cygnus A can be picked up by amateur astronomers with the simplest of backyard equipment. This means that Cygnus A must be one of the most luminous radio sources in the sky. In fact, its radio luminosity is 10^7 times as great as that of an ordinary galaxy like the Milky Way. The object that creates the Cygnus A radio emission must be something quite extraordinary.

Cygnus A was not the only curious radio source to draw the attention of astronomers. In 1960, Allan Sandage used the 200-inch telescope at Palomar to discover a "star" at the location of a radio source designated 3C 48. (The 3C refers to the *Third Cambridge Catalogue,* a compendium of radio sources.) Ordinary stars are not strong sources of radio emission, so 3C 48 must be something unusual.

Although 3C 48 and 3C 273 were clearly oddballs, many astronomers thought they were just strange stars in our Galaxy. A breakthrough occurred in 1963, when Maarten Schmidt at Caltech took a more careful look at the spectrum of 3C 273. Schmidt determined that 3C 273 has a recessional velocity of 100,000 miles per hour, nearly 15% of the speed of light. No star could be moving this fast and remain within our Galaxy for very long. Hence, Schmidt concluded that 3C 273 could not be a nearby star, but must lie outside the Milky Way. To be detected at such distances, 3C 273 must be an extraordinarily powerful source of both visible and radio light.

Because of their strong radio emission and starlike appearance, 3C 48 and 3C 273 were dubbed *quasi-stellar radio sources,* a term soon shortened to **quasars.** After the first quasars were discovered by their radio emission, many similar ultradistant, starlike objects were found that emit little or no radio wavelength light. These "radio-quiet" quasars were originally called *quasi-stellar objects,* or *QSOs*, to distinguish them from radio emitters. Today, however, the term "quasar" is often used to include both types.

ConceptCheck 14-15: What makes Cassiopeia A and Sagittarius A so unusual?

Quasars Are the Ultraluminous Centers of Distant Galaxies

What are these strange, brilliant, and distant objects? Quasars cannot simply be very large and luminous galaxies, because their spectra are totally different. Quasars turn out to be ultraluminous objects located at the centers of remote galaxies.

In the 1970s, observations showed that quasars are associated with some of the most remote galaxies. Long-exposure images of quasars showed that they are often found in groups or clusters of galaxies, and that the distance to these quasars is essentially the same as the distances to the galaxies that surround them. A quasar's spectra can indicate its distance from Earth, as do the distances of galaxies.

It is very difficult to observe the "host galaxy" in which a quasar is located because the quasar's light overwhelms light from the galaxy's individual stars (**Figure 14-30*a***). Nevertheless, improved technology and painstaking observations have revealed some basic properties of these host galaxies (Figure 14-30*b*). Relatively nearby radio-quiet quasars tend to be located in spiral galaxies, whereas radio-loud quasars as well as more distant radio-quiet quasars tend to be located in ellipticals (Figure 14-30*c*). However, a large percentage of these host galaxies have distorted shapes or are otherwise peculiar. Many have nearby companion galaxies, suggesting a link between collisions or

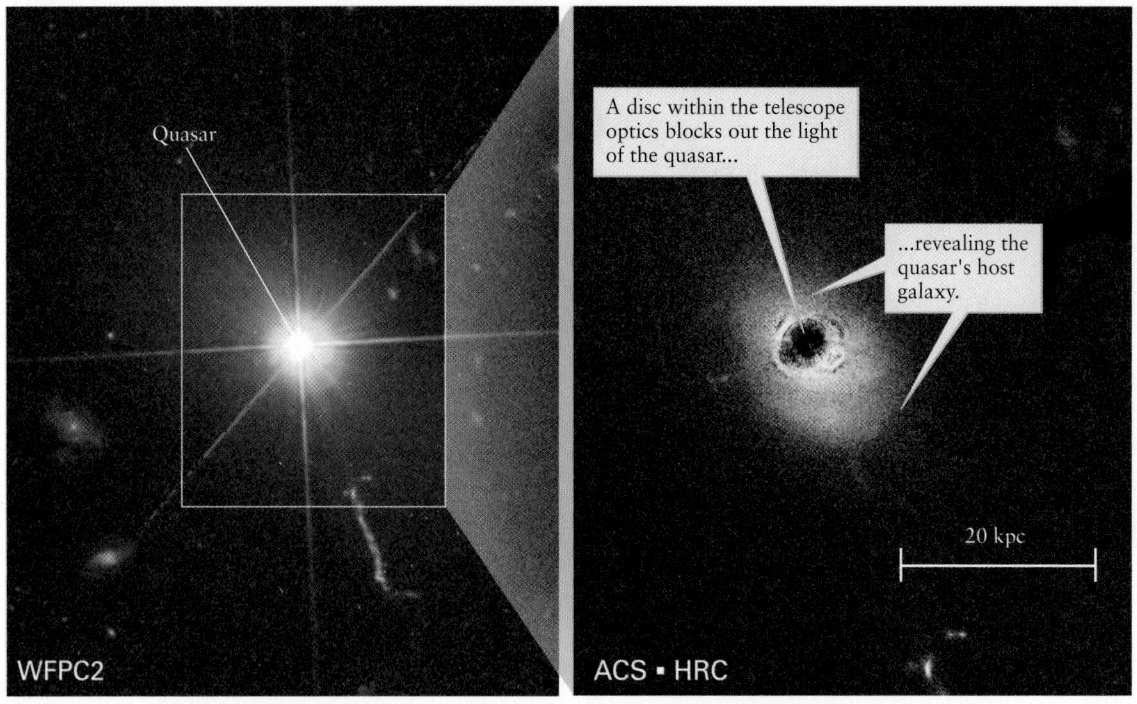

(a) 3C 273

(b) The host galaxy of 3C 273

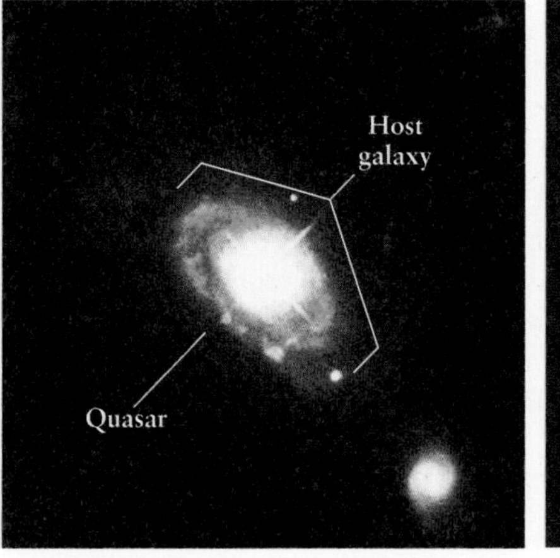

(c) PG 0052+251 and its host galaxy

(d) PG 1012+008 and its host galaxy

FIGURE 14-30 R I V U X G **Quasars and Their Host Galaxies** (a) In this 1994 image from the Hubble Space Telescope (HST), the glare of quasar 3C 273 hides its host galaxy. (b) This 2002 image that reveals the host galaxy was made using an upgraded camera aboard HST. (c) Quasar PG 0052+251 is located at the center of an apparently normal spiral galaxy. Other quasars are found at the centers of ordinary looking elliptical galaxies. (d) The galaxy that hosts quasar PG 1012+008 is in the process of merging with a second luminous galaxy. The wispy material surrounding the quasar may have been pulled out of the galaxies by tidal forces (see Figure 14-24). The two merging galaxies are just 31,000 ly apart. Another small galaxy to the left of the quasar may also be merging with the others. *(a: NASA and J. Bahcall [Institute for Advanced Study]; b: NASA, A. Martel [JHU], H. Ford [JHU], M. Clampin [STScI], G. Hartig [STScI], G. Illingworth [UCO/Lick Observatory], the ACS Science Team and ESA; c, d: J. Bahcall [Institute for Advanced Study], M. Disney [University of Wales], and NASA)*

mergers and the quasar itself (Figure 14-30*d*). While observations like those in Figure 14-30 show that quasars lie at the centers of galaxies, they do not really explain what quasars are.

> **CalculationCheck 14-1: If two quasars have identical energy output, but one is 3 times farther from us than the other, what is the difference in brightness as seen from Earth?**

Active Galaxies and AGNs

Quasars are members of a diverse group of very distant, superluminous objects now collectively known as **active galaxies.** The activity of such a galaxy comes from an energy source at its center. Hence, astronomers say that these galaxies possess **active galactic nuclei,** or **AGNs.**

One of the features that distinguishes one type of AGN from another is the width of the emission lines in their

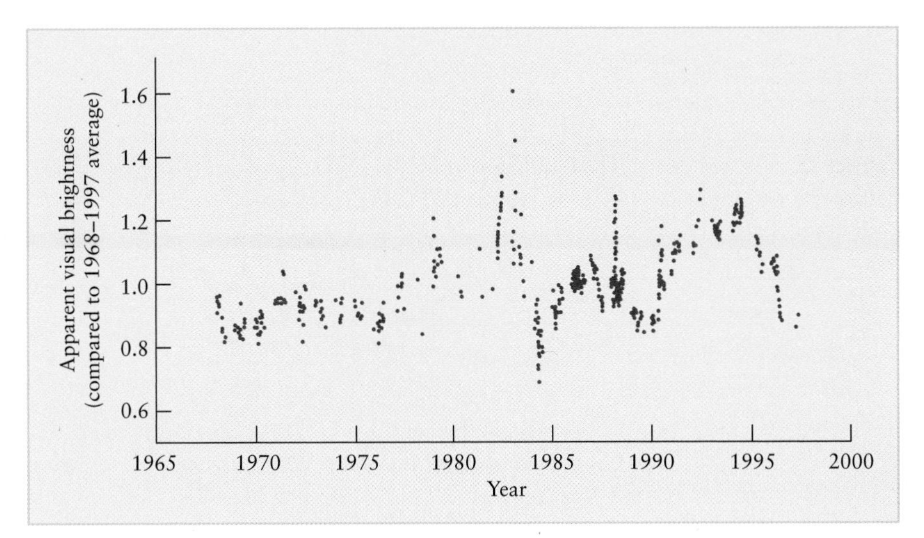

FIGURE 14-31 Brightness Variations of an AGN This graph shows variations over a 29-year period in the apparent brightness of the quasar 3C 273 (see Figure 14-30). Note the large outburst in 1982–1983 and the somewhat smaller ones in 1988 and 1992. *(Adapted from M. Türler, S. Paltani, and T. J.-L. Courvoisier)*

spectra. The widths of these emission lines indicate that individual light-emitting gas clouds are moving within the quasar at very high speeds (around 20 million miles per hour or 9000 km/s). Some of the clouds are moving toward us, causing the emitted light to have a shorter wavelength and higher frequency, while other clouds are moving away from us and emit light with longer wavelength and lower frequency.

One characteristic that is common to *all* types of active galactic nuclei is variability. For example, **Figure 14-31** shows brightness fluctuations of the quasar 3C 273 as determined from 29 years of observations. The brightness of 3C 273 increased by 60% from the beginning to the end of 1982, then declined to the starting value in just five months. Other AGNs undergo greater fluctuations in brightness (by a factor of 25 or more) that occur even more rapidly (X-ray observations reveal that some objects vary in brightness over time intervals as short as 3 hours).

These fluctuations in brightness allow astronomers to place strict limits on the maximum size of a light source. An object cannot vary in brightness faster than light can travel across that object. For example, an object that is 1 ly in diameter cannot vary significantly in brightness over a period of less than 1 year.

To understand this limitation, imagine an object that measures 1 ly across, as in **Figure 14-32.** Suppose the entire

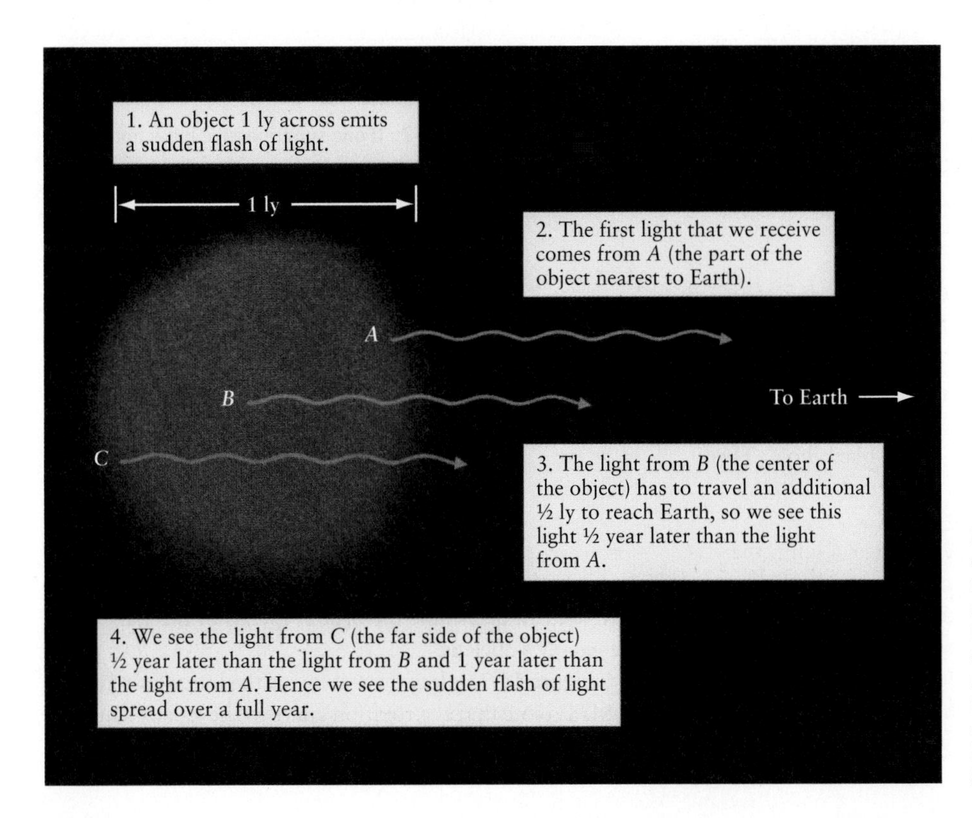

FIGURE 14-32 A Limit on the Speed of Variations in Brightness The rapidity with which the brightness of an object can vary significantly is limited by the time it takes light to travel across the object. If the object 1 ly in size emits a sudden flash of light, the flash will be observed from Earth to last a full year. If the object is 2 ly in size, brightness variations will last at least 2 years as seen from Earth, and so on.

object emits a brief flash of light. Photons from that part of the object nearest Earth arrive at our telescopes first. Photons from the middle of the object arrive at Earth 6 months later. Finally, light from the far side of the object arrives a year after the first photons. Although the object emitted a sudden flash of light, we observe a gradual increase in brightness that lasts a full year. In other words, the flash is stretched out over an interval equal to the difference in the light travel time between the nearest and farthest observable regions of the object.

The rapid flickering exhibited by active galactic nuclei means that they emit their energy from a small volume, possibly less than 1 light-day across. In other words, a region no larger than our solar system can emit more energy per second than a thousand galaxies! Astrophysicists therefore face the challenge of explaining how so much energy can be produced in such a very small volume.

ConceptCheck 14-16: Why do astronomers think the energy source of an AGN has a relatively small diameter?

14-8 Supermassive black holes may be the "central engines" that power active galaxies

In 1968, the British astronomer Donald Lynden-Bell pointed out that a black hole lurking at the center of a galaxy could be the "central engine" powering an active galactic nucleus. Lynden-Bell theorized that as gases fall onto a black hole, their gravitational energy would be converted into radiation. To produce as much radiation as is seen from active galactic nuclei, the black hole would have to be very massive indeed.

The Eddington Limit and Black Hole Sizes

How large a black hole would be needed to power an active galactic nucleus? You might think that what really matters is not the size of the black hole, but rather the amount of gas that falls onto it and releases energy. However, there is a natural limit to the luminosity that can be radiated by accretion onto a compact object like a black hole. This is called the **Eddington limit,** after the British astrophysicist Sir Arthur Eddington.

If the luminosity exceeds the Eddington limit, there is so much *radiation pressure*—the pressure produced by photons streaming outward from the infalling material—that the surrounding gas is pushed away rather than falling inward onto the black hole. Without a source of gas to provide energy, the luminosity naturally decreases to below the Eddington limit, at which point gas can again fall inward. This limit allows us to calculate the minimum mass of an active galactic nucleus.

Numerically, the Eddington limit is:

The Eddington limit

$$L_{\text{Edd}} = 30{,}000\left(\frac{M}{M_{\odot}}\right)L_{\odot}$$

L_{Edd} = maximum luminosity that can be radiated by accretion onto a compact object
M = mass of the compact object
$M_{\odot}$ = mass of the Sun
$L_{\odot}$ = luminosity of the Sun

What this means is that the mass of the black hole must also be quite large because the tremendous luminosity of an active galactic nucleus must be less than or equal to its Eddington limit.

Astronomers have indeed found evidence for such **supermassive black holes** at the centers of many nearby normal galaxies. As we saw in section 13-6, at the center of our own Milky Way Galaxy lies what is almost certainly a black hole of about 3.7×10^6 solar masses—supermassive in comparison to a star, but less than 1% the mass of the behemoth black hole at the center of 3C 273.

Theory suggests that unlike stellar-mass black holes, which require a supernova to produce them, supermassive black holes can be produced without extreme pressures or densities. This may help to explain why they appear to be a feature of so many galaxies.

ConceptCheck 14-17: What keeps a supermassive black hole from having an unlimited luminosity?

CalculationCheck 14-2: If one supermassive black hole has twice the mass of a second supermassive black hole, how is its Eddington limit different?

Measuring Black Hole Masses in Galaxies

One galaxy that probably has a black hole at its center is the Andromeda Galaxy (M31), shown in Figure 14-2. M31 is only 2.5 million ly from Earth, close enough that details in its core as small as 3 ly across can be resolved under the best seeing conditions.

In the mid-1980s, astronomers made high-resolution spectroscopic observations of M31's core. By measuring the Doppler shifts of spectral lines at various locations in the core, we can determine the orbital speeds of the stars about the galaxy's nucleus.

Figure 14-33 plots the results for the innermost 1000 ly of M31. Note that the rotation curve in the galaxy's nucleus does not follow the trend set in the outer core. Rather, there

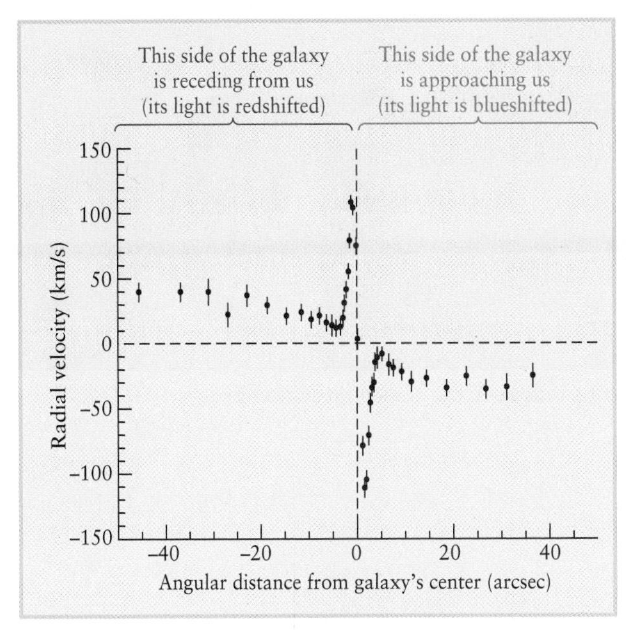

FIGURE 14-33 The Rotation Curve of the Core of M31 This graph plots radial velocity of matter in the core of M31 versus the angular distance from the galaxy's center. Note the sharp peaks, one blueshifted and one redshifted, within 5 arcsec of the galaxy's center. This indicates the presence of a compact, very massive object at the center of the galaxy. At the distance of M31, 1 arcsec corresponds to 12 ly. *(Adapted from J. Kormendy)*

are sharp peaks—one on the approaching side of the galaxy and the other on the receding side—within 5 arcsec of the galaxy's center.

The most straightforward interpretation is that the peaks are caused by the orbital motions of stars around M31's center. Stars on one side of the galaxy's center are approaching us while stars on the other side are receding from us.

The high speeds of stars orbiting close to M31's center indicate the presence of a massive central object. We can use Newton's form of Kepler's third law and our knowledge of these stars' orbital speeds to calculate the mass of this object. Similarly, this is the same method we use to calculate the mass of the supermassive black hole at the center of our Milky Way Galaxy. The difference is that we can track individual stars at the center of our Galaxy, while the data in Figure 14-33 come from the combined light of many stars in M31. Such calculations show that there must be about 3×10^7 solar masses within 16 ly of the center of M31. That much matter confined to such a small volume strongly suggests the presence of a supermassive black hole. Observations of M31 with the Chandra X-ray Observatory are consistent with this picture.

By applying high-resolution spectroscopy to the cores of other nearby galaxies, astronomers have discovered a number of supermassive black holes like the one in M31. Unfortunately, this technique for identifying black holes is difficult to apply to quasars, which are very distant and have small angular sizes. The evidence for supermassive black holes in quasars is therefore circumstantial, yet compelling: No other known energy source could provide enough power to sustain a quasar's intense light output.

ConceptCheck 14-18: **How is the evidence for supermassive black holes in nearby galaxies different from the evidence for supermassive black holes in distant quasars?**

14-9 Galaxies may have formed from the merger of smaller objects

How do galaxies form and how do they evolve? Astronomers can gain important clues about galactic evolution simply by looking deep into space. The more distant a galaxy is, the longer its light takes to reach us. The Andromeda Galaxy is 2.2 million ly away from us, which means that the light we see from it tonight left Andromeda 2.2 million years ago. We are in effect looking into Andromeda's past. As we examine galaxies that are at increasing distances from Earth, we are actually looking further and further back in time. By looking into the past, we can see galaxies in their earliest stages.

Building Galaxies from the "Bottom Up"

The Hubble Space Telescope images in **Figure 14-34** provide a glimpse of galaxy formation in the early universe. Figure 14-34*a* shows a number of galaxylike objects some 11 billion ly away and are thus seen as they were 11 billion years ago. These objects are smaller than even the smallest galaxies we see in the present-day universe and have unusual, irregular shapes (Figure 14-34*b*). Furthermore, these objects are scattered over an area only 2 million ly across—less than the distance between the Milky Way Galaxy and M31—making it quite probable that they would collide and merge with each other. These collisions would be aided by the dark matter associated with each subgalactic object, which increases the object's mass and, hence, the gravitational forces pulling the objects together. Such mergers would eventually give rise to a normal-sized galaxy.

Images such as those in Figure 14-34 lead astronomers to conclude that galaxies formed "from the bottom up"—that is, by the merger of smaller objects like those in Figure 14-34*b* to form full-size galaxies. (These same images rule out an older idea that galaxies formed from the "top down"—that is, directly from immense, galaxy-sized clouds of material.) The blue color of the objects in Figure 14-34*b* indicates the presence of young stars. Observations indicate that the very first stars formed about 13.5 billion years ago, when the universe was only about 200 million years old.

(a) A portion of the constellation Hercules

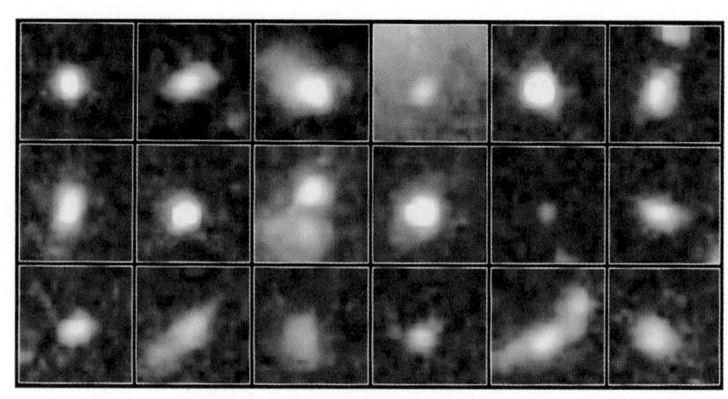

(b) Close-up images of the numbered objects in (a)

FIGURE 14-34 R I V U X G **The Building Blocks of Galaxies** (a) In this Hubble Space Telescope image, the objects outlined by boxes are about 3400 Mpc (11 billion ly) from Earth and are only 2000 to 3000 ly across—larger than a star cluster but smaller than even dwarf elliptical galaxies like that shown in Figure 14-9. The black squares at the top right are due to the shape of the camera. (b) Close-up images of 18 of the smallest galaxies from the constellation of Hercules. If these objects were to merge, the result would be a full-sized galaxy such as we see in the nearby universe today. *(Rogier Windhorst and Sam Pascarelle, Arizona State University; NASA)*

ConceptCheck 14-19: **Did today's galaxies form from combining smaller galaxies or from the separation of larger galaxies?**

Forming Spirals, Lenticulars, and Ellipticals

Go to Video 14-6

Once a number of subgalactic units combine, they make an object called a *protogalaxy*. The rate at which stars form within a protogalaxy may determine whether this protogalaxy becomes a spiral or an elliptical. If stars form relatively slowly, the gas surrounding them has enough time to settle by collisions to form a flattened disk, much as happened on a much smaller scale in the solar nebula during the formation of our Sun and solar system. Star formation continues because the disk contains an ample amount of hydrogen from which to make new stars. The result is a spiral galaxy (**Figure 14-35*a***). But if stars initially form in the protogalaxy at a rapid rate, virtually all of the available gas is used up to make stars before a disk can form. In this case what results is an elliptical galaxy (Figure 14-35*b*).

Galaxies change shape when gravitationally distorting other galaxies.

Figure 14-35*c* compares the stellar birthrate in the two types of galaxies. This graph helps us understand some of the differences between spiral and elliptical galaxies that we described earlier. Protogalaxies are thought to have been composed almost exclusively of hydrogen and helium gas, so the first stars were Population II stars with hardly any metals (that is, heavy elements). As stars die and form planetary nebulae or supernovae, they eject gases rich in metals into the interstellar medium. In a spiral galaxy, there is ongoing star formation in the disk, so these metals are incorporated into new generations of stars, making relatively metal-rich Population I stars like the Sun. By contrast, an elliptical galaxy has a single flurry of star formation when it is young, after which star formation ceases. Elliptical galaxies therefore contain only metal-poor Population II stars.

Figure 14-35*c* shows that both elliptical and spiral galaxies form stars most rapidly when they are young. This idea is borne out by the observation that very distant galaxies tend to be blue, which means that galaxies were bluer in the distant past than they are today. (Note the very blue colors of the distant, gravitationally lensed galaxies shown in Figure 14-27*c*, as well as those of the subgalactic objects shown in Figure 14-34*b*.) Spectroscopic studies of such galaxies in the 1980s by James Gunn and Alan Dressler demonstrated that most owe their blue color to vigorous star formation, often occurring in intense, episodic bursts. The hot, luminous, and short-lived O and B stars produced in these bursts of star formation give blue galaxies their characteristic color.

ConceptCheck 14-20: **Why do elliptical galaxies contain only Population II stars?**

(a) Formation of a spiral galaxy

(b) Formation of an elliptical galaxy

In an elliptical galaxy, there is a brief, intense burst of star formation, when the galaxy is young.

In a spiral galaxy, star formation continues at a more leisurely pace that extends over billions of years.

Stellar birthrate ($M_\odot$/ year)

100 75 50 25

1 3 5 10

Billions of years

(c) The stellar birthrate in galaxies

FIGURE 14-35 The Formation of Spiral and Elliptical Galaxies (a) If the initial star formation rate in a protogalaxy is low, it can evolve into a spiral galaxy with a disk. (b) If the initial star formation rate is rapid, no gas is left to form a disk. The result is an elliptical galaxy. (c) This graph shows how the rate of star birth (in solar masses per year) varies with age in spiral and elliptical galaxies.

An Evolving Universe of Galaxies

In addition to changes in galaxy colors, the character of the galactic population has also changed over the past several billion years. In nearby rich clusters, only about 5% of the galaxies are spirals. But observations of rich clusters at distances corresponding to looking about 4 billion years into the past show that about 30% of their galaxies were spirals.

Why were spiral galaxies more common in rich clusters in the distant past? Galactic collisions and mergers are probably responsible. During a collision, interstellar gas in the colliding galaxies is vigorously compressed, triggering a burst of star formation. A succession of collisions produces a series of star-forming episodes that create numerous bright, hot O and B stars, which become dispersed along arching spiral arms by the galaxy's rotation. Eventually, however, the gas is used up; star formation then ends and the spiral arms become less visible. Furthermore, tidal forces tend to disrupt colliding galaxies, strewing their stars across intergalactic space until the galaxies are completely disrupted (see Figure 14-24).

A full description of galaxy formation and evolution must include the effects of dark matter. As we have seen, only about 10% of the mass of a galaxy—its stars, gas, and dust—emits detectable light of any kind. As yet we have no idea what the remaining 90% looks like or what it is made of. The dilemma of dark matter is one of the most challenging problems facing astronomers today.

ConceptCheck 14-21: **What happens to spiral galaxies during galaxy mergers?**

KEY IDEAS AND TERMS

14-1 When galaxies were first discovered, it was not clear that they lie far beyond the Milky Way until their variable stars were carefully observed

- Debates about the nature of and distance to so-called spiral nebulae were ongoing until about 100 years ago.
- Spiral nebulae were confirmed to be distant galaxies when Cepheid stars were identified in them, allowing distance to be measured.

14-2 Hubble devised a system for classifying galaxies according to their appearance

- Galaxies can be grouped into four major categories organized on a **tuning fork diagram** using the **Hubble classification.**
- The disks of **spiral galaxies** and **barred spiral galaxies** are sites of active star formation.

spiral galaxies. How exactly would you design the positions of the band members on the field to represent the different spiral galaxies of classes Sa, Sb, and Sc? Create two columns on your paper by drawing a line from top to bottom, drawing sketches in the left-hand column, and writing a description of each sketch in the right-hand column. Also include what the band's opening formation and final formation should be.

Observing Questions

1. Use *Starry Night*™ to visit a variety of galaxies and determine whether they are spiral, barred spiral, elliptical, or irregular. Click on **Home** to see the sky from your home location. Click on the **Options** tab, expand the **Deep Space** layer and click **Off** all images except **Messier Objects** and **Bright NGC Objects.** Type **Ctrl-H** (**Cmd-H** on a Mac) or select **View > Hide Horizon** from the menu to remove the horizon. Also select **View > Hide Daylight** to remove daylight from the view. Use the **Find** pane to visit each of the galaxies listed below. First click the icon in the search box of the **Find** pane and choose **Search All** from the menu. Then, for each object, type its name in the search box of the **Find** pane and press the **Enter** key. (*Hint:* To go to the galaxy without slewing, press the spacebar.) Use the **Zoom** buttons to examine each galaxy in detail and then classify it as a spiral (S), barred spiral (SB), elliptical (E), or irregular (Irr), and the subclassification of each galaxy (e.g., Sa, E5).
   a) M33
   b) M58
   c) M74
   d) M81
   e) M83
   f) M94
   g) M109
   h) Large Magellanic Cloud
   i) Small Magellanic Cloud
   j) NGC1232
   k) M84
   l) M86
   m) M59
2. Clusters of galaxies contain different numbers and distributions of galaxies and harbor significant amounts of the mysterious dark matter. In this exercise, you can use *Starry Night*™ to compare a few of these groupings and see the gravitational effect of dark matter. You can start by looking at one of the largest galaxy clusters, the Virgo cluster. Select **Favourites > Explorations > Virgo Cluster-Milky Way** from the menu. You are looking at this group of galaxies from a very large distance out in space, at about 66 Mly from the Sun. Our Milky Way Galaxy is labeled at the bottom left of the view, across a void in space from this cluster. Use the location scroller to move around the Virgo cluster and consider its overall shape and its relationship to neighboring galaxies.
   a) What is the general shape of the Virgo cluster? **Zoom in** toward this cluster until individual galaxies are shown and use the location scroller to help you to identify each classification of galaxy (elliptical, spiral and irregular) within the group. Select **File > Revert** to return to the original view and identify several other groups of galaxies. Select one or two clusters of galaxies in turn, move the cursor over a galaxy within the selected group and right-click to open the object contextual menu and select the **Highlight** option to identify this group. You can select the **Centre** option to move the selected cluster to the center of the view and examine the cluster's extent across space. Again, use the location scroller and **Zoom in** to examine this cluster from various viewpoints.
   b) Describe the distributions of the galaxies within the clusters, compared to that in the Virgo Cluster. How do the shapes and relative sizes of these clusters compare to each other and to the Virgo cluster? See if you can recognize the walls of galaxies surrounding large voids in space that link these concentrated regions of galaxies.
   c) Recent studies have revealed the presence within clusters of galaxies of mysterious dark matter, detected only by its gravitational effect on light. You can examine one example of this effect in a Hubble Space Telescope image of such a cluster. Click on the **Home** button to return to your sky. Click on the **Find** tab and ensure that the search box is empty. Click on the magnifying glass icon in the search box to open a dropdown list and click on **Hubble Images.** In the list of Hubble images, click on **Gravitational Lens** to center on this image of a cluster of galaxies known as CL0024+1654. **Zoom in** to a field of view of about 1 arcminute. This Hubble Space Telescope image shows a rich cluster of ordinary-looking yellowish galaxies surrounded by blue arcs of light. These arcs are multiple images of a very distant galaxy, as seen through the gravitational "lens" of dark matter pervading the cluster of galaxies. The blue color of these images suggests that this distant galaxy is composed mostly of young, blue stars. The distribution of this mysterious substance within galactic clusters can be inferred from these types of images of distant clusters of galaxies.
3. Use the *Starry Night*™ program to investigate the large-scale structure of the universe. Open **Favourites > Explorations > Large Scale Structure,** and select the location scroller tool. The main window shows a view looking toward Earth from a location 311 million ly away. (For comparison, the Andromeda Galaxy is only about 2 million ly away from Earth.) Each dot on the screen is a galaxy. If you position the cursor anywhere on the screen and click and hold the mouse button (left mouse button on a two-button mouse), a small, circular arrow appears near the bottom center of the screen. The Milky Way Galaxy, with Earth and the Sun in it, is near the center of this circular arrow, although Earth and the Sun are too small to be visible in this view.
   a) What is the overall distribution of galaxies on the screen (uniformly distributed, or grouped into isolated clusters with empty space between clusters, or many galaxies grouped into clusters and many not, with several empty regions, or voids)?
   b) Use the location scroller to move your viewpoint around this part of the universe (it looks as though you are rotating the universe). The apparent motion of the galaxies relative to each other gives a three-dimensional effect to the view on the screen. What is the overall distribution of galaxies in the universe, as suggested by this 3-D effect (e.g., uniformly distributed throughout space, or grouped into clusters that are in turn grouped into superclusters with empty space between clusters and between superclusters, or grouped into clusters that are linked by lines of galaxies to form superclusters with empty regions, or voids, between them)?
4. Use the *Starry Night*™ program to observe a peculiar galaxy, a galaxy that shows features not included in the Hubble classification scheme. Click the **Home** button in the toolbar to place yourself at your home location at the present time and click the **Stop** button. Use the **View** menu or button bar to hide the horizon and hide daylight.
   a) Open the **Find** pane, click the icon in the search box and select **Search All** from the list. Then type NGC 4314 in the search box, click the icon next to the name in the list box, and choose **Magnify.** The image is a Hubble Space Telescope view of only the innermost part of this barred spiral galaxy. The bar and the

spiral arms are located beyond the edge of the image seen here. Although NGC 4314 is billions of years old, the image shows a ring of intense star formation that has occurred only in the last few million years, close to the galaxy's small core. In which constellation is NGC 4314 located?

b) What name could you apply to this galaxy based on the peculiar features visible here?

ANSWERS

ConceptChecks

ConceptCheck 14-1: If spiral nebulae are closer than some of the stars of our Galaxy, then the evidence presented supports Shapley's argument that spiral nebulae are within our Galaxy, and not in a large galaxy very far away.

ConceptCheck 14-2: Cepheid variables identified in other galaxies have well-known luminosities, and comparing luminosity to brightness reveals distances to other galaxies.

ConceptCheck 14-3: Sc galaxies have the most active star formation so a sketch would have a smaller central bulge and a relatively large star-forming disk with considerable gas and dust.

ConceptCheck 14-4: In both spiral and barred spirals, the designation *a* is used for tightly wound arms and *c* for loosely wrapped arms.

ConceptCheck 14-5: Elliptical galaxies have almost no gas or dust available for the formation of stars.

ConceptCheck 14-6: If the Type Ia supernova appeared to be dimmer because of intervening dust, then astronomers would mistakenly believe that the supernova was farther away and, subsequently, that the galaxy was farther away than it really is.

ConceptCheck 14-7: A galaxy that has almost no variability in the hydrogen light emitted is assumed to be rotating slowly because it has very few stars and must be quite small and have a low luminosity. However, in this case, if it appears bright, then it must be very close to our Galaxy.

ConceptCheck 14-8: Because the base of the distance ladder depends on parallax, all rungs of the ladder are completely dependent on an accurate understanding of brightness as determined by parallax.

ConceptCheck 14-9: Regular clusters have a spherical distribution and are observed to contain significantly more galaxies, most of which lack spiral structures, as compared to irregular clusters that are more spread out, less densely packed but have a wider variety of galaxy shapes included.

ConceptCheck 14-10: No, although on the largest scales, galaxies appear in every direction. Observations of superclusters show they are clumped into uneven groups and into long filaments.

ConceptCheck 14-11: The 54-member Local Group, dominated by the Milky Way Galaxy and the Andromeda Galaxy, contains mostly dwarf ellipticals.

ConceptCheck 14-12: Collisions between gas and dust in colliding galaxies will warm gas, causing it to glow in X-rays.

ConceptCheck 14-13: Simulations suggest that a single galaxy will form with many newly formed stars.

ConceptCheck 14-14: The best evidence is that the rotation curves in galaxies cannot be accounted for by the observed mass and the ability of galaxies and galaxy clusters to gravitationally lens more distant sources of light.

ConceptCheck 14-15: These objects are quite distant, yet profoundly luminous and exhibit emission lines rather than absorption lines.

ConceptCheck 14-16: An object that has a very short period of brightening and dimming can only do so if quite small in diameter, otherwise we would see a gradual brightening and dimming as the light from the closest regions arrives first and the light from the most distant regions arrives later.

ConceptCheck 14-17: Material falling inward toward a supermassive black hole can become so bright that its emitted energy (radiation pressure) can actually prevent more material from approaching the supermassive black hole.

ConceptCheck 14-18: We can observe the rotation curves for nearby galaxies, whereas for quasars, we infer that only a compact supermassive black hole can account for the energy we observe being emitted.

ConceptCheck 14-19: Earlier galaxies appear to have been much smaller, suggesting that today's large galaxies formed by collisions and combinations of smaller galaxies.

ConceptCheck 14-20: Elliptical galaxies create most of their stars during the initial formation of the galaxy. Metal-rich, Population I stars can only form from the remains of Population II stars, which can only occur in later generations of star formation, which do not occur in elliptical galaxies.

ConceptCheck 14-21: The arms become distorted during collisions, resulting in astronomers observing far fewer spiral galaxies today than in the past.

CalculationChecks

CalculationCheck 14-1: Using the inverse-square law, if one object is 3 times farther away, it must be $1/3^2$ dimmer, or only 1/9 as bright.

CalculationCheck 14-2: If $L_{Edd} = 30{,}000\ (M \div M_\odot)L_\odot$, then doubling the mass doubles the Eddington limit.

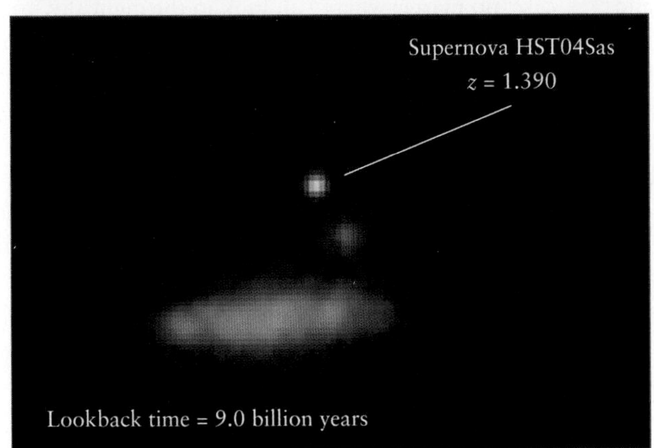

R I V U X G Very distant supernovae—which we see as they were billions of years ago—help us understand the evolution of the universe. *(NASA, ESA, and A. Riess [STScI])*

15

Observing the Evolution of the Universe

CHAPTER LEARNING OBJECTIVES By studying the sections of this chapter, you will learn:

15-1 The darkness of the night sky tells us about the nature of the universe

15-2 Our observations show us that the universe is expanding

15-3 The expanding universe emerged from a cataclysmic event called the Big Bang

15-4 The microwave radiation that fills all space is compelling evidence of a hot Big Bang

15-5 The universe was a rapidly expanding, hot, opaque plasma during its first 300,000 years and has slowly cooled

15-6 The shape of the universe indicates its matter and energy content

15-7 Observations of distant supernovae indicate that we live in an accelerating universe

So far in this book we have cataloged the contents of the universe. Our scope has ranged from submicroscopic objects, such as atomic nuclei, to superclusters of galaxies hundreds of millions of light-years across. In between, we have studied planets, moons, and stars.

But now we turn our focus beyond what we find in the universe to the nature of the universe itself—the subject of the science called *cosmology*. How large is the universe? What is its structure? How long has it existed, and how has it changed over time?

In this chapter we will see that the universe is expanding. This expansion began with an event at the beginning of time called the Big Bang. We will see direct evidence of the Big Bang in the form of microwave radiation from space. This radiation is the faint afterglow of a primordial fireball that filled all space shortly after the beginning of the universe.

Will the universe continue to expand forever, or will it eventually collapse back on itself? We will find that to predict the future of the universe, we must first understand what happened in the remote past. To this end, astronomers study luminous supernovae like the example shown in the photos that open this chapter. Supernovae can be seen across billions of light-years and can tell us about conditions in the universe billions of years ago. We will see how recent results from such supernovae, as well as from studies of the Big Bang's afterglow, have revolutionized our understanding of cosmology and given us new insights into our place in the cosmos. ■

15-1 The darkness of the night sky tells us about the nature of the universe

When meeting a person for the first time, how often do you ask, "Where do you live?" Indeed, our sense of place in terms of where we have been, where we are now, and where we are going are important aspects of who we are as human beings. So, in the context of astronomy, where we have been, where we are now, and where we are going are equally important questions about our universe. **Cosmology** is the science concerned with the structure and evolution of the universe as a whole. One of the most profound and basic questions in cosmology may at first seem foolish: Why is the sky dark at night? This question was brought to public attention in the early 1800s by the German amateur astronomer Heinrich Olbers.

Olbers's Paradox and Newton's Static Universe

Olbers and his contemporaries pictured our universe as stars scattered more or less randomly throughout infinite space. The physicist Isaac Newton thought that no other conception made sense. The gravitational forces between any *finite* number of stars, he argued, would in time cause them all to fall together, and the universe would soon be a compact blob.

Obviously, this has not happened. Newton concluded that we must be living amid a static, infinite expanse of stars. In this model, the universe is infinitely old, and it will exist forever without major changes in its structure. Olbers noticed, however, that a static, infinite universe presents a major puzzle.

If space goes on forever, with stars scattered throughout it, then any line of sight must eventually hit a star. In this case, no matter where you look in the night sky, you should ultimately see a star. As a result, the entire sky should be as bright as an average star, so, even at night, the sky should blaze like the surface of the Sun. **Olbers's paradox** is that the night sky is actually *dark* (**Figure 15-1**).

Olbers's paradox suggests that something is wrong with Newton's infinite, static universe. According to the classical, Newtonian picture of reality, space is like a gigantic flat sheet of inflexible, rectangular graph paper. (Space is actually three-dimensional, but it is easier to visualize just two of its three dimensions. In a similar way, an ordinary map represents the three-dimensional surface of Earth, with its hills and valleys, as a flat, two-dimensional surface.) This rigid, flat, Newtonian space stretches on and on, totally independent of stars or galaxies or anything else. As it turns out, this is not how our universe is structured, and the resolution to Olbers's paradox about the nature of space came from a surprising discovery by a young astronomer named Edwin Hubble.

> ConceptCheck 15-1: **Why does Olbers's paradox suggest that the night sky should actually be light instead of dark?**

FIGURE 15-1 R I V U X G **The Dark Night Sky** If the universe were infinitely old and filled uniformly with stars that were fixed in place, the night sky would be ablaze with light. In fact, the night sky is dark, punctuated only by the light from isolated stars and galaxies. Hence, this simple picture of an infinite, static universe cannot be correct. *(NASA, ESA, and The Hubble Heritage Team [STScI/AURA])*

15-2 Our observations show us that the universe is expanding

During the 1920s, Edwin Hubble, working with Milton Humason, photographed the spectra of many galaxies with the 100-in. (2.5-m) telescope on Mount Wilson in California. To their great surprise, they found that the measured spectral line positions were often not located exactly where they expected them to be. Rather, they found that spectral line positions were often shifted toward the red end of the spectrum. Recall from studying the Doppler effect that spectral lines from glowing objects that are moving away from Earth exhibit a redshift toward longer wavelength positions. Observing the redshifted spectral lines from the galaxies, they found that most galaxies are moving away from Earth.

A second set of observations added to the story line. Hubble made measurements of the fluctuating apparent brightnesses and pulsation periods of Cepheid variables in these galaxies. Using the Cepheid variables as standard candles allowed Hubble and Humason to estimate the distance to each galaxy. To their great surprise, they found a direct correlation between the distance to a galaxy and its redshift. This general rule is one of the most important in astronomy and can be summarized as

The more distant a galaxy, the greater its redshift and the more rapidly the distance between us is increasing.

In other words, nearby galaxies are moving away from us slowly, and more distant galaxies appear to be rushing away from us much more rapidly. **Figure 15-2** shows this relationship for five representative elliptical galaxies. These observations lead us to an inescapable conclusion—our universe is expanding.

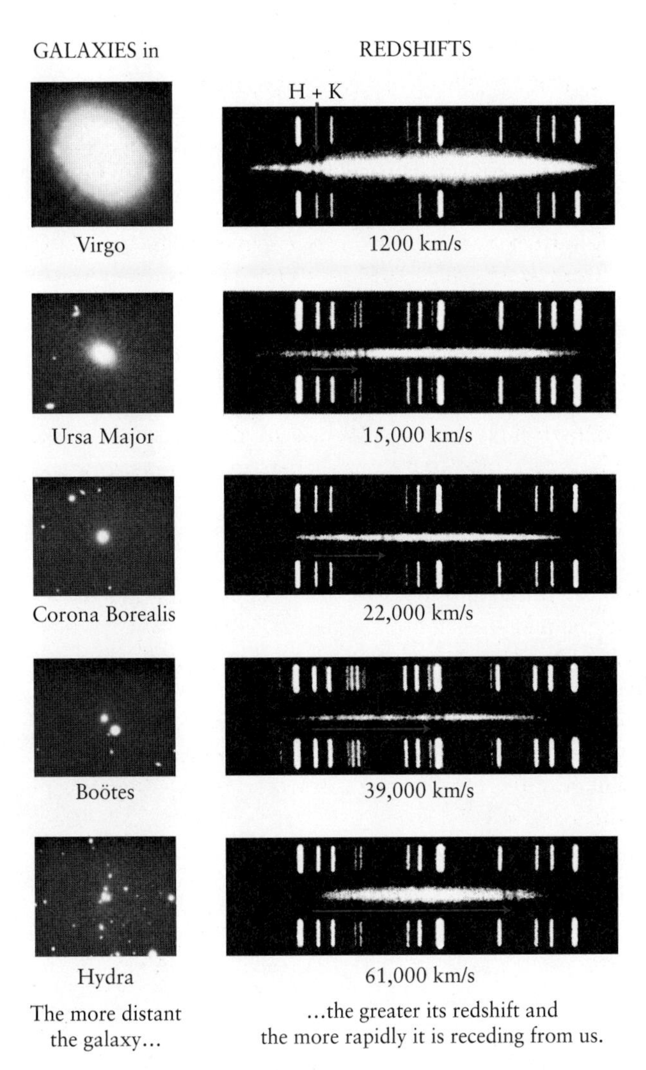

FIGURE 15-2 RIVUXG **Relating the Distances and Redshifts of Galaxies** These five galaxies are arranged, from top to bottom, in order of increasing distance from us. All are shown at the same magnification. Each galaxy's spectrum is a bright band with dark absorption lines; the bright lines above and below it are a comparison spectrum of a light source at the observatory on Earth. The horizontal red arrows show how much the H and K lines of singly ionized calcium are redshifted in each galaxy's spectrum. Below each spectrum is the recessional velocity calculated from the redshift. The more distant a galaxy is, the greater its redshift. *(The Observatories of the Carnegie Institution of Washington)*

ANALOGY What does it actually mean to say that the universe is expanding? According to general relativity, space itself is not rigid. The amount of space between widely separated locations changes over time. A good analogy is that of baking a chocolate chip cake, as in **Figure 15-3.** As the cake expands during baking, the amount of space between the chocolate chips gets larger and larger. In the same way, as the universe expands, the amount of space between widely separated galaxies increases. The expansion of the universe *is* the expansion of space.

CAUTION It is important to realize that the expansion of the universe occurs primarily in the vast spaces that separate clusters of galaxies. Just as the chocolate chips in Figure 15-3 do not expand as the cake expands during baking, galaxies themselves do not expand. Einstein and others have established that an object that is held together by its own gravity, such as a galaxy or a cluster of galaxies, is always contained within a patch of nonexpanding space. A galaxy's gravitational field produces this nonexpanding region, which is indistinguishable from the rigid space described by Newton. Thus, Earth and your body, for example, are not getting any bigger. Only the distance between widely separated galaxies increases with time. **Cosmic Connections: "Urban Legends" about the Expanding Universe** has more to say about several misconceptions concerning the expanding universe.

FIGURE 15-3 The Expanding Chocolate Chip Cake Analogy The expanding universe can be compared to what happens inside a chocolate chip cake as the cake expands during baking. (The cake is floating weightlessly inside the oven of an orbiting spacecraft crewed by hungry astronauts.) All of the chocolate chips in the cake recede from one another as the cake expands, just as all the galaxies recede from one another as the universe expands.

COSMIC CONNECTIONS "Urban Legends" about the Expanding Universe

There are a number of common misconceptions or "urban legends" about what happens as the universe expands. The illustrations below depict the myth and the reality for three of these "urban legends."

Urban Legend #1:
The expansion of the universe means that as time goes by, galaxies move away from each other through empty space. In this picture, space is simply a background upon which the galaxies act out their parts.

Reality:
The expansion of the universe means that as time goes by, *space itself* expands. As it expands, it carries the galaxies along with it.

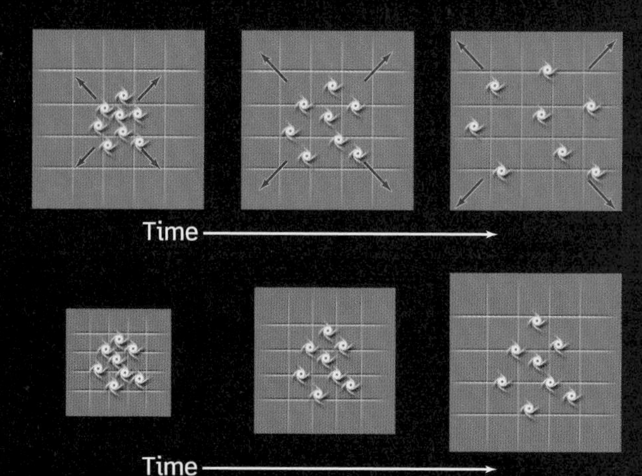

Urban Legend #2:
The redshift of light from distant galaxies is a Doppler shift. It occurs because these galaxies are moving away from us rapidly.

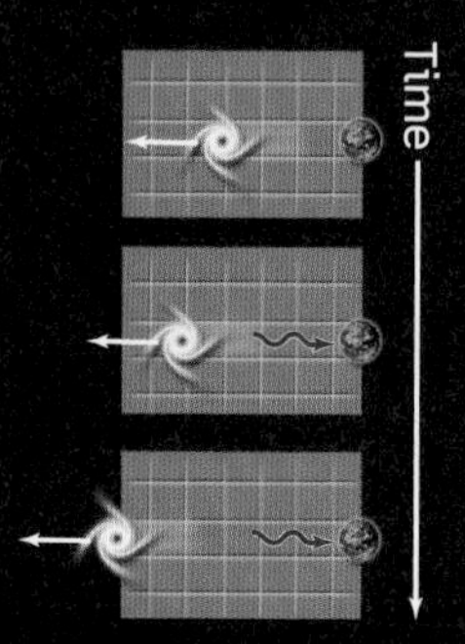

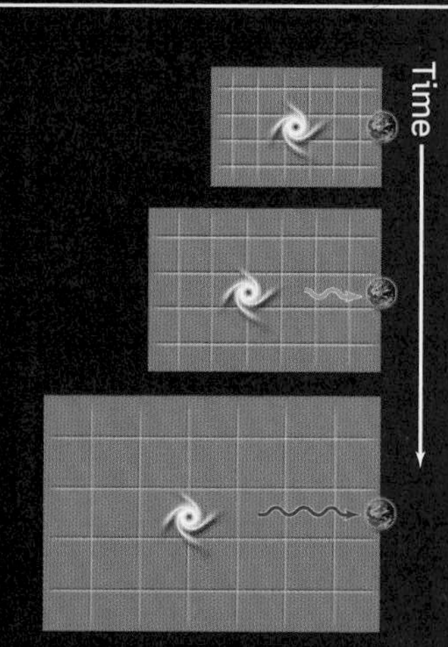

Reality:
As a photon travels through intergalactic space, its wavelength expands as the space through which it is traveling expands. This is called a *cosmological* redshift.

Urban Legend #3:
As the universe expands, so do objects within the universe. Hence galaxies within a cluster are now more spread out than they were billions of years ago.

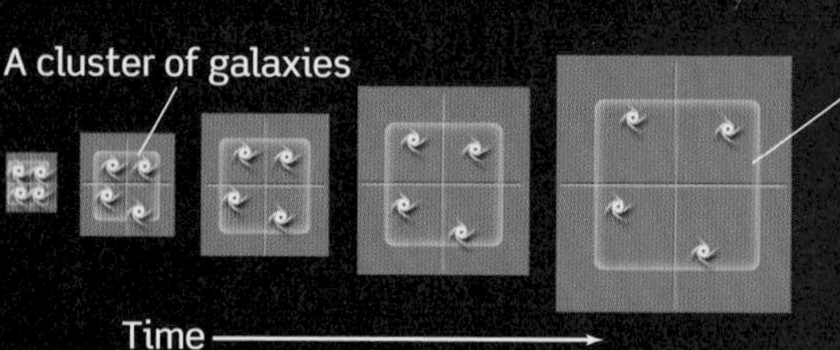

In this picture, the cluster expands as the universe expands.

Reality:
At first the expansion of the universe tends to pull the galaxies of a cluster away from each other. But the force of gravitational attraction binds the members of the clusters together, so the cluster stabilizes at a certain size.

The cluster has stabilized. The size no longer changes within the cluster, even as the clusters are farther from other clusters as the universe expands.

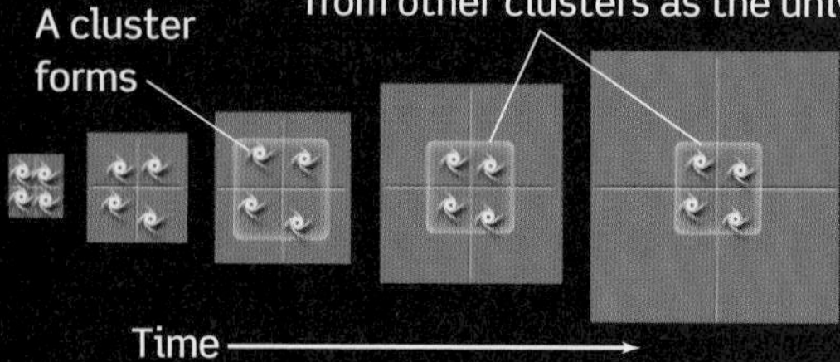

This notion has critically important implications for Olbers's paradox presented in the first section of this chapter. If the universe is expanding, then some stars and galaxies are so far away that their light has not yet had time to travel through space to get here!

ConceptCheck 15-2: Although nearly all distant galaxies have measureable redshifts, the relatively nearby Andromeda Galaxy exhibits an overall blue shift. What does this mean about the Andromeda Galaxy's movement?

Redshift of a Receding Object

Hubble estimated the distances to a number of galaxies and the redshifts of those galaxies. We can write a straightforward mathematical relationship between how much a spectral line position has shifted compared to its expected position. We call this ratio a galaxy's **redshift,** denoted by the symbol z. Using a calculator, this is found by taking the wavelength (λ) observed for a given spectral line, subtracting from it the ordinary, unshifted wavelength of that line (λ_0) to get the wavelength difference ($\Delta\lambda$), and then dividing that difference by λ_0:

$$z = \frac{\lambda - \lambda_0}{\lambda_0} = \frac{\Delta\lambda}{\lambda_0}$$

z = **redshift of an object**
λ_0 = **ordinary, unshifted wavelength of a spectral line**
λ = **wavelength of that spectral line that is actually observed from the object**

Astronomers often express the distance to a remote galaxy simply in terms of its redshift z (which can be measured very accurately). For example, an astronomer might simply say that a certain galaxy is "at $z = 0.128$" by making use of the following general rule:

The greater the redshift of a distant galaxy, the greater its distance.

This relationship between the distances to galaxies and their redshifts was one of the most important astronomical discoveries of the twentieth century. In 1929, Hubble published this discovery, which is now known as the **Hubble law.** The Hubble law is that there is a direct relationship between a galaxy's distance and velocity—distant objects are seen to be moving away faster—and is easily stated as a formula:

The Hubble law

$$v = H_0 d$$

v = **recessional velocity of a galaxy**
H_0 = **Hubble constant**
d = **distance to the galaxy**

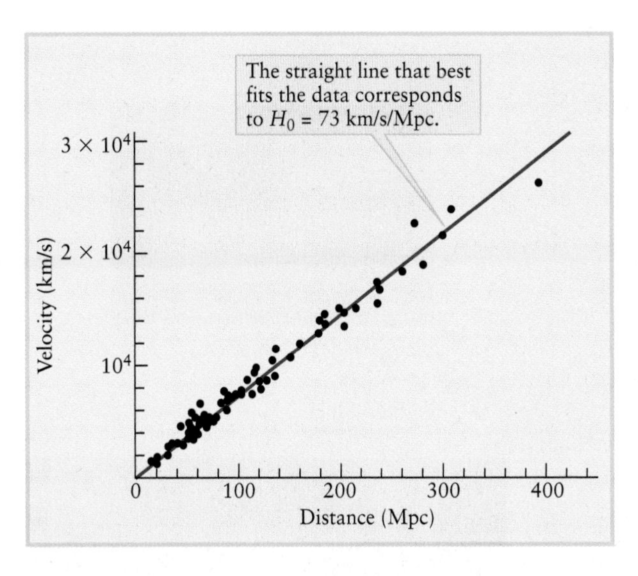

FIGURE 15-4
The Hubble Law This graph plots the distances and recessional velocities of a sample of galaxies. The straight line is the best fit to the data. This linear relationship between distance and recessional velocity is called the Hubble law.

This formula is the equation for the straight line displayed in **Figure 15-4,** and the **Hubble constant** H_0 is the slope of this straight line. To make the calculations easier, we often measure speeds in kilometers per seconds, rather than miles per hour, and distances in megaparsecs, rather than millions of light years—recall that a megaparsec is 3.26 million ly. From the data plotted on this graph we find that H_0 = 73 km/s/Mpc (say "73 kilometers per second per megaparsec"). In other words, for each million parsecs to a galaxy, the galaxy's speed away from us increases by 73 km/s. For example, a galaxy located 100 million parsecs from Earth should be rushing away from us with a speed of 7300 km/s. (In other books you may see the units of the Hubble constant written with exponents: 73 km s^{-1} Mpc^{-1}.)

Determining the value of H_0 has been an important task of astronomers for a very simple reason: The Hubble constant is one of the most important numbers in all astronomy. It expresses the rate at which the universe is expanding and helps estimate the age of the universe. Furthermore, the Hubble law can be used to determine the distances to extremely remote galaxies. If the redshift of a galaxy is known, the Hubble law can be used to determine its distance from Earth.

Go to Video 15-1

The Hubble law is a direct proportionality—that is, a galaxy twice as far away is receding from us twice as fast. This is just what we would expect in an expanding universe. To see why this is so, imagine a grid of parallel lines (as on a piece of graph paper) crisscrossing the universe. **Figure 15-5*a*** shows a series of such gridlines 100 Mpc (326 million ly) apart, along with five galaxies labeled A, B, C, D, and E that happen to lie where gridlines cross. As the universe expands in all directions, the gridlines and the attached galaxies spread apart. (This is just what would happen if the universe were a two-dimensional rubber sheet that was being pulled equally on all sides. Alternatively, you can imagine that Figure 15-5*a* depicts a very small portion of the

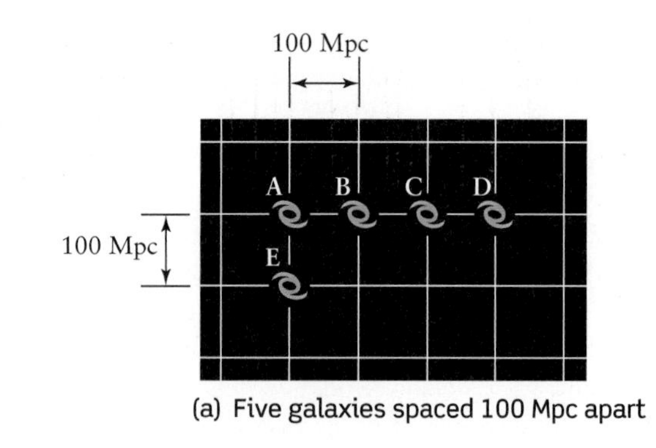

(a) Five galaxies spaced 100 Mpc apart

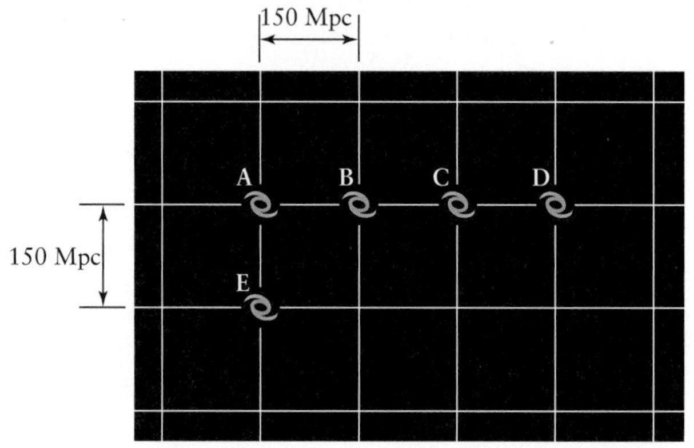

(b) The expansion of the universe spreads the galaxies apart

	Original distance (Mpc)	Later distance (Mpc)	Change in distance (Mpc)
A–B	100	150	50
A–C	200	300	100
A–D	300	450	150
A–E	100	150	50

FIGURE 15-5 The Expanding Universe and the Hubble Law (a) Imagine five galaxies labeled A, B, C, D, and E. At the time shown here, adjacent galaxies are 100 Mpc apart, or 326 million ly. (b) As the universe expands, by some later time the spacing between adjacent galaxies has increased to 150 Mpc. The table shows that the greater the original distance between galaxies, the greater the amount that distance has increased. This agrees with the Hubble law.

chocolate chip cake in Figure 15-3, with galaxies taking the place of chocolate chips.)

The greater the redshift of a distant galaxy, the greater its distance.

Figure 15-5*b* shows the universe at a later time, when the gridlines are 50% farther apart (150 Mpc) and all the distances between galaxies are 50% greater than in Figure 15-5*a*. Imagine that A represents our Galaxy, the Milky Way. The table accompanying Figure 15-5 shows how far each of the other galaxies has moved away from us during the expansion: Galaxies A and B and galaxies A and E were originally 100 Mpc apart, and have moved away from each other by an additional 50 Mpc; galaxies A and C, which were originally 200 Mpc apart, have increased their separation by an additional 100 Mpc; and the distance between galaxies A and D, originally 300 Mpc, has increased by an additional 150 Mpc. In other words, the increase in distance between any pair of galaxies is in direct proportion to the original distance; if the original distance is twice as great, the increase in distance is also twice as great.

CAUTION It may seem that if the universe is expanding, and if we see all the distant galaxies rushing away from us, then we must be in a special position at the very center of the universe. In fact, the expansion of the universe looks the same from the vantage point of *any* galaxy. For example, as seen from galaxy D in Figure 15-5, the initial distances to galaxies A, B, and C are 300 Mpc, 200 Mpc, and 100 Mpc, respectively. Between parts *a* and *b* of the figure, these distances increase by 150 Mpc, 100 Mpc, and 50 Mpc, respectively. So, as seen from galaxy D as well, the recessional velocity increases in direct proportion with the distance, and in the same proportion as seen from galaxy A. In other words, no matter which galaxy you call home, you will see all the other galaxies receding from you in accordance with the same Hubble law (and the same Hubble constant) that we observe from Earth.

Figure 15-3 also shows that the expansion of the universe looks the same from one galaxy as from any other. An insect sitting on any one of the chocolate chips would see all the other chips moving away. If the cake were infinitely long, it would not actually have a center; as seen from any chocolate chip within such a cake, the cake would extend off to infinity to the left and to the right, and the expansion of the cake would appear to be centered on that chip. Likewise, because every point in the universe appears to be at the center of the expansion, it follows that our universe has no center at all. (Later in this chapter we will see evidence that the universe, like our imaginary cake, is indeed infinite.)

CAUTION "If the universe is expanding, what is it expanding into?" This commonly asked question arises only if we take our chocolate chip cake analogy too literally. In Figure 15-3, the cake (representing the universe) expands in three-dimensional space into the surrounding air. But the actual universe includes *all* space; there is nothing "beyond" it, because there is no "beyond." Asking "What lies beyond the universe?" is as meaningless as asking "Where on Earth is north of the North Pole?"

The ongoing expansion of space explains why the light from remote galaxies is redshifted. Imagine a photon coming toward us from a distant galaxy. As the photon travels through space, the space is expanding, so the photon's wavelength becomes stretched. When the photon reaches our eyes, we see an increased wavelength: The photon has been redshifted. The longer the photon's journey, the more its wavelength will have been stretched. Thus, photons from distant galaxies have larger redshifts than those of photons from nearby galaxies, as expressed by the Hubble law. To be clear, a redshift caused by the expansion of the universe is properly called a **cosmological redshift.** It is *not* the same as a Doppler shift. Doppler shifts are caused by an object's *motion through space,* whereas a cosmological redshift is caused by the *expansion of space* (**Figure 15-6**).

CAUTION A common misconception about the Hubble law is that *all* galaxies are moving away from the Milky Way. The reality is that galaxies have their own motions relative

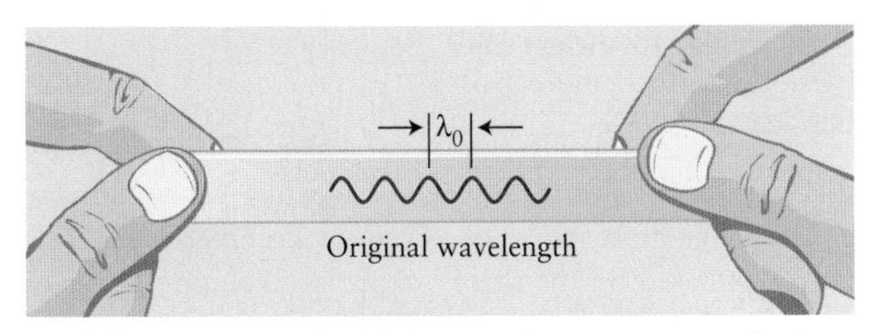

(a) A wave drawn on a rubber band ...

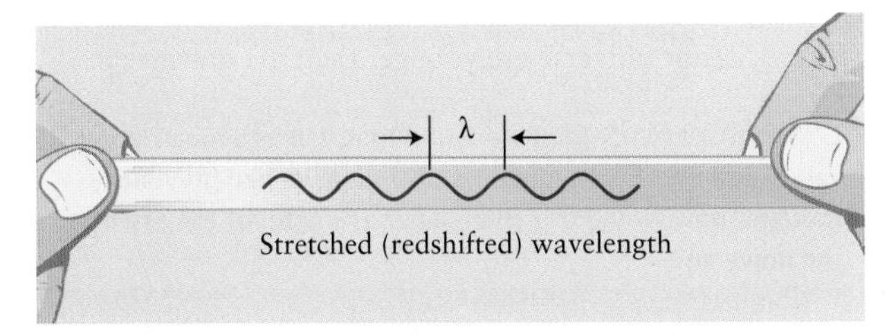

(b) ... increases in wavelength as the rubber band is stretched.

FIGURE 15-6 Cosmological Redshift A wave drawn on a rubber band stretches along with the rubber band. In an analogous way, a light wave traveling through an expanding universe "stretches," that is, its wavelength increases.

to one another, thanks to their mutual gravitational attraction. For nearby galaxies, the speed of the Hubble flow is small compared to these intrinsic velocities. Hence, some of the nearest galaxies, including M31 (shown in Figure 14-2), are actually approaching us and have blueshifts rather than redshifts. But for distant galaxies, the Hubble speed $v = H_0 d$ is much greater than any intrinsic motion that the galaxies might have. Even if the intrinsic velocity of such a distant galaxy is toward the Milky Way, the fast-moving Hubble flow sweeps that galaxy away from us.

ConceptCheck 15-3: **What two observations about galaxies most strongly suggest that our universe is expanding?**

CalculationCheck 15-1: **What is the redshift *z*-value for a galaxy that has a galaxy spectral line shifted to 725.6 nm when a stationary object would emit the line at 656.3 nm?**

CalculationCheck 15-2: **What is the distance to a galaxy that is observed to have a recessional velocity of 10,000 km/s?**

The Cosmological Principle

These ideas about the expanding universe demonstrate the central philosophy of cosmology. In cosmology, unlike other sciences, we cannot carry out controlled experiments or even make comparisons: There is only one universe that we can observe. To make progress in cosmology, we must accept certain philosophical assumptions or abandon hope of understanding the nature of the universe. The Hubble law provides a classic example. It could be interpreted to mean that we are at the center of the universe. We reject this interpretation, however, because it violates a cosmological extension of Copernicus's belief that we do not occupy any particular special or privileged location in space.

When Einstein began applying his thinking to cosmology about a century ago, he made a daring assumption: Over very large distances the universe is **homogeneous,** meaning that every region is the same as every other region, and **isotropic,** meaning that the universe looks the same in every direction. In other words, if you could stand back and look at a very large region of space, any one part of the universe would look basically the same as any other part, with the same kinds of galaxies distributed through space in the same way. The assumption that the universe is homogeneous and isotropic constitutes the **cosmological principle.** It gives precise meaning to the idea that we do not occupy a special location in space.

Models of the universe based on the cosmological principle have proven remarkably successful in describing the structure and evolution of the universe and in interpreting observational data. All the discussion about the universe in this book assumes that the universe is homogeneous and isotropic on the largest scale.

ConceptCheck 15-4: **If we notice in the night sky that there are more stars along the Milky Way than in other regions of the sky, is this consistent or inconsistent with the cosmological principle?**

15-3 The expanding universe emerged from a cataclysmic event called the Big Bang

The universe has been expanding for billions of years. This means that in the past the matter in the universe must have been closer together and therefore denser than it is today. If we look far enough into the very distant past, there must have been a time when the density of matter was almost inconceivably high. This leads us to conclude that some sort of tremendous event caused ultradense matter to begin the expansion that continues to the present day. This event, which we have named the **Big Bang,** marks the creation of the universe.

CAUTION It is not correct to think of the Big Bang as an explosion. When a bomb explodes, pieces of debris fly off *into space* from a central location. If you could trace all the pieces back to their origin, you could find out exactly where the bomb had been. This process is not possible with the universe, however, because the universe itself always has and always will consist of all space. As we have seen, the universe logically cannot have an edge.

Estimating the Age of the Universe

How long ago did the Big Bang take place? To estimate an answer, imagine two galaxies that today are separated by a distance d and receding from each other with a velocity v. A movie of these galaxies would show them flying apart. If you were to run the movie backward, you would see the two galaxies approaching each other as time runs in reverse. We can calculate the time T_0 it would take for the galaxies to collide by using a version of the familiar *distance* = *rate* × *time* equation as:

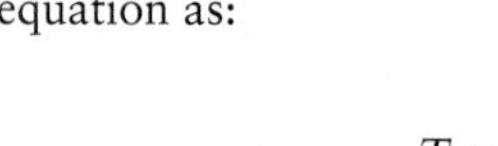

$$T_0 = \frac{d}{v}$$

This says that the time to travel a distance d at velocity v is equal to the ratio d/v. (As an example, to travel a distance of 360 km at a velocity of 90 km/h takes (360 km)/(90 km/h) = 4 hours.) If we use the Hubble law, $v = H_0 d$, to replace the velocity v in this equation, we get an even more powerful and easy-to-use formula:

$$T_0 = \frac{d}{H_0 d} = \frac{1}{H_0}$$

If galaxies are moving apart at a known rate, then we can calculate how long it has been since they were all together.

Note that the distance of separation, d, has canceled out and does not appear in the final expression. This means that T_0 is the same for *all* galaxies. This is the time in the past when all galaxies were crushed together, the time back to the Big Bang. In other words, the reciprocal of the Hubble constant H_0 gives us an estimate of the age of the universe, which is one reason why H_0 is such an important quantity in cosmology.

The most recent observations suggest that H_0 = 73 km/s/Mpc to within a few percent, and this is the value we choose as our standard. Using this value, our estimate for the age of the universe is

$$T_0 = \frac{1}{73 \text{ km/s/Mpc}}$$

To convert this into units of time, we simply need to remember that 1 Mpc equals 3.09×10^{19} km and 1 year equals 3.156×10^7 seconds. Converting units, we get

$$T_0 = \frac{1 \text{ Mpc s}}{73 \text{ km}} \times \frac{3.09 \times 10^{19} \text{ km}}{1 \text{ Mpc}} \times \frac{1 \text{ year}}{3.156 \times 10^7 \text{ s}}$$
$$= 1.34 \times 10^{10} \text{ years} = 13.4 \text{ billion years}$$

This calculation shows our universe is nearly 14 billion years old. By comparison, the age of our solar system is only 4.56 billion years, or about one-third the age of the universe. Thus, the formation of our home planet is a relatively recent event in the history of the cosmos.

The value of H_0 has an uncertainty of about 5%, so our simple estimate of the age of the universe is likewise uncertain by at least 5%. Furthermore, the formula $T_0 = 1/H_0$ is at best an approximation, because in deriving it we assumed that the universe expands at a constant rate, which may or may not be true. When these factors are taken into consideration, we find that the age of the universe is 13.7 billion years, with an uncertainty of about 0.2 billion years. This is remarkably close to our simple estimate.

Go to Video 15-2

Whatever the true age of the universe, it must be at least as old as the oldest stars. The oldest stars that we can observe readily lie in the Milky Way's globular clusters. The most recent observations, combined with calculations based on the theory of stellar evolution, indicate that these stars are about 13.6 billion years old (with an uncertainty of about 10%). Encouragingly enough, this is (slightly) less than the calculated age of the universe: The oldest stars in our universe are younger than the universe itself!

ConceptCheck 15-5: If we found that H_0 was a much larger value, due to galaxies moving faster than we had previously thought, how would this change our estimate for the age of the universe?

Our Observable Universe and the Dark Night Sky

The Big Bang helps to even further resolve Olbers's paradox posited at the beginning of this chapter. We know that the universe had a definite beginning, and thus its age is finite (as opposed to infinite). If the universe is 13.7 billion years old, then the most distant objects that we can see are those whose light has traveled 13.7 billion years to reach us—due to the expansion of the universe, these objects are now more than 13.7 billion light-years away. As a result, we can only see objects that lie within an immense sphere centered on Earth (**Figure 15-7**). This is true even if the universe is infinite, with galaxies scattered throughout its limitless expanse.

Our entire **observable universe** is located inside an imaginary sphere, as depicted in Figure 15-7. We cannot see anything beyond the edge of this imaginary sphere, because the time required for light to reach us from these incredibly remote distances is greater than the present age of the universe. As time goes by, light from more distant parts of the universe reaches us for the first time, and the size of our observable universe increases. Galaxies are distributed sparsely enough in our observable universe that there are no stars along most of our lines of sight. This helps explain why the night sky is dark.

Besides the finite age of the universe, a second effect also contributes significantly to the darkness of the night sky—the redshift. According to the Hubble law, the greater the distance to a galaxy, the greater the redshift. When a photon is redshifted, its wavelength becomes longer, and its energy—which is inversely proportional to its wavelength—decreases. Consequently, even though there are many galaxies far from Earth, they have large redshifts and their light does not carry much energy. A galaxy nearly at the edge of our observable universe has a nearly infinite redshift, meaning that the light we receive from that galaxy carries practically no energy at all. This decrease in photon energy because of the expansion of the universe decreases the brilliance of remote galaxies, helping to make the night sky dark.

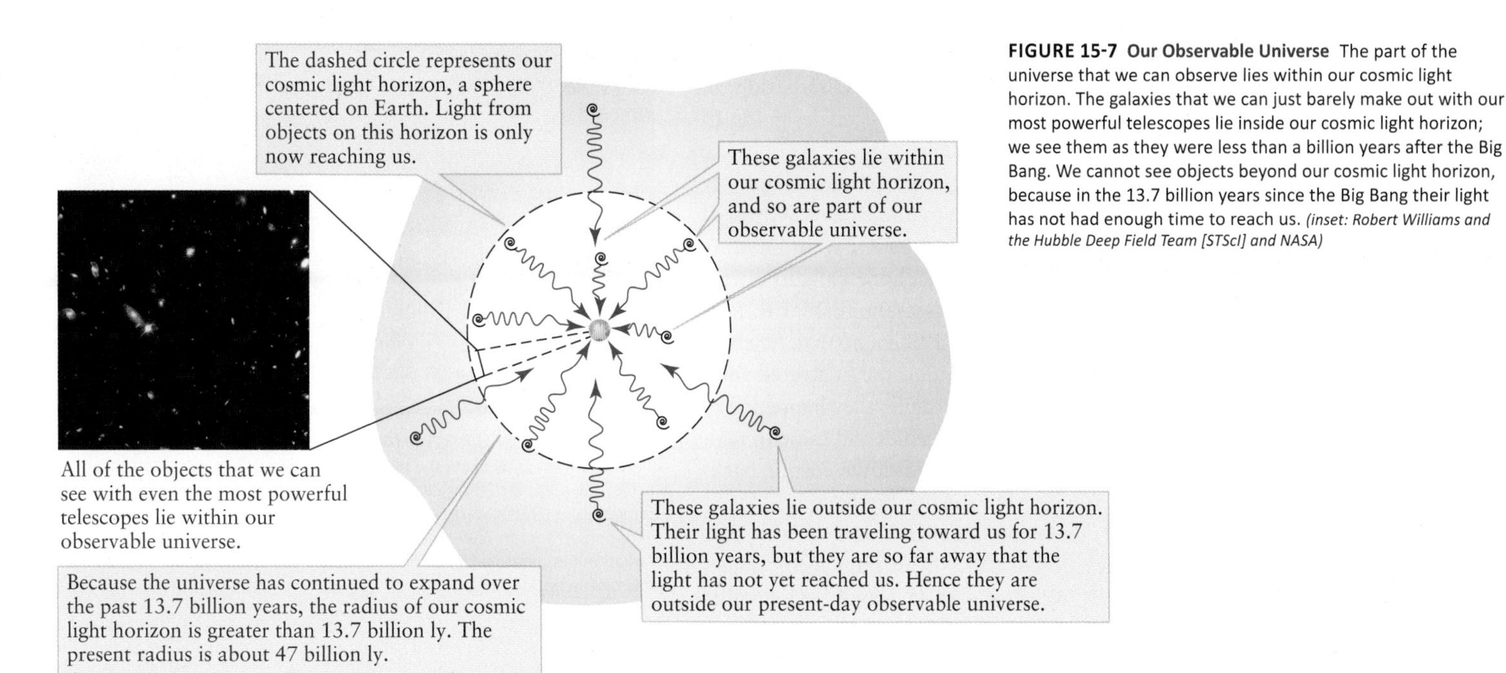

FIGURE 15-7 Our Observable Universe The part of the universe that we can observe lies within our cosmic light horizon. The galaxies that we can just barely make out with our most powerful telescopes lie inside our cosmic light horizon; we see them as they were less than a billion years after the Big Bang. We cannot see objects beyond our cosmic light horizon, because in the 13.7 billion years since the Big Bang their light has not had enough time to reach us. *(inset: Robert Williams and the Hubble Deep Field Team [STScI] and NASA)*

The concept of a Big Bang origin for the universe is a straightforward, logical consequence of having an expanding universe. If you can just imagine far enough back into the past, you can arrive at a time 13.7 billion years ago, when the density throughout the universe was infinite. As a result, throughout the universe space and time were completely jumbled up in a condition of infinite curvature similar to that at the singularity found at the center of a black hole. For this reason, a better name for the Big Bang might be the cosmic singularity. Thanks to the infinite curvature, the usual laws of physics do not tell us exactly what happened at the moment of the Big Bang.

ConceptCheck 15-6: Why does our observable universe get larger over time?

15-4 The microwave radiation that fills all space is compelling evidence of a hot Big Bang

One of the major advances in twentieth-century astronomy was the discovery of the origin of the heavy elements. We know today that essentially all the heavy elements are created by thermonuclear reactions at the centers of stars and in supernovae. The starting point of all these reactions is the fusion of hydrogen into helium. But as astronomers began to understand the details of thermonuclear synthesis in the 1960s, they were faced with a dilemma: There is far more helium in the universe than could have been created by hydrogen fusion in stars.

For example, the Sun consists of about 74% hydrogen and 25% helium by mass, leaving only 1% for all the remaining heavier elements combined. This 1% can be understood as material produced inside earlier generations of massive stars that long ago cast these heavy elements out into space when they became supernovae. Some freshly made helium, produced by the thermonuclear fusion of hydrogen within the stars, certainly accompanied these heavy elements. But calculations showed that the amount of helium produced in this way was not nearly enough to account for one-quarter of the Sun's mass. Because it was thought that the universe originally contained only hydrogen—the simplest of all the chemical elements—the presence of so much helium posed a major dilemma.

A Hot Big Bang and the Cosmic Microwave Background

Today, it seems obvious that the universe immediately following the Big Bang must have been so incredibly hot that thermonuclear reactions occurred everywhere throughout space. In the 1960s, the Princeton University physicists Robert Dicke and P. J. E. Peebles discovered that they could indeed account for today's high abundance of helium by assuming that the early universe had been at least as hot as the Sun's center, where helium is currently being produced. The hot early universe must therefore have been filled with many high-energy, short-wavelength photons.

The universe has expanded so much since those ancient times that all those short-wavelength photons have had their wavelengths stretched by a tremendous factor. As a result, they have become low-energy, long-wavelength photons. The temperature of this cosmic energy field is now only a few degrees above absolute zero. By Wien's law (described in Section 2-3), the majority of light emitted at such a low temperature should have its peak intensity at microwave wavelengths of approximately 1 millimeter. Hence, this energy field, which fills all of space, is called the **cosmic microwave background** or **cosmic background radiation.** In the early 1960s, Dicke and his colleagues began designing an antenna to detect this microwave light.

Meanwhile, scientists working on something completely unrelated to astronomy, Arno Penzias and Robert Wilson of Bell Telephone Laboratories in New Jersey, were building a new antenna designed to relay telephone calls to Earth-orbiting communications satellites (see Figure 15-13). Penzias and Wilson were deeply puzzled when, no matter where in the sky they pointed their antenna, they detected faint background noise. Thanks to a colleague, they happened to learn about the work of Dicke and Peebles and came to realize that they had discovered the cooled-down cosmic background radiation left over from the hot Big Bang.

The first high-precision measurements of the cosmic microwave background came from the Cosmic Background Explorer (COBE, pronounced "coe-bee") satellite (see Figure 15-13). Data from COBE's spectrometer, shown in **Figure 15-8,** demonstrate that this ancient microwave light has the spectrum of a blackbody with a temperature of 2.725 K—just a few degrees above absolute zero as predicted.

An important feature of the microwave background is that its intensity is almost perfectly isotropic, that is, the same in all directions. In other words, we detect nearly the same background intensity from all parts of the sky. However, extremely accurate measurements first made from high-flying airplanes, and later from high-altitude balloons and from COBE, reveal a very slight variation in temperature across the sky. The microwave background appears slightly warmer than average toward the constellation of Leo and slightly cooler than average in the opposite direction toward Aquarius. Between the warm spot in Leo and the cool spot in Aquarius, the background temperature declines smoothly across the sky. **Figure 15-9** is a map of the microwave sky showing this variation.

This apparent variation in temperature is caused by Earth's motion through the cosmos. If we were at rest with respect to the microwave background, the microwave light

Warmer radiation
Toward Aquarius
Toward Leo
Cooler radiation

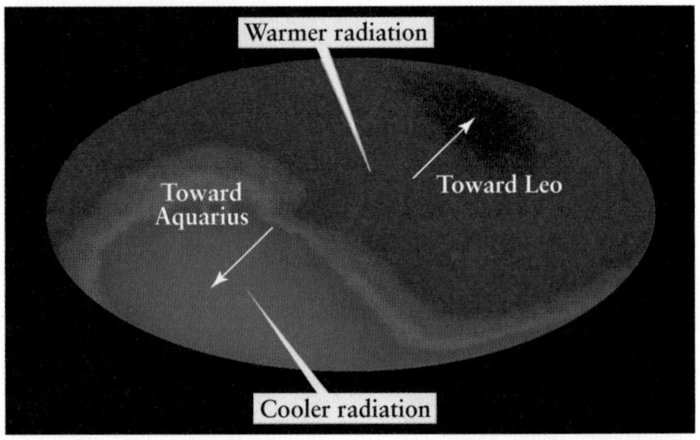

FIGURE 15-9 R I V U X G **The Microwave Sky** In this map of the entire sky made from COBE data, the plane of the Milky Way runs horizontally across the map, with the galactic center in the middle. Color indicates temperature—red is warm and blue is cool. The small temperature variation across the sky—only 0.0033 K above or below the average radiation temperature of 2.725 K—is caused by Earth's motion through the microwave background. *(NASA)*

observed would be even more nearly isotropic. Because we are moving through this field of microwave light, however, we see a Doppler shift. Specifically, we see shorter-than-average wavelengths in the direction toward which we are moving, as drawn in **Figure 15-10.** A decrease in wavelength corresponds to an increase in photon energy and thus an increase in the temperature. The slight temperature excess observed, about 0.00337 K, corresponds to a speed of 371 km/s, which is about 830,000 mi/h. Conversely, we

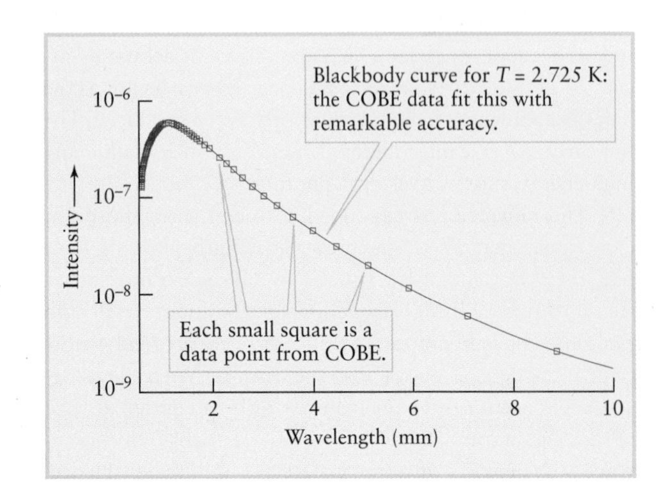

FIGURE 15-8 Spectrum of the Cosmic Microwave Background A blackbody curve gives an excellent match to the Cosmic Background Explorer (COBE) satellite data measuring the spectrum and angular distribution of the cosmic microwave background over a wavelength range of photons from 1 μm to 1 cm. *(Data from E. Cheng/NASA COBE Science Team)*

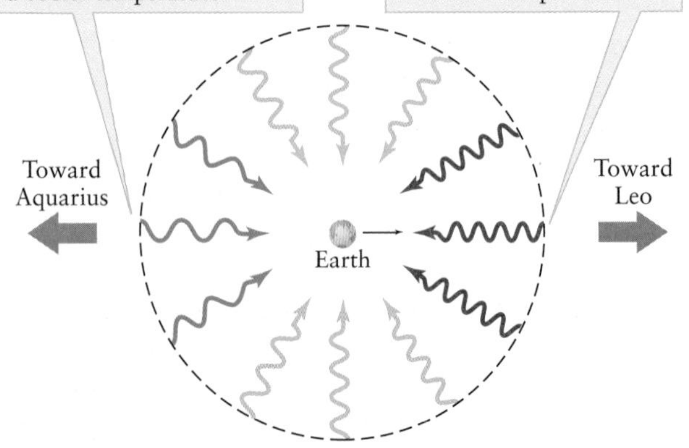

FIGURE 15-10 Our Motion Through the Microwave Background Because of the Doppler effect, we detect shorter wavelengths in the microwave background and a higher temperature of radiation in that part of the sky toward which we are moving. This part of the sky is the area shown in red in Figure 15-9. In the opposite part of the sky, shown in blue in Figure 15-9, the microwave radiation has longer wavelengths and a cooler temperature.

see longer-than-average wavelengths in that part of the sky from which we are receding. An increase in wavelength corresponds to a decline in photon energy and, hence, a decline in temperature.

Our solar system is thus traveling away from Aquarius and toward Leo at a speed of 371 km/s. Taking into account the known velocity of the Sun around the center of our Galaxy, we find that the entire Local Group of galaxies, including our Milky Way Galaxy, is moving at about 620 km/s toward the Hydra-Centaurus supercluster, which is about 1.4 million mi/h. Observations show that thousands of other galaxies are being carried in this direction, as is the Hydra-Centaurus supercluster itself. This tremendous flow of matter is thought to be due to the gravitational pull of an enormous collection of visible galaxies and dark matter lying in that direction. The existence of such concentrations of mass, as well as the existence of superclusters of galaxies, shows that the universe is rather "lumpy" on scales of 100 Mpc (326 million ly) or smaller. It is only on larger scales that the universe is homogeneous and isotropic.

> ConceptCheck 15-7: **If the early universe was filled with high-energy, short-wavelength photons, why are these observed today to be low-energy, long-wavelength microwave photons?**

15-5 The universe was a rapidly expanding, hot, opaque plasma during its first 300,000 years and has slowly cooled

Go to Video 15-3

Everything in the universe falls into one of two categories—energy or matter. One form of energy is radiation, that is, photons. There are many photons of starlight traveling across space, but the vast majority of photons in the universe belong to the cosmic microwave background. The matter in the universe is contained in such luminous objects as stars, planets, and galaxies, as well as in nonluminous dark matter. A natural question to ask is this: Which plays a more important role in the universe, radiation or matter? As we will see, the answer to this question is different for the early universe from the answer for our universe today.

Radiation and Matter in the Universe

Most of outer space is seemingly empty. Although the average density of matter in the universe is tiny by Earth standards, it turns out that it is thousands of times larger than the mass density of radiation in the universe. However, this was not always the case. Matter prevails over radiation today only because the energy now carried by microwave photons is so small. Nevertheless, the number of photons in the microwave background is astounding. Today, there are 410 million (4.1×10^8) photons in every cubic meter of space. In other words, the photons in space outnumber atoms by roughly a billion (10^9) to one. In terms of total number of particles, the universe thus consists almost entirely of microwave photons. This radiation field no longer has much "clout," however, because its photons have been redshifted to long wavelengths and low energies after 13.7 billion years of being stretched by the expansion of the universe.

Light remaining from the formation of the universe is seen everywhere in the form of microwave photons.

In contrast, think back toward the Big Bang. The universe becomes increasingly compressed, and thus the density of matter increases as we go back in time. The photons in the background radiation also become more crowded together as we go back in time. But, in addition, the photons become less redshifted and thus have shorter wavelengths and higher energy than they do today. Because of this added energy, the mass density of radiation increases more quickly as we go back in time than does the average density of matter. In fact, as **Figure 15-11** shows, there was a time in the ancient past when these two quantities were equal. Before this time, radiation held sway over matter. This transition from a radiation-dominated universe to a matter-dominated universe occurred about 70,000 years after the Big Bang.

> ConceptCheck 15-8: **At the time Earth formed, was the universe dominated by energy or by matter?**

When the First Atoms Formed

The nature of the universe changed again in a fundamental way about 380,000 years after the Big Bang, when the temperature of the radiation background was about $1100 \times 2.725\ \text{K} = 3000\ \text{K}$. To see the significance of this moment in cosmic history, recall that hydrogen is by far the most abundant element in the universe—hydrogen atoms outnumber helium atoms by about 12 to 1. A hydrogen atom consists of

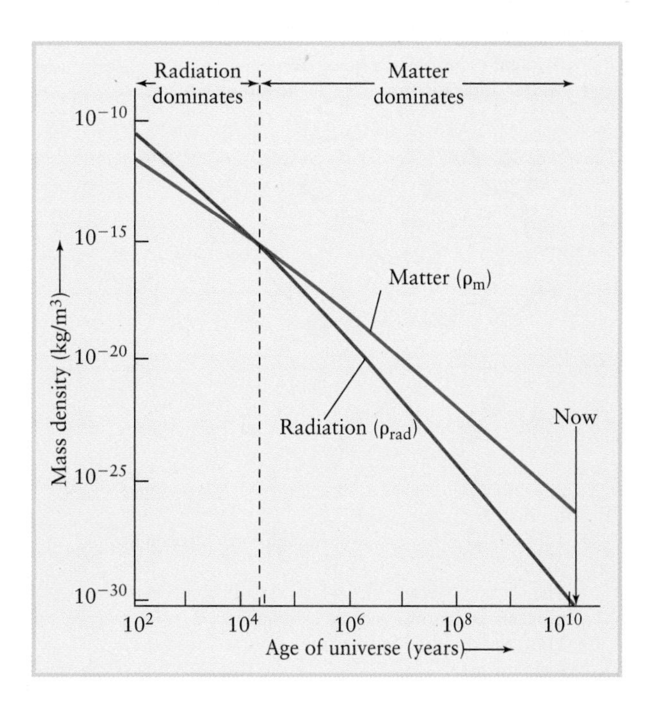

FIGURE 15-11 The Evolution of Density For approximately 24,000 years after the Big Bang, the mass density of radiation (ρ_{rad}, shown in red) exceeded the matter density (ρ_m, shown in blue), and the universe was dominated by radiation. Later, however, continued expansion of the universe caused ρ_{rad} to become less than ρ_m, at which point the universe became dominated by matter. *(Data from Clem Pryke, University of Chicago)*

a single proton orbited by a single electron, and it takes relatively little energy to knock the proton and electron apart. In fact, ultraviolet radiation warmer than about 3000 K easily ionizes hydrogen. Thus, hydrogen atoms could not survive in the universe that existed in the first 380,000 years after the Big Bang, when the background photons had energies great enough to prevent electrons and protons from binding to form hydrogen atoms (**Figure 15-12**). Only since then have the energies of these photons been low enough to permit hydrogen atoms to exist (Figure 15-12*b*).

The epoch when atoms first formed at t = 380,000 years is called the **era of recombination.** This refers to electrons "recombining" to form atoms. (The name is a bit misleading, because the electrons and protons had never *before* combined into atoms.)

Prior to t = 380,000 years, the universe was completely filled with a shimmering expanse of high-energy photons colliding vigorously with protons and electrons. This state of matter, called a **plasma,** is opaque, just like the glowing gases inside a discharge tube (like a neon advertising sign).

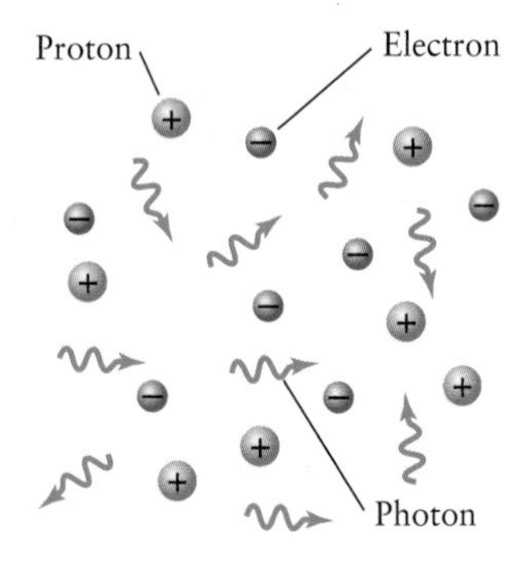

(a) Before recombination:
- Temperatures were so high that electrons and protons could not combine to form hydrogen atoms.
- The universe was opaque: Photons underwent frequent collisions with electrons.
- Matter and radiation were at the same temperature.

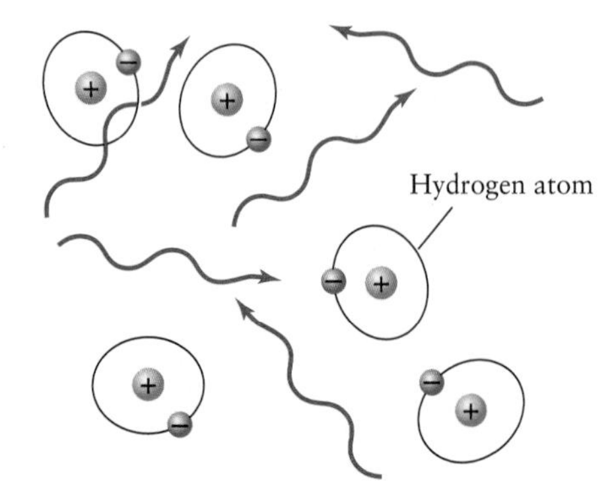

(b) After recombination:
- Temperatures became low enough for hydrogen atoms to form.
- The universe became transparent: Collisions between photons and atoms became infrequent.
- Matter and radiation were no longer at the same temperature.

FIGURE 15-12
The Era of Recombination
(a) Before recombination, the energy of photons from the cosmic background was high enough to prevent protons and electrons from forming hydrogen atoms. (b) Some 380,000 years after the Big Bang, the energy of the cosmic background radiation became low enough that hydrogen atoms could survive.

The surface and interior of the Sun are also a hot, glowing, opaque plasma.

After t = 380,000 years, the photons no longer had enough energy to keep the protons and electrons apart. As soon as the temperature of the field fell below about 3000 K, protons and electrons began combining to form hydrogen atoms. These atoms do not absorb low-energy photons, and so space became transparent! All the photons that heretofore had been vigorously colliding with charged particles could now stream unimpeded across space. Today, these same photons constitute the microwave background.

Before recombination, matter and the energy field had the same temperature, because photons, electrons, and protons were all in continuous interaction with one another. After recombination, photons and atoms hardly interacted at all, and thus the temperature of matter in the universe was no longer the same as the temperature. Thus, T = 2.725 K is the temperature of the present-day background radiation field, *not* the temperature of the matter in the universe. Note that while the temperature of the background is very uniform, the temperature of matter in the universe is anything but: It ranges from hundreds of millions of kelvins in the interiors of giant stars to a few tens of kelvins in the interstellar medium.

ANALOGY A good analogy is the behavior of a glass of cold water. If you hold the glass in your hand, the water will get warmer and your hand will get colder until both the water and your hand are at the same temperature. But if you set the glass down and do not touch it, so that the glass and your hand do not interact, their temperatures are decoupled: The water will stay cold and your hand will stay warm for much longer.

Because the universe was opaque prior to t = 380,000 years, we cannot see any further into the past than the era of recombination. In particular, we cannot see back to the era when the universe was radiation-dominated. The microwave background contains the most ancient photons we will ever be able to observe.

ConceptCheck 15-9: **If the universe had cooled more slowly, would the first atoms have appeared more quickly or more slowly?**

Nonuniformities in the Early Universe and the Origin of Galaxies

Careful analysis of COBE data showed that the microwave photons in the cosmic background are not completely isotropic. Even when the effects of Earth's motion are accounted for, there remain variations in the temperature of this energy field of about 100 μK (100 microkelvins, or 10^{-4} K) above or below the average 2.725 K temperature. These tiny temperature variations indicate that the matter and energy in the universe were not totally uniform at the moment of recombination. Regions that were slightly denser than average were also slightly cooler than average; less dense regions were slightly warmer. When energy decoupled from matter at the time of recombination, the radiation preserved a record of these variations in temperature and density.

Astronomers place great importance on studying temperature variations in the cosmic background. The reason is that concentrations of mass in our present-day universe, such as superclusters of galaxies, are thought to have formed from the denser regions in the early universe. Within these immense concentrations formed the galaxies, stars, and planets. Thus, by studying these nonuniformities, we are really studying our origins.

Unfortunately, the detectors on board COBE had a relatively coarse angular resolution of 7° and thus could not give a detailed picture of these temperature variations. In 1998 two balloon-borne experiments, BOOMERANG and MAXIMA, carried new, high-resolution telescopes aloft to study the cosmic background radiation with unprecedented precision. The best all-sky coverage of the background radiation has come from the state-of-the-art detectors on board the Wilkinson Microwave Anisotropy Probe (WMAP). Shown in **Figure 15-13,** the spacecraft is named for the late David Wilkinson of Princeton University, who was a pioneer in studies of the cosmic background radiation. This figure also shows a map of the sky based on data taken by the WMAP detectors. This map shows us the state of the universe when it was less than 0.003% of its present age.

ConceptCheck 15-10: What does the WMAP data show and how does it compare to earlier observations?

The Newborn Universe Underwent a Brief Period of Vigorous Expansion

With the discovery of the cosmic microwave background, astronomers had direct evidence to support the expansion observations made by Hubble that the universe began with a hot Big Bang. In the early 1980s, working independently, Alexei Starobinsky at the L. D. Landau Institute of Theoretical Physics in Moscow and Alan Guth at Stanford University suggested that the universe might have experienced a brief period of extremely rapid expansion just after the Big Bang started. Their theories provided a way of getting around some problems in earlier thinking about how exactly the Big Bang occurred. During this **inflationary epoch,**

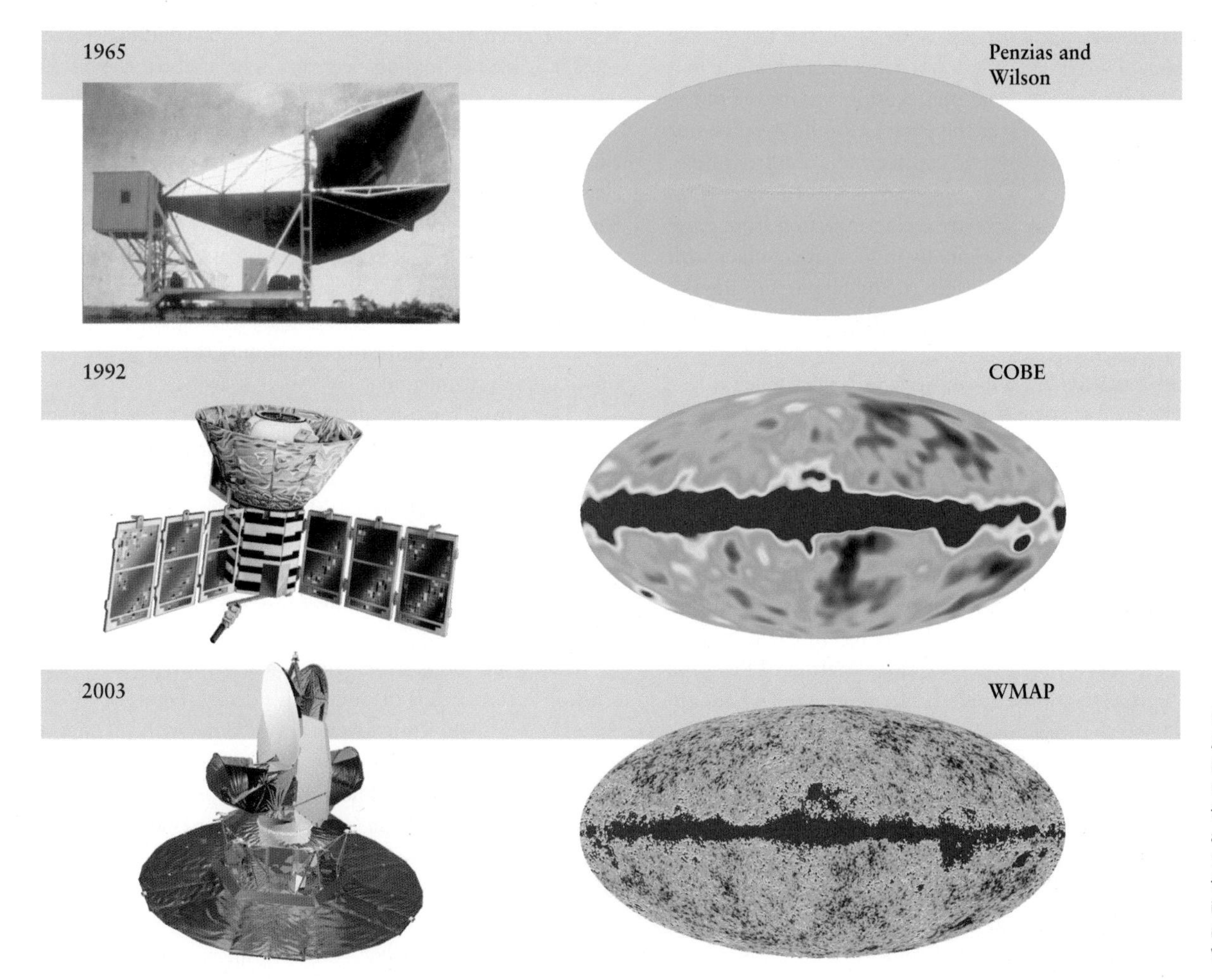

FIGURE 15-13 Looking for Ancient Background Photons From the ground on Earth, Nobel Prize winners Penzias and Wilson first observed the remnant afterglow from the Big Bang in 1965. Then, from space, COBE first discovered the patterns in the afterglow that was later more accurately observed by WMAP. *(NASA/WMAP Science Team)*

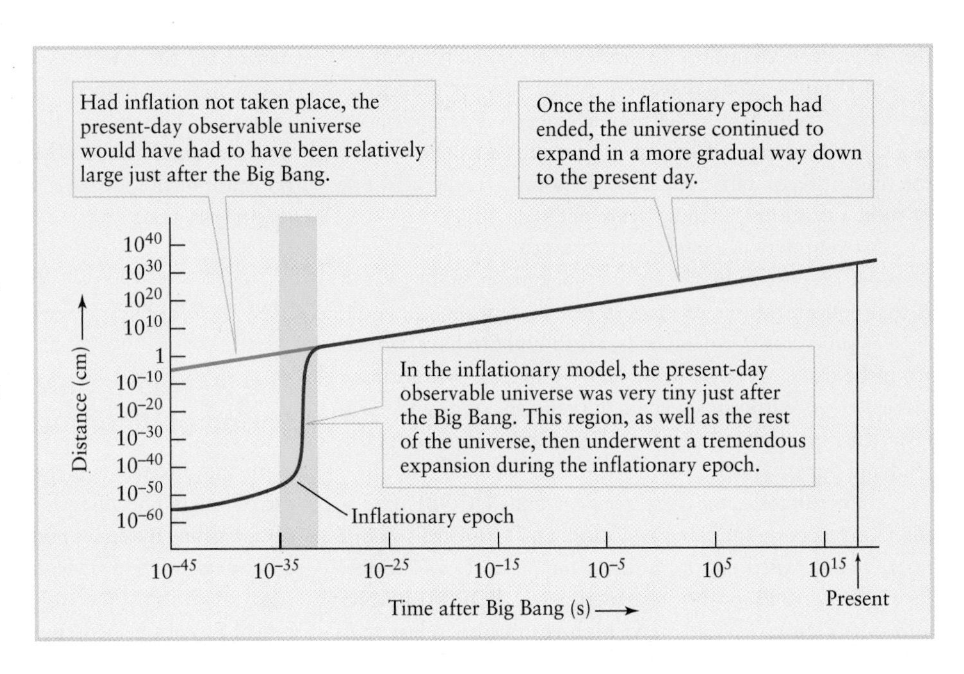

FIGURE 15-14 The Observable Universe With and Without Inflation According to the inflationary model (shown in purple), the universe expanded by a factor of about 10^{50} shortly after the Big Bang. This growth in the size of the present-day observable universe—that portion of the universe that lies within our present cosmic light horizon—occurred during a very brief interval, as indicated by the vertical shaded area on the graph. The blue line shows the projected size of the present-day observable universe soon after the Big Bang if inflation had not taken place. *(Adapted from A. Guth)*

the universe expanded outward in all directions by a factor of about 10^{50}. This epoch of dramatic expansion may have lasted only about 10^{-32} second (**Figure 15-14**), but would have been enough to account for the observations we are making today.

Inflation accounts for the isotropy of the microwave background. During the inflationary epoch, much of the material that was originally near our location was moved out to tremendous distances. Over the past 13.7 billion years, our view has expanded so that we can see radiation from these distant regions. Hence, when we examine microwaves from opposite parts of the sky, we are seeing radiation from parts of the universe that were originally in intimate contact with one another. This common origin is why all parts of the sky have almost exactly the same temperature.

ConceptCheck 15-11: Is the universe expanding today at the same rate it did right after the Big Bang?

The Fundamental Forces of Nature

If the universe went through an episode of extreme inflation, what could have triggered it? Our understanding is that inflation was one of a sequence of remarkable events during the first 10^{-12} second after the Big Bang. In each of these events there was a fundamental transformation of the basic physical properties of the universe. To understand what happened during that brief moment of time, when the universe was a hot, dense sea of fast-moving particles and energetic photons colliding with each other, we must first understand how particles interact at very high energies.

Just *four* fundamental forces—gravitation, electromagnetism, and the strong and weak forces—explain the interactions of everything in the universe. Of these forces, gravitation is the most familiar (**Figure 15-15*a***). It is a long-range force that dominates the universe over astronomical distances. The electromagnetic force is also a long-range force, but it is intrinsically much stronger than the gravitational force. For example, the electromagnetic force between an electron and a proton is about 10^{39} times stronger than the gravitational force between those two particles. That is why the electromagnetic force, not the gravitational force, is responsible for holding electrons in orbit about the nuclei in atoms.

We do not generally observe longer-range effects of the electromagnetic force, because there is usually a negative electric charge for every positive charge and a south magnetic pole for every north magnetic pole. Thus, over great volumes of space the effects of electromagnetism effectively cancel out. No similar canceling occurs with gravity because there is no equivalent "negative mass." This explains why the force that holds Earth in orbit around the Sun is gravitational, not electromagnetic.

The **strong force** holds protons and neutrons together inside the nuclei of atoms (Figure 15-15*b*). It is said to be a *short-range force:* Its influence extends only over distances less than the diameter of a proton, about 10^{-15} m. Without the strong force, nuclei would disintegrate because of the electromagnetic repulsion of the positively charged protons. In fact, the strong force overpowers the electromagnetic forces inside nuclei.

The **weak force,** which also is a short-range force, is at work in certain kinds of radioactive decay (Figure 15-15*c*). An example is the transformation of a neutron (n) into a proton (p), in which an electron (e^-) is released along with a nearly massless particle called an *antineutrino* ($\bar{\nu}$).

In the 1970s, Sheldon Glashow, Howard Georgi, Jogesh Pati, and Abdus Salam proposed **grand unified theories** (or **GUTs**), which predict that the strong force becomes unified with the weak and electromagnetic forces at high energies and temperature. In other words, in special situations like the early universe, the strong, weak, and electromagnetic

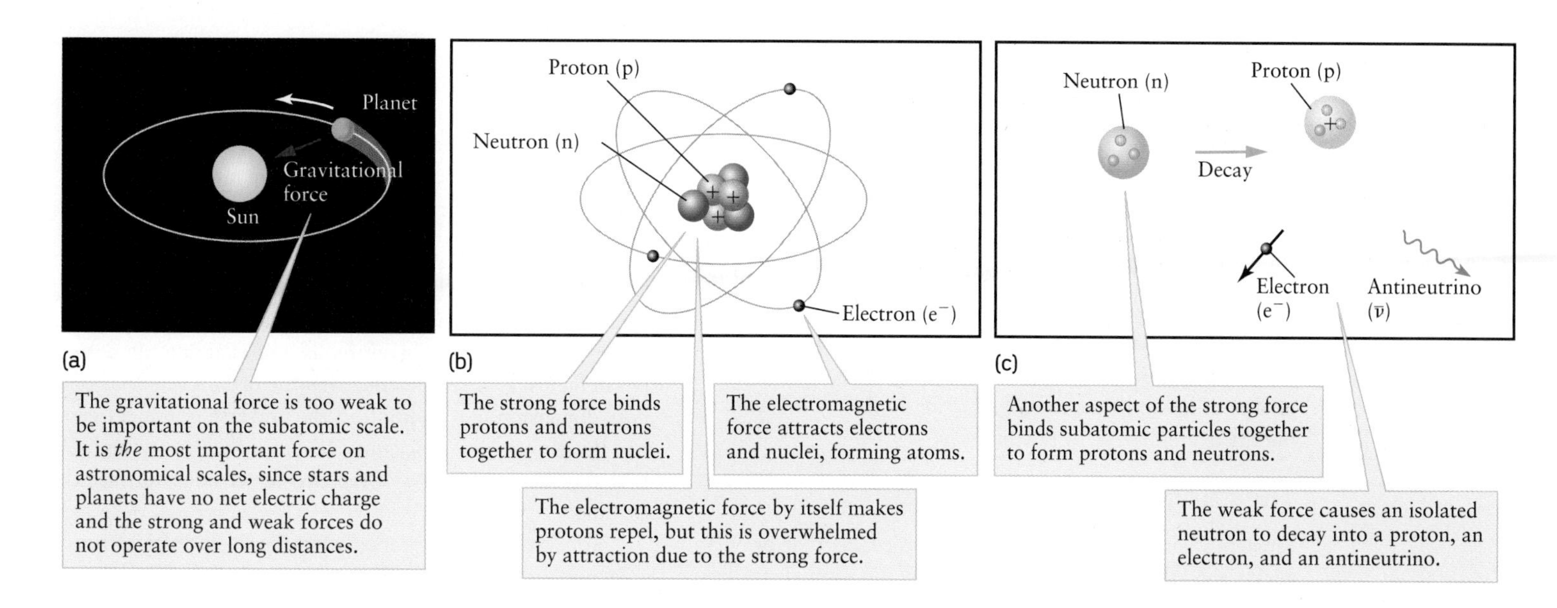

interactions would all be long-range forces and would be indistinguishable from each other.

Many physicists suspect that all four forces may be unified at tremendously high energies (**Figure 15-16**). That is, if particles were to collide at these colossal energies, such as occurred in the very young universe during the Big Bang, there would be no difference between the gravitational, electromagnetic, and nuclear forces. However, no one has yet succeeded in working out the details of such a supergrand unified theory, which is sometimes called a **theory of everything** (or TOE).

ConceptCheck 15-12: Which of the four fundamental forces is the most influential between two protons?

Force	Relative strength	Range	Example
Strong	1	10^{-15} m	holding protons, neutrons, and nuclei together
Electromagnetic	1/137	infinite	holding atoms together
Weak	10^{-4}	10^{-16} m	radioactive decay
Gravitational	6×10^{-39}	infinite	holding the solar system together

FIGURE 15-15 The Four Forces (a) Gravitation is dominant on the scales of planets, star systems, and galaxies, while (b, c) the strong, electromagnetic, and weak forces hold sway on the scale of atoms and nuclei. The table lists the relative strength of each force and provides examples of the domains in which each force dominates.

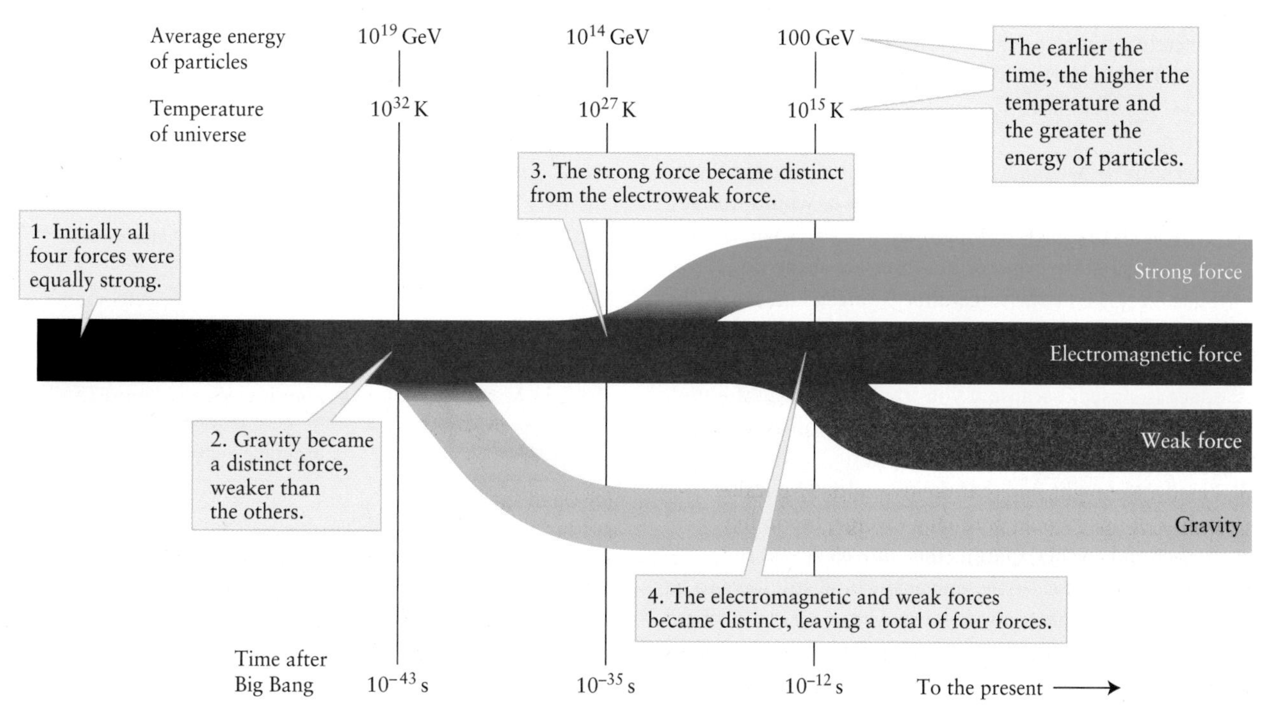

FIGURE 15-16 Unification of the Four Forces The strength of the four fundamental forces depends on the energy of the particles that interact. As shown in this schematic diagram, the higher the energy, the more the forces resemble each other. Also included here are the temperature of the universe and the time after the Big Bang when the strengths of the forces are thought to have been equal.

15-6 The shape of the universe indicates its matter and energy content

Go to Video 15-4

We have seen that by following the mass densities of radiation and matter, we can learn about the evolution of the universe. But it is equally important to know the combined mass density of all forms of matter and energy. (In an analogous way, an accountant needs to know the overall financial status of a company, not just individual profits or losses.) Remarkably, we can do this by investigating the overall shape of the universe.

The Curvature of the Universe

Einstein's general theory of relativity explains that gravity curves the fabric of space. Furthermore, the equivalence between matter and energy, expressed by Einstein's equation $E = mc^2$, tells that either matter or energy produces gravity. Thus, the matter and energy scattered across space should give the universe an overall curvature. The degree of curvature depends on the combined average mass density of *all* forms of matter and energy. Thus, by measuring the curvature of space, we should be able to learn about the content of the universe as a whole.

To see what astronomers mean by the curvature of the universe, imagine shining two powerful laser beams out into space so that they are perfectly parallel as they leave Earth. Furthermore, suppose that nothing gets in the way of these two beams, so that we can follow them for billions of light-years as they travel across the universe and across the space whose curvature we wish to detect. There are only three possibilities:

1. We might find that our two beams of light remain perfectly parallel, even after traversing billions of light-years. In this case, space would not be curved: The universe would have zero curvature, and space would be **flat.**
2. Alternatively, we might find that our two beams of light gradually converge. In such a case, space would not be flat. Recall that lines of longitude on Earth's surface are parallel at the equator but intersect at the poles. Thus, in this case the three-dimensional geometry of the universe would be analogous to the two-dimensional geometry of a spherical surface. We would then say that space is **closed,** because if you travel in a straight line in any direction in such a universe, you will eventually return to your starting point.
3. Finally, we might find that the two initially parallel beams of light would gradually diverge, becoming farther and farther apart as they moved across the universe. In this case, the universe would still have to be curved, but in the opposite sense. In the same way that a sphere is a positively curved two-dimensional surface, a saddle is a good example of a negatively curved two-dimensional surface. Parallel lines drawn on a sphere always converge, but parallel lines drawn on a saddle always diverge. Such a universe is called **open** because if you were to travel in a straight line in any direction, you would never return to your starting point.

Figure 15-17 illustrates the three cases of flat, closed, and open. Real space is three-dimensional, but we have drawn the three cases as analogous, more easily visualized two-dimensional surfaces. Therefore, as you examine the drawings in Figure 15-17, remember that the real universe has one more dimension. For example, if the universe is in fact open, then the geometry of space must be the (difficult-to-visualize) three-dimensional analog of the two-dimensional surface of a saddle.

Note that in accordance with the cosmological principle, none of these models of the universe has an "edge" or a "center." This is clearly the case for both the flat and open universes, because they are infinite and extend forever in all directions. Alternatively, a closed universe is finite, but it also lacks a center and an edge. You could walk forever around the surface of a closed spherical shape (like the surface of Earth if you could walk on the oceans) without ever finding a center or an edge.

The shape of the universe is determined by the value of a combined density of mass and energy for the universe. If this value is large, then the universe is closed. Alternatively, if it is small, then the universe is open. In the special case that the combined density of mass in the universe is "just

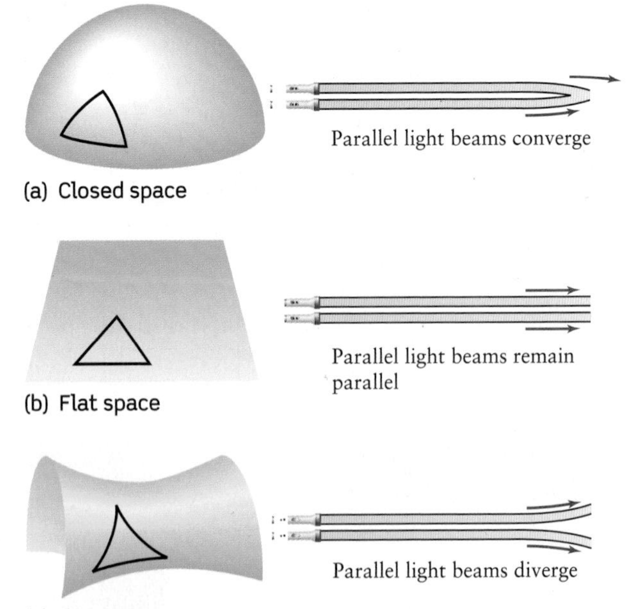

FIGURE 15-17 The Geometry of the Universe The shape of the universe is either (a) closed, (b) flat, or (c) open. The curvature depends on whether the combined density of all mass and energy is greater than, equal to, or less than a critical value that is "just right." In theory, such a curvature could be determined by seeing whether two laser beams initially parallel to each other would converge, remain parallel, or spread apart.

right," then the universe is flat. To help you get a sense for what would be "just right," a sample of hydrogen gas with this perfect density would contain just 6 hydrogen atoms per cubic meter—that's not much!

A way to determine the curvature of the universe that is both practical and precise is to see if light rays bend toward or away from each other, as shown in Figure 15-17. The greater the distance a pair of light rays has traveled, and, hence, the longer the time the light has been in flight, the more pronounced any such bending should be. Therefore, astronomers test for the presence of such bending by examining the oldest radiation in the universe: the cosmic microwave background.

When carefully observing the cosmic microwave background, astronomers have discovered that there are localized "hot spots" due to density variations in the early universe. The apparent size of these hot spots depends on the curvature of the universe (**Figure 15-18**). If the universe is closed, the bending of light rays from a hot spot will make the spot appear larger (Figure 15-18*a*); if the universe is open, the light rays will bend the other way and the hot spots will appear smaller (Figure 15-18*c*). Only in a flat universe will the light rays travel along straight lines, so that the hot spots appear with their true size (Figure 15-18*b*).

By calculating what conditions were like in the early universe, astrophysicists find that in a flat universe, the dominant "hot spots" in the cosmic background radiation should have an angular size of about 1°. This is just what the BOOMERANG and MAXIMA experiments observed, and what the WMAP observations have confirmed (see Figure 15-13). Hence, the curvature of the universe must be very close to zero, and it appears that our universe must either be flat or very nearly so.

This brings us to a curious problem. If we add the total amount of known energy and matter in the universe, including dark matter, together they account for only 24% of the total amount in the observed universe! The dilemma is this: What could account for the rest? The source of the missing amount must be some unexpected form of energy that we cannot detect from its gravitational effects (the technique astronomers use to detect dark matter). It must also not emit detectable light of any kind. We refer to this mysterious energy as **dark energy.** We do not know what it is, but whatever dark energy is, it accounts for 76% of the contents of the universe!

The concept of dark energy is actually due to Einstein. When he proposed the existence of a cosmological constant, he was suggesting that the universe is filled with a form of energy that by itself tends to make the universe expand. Unlike gravity, which tends to make objects attract, the energy associated with a cosmological constant would provide a form of "antigravity." Hence, it would not be detected in the same way as matter. These ideas concerning dark energy are extraordinary, and extraordinary claims require extraordinary evidence to confirm them. As we will see in the next section, a crucial test is to examine how the rate of expansion of the universe has evolved over the eons.

ConceptCheck 15-13: If you could start walking in a straight line around planet Earth and return to your exact starting point, would you categorize Earth as a flat, closed, or open world?

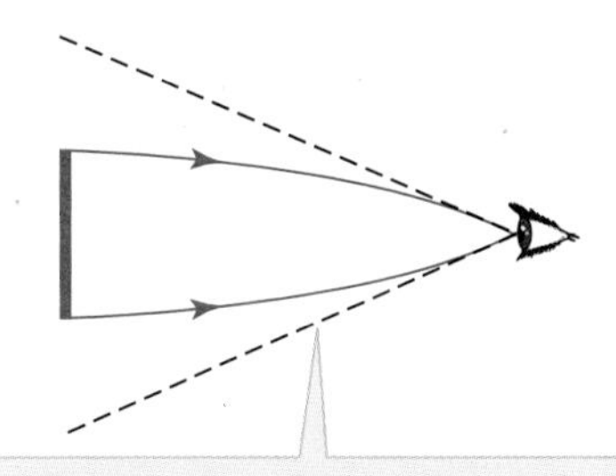

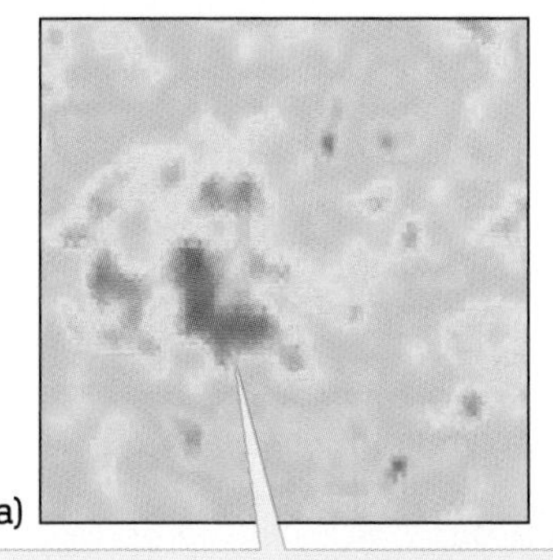

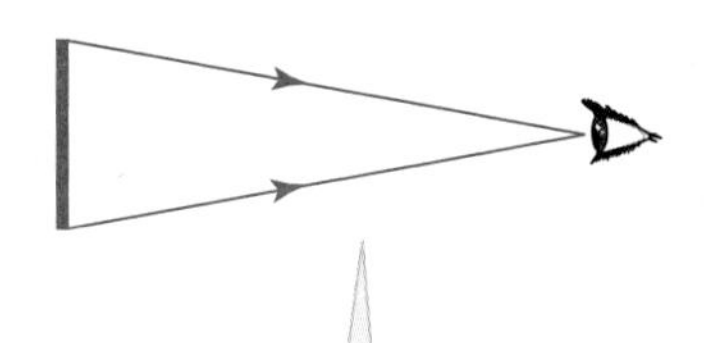

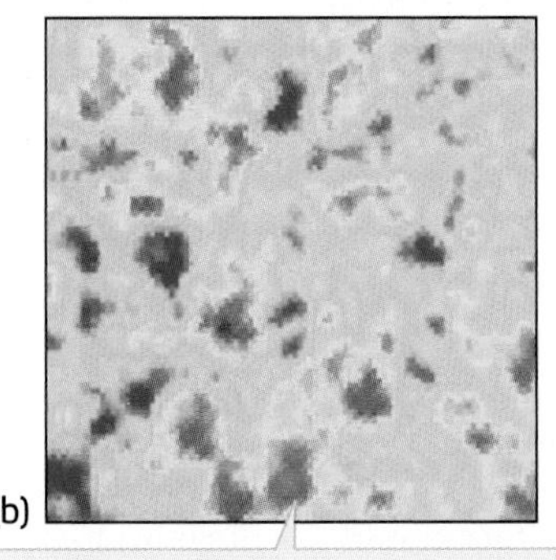

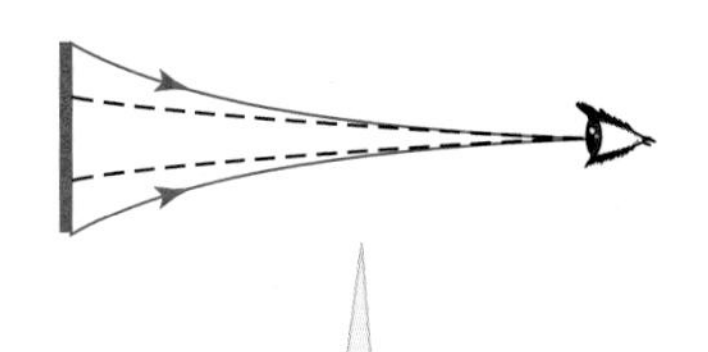

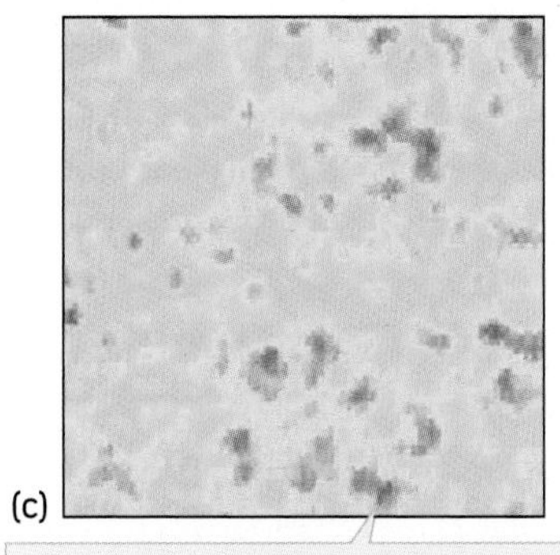

FIGURE 15-18 RIVUXG **The Cosmic Microwave Background and the Curvature of Space** Temperature variations in the early universe appear as "hot spots" in the cosmic microwave background. The apparent size of these spots depends on the curvature of space. *(The Boomerang Collaboration)*

15-7 Observations of distant supernovae indicate that we live in an accelerating universe

At great distances, the universe's expansion seems to be accelerating.

We have seen that the universe is expanding. But does the rate of expansion stay the same? Because there is matter in the universe, and because gravity tends to pull the bits of matter in the universe toward one another, we would expect that the expansion should slow down with time. (In the same way, a cannonball shot upward from the surface of Earth will slow down as it ascends because of Earth's gravitational pull.) If there is a cosmological constant, however, its associated dark energy will exert an outward pressure that tends to accelerate the expansion. Which of these effects is more important?

Modeling the Expansion History of the Universe

To answer this question, astronomers study the relationship between redshift and distance for extremely remote galaxies. We see these galaxies as they were billions of years ago. If the rate of expansion was the same in the distant past as it is now, the same Hubble law should apply to distant galaxies as to nearby ones. But if the rate of expansion has either increased or decreased, we will find important deviations from the Hubble law.

To see how astronomers approach this problem, first imagine two different parallel universes. Both Universe #1 and Universe #2 are expanding at constant rates, so for both universes there is a direct proportion between recessional velocity v and distance d as expressed by the Hubble law $v = H_0d$. Hence, a graph of distance versus recessional velocity for either universe is a straight line, as **Figure 15-19*a*** shows. The only difference is that Universe #1 is expanding at a slower rate than Universe #2. Hence, a galaxy at a certain distance from Earth in Universe #1 will have a slower recessional velocity than a galaxy at the same distance from Earth in Universe #2. As a result, the graph of distance versus recessional velocity for slowly expanding Universe #1 (shown in blue) has a steeper slope than the graph for rapidly expanding Universe #2 (shown in green). Keep this observation in mind: A slower expansion means a steeper slope on a graph of distance versus recessional velocity.

Now consider *our* universe and allow for the possibility that the expansion rate may change over time. If we observe very remote galaxies, we are seeing them as they were in the remote past. If the expansion of the universe in the remote past was slower or faster than it is now, the slope of the graph of distance versus recessional velocity will be different for those remote galaxies. If the expansion was slower, then the slope will be steeper for distant galaxies (shown in blue in Figure 15-19*b*); if the expansion was faster, the slope will be shallower for distant galaxies (shown in green in Figure 15-19*b*). In either case, there will be a deviation from the straight-line Hubble law (shown in purple in Figure 15-19*b*).

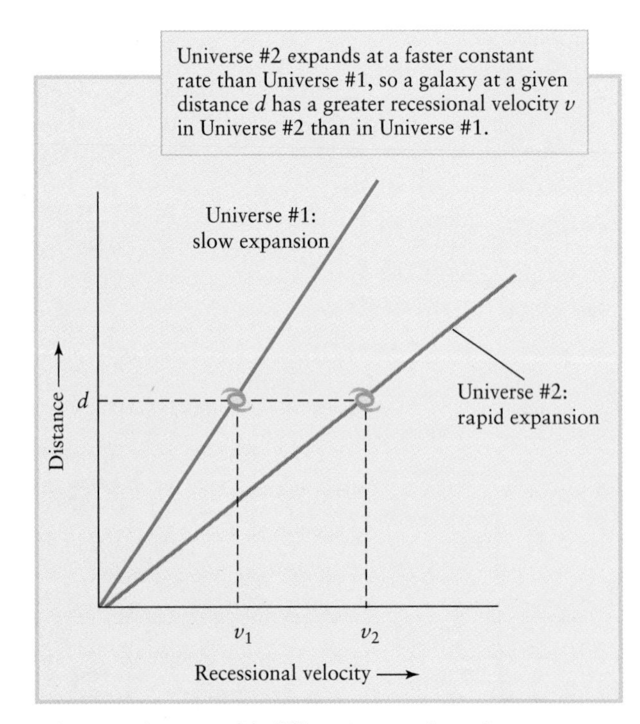

(a) Two universes with different expansion rates

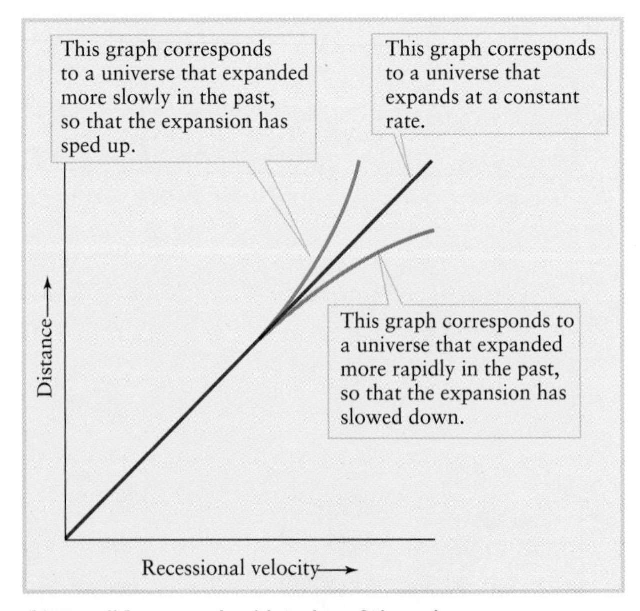

(b) Possible expansion histories of the universe

FIGURE 15-19 Varying Rates of Cosmic Expansion (a) Imagine two universes, #1 and #2. Each expands at its own constant rate. For a galaxy at a given distance, the recessional velocity will be greater in the more rapidly expanding universe. Hence, the graph of distance *d* versus recessional velocity *v* will have a shallower slope for the rapidly expanding universe and a steeper slope for the slowly expanding one. (b) If the rate of expansion of our universe was more rapid in the distant past, corresponding to remote distances, the graph of *d* versus *v* will have a shallower slope for large distances (green curve). If the expansion rate was slower in the distant past, the graph will have a steeper slope for large distances (blue curve).

ConceptCheck 15-14: In a universe that is expanding quickly, will the slope of a Hubble graph be steeper or shallower?

Measuring Ancient Expansion with Type Ia Supernovae

Which of the possibilities shown in Figure 15-19*b* represents the actual history of the expansion of our universe? In Section 15-2, we looked at the observed relationship between distance and recessional velocity for galaxies. Figure 15-4 is a plot of some representative data. The data points appear to lie along a straight line, suggesting that the rate of cosmological expansion has not changed. (Figure 15-4 is actually a graph of recessional velocity versus distance, not the other way around. But a straight line on one kind of graph will be a straight line on the other, because in either case there is a direct proportion between the two quantities being graphed.) However, the graph in Figure 15-4 was based on measurements of galaxies no farther than 400 Mpc (1.3 billion ly) from Earth, which means we are looking only 1.3 billion years into the past. The straightness of the line in Figure 15-4 means only that the expansion of the universe has been relatively constant over the past 1.3 billion years—only 10% of the age of the universe, and a relatively brief interval on the cosmic scale.

Now suppose that you were to measure the redshifts and distances of galaxies *several* billion light-years from Earth. The light from these galaxies has taken billions of years to arrive at your telescope, so your measurements will reveal how fast the universe was expanding billions of years ago. To do this, we need a technique that will allow us to find the distances to these very remote galaxies. One way to do this is to identify Type Ia supernovae as standard candles in such very distant galaxies. These supernovae are among the most luminous objects in the universe, and hence can be detected even at extremely large distances (see Figure 14-12). The maximum brightness of a supernova tells astronomers its distance through the inverse-square law for light, and the redshift of the supernova's spectrum tells them its recessional velocity. Taken together, we can calculate the distance to these galaxies.

ConceptCheck 15-15: What assumption must one make about Type Ia supernovae in order to use them to measure distances to galaxies?

An Accelerating Universe Filled with Dark Energy

In 1998, two prominent research groups—the Supernova Cosmology Project, led by Saul Perlmutter of Lawrence Berkeley National Laboratory, and the High-Z Supernova Search Team, led by Brian Schmidt of the Mount Stromlo and Siding Springs Observatories in Australia—reported their results from a survey of Type Ia supernovae in galaxies at redshifts of 0.2 or greater, corresponding to distances beyond 750 Mpc (2.4 billion ly). **Figure 15-20** shows some of their data, along with more recent observations, on a graph of apparent magnitude versus redshift. A greater apparent magnitude corresponds to a dimmer supernova, which means that the supernova is more distant. A greater redshift implies a greater recessional velocity. Hence, this graph is basically the same as those in Figure 15-19.

What we see here is surprising indeed: When the curve lies in the blue region of Figure 15-20, dark energy has made the expansion of the universe speed up over time. Hence, the expansion of the universe was slower in the distant past, which means that we live in an *accelerating* universe. Just like the blue curve in Figure 15-19*b*, the data in Figure 15-20 show that supernovae of a certain brightness (and hence a given distance) have smaller redshifts (and hence smaller recessional velocities) than would be the case if the expansion rate had always been the same. These data provide compelling evidence of the existence of dark energy.

Roughly speaking, the data in Figure 15-20 indicate the relative importance of dark energy (which tends to make the expansion speed up) and gravitational attraction between galaxies (which tends to make the attraction slow down). Taken together and combined with other observations, this points to a radically different conclusion about

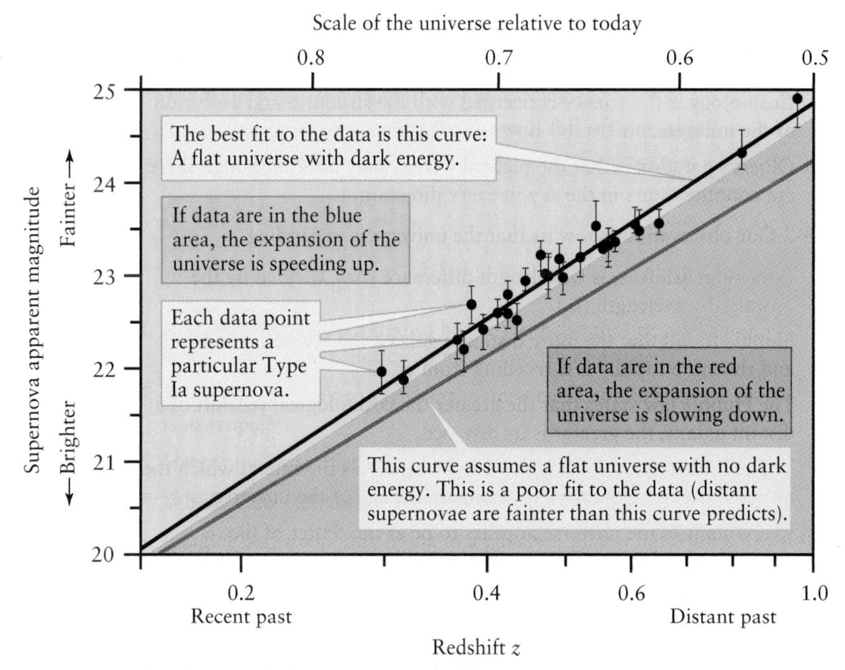

FIGURE 15-20 The Hubble Diagram for Distant Supernovae This graph shows apparent magnitude versus redshift for supernovae in distant galaxies. The greater the apparent magnitude, the dimmer the supernova and the greater the distance to it and its host galaxy. If the expansion of the universe is speeding up, the data will lie in the blue area; if it is slowing down, the data will lie in the red area. The data show that the expansion is in fact speeding up. *(Data from The Supernova Cosmology Project/ R. A. Knop et al.)*

Web Chat Questions

1. How can astronomers be certain that the cosmic microwave background fills the entire cosmos, not just the vicinity of Earth?
2. How does the evidence for the Big Bang confirm or conflict with religious or spiritual views of the beginning of time?
3. Some GUTs predict that the proton is unstable, although with a half-life far longer than the present age of the universe. What would it be like to live at a time when protons were decaying in large numbers?

Collaborative Exercises

1. As a group, create a four- to six-panel cartoon strip showing a discussion between two individuals describing why the sky is dark at night.
2. Imagine your firm, Creative Cosmologists Coalition, has been hired to create a three-panel, folded brochure describing the principal observations that astronomers use to infer the existence of a Big Bang. Create this brochure on an 8½ × 11 piece of paper. Be sure each member of your group supervises the development of a different portion of the brochure and that the small print acknowledges who in your group was primarily responsible for which portion.
3. The three potential geometries of the universe are shown in Figure 15-17. To demonstrate this, ask one member of your group to hold a piece of paper in one of the positions while another member draws two parallel lines that never change in one geometry, eventually cross in another geometry, and eventually diverge in another.
4. The four fundamental forces of nature are the strong force, the weak force, the gravitational force, and the electromagnetic force. List four things at your school that rely on one of these fundamental forces, and explain how each thing is dependent on one of the fundamental forces.
5. Consider the following hypothetical scenario adapted from a daytime cable television talk show. Chris states that Pat borrowed Chris's telescope without permission. Tyler purchased balloons and a new telescope eyepiece without telling Chris. Sean borrowed star maps from the library, with the library's permission, but without telling Pat. Eventually, when the four met on Sunday evening, Chris was crying and speechless. Can you create a "grand unified theory" that explains this entire situation?

Observing Questions

1. In an attempt to explore the far reaches of the universe, the Hubble Space Telescope (HST) took long-exposure images of very dark regions of space that appear to contain no bright stars or galaxies. These images, known as the Hubble Deep Field and Hubble Ultra Deep Field images, reveal very rich fields of faint and very distant galaxies. The light now arriving at Earth from some of these galaxies has traveled for more than 13 billion years and was collected by the HST at a rate of a few photons per minute! This light was emitted very early in the life of the universe, only a few 100 million years after the Big Bang. You can examine and measure these two images.
   a) In *Starry Night*™, open the **Options** pane and ensure that the **Hubble Images** option is checked in the **Deep Space** layer. Open the **Find** pane, ensure that the search edit box is empty, and click on the icon in this box to display a list of image sources. Click on **Hubble Images** and double-click on **Hubble Deep Field** to center the view on this dark region of space. Note its position with respect to the Big Dipper. (*Note:* If you cannot identify this region of the northern sky, click on **View > Constellations > Asterisms** and **View > Constellations > Labels.** Remove these indicators after you have identified the region.) **Zoom** in to a field of view about 3° wide and note that the region still appears to be devoid of objects. **Zoom** in again until the **Hubble Deep Field** (HDF) fills the view. One-quarter of the full HDF, with dimensions of 1.15′ × 1.15′, is displayed in *Starry Night*™. The bright object with spikes radiating from it is a star in our own Galaxy (the spikes are caused by diffraction by the supports for Hubble's secondary mirror), but all of the other objects that appear on this long-exposure image are galaxies containing millions of stars. Examine this image carefully and attempt to identify some examples of each kind of galaxy—spiral, barred spiral, elliptical, and irregular—in this field. Choose 5 or 6 of the largest galaxies in this field, record their shapes and galaxy types, and use the angular separation tool to measure carefully and record their angular dimensions.
   b) Click the **Zoom** panel in the toolbar and select **90°** from the dropdown menu. Return to the **Find** pane and the list of **Hubble Images** and double-click on **Hubble Ultra Deep Field** (HUDF) to center the view of this "dark" region of the sky. **Zoom** in on this region and note that, even at a field of view as small as 2°, no objects can be seen in the position of this long-exposure image. **Zoom** in further until the full HUDF, with dimensions of 3.3′ × 3.3′, fills the field of view to see this rich field of faint and very distant galaxies. Examine this image carefully and attempt to identify each kind of galaxy—spiral, barred spiral, elliptical, and irregular—in this field. Again, select 5 or 6 of the largest galaxies in this field, record their shapes and galaxy types, and use the angular separation tool to measure their dimensions.
   c) Consider the mix of different kinds of galaxies and assess whether the proportions of different kinds are the same in these two images. Compare the angular sizes of the largest galaxies in these two images.
2. Use the *Starry Night* ™ program to examine a very distant pair of quasars. Most quasars are starlike in appearance but their spectra show very large redshifts. If these redshifts are cosmological (i.e., caused by the expansion of the universe), then quasars obey Hubble's law, and their large redshifts show that they must be very far away. Since they appear bright in our sky, they must be very luminous, with outputs equivalent to that of billions and even trillions of Suns. They are part of a group of objects known as Active Galactic Nuclei. The source of energy output of these very luminous objects is probably a supermassive black hole at the object's center. Select **Favourites > Explorations > Quasar Pair** and **Zoom** in to a field of view about 30 arcseconds wide. Open the **Info** pane for this pair of quasars and note the **Distance from observer** as given in the **Position in Space** layer.
   a) What is the distance from Earth of these quasars?
   b) Using a Hubble constant of 73 km/s/Mpc, calculate the expected recessional velocity of this quasar pair. (*Note:* 1 pc = 3.26 ly.)
   c) What is the ratio of the recessional velocity of these quasars to the speed of light?

ANSWERS

ConceptChecks

ConceptCheck 15-1: Because if the universe is infinitely large, then there should be stars in every possible direction you look, resulting in a light night sky instead of a dark night sky.

ConceptCheck 15-2: The spectral shift toward the shorter blue wavelengths means that the Andromeda Galaxy is moving toward our Milky Way Galaxy.

ConceptCheck 15-3: The two most important observations are that galaxies have redshifted spectra and that the most distant galaxies are moving away from us fastest.

ConceptCheck 15-4: The cosmological principle applies over very large regions of outer space; it does not apply to the distribution of stars in objects as small as a galaxy.

ConceptCheck 15-5: The age of the universe, T_0, is calculated as $1/H_0$; so, if H_0 is larger, then T_0 must be smaller.

ConceptCheck 15-6: The observable universe becomes larger as more light from never-before-seen distant galaxies has sufficient time to reach us.

ConceptCheck 15-7: The expansion of the universe over the past 13.7 billion years has stretched and redshifted these photons to such a great degree that they are now low-energy, long-wavelength microwave photons.

ConceptCheck 15-8: Earth formed about 4.5 billion years ago, when the universe was about 9 billion years old, and the universe has been dominated by matter for most of its history.

ConceptCheck 15-9: The universe had to be sufficiently cool for atoms to form, so if the early universe was hotter for longer, the first atoms would have appeared more slowly.

ConceptCheck 15-10: The WMAP data shows slight differences in temperature at various locations from which astronomers think were the locations for the initial formation of galaxies and is significantly more detailed than earlier observations.

ConceptCheck 15-11: No. The universe underwent a very brief but dramatic expansion, right after the Big Bang.

ConceptCheck 15-12: The strong nuclear force dominates in the short-distance realm of atomic nuclei.

ConceptCheck 15-13: In a closed universe, any beam of light eventually comes back to its starting point, so in this geometrical sense, Earth is a *closed* universe.

ConceptCheck 15-14: The slope of the line with value H_0 will be shallower for a rapidly expanding universe.

ConceptCheck 15-15: Astronomers must make the assumption that all Type Ia supernovae have identical luminosities, regardless of the galaxy in which they are found.

ConceptCheck 15-16: Today's universe is dominated by dark energy, followed by matter, and lastly by radiation energy.

CalculationChecks

CalculationCheck 15-1: Because z = the change in wavelength ($\lambda - \lambda_0$) divided by the original λ_0, $z = (725.6 \text{ nm} - 656.3 \text{ nm}) \div 656.3 \text{ nm} = 0.11$.

CalculationCheck 15-2: Hubble's Law, $v = H_0 d$, can be rearranged as $d = v \div H_0$. Using $H_0 = 73$ km/s/Mpc, $d = v \div H_0 = 10{,}000 \text{ km/s} \div 73 \text{ km/s/Mpc} = 134$ Mpc (million parsecs).

APPENDICES

Much of the astronomical data presented in these appendices use international metric units. To convert from kilometers to miles, multiply the number of kilometers × 0.621. To convert mass in kilograms to pounds-mass, multiple the number of kilograms × 2.20.

Appendix 1 The Planets: Orbital Data

Planet	Semimajor axis (10^6 km)	Semimajor axis (AU)	Sidereal period (years)	Sidereal period (days)	Synodic period (days)	Average orbital speed (km/s)	Orbital eccentricity	Inclination of orbit to ecliptic (°)
Mercury	57.9	0.387	0.241	87.969	115.88	47.9	0.206	7.00
Venus	108.2	0.723	0.615	224.70	583.92	35.0	0.007	3.39
Earth	149.6	1.000	1.000	365.256	—	29.79	0.017	0.00
Mars	227.9	1.524	1.88	686.98	779.94	24.1	0.093	1.85
Jupiter	778.3	5.203	11.86		398.9	13.1	0.048	1.30
Saturn	1429	9.554	29.46		378.1	9.64	0.053	2.48
Uranus	2871	19.194	84.10		369.7	6.83	0.043	0.77
Neptune	4498	30.066	164.86		367.5	5.5	0.010	1.77

Appendix 2 The Planets: Physical Data

Planet	Equatorial diameter (km)	Equatorial diameter (Earth = 1)	Mass (kg)	Mass (Earth = 1)	Average density (kg/m^3)	Rotation period* (solar days)	Inclination of equator to orbit (°)	Surface gravity (Earth = 1)	Albedo	Escape speed (km/s)
Mercury	4880	0.383	3.302×10^{23}	0.0553	5430	58.646	0.5	0.38	0.12	4.3
Venus	12,104	0.949	4.868×10^{24}	0.8149	5243	243.01^{R}	177.4	0.91	0.59	10.4
Earth	12,756	1.000	5.974×10^{24}	1.000	5515	0.997	23.45	1.000	0.39	11.2
Mars	6794	0.533	6.418×10^{23}	0.107	3934	1.026	25.19	0.38	0.15	5.0
Jupiter	142,984	11.209	1.899×10^{27}	317.8	1326	0.414	3.12	2.36	0.44	59.5
Saturn	120,536	9.449	5.685×10^{26}	95.16	687	0.444	26.73	1.1	0.47	35.5
Uranus	51,118	4.007	8.682×10^{25}	14.53	1318	0.718^{R}	97.86	0.92	0.56	21.3
Neptune	49,528	3.883	1.024×10^{26}	17.15	1638	0.671	29.56	1.1	0.51	23.5

*For Jupiter, Saturn, Uranus, and Neptune, the internal rotation period is given. A superscript R means that the rotation is retrograde (opposite the planet's orbital motion).

Appendix 3 Satellites of the Planets

Planet	Satellite	Date of discovery	Average distance from center of planet (km)	Orbital (sidereal) period (days)*	Orbital eccentricity	Size of satellite (km)**	Mass (kg)
EARTH	Moon	—	384,400	0027.322	0.0549	3476	7.349×10^{22}
MARS	Phobos	1877	9378	0.319	0.01	28 × 23 × 20	1.1×10^{16}
	Deimos	1877	23,460	1.263	0.00	16 × 12 × 10	1.8×10^{15}
JUPITER	Metis	1979	128,000	0.295	0.0012	44	1.2×10^{17}
	Adrastea	1979	129,000	0.298	0.0018	24 × 20 × 16	7.5×10^{15}
	Amalthea	1892	181,400	0.498	0.0031	270 × 200 × 155	2.1×10^{18}
	Thebe	1979	221,900	0.675	0.0177	98	1.5×10^{18}
	Io	1610	421,600	1.769	0.0041	3642	8.932×10^{22}
	Europa	1610	670,900	3.551	0.0094	3120	4.791×10^{22}
	Ganymede	1610	1,070,000	7.155	0.0011	5268	1.482×10^{23}
	Callisto	1610	1,883,000	16.689	0.0074	4800	1.077×10^{23}
	Themisto	2000	7,284,000	130.02	0.2426	9	6.9×10^{14}
	Leda	1974	11,165,000	240.92	0.1636	18	1.1×10^{16}
	Himalia	1904	11,461,000	250.56	0.1623	184	6.7×10^{18}
	Lysithea	1938	11,717,000	259.20	0.1124	38	6.3×10^{16}
	Elara	1905	11,741,000	259.64	0.2174	78	8.7×10^{17}
	Kallichore	2000	12,555,000	456.10	0.2480	4	1.5×10^{13}
	S/2003 J12	2003	15,912,000	489.52^{R}	0.6056	1	1.5×10^{12}
	Carpo	2003	16,989,000	456.10	0.4297	3	4.5×10^{13}
	Euporie	2001	19,304,000	550.74^{R}	0.1432	2	1.5×10^{13}
	S/2003 J3	2003	20,221,000	583.88^{R}	0.1970	2	1.5×10^{13}
	S/2003 J18	2003	20,514,000	596.59^{R}	0.0148	2	1.5×10^{13}
	Orthosie	2001	20,720,000	622.56^{R}	0.2808	2	1.5×10^{13}
	Euanthe	2001	20,797,000	620.49^{R}	0.2321	3	4.5×10^{13}
	Harpalyke	2000	20,858,000	623.31^{R}	0.2268	4	1.2×10^{14}
	Praxidike	2000	20,907,000	625.38^{R}	0.2308	7	4.3×10^{14}
	Thyone	2001	20,939,000	627.21^{R}	0.2286	4	9.0×10^{13}
	S/2003 J16	2003	20,963,000	616.36^{R}	0.2245	2	1.5×10^{13}
	Iocaste	2000	21,061,000	631.60^{R}	0.2160	5	1.9×10^{14}
	Mneme	2003	21,069,000	620.04^{R}	0.2273	2	1.5×10^{13}
	Hermippe	2001	21,131,000	633.90^{R}	0.2096	4	9.0×10^{13}
	Thelxinoe	2003	21,162,000	628.09^{R}	0.2206	2	1.5×10^{13}
	Helike	2003	21,263,000	634.77^{R}	0.1558	4	9.0×10^{13}
	Ananke	1951	21,276,000	629.77^{R}	0.2435	28	3.0×10^{16}
	S/2003 J15	2003	22,627,000	689.77^{R}	0.1916	2	1.5×10^{13}
	Eurydome	2001	22,865,000	717.33^{R}	0.2759	3	4.5×10^{13}
	Arche	2002	22,931,000	723.90^{R}	0.2588	3	4.5×10^{13}
	S/2003 J17	2003	23,001,000	714.47^{R}	0.2379	2	1.5×10^{13}
	Pasithee	2001	23,004,000	719.44^{R}	0.2675	2	1.5×10^{13}
	S/2003 J10	2003	23,042,000	716.25^{R}	0.4295	2	1.5×10^{13}
	Chaldene	2000	23,100,000	723.70^{R}	0.2519	4	7.5×10^{13}
	Isonoe	2000	23,155,000	726.25^{R}	0.2471	4	7.5×10^{13}

Appendix 3 Satellites of the Planets

Planet	Satellite	Date of discovery	Average distance from center of planet (km)	Orbital (sidereal) period (days)*	Orbital eccentricity	Size of satellite (km)**	Mass (kg)
JUPITER (continued)	Erinome	2000	23,196,000	728.51^R	0.2665	3	4.5×10^{13}
	Kale	2001	23,217,000	729.47^R	0.2599	2	1.5×10^{13}
	Aitne	2001	23,229,000	730.18^R	0.2643	3	4.5×10^{13}
	Taygete	2000	23,280,000	732.41^R	0.2525	5	1.6×10^{14}
	S/2003 J9	2003	23,384,000	733.29^R	0.2632	1	1.5×10^{12}
	Carme	1938	23,404,000	734.17^R	0.2533	46	1.3×10^{17}
	Sponde	2001	23,487,000	748.34^R	0.3121	2	1.5×10^{13}
	Megaclite	2000	23,493,000	752.88^R	0.4197	6	2.1×10^{14}
	S/2003 J5	2003	23,495,000	738.73^R	0.2478	4	9.0×10^{13}
	S/2003 J19	2003	23,533,000	740.42^R	0.2557	2	1.5×10^{13}
	S/2003 J23	2003	23,563,000	732.44^R	0.2714	2	1.5×10^{13}
	Kalyke	2000	23,566,000	742.03^R	0.2465	5	1.9×10^{14}
	S/2003 J14	2003	23,614,000	779.23^R	0.3439	2	1.5×10^{13}
	Pasiphaë	1908	23,624,000	743.63^R	0.4090	58	3.0×10^{17}
	Eukelade	2003	23,661,000	746.39^R	0.2721	4	9.0×10^{13}
	S/2003 J4	2003	23,930,000	755.24^R	0.3618	2	1.5×10^{13}
	Sinope	1914	23,939,000	758.90^R	0.2495	38	7.5×10^{16}
	Hegemone	2003	23,947,000	739.60^R	0.3276	3	4.5×10^{13}
	Aoede	2003	23,981,000	761.50^R	0.4322	4	9.0×10^{13}
	Kallichore	2003	24,043,000	764.74^R	0.2640	2	1.5×10^{13}
	Autonoe	2000	24,046,000	760.95^R	0.3168	4	9.0×10^{13}
	Callirrhoe	1999	24,103,000	758.77^R	0.2828	7	8.7×10^{14}
	Cyllene	2003	24,349,000	751.91^R	0.3189	2	1.5×10^{13}
	S/2003 J2	2003	29,541,000	979.99^R	0.2255	2	1.5×10^{13}
SATURN	Pan	1981	133,580	0.575	0	$35 \times 35 \times 23$	4.9×10^{15}
	Daphnis	2005	136,500	0.594	0	7	3×10^{14}
	Atlas	1980	137,670	0.602	0.0012	$46 \times 38 \times 19$	6.6×10^{15}
	Prometheus	1980	139,380	0.613	0.0022	$119 \times 87 \times 61$	1.6×10^{17}
	Pandora	1980	141,720	0.629	0.0042	$103 \times 80 \times 64$	1.4×10^{17}
	Epimetheus	1980	151,410	0.694	0.0098	$135 \times 108 \times 105$	5.3×10^{17}
	Janus	1980	151,460	0.695	0.0068	$193 \times 173 \times 137$	1.9×10^{18}
	Mimas	1789	185,540	0.942	0.0196	397	3.8×10^{19}
	Methone	2004	194,440	1.01	0.0001	3	2×10^{13}
	Pallene	2004	212,280	1.154	0.004	4	4×10^{13}
	Enceladus	1789	238,040	1.37	0.0047	504	1.1×10^{20}
	Tethys	1684	294,670	1.888	0.0001	1066	6.2×10^{20}
	Telesto	1980	294,710	1.888	0.0002	$29 \times 22 \times 20$	8×10^{15}
	Calypso	1980	294,710	1.888	0.0005	$30 \times 23 \times 14$	5×10^{15}
	Polydeuces	2004	377,200	2.737	0.0192	3.5	3×10^{13}
	Dione	1684	377,420	2.737	0.0022	1123	1.1×10^{21}
	Helene	1980	377,420	2.737	0.0071	$36 \times 32 \times 30$	2×10^{16}
	Rhea	1672	527,070	4.518	0.001	1528	2.3×10^{21}

Appendix 3 Satellites of the Planets

Planet	Satellite	Date of discovery	Average distance from center of planet (km)	Orbital (sidereal) period (days)*	Orbital eccentricity	Size of satellite (km)**	Mass (kg)
SATURN (continued)	Titan	1655	1,221,870	15.95	0.0288	5150	1.34×10^{23}
	Hyperion	1848	1,500,880	21.28	0.0274	360 × 280 × 225	5.7×10^{18}
	Iapetus	1671	3,560,840	79.33	0.0283	1472	2.0×10^{21}
	Kiviuq	2000	11,111,000	449.22	0.3288	16	3×10^{15}
	Ijiraq	2000	11,124,000	451.43	0.3163	12	1×10^{15}
	Phoebe	1898	12,947,780	550.31^{R}	0.1635	230 × 220 × 210	8.3×10^{18}
	Paaliaq	2000	15,200,000	686.93	0.3631	19	4×10^{15}
	Skathi	2000	15,541,000	728.21^{R}	0.2701	6	2×10^{14}
	Albiorix	2000	16,182,000	783.46	0.477	32	3×10^{16}
	S/2007 S2	2007	16,560,000	792.96^{R}	0.2418	6	1.5×10^{14}
	Bebhionn	2004	17,153,520	838.77	0.4691	6	2×10^{14}
	Erriapo	2000	17,343,000	871.18	0.4724	10	9×10^{14}
	Siarnaq	2000	17,531,000	895.55	0.2961	40	4×10^{16}
	Skoll	2006	17,473,800	862.37	0.418	6	2×10^{14}
	Tarqeq	2007	17,910,600	894.86	0.1081	7	2.3×10^{14}
	Tarvos	2000	17,983,000	926.23	0.5305	15	3×10^{15}
	Greip	2006	18,065,700	906.56	0.374	6	2×10^{14}
	S/2004 S19	2004	18,217,125	912^{R}	0.36	8	8×10^{14}
	S/2004 S13	2004	18,403,000	933.45^{R}	0.2586	6	2×10^{14}
	Jarnsaxa	2006	18,556,900	943.78	0.192	6	2×10^{14}
	Mundilfari	2000	18,685,000	952.67^{R}	0.21	7	3×10^{14}
	S/2006 S1	2006	18,930,200	972.41	0.13	6	2×10^{14}
	Narvi	2003	19,007,000	1003.93^{R}	0.4309	7	3×10^{14}
	S/2004 S15	2004	19,338,000	1005.93^{R}	0.1428	6	2×10^{14}
	S/2004 S17	2004	19,447,000	1014.7^{R}	0.1793	4	4×10^{13}
	Suttungr	2000	19,459,000	1016.67^{R}	0.114	7	3×10^{14}
	S/2004 S14	2004	19,856,000	1038.67^{R}	0.3715	6	2×10^{14}
	S/2004 S12	2004	19,878,000	1046.16^{R}	0.3261	5	2×10^{14}
	S/2004 S18	2004	20,129,000	1083.57^{R}	0.5214	7	3×10^{14}
	S/2004 S9	2004	20,390,000	1086.1^{R}	0.2397	5	2×10^{14}
	Thrymr	2000	20,474,000	1094.23^{R}	0.4652	7	3×10^{14}
	S/2007 S3	2007	20,518,500	1100^{R}	0.130	5	9×10^{14}
	S/2004 S10	2004	20,735,000	1116.47^{R}	0.252	6	2×10^{14}
	S/2004 S7	2004	20,999,000	1140.28^{R}	0.5299	6	2×10^{14}
	S/2006 S3	2006	21,076,300	1142.37	0.471	6	2×10^{14}
	Surtur	2006	22,288,916	1242.36	0.368	6	2×10^{14}
	Kari	2006	22,321,200	1245.06	0.341	7	3×10^{14}
	S/2004 S16	2004	22,453,000	1260.28^{R}	0.1364	4	4×10^{13}
	Loge	2006	22,984,322	1300.95^{R}	0.1390	6	1.5×10^{14}
	Ymir	2000	23,040,000	1315.21^{R}	0.335	18	3×10^{15}
	S/2006 S5	2006	23,190,000	1314^{R}	0.139	6	2×10^{14}
	S/2004 S8	2004	25,108,000	1490.87^{R}	0.2064	6	2×10^{14}

Appendix 3 Satellites of the Planets

Planet	Satellite	Date of discovery	Average distance from center of planet (km)	Orbital (sidereal) period (days)*	Orbital eccentricity	Size of satellite (km)**	Mass (kg)
URANUS	Cordelia	1986	49,800	0.335	0.0003	50 × 36	4.4×10^{16}
	Ophelia	1986	53,800	0.376	0.0099	54 × 38	5.3×10^{16}
	Bianca	1986	59,200	0.435	0.0009	64 × 46	9.2×10^{16}
	Cressida	1986	61,800	0.464	0.0004	92 × 74	3.4×10^{17}
	Desdemona	1986	62,700	0.474	0.0001	90 × 54	1.8×10^{17}
	Juliet	1986	64,400	0.493	0.0007	150 × 74	5.6×10^{17}
	Portia	1986	66,100	0.513	0.0001	156 × 126	1.7×10^{18}
	Rosalind	1986	69,900	0.558	0.0001	72	2.5×10^{17}
	Cupid	2003	74,800	0.618	0.0013	18	3.8×10^{15}
	Belinda	1986	75,300	0.624	0.0001	128 × 64	3.6×10^{17}
	Perdita	1986	76,420	0.638	0.003	26	1.3×10^{16}
	Puck	1985	86,000	0.762	0.0001	162	2.9×10^{18}
	Mab	2003	97,734	0.923	0.0025	24	1.0×10^{16}
	Miranda	1948	129,900	1.413	0.0013	471	6.59×10^{19}
	Ariel	1851	190,900	2.52	0.0012	1158	1.35×10^{21}
	Umbriel	1851	266,000	4.144	0.0039	1169	1.2×10^{21}
	Titania	1787	436,300	8.706	0.0011	1578	3.53×10^{21}
	Oberon	1787	583,500	13.46	0.0014	1522	3.01×10^{21}
	Francisco	2001	4,276,000	266.56	0.1459	22	1.4×10^{15}
	Caliban	1997	7,231,000	579.73^{R}	0.1587	72	7.4×10^{17}
	Stephano	1999	8,004,000	677.36^{R}	0.2292	32	6.0×10^{15}
	Trinculo	2001	8,504,000	749.24^{R}	0.22	18	7.5×10^{14}
	Sycorax	1997	12,179,000	1288.3^{R}	0.5224	150	5.4×10^{18}
	Margaret	2003	14,345,000	1687.01	0.6608	20	1.0×10^{15}
	Prospero	1999	16,256,000	1978.29^{R}	0.4448	25	2.1×10^{16}
	Setebos	1999	17,418,000	2225.21^{R}	0.5914	24	2.1×10^{16}
	Ferdinand	2003	20,901,000	2887.21^{R}	0.3682	21	4.4×10^{16}
NEPTUNE	Naiad	1989	48,227	0.294	0.0004	96 × 60 × 52	1.9×10^{17}
	Thalassa	1989	50,075	0.311	0.0002	108 × 100 × 52	3.5×10^{17}
	Despina	1989	52,526	0.335	0.0002	180 × 150 × 130	2.1×10^{18}
	Galatea	1989	61,953	0.429	0	204 × 184 × 144	2.1×10^{18}
	Larissa	1989	73,548	0.555	0.0014	216 × 204 × 164	4.2×10^{18}
	Proteus	1989	117,647	1.122	0.0005	440 × 416 × 404	4.4×10^{19}
	Triton	1846	354,800	5.877^{R}	0	2706	2.15×10^{22}
	Nereid	1949	5,513,400	360.14	0.7512	340	3.1×10^{19}
	S/2002 N1	2002	15,728,000	1879.71^{R}	0.5711	62	1.8×10^{17}
	S/2002 N2	2002	22,422,000	2914.07	0.2931	44	6.3×10^{16}
	S/2002 N3	2002	23,571,000	3167.85	0.4237	42	5.5×10^{16}
	Psamathe	2003	46,695,000	9115.91^{R}	0.4499	24	1.0×10^{16}
	S/2002 N4	2002	48,387,000	9373.99^{R}	0.4945	60	1.6×10^{17}

This table was compiled from data provided by the Jet Propulsion Laboratory.

*A superscript R means that the satellite orbits in a retrograde direction (opposite to the planet's rotation).

**The size of a spherical satellite is equal to its diameter.

Appendix 4 The Nearest Stars

Name	Parallax (arcsec)	Distance (parsecs)	Distance (light-years)	Spectral Type	Proper motion (arcsec/yr)	Apparent visual magnitude	Absolute visual magnitude	Mass (Sun = 1)
Proxima Centauri	0.772	1.30	4.22	M5.5 V	3.853	+11.09	+15.53	0.107
Alpha Centauri A	0.747	1.34	4.36	G2 V	3.710	−0.01	+4.36	1.144
Alpha Centauri B	0.747	1.34	4.36	K0 V	3.724	+1.34	+5.71	0.916
Barnard's Star	0.547	1.83	5.96	M4.0 V	10.358	+9.53	+13.22	0.166
Wolf 359	0.419	2.39	7.78	M6.0 V	4.696	+13.44	+16.55	0.092
Lalande 21185	0.393	2.54	8.29	M2.0 V	4.802	+7.47	+10.44	0.464
Sirius A	0.380	2.63	8.58	A1 V	1.339	−1.43	+1.47	1.991
Sirius B	0.380	2.63	8.58	white dwarf	1.339	+8.44	+11.34	0.500
UV Ceti	0.374	2.68	8.73	M5.5 V	3.368	+12.54	+15.40	0.109
BL Ceti	0.374	2.68	8.73	M6.0 V	3.368	+12.99	+15.85	0.102
Ross 154	0.337	2.97	9.68	M3.5 V	0.666	+10.43	+13.07	0.171
Ross 248	0.316	3.16	10.32	M5.5 V	1.617	+12.29	+14.79	0.121
Epsilon Eridani	0.310	3.23	10.52	K2 V	0.977	+3.73	+6.19	0.850
Lacailee 9352	0.304	3.29	10.74	M1.5 V	6.896	+7.34	+9.75	0.529
Ross 128	0.299	3.35	10.92	M4.0 V	1.361	+11.13	+13.51	0.156
EZ Aquarii A	0.290	3.45	11.27	M5.0 V	3.254	+13.33	+15.64	0.105
EZ Aquarii B	0.290	3.45	11.27	—	3.254	+13.27	+15.58	0.106
EZ Aquarii C	0.290	3.45	11.27	—	3.254	+14.03	+16.34	0.095
Procyon A	0.286	3.50	11.40	F5 IV-V	1.259	+0.38	+2.66	1.569
Procyon B	0.286	3.50	11.40	white dwarf	1.259	+10.70	+12.98	0.500
61 Cygni A	0.286	3.50	11.40	K5.0 V	5.281	+5.21	+7.49	0.703
61 Cygni B	0.286	3.50	11.40	K7.0 V	5.172	+6.03	+8.31	0.630
GJ725 A	0.283	3.53	11.53	M3.0 V	2.238	+8.90	+11.16	0.351
GJ725 B	0.283	3.53	11.53	M3.5 V	2.313	+9.69	+11.95	0.259
GX Andromedae	0.281	3.56	11.62	M1.5 V	2.918	+8.08	+10.32	0.486
GQ Andromedae	0.281	3.56	11.62	M3.5 V	2.918	+11.06	+13.30	0.163
Epsilon Indi A	0.276	3.63	11.82	K5 V	4.704	+4.69	+6.89	0.766
Epsilon Indi B	0.276	3.63	11.82	T1.0	4.823			0.044
Epsilon Indi C	0.276	3.63	11.82	T6.0	4.823			0.028
DX Cancri	0.276	3.63	11.83	M6.5 V	1.290	+14.78	+16.98	0.087
Tau Ceti	0.274	3.64	11.89	G8 V	1.922	+3.49	+5.68	0.921
RECONS 1	0.272	3.68	11.99	M5.5 V	0.814	+13.03	+15.21	0.113
YZ Ceti	0.269	3.72	12.13	M4.5 V	1.372	+12.02	+14.17	0.136
Luyten's Star	0.264	3.79	12.37	M3.5 V	3.738	+9.86	+11.97	0.257
Kapteyn's Star	0.255	3.92	12.78	M1.5 V	8.670	+8.84	+10.87	0.393
AX Microscopium	0.253	3.95	12.87	M0.0 V	3.455	+6.67	+8.69	0.600

This table, compiled from data reported by the Research Consortium on Nearby Stars, lists all known stars within 4.00 parsecs (13.05 light-years).
*Stars that are components of multiple star systems are labeled A, B, and C.

Appendix 5 The Visually Brightest Stars

Name	Designation	Distance (parsecs)	Distance (light-years)	Spectral type	Radial velocity (km/s)*	Proper motion (arcsec/year)	Apparent visual magnitude	Apparent visual brightness (Sirius = 1)**	Absolute visual magnitude
Sirius A	α CMa A	2.63	8.58	A1 V	−7.6	1.34	−1.43	1.000	+1.46
Canopus	α Car	95.9	313	F0 II	+20.5	0.03	−0.72	0.520	−5.63
Arcturus	α Boo	11.3	36.7	K1.5 III	−5.2	2.28	−0.04	0.278	−0.30
Alpha Centauri A	α Cen A	1.34	4.36	G2 V	−25	3.71	−0.01	0.270	+4.36
Vega	α Lyr	7.76	25.3	A0 V	−13.9	0.35	+0.03	0.261	+0.58
Capella	α Aur	12.9	42.2	G5 III	+30.2	0.43	+0.08	0.249	−0.48
Rigel	α Ori A	237	773	B8 Ia	+20.7	0.002	+0.12	0.240	−6.75
Procyon	α CMi A	3.50	11.4	F5 IV-V	−3.2	1.26	+0.34	0.196	+2.62
Achernar	α Eri	44.1	144	B3 V	+16	0.10	+0.50	0.169	−2.72
Betelgeuse	α Ori	131	427	M1 Iab	+21	0.03	+0.58	0.157	−5.01
Hadar	β Cen	161	525	B1 III	+5.9	0.04	+0.60	0.154	−5.43
Altair	α Aql	51.4	168	A7 V	−26.1	0.66	+0.77	0.132	−2.79
Aldebaran	α Tau A	20.0	65.1	K5 III	+54.3	0.20	+0.85	0.122	−0.65
Spica	α Vir	80.4	262	B1 III-IV	+1	0.05	+1.04	0.103	−3.49
Antares	α Sco A	185	604	M1.5 Iab	−3.4	0.03	+1.09	0.098	−5.25
Pollux	β Gem	10.3	33.7	K0 IIIb	+3.3	0.63	+1.15	0.093	+1.08
Fomalhaut	α PsA	7.69	25.1	A3 V	+6.5	0.37	+1.16	0.092	+1.73
Deneb	α Cyg	990	3230	A2 Ia	−4.5	0.002	+1.25	0.085	−8.73
Mimosa	β Cru	108	353	B0.5 IV	+15.6	0.05	+1.297	0.081	−3.87
Regulus	α Leo A	23.8	77.5	B7 V	+5.9	0.25	+1.35	0.077	−0.53

Data in this table were compiled from SIMBAD database operated at the Centre de Données Astronomiques de Strasbourg, France.

*A positive radial velocity means that the star is receding; a negative radial velocity means that the star is approaching.

**This is a ratio of the star's apparent brightness to that of Sirius, the brightest star in the night sky.

Note: Acrux, or α Cru (the brightest star in Crux, the Southern Cross) appears to the naked eye as a star of apparent magnitude +0.87, the same as Aldebaran. However, it does not appear in this table because Acrux is actually a binary star system. The blue-white component stars of this binary system have apparent magnitudes of +1.4 and +1.9, and so they are dimmer than any of the stars listed here.

Appendix 6 Some Useful Mathematics

Area of a rectangle of sides a and b	$A = ab$
Volume of a rectangular solid of sides a, b, and c	$V = abc$
Hypotenuse of a right triangle whose other sides are a and b	$c = \sqrt{a^2 + b^2}$
Circumference of a circle of radius r	$C = 2\pi r$
Area of a circle of radius r	$A = \pi r^2$
Surface ares of a sphere of radius r	$A = 4\pi r^2$
Volume of a sphere of radius r	$V = 4\pi r^3/3$
Value of π	$\pi = 3.1415926536$

granule A convective cell in the solar photosphere. (Chapter 9)

gravitational lens A massive object that deflects light rays from a remote source, forming an image much as an ordinary lens does. (Chapter 14)

Great Red Spot A prominent high-pressure system in Jupiter's southern hemisphere. (Chapter 7)

greatest eastern elongation The configuration of an inferior planet at its greatest angular distance east of the Sun. (Chapter 3)

greatest western elongation The configuration of an inferior planet at its greatest angular distance west of the Sun. (Chapter 3)

greenhouse effect The trapping of infrared radiation near a planet's surface by the planet's atmosphere. (Chapter 5)

greenhouse gas A substance whose presence in a planet's atmosphere enhances the greenhouse effect. (Chapter 5)

group (of galaxies) A gravitationally bound collection of about twenty galaxies. (Chapter 14)

GUT See g*rand unified theory.*

halo (of a galaxy) A spherical distribution of globular clusters and Population II stars that surround a spiral galaxy. (Chapter 13)

heliocentric model A Sun-centered theory of the universe. (Chapter 3)

helioseismology The study of the vibrations of the Sun as a whole. (Chapter 9)

helium flash The nearly explosive beginning of helium fusion in the dense core of a red giant star. (Chapter 11)

Hertzsprung-Russell (H-R) diagram A plot of the luminosity (or absolute magnitude) of stars against their surface temperature (or spectral type). (Chapter 10)

highlands (on Mars) See *southern highlands (on Mars).*

homogeneous Having the same property in one region as in every other region. (Chapter 15)

horizontal branch An evolutionary track on the H-R diagram on which are plotted stars with helium-fusing cores surrounded by hydrogen-fusing shells; because luminosity changes relatively little during this stage, the track is almost horizontal. (Chapter 11)

Hubble classification A method of classifying galaxies as spirals, barred spirals, ellipticals, or irregulars according to their appearance. (Chapter 14)

Hubble constant (H_0) In the Hubble law, the constant of proportionality between the recessional velocities of remote galaxies and their distances. (Chapter 15)

Hubble law (Chapter 15) The relationship between a galaxy's distance from Earth and its recessional velocity

hydrostatic equilibrium A balance between the weight of a layer in a star and the pressure that supports it. (Chapter 9)

hypothesis An idea or collection of ideas that seems to explain a specified phenomenon; a conjecture. (Chapter 1)

ice rafts (Europa) Segments of Europa's icy crust that have been moved by tectonic disturbances. (Chapter 6)

ices Frozen substances heavier than hydrogen and helium, including solidified water. (Chapter 4)

impact crater A circular depression on a planet or satellite caused by the impact of a meteoroid. (Chapter 6)

inferior conjunction The configuration in which an inferior planet is between the Sun and Earth. (Chapter 3)

inflationary epoch A brief period shortly after the Big Bang during which the scale of the universe increased very rapidly. (Chapter 15)

infrared light (IR) Electromagnetic radiation of wavelength longer than visible light but shorter than microwaves or radio waves; covers the range from about 700 nm to 1 mm. (Chapter 2)

inner core (of Earth) The solid innermost portion of Earth's iron-rich core. (Chapter 5)

interstellar medium Gas and dust in interstellar space. (Chapter 11)

intracluster gas Superheated, ionized gas in the space between nearby galaxies. (Chapter 14)

inverse-square law The statement that the apparent brightness of a light source varies inversely with the square of the distance from the source. (Chapter 10)

irregular cluster (of galaxies) A sprawling collection of galaxies whose overall distribution in space does not exhibit any noticeable spherical symmetry. (Chapter 14)

irregular galaxy An asymmetrical galaxy having neither spiral arms nor an elliptical shape. (Chapter 14)

isotropic Having the same property in all directions. (Chapter 15)

joule (J) The amount of energy expended in applying the force of one newton over a distance of one meter. (Chapter 2)

Jovian (planets) Low-density planets composed primarily of hydrogen and helium, including Jupiter, Saturn, Uranus, and Neptune. (Chapter 4)

Kepler's first law The statement that each planet moves around the Sun in an elliptical orbit with the Sun at one focus of the ellipse. (Chapter 3)

Kepler's second law The statement that a planet sweeps out equal areas in equal times as it orbits the Sun; also called the law of equal areas. (Chapter 3)

Kepler's third law A relationship between the period of an orbiting object and the semimajor axis of its elliptical orbit. (Chapter 3)

Kirchoff's laws A series of predictions describing how light from glowing gases interacts with matter. (Chapter 2)

Kuiper belt A region that extends from around the orbit of Pluto to about 500 AU from the Sun where many icy objects orbit the Sun. (Chapter 4)

law of equal areas See *Kepler's second law.*

law of universal gravitation A formula deduced by Isaac Newton that expresses the strength of the force of gravity that two masses exert on each other. (Chapter 3)

lenticular galaxy A galaxy with a central bulge and a disk but no spiral structure; an S0 galaxy. (Chapter 14)

light curve A graph that displays how the brightness of a star or other astronomical object varies over time. (Chapter 12)

light scattering The physical process of transmission and absorption that occurs when light interacts with matter. (Chapter 2)

light-gathering power A measure of the amount of radiation brought to a focus by a telescope. (Chapter 2)

light-year (ly) The distance light travels in a vacuum in one year. (Chapter 3)

line of nodes The line where the plane of Earth's orbit intersects the plane of the Moon's orbit. (Chapter 1)

lithosphere The solid, upper layer of Earth; essentially Earth's crust. (Chapter 5)

Local Group The cluster of galaxies of which our Galaxy is a member. (Chapter 14)

lowlands (on Mars) See *northern lowlands (on Mars).*

luminosity The rate at which electromagnetic radiation is emitted from a star or other object. (Chapter 9, Chapter 10)

luminosity class A classification of a star of a given spectral type according to its luminosity. (Chapter 10)

lunar eclipse An eclipse of the Moon by Earth; a passage of the Moon through Earth's shadow. (Chapter 1)

lunar month See *synodic month.*

lunar phase The appearance of the illuminated area of the Moon as seen from Earth. (Chapter 1)

magnetic-dynamo model A theory that explains the solar cycle as a result of the Sun's differential rotation acting on the Sun's magnetic field. (Chapter 9)

magnetosphere The region around a planet occupied by its magnetic field. (Chapter 5)

magnification The factor by which the apparent angular size of an object is increased when viewed through a telescope. (Chapter 2)

magnifying power See *magnification.*

magnitude scale A system for rating the brightness of objects, using the largest numbers for the dimmest objects. (Chapter 10)

main sequence A category of stable stars whose primary energy source is the conversion of hydrogen into helium. (Chapter 10)

mantle (of a planet) That portion of a terrestrial planet located between its crust and core. (Chapter 5)

March equinox One of two times each year that the ecliptic and the celestial equator intersect and day and night are each 12 hours long. See also *September equinox.* (Chapter 1)

mass A measure of the total amount of material in an object. (Chapter 3)

massive compact halo object (MACHO) The general name for any kind of object with mass that emits little or no light and which might explain the apparent presence of dark matter surrounding galaxies. (Chapter 13)

mass-luminosity relation A relationship between the masses and luminosities of main-sequence stars. (Chapter 10)

medium, interstellar See *interstellar medium.*

mesosphere A layer in Earth's atmosphere above the stratosphere. (Chapter 5)

metal-poor Describes a star that, compared to the Sun, is deficient in elements heavier than helium; also called a Population II star. (Chapter 11)

metal-rich Describes a star whose abundance of heavy elements is roughly comparable to that of the Sun; also called a Population I star. (Chapter 11)

metals Specific to astronomy, metals are any elements heavier than hydrogen or helium. (Chapter 10)

microlensing The apparent brightening of a star caused by an object's passing between the star and an observer. (Chapter 4)

microwaves Short-wavelength radio waves, roughly 1mm to 10 cm. (Chapter 2)

Milky Way Our Galaxy; the band of faint stars seen from Earth in the plane of our Galaxy's disk. (Chapter 13)

minor planet See *asteroid.*

moon Any naturally occurring celestial body orbiting a planet, dwarf planet, or asteroid. (Chapter 4)

nanometer A unit of measurement that is one-billionth of a meter. (Chapter 2)

near-infrared A part of the electromagnetic spectrum with wavelengths between 700 nm and 1400 nm. (Chapter 13)

nebula (*plural* **nebulae**) A cloud of interstellar gas and dust. (Chapter 4, Chapter 11)

nebular hypothesis The idea that the Sun and the rest of the solar system formed from a cloud of interstellar material. (Chapter 4)

neutrino A subatomic particle with no electric charge and very little mass, yet one that is important in many nuclear reactions. (Chapter 9)

neutron star A very compact, dense star composed almost entirely of neutrons. (Chapter 12)

new moon The phase of the Moon when the dark hemisphere of the Moon faces Earth. (Chapter 1)

Newton's first law of motion The statement that a body remains at rest, or moves in a straight line at a constant speed, unless acted upon by a net outside force; the law of inertia. (Chapter 3)

Newton's laws of motion A series of descriptions governing how objects move based on the net forces applied. (Chapter 3)

Newton's second law of motion A relationship between the acceleration of an object, the object's mass, and the net outside force acting on the mass. (Chapter 3)

Newton's third law of motion The statement that whenever one body exerts a force on a second body, the second body exerts an equal and opposite force on the first body. (Chapter 3)

node See *line of nodes.*

north celestial pole The point directly above Earth's north pole where Earth's axis of rotation, if extended, would intersect the celestial sphere. (Chapter 1)

northern lights See *aurora borealis.*

northern lowlands (on Mars) Relatively young and crater-free terrain in the Martian northern hemisphere. (Chapter 6)

northern solstice The point on the ecliptic where the Sun is as far south of the celestial equator as it can get. Also used to refer to the date on which the Sun passes through this point. (Chapter 1)

spiral galaxy A flattened, rotating galaxy with pinwheel-like spiral arms winding outward from the galaxy's nucleus. (Chapter 14)

standard candle A class of objects whose distances are determined by comparing their known energy output and their observed brightness. (Chapter 14)

starburst galaxy A galaxy that is experiencing an exceptionally high rate of star formation. (Chapter 14)

Stefan-Boltzmann law A relationship between the temperature of a blackbody and the rate at which it radiates energy. (Chapter 2)

stellar parallax The apparent displacement of a star due to Earth's motion around the Sun. (Chapter 10)

stratosphere A layer in Earth's atmosphere directly above the troposphere. (Chapter 5)

strong force The force that binds protons and neutrons together in nuclei. (Chapter 15)

subduction zone A location where colliding tectonic plates cause Earth's crust to be pulled down into the mantle. (Chapter 5)

sunspot A temporary cool region in the solar photosphere. (Chapter 9)

sunspot cycle The semiregular 11-year period with which the number of sunspots fluctuates. (Chapter 9)

sunspot maximum/minimum That time during the sunspot cycle when the number of sunspots is highest/lowest. (Chapter 9)

supercluster A collection of clusters of galaxies. (Chapter 14)

supergiant A very large, extremely luminous star of luminosity class I. (Chapter 10, Chapter 12)

supergrand unified theory A complete description of all forces and particles, as well as the structure of space and time; a "theory of everything" (TOE). (Chapter 15)

superior conjunction The configuration in which a planet is behind the Sun as viewed from Earth. (Chapter 3)

supermassive black hole A black hole with a mass of a million or more Suns. (Chapter 12, Chapter 14)

supernova remnant The gases ejected by a supernova. (Chapter 11, Chapter 12)

surface wave A type of seismic wave that travels only over Earth's surface. (Chapter 5)

synchronous rotation The rotation of a body with a period equal to its orbital period. (Chapter 1)

synodic month The period of revolution of the Moon with respect to the Sun; the length of one cycle of lunar phases; also called the lunar month. (Chapter 1)

synodic period The interval between successive occurrences of the same configuration of a planet. (Chapter 3)

T Tauri stars Young variable stars associated with interstellar matter that show erratic changes in luminosity. (Chapter 11)

T Tauri wind A flow of particles away from a T Tauri star. (Chapter 4)

terrestrial (planets) High-density worlds with solid surfaces, including Mercury, Venus, Earth, and Mars. (Chapter 4)

theory A hypothesis that has withstood experimental or observational tests. (Chapter 1)

theory of everything (TOE) See *supergrand unified theory.*

thermal equilibrium A balance between the input and outflow of heat in a system. (Chapter 9)

thermal pulse A brief burst in energy output from the helium-fusing shell of an aging low-mass star. (Chapter 11)

thermonuclear fusion The combining of nuclei under conditions of high temperature in a process that releases substantial energy. (Chapter 9)

thermosphere A region in Earth's atmosphere between the mesosphere and the exosphere. (Chapter 5)

third quarter moon The phase of the Moon that occurs when the Moon is 90° west of the Sun. (Chapter 1)

TOE See *supergrand unified theory.*

total lunar eclipse A lunar eclipse during which the Moon is completely immersed in Earth's umbra. (Chapter 1)

total solar eclipse An event in which the Moon completely blocks the Sun's disk from being observed from a particular location on Earth. (Chapter 1)

totality (lunar eclipse) The period during a total lunar eclipse when the Moon is entirely within Earth's umbra. (Chapter 1)

totality (solar eclipse) The period during a total solar eclipse when the disk of the Sun is completely hidden. (Chapter 1)

transit method The process of identifying extrasolar planets orbiting other stars by monitoring a drop in a star's brightness that is due to an intervening planet. (Chapter 4)

trans-Neptunian object (TNO) Any small body of rock and ice that orbits the Sun within the solar system, but beyond the orbit of Neptune. (Chapter 4)

triple alpha process A sequence of two thermonuclear reactions in which three helium nuclei combine to form one carbon nucleus. (Chapter 11)

Tropic of Cancer A circle of latitude 23½° north of Earth's equator. (Chapter 1)

Tropic of Capricorn A circle of latitude 23½° south of Earth's equator. (Chapter 1)

troposphere The lowest level in Earth's atmosphere. (Chapter 5)

Tully-Fisher relation A correlation between the width of the 21-cm line of a spiral galaxy and the total luminosity of that galaxy. (Chapter 14)

tuning fork diagram A diagram that summarizes Edwin Hubble's classification scheme for spiral, barred spiral, and elliptical galaxies. (Chapter 14)

22-year solar cycle The semiregular 22-year interval between successive appearances of sunspots at the same latitude and with the same magnetic polarity. (Chapter 9)

Type I supernovae Supernovae whose spectra lack hydrogen lines. Type I supernovae are further classified as Type Ia, Ib, or Ic. (Chapter 12)

Type II supernovae Supernovae with hydrogen emission lines in their spectra, caused by the explosion of a massive star. (Chapter 12)

ultraviolet light (UV) A subset of the electromagnetic spectrum of light ranging from 400 nm to 10 nm. (Chapter 2)

umbra (of a shadow) (*plural* **umbrae**) The central, completely dark portion of a shadow. (Chapter 1)

universal constant of gravitation (*G*) An empirical constant relating force to mass and distance in Newton's law of gravitation equations. Chapter 3

velocity The speed and direction of an object's motion. (Chapter 3)

visible light Electromagnetic radiation detectable by the human eye. (Chapter 2)

void A large volume of space, typically 30 Mpc to 120 Mpc (100 million ly to 400 million ly) in diameter, that contains very few galaxies. (Chapter 14)

waning crescent moon The phase of the Moon that occurs between third quarter and new moon. (Chapter 1)

waning gibbous moon The phase of the Moon that occurs between full moon and third quarter. (Chapter 1)

water cycle The process by which water circulates in different phases around an object. (Chapter 5)

water hole A range of frequencies in the microwave spectrum suitable for interstellar radio communication. (Chapter 8)

watt (W) A unit of measure describing the transfer of energy per second. (Chapter 2)

wavelength The distance between two successive wave crests. (Chapter 2)

waxing crescent moon The phase of the Moon that occurs between new moon and first quarter. (Chapter 1)

waxing gibbous moon The phase of the Moon that occurs between first quarter and full moon. (Chapter 1)

weak force The short-range force that is responsible for transforming certain particles into other particles, such as the decay of a neutron into a proton. (Chapter 15)

weakly interacting massive particles (WIMPs) (Chapter 13) The general name for particles that interact only through gravity and the weak force and which might explain the apparent presence of dark matter surrounding galaxies.

white dwarf A low-mass star that has exhausted all its thermonuclear fuel and contracted to a size roughly equal to the size of Earth. (Chapter 10, Chapter 11)

Wien's law A relationship between the temperature of a blackbody and the wavelength at which it emits the greatest intensity of radiation. (Chapter 2)

X-ray burster A nonperiodic X-ray source that emits powerful bursts of X-rays. (Chapter 12)

X-rays Electromagnetic radiation whose wavelength is between that of ultraviolet light and gamma rays. (Chapter 2)

Zeeman effect A splitting or broadening of spectral lines due to a magnetic field. (Chapter 9)

zenith The point on the celestial sphere directly overhead an observer. (Chapter 1)

zone A light-colored band in Jupiter's atmosphere. (Chapter 7)

INDEX

Page numbers in *italics* indicate figures and captions.

STAR CHARTS

The following set of star charts, one for each month of the year, are useful in the northern hemisphere only. To use these charts, first select the chart that best corresponds to the date and time of your observations. Hold the chart vertically as shown in the above illustration and turn it so that the direction you are facing is shown at the bottom.

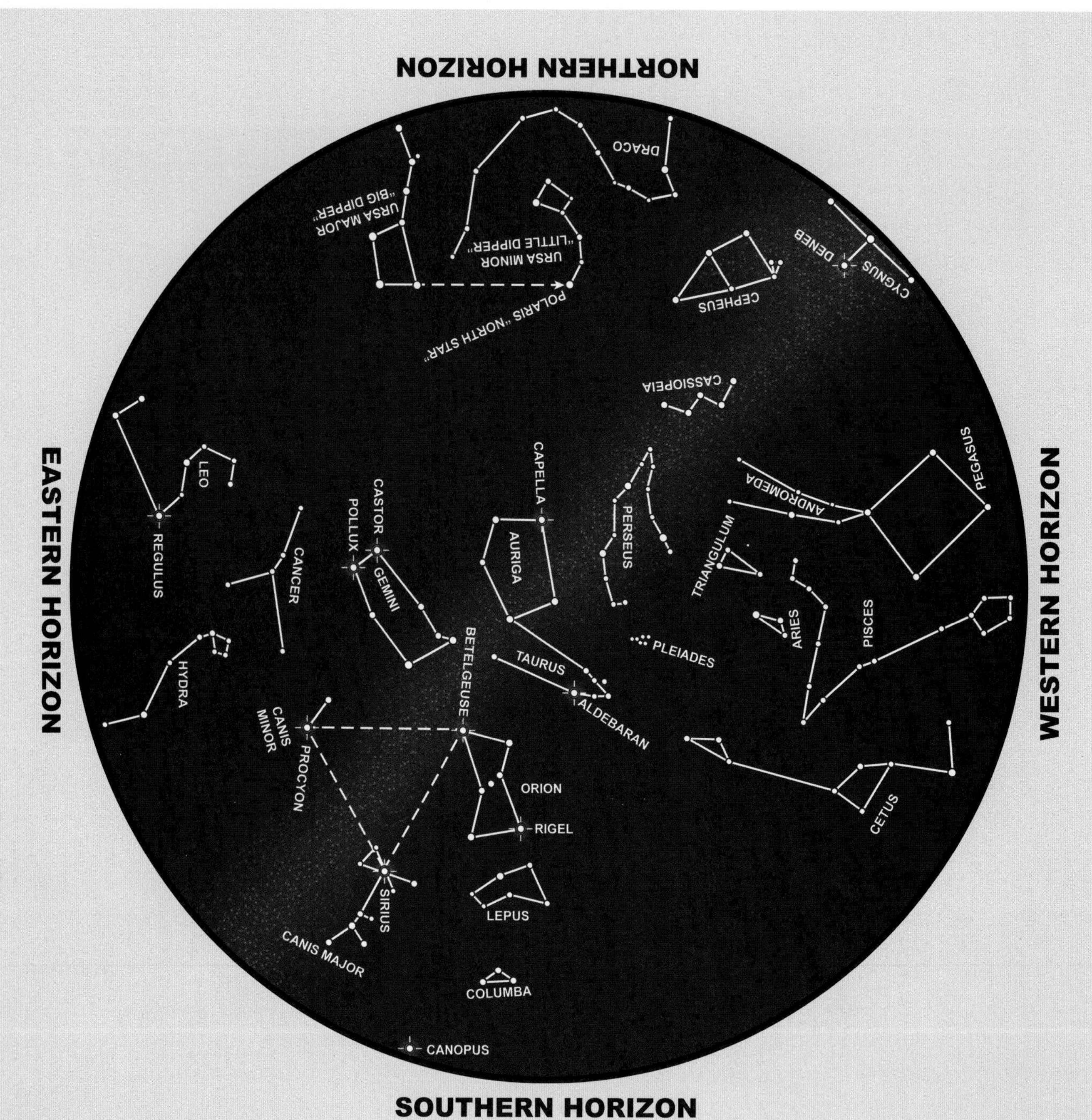

THE NIGHT SKY IN JANUARY

Chart time (Local Standard Time):
10 pm: First of January
9 pm: Middle of January
8 pm: Last of January

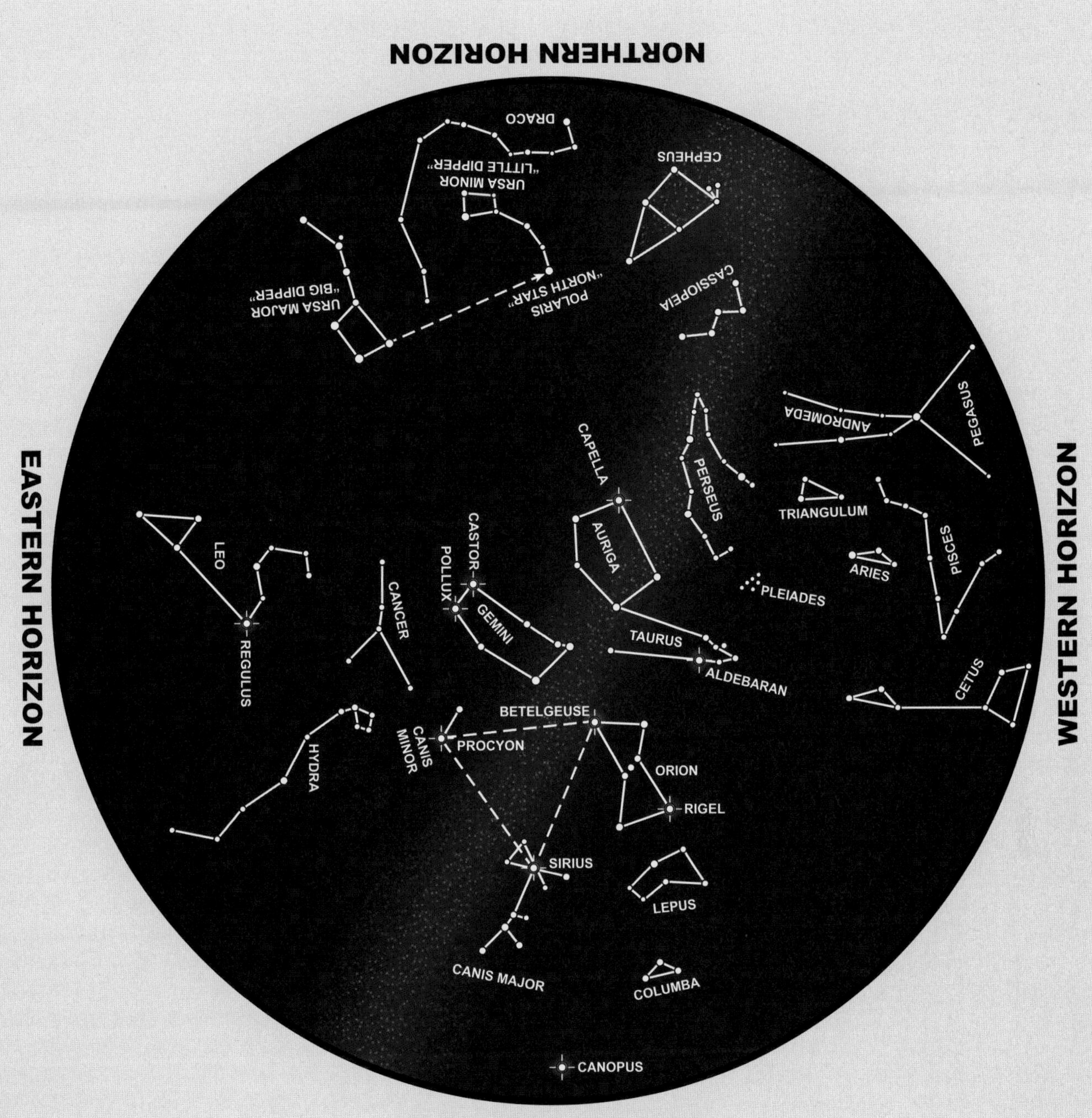

THE NIGHT SKY IN FEBRUARY

Chart time (Local Standard Time):
10 pm: First of February
9 pm: Middle of February
8 pm: Last of February

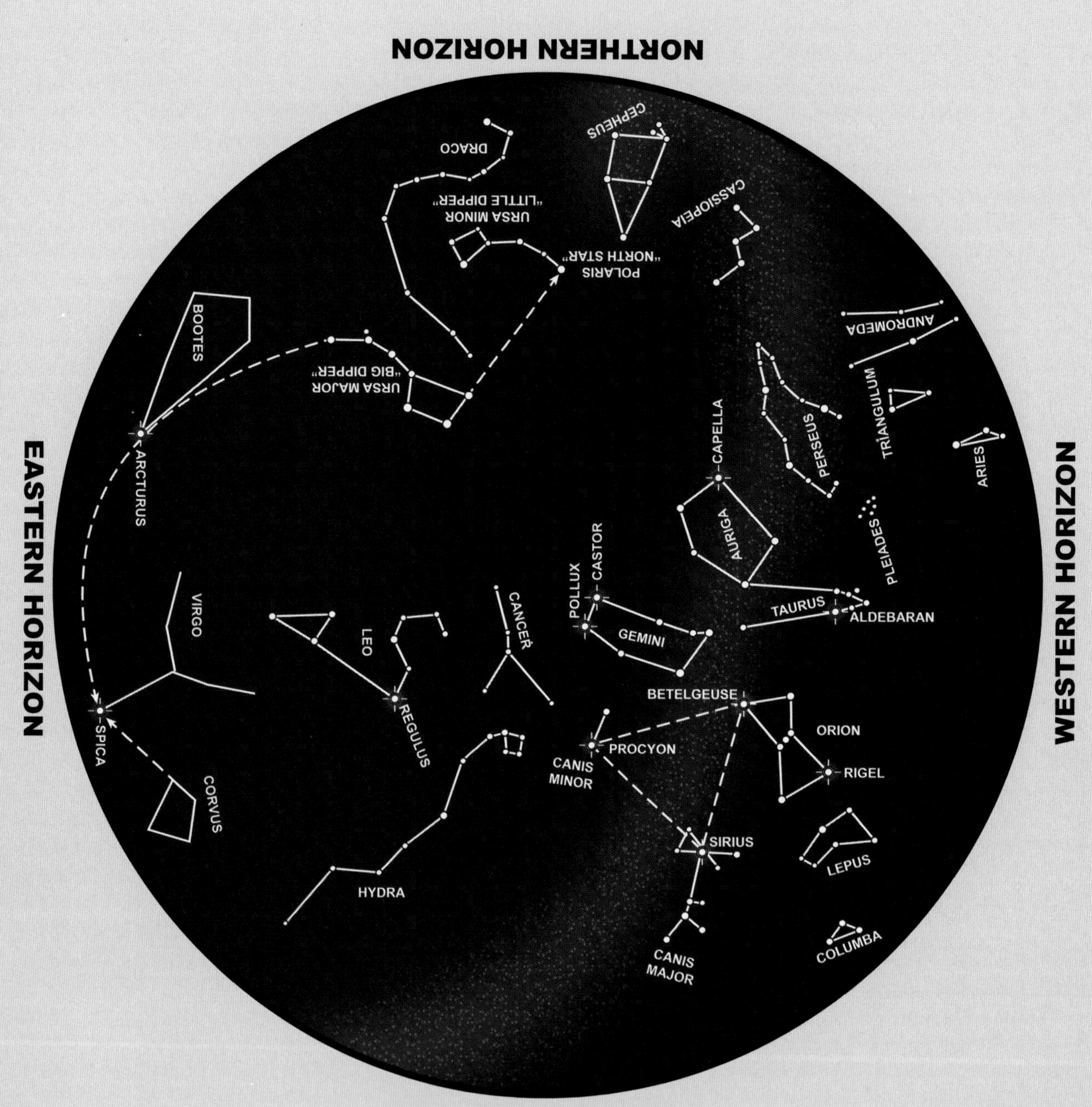

THE NIGHT SKY IN MARCH

Chart time (Local Standard Time):
10 pm: First of March
9 pm: Middle of March
8 pm: Last of March

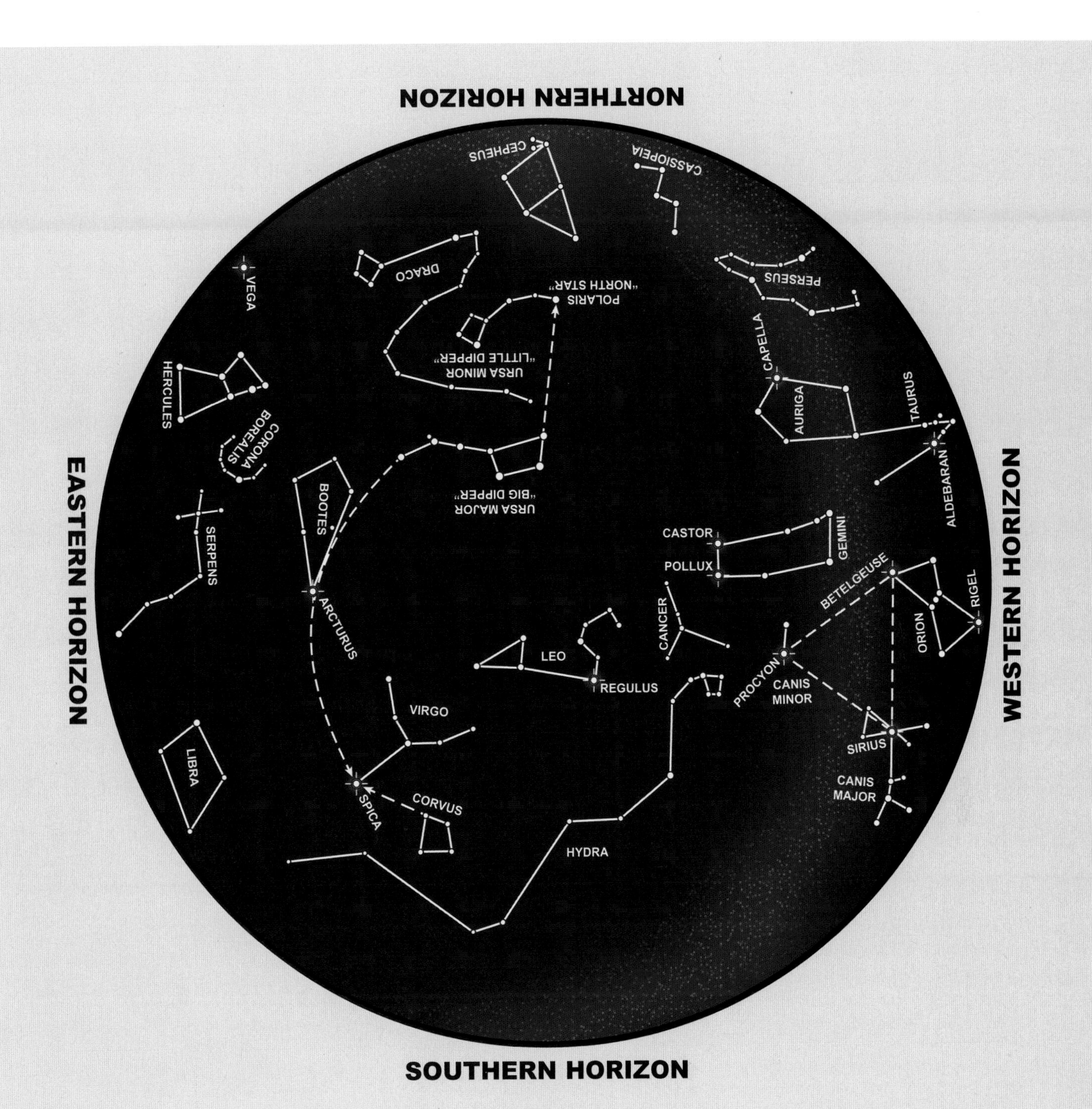

THE NIGHT SKY IN APRIL

Chart time (Daylight Saving Time):
11 pm: First of April
10 pm: Middle of April
9 pm: Last of April

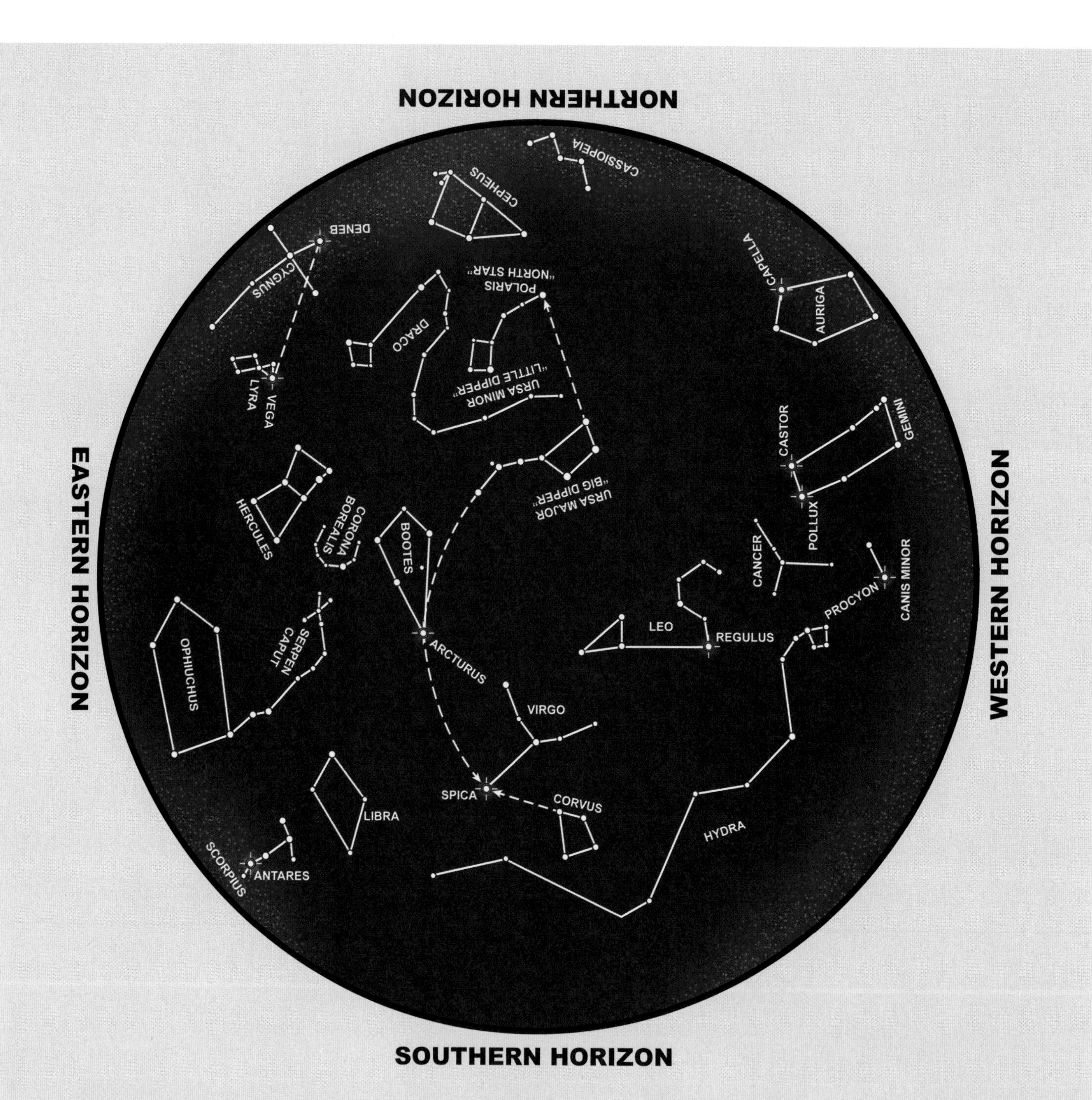

THE NIGHT SKY IN MAY

Chart time (Daylight Saving Time):
11 pm: First of May
10 pm: Middle of May
9 pm: Last of May

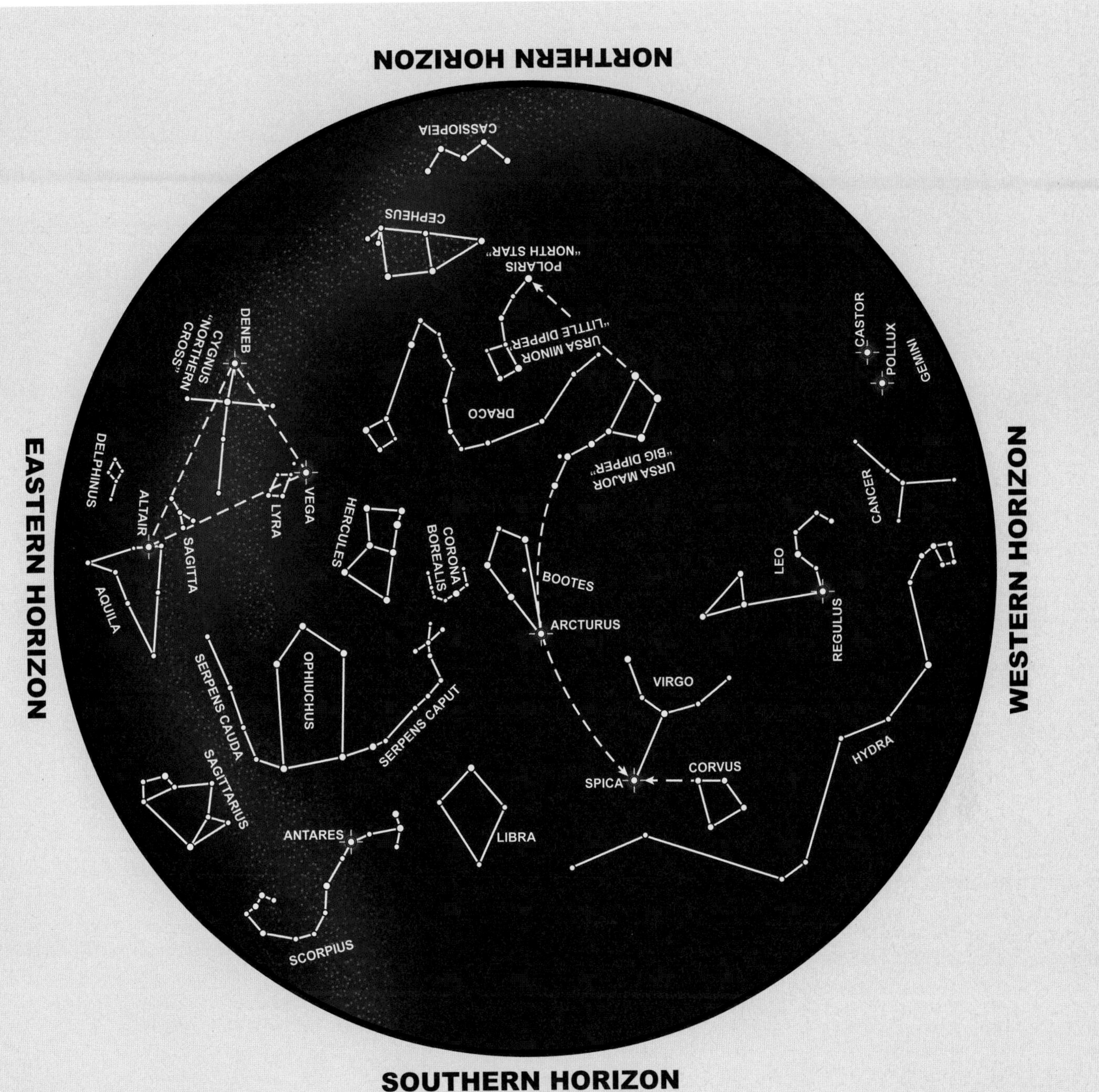

THE NIGHT SKY IN JUNE

Chart time (Daylight Saving Time):
11 pm: First of June
10 pm: Middle of June
9 pm: Last of June

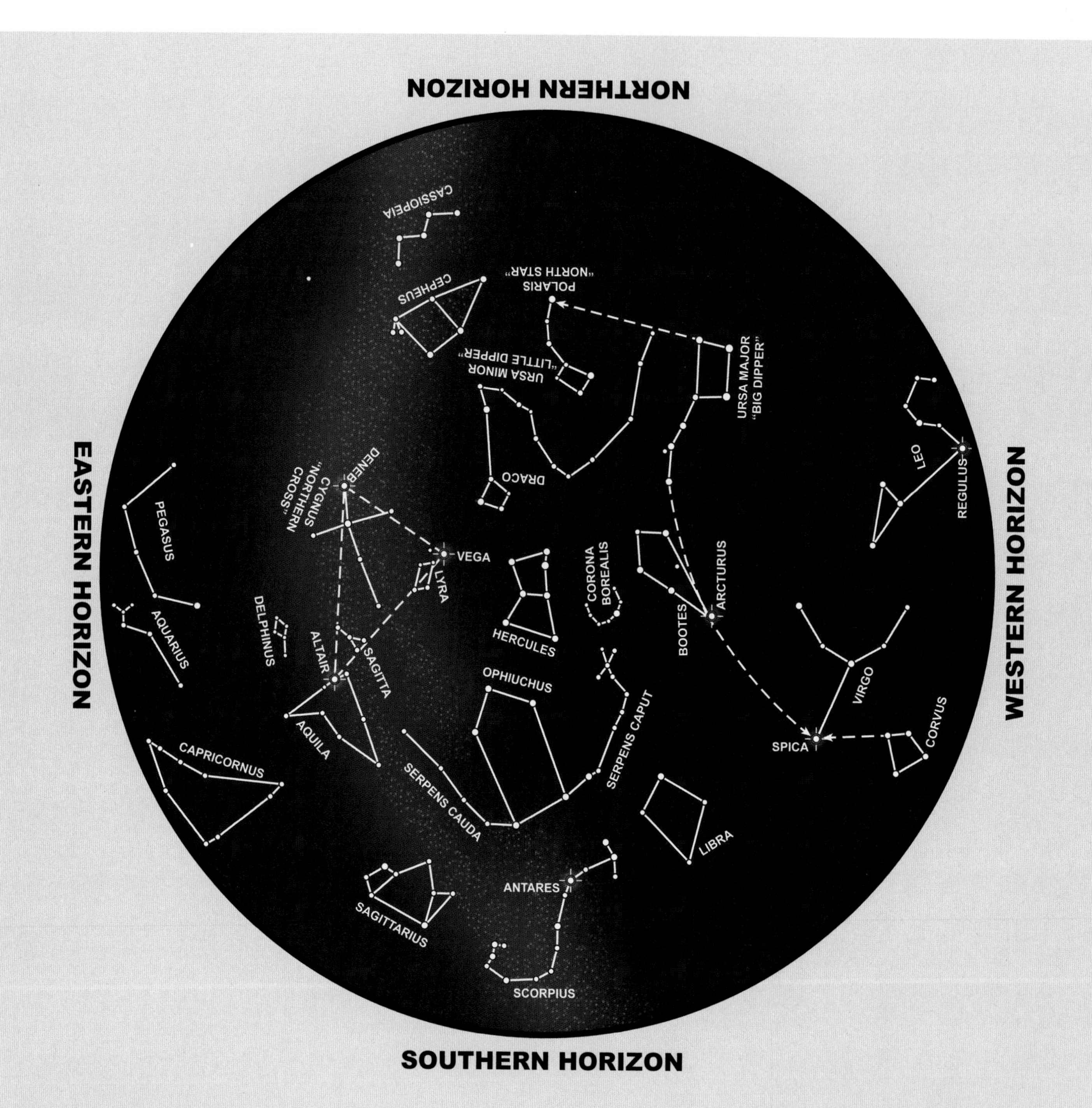

THE NIGHT SKY IN JULY

Chart time (Daylight Saving Time):
11 pm: First of July
10 pm: Middle of July
9 pm: Last of July

THE NIGHT SKY IN AUGUST

Chart time (Daylight Saving Time):
11 pm: First of August
10 pm: Middle of August
9 pm: Last of August

THE NIGHT SKY IN SEPTEMBER

Chart time (Daylight Saving Time):
11 pm: First of September
10 pm: Middle of September
9 pm: Last of September

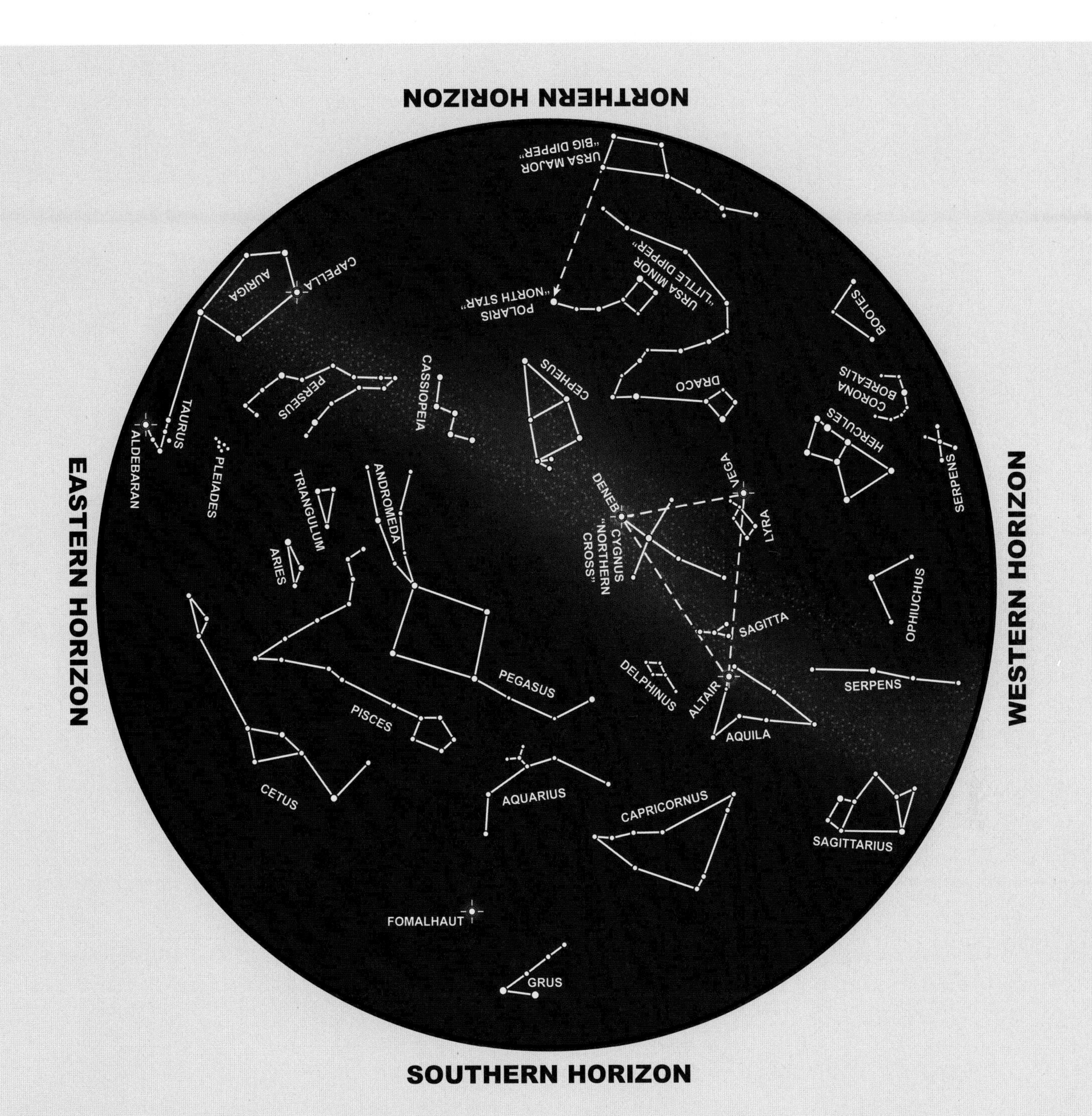

THE NIGHT SKY IN OCTOBER

Chart time (Daylight Saving Time):
11 pm: First of October
10 pm: Middle of October
9 pm: Last of October

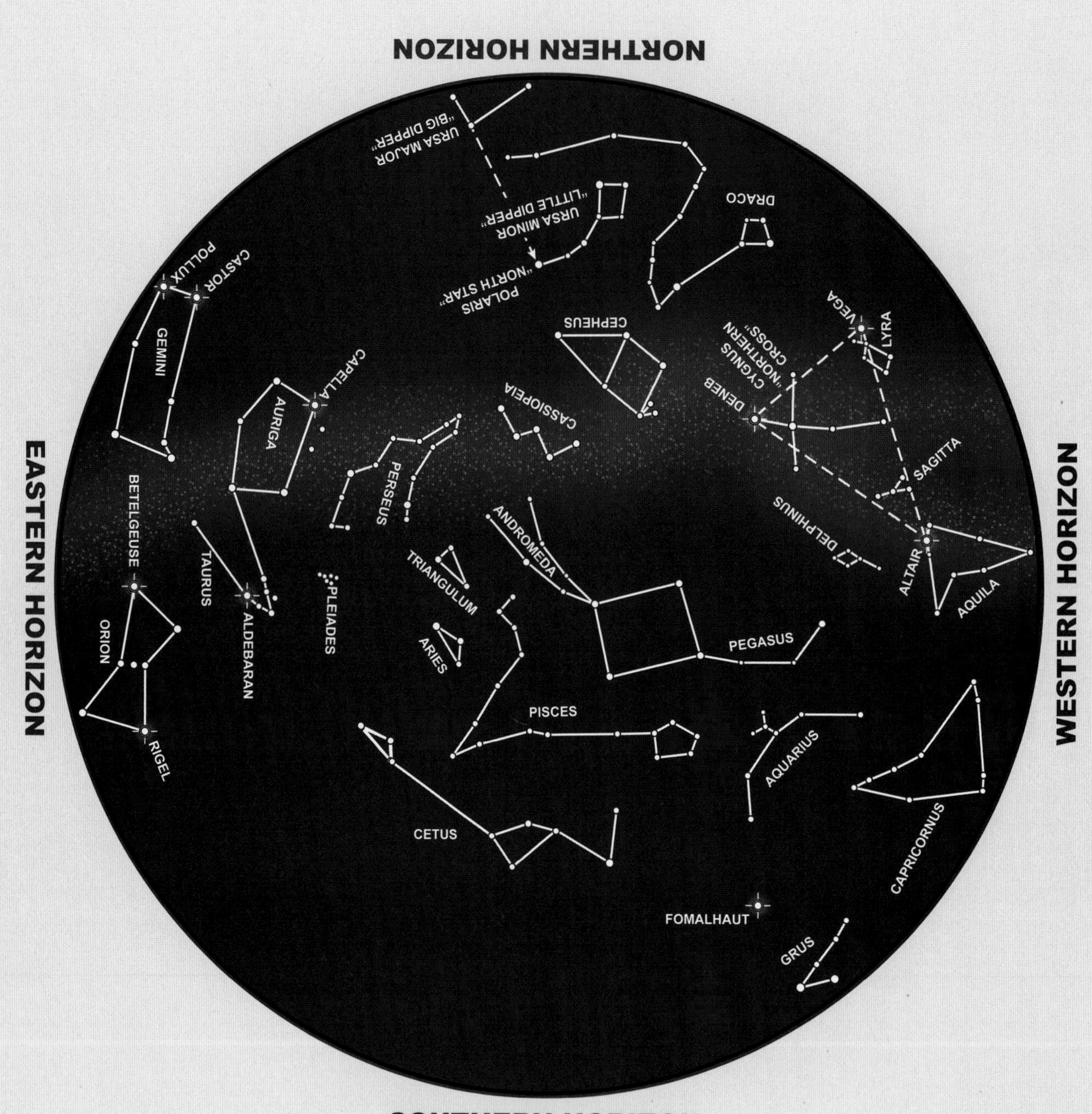

THE NIGHT SKY IN NOVEMBER

Chart time (Local Standard Time):
10 pm: First of November
9 pm: Middle of November
8 pm: Last of November

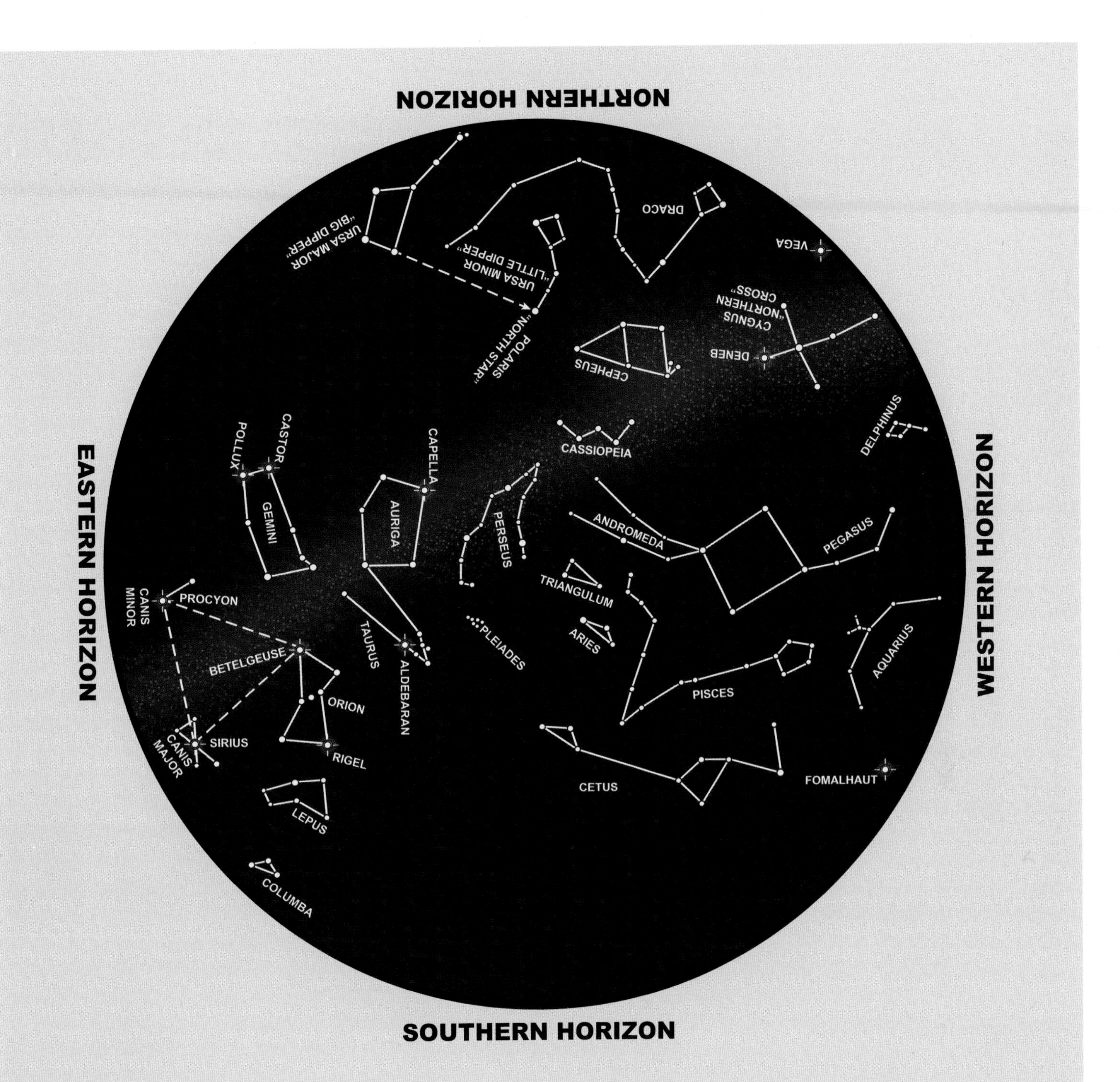

THE NIGHT SKY IN DECEMBER

Chart time (Local Standard Time):
10 pm: First of December
9 pm: Middle of December
8 pm: Last of December